3rd Southeast Asian Conference on Geophysics

IOP Conference Series: Earth and Environmental Science
Volume 873

Online
3-5 November 2020

ISBN: 978-1-7138-4855-4
ISSN: 1755-1307

Printed from e-media with permission by:

Curran Associates, Inc.
57 Morehouse Lane
Red Hook, NY 12571

Some format issues inherent in the e-media version may also appear in this print version.

Institute of Physics
Dirac House, Temple Back
Bristol BS1 6BE UK

Phone: 44 1 17 929 7481
Fax: 44 1 17 920 0979

techtracking@iop.org

Additional copies of this publication are available from:

Curran Associates, Inc.
57 Morehouse Lane
Red Hook, NY 12571 USA
Phone: 845-758-0400
Fax: 845-758-2633
Email: curran@proceedings.com
Web: www.proceedings.com

TABLE OF CONTENTS

IOP Conf. Series: Earth and Environmental Science **873** (2021) 011001 doi:10.1088/1755-1315/873/1/011001

The 3rd Southeast Asian Conference on Geophysics (SEACG) 2020: Future Challenges and Opportunities in Geophysics (Preface)

Zulfakriza[1*], Andri Hendriyana[1], Erdinc Saygin[2], Chanmaly Chhun[3]

[1]Department of Geophysical Engineering, Faculty of Mining and Petroleum Engineering, Institut Teknologi Bandung, Indonesia.
[2]CSIRO Australian Resources Research Centre 26 Dick Perry Avenue, Kensington WA 6151 Australia.
[3]Kyushu University, Japan,

*email: zulfakriza@itb.ac.id

The issue of IOP Conference Series: Earth and Environmental Science contains the outstanding papers presented at Southeast Asian Conference on Geophysics (SEACG) 2020. Due to Covid-19 Pandemic, the 3rd SEACG 2020 Conference was held online on 3 – 5 November 2020.

The Southeast Asian Conference on Geophysics (SEACG) is a biannual event held by the Geophysical Engineering Department, Faculty of Mining and Petroleum Engineering, Bandung Institute of Technology, Bandung, Indonesia. The 3rd SEACG 2020, held in collaboration with PEER-USAID Project, Institute of Research and Community Services – ITB, Center for Earthquake Science, and Technology (CEST) -ITB, and Center for CO2 & Flared Gas Utilization - ITB.

The conference's theme of SEACG 2020 was "Future Challenges and Opportunities in Geophysics". Some new geophysical insights and techniques were introduced and delivered by the invited and contributed speakers. Energies, Environments and infrastructures are the objects that their integrity and sustainability must be contemplated and considered by all earth scientists and engineers, in order to make sure the sustainability of human security in the future.

The SEACG 2020 created an opportunity for academics, professionals, and students to promote and discuss the scientific results related to recent development in geophysical methods and related science. The session topics during the conference of SEACG 2020 include rock and paleomagnetism; near-surface geophysics; geophysical approaches in hydrocarbon and geothermal exploration; microseismic for new and renewable energy exploration; environmental geophysics; volcanology; earthquake seismology; exploration seismology; geohazard mitigation; big data and machine learning in geosciences; and current developments on potential method in geoscience.

There were 13 invited international speakers, 151 contributed speakers with 5 panel discussions and 30 parallel sessions. The interesting topics discussed in the panel discussion with invited speakers from Bandung Institute of Technology Indonesia, Australian National University, CSIRO – Australia, Kyushu University – Japan, Karlsruhe Institute of Technology – Germany, and ISTerre Greanoble University, France.

The keynote and plenary session were held in the beginning of each day (3, 4, and 5 November 2020). For the 1st day (3 November 2020) Prof. Dr. Djoko Santoso from Bandung Institute of Technology became a keynote speaker and talked about Toward Decarbonization Society through Geophysics.

Then, the panel discussion with topic about Challenges and Opportunities of Geophysical Exploration to Meet Future Energy Demands that was delivered by Prof. Dr. Sigit Sukmono (Bandung Institute of Technology) and Prof. Takeshi Tsuji, PhD (Kyushu University, Japan). Special Session of SEACG 2020 in the 1st day was a Collaborative Workshop between Bandung Institute of Technology and Karlsruhe Institute of Technology, Germany. The speakers of this session were Dr. Wahyu Triyoso (Bandung Institute of Technology), Prof. Dr. Andreas Rietbrock (Karlsruhe Institute of Technology, Germany) and 5 (five) students from Bandung Institute of Technology and Karlsruhe Institute of Technology, Germany.

For 2nd day (4 November 2020), the panel discussion with topic of Future Perspective on Seismology and Geohazard Mitigation, the speakers for this session were Prof. Phil Cummins, PhD (Australian National University), Prof. Sri Widiyantoro, PhD (Bandung Institute of Technology) and Prof. Dr. Andri Dian Nugraha (Bandung Institute of Technology). The special session for 2nd day was about Volcano Seismology delivered by Dr. Devy Kamil Syahbana (CVGHM, Indonesia) and Dr. Corentin Caudron (ISTerre Greanoble University, France).

Finally, the 3rd panel discussion, held on 5 November 2020 with the topic of Open Forum in Applied Geophysics. The speakers of this session were Prof. Dr. Wawan Gunawan A. Kadir (Bandung Institute of Technology), Prof. Dr. Satria Bijaksana (Bandung Institute of Technology) and Dr. Erdinc Saygin (CSIRO, Australia).

Documentation:

SEACG 2020 Highlights

Day-1

Tuesday, 3 November

SEACG 2020 OPENING
Opening and Welcome Speech

Dr. Zulfakriza *as SEACG 2020 Chairman*
Dr. Ir. Warsa *as Head of Geophysical Engineering Bachelor Degree Program*
Dr. Fatkhan *as Head of Geophysical Engineering Magister & Doctorate Degree Program*

TIME	08.00 – 08.20 (UTC+7)
ROOM	*ZOOM 7*

KEYNOTE SESSION I
"Toward Decarbonization Society through Geophysics"

TIME	08.20 – 08.40 (UTC+7)
ROOM	*ZOOM 7*
MODERATOR	Sonny Winardhie, Ph.D
SPEAKERS	Prof. Dr. Djoko Santoso *Institut Teknologi Bandung, Indonesia*

PANEL DISCUSSION I
"Challenges and Opportunities of Geophysical Exploration to Meet Future Energy Demands."

TIME	08.40 – 10.00 (UTC+7)
ROOM	*ZOOM 7*
MODERATOR	Sonny Winardhie, Ph.D
SPEAKERS	Prof. Dr. Sigit Sukmono *Institut Teknologi Bandung, Indonesia* Prof. Takeshi Tsuji *Kyushu University, Japan*

SCIENTIFIC PARALLEL SESSION I

TIME	10.10 – 11.50 (UTC+7)
Room	Topics
ZOOM 1	Reservoir Geophysics-1
ZOOM 2	Resistivity Methods and Application
ZOOM 3	Geomagnetic Methods and Application
ZOOM 4	Microtremor and HVSR Analysis
ZOOM 5	Volcanology (1)
ZOOM 6	Earthquakes Seismology (1)

<< 11.50 - 13.00 SESSION BREAK >>

SCIENTIFIC PARALLEL SESSION II

TIME	13.10 – 14.50 (UTC+7)
Room	Topics
ZOOM 1	Reservoir geophysics-2
ZOOM 2	Gravity Methods and Application (1)
ZOOM 3	Seismic Hazard (1)
ZOOM 4	Landslide and Liquefaction Investigation
ZOOM 5	Volcanology (1)
ZOOM 6	Earthquakes Seismology (2)

SEACG 2020 SPECIAL SESSION I
"KIT-ITB Collaborative Workshop"

TIME	15.00 – 17.35 (UTC+7)
ROOM	*ZOOM 7*
MODERATOR	Dr. Moh. Rachmat Sule
SPEAKERS	Dr. Wahyu Triyoso *Institut Teknologi Bandung, Indonesia* Prof. Andreas Rietbrock *Karlsruhe Institute of Technology, Germany*
STUDENT PRESENTATION	I Made Kris Adi Astra Lucky Kriski Muhtar Sonia Sortan Aria Widhi Baskara V. Reddy

IOP Conf. Series: Earth and Environmental Science **873** (2021) 011001 doi:10.1088/1755-1315/873/1/011001

Day-2

Wednesday, 4 November

PANEL DISCUSSION II

"Future Perspective on Seismology and Geohazard Mitigation"

TIME	**08.00 – 09.35 (UTC+7)**
ROOM	*ZOOM 7*
MODERATOR	**Dr. Endra Gunawan**
SPEAKERS	**Prof. Dr. Phil Cummins** *Australian National University*
	Prof. Sri Widiyantoro, Ph.D *Institut Teknologi Bandung / Maranatha Christian University*
	Prof. Dr. Andri Dian Nugraha *Institut Teknologi Bandung, Indonesia*

SCIENTIFIC PARALLEL SESSION III

TIME	**09.45 – 11.25 (UTC+7)**
Room	**Topics**
ZOOM 1	**Seismic Processing**
ZOOM 2	**Gravity Methods and Application (2)**
ZOOM 3	**Seismic Hazard (2)**
ZOOM 4	**GPS Analysis and Crustal Deformation**
ZOOM 5	**Seismic Tomography (1)**
ZOOM 6	**Earthquakes Seismology (3)**

<< 11.25 - 13.00 SESSION BREAK >>

SCIENTIFIC PARALLEL SESSION IV

TIME	**13.10 – 14.50 (UTC+7)**
Room	**Topics**
ZOOM 1	**Rock Physics and Shale Gas Investigation**
ZOOM 2	**Near-surface geophysics**
ZOOM 3	**Tsunami and Hazard Mitigation**
ZOOM 4	**Geotechnics & Environmental Geophysics**
ZOOM 5	**Seismic Tomography (2)**
ZOOM 6	**Earthquakes Seismology (4)**

SEACG 2020 SPECIAL SESSION II

"Volcanic Seismology"

TIME	**15.00 – 17.35 (UTC+7)**
ROOM	*ZOOM 7*
MODERATOR	**Dr. Zulfakriza**
SPEAKERS	**Dr. Devy K. Syahbana** *CVGHM, Indonesia*
	Dr. Corentin Caudron *ISTerre, Grenoble, France*

Day-3

Thursday, 5 November

PANEL DISCUSSION III

"Open Forum in Applied Geophysics"

TIME	**08.00 – 09.35 (UTC+7)**
ROOM	*ZOOM 7*
MODERATOR	**Dr. David Prambudi Sahara**
SPEAKERS	**Prof. Dr. Wawan G.A. Kadir** *Institut Teknologi Bandung*
	Prof. Satria Bijaksana *Institut Teknologi Bandung*
	Dr. Erdinc Saygin *CSIRO - Australia*

SCIENTIFIC PARALLEL SESSION V

TIME	**09.45 – 11.25 (UTC+7)**
Room	**Topics**
ZOOM 1	**Machine Learning in Geophysics**
ZOOM 2	**Active Fault Identification**
ZOOM 3	**Earthquakes Source Investigation**
ZOOM 4	**Land subsidence & Liquefaction**
ZOOM 5	**General Seismology (1)**
ZOOM 6	**General Seismology (2)**

CLOSING

TIME	**11.30 – 12.00 (UTC+7)**

SEACG
Southeast Asian Conference on Geophysics

The 3rd Southeast Asian Conference on Geophysics
Online Conference
03-05 November 2020
seacg.fttm.itb.ac.id

Andri Dian N.
SEACG 7
Warsa
Zulfakriza Z.
SEACG
Southeast Asian Conference on Geophysics
The 3rd Southeast Asian Conference on Geophysics
"Future Challenges and Opportunities in Geophysics"
Online Conference
03-05 November 2020
Stop Video
Participants
Chat
Share Screen
Record
Reactions

SEACG
Southeast Asian Conference on Geophysics
The 3rd Southeast Asian Conference on Geophysics
"Future Challenges and Opportunities in Geophysics"
Online Conference
03-05 November 2020
seacg.fttm.itb.ac.id
Day-2
Wednesday, 4 November
PANEL DISCUSSION II
"Future Perspective on Seismology and Geohazard Mitigation"
TIME 08.00 – 09.35 (UTC+7)
ROOM ZOOM 7
MODERATOR Dr. Endra Gunawan
SPEAKERS Prof. Dr. Phil Cummins
Australian National University
Prof. Sri Widiyantoro, Ph.D
Institut Teknologi Bandung / Maranatha Christian University
Prof. Dr. Andri Dian Nugraha
Institut Teknologi Bandung, Indonesia
Panel Discussion will be started at 08.00 UTC +7
11:04
Phil Cummins
Rahma Hanifa
Endra Gunawan
SEACG 7
Zulfakriza Z.
Start Video
Participants
Chat
Share Screen
Record
Reactions

IOP Conf. Series: Earth and Environmental Science **873** (2021) 011001 doi:10.1088/1755-1315/873/1/011001

IOP Conf. Series: Earth and Environmental Science **873** (2021) 011001 doi:10.1088/1755-1315/873/1/011001

The 3rd Southeast Asian Conference on Geophysics
03 - 05 November 2020

Geophysical Engineering
Faculty of Mining and Petroleum Engineering

Partnership for Enhanced Engagement in Research (PEER) Science

INDONESIA CENTER OF EXCELLENCE CARBON CAPTURE STORAGE

CEST
CENTER FOR EARTHQUAKE SCIENCE AND TECHNOLOGY
RESEARCH CENTER FOR DISASTER MITIGATION

IOP Conf. Series: Earth and Environmental Science **873** (2021) 011002 doi:10.1088/1755-1315/873/1/011002

The 3rd Southeast Asian Conference on Geophysics (SEACG) 2020

Peer Review Declaration

All papers published in this volume of IOP Conference Series: Earth and Environmental Science have been peer reviewed through processes administered by the editors. Reviews were conducted by expert referees to the professional and scientific standards expected of a proceedings journal published by IOP Publishing.

- **Type of peer review:** Single blind
- **Conference submission management system:** Conference web submission system
- **Number of submission Received**: 151 manuscripts received
- **Number of submission sent for review**: 110 manuscripts sent for review
- **Number of submission accepted**: 102 manuscript accepted by editor
- **Acceptance Rate (Number of Submissions Accepted / Number of Submissions Received X 100):** (102/151) x 100 = 67.5 %
- **Average number of reviews per paper:** Average 2 reviewers of each paper
- **Total number of reviewers involved:** 45 reviewers
- **Any additional info on review process:**
  - We did plagiarism check for each paper using licenced Turnitin online software. And we accepted the level of plagiarism each manuscript below 20 %.
  - The review process was conducted using google form that integrated with conference secretariat email. We sent a review invitation to reviewer by official email of SEACG 2020 and the reviewers conducted review process based on 4 sections:
    - Technical Criteria
      - Scientific merit of the manuscript: notably scientific rigour, accuracy and correctness.
      - Clarity of expression; communication of ideas; readability and discussion of concepts.

IOP Conf. Series: Earth and Environmental Science **873** (2021) 011002 doi:10.1088/1755-1315/873/1/011002

        - Sufficient discussion of the context of the work, and suitable referencing.
    - Quality Criteria
        - Originality: Is the work relevant and novel?
        - Motivation: Does the problem considered have a sound motivation? All papers should clearly demonstrate the scientific interest of the results.
        - Repetition: Have significant parts of the manuscript already been published?
        - Length: Is the content of the work of sufficient scientific interest to justify its length?
    - Presentation Criteria
        - Title: Is it adequate and appropriate for the content of the article?
        - Abstract: Does it contain the essential information of the article? Is it complete? Is it suitable for inclusion by itself in an abstracting service?
        - Diagrams, figures, tables and captions: Are they essential and clear?
        - Text and mathematics: Are they brief but still clear? If you recommend shortening, please suggest what should be omitted.
        - Conclusion: Does the paper contain a carefully written conclusion, summarizing what has been learned and why it is interesting and useful?
    - Decision
        - Reviewer comment
        - Suggestion
            - Ready to publish
            - Minor correction
            - Major correction
            - Reject
  - All the comments by reviewers were sent to the author to make the correction and revision within 20 days. The editor will check each revised manuscript carefully and make sure the authors have addressed the questions and comments from reviewer. If the manuscript need to do revision for 2nd circle, the editor would send it back to the authors.

- **Contact person for queries:**

Dr. Zulfakriza
Geophysical Engineering Dept.
Faculty of Mining and Petroleum Engineering
Bandung Institute of Technology, Indonesia
zulfakriza@itb.ac.id

IOP Conf. Series: Earth and Environmental Science **873** (2021) 012001 doi:10.1088/1755-1315/873/1/012001

Analysis of the Mw 6.5 Ambon Earthquake (September 26, 2019) based on the aftershocks hypocenter relocation

P Supendi[1,2], A D Nugraha[2,3], S Widiyantoro[3,4], C I Abdullah[5], Daryono[1], D Karnawati[1], K H Palgunadi[6], S Rohadi[1], M T Gunawan[1], S Rosalia[2,3]

[1] Agency for Meteorology, Climatology and Geophysics (BMKG), Jakarta 10720, Indonesia
[2] Center for Earthquake Science and Technology (CEST), Research Center for Disaster Mitigation, Institut Teknologi Bandung, Bandung 40132, Indonesia
[3] Global Geophysics Research Group, Faculty of Mining and Petroleum Engineering, Institut Teknologi Bandung, Bandung 40132, Indonesia
[4] Faculty of Engineering, Maranatha Christian University, Bandung 40164, Indonesia
[5] Geodynamic and Sedimentology Research Group, Faculty of Earth Sciences and Technology, Institut Teknologi Bandung, Bandung 40132, Indonesia
[6] Physical Science and Engineering, King Abdullah University of Science and Technology, Thuwal, Saudi Arabia

E-mail: pepen_geophysics@yahoo.com

Abstract. On September 26, 2019, an earthquake (Mw 6.5) occurred in the northern part of Ambon Island, Molucca, East Indonesia. The National Authority show this earthquake caused infrastructures damage and 28 fatalities. Up to October 31, 2019, the Indonesian Agency for Meteorology, Climatology, and Geophysics (BMKG) seismic network had recorded 479 aftershock events. We have relocated 463 out of the 479 aftershocks using the double-difference method. Our results show that some improvements in the hypocenter locations, where the focal depths of initial earthquakes fixed at 10 km have been updated to ~9.8 km depth on average. There are two aftershock clusters in a North-South direction of ~35 km length between Ambon Island and Haruku Island and in a West-East direction of ~30 km length in the Ambon Island, each with a width of ~8 km.

1. Introduction

Ambon Island, north of the Banda Arc, eastern Indonesia, was shaken by an earthquake that occurred on September 26, 2019. According to the BMKG report, this earthquake occurred at 23:46:45 UTC with a magnitude (Mw) 6.5, the epicenter is located at 3.38° S; 128.43° E with a depth of 10 km. The focal mechanism solution of the mainshock shows a strike-slip fault (Figure 1). The analysis of the BMKG shakemap indicate that this earthquake was moderate to strong level (V to VI) on the Modified Mercalli Intensity (MMI) scale in Ambon Island and its vicinity. The authority reported that there were at least 28 fatalities, 150 people were seriously injured, and 240,000 people were evacuated. About minutes after the mainshock, a sequence of aftershocks hit this area for a few weeks.

The aim of this research is to relocate the location of aftershocks to explain the relationship between the events and the existing fault in the research area. Based on historical seismicity, this area had been hit by several major events in the past, e.g., the 1674 tsunami earthquake (Mw 8.1) on the northern part of Ambon [1] and the 1950 (Mw 7.3) earthquake southern part of Ambon Island [2].

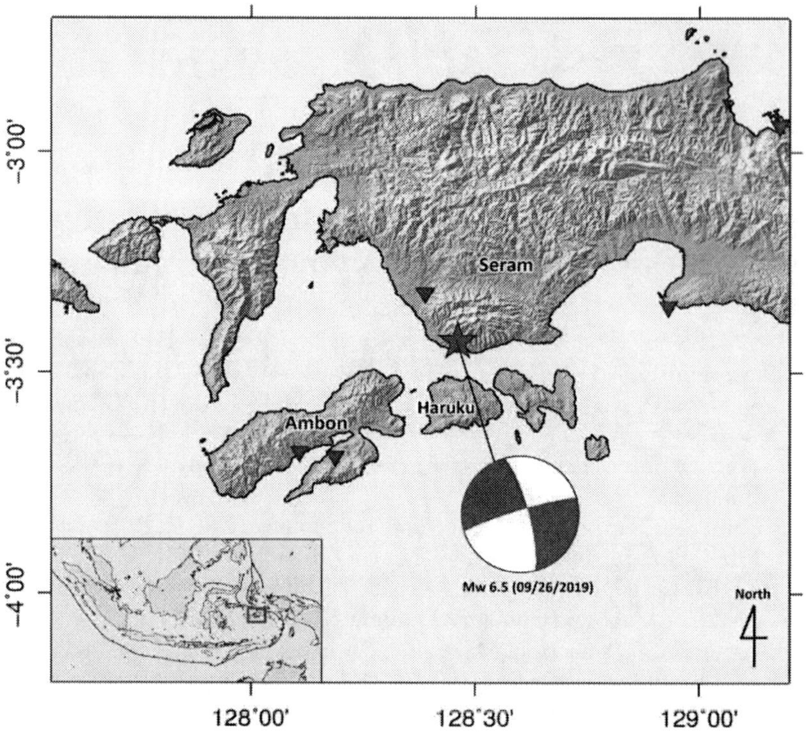

Figure 1. Map view of the research area. Inset depicts the Indonesian region. The mainshock (red star) is taken from BMKG and the focal mechanism solution extracted from the Global Centroid Moment Tensor (GCMT). Blue inverted triangles depict the BMKG seismic stations.

2. Data and Method

We used the aftershocks arrival-time data from the BMKG earthquake catalog for 36 days (26 September to 31 October 2019). During the time, there were 479 events with depths of less than 30 km for local magnitude (M_L) 1.5 to 5.0 with 3897 and1482 of P-and S-phase arrival times, respectively. The BMKG hypocenter locations generated by the SeisComP3 Program (GFZ) used the IASP91 velocity model [3]. For relocating the hypocenter of the aftershocks, we used the HypoDD program [4] which utilizes the Double-Difference method [5]. The method assumes that ray paths of two or more hypocenters to the station are considered to be similar and propagate passing the same medium if the hypocenters distance pair is smaller than the hypocenters and station distance. Residuals between observed and calculated travel-time differences were minimized by HypoDD in an iterative procedure. We used the AK135 velocity model [6] for this research. To estimate the relocation uncertainty, we performed a Bootstrap method [7, 8].

3. Results and Discussion

Totally, 463 out of 479 aftershocks have been relocated (Figure 2). Our relocation result shows there are patterns in N-S and E-W headings of ~ 35 km and ~ 30 km lengths, respectively, and each with a width of ~ 8 km. Average horizontal and vertical of hypocenter uncertainty are 0.9 and 1.04 km (Figure 3). Based on the previous study [9], the fault plane of the mainshock is in the N-S direction. Our interpretation that the second trend of the aftershock in the S-W direction probably was related to triggered seismicity of the active fault by the mainshock.

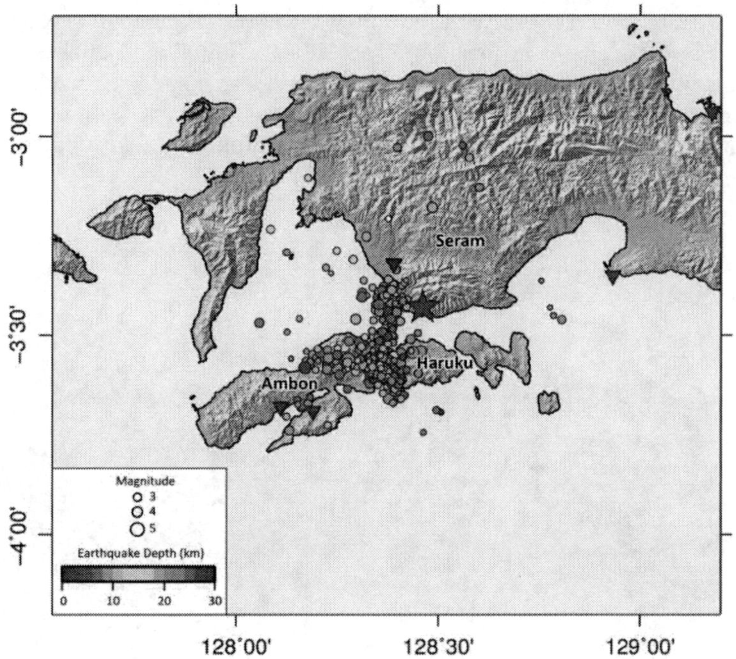

Figure 2. Map view for relocated aftershocks depicted in red to green dots. The red star is the epicenter of the mainshock from BMKG. Blue inverted triangles represent the BMKG seismic stations.

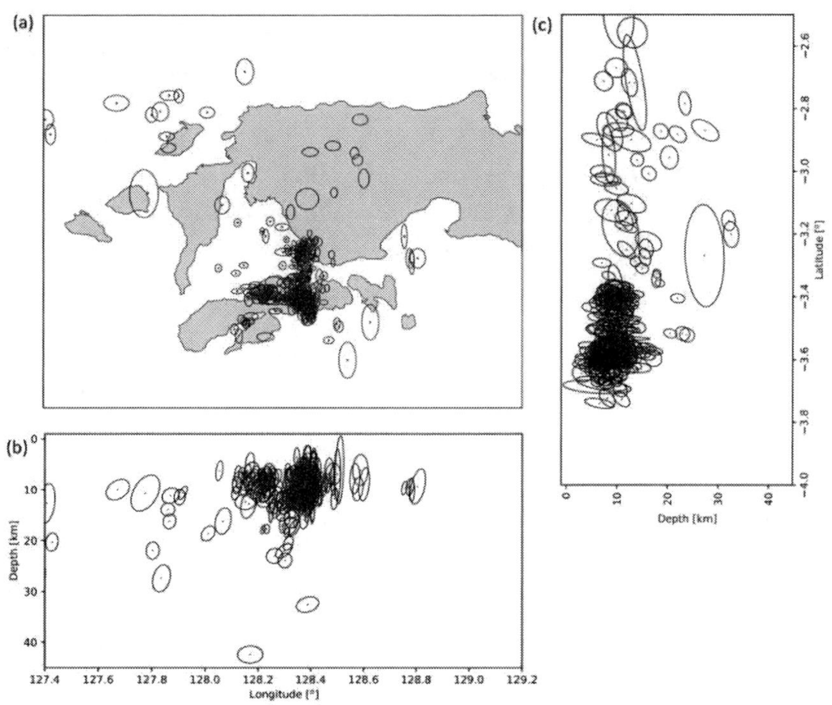

Figure 3. (a) Map view of relative location uncertainties for the 463 aftershocks; (b) depth view along longitude; and (c) depth view along latitude.

We compare the relocated aftershocks with the initial locations (Figure 4). Hypocenter relocation of aftershocks indicates improvement in location, especially the initial earthquakes at fixed depth have been updated. Generally, the focal depths of relocated aftershocks in the N-S trending is less than 10 km on average. Tomography study in the area [10] also shows a significant feature at 10 km depth with a contrast velocity underlying the area between Ambon and Haruku Islands, suggesting the existence of an active fault line.

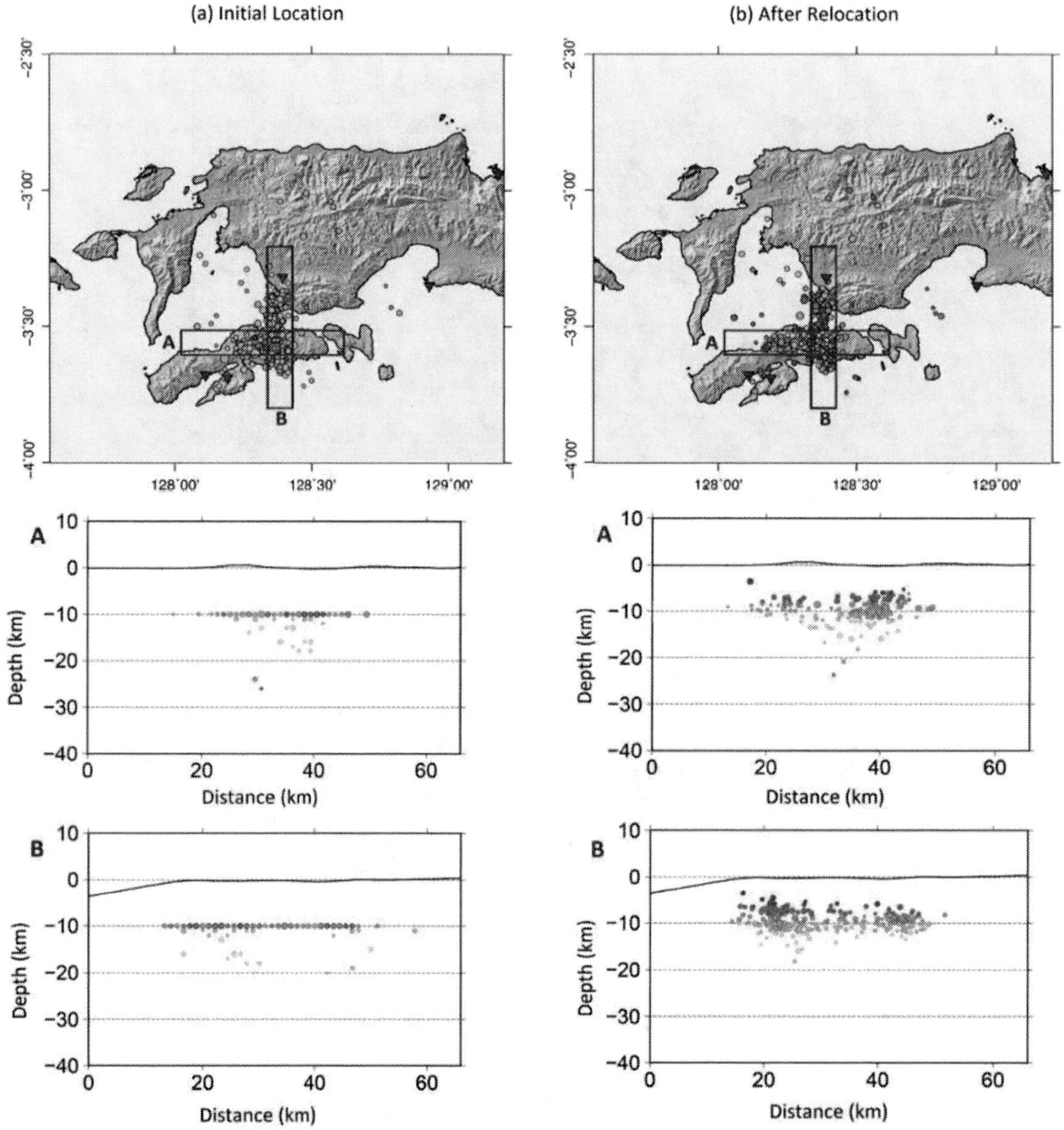

Figure 4. Distribution of the aftershocks and vertical cross-sections for (a) BMKG locations, and (b) hypoDD relocation, 463 events repectively. The red to the green dots depict the epicenter of earthquakes based on depth in km.

4. Concluding Remarks

We relocated 463 aftershocks of the Ambon earthquake from 26 September to 31 October 2019. The initial locations fixed at 10 km depth have been updated. There are two aftershock clusters in North-South and West-East directions. The N-S trend is the main fault plane located between Ambon Island and Haruku Island, whereas the S-W trend is probably related to triggered seismicity of another fault plane in Ambon Island.

Acknowledgements

We thank BMKG for the earthquake data. We made all figures using the Generic Mapping Tools [11].

References

[1]. Løvholt F, Ku¨hn D, Bungum H, Harbitz C B, and Glimsdal S 2012 Historical tsunamis and present tsunami hazard in eastern Indonesia and the southern Philippines *J. Geophys. Res. Solid Earth* **117** (B9) B09310 https://doi.org/10.1029/2012JB009425.

[2]. Latief H, Kodijat A, Ismoyo D, Bustamam B, Adyasar D, Nurbandika N, and Rahayu H 2016 Air turun naik di tiga negeri *United Nations Educational, Scientific, and Cultural Organization*, Office Jakarta - Indian Ocean Tsunami Information Centre (in Indonesian).

[3]. Kennett B L N and Engdahl E R 1991 Travel times for global earthquake location and phase association *Geophys. J. Int.* **105** 429-465.

[4]. Waldhauser F 2001 HypoDD – A Program to Compute Double-Difference Hypocenter Locations *United States Geological Survey.*

[5]. Waldhauser F and Ellsworth W L 2000 A double-difference earthquake location algorithm: Method and application to the Northern Hayward fault California *Bull. Seismol. Soc. Am.* **90** 1353–1368.

[6]. Kennett B L N, Engdahl E R, and Buland R 1995 Constraints on seismic velocities in the Earth from traveltimes *Geophys J. Int.*, 122, 108-124.

[7]. Efron B 1982 The Jackknife, the Bootstrap and other Resampling Plans *J. Soc. Ind. Appl. Math.*

[8]. Billings S D 1994 Simulated annealing for earthquake location *Geophys. J. Int.* **118** 680–692.

[9]. Sahara D P, Nugraha A D, Muhari A, Rusdin A A, Rosalia S, Priyono A, Zulfakriza Z, Widiyantoro S, Puspito N T, Lesmana A, Kusumawati D, Ardianto, Baskara A W, Halauwet Y, Shiddiqi H A, Pradisti R, Mozef P W, Tuakia M Z, Elly E 2021 Analysis of the Source Mechanism of the Mw 6.5 Ambon, Indonesia Earthquake and Its Impact to the Surrounding Fault Network Using Local Seismic Network Data *Tectonophysics* **779** 228709.

[10]. Nugraha A D, Supendi P, Rosalia S, Sahara D P, Muhari A, Priyono A, Zulfakriza Z, Widiyantoro S, Puspito N T, Harisandi A A 2020 Local Earthquake Tomography to Investigate the Fault Geometry Causing the Mw 6.5 Ambon, Indonesia, 26 September 2019, *J. Geophys. Res. Lett.* (submitted).

[11]. Wessel P and Smith W H F 1998 New, Improved Version of Generic Mapping Tools Released. EOS Trans AGU **79** (47) p. 579.

IOP Conf. Series: Earth and Environmental Science **873** (2021) 012002 doi:10.1088/1755-1315/873/1/012002

Earthquake Swarm Analysis around Mt. Salak, West Java, Indonesia, Using BMKG Data from August 10 to November 24, 2019

Pepen Supendi[1,2], Nanang T. Puspito[2,3], Andri Dian Nugraha[2,3], Sri Widiyantoro[2,3,4], Chalid Idham Abdullah[5], Daryono[1], Dwikorita Karnawati[1], Supriyanto Rohadi[1], Zulfakriza[2,3], David P. Sahara[2,3]

[1] Agency for Meteorology, Climatology and Geophysics (BMKG), Jakarta, Indonesia
[2] Center for Earthquake Science and Technology (CEST), Research Center for Disaster Mitigation, Institut Teknologi Bandung, Bandung 40132, Indonesia
[3] Global Geophysics Research Group, Faculty of Mining and Petroleum Engineering, Institut Teknologi Bandung, Bandung 40132, Indonesia
[4] Faculty of Engineering, Maranatha Christian University, Bandung 40164, Indonesia
[5] Geodynamic and Sedimentology Research Group, Faculty of Earth Sciences and Technology, Institut Teknologi Bandung, Bandung 40132, Indonesia

pepen_geophysics@yahoo.com

Abstract. Earthquake swarms commonly come approximately active tectonic and volcanic area. Interestingly, the swarm events occurred ~23 km southwest from Mt. Salak-Bogor, West Java, Indonesia, from August 10 to November 24, 2019, and were recorded by local/regional network of the Indonesian Agency for Meteorology, Climatology, and Geophysics (BMKG). Our previous study showed that in this area a destructive M_L 4.6 earthquake with thrust faulting occurred on September 8, 2012. The double-difference method was applied to update the hypocenter locations from the BMKG data. In the time period of ~3.5 months, we relocated 79 swarm events with ~9.4 km depth average for local magnitude (M_L) 2.2 to 4.2. The source mechanism result for selected events shows a strike-slip faulting. Our interpretation is that these swarm events are probably related to stress change due to volcano-tectonic activity.

1. Introduction

On August 10, 2019, there was a swarm activity around 23 km in the SW direction from Mt. Salak-Bogor, West Java. Up to November 24, 2019, BMKG recorded 92 swarm earthquakes in this area. Several events were weak to light shaking and caused slight damage to several houses in Nanggung sub-district, Bogor. Previous studies show that West Java has a high tectonic activity due to active faults on land, that consists of Baribis, Lembang, Cimandiri, and Garsela Faults [1] that are adjacent to high population density area. Interestingly, these swarm events are not located on these faults. The seismic sequences types have been classified by Mogi [2] into three types, i.e., foreshocks-mainshock-aftershocks; mainshock-aftershock; and swarm. The swarm earthquake is characterized by the continuous seismic activity with a relatively small magnitude without any major earthquakes. Some

IOP Conf. Series: Earth and Environmental Science **873** (2021) 012002 doi:10.1088/1755-1315/873/1/012002

areas in Indonesia that have occurred of the earthquake swarms i.e., in Madiun, East Java [3]; Jailolo, around Bekancan, North Sumatra [4]; and West Halmahera, North Maluku [5]. This research aims to obtain a more precise hypocenter location and to analyze the source mechanism of the earthquakes swarm.

2. Data and Method

We used the BMKG arrival-times data from 10 August to 24 November 2019. There were 92 events, consist of 654 P-phase and 592 S-phase arrival times; with focal depths of less than 30 km and local magnitude (M_L) 2.2 to 4.2. We applied Double-Difference method [6] to relocate hypocenters using the HypoDD program [7]. We processed the waveform data with frequency 0.04 Hz to 0.09 Hz from BMKG seismic stations in Western Java and its vicinity to conduct moment tensor inversion by using ISOLA package [8]. We utilized the AK135 velocity model [9] in the research.

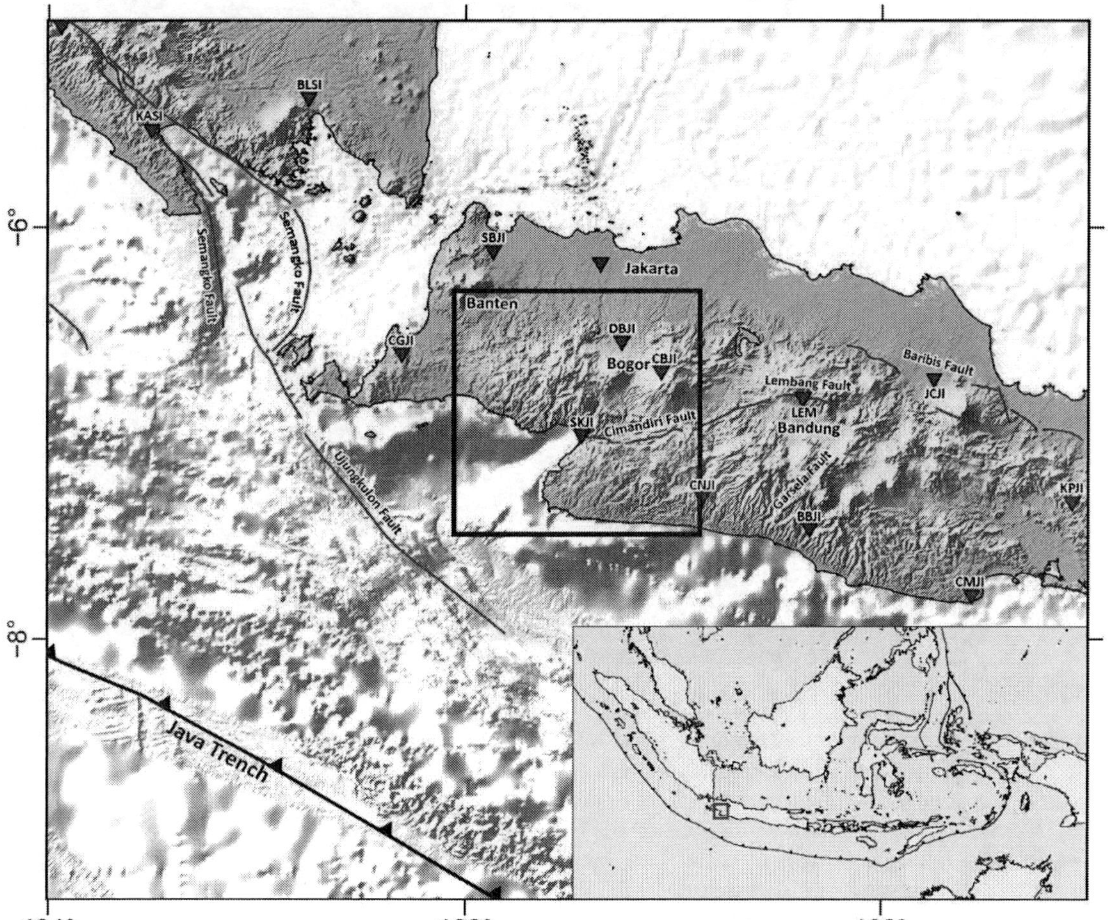

Figure 1. Map view of the research area (inside the black box), inset shows the Indonesian region. Red lines depict crustal faults extracted from Irsyam et al. [1]. The BMKG seismic stations is depicted by the blue inverted triangles.

3. Results and Discussion

Generally, 73 out of 92 relocated swarm earthquakes around southwest of Mt. Salak, Bogor (Figure 2) shows there are two clusters (dashed blue circles in Figure 2) that probably related to different sources, namely cluster 1 is probably caused by volcano-tectonic activity in the southwest of Mt. Salak. The M_L 4.6 Bogor crustal earthquake (September 8, 2012) with thrust faulting was occurred in the cluster 1 area and caused buildings damage [10].

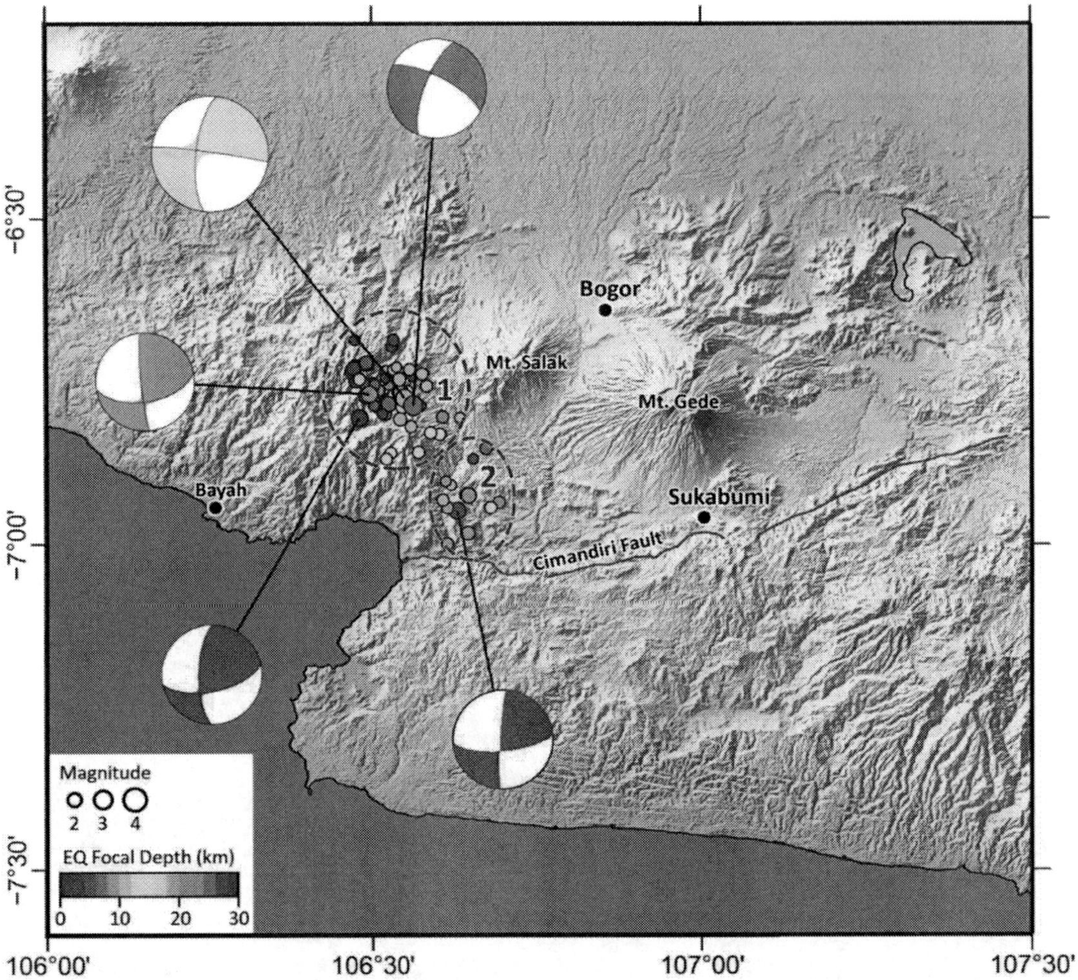

Figure 2. Relocated swarm earthquakes (red to green circles) and the focal mechanism solutions for selected events. Dashed blue circles represent two clusters of earthquakes in the research area. Red traces correspond to the major crustal faults in the region extracted from [1].

Based on the geological map by Effendi et al. [11], the Bogor zone is composed of volcanic and intrusive rocks. Subsurface structure of Mt. Salak and its surrounding in West Java as part of a modern volcanic arc is dominated by heat source beneath the surface manifested as hot springs in the area. Stress changes due to the increase of hot material probably generated to activate a local active fault in cluster 1. On the other hand, cluster 2 is probably caused by another active fault (Citarik fault zone?) close to the Cimandiri Fault. Our vertical cross-sections show that the fixed depth has been revised (Figure 3). Generally, the focal depth in each cluster is less than 20 km, on average 9.4 km depth. The focal mechanism solution for the selected event ($M_L > 3.5$) shows a strike-slip fault type.

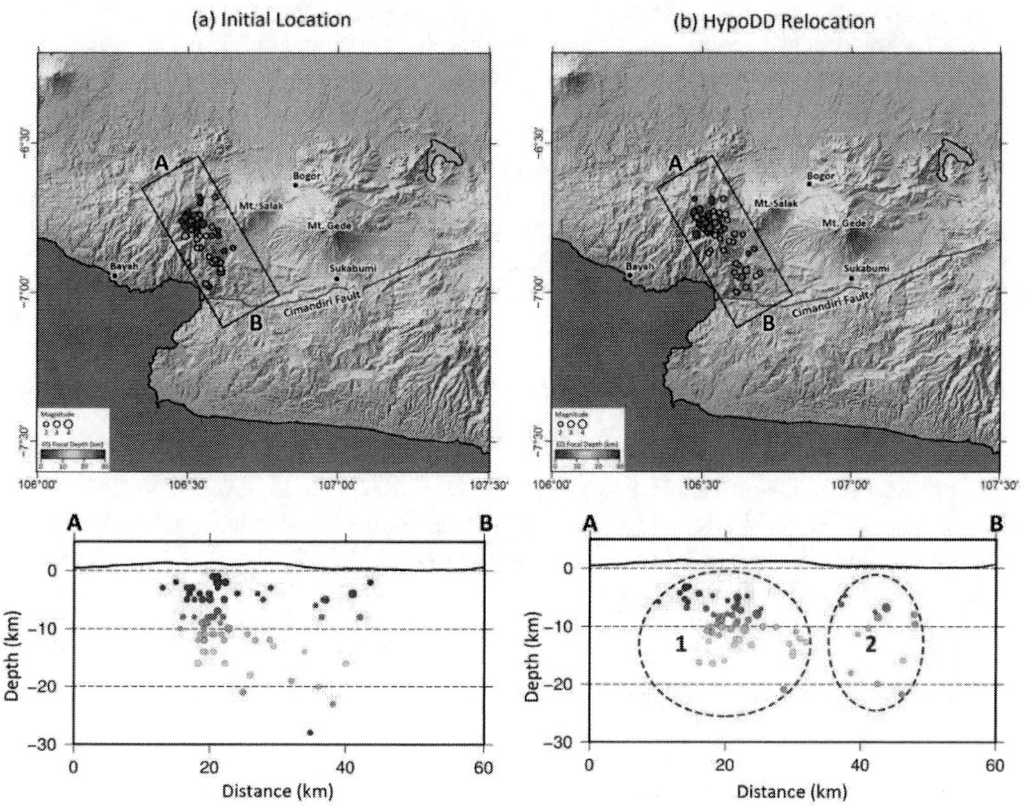

Figure 3. Map view and cross-section of the swarm earthquakes (red to green dots) around Mt. Salak-Bogor. (a) Initial locations from BMKG catalog. (b) After relocation; 73 events respectively.

4. Concluding Remarks

Hypocenter relocation of swarm earthquakes indicates an improvement in hypocentre locations, Initial earthquakes at a fixed depth have been updated. There are swarm earthquakes around and beneath the Bogor area, which are probably caused by both tectonic (local fault) and volcanic activities.

Acknowledgements

We are grateful to BMKG for the earthquake catalog data. We have used the Generic Mapping Tools [13] to present all figures.

References

[1]. Irsyam M, Cummins P R, Asrurifak M, Faizal L, Natawidjaja D H, Widiyantoro S, Meilano I, Triyoso W, Rudiyanto A, Hidayati S, Ridwan M, Hanifa N R and Syahbana A J 2020 Development of the 2017 national seismic hazard maps of Indonesia *Earthquake Spectra* 875529302095120

[2]. Mogi K 1963 Some discussion on aftershocks, foreshocks and earthquake swarms – the fracture of semi-infinite body caused by an inner stress origin and its relation to the earthquake phenomena *Bull Earthquake Res. Inst.* 41:615–658

[3]. Santoso D, Wahyudi E J, Kadir W G A, Alawiyah S, Nugraha A D, Supendi P and Parnadi W W 2018 Gravity Structure around Mt. Pandan, Madiun, East Java, Indonesia and Its Relationship to 2016 Seismic Activity *Open Geosciences* **10** 882–8

[4]. Nugraha A D, Supendi P, Widiyantoro S, Daryono and Wiyono S 2018 Earthquake swarm analysis around Bekancan area, North Sumatra, Indonesia using the BMKG network data: Time periods of February 29, 2015 to July 10, 2017 *IOP Conference Proceedings* p 020092

[5]. Nugraha A D, Supendi P, Widiyantoro S, Daryono and Wiyono S 2018 Hypocenter relocation of earthquake swarm around Jailolo volcano, North Molucca, Indonesia using the BMKG network data: Time periods of September 27-October 10, *IOP Conference Proceedings* p 020093

[6]. Waldhauser F and Ellsworth W L 2000 A double-difference earthquake location algorithm: Method and application to the Northern Hayward fault California *Bull. Seismol. Soc. Am.* **90** 1353–1368.

[7]. Waldhauser F 2001 HypoDD – A Program to Compute Double-Difference Hypocenter Locations *United States Geological Survey.*

[8]. Sokos E N and Zahradnik J 2008 ISOLA a Fortran code and a Matlab GUI to perform multiple-point source inversion of seismic data *Computers & Geosciences* **34** 967–977

[9]. Kennett B L N, Engdahl E R and Buland R 1995 Constraints on seismic velocities in the Earth from traveltimes *Geophysical Journal International* **122** 108–124

[10].Supendi P, Nugraha A D, Puspito N T, Widiyantoro S and Daryono D 2018 Identification of active faults in West Java, Indonesia, based on earthquake hypocenter determination, relocation, and focal mechanism analysis *Geoscience Letters* **5,** https://doi.org/10.1186/s40562-018-0130-y

[11].Effendi AC, Kusuma, and Hermanto B 1998 Peta geologi lembar Bogor, Jawa. Pusat Penelitian dan Pengembangan Badan Geologi (in Indonesian)

[12].Lomax A and Michelini A 2009 Mwpd: A duration–amplitude procedure for rapid determination of earthquake magnitude and tsunamigenic potential from P waveforms *Geophysical Journal International* **176** 200–214

[13].Wessel P and Smith W H F 1998 New, improved version of generic mapping tools released *Eos, Transactions American Geophysical Union* **79** 579–579

3D Seismic Waves Velocity Structure of West Sumatera Seismic Waves Using Earthquake Data from January 2010-December 2017

Rizki Wulandari[1], Tedi Yudistira[2], Erlangga Ibrahim Fattah[1], Atin Nur Aulia[3] and Adhi Wibowo[4]

[1] Geophysical Engineering, Department of Manufacturing and Geo-Engineering, Sumatra Institute of Technology, Way Huwi, South Lampung, Lampung 35365
[2] Geophysical Engineering, Faculty of Mining and Petroleum Engineering, Bandung Institute of Technology, Jalan Tamansari 64, Bandung, West Java 40116
[3] Geotechnology Research Center, Indonesian Institute of Sciences, Jalan Sangkuriang, Kompleks LIPI, Bandung, West Java 40136
[4] Meteorology, Climatology and Geophysics Agency, Kotabumi North Lampung Geophysics Station, Jalan Raden Intan No. 219, Kotaalam, Kotabumi Selatan - North Lampung 34519

E-mail: rizki.12116060@student.itera.ac.id

Abstract. This study analyses the P wave velocity (Vp) and Vp/Vs based on 3D seismic tomography in West Sumatra. Seismic tomography is a method of reconstructing the image of the subsurface structure of the earth using travel time data. This research uses secondary data obtained from the Incorporated Research Institutions for Seismology (IRIS) from January 2010 to December 2017. The data obtained were 472 earthquake events and 21 seismic stations that recorded the earthquake events. This research consists of are the hypocenter relocation, which will simultaneously renew the 1D velocity model using VELEST software, and tomographic inversion using SIMULPS12 software. The minimum anomaly of Vp/Vs value is around 1.39, while the maximum anomaly Vp/Vs value is around 2.05. The Vp distribution results have low and high Vp/Vs values around Sumatra Fault Zone and Mentawai Fault Zone. Anomaly results from the tomogram of these areas have an association with saturated pressure sedimentary areas and the presence of fractures which will further contribute to earthquake events.

Keywords: seismic tomography, Vp, Vp/Vs

1. Introduction

Indonesia is an archipelago located on the confluence of four plates, namely Eurasian, Indo-Australian, Pacific, and the Philippine Sea[1]. The convergence of these plates resulted in the formation of complex tectonic arrangements, such as the formation of archipelago arcs with oceanic trenches, non-volcanic arcs composed of accretionary prisms, volcanic arcs, and back-arc basins[2]. The friction between the plates causes the release of energy, where this energy causes an elastic wave effect so that vibrations can be felt on the surface or are called earthquakes.

IOP Conf. Series: Earth and Environmental Science **873** (2021) 012003 doi:10.1088/1755-1315/873/1/012003

One of the regions near the plate convergence zone has a complex geological structure in West Sumatra. This area has a high potential for earthquake hazards. Apart from the subduction area between the Indo-Australian and Eurasian tectonic plates, there are two active faults: the Mentawai Fault Zone (MFZ) and the Sumatra Fault Zone (SFZ). This is evidenced by several earthquakes that damaged this area, including the 2007 Padang Panjang earthquake with Mw 6.4 and 6.3, the Pariaman earthquake in 2009 Mw 7.9, and the Pagai Selatan earthquake in the Mentawai Islands in 2010 with Mw 7,2.

These events caused damage and casualties. Optimal mitigation efforts need to be done to minimize the impact of the disaster that occurs. Efforts can be made to identify active seismic zones by using the imaging method of subsurface structures. One of these methods is seismic tomography, and this method reconstructs the subsurface structure of the earth using travel time data. In theory, the seismic tomography will present an image of the earth's subsurface in the velocity domain[3].

2. Data and Methodology

The data in this study was used from the Incorporated Research Institutions for Seismology (IRIS) from January 2010 to December 2017. The data obtained is an earthquake catalog which contains information about the time of the earthquake, time of origin, arrival time of P and S waves, earthquake hypocenter earthquake (latitude, longitude, and depth), 1D seismic wave velocity model, and observation station data.

The data used for the computation of tomographic inversion were 295 events with 21 recording stations. The data used are earthquake event data with a minimum number of recording stations of 4 stations and azimuth GAP value <220°. Furthermore, model parameterization is carried out by determining the grid with an area of 1000 km×1000 km and a depth of 70 km, the model was built with dimensions of 11×11×7 blocks, as shown in Figure 1. The initial velocity model used in the tomographic inversion process is shown in Table 1. The distribution of the seismic station density, the position of the epicenter, and the target area of the study are the primary consideration factor determining the grid size in the area.

Table 1. The initial velocity model[4]

Depth (km)	Velocity (km/s)
0	5.53
20	6.86
27.5	7.74
35	8.04
50	8.15
70	8.23

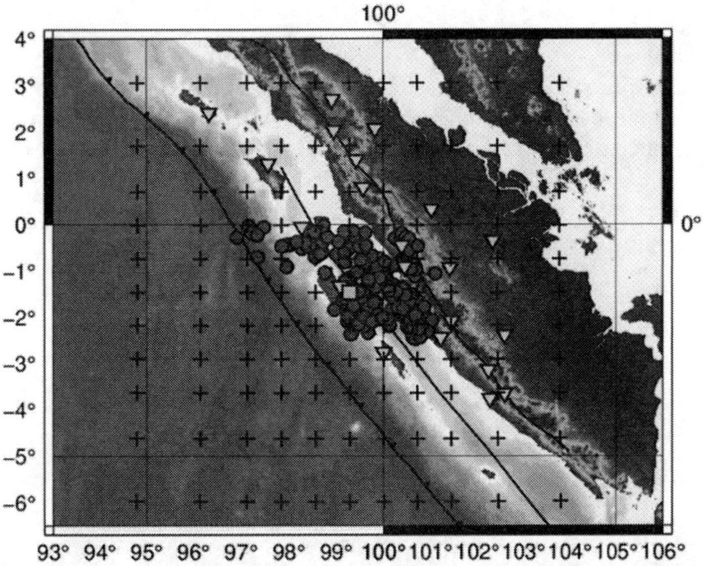

Figure 1. Grid configuration used for travel time seismic tomography calculations with epicenter distribution and earthquake recording stations. The epicenter used is the result of hypocenter relocation using the Coupled Velocity Hypocenter method[5]

In this study, a damping test was carried out using a trade-off curve analysis to obtain the optimum damping value. In this study, Vp's optimum damping value was 40 shown in Figure 2.a, and Vp/Vs was 85 in Figure 2.b.

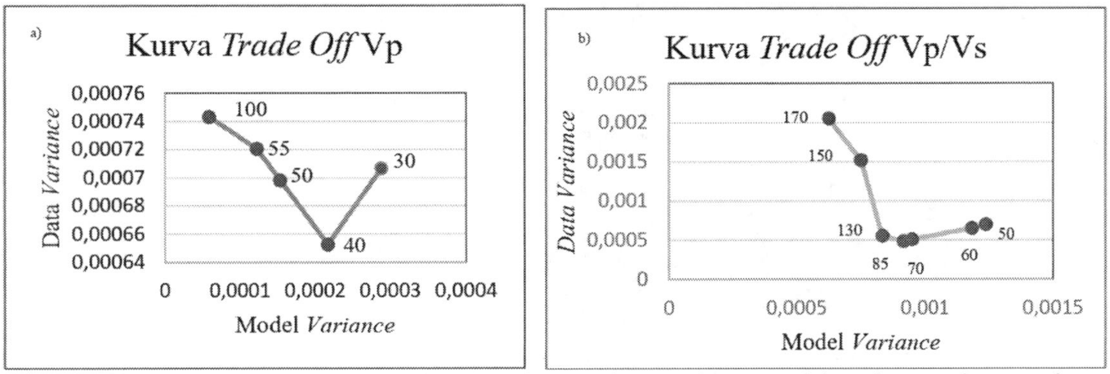

Figure 2. The trade-off curve was showing data variance with a model variance where optimum damping a). Vp 40 and b). Vp/Vs 85

After the tomographic inversion stage, a resolution test is carried out, aiming to see the resulting resolution, and is carried out before the interpretation of the tomogram. In this study, two resolution testing methods were carried out, namely the Checkboard Resolution Test (CRT) and the Derivative Weight Sum (DWS). The CRT model is tested using the forward modeling method. If the results obtained from the CRT process are the same or close to the initial model, the tomographic inversion results have a good resolution. Furthermore, DWS helps test the resolution of the tomographic results based on each node's ray distribution.

IOP Conf. Series: Earth and Environmental Science **873** (2021) 012003 doi:10.1088/1755-1315/873/1/012003

3. Results and Discussion

3.1 Hypocenter Relocation

As a validation of the hypocenter relocation study, a comparison between the residual values' frequency before and after the hypocenter relocation was made in Figure 3. This validation was carried out to test the relocation results' accuracy, and the relocation results were said to be good if the residual time value was close to zero. The residual RMS value obtained after relocating was smaller than before the relocation. This is because the velocity model used is a 3D velocity model so that the relocation results obtained are better as shown in Figure 4.

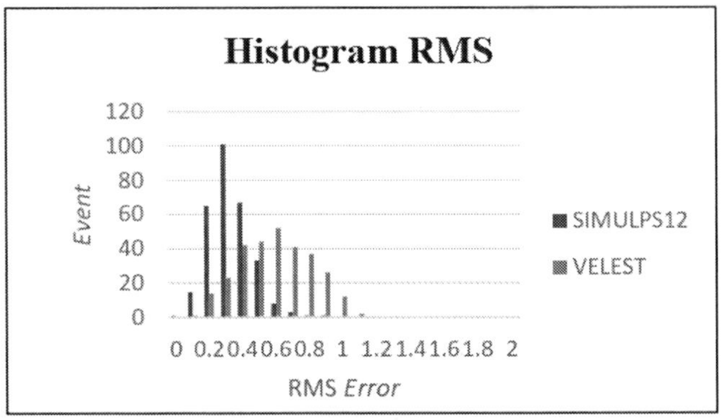

Figure 3. Histogram RMS error, where orange color is the result of VELEST software and blue color is the result of relocation with SIMULPS12 software

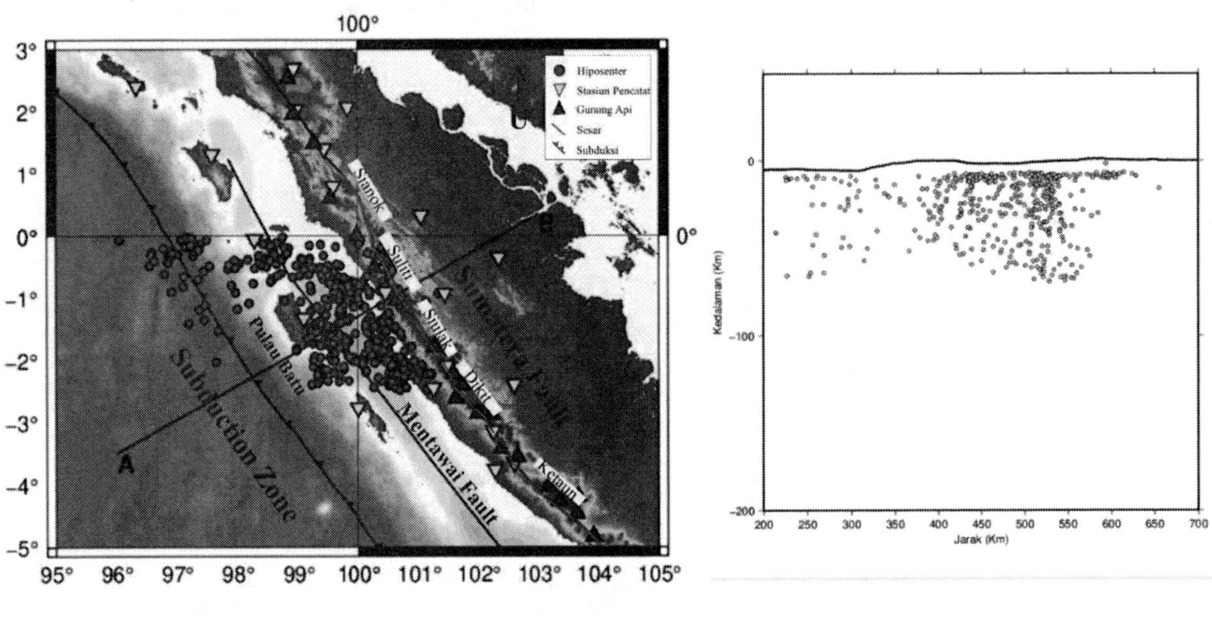

Figure 4. The result after relocation using SIMULPS12

IOP Conf. Series: Earth and Environmental Science **873** (2021) 012003 doi:10.1088/1755-1315/873/1/012003

3.2 Resolution Test and Seismic Tomographic Image

The difference in quality and amount of arrival time data between Vp and Vs will result in different artifacts on each model. To avoid this, Vp is modeled directly from the arrival time of P-wave, but for Vs model is calculated by inversion of Vp/Vs with Ts-Tp data[6]. So the resolution test is done on the inversion results of Vp and Vp/Vs tomography.

As can be seen in Figure 5.a and Figure 6.a are DWS Vp's results, the blocks that have good resolution are those that pass a lot of rays marked with gray blocks. In this study, the minimum DWS Vp score is 0, and the maximum value of 12,665. In this study, an average DWS Vp score of 400 was obtained, so a DWS value greater than 400 is chosen to limit the area where a lot of light passes so that changes are visible, which are marked with dotted black lines. Figure 5.b and Figure 6.b are a resolution test with CRT Vp which shows well-resolution blocks seen in the checkerboard models, namely horizontal slices with a depth of 20, 30, and 50 km marked with a dotted black line. Figure 5.c and Figure 6.c are the result of tomography inversion, high-velocity anomalies are characterized by blue gradations with tomographic results ranging from +5% to +10%, which are relatively high compared to the surroundings. In comparison, low-velocity anomalies are characterized by red gradations ranging from -5% to -10%, which are lower than the surrounding area.

Similar to the P wave, the Vp/Vs also get the results of the resolution test shown in Figure 7.a and Figure 8.a. This study obtained a minimum and a maximum Vp/Vs DWS of 0 and 1,172. The average Vp/Vs DWS value of 125, so a value greater than 125 is chosen to limit the area where a lot of light passes so that changes are visible, which are marked with black dotted lines. Figure 7.b and Figure 8.b are a resolution test result with Vp/Vs CRT showing well-resolution blocks seen in reconstructed areas such as chessboard models, i.e. horizontal slices with depths of 20, 30, and 50 km. There is little difference between resolution test results on Vp and Vp/Vs, where P wave CRT results show better chessboard patterns. This can be due to the data quality and the number of more P wave phases used in the tomography inversion process. Figure 7.c and Figure 8.c are the result of Vp/Vs. The initial Vp/Vs model used in this study was 1.73. In this study, the minimum and maximum Vp/Vs were 1.39 and 2.05.

Dark blue gradations show maximum positive velocity anomalies, while deep red indicates maximum negative velocity anomalies. The positive velocity anomaly suggests that the resulting velocity is greater than that of the initial model velocity in the area and indicates harder and denser rocks in the area. Negative velocity anomalies indicate that the resulting velocity is smaller than the average velocity of the site. The lower P wave velocity value observed in rocks with high porosity and the presence of boulders will also reduce seismic velocity compared to intact rock[5].

IOP Conf. Series: Earth and Environmental Science **873** (2021) 012003 doi:10.1088/1755-1315/873/1/012003

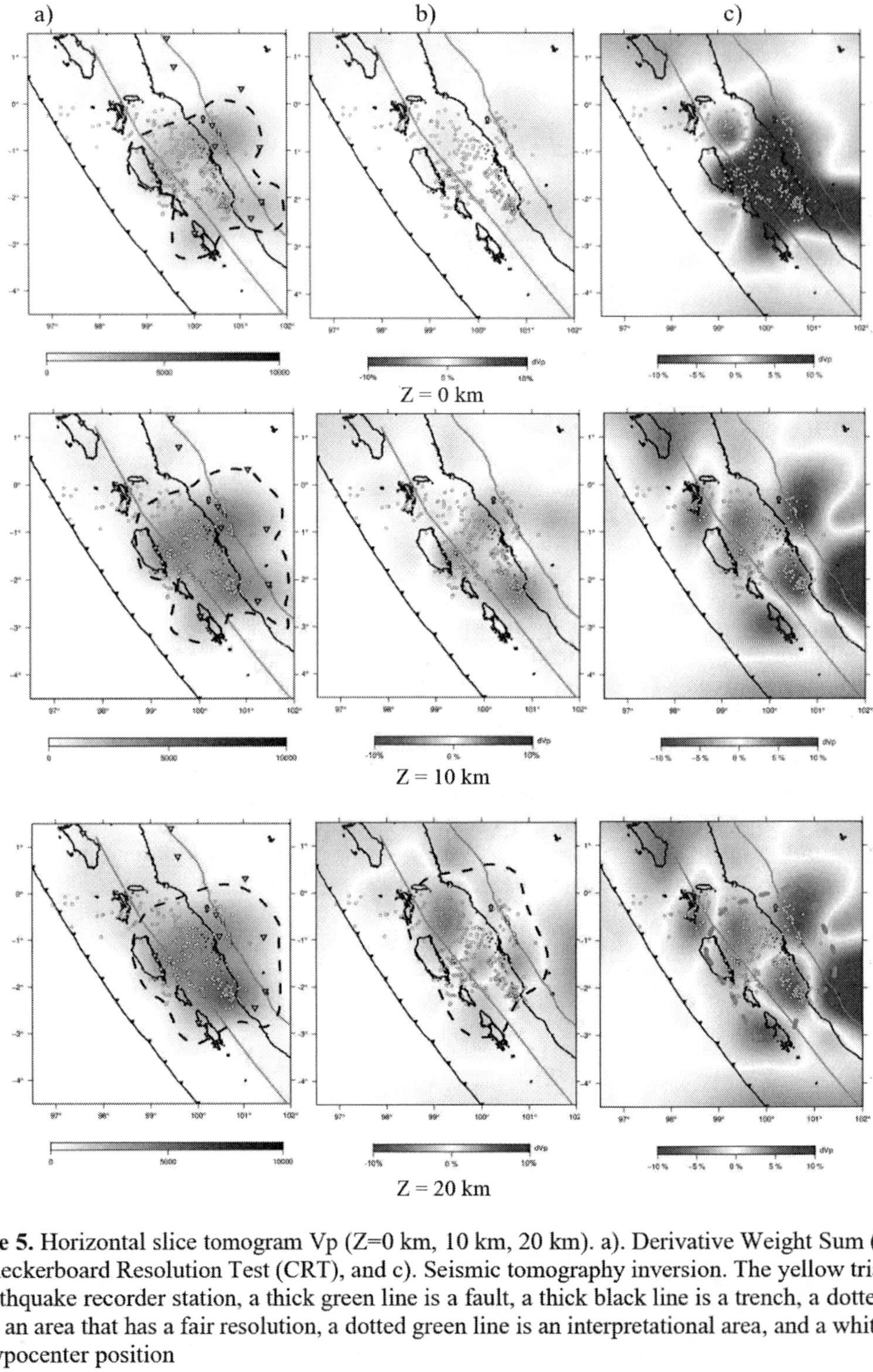

Figure 5. Horizontal slice tomogram Vp (Z=0 km, 10 km, 20 km). a). Derivative Weight Sum (DWS), b). Checkerboard Resolution Test (CRT), and c). Seismic tomography inversion. The yellow triangle is an earthquake recorder station, a thick green line is a fault, a thick black line is a trench, a dotted black line is an area that has a fair resolution, a dotted green line is an interpretational area, and a white circle is a hypocenter position

IOP Conf. Series: Earth and Environmental Science **873** (2021) 012003 doi:10.1088/1755-1315/873/1/012003

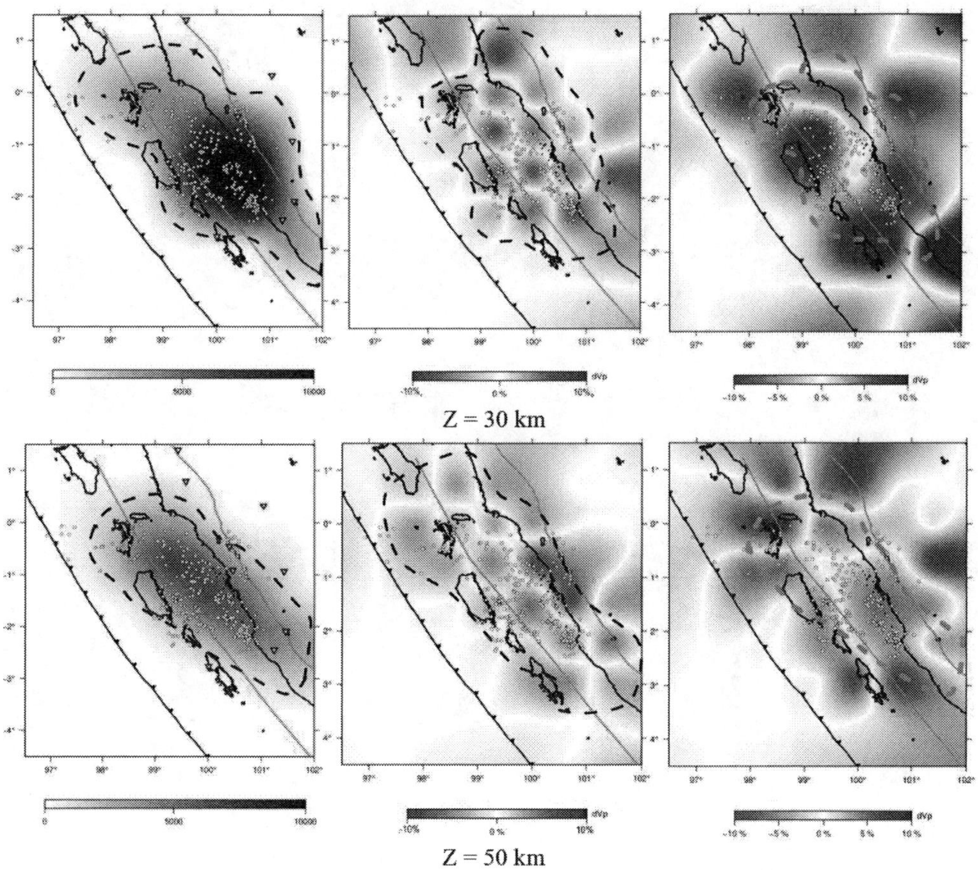

Figure 6. Horizontal slice tomogram Vp (Z=30 km and 50 km). a). Derivative Weight Sum (DWS), b). Checkerboard Resolution Test (CRT), and c). Seismic tomography inversion. The yellow triangle is an earthquake recorder station, a thick green line is a fault, a thick black line is a trench, a dotted black line is an area that has a fair resolution, a dotted green line is an interpretational area, and a white circle is a hypocenter position

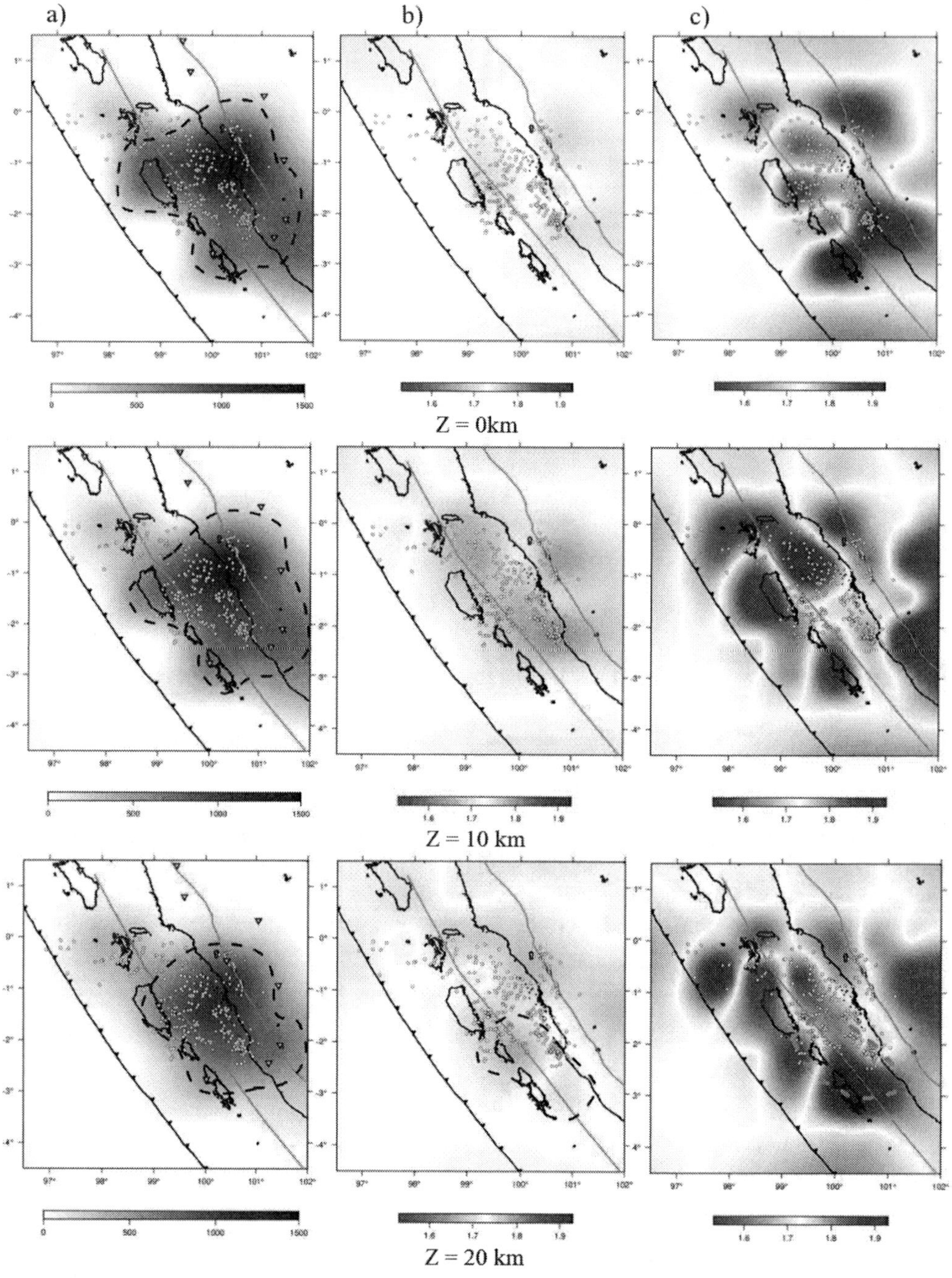

Figure 7. Horizontal slice tomogram Vp/Vs (Z=0 km, 10 km, and 20 km). a). Derivative Weight Sum (DWS), b). Checkerboard Resolution Test (CRT), and c). Seismic tomography inversion. The yellow triangle is an earthquake recorder station, a thick green line is a fault, a thick black line is a trench, a dotted black line is an area that has a fair resolution, a dotted green line is an interpretational area, and a white circle is a hypocenter position

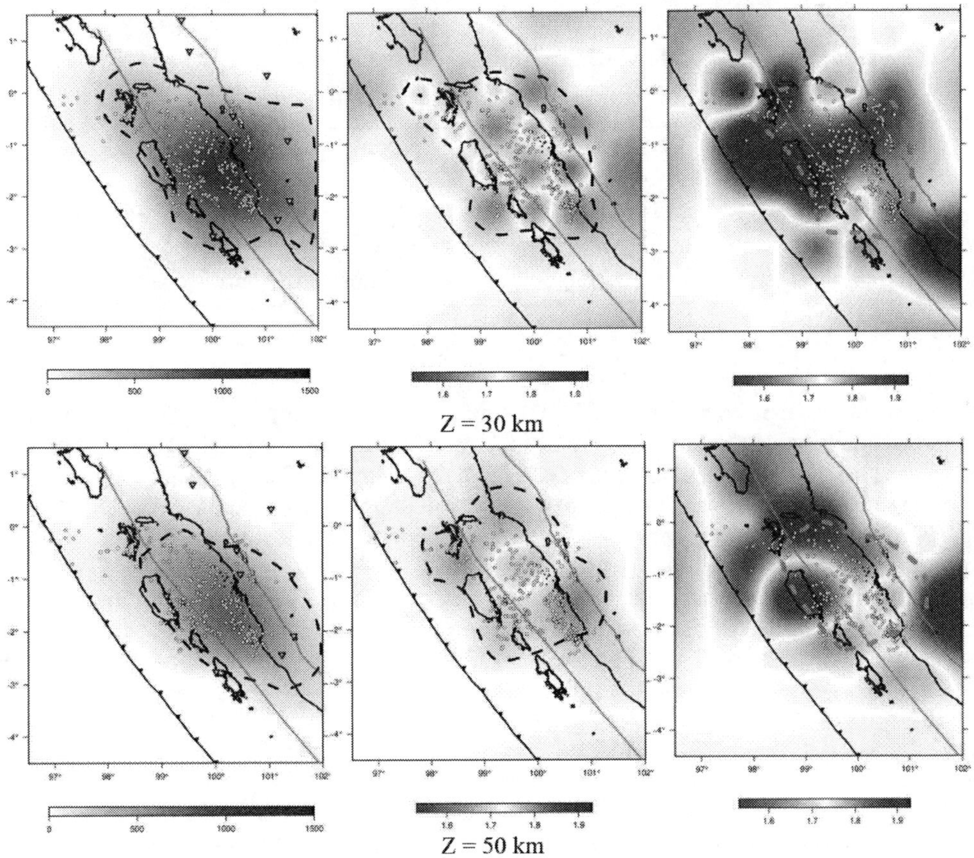

Figure 8. Horizontal slice tomogram Vp/Vs (Z=30 km and 50 km). a). Derivative Weight Sum (DWS), b). Checkerboard Resolution Test (CRT), and c). Seismic tomography inversion. The yellow triangle is an earthquake recorder station, a thick green line is a fault, a thick black line is a trench, a dotted black line is an area that has a fair resolution, a dotted green line is an interpretational area, and a white circle is a hypocenter position

In general, horizontal slices on tomograms with a depth of 20 km are seen that the earthquake event is in the forearc area and Mentawai fault zone. This area appears to have a low P wave velocity anomaly except in the Pulau Batu. Simultaneously, for Vp/Vs tomograms, it is obtained a relatively high Vp/Vs value (1.73–2.05). Also, along the Sumatra Fault Zone area is seen to have a low P wave velocity anomaly characterized by a red anomaly. However, the Vp/Vs tomogram results have a high Vp/Vs value in the Siulak segment, Dikit segment, and Ketaun segment. This can identify that the third area of the segment is an area of high porosity. At the same time, the Sianok segment has a low Vp/Vs value (1.29-1.73). This indicates that this area has hard rocks and is denser than the surrounding area.

At a horizontal slice with a depth of 30 km, the Sumatra Fault Zone area has a low Vp anomaly and high Vp/Vs value. Mentawai Fault Zone has a low Vp anomaly, and the Vp/Vs tomogram result obtained a relatively high Vp/Vs value. Tomogram results at 50 km are seen that in Sumatra Fault Zone area has low Vp anomalies except in Ketaun segment while obtained low Vp/Vs value. There are also differences in Vp and Vp/Vs anomalies in Siberut island areas where Vp is high while the Vp/Vs value is low. Areas in the forearc were observed to have low Vp values and high Vp/Vs values were interpreted as fracture areas or sediment pressure saturated. In general, the research results are shown in Table 2.

Table 2. Catalog of interpretations

Vp	Vp/Vs	Interpretation
Low	High	Presence of damage or saturated pressurized sedimentary areas
Low	Low	The presence of weak areas such as faults or sub-surface structures resulting from tectonic activity
High	Low	The conduit of magma intrusion or hard rock area

4. Conclusion

The result is there are variations of Vp and Vp/Vs anomalies distribution around the Sumatra Fault Zone and Mentawai Fault Zone. The low Vp anomaly and high Vp/Vs value are associated with high porosity and fractures areas. In the horizontal slice (Z=50 km), the Mentawai Fault Zone has a high Vp and low Vp/Vs values are areas harder and denser than the surrounding rocks. Whereas areas that have low Vp and Vp/Vs values are fracturing areas. The areas are associated with saturated pressure will affect the pore pressure, which will contribute to earthquakes.

Some parts of the study area cannot be interpreted because of the lack of earthquake events and uneven station distribution. Vs also have less resolution because the S-phase number is minimal, making Vp/Vs interpretation more difficult. For further study, the S-phase should be re-picked to get more reliable data.

Acknowledgments

Say thanks to the supervisor, Mr. Dr. Tedi Yudistira S.Si., M.Si., Mr. Erlangga Ibrahim Fattah S.Si., M.T., Mrs. Atin Nur Aulia, and Mr. Adhi Wibowo has provided direction and guidance to the author in completing this research. As well as thanks to the Geophysical Engineering lecturers at the Sumatra Institute of Technology.

References

[1] Hall R 2002 Cenozoic geological and plate tectonic evolution of SE Asia and the SW Pacific: Computer-based reconstructions, model and animations *J. Asian Earth Sci.* 20(4) pp 353–431 doi: 10.1016/S1367-9120(01)00069-4.

[2] Hamilton W 1973 Tectonics of the Indonesian Region *Bull. Geol. Soc. Malaysia* 6 pp. 3–10 doi: 10.7186/bgsm06197301.

[3] Darman H, Sidi F H and Indonesia I A G 2000 *An outline of the geology of Indonesia* Jakarta: Indonesian Association of Geologists.

[4] Shelawisata H G, Warnana D D and Nur Rochman J P G 2018 Pemodelan 3D Kecepatan Gelombang P Dengan Metode Double Difference Dan Tomografi Travel Time Untuk Identifikasi Patahan (Studi Kasus Patahan Palu- Koro) *J. Geosaintek* 4(2) pp 51 doi: 10.12962/j25023659.v4i2.4295.

[5] Collings R *et al* 2012 Structure and seismogenic properties of the Mentawai segment of the Sumatra subduction zone revealed by local earthquake traveltime tomography *J. Geophys. Res. Solid Earth* 117(1) pp 1–23 doi: 10.1029/2011JB008469.

[6] Kissling E, Ellsworth W L, Eberhart-Phillips D and Kradolfer U 1994 Initial reference models in local earthquake tomography *J. Geophys. Res.* 99(B10) doi: 10.1029/93jb03138.

IOP Conf. Series: Earth and Environmental Science **873** (2021) 012004 doi:10.1088/1755-1315/873/1/012004

Identifying Subsurface Structures beneath the Sumani and Sianok Segments of the Great Sumatran Fault using Combined Geophysical Methods

Harman Amir[1,2,*], Satria Bijaksana[1], Darharta Dahrin[1], Andri Dian Nugraha[1], Ilham Arisbaya[1,3]

[1]Faculty of Mining and Petroleum Engineering, Institut Teknologi Bandung, Jalan Ganesa 10, Bandung 40132, Indonesia

[2]Faculty of Mathematics and Natural Sciences, Padang State University, Jalan Prof. Dr. Hamka, Padang 25171, Indonesia

[3]Research Center for Geotechnology, Indonesia Institute of Sciences (LIPI), Jalan Sangkuriang, Bandung 40135, Indonesia

*Corresponding author: harman_unp@yahoo.com

Abstract. The 1900 km long Great Sumatran Fault is a complex active fault system that is divided into segments that include the Sumani and Sianok segments in a rather densely populated area of the West Sumatra Province. Major earthquakes have occurred in these two segments that include the March 2007 Sumatra earthquake. Mitigating future risks requires a better understanding of these complex segments. To identify the subsurface structures beneath the Sumani and Sianok segments, we are conducting combined geophysical study that include gravity and magnetic. Gravity data were obtained from the published regional Bouguer anomaly map of the area around these two segments. The measurements from which the map was derived were rather sparse. Thus, more detailed magnetic measurements were carried out in this study. Magnetic measurements were also expected to be more sensitive as the predominant rocks in the study area are volcanic as well as other type of intrusive rocks. These gravity and magnetic analyses were complemented by seismicity data that include relocated seismicity data that will enhance the modelling of subsurface structures. Progress of this study will be reported. Challenges and obstacles will also be presented.

1. Introduction

Located in Indonesia, the Great Sumatran Fault (GSF) is a 1900 km long fault system that is still very active. Studies have shown that this fault system is caused by subduction of Indo-Australian plate beneath the Eurasian plate [1-4]. GSF could be divided into 19 segments [5] and four of these segments (Sumani, Sianok, Siulak and Sumpur) passed through the province of West Sumatera. Two segments, i.e., Sumani and Sianok are particularly intriguing because of their history of damaging earthquakes as well as their proximity to major population centers in the province of West Sumatra [6-8]. Moreover, there were records of damaging doublet earthquakes in the vicinity of Sumani [9-10].

Despite of their geological risks, the structures underneath these two segments are not well understood.

To resolve these questions, an integrated geophysical study was proposed. The study area is located within these coordinates (99°50' to 100°45'East and 0°05' to 0°50'South) covering an area of about 103 km times 85 km. The main methods are passive or potential methods, i.e., magnetic and gravity [11-12]. These two methods are very suitable for resolving the geometry of underneath structures [13-15]. These two methods are complementary as each of them relying on different physical properties: magnetic susceptibility for magnetic method and density for gravity method [16-17]. Earlier, this area was also subjected to qualitative gravity interpretation [18] based only on Bouguer anomaly map produced by *Pusat Survey Geology* (PSG or The Center for Geological Survey) of Indonesia. It is expected that the proposed study will produce more comprehensive results but qualitatively as well as quantitatively.

2. Geological Setting

Regionally, the research area are segments of GSF caused by oblique subduction between Indo-Australia and Eurasia plates causing significant lateral force [1]. Based on combined geology and GPS surveys, this lateral force along GSF creates slip rate of about 14-15 mm/year [19] causing seismically active regions that include the Sumani and Sianok segments of the GSF [20]. These two segments are show in Figure 1.

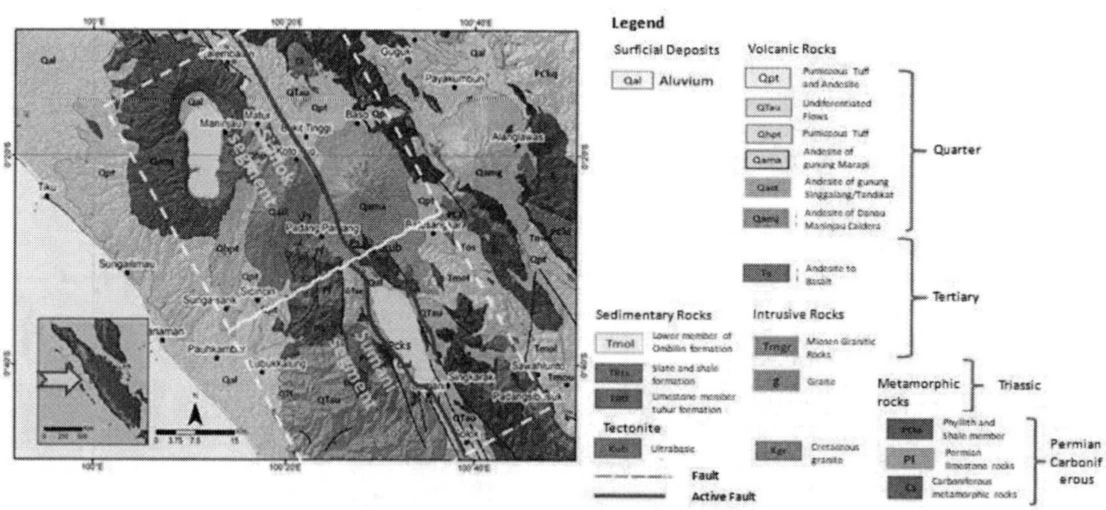

Figure 1. Simplified geological map of Sianok and Sumani segments of the GSF. The active faults are shown in red bold lines (modified from [21-22]).

Based on the regional geological maps [21-22], the research area shown in Figure 1 comprises of various type of rocks ranging from Carboniferous and Permian metamorphic rocks, Tertiary intrusive rocks to quaternary volcanic rocks and sediments. Around Lake Singkarak, Permian metamorphic rocks present along with Cretaceous granite and ultrabasic rocks. Permian rocks also appear in this area as phyllite and shale. Around these two segments, Tertiary rocks appear as andesite to basalt near Lubuk Sikaping, Pasaman while in the foothill of Mt. Singgalang, there are Miocene granitic rocks. Other Tertiary rocks appear as limestone as well as slate and shale.

Younger Quaternary rocks (Holocene to Pleistocene in age) appear in several places that include the Marapi andesite (near Mt. Marapi) and the Singgalang Tandikat andesite (around Mt. Singgalang-Tandikat. Around Lake Maninjau, there are Maninjau andesite as well as pumiceous tuff and andesite. Around Lake Singkarak, there are lahar, conglomerate, and undifferentiated flows. As shown in Figure

1, the predominant rocks around Sianok and Sumani segments are relatively Quaternary volcanic rocks followed by Tertiary and Permian rocks.

3. Methods

The main method in this study is geomagnetic survey carried out using GEM Proton Precision Magnetometer GSM-19T (GEM system, Markham, Canada) in the following cities and districts in the province of West Sumatra (Tanah Datar, Solok, Padang Pariaman, Agam, Pasaman Timur, Padang, Padang Panjang, Batusangkar and Bukittinggi). The measurements were carried out along highways, streets and dirt roads that are accessible either by car or motorcycle. The density of survey points varies between 500 m to 1000 m depending on their distance to the suspected fault lines (based on geological maps). The density is higher in the vicinity of fault lines. The coordinates of survey points were determined using Garmin GPS system.

Field magnetic data were then subjected to various corrections that include diurnal correction and IGRF (International Geomagnetic Reference Field) correction to obtain the magnetic intensity values. To complement magnetic data, gravity data in the form of Bouguer anomaly were used based on the anomaly maps for Solok and Padang regions derived from ground measurements with station spacing of about 5 km [23-24]. The gravity data were digitized so that they could be integrated seamlessly with magnetic data. The magnetic and gravity data were then processed further using GM-SYS Oasis Montaj (Geosoft Inc., Toronto, Canada). The processing includes gridding (with a minimum curvature approach) due to the relatively equal distance of the measurement stations and that is followed by extrapolation processing through adjustment of blanking distance. The values of the emptying distance were set at 5000 because of these values the magnetic anomaly is easier to be interpreted. The data were then filtered using Gaussian Regional/Residual filter. This type of filter was chosen as it produces better magnetic anomaly pattern compared to that of other filters such as Bandpass filter. This filtering process was carried out using the inclination and declination values of respectively -18° and -0.5° representing that of this area.

4. Progress Report

So far, 798 survey points have been measured in Solok, Tanah Datar, Padang Panjang, Agam and Padang Pariaman (see Figure 2). The points are in the area around Sumani which constitute the first phase of this research. Interim results from the measurements of the Sumani segment have been submitted to the journal Tectonophysics. The surveys were carried out in 23 days using multiple magnetometers. The geomagnetic surveys in Sianok segments were initially planned in 2020. However, due to COVID-19 pandemic, the surveys would be postponed to 2021. The survey lines in Sianok segments are 437 km long with about 637 survey points (shown in Figure 3). Like that in Sumani segment the distance between each survey point is about 500-1000 m.

The condition in the research area is not always easy as the survey lines often must follow undulated terrain along the mountainous Bukit Barisan Range. Some of the survey points are located around 850 m above sea level. The roads are also often impassable to car so that often motorcycle must be used to access the survey points. Some points are also located in Lake Singkarak requiring a special modification on sampan (small boats) that are available on the lake shores.

5. Conclusions

Identification of the subsurface structures beneath the Sumani and Sianok Segments of the GSF is currently underway employing magnetic and gravity data. The magnetic surveys have been completed in the Sumani segment but were delayed in the Sianok segment due to COVID-19 pandemic. In total 798 survey points have been measured and 637 points are to be measured as soon as the pandemic is over. Interpretation of the magnetic anomaly maps of Sumani segments is underway.

Acknowledgments
This research is financially supported by the research grants to Darharta Dahrin (PDUPT: *Penelitian Dasar Unggulan Perguruan Tinggi*) and to Satria Bijaksana (PDD: *Penelitian Disertasi Doktor*). The authors thank the Ministry of Education and Culture of the Republic of Indonesia for awarding these grants. Harman Amir is the recipient of doctoral scholarship from Padang State University and is thankful to its Rector. The authors also thank the West Sumatra Provincial Government for its permission to carry out this research in that province.

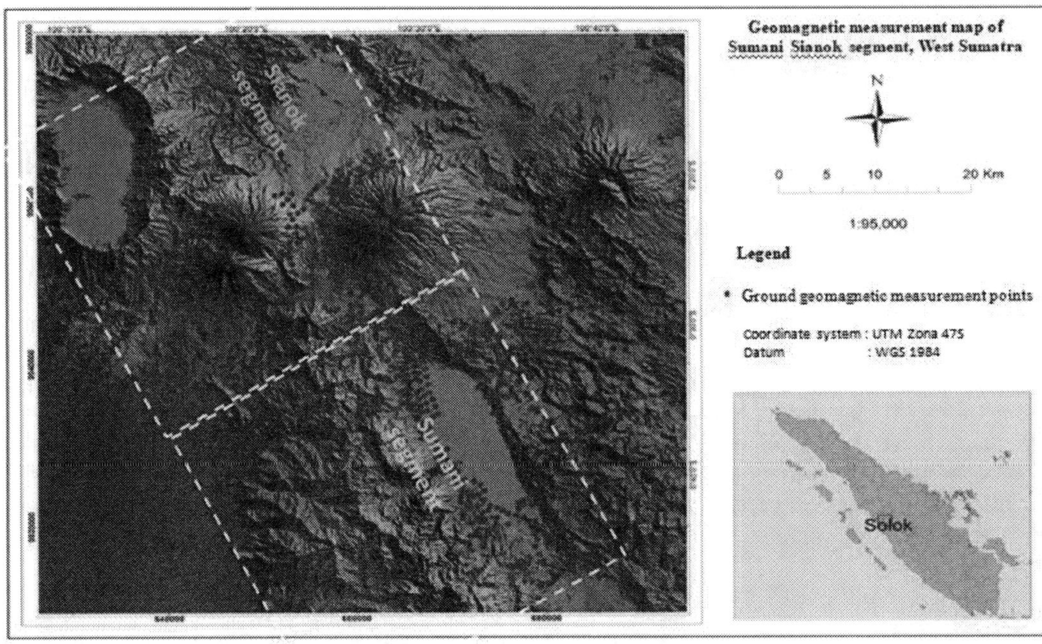

Figure 2. Locations of geomagnetic survey points (shown as red dots) in the Sumani and Sianok segments of GSF. The SRTM satellite image for Sumatra was obtained from the following site: https://www.earthdata.nasa.gov (downloaded on March 5, 2019).

IOP Conf. Series: Earth and Environmental Science **873** (2021) 012004 doi:10.1088/1755-1315/873/1/012004

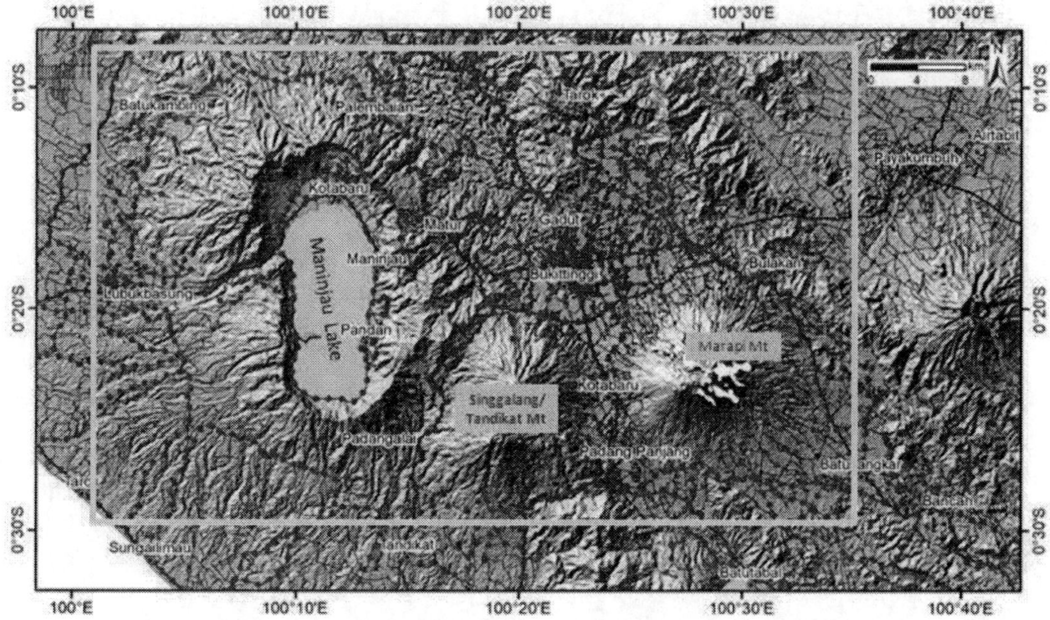

Figure 3. Locations of the proposed geomagnetic survey points (shown as red dots) in Sianok segment of GSF. In total there are about 637 survey points that will be measured.

References

[1] Hamilton W 1979 *Tectonic of the Indonesian Region, Geological Survey Professional Paper 1078* (Washington: US Government Printing Office)
[2] Posavec M, Taylor D, Leeuwen,T V and Spector A 1973 *Geol. Soc. Malaysia* **6**, 43-60
[3] Curray J R 2005 *J Asian Earth Sci.***25**, 187–232
[4] Barber A J, M J Crow and J S Milson 2005 *(eds) Sumatra: geology, resources and tectonic evolution*(London: Geological Society)
[5] Sieh K and Natawidjaja D H 2000 *J. Geophys. Res.***105**, 28,295–28,326
[6] Natawidjaja D H and Triyoso W 2007 *Hazard*, **1**, 2147
[7] Sahara D P, Widiyantoro S, and Irsyam M 2018 *J Asian Earth Sci.***164**, 1-8.
[8] Nugraha A D, Shiddiqi H A, Widiyantoro S, Thurber C H, Pesicek J D, Zhang H, Wiyono S H, Ramdhan M, Wandono and Irsyam M 2018 *Seismol. Res. Lett.***XX**.
[9] Nakano M, Kumagai H, Toda S, Ando R, Yamashina T, Inoue H and Sunarjo 2010 *Geophys. J. Int.***181**, 141-153
[10] Daryono M R, Natawidjaja D H and Sieh K 2012 *B. Seismol. Soc. Am.***102**, 2356-2367.
[11] Reynold J M 1997 *An Introduction to Applied and Environmental Geophysics* (England: Jhon Wiley & Sons)
[12] Mariita N O 2009 *The Magnetic Method, Presented at Short Course IV on Exploration for Geothermal Resources* (Kenya: UNU-GTP)
[13] Telford W M, Geldart L P and Sheriff R E 1990 *Applied Geophysics* (New York: Cambridge University Press)
[14] Chen G, Liu T, Sun J, Cheng Q, Sahoo B, Zhang Z and Zhang H 2015 *J. Appl. Geophys.* **120**, 14–25.
[15] Jiang S, Gao S, Li S, Zhang H and Zhao F 2007 *Earth Sci.***24**, 152-170.
[16] Zubaidah T, Korte M, Mandae M and Hamoudi M 2014 *J Asian Earth Sci.***80**, 172-184.

IOP Conf. Series: Earth and Environmental Science **873** (2021) 012004 doi:10.1088/1755-1315/873/1/012004

[17] Araffa S A, El-bohoti M, Mekkawi M, Ismail E, Khlail A, El- Razek and E M A 2018 *NRIAG J Astro Geophys*, **7**, 162–174.
[18] Saragih RDandBrotopuspito KS 2018 *Journal of Physics*: *Conference Series*, **1011**.
[19] Bradley K E, Feng L, Hill E M, Natawidjaja D H and Sieh K 2017 *J Geophys Res***122**, 572-591
[20] Irsyam M, Widiyantoro S, Natawidjaja D H, Meilano I, Rudyanto A, Hidayati S, Triyoso W, Hanifa N R, Djarwadi D, Faizal L and Sunardjito 2017 *Peta Sumber dan BahayaGempa Indonesia*, (Bandung: Pusat Penelitian dan Pengembangan Perumahan dan Permukiman, Kementerian PekerjaanUmum dan Perumahan Rakyat)
[21] Kastowo, Leo G H, Gafoer S and Amin T C 1996 *Geological Map of the Padang Quadrangle, Sumatera* (Bandung: Geological Research and Development Center)
[22] Silitonga P H and Kastowo 1995 *Geological Map of the Solok Quadrangle, Sumatera* (Bandung: Geological Research and Development Center)
[23] Buyung N, Subagio and Walker A S D 1992 *Bouguer Anomaly Map of The Padang Quadrangle, Sumatera* (Bandung: Geological Research and Development Center)
[24] Buyung N, Mirnanda E and Walker A S D 1992 *Bouguer Anomaly Map of The Solok Quadrangle, Sumatera* (Bandung: Geological Research and Development Center))

Seismic Vulnerability Index Calculation for Mitigation Purposes at Cilacap District

Robert Edison,[1*] Supriyanto Rohadi,[1] Yusuf Perdana,[1] Nelly Florida Riama,[2] Dwikorita Karnawati

[1,2,3] Badan Meteorologi, Klimatologi, dan Geofisika. Jl. Angkasa No. 2, Indonesia

rkolibu@gmail.com

Abstract. Seismic vulnerability index is one of the key factors in mitigation that shows the vulnerability of the soil layer beneath when passed through with a wave, the more vulnerable the soil layer, the more damage it done when an earthquake happens. Seismic vulnerability is calculated using two variables, that is dominant frequencies and amplification that are obtained by analyzing HVSR curve. HVSR are used to determine the dominant frequency by determining the maximum amplification in that area. HVSR curve is obtained by measuring microbemor data in 163 spots with 30 minutes-minimum duration in Cilacap with a portable seismograph. Mierotremor is a natural vibration that is caused by continuous vibration that come from beneath the surface, sometimes mixed up by the vibrabon that is caused by human activities such as pipe-flow, vehicles, etc. Thus, the purposes of this research are to determine which area is more vulnerable than others, based on the seismic vulnerability index, so it could be a reference for regional development to classified is it safe or unsafe to build in that area, remembering Cilacap is one of the most developed Districts in Central Java.

1. Introduction

In a high seismic prone area, such as Cilacap, human and economic loss are more likely due to construction or not seismically designed. When an earthquake happens in the surface, the buildings are being hit by seismic wave that could be separated as horizontal and vertical component. The vertical component isn't threatening life because the array is in line with gravity, meanwhile the horizontal component threatening because it creates a movement that caused the building above it to swing. This condition could cause the building to collapse [6].

Because Cilacap is one of the developed District at Java, it is necessary to make a mitigation attempt so it could be used a reference to understand which area is more vulnerability when passed by a seismic wave than the other. One of the methods that could be used to understand this is by mapping the seismic vulnerability index, which gotten by measuring microtremor in certain place. If compared to Jakarta which consist of Basin, it is assumed that the seismic vulnerability index in Cilacap would look alike in Jakarta [4].

2. Data and Methods

2.1. Data

The data in this research is the results of measuring microtremor wave in area which shown below:

IOP Conf. Series: Earth and Environmental Science **873** (2021) 012005 doi:10.1088/1755-1315/873/1/012005

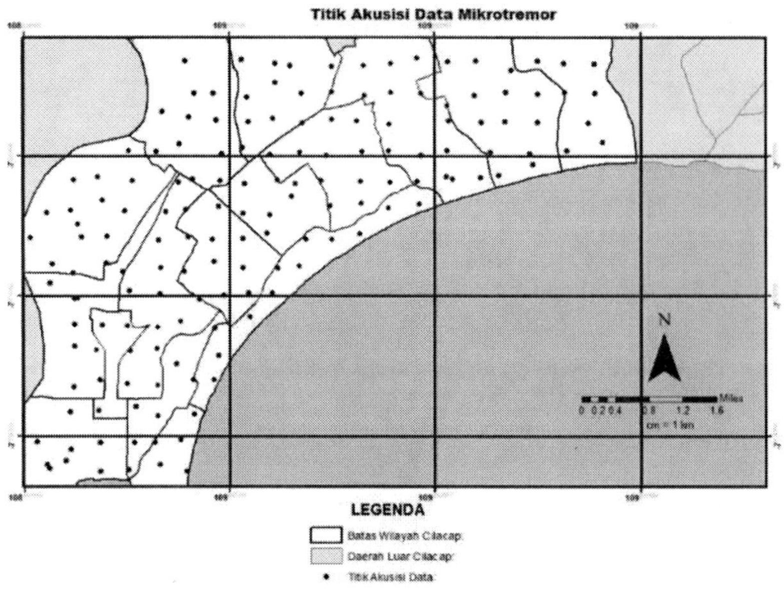

Figure 1. Microtremor measurement at Cilacap

Microtremor wave are measured using TDS-303 Portable Seismograph for a 30 minutes-minimum of data recording. A_0 and f_0 then determined using Nakamura technique [7] on Geopsy [9].

2.2. Methods

The list of authors should be indented 25 mm to match the abstract. The style for the names is initials then surname, with a comma after all but the last two names, which are separated by 'and'. Initials should not have full stops—for example **A J Smith** and *not* **A. J. Smith**. First names in full may be used if desired. If an author has additional information to appear as a footnote, such as a permanent address or to indicate that they are the corresponding author, the footnote should be entered after the surname.

2.2.1. Estimating Soil classification based on dominant period

Soil classification can be determined using dominant period gained from the microtremor using equation:

$$T_0 = \frac{1}{f_0}$$

Where f_0 *is* dominant frequency of HVSR wave

2.2.2. Calculating Seismic Vulnerability Index.

Seismic Vulnerability Index can be calculated using:

$$K_g = \frac{A_0}{{f_0}^2}$$

Where:
A_0 = maximum aplitude of HVSR wave
f_0 = dominant frequencies of HVSR wave.

2.2.3. Flow Chart

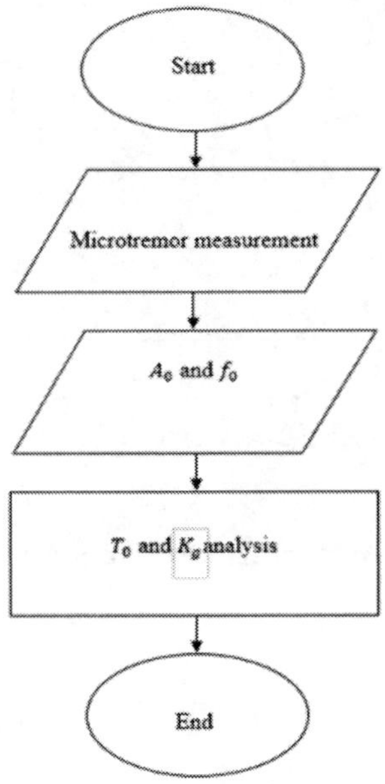

Figure 2. Flow Chart

3. Results

A_0 and f_0 estimated using HVSR data from processing microtremor data using Geopsy [9], as below:

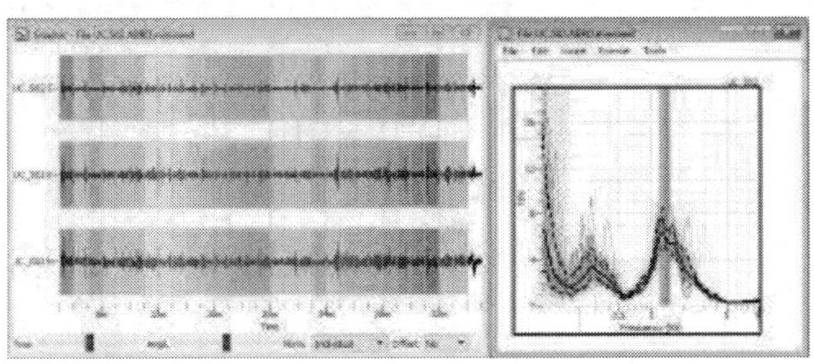

Figure 3. Microtremor data processing in *Geopsy*

The data that has been recorded using portable seismograph then processed with Geopsy to obtain the frequency based on the amplification value. Then, the value of dominant period is calculated to classify the soil. Meanwhile the seismic vulnerability index is calculated with the frequencies and the amplification value.

3.1. Dominant period estimation

This research is using microtremor data that measured in 162 spots in Cilacap, which varied from 0.345 — 3.809 s that can be classified into Type III (sand, clay that categorized as alluvium) and Type IV (delta sedimentation) soil which is shown below:

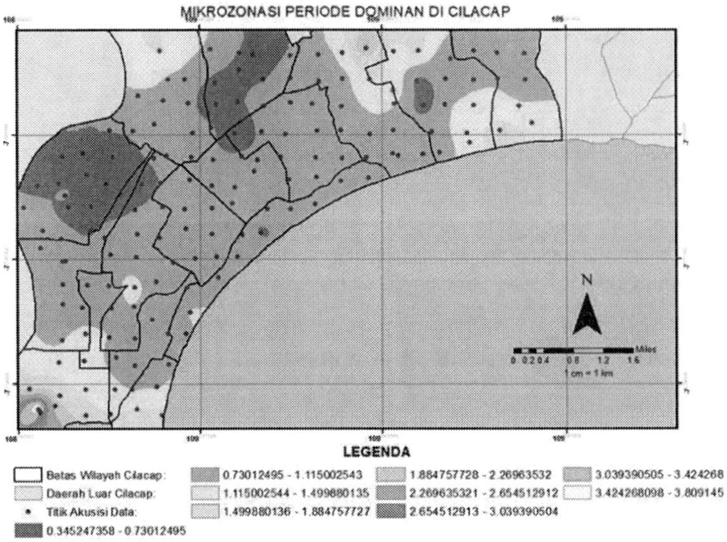

Figure 4. Dominant period microzonation

3.2. Amplification estimation

This research is using microtremor data that measured in 162 spots in Cilacap, which varied from 1.95 — 6.159 that was obtained from peak amplification on a HVSR curve on *Geopsy,* which is shown below:

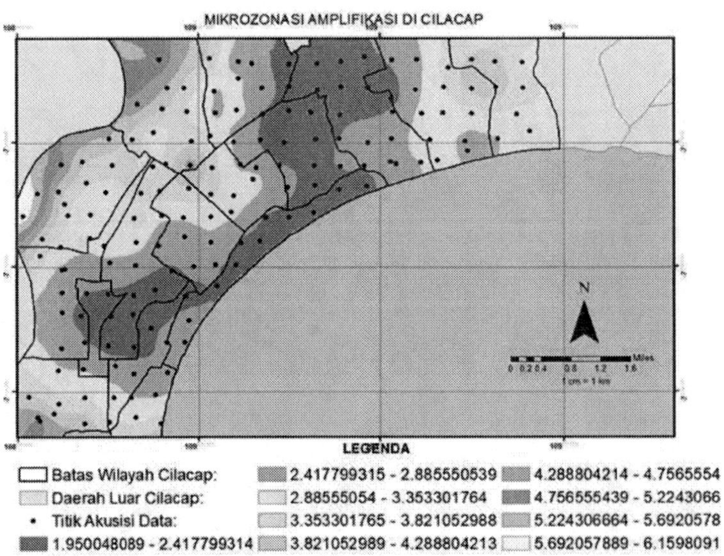

Figure 5. Amplification microzonation

3.3. Seismic vulnerability index calculation

The seismic vulnerability index is showing which area are more prone due to seismic activity based on the geological condition. The area that's way more prone are shown in red, while the safer area is shown in green. The seismic vulnerability index is varied from 3.184-56.827 as shown in figure below:

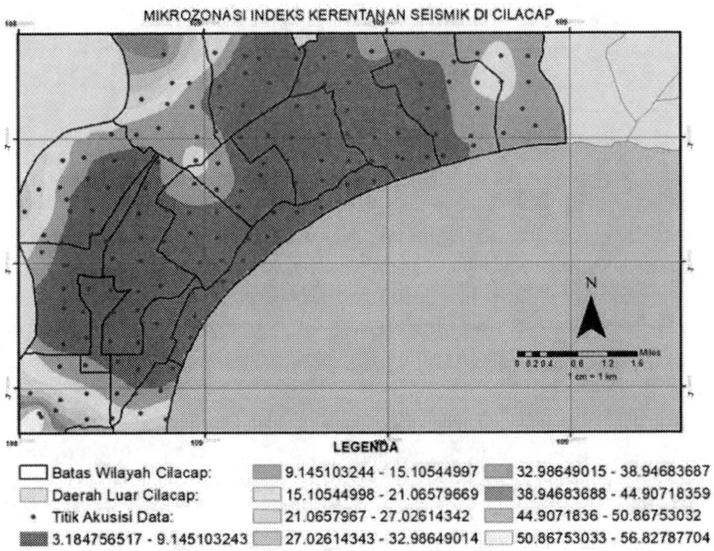

Figure 6. Seimic vulnerability index microzonation

4. Discussion

Based on dominant period calculation, Cilacap is composed of Type III and Type IV soil based on Kanai and Tanaka's Soil Classification [1], where Type III soil is consists of sand, clayclay, and alluvium soil, while Type IV soil consist of delta's sedimentation, topsoil, mud, humushumus, or soft soil with 30 m depths. The Type III soil are seen in the South-Westnorthern area of Cilacap, while the rest of Cilacap are covered in Type IV soil. According to Elnashai & Sarno (2008) [4], the importance of geological surface condition effects soil vibration. Site effect had important role because of the geological difference between every area that determines the amplification and dominant frequencies.

Calculation of seismic vulnerability index are depended by the dominant frequency and the amplification. In this study, even though the seismic vulnerability index is varied, if compared to the geological map [2] are categorized as Qac (coastal deposit) that generally consists of well-medium sorted sands, very loose and categorized as shallow, with highest seismic vulnerability index located in southern Cilacap. Based on the calculation of seismic vulnerability index, the South-West area has seismic vulnerability index higher than the rest of Cilacap, it is estimated during earthquake the South-West area of Cilacap to have more destruction during earthquake. While the seismic vulnerability index microzonation is compared to the dominant period microzonation, it can be assumed that the higher the dominant period of the area, the higher the seismic vulnerability index. This can be explained in softer soil, the earthquake wave that traveled through the soil is being reflected and refracted in the soil because of the different length between soil particle in soft soil, while in harder soil the earthquake wave could traveled through the soil without being much reflected and refracted because of the rigidity of the soil. This caused the building in soft soil to shake more than the building in hard soil [5].

Variation of the seismic vulnerability index could be correlated to level of destruction of an earthquake, region with higher seismic vulnerability index tend to have more destruction when an earthquake happens. It is important to conduct a seismic risk evaluation based on past earthquake event and planning mitigation plans are required to minimize the damage of the earthquake. Mitigation plans that could be approached for this area such as: foundation engineering, building a better structure, or even avoiding building houses and offices in that area. This study also shows the important need of local geological study (not only on the surface), but also the thickness of the sediments that have an impact when an earthquake happens.

5. Conclusions

According to microzonation, seismic vulnerability index is higher in area with harder soil. While seismic vulnerability index at the South-West area of microzonation area in Cilacap is the highest than other area (56,827).

References

[1] Arifin S S, Mulyatno B S, Marjiyono and Setianegara 2013 Penentuan Zona Rawan Guncangan Gempa Bumi Berdasarkan Analisis Nilai Amplifikasi HVSR Mikrotremor dan Analisis Periode Dominan Daerah Liwa dan Sekitarnya *Jurnal Geofisika Eksplorasi* **2** 30-40.
[2] Asikin S, Handoyo A, Prasistho B and Gafoer S 1992 *Peta Geologi Lembar Banyumas, Jawa Barat* Bandung: Pusat Penelitian dan Pengembangan Geologi.
[3] Cipta A, Cummins P, Dettmer J, Saygin E, Irsyam M, Rudyanto A and Murjaya J 2018 Seismic Velocity Structure of the Jakarta Basin, Indonesia, using Trans-dimensional Bayesian Inversion of Horizontal-to-Vertical Spectral Ratios *Geophysics Journal International.*
[4] Elnashai A S and Sarno L D 2008 *Fundamentals of Earthquake Engineering* United Kingdom: John Wiley & Son.
[5] Gadallah R M and Fisher R 2009 *Exploration Geophysics* Berlin: Springer.
[6] Lermo J and Chavez-Garcia J 1994 Site Effect Evaluation using Spectral Ratios with only One Station *Bulletin of the Seismic Society of America* **83** 1574-1594.
[7] Nakamura Y 1989 A Method for Dynamic Characteristics Estimation of Subsurfaces Using Microtremor on the Ground Surface *QR of RTRI* **30** 1 25-33.
[8] Nakamura Y 2000 Clear Identification of Fundamental Idea of Nakamura's Technique and its Applications *12WCEE 2656.*
[9] Project S E 2004 *SESAME: Guidelines for The Implementation of The H/V Spectral Ratio Technique on Ambient Vibrations* European Commission-Research General Directorate.
[10] Widia R, Isburhan P, Nuraeni G, Ry R V, Yudistira T, Cipta A and Cummins P 2019 Horizontal-to-Vertical Spectral Ratio (HVSR) Method for Earthquake Risk Determination of Jakarta City with Microtremor Data *2nd Southeast Asian Conference on Geophysics*.

IOP Conf. Series: Earth and Environmental Science **873** (2021) 012006 doi:10.1088/1755-1315/873/1/012006

Network of Radon Gas Concentration Monitoring of Research and Development Centre – BMKG for Earthquake Precursor Research in Indonesia

Agustya Adi Martha [1], Angga Setiyo Prayogo[1], Jimmi Nugraha[1], Suliyanti Pakpahan[1], Nelly Florida Riama[1]

[1] Agency for Meteorology, Climatology and Geophysics (BMKG), Jakarta, Indonesia

E-mail: agustyaadi@gmail.com

Abstract. The geographical position of Indonesia, which is flanked by several subduction zones and the presence of active faults in the sea and land make Indonesian territory prone to earthquakes and tsunamis which can result in many deaths and damaged. There are several efforts we can do to minimize the occurrence of earthquakes, including developing earthquake resistant buildings, increasing the ability/capacity of the community, and predicting earthquakes or better known as earthquake precursors. The BMKG Research Centre began conducting research on earthquake prediction using several methods, including the Radon monitoring method. Monitoring of Radon gas concentrations for earthquake precursors has several advantages, including the presence of radioactive gas which is abundant in ground water that has a half-life of 3.2 days. Radon is the result of decay of uranium 278U which is abundant in the earth's crust rock so that when rock friction occurs, the Radon gas can be detached. Based on the results of Radon monitoring at Tadulako and Palolo stations - Southeast Sulawesi province, there was a change in the pattern of radon gas concentration and water level rising up and down drastically and a gradual decrease in ground water temperature before the earthquake on 28 September 2018. In addition to Central Sulawesi, since 2012 the Centre for Research and Development of BMKG has been conducting research to monitor radon gas concentrations in the DI Yogyakarta region precisely in Piyungan and Pundong districts with the aim of monitoring radon gas concentrations in the Opak fault. In 2021, the BMKG Research and Development Centre added a new radon gas monitoring network in the active fault areas of Cimandiri and Lembang in the West Java province. There are 1 station in the Cimandiri fault segment and 2 stations in the Lembang fault section. It is hoped that in the future the results of monitoring can reduce the impact caused by the earthquake disaster in Indonesia.

1. Introduction

Geographically, Indonesia is located in an area of several active tectonic plates. This has resulted in the existence of many faults on land that can cause earthquakes throughout Indonesia. Figure 1 shows the earthquake events detected by BMKG that occurred in the period January to September 2020, with a magnitude of 1 to 7. In this period no less than 3500 earthquakes occurred in Indonesia's area.

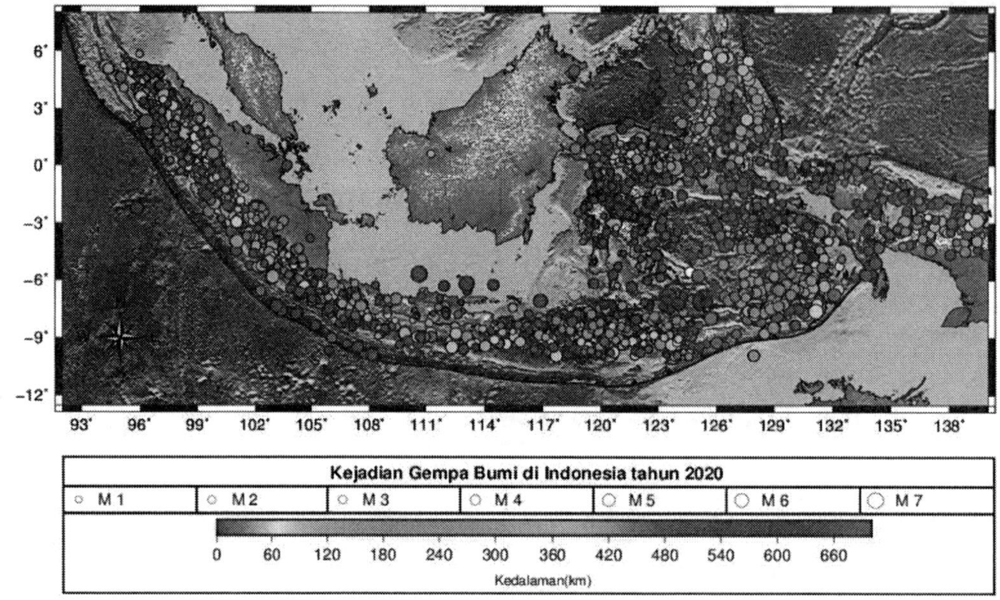

Figure 1. Earthquake Epicenter January - September 2020

Research and Development Center of BMKG has initiated efforts to predict the occurrence of earthquakes since the beginning of 2010 until now. Various methods have been applied and developed, ranging from geophysical, geo-atmosphere and geochemical anomalies. This year we are concentrating on improving the performance of monitoring the geochemical anomalies from the changes in radon gas concentrations before earthquakes occur. Radon gas (222Rn) is a radioactive gas resulting from the decayment of radium-226 which is found in soil, rock, air and water with varying concentrations with a half-life of about 3.8 days [1, 2, 3, 4].

Several studies have shown anomalous changes in radon gas concentrations prior to the earthquake, including changes in ground water and radon gas concentrations before the Kobe - Japan echo [5], increased radon gas anomalies in San Andreas faults [6] and the Nepal region [7]. Efforts to improve monitoring and analysis of radon data continue to be made, as well as methodologies and mathematical modeling, to strengthen the link between radon gas anomalies and geodynamic processes [8]. Precursor research based on the anomaly monitoring of radon gas concentrations in a fault is promising to be applied and developed in Indonesia. The Center for Research and Development of BMKG as an institution that has the task of carrying out research, assessment, and development in the field of Geophysics, has been started conducting pre-surgery earthquake research using this method since 2015. This year BMKG has installed radon monitoring stations in the regions of Central Sulawesi, Jogjakarta, and West Java.

2. Radon Monitoring Station System

In 2015, the Research and Development Center of BMKG began to intensify precursor research using the anomaly monitoring method for radon gas concentrations in the soil [9]. In addition to radon monitoring equipment, we integrate water level monitoring instruments and Geo Temperature (Figure 2). The integrationis based on the success of previous researchers in predicting earthquake events using this method [5, 10, 11, 12].

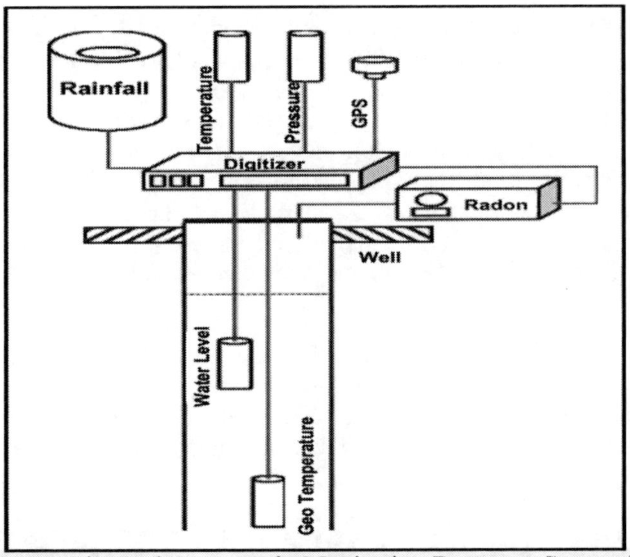

Figure 2. The Radon Anomaly Monitoring Precursor System [13]

The Research and Development Center of BMKG for the first time built a radon monitoring station that is integrated with water level and geo-temperature monitoring in the DI Yogyakarta region (Figure 3). We use a monitoring system developed by TAIDE, where this system is integrated between radon gas concentration information, water level, geo-temperature in the well with meteorological information such as air pressure, temperature and rainfall. All this information is sent in real time to the Central BMKG Research and Development Center.

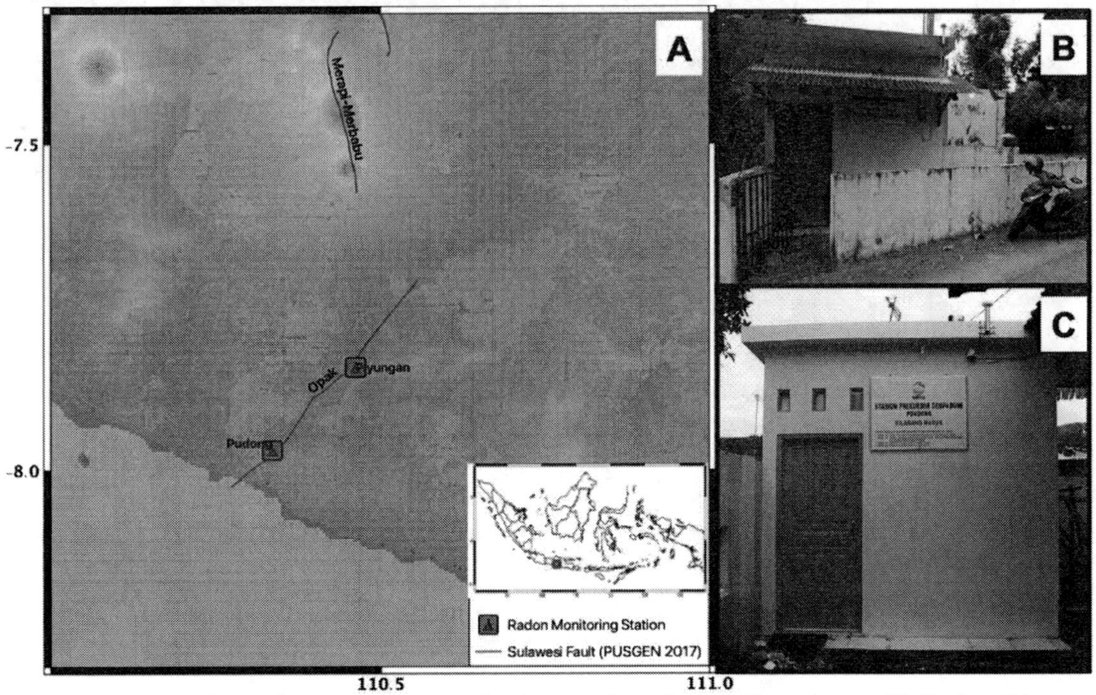

Figure 3.A) Location of two Radon Monitoring Stations in DI Yogyakarta, B) Radon Piyungan monitoring stations, C) Radon Pundong monitoring stations

We chose to install 2 radon stations in Jogjakarta by considering the presence of an active Opak fault which caused a devastating earthquake in 2006. The first station is located in the northern part of the Yogyakarta region, precisely in Piyungan District. The second station is located in the southern part of the DIY region, precisely in Pundong District. These stations are located close to the Opak fault, in the hope that before the earthquake occurs, radon that is released around the fault can be immediately detected by these stations. In 2016, the BMKG Research and Development Center built two radon stations with the same system in the Central Sulawesi region. The first station is located at Tadulako University which is adjacent to the Palukoro active fault and the second station is installed at a location adjacent to the Palolo fault (Figure 4). The second station is installed at a location adjacent to the Palolo fault.

The installation of a radon monitoring station in the same area but in different fault segments has the objective of monitoring radon anomalies in the same segment and different segments. It is hoped that we will be able to find out the effect of radon, water level and geotemperature anomaly before the earthquake occurs on the closest active fault and distant segments of the negative fault.

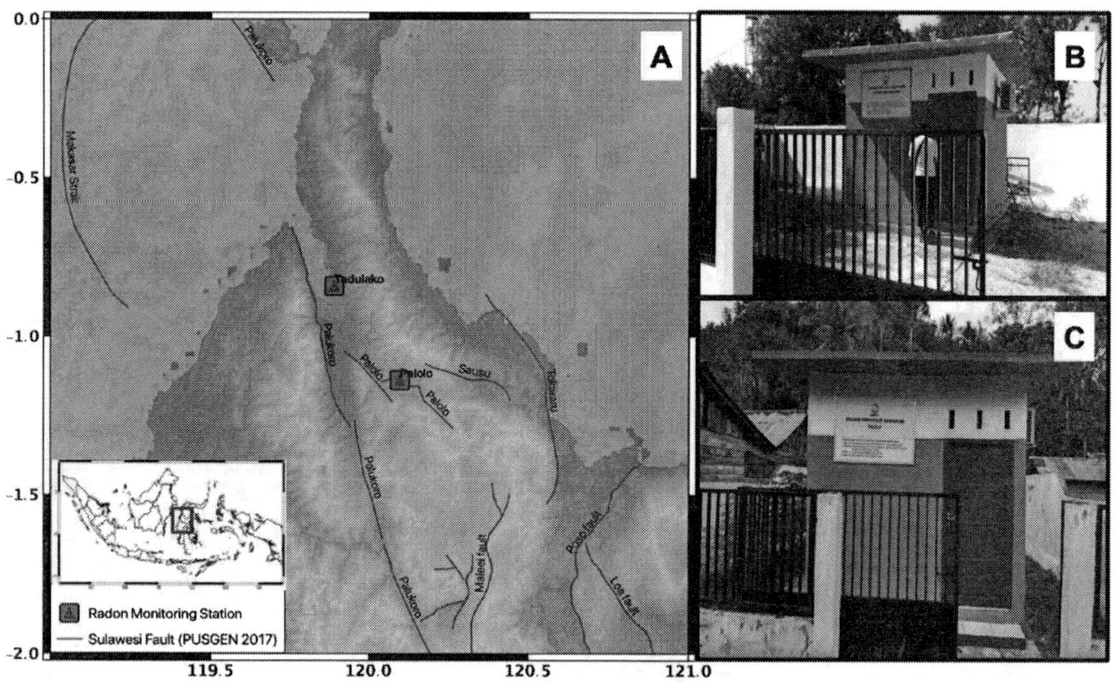

Figure 4.A) Location of two Radon Monitoring Stations in Central Sulawesi, B) Radon Tadulako monitoring stations, C) Radon Palolo monitoring stations

3. Monitoring of Yogyakarta Earthquake Precursors

Parameters observed in radon-based earthquake precursor monitoring in Yogyakarta include radon gas, subsurface temperature, water level, air pressure, surface temperature and rainfall. Figures 5 and 6 show the results of monitoring of precursor parameters at Pundong and Piyungan stations and maps of the surrounding epicenters for January - February 2019.

IOP Conf. Series: Earth and Environmental Science **873** (2021) 012006 doi:10.1088/1755-1315/873/1/012006

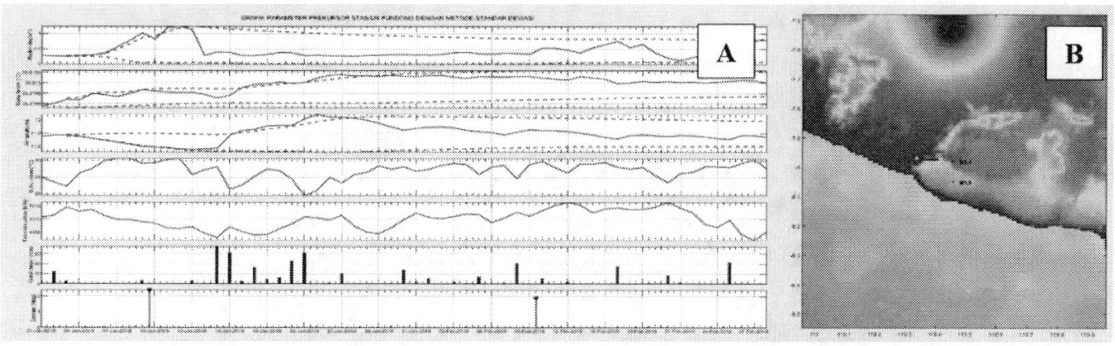

Figure 5. A) Graph of the observation of earthquake precursors at Pundong Station in January - February 2019 and their standard deviation, B) Map of earthquake epicenters around Pundong station for January - February 2019 (Yellow triangle is Radon Station, red circles are Earthquake Epicenter).

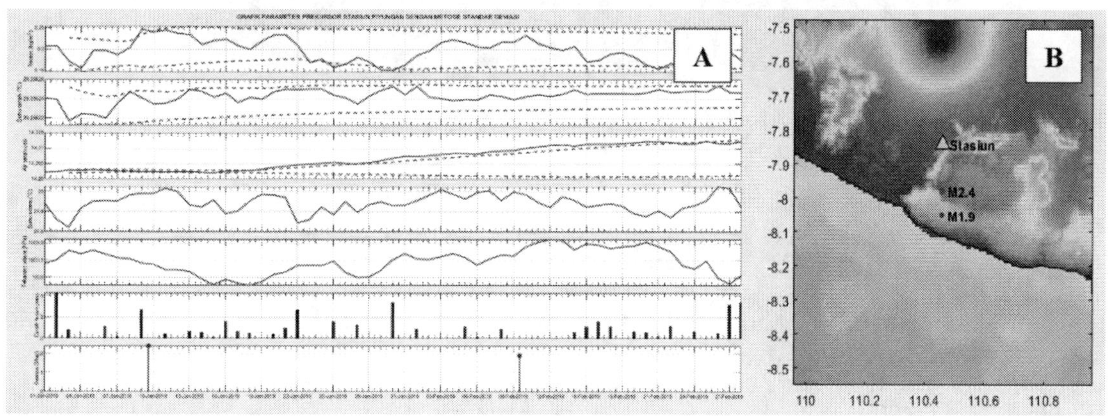

Figure 6. A) Graph of earthquake precursor observations at Piyungan station from January to February 2019 and their standard deviation. B) Map of earthquake epicenters around Piyungan station for January - February 2019 (Yellow triangle is Radon Station, red circles are Earthquake Epicenter).

There were 2 earthquakes around the Pundong and Piyungan precursor stations. The first earthquake occurred on January 9, 2019 with a magnitude of M2.4 at a distance of 17 km from Pundong Station and 18 km from Piyungan Station. The second earthquake occurred on 9 February 2019 with a magnitude of M1.9 at a distance of 19 km from Pundong Station and 25 km from Piyungan Station (Figures 6).

Figure 5 and Figure 6 show that the January 9 earthquake caused anomalies at Pundong and Piyungan Stations. At the Pundong station, the precursor to the 9 January 2019 earthquake, there was an increase in radon gas and a decrease in groundwater since 5 days before the incident, while at Piyungan station the increase in radon gas concentrations occurred 6 days before the incident but was not significant. The 9 February 2019 earthquake only caused radon gas anomaly at Piyungan Station. An insignificant increase in radon gas concentrations occurred at Piyungan Station since 9 days before the 9 February 2019 earthquake.

4. Monitoring Network of Lembang Radon Station - Cimandiri 2020

Figure 7. Location of Radon Monitoring Station in West Java Region (Left) and Radon station above ground water well

At the end of 2020 we started reinstalling 3 radon stations in the West Java region (Figure 7). We use the Rad7 instrumentation to measure radon and thoron values in groundwater wells. We installed 1 station in the Cimandiri fault segment right at the BMKG Earth Magnet observatory station, and 2 stations in the Lembang fault segment to be precise at the earthquake observation station and BMKG official residence. In this monitoring network 2 stations are placed in close proximity to each other around the Lembang fault segment in order to determine the anomalous effect of radon gas concentrations before the earthquake occurred at stations that have different distances from the source (Lembang Fault). All data on this network will be sent in real time to the Research and Development Center of BMKG, in order to monitor activities before an earthquake occurs in the research area.

5. Concluding Remarks

Monitoring of radon gas concentration anomalies for research efforts on earthquake precursors, we value is quite promising to be applied around active faults that are scattered throughout Indonesia. In the period from January to Fenruari there were 2 earthquakes in the Jogjakarta area which were adjacent to the Pundong Radon Station. The monitoring results at the Pundong Station showed a significant anomaly 4 days before the earthquake occurred, as well as the piyungan sensor which has a further distance from the epicenter of the earthquake, although the detected anomaly was not as significant as the Pundong station. Based on our observations in the Central Sulawesi and Yogyakarta regions, anomalies of radon gas concentration as an earthquake precursor can be detected depending on the distance, magnitude and length of the active fault that releases energy. We are currently starting to develop an algorithm for the relationship between the anomalous magnitude of radon gas concentration, water level and geo temperature to the magnitude and hypocenter of the earthquake that occurred.

References

[1] Timothy M K 2003 Geological Hazards : A Sourcebook Greenwood Press
[2] Oh Y H and Kim G 2015 A radon-thoron isotope pair as a reliable earthquake precursor www.nature.com/scientificreport
[3] Kuo T, Lin C, Chang G, Fan K, Cheng W and Lewis C 2010 Estimation of aseimic crustalstrain using radon repetitive radon precursors of the 2003 M 6.8, 2006 M 6.1, and 2008 M 5.0 earthquakes in eastern Taiwan *Nat. Hazards* **53** 219 - 228 DOI: 10.1007/s11069-009-9423-y

[4] Yasuoka Y, Kawada Y, Nagahama H, Omori Y, Ishikawa T, Tokonami S and Shinogi M 2009 Preseismic changes in atmospheric radon concentration and crustal strain *Phys. Chem. Earth* **34** 431 - 434 DOI: 10.1016/j.pce.2008.06.005

[5] Igarashi G, Saeki S and Takahara N 1995 Ground-water radon anomaly before the Kobe earthquake in Japan *Nature* **269** 60 – 61

[6] King C 1978 Radon emanation on San Andreas Fault *Nature* **271** 516–519

[7] Deb A, Gazi M and Barman C 2016 Anomalous soil radon fluctuations – signal of earthquakes in Nepal and eastern India regions *Journal earth system science* **125** 1657–1665 DOI: 10.1007/s12040-016-0757-z

[8] Imme G and Morelli D 2012 Radon as Earthquake Precursor, Universita di Catania - INFN Sezione di Catania, Italy

[9] Sunardi B, Rohadi S, Muslim B, Ngadmanto D, Susilanto P, Sulastri, Pakpahan S, Kurniawan T and Prayogo A S 2018 Radon Gas Variation and Seismic Activities around Opak Fault *Journal of Environtment and Geological Hazards* **9** 11-20

[10] Panda S K, Choudhury S, Saraf A K, Das J D 2007 MODIS land surface temperature data detects thermal anomaly preceding 08 October 2005 Kashmir earthquake *International Journal of Remote Sensing* **28** 4587-4596

[11] Roeloffs E A 1996 Poroelastic techniques in the study of earthquake related hydrological phenomena *Adv. Geophys* **37** 135 – 195

[12] Roeloffs E A 1998 Persistent water level changes in a well near Parkfield, California due to local and distant earthquakes *J. Geophys. Res* **103** 869 – 889

[13] Taide 2014 ADP seismic premonition observation system *TAIDE Enterprise Co., Ltd, China*. DOI: 10.1016/S0264-3707(00)00022-3.

Determination of Hypocenter Using Geiger Method in Sinabung Volcano, April-July 2016 Period

Y Annisa[1], G C Astriyan[1], S Wahyunia[1], N Indrastuti[2], M F I Massinai[1]

[1] Geophysics Department, Faculty of Mathematics and Natural Science, Hasanuddin University, Makassar, Indonesia
[2] Center for Volcanology and Geological Hazard Mitigation, Bandung, Indonesia

yusrinannisa127@gmail.com

Abstract. Sinabung is a volcano located in the Karo Highlands, Karo District, North Sumatra, Indonesia, with the highest peak of 2460 meters mean sea level. Volcanic earthquake is an earthquake that occurs due to volcanic activity. This is caused by the movement of magma upwards in the volcano. This study aims to determine the type of earthquake, hypocenter position and epicenter of volcanic earthquakes in Sinabung volcano in April-July 2016. The principle of this study was carried out by analyzing volcanic earthquake data in Sinabung volcano in April-July 2016. The data is recorded data (seismogram) or in other words is secondary data from Sinabung volcano on 7 seismometer stations namely Sukanalu, Lau Kawar, Sigarang-Garang, Mardinding, Gamber, Sibayak, and Kebayaken stations. Earthquake data in April-July 2016 revealed that there were 24 earthquake events in a period of 3 months which were the results of picking up the P and S waves, where volcanic earthquakes were obtained only in the form of volcanic earthquake type A and type B volcanic earthquake. Sinabung volcano has an earthquake activity that high enough so that the status of Sinabung volcano is still at level III (standby) status. Based on the hypocenter of several VA and VB earthquakes that occurred in April-July 2016, it can be concluded that the distribution of the hypocenter of the volcanic earthquake shows that the maximum depth of the volcanic earthquake is 10.000 meters and the position of the earthquake is spread at the point between Sinabung volcano and Mount Sibayak.

1. Introduction

Volcanic earthquake (volcano) is an earthquake that happen because of volcano activity, either it is magma movement reaching earth surface or eruption/gas wind erupted from inside the volcano [1].
Indonesia is one of country located at intersection of tectonic plates, those are Eurasia Plate, Indo-Australia Plate and Pacific Plate [2]. These plates boundaries are subduction zone thus forming mountain ranges and compression structures [3]. Because of these, Indonesia is known to have the most active volcanoes in the world, 127 (30%) of them are located in Indonesia [4].
One of the active volcanoes located in North Sumatera is Sinabung volcano. Sinabung volcano is at Karo Regency, 80 kilometers away from Medan, the capital city of North Sumatera [4]. This mountain which 2460 meters mean sea level was formed at northwest edges of fractured basin Toba Tua [5]. Seismic activities note that since 1600-2000s, the volcano never erupted, but then suddenly became active in 2010 [2]. Sinabung volcano erupts volcanic materials reaching 7 to 8 kilometers height [6].

Besides, Sinabung volcano has four main craters including stratovolcano or layered volcano, alternating between lava melting and pyroclastic in every single eruption period [7] [8].
This study analyze Sinabung volcano to reveal earthquake capacity happening either it is shallow volcanic earthquake or deep volcanic earthquake by using GAD method (Geiger's method with Adaptive Damping) [9] [10]. Geiger Method is iteration steps by optimizing least square in order to determine hypocenter location [11] [12]. The basic of this method are minimizing travel time residuals between observation data and calculated travel times by calculating using distant seismic networks by employing spherical earth travel time [13]. Geiger Method needs input such as arrival time parameter, velocity model and station location on Cartesian coordinate [14] [15]. This method uses arrival time of P wave and/or S wave as input. The equation is solved by using damped least square with adaptive damping. The damping constant is adapted with previous iteration results, thus stable solution may be obtained [16]. Because GAD method uses arrival time of P and S wave, researcher do pick of arrival time from waveform which resulted from Sinabung volcano. From this research, by knowing hypocenter and magma movement prediction, recommended disaster mitigation information can be constructed for society living nearby Sinabung volcano.

2. Data and Methodology

Seismic records of earthquakes at Sinabung volcano is processed using Software Swarm 2.7.2 and GAD. Software Swarm is used for determining earthquake type and arrival time picking of P and S wave, while GAD program is used for determining hypocenter and epicenter from the earthquakes.
The data used include seismic records from April to July 2016 measured by 7 seismometers at Sinabung volcano stations. But the proceeded data was chosen only in April, May and July. This is because error of data records in June thus cannot be proceeded.
Earthquake or seismic activity observations was done by using data records (seismograph) transmitted from observation stations to Volcanology Observation Center. Nearby Sinabung volcano, there are 7 stations used: Sukanalu, Lau Kawar, Sigarang-Garang, Mardinding, Gamber, Sibayak and Kebayaken. Coordinate data of the stations are provided in **Table 1.** below:

Table 1. Stations Coordinate of Sinabung Volcano

No.	Station Name	Easting	Northing
1.	SKN (Sukanalu)	434756.00	351016.00
2.	LKW (Lau Kawar)	431671.00	352796.00
3.	SGR (Sigarang-Garang)	434243.00	352275.00
4.	MMD (Mardinding)	429590.69	349224.74
5.	GBR (Gamber)	437352.89	348700.42
6.	SBY (Sibayak)	444684.00	358076.42
7.	KBY (Kebayaken)	436267.00	356063.00

The following is flowchart of Sinabung volcano earthquakes data processing:

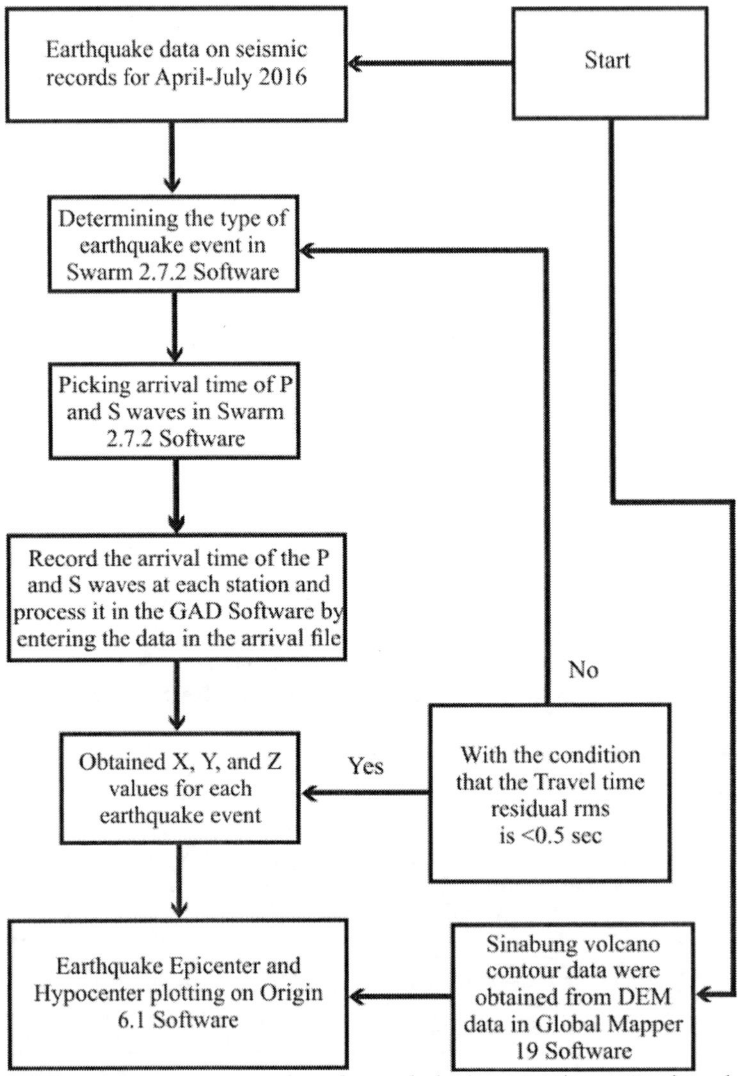

Figure 1. Data Processing Flowchart of Sinabung Volcano Earthquakes

3. Result and Discussion

3.1. Result

The results provide 24 earthquakes in period of 3 months, where volcanic earthquakes obtained are volcanic earthquake type A and type B. Besides, there are another kind of earthquake recorded: far tectonic earthquake, low frequency earthquake and wind earthquake. From all of these obtained earthquakes, only VTA and VTB were proceeded.

Station locations and Sinabung volcano peak using Google Earth are visualized in **Figure 2**, while earthquake plotting against epicenter is visualized in **Figure 3.** Hypocenter seen from North-South and East-West is visualized in **Figure 4.** and **5.**

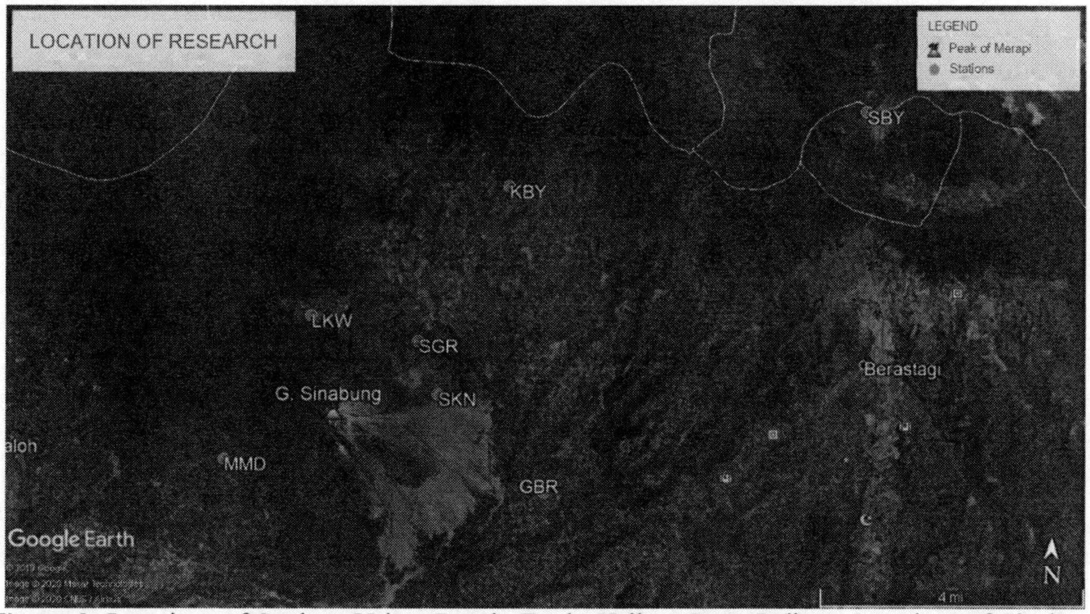

Figure 2. Locations of Stations Using Google Earth. Yellow Spots Indicate Locations of Sinabung Volcano Observation Stations, Orange Triangle is Sinabung Volcano Peak.

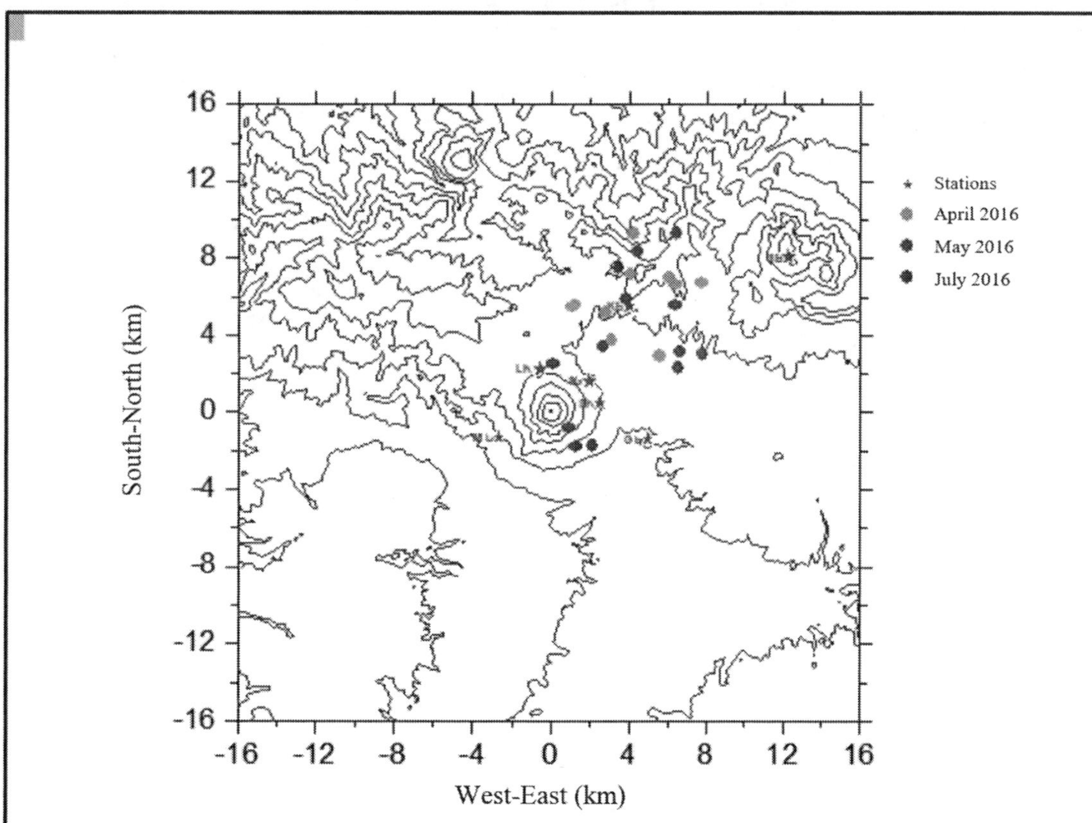

Figure 3. Earthquake Type A and B Positions in Sinabung Volcano in Period of April-July 2016

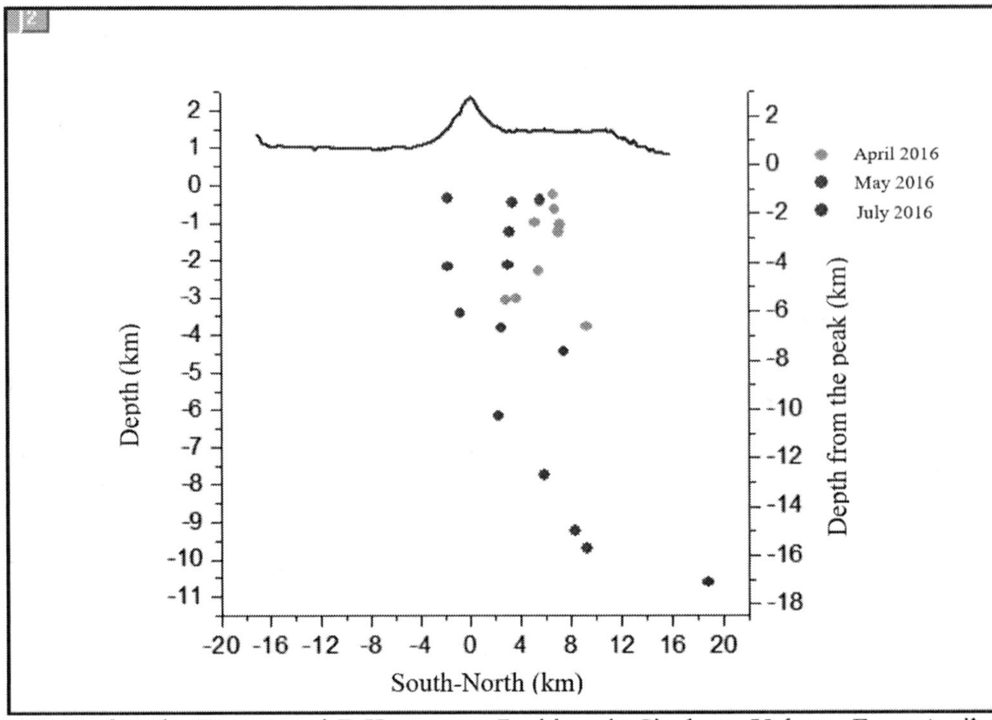

Figure 4. Earthquake Type A and B Hypocenter Positions in Sinabung Volcano From April to July 2016 (South-North Based).

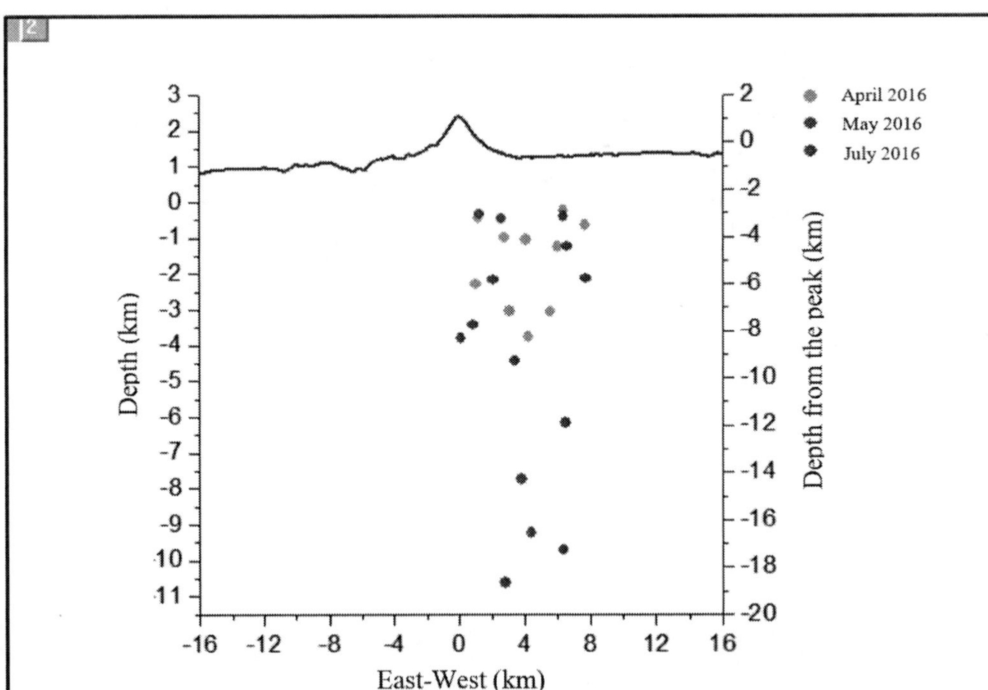

Figure 5. Earthquake Type A and B Hypocenter Positions in Sinabung Volcano From April to July 2016 (East-West Based).

3.2. Discussion
Data processing was done to obtain volcanic earthquake hypocenter and epicenter positions in period of April-July 2016. Processing was done by using GAD method which use arrival time of P and S wave. After identification of earthquakes in Sinabung volcano, 27 wind earthquakes, 6 low frequency earthquakes (LF), 10 far tectonic earthquakes, 14 VTA earthquakes and 10 VTB earthquakes were obtained, resulting 24 volcanic earthquakes in total from April-July 2016. These indicate earthquake activity of Sinabung volcano is high enough thus Sinabung volcano status is at level III (standby).
From Origin 7 Program, it can be seen the volcanic earthquakes type A and B epicenter and hypocenter locations, where maximum value of volcanic earthquake depth in the hypocenter distributions is 10.000 meters and earthquakes spots spread between Sinabung volcano and Mount Sibayak. Volcanic earthquake hypocenter seen indicate that during period of April-July 2016, earthquake locations dominate at around lower parts of Sinabung volcano. These are because of magma activity and volcanic eruption in 2016.

4. Conclusions
From data processing of volcanic earthquakes type A (VA) and B (VB) in Sinabung volcano, it can be concluded that types of recorded earthquake in Sinabung volcano are Shallow Volcanic-Tectonic Earthquake (VTB), Deep Volcanic-Tectonic Earthquake (VTA), Far Tectonic Earthquake, Low Frequency Earthquake and Wind Earthquake. The amount of identified volcanic earthquake in Sinabung volcano are 24. Sinabung volcano hypocenter distributions are clustered under the volcano reaching 0,5-10 kilometers depth. Based on the hypocenter distributions shown in the previous visualization, the distributions are seen from about North to Southeast of Sinabung volcano reaching 10 kilometers depth.

5. Acknowledgments
Researcher give great thanks to Allah SWT, who has given health and strength so that researcher can finish this research. Researcher also would like to thank parents, supervisors and all friends in Geophysics Department. We also would like to thank Center for Volcanology and Geological Hazard Mitigation (PVMBG) for granting us to publish final data, and also to Hasanuddin University for giving financial support for this research.

References
[1] Hidayati S 2010 *Pengenalan Seismologi Gunungapi* (Bandung: Diklat Pelaksana Pemula Gunungapi baru, Pusat Vulkanologi dan Mitigasi Bencana Geologi)
[2] Ismail N, Yanis M, Aramanda L and Surbakti M S 2019 Analisis Kerentanan Situs Heritage Terhadap Ancaman Letusan Gunung Api Sinabung *J.Geografi* **11(1)** 76-85
[3] Nugroho H, Widiyantoro S and Ibrahim G 2017 Penentuan Posisi Hiposenter Gempabumi dengan Menggunakan Metoda Guided Grid Search dan Model Struktur Kecepatan Tiga Dimensi *J. Meteorologi dan Geofisika* **8(1)** 48-60
[4] Sukarman and Suprapto 2015 Sebaran dan Karakteristik Material Vulkanik Hasil Erupsi Gunung Sinabung di Sumatera Utara *J. Tanah dan Iklim* **39(1)** 9-18
[5] Suprapto, Nurmasari R and Rosyida A 2015 Data Bencana Indonesia 2014 (Jakarta: Pusat Data, Informasi dan Hubungan Masyarakat Badan Nasional Penanggulangan Bencana)
[6] BNPB 2012 Peraturan Kepala Badan Nasional Penanggulangan Bencana No. 02 Tahun 2012 Tentang Pemodan Umum Pengkajian Risiko Bencana (Jakarta: BNPB)
[7] Santoso M S 1992 G. Sinabung (B) Berita Berkala Vulkanologi Edisi Khusus (Bandung: Direktorat Vulkanologi)
[8] Indrastuti N 2014 *Studi Kegempaan dan Seismik Tomografi Gunung Sinabung* (Bandung: Program Studi Sains Kebumian. Institut Teknologi Bandung)
[9] Massinai M A and Massinai M F I 2018 Determination Hypocentre and Focal Mechanism Earthquake of Oct 31, 2016 in Bone, South Sulawesi *IOP Conf. Series: J. of Phys.* **979** 012045

[10] Lahr J C 1979 HYPOELLIPSE: A Computer Program for Determining Local Earthquake Hypocentral Parameters, Magnitude and First Motion Pattern Open-le Report 79-431 US Geological Survey Menlo Park California 310

[11] Rasmid and Ramdhan M I 2014 Interpretasi Episenter dan Hiposenter Sesar Lembang *J. of Physics* **1(1)** 1-7

[12] Geiger L 1912 Probability Method for the Determination of Earthquake Epicenters from the Arrival Time Only *Bull. St. Louis Univ* **8** 56–71

[13] Lahr J C 1999 HYPOELLIPSE: A Computer Program for Determining Local Earthquake Hypocentral Parameters, Magnitude and First Motion Pattern (Y2K Compliant Version) Open-File Report 99-023 (US Geological Survey Golden Colorado paper and on-line editions) 112 pp

[14] Massinai M F I, Astuti A P, Kiraman M R B, Massinai M A and Ramdhan M 2019 Hypocenter determination and focal mechanism solution of May 29 2017 earthquake around Poso, Central Sulawesi, Indonesia *J. of Phys.* **1341** 082017

[15] Nugraha A K, Maryanto S and Triastuty H 2017 *J. Neutrino* **9(2)** 44–51 doi: 10.18860/neu.v9i2.4103

[16] Nishi K 2005 *Hypocenter Calculation Software GAD (Geiger's method with Adaptive Damping)* (Silver Expert JICA Indonesia version 1)

IOP Conf. Series: Earth and Environmental Science **873** (2021) 012008 doi:10.1088/1755-1315/873/1/012008

Gravity Data Analysis Based on Optimum Upward Continuation Filter and 3D Inverse Modelling (Case Study at Sedimentary Basin in Volcanic Region Malang and Its Surrounding Area, East Java)

I Setiadi[1], Marjiyono[2] and T B Nainggolan[1]
[1]Marine Geological Institute, Jl .Dr. Djundjunan, No. 236, Bandung
[2]Centre for Geological Survey, Jl. Diponegoro No. 57, Bandung

Email : setiadi_i@yahoo.com

Abstract. The study on the fore-arc sedimentary basin for hydrocarbon exploration is rare because of the more complicated geological structures, and conventional seismic methods cannot optimally penetrate the rock layers as there are many volcanic and limestone rocks. One of the natural resources potential in the Southern part of the East Java region, especially in Malang and its surrounding areas is the possibility of hydrocarbons in the fore-arc basin, so research is needed to know the existence of these sedimentary basins. The gravity method is one of the geophysical methods used to assess sedimentary basins based on physical parameters of mass density. The aims of this research are to delineate the sedimentary sub-basin, to find out its structure pattern, interpret subsurface geological and basement configuration. The data analysis approach used in this study involves spectral analysis, upward continuation filter, and 3D inverse modeling. The maximum height for the optimum upward continuation filter is 3000 m, which results in regional and residual anomalies. There were five sedimentary sub-basins identified based on residual gravity anomaly, and the gravity anomalies can also detect structure patterns such as basement high, lineament, and fault pattern. The bedrock is supposed as an intermediate igneous rock with a mass density of around 2.7 gr/cc according to the results of 3D inverse modeling. Deposition from bottom to upward is Mandalika, Nampol, and Wonosari Formations and completed by the uppermost are quaternary volcanic rocks. The inversion modeling results show that the Malang and surrounding areas have thick sedimentary rocks covered by volcanic deposits, which is impressive for further investigation to explore the possibility of the hydrocarbon existence in these areas.

1. Introduction

The Indo-Australian plate's subduction activities influenced the Geological setting of the southern part of East Java. The subduction activities resulted in geological structures that have implications for the potential of natural resources and disasters in this area. One of the Southern parts of East Java's potentials resources is the possibility of hydrocarbons occurrence in the fore-arc basin. The presence of oil and gas seepage on the surface at the volcanic area in the southern part of Java indicates an active petroleum system where volcanic rocks covered the sedimentary basin [1]. The Conventional seismic method cannot optimally penetrate the rock layers as there are many volcanic rocks and limestone. Information about basin boundaries, geological structure trends, and bedrock configuration in the fore-arc basin are essential to be known as initial information for hydrocarbon exploration. One of the geophysical methods that can be used to delineate and interpret sedimentary basins based on physical parameters of mass density is the gravity method. However, some techniques that can be used to separate regional and residual anomalies have been carried out by previous researchers include the use of upward continuation filter and trend surface analysis [2], application of polynomial filters on gravity data [3]. The purposes of this study are to delineate sedimentary sub-basins, interpret geological structure trends, basement configuration, and subsurface geological model due to gravity data analysis using the best upward continuation filter and 3D inverse modeling. This paper will do separation regional and residual anomalies from Bouguer gravity anomalies using the optimum upward continuation technique as proposed by [4]. The Optimum upward continuation filter is carried out by correlation between upward continuation at two successive heights. The maximum deflection results from the cross-correlation of two successive heights will be used to determine the optimum upward continuation filter then applied to calculate regional and residual anomalies. Modeling was done by 3D inversion using grablox software, and geological information is used as a constraint to interpret the subsurface geological model. The study's benefits include designing a 3D model to know subsurface geological conditions based on the physical parameter of density and offering an alternative approach to obtaining the best continuation elevation value in obtaining regional anomalies.

2. Data and Method

This research used gravity data from field measurement carried out by the sedimentary basin research team at the Center for Geological Survey (CGS) in 2017 and geological information as a constraint for analyzing and interpreting of the gravity model. This study used the optimum upward continuation filter and the subsurface geological model using 3D inversion modeling. The step of translating potential field data calculated at a surface level into data at a higher surface level is known as the upward continuation. The basic principle of this method is formulated by [5] as in the equation below:

$$U(x,y,z0-\Delta z) = \frac{\Delta z}{2\pi}\int_{-\infty}^{\infty}\int_{-\infty}^{\infty}\left(\frac{U(x',y',z0)}{[(x-x')^2+(y-y')^2+(\Delta z)^2]^{3/2}}\right)dx'dy' \quad \ldots\ldots\ldots\ldots\ (1)$$

$\Delta z > 0$, with $U(x,y,z0-\Delta z)$ is the total potential field in the upward continuation plane at the height h to the surface area (measurement plane), while the magnitude field $U(x',y',z0)$ was known.
The main problem with this method is to determine the height level of the continuation. When using the upward continuation method, the continuation height is determined subjectively by comparing the level of continuation heights, while the continuation height affects the result of the regional anomaly. It is necessary to find out the optimum upward continuation height value so that the result of the regional anomaly is optimum. One of the possible methods to estimate the optimum upward continuation height is written by Zeng et al. (2007) namely by using cross-correlation to two successive continuation heights. The result of cross-correlation at successive heights will pass the maximum deflection at a certain height which is used to estimate the optimum upward continuation height. The cross-correlation r between the upward continuation anomaly g1 and g2 can be calculated using the following equation (Zeng, et al., 2007).

$$r_{g1g2} = \frac{\sum_{i=1}^{M}\sum_{j=1}^{N} g1(xi,yj)g2(xi,yj)}{\sqrt{\sum_{i=1}^{M}\sum_{j=1}^{N} g1^2(xi,yj)\sum_{i=1}^{M}\sum_{j=1}^{N} g2^2(xi,yj)}} \quad \ldots\ldots\ldots\ldots\ (2)$$

The anomaly still consists of a mixture of regional and residual anomalies when the level height of upward continuation is less than the optimal value. The residual anomaly will decrease and almost disappear if the height of the upward continuation is greater than the optimal value; this condition will create a regional anomaly pattern [4]. Inversion modeling was done by optimizing the Singular Value Decomposition approach to gravity anomalies. The method is based on the linear inversion technique. Singular Value Decomposition (SVD) is a matrix factoring method that is closely related to the singular value of the matrix, which is a well-known numerical analysis technique for diagonalizing the matrix. Other than SVD, the Occam inversion method which consists of Occam density and Occam height is also carried out at this inversion. Occam density is used to optimize or maximize the block density value, resulting in a smoother model, while Occam h (Occam heights) is used to optimize block height or layer depth. The result of this inversion modeling is the mass density distribution of the rock in three dimensions. Inversion modeling is carried out with Grablox software [6], then plotted using Matlab.

3. Result and Discussion

Gravity data that has been processed entirely as well as gridding and contouring produce a gravity anomaly pattern and measurement point distribution as shown in Figure 1. From the figure, it can be seen that the distribution of measurement points is relatively even with a spacing of about 2 Km. The gravity anomaly value between (35 to 121) mGal, the anomaly pattern is divided into three different anomalies patterns, the first is a high anomaly with anomaly range between (100 to 121) mGal, moderate anomaly ranges from (60 to 99) mGal, and Low anomaly has values between (35 to 59) mGal.

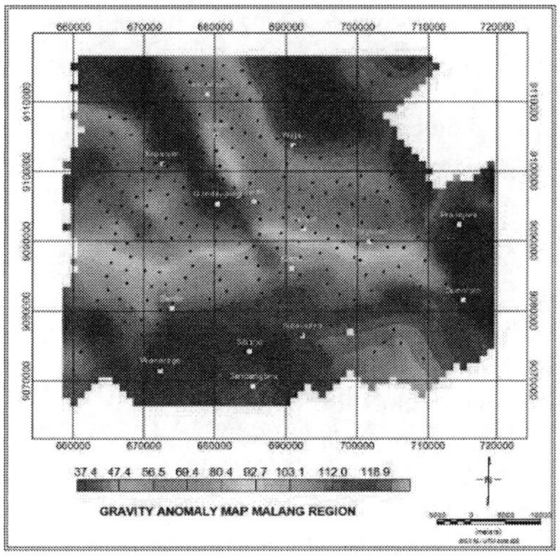

Figure 1. Gravity anomaly map at Malang and its surrounding area.

High gravity anomaly characterized by red color occupies in the southern part and covering area around Sidomulyo, Sitiarjo, Tambakasri, Purwodadi, and Pujiharjo, possibly related to the andesitic basement a mixture of oceanic and continental crust which is in some places appears as an intrusion on the surface. The moderate anomaly is characterized in green color and probably related to the basin's boundaries of the Malang area. Meanwhile, the low anomaly marked in blue color is probably the deepest basin of the Malang area Southern part of East Java. The obtained gravity anomaly pattern was then subjected to an Upward Continuation filter at different heights to determine the best regional anomaly.

IOP Conf. Series: Earth and Environmental Science **873** (2021) 012008 doi:10.1088/1755-1315/873/1/012008

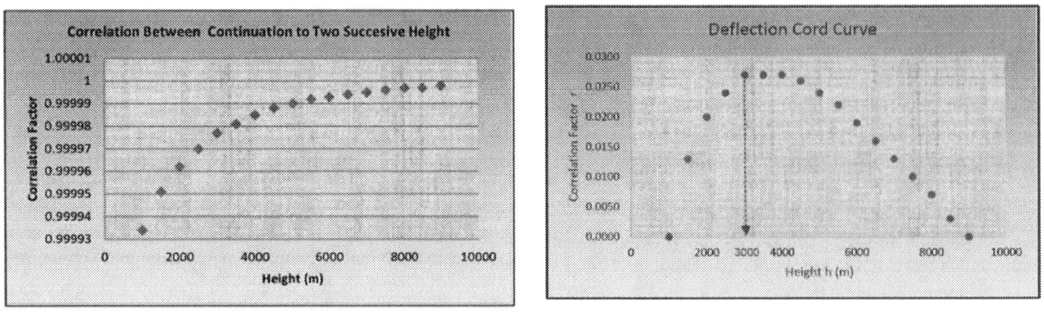

(a) (b)

Figure 2. (a) Cross-correlation upward continuation to two successive height and (b) maximum deflection curve

The upward continuation heights were used starting from 500 m to 9000 m with an interval height of 500 m. The upward continuation was calculated on each height map using a grid size (2 x 2) Km using Geosoft Oasis Montaj software. Cross-correlation was carried out at successive height with an interval of 500 m using Matlab. Figure 2 shows the cross-correlation effects of continuation to two successive heights, as well as the deflection curve resulting from the correlation between two successive heights. The cross-correlation results between two successive heights are then calculated to determine the maximum deflection so that the optimum height of upward continuation can be found and will be used to calculate the residual anomaly. The height associated with the maximum deflection is used to estimate the optimum upward continuation height. The maximum deflection at 3000 m upward continuation height is shown in Figure 2b. When the upward continuation height is less than the optimum height, the upward continuation anomaly has two elements: regional and residual anomalies. As the upward continuation height reaches the optimum value, the residual anomaly will lower and almost vanish. Meanwhile, the residual anomaly will attenuate and reach zero when the upward continuation height is greater than the optimum height. The maximum deflection value indicates the optimum residual variable and has not been attenuated when the upward continuation height is at its optimal value **[7]**. The regional anomaly is obtained from the Bouguer Anomaly from the result of the optimum upward continuation filter at 3000 m height. At the same time the residual anomaly is obtained by subtracting Bouguer anomaly and Regional anomaly. The interpretation of gravity anomaly will be made by qualitative and quantitative. The qualitative interpretation was done to interpret the lateral anomaly changes based on the residual anomaly that has been obtained from the optimum upward continuation filter results.

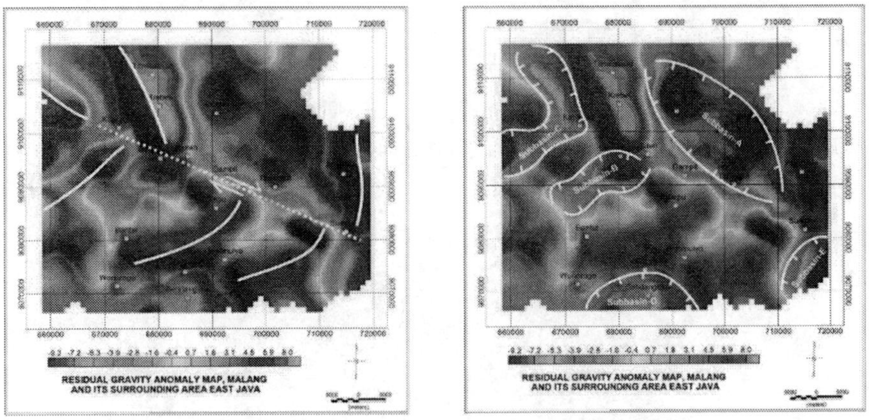

(a) (b))

Figure 3. Qualitative interpretation based on residual anomaly map, (a) lineament patterns and structural trend, (b) sub-basin delineation of the research area

Qualitative interpretation is used to determine the geological structure patterns, alignment, and delineation of the sub-basin. In general, the alignment of the structural geological patterns has a relative southwest-northeast direction as shown in Figure 3a, this is following the regional pattern, namely a Meratus trend which is probably due to the influence of subduction activity in this area. The sub-basin patterns that can be delineated from the residual anomaly consist of 5 sub-basins as shown in Figure 3b. The residual gravity anomaly indicates that the low anomaly (blue color) is supposed as a graben filled by sedimentary rock with low-density values. The presence of the thick sedimentary rocks makes this area interesting for further research, especially regarding the petroleum system of geology, such as to determine the source rock of hydrocarbons, reservoir rock, and structures where hydrocarbons are trapped. The purposes of quantitative interpretation are to interpret the subsurface geological model such as the dimensions, the type of rock based on the physical parameters (density), shape and size of the model. The quantitative interpretation results are expected to know the depth of the bedrock and the composition of the sedimentary rock above it. The quantitative interpretation was carried out by doing 3D inversion modeling using the grablox program [6] which is then plotting using the Matlab program. The results of the 3D inversion and the cross-section direction of the 2D model, the spread of the sedimentary rock based on the 3D inversion model, and the subsurface cross-sections models of the AA' and BB' can be seen in Figure 4. The inversion results show that the rocks density variation ranges between (2.4 to 3.1) gr/cc represent sub-surface density values of the Malang and its surrounding area (Figure 4a). The distribution of sedimentary rocks based on 3D inversion results has density values between (2.4 to 2.55) gr/cc and can be seen in (Figure 4b). The spreading of sedimentary rocks is in the center, northeast, and southwest part of the research areas. The AA' section (Figure 4c) has a relative North-South direction, the model on the AA' section shows that the basement may have a density value of about 2.7 gr/cc which is interpreted as an intermediate igneous rock of the Mandalika Formation (Old Andesite Formation), then above it is Nampol Formation consists of sandstones, marl, clay, and tuff. Above the Nampol Formation is the Wonosari Formation, which is consists of limestone, and the uppermost layer is quaternary volcanic deposits.

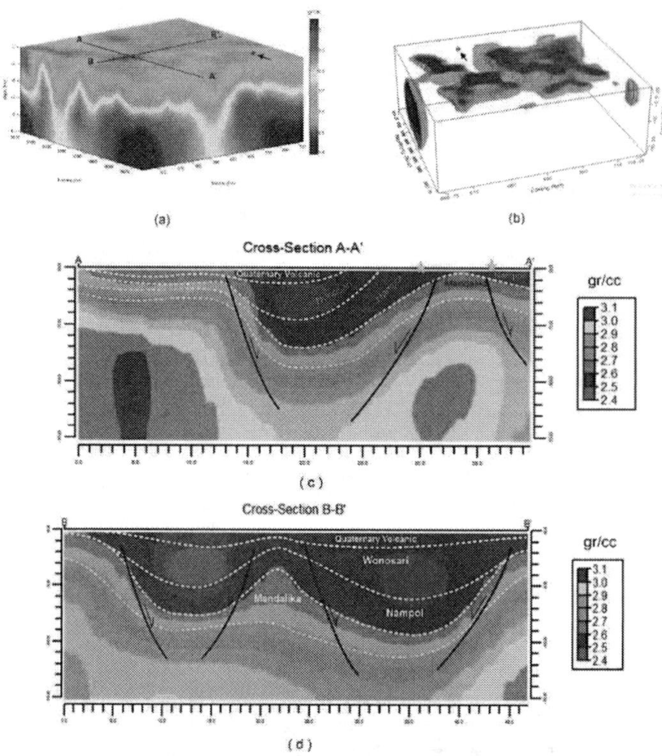

Figure 4. Subsurface geological interpretation based on 3D gravity inversion, (a) 3D inverse modeling result, (b) spreading of sedimentary rocks based on 3D inverse modeling, (c) 2D cross-section AA' and (c) 2D cross-section BB'

Figure 4d shows the cross-section of BB' under the surface of the 3D inversion result. BB's cross-section has a West-East relative direction through the Wonosari Formation and volcanic rocks on the surface. The model results in this section show that the basement undulation which consists of andesite rock is located at a depth between (2-7) km with a density value of about 2.7 gr/cc. Above the basement is the Nampol Formation consists of sandstones, marl, clay, and tuff; above Nampol Formation is the Wonosari Formation which consists of limestone, uppermost layer is the quaternary rock. Based on geological information, the outcrop in the study area is quaternary volcanic rocks, only a small part of the outcrops occupied by sedimentary rocks, namely in the central and southern regions. The gravity analysis results show that in several places where volcanic rocks are exposed on the surface, gravity anomalies and the inversion model show some low anomalies and deep basins. This shows that there are some thick sedimentary rocks which are covered by quaternary volcanic rocks in this location. The presence of thick sedimentary rocks below the volcanic layer is interest for further research, especially the petroleum system that develops in this area.

4. Conclusion

The structure pattern in the study area due to the residual anomalies analysis results has trend in a southwest-northeast direction, this follows the regional pattern namely a Meratus trend, which is probably caused by the influence of subduction activity from the south movement. The optimum upward continuation filter results show that the optimum height is at 3000 m height; this result was used to determine regional and residual anomalies. Regional anomaly shows anomaly with long wavelengths, which is interpreted as regional structures while residual anomaly reflecting anomaly with shorter wavelengths and be interpreted as localized (shallow) structures. The number of sedimentary sub-basins that can be delineated based on the gravity data analysis is five sedimentary sub-basins. The basement in the study area is interpreted to be andesite rocks with a mass density of about 2.7 gr/cc. The gravity analysis results are useful for knowing geological structure, fault, graben patterns, and basement high which are interesting for further research to find more detail of the petroleum system in this area.

References

[1] Satyana, A.H., 2015. *Subvolcanic hidrocarbon prospectivity of Java : Opportunities and challenges*, Proceedings Indonesian Petroleum Association (IPA), Thirty-Ninth Annual Convention and Exhibition, May 2015.

[2] Kebede, H., Alemu, A., and Fisseha, S., 2020. *Upward Continuation and Polynomial Trend Analysis as a Gravity Data Decomposition, case study at Ziway-Shala Basin, Central Main Ethiopian Rift*. Heliyon, vol.6. https://doi.org/10.1016/j.heliyon.2020.203292.

[3] Okiwelu, A.A., Osazua, B.I., and Lawal, K.M., 2010. Isolation of Residuals Determined from Polynomial Fitting to Gravity Data of Calabar Flank, South eastern Nigeria. *The Pacific Journal of Science and Technology*, Volume 11. Number 1, pp. 576-585

[4] Zeng, H., Xu, D., and Tan, H., 2007. *A model study for estimating optimum upward-continuation height for gravity separation with application to a Bouguer gravity anomaly over a mineral deposit, Jilin province, northeast China*. Geophysics Vol. 72, No. 4 July-August 2007; P. I45–I50, 9 Figs., 1tABe.10.1190/1.2719497.

[5] Blakely, R. J., 1996, *Potential Theory in Gravity and Magnetic Applications*, Cambridge University Press, Cambridge.

[6] Pirttijarvi, M., 2008, *Gravity interpretation and modeling software based on 3-D block models. User's guide to version 1.6b*. Department of Physics Sciences. University of Oulu. Finlandia.

[7] Muharina, G.N., Setiadi, I., Santoso, B., and Supriyana, E., 2019. *Delineation of sedimentary sub-basin and basement configuration based on the upward continuation filter and 2d and 3d modeling of gravity data (a case study: Rembang Area and its surroundings)*, IOP Conf. Series: Earth and Environmental Science 311 (2019) 01206, Padjadjaran Earth Dialogues, International Symposium on Geophysical Issues.

IOP Conf. Series: Earth and Environmental Science **873** (2021) 012009 doi:10.1088/1755-1315/873/1/012009

The Estimation of Seismic Site Amplification of Bukittinggi City, West Sumatera, Indonesia

Marjiyono[1], I Setiadi[2] J Setiawan[1]
[1]Centre for Geological Survey, Energy and Mineral Resources Ministry
Jl. Diponegoro No. 57, Bandung
[2]Marine Geological Institute, Energy and Mineral Resources Ministry
Jl .Dr. Djundjunan, No. 236, Bandung

Email : marjiyono@esdm.go.id

Abstract
Bukittinggi city area is exposed substantial risk of both, close to seismic source line and site amplification due to soft surface sediment / soil layer. The Great Sumateran Fault (GSF) which is crossing this area, notorious as a very active seismic source. Its responsible to some major earthquake in the vicinity of the fault line. Generally, Bukittinggi area is covered by volcanic product as pumiceous tuff. We applied combined array and single station microtremor measurement to characterize near surface sediment in this area. Based on analysis microtremor single station and array using spatial autocorrelation (SPAC) method, weathered surface layer thickness is in the range of 0-108 m with shear wave velocity in the range of 62- 190 m/s, while the fresh pumiceous tuff is in the range of 375-629 m/s. The estimation of site amplification in this area which is indicated by mean amplification is in the range of 1-1.95. The high amplification zone is occupies the northernmost area and some area locally in the southern part of Bukittinggi city.

1. Introduction

As a main tourism destination in Sumatera Island, Bukittinggi city is undergoing rapid development. The 25.239 km^2 of Bukittinggi area inhabited by 130.773 populations with annual growth rate 1.55 % [1]. Tectonically, this area is crossed by very active fault structure Sianok Segment of the Great Sumateran Fault (GSF). Some major earthquakes which were occurred due to the activity of the GSF in West Sumatera area in the last century; Padang Panjang (1926), Singkarak (1943), Pasaman (1977), Agam (2003) and Batusangkar Earthquake (2007). The last event was caused 70 casualties and more than 5 thousands building damaged.

The intensity of ground shaking during an earthquake at a specific site is affected not only by the magnitude of the earthquake and its distance to the hypocenter, but also by the local surface sediment. As compared to nearby hard rock sites, soft sediment / soil layers in the surface will amplify seismic waves and cause more ground shaking. Based on geological mapping [2], the surface lithology in this area dominated by pumiceous tuff which is predicted as product of Maninjau eruption. The fresh of this volcanic clastics is outcropped in the bottom of Sianok canyon as a massive ignimbrite tuff, while top layer in the research area generally are soil, weathered of the volcanic clastics. Physically, the soil is as soft layer, therefore its may amplify seismic wave during earthquake occurred. In order to provide the level of site amplification data for spatial planning of future urban growth, it is essential to characterize the surface soft sediment / soil in this region.

IOP Conf. Series: Earth and Environmental Science **873** (2021) 012009 doi:10.1088/1755-1315/873/1/012009

2. Data and Method

The site amplification is known as the amplitude peak of the spectrum ratio between the ground surface and the base layer, and it is influenced by several factors such as the shear wave velocity of the surface sediment and the base layer, the density of the sediment layer, and the internal damping of each sediment layer. According to Kokusho & Sato [3] the ratio of shear wave velocity between the base rock and the surface soft sediment layer is the most important influencing factor. Theoretically, the amplification in a multilayer sediment structure (figure 1) can be expressed as:

$$A = \sum_{i=0}^{N}\left(\frac{\rho_i V_{Si}}{\rho_{(i+1)} V_{S(i+1)}}\right) exp\left(-\frac{\pi f_{oi} h_i}{Q_i V_{Si}}\right) \qquad (1)$$

where ρ_i : density of i-th sediment layer
V_{si}: shear wave velocity of i-th sediment layer
h_i : thickness of i-th sediment layer
f_{oi}: resonant frequency of i-th sediment layer
Q_i : quality factor of i-th sediment layer

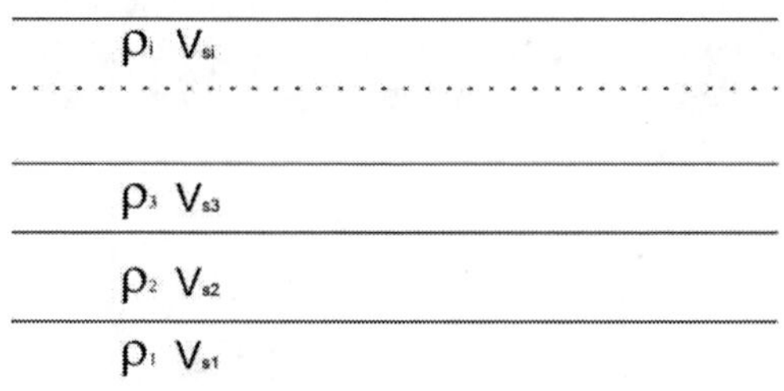

Figure 1. Multilayer of subsurface model for site amplification calculation, the site amplification in the surface is superposition of all individual layer.

In this paper, we approach the estimation of amplification by simple 2 layer subsurface model, i.e. surface soft sediment / soil and a base layer, therefore equation (1) will be :

$$A_2 = \left(\frac{\rho_b V_{Sb}}{\rho_S V_{SS}}\right) exp\left(-\frac{\pi f_o h_S}{Q_S V_{SS}}\right) \qquad (2)$$

where ρ_b : base layer density
ρ_s : surface soft sediment / soil density
V_{sb}: shear wave velocity of base layer
V_{ss}: shear wave velocity of surface sediment / soil layer
h_s : thickness of surface sediment / soil layer
f_o: resonant frequency of surface sediment / soil layer
Q_s : quality factor of surface sediment / soil layer

We applied array and single station microtremor measurement in Bukittinggi area. Array microtremor recording has been shown to provide a fair estimate of the shear wave velocity structure [4]. The spatial autocorrelation (SPAC) method was applied to array data of microtremors for determining shear wave velocity structures. This approach is one of the most effective for extracting surface wave phase velocities from ambient seismic noise [5]. To determine the geometry of the soil layer, it is important to combine the shear wave velocity profiles obtained by the SPAC method with observed resonance frequencies on horizontal to vertical spectral ratio (HVSR) results.

Array microtremor measurements were carried out in eight locations and its distributed evenly in order represent near surface properties in Bukittinggi area, generally. All of 1D velocity profiles which were inverted from the phase velocity dispersion of surface waves then interpolated to provide velocity structure, laterally.
Single station microtremor measurements were carried out in 96 points. The horizontal to vertical spectrum ratio (HVSR) is used to calculate the soft surface sediment layer's resonance frequency. Ibs-von Seht [6] found that the frequency of the main peak of the HVSR are well correlated to the sediment thickness at the site. According to another researcher, HVSR observations provide a reliable estimate of the fundamental frequency of soft sediments resonance over hard rock [7]. Mathematically, the relation between resonant frequency (f_o), the thickness of and the shear wave velocity (V_{ss}) soil layer can be expressed as :

$$\frac{1}{f_o} = \frac{4h_s}{V_{ss}} \tag{3}$$

Based on resonant frequency data which is obtained from single station microtremor measurements and the shear velocity distribution the result of interpolation of shear wave velocity profiles from array microtremor measurements, then the thickness of soil layer can be mapped for all area. The combined-array and single station microtremor measurements allow to derive all parameter (Figure 2) which is needed for site amplification calculation.

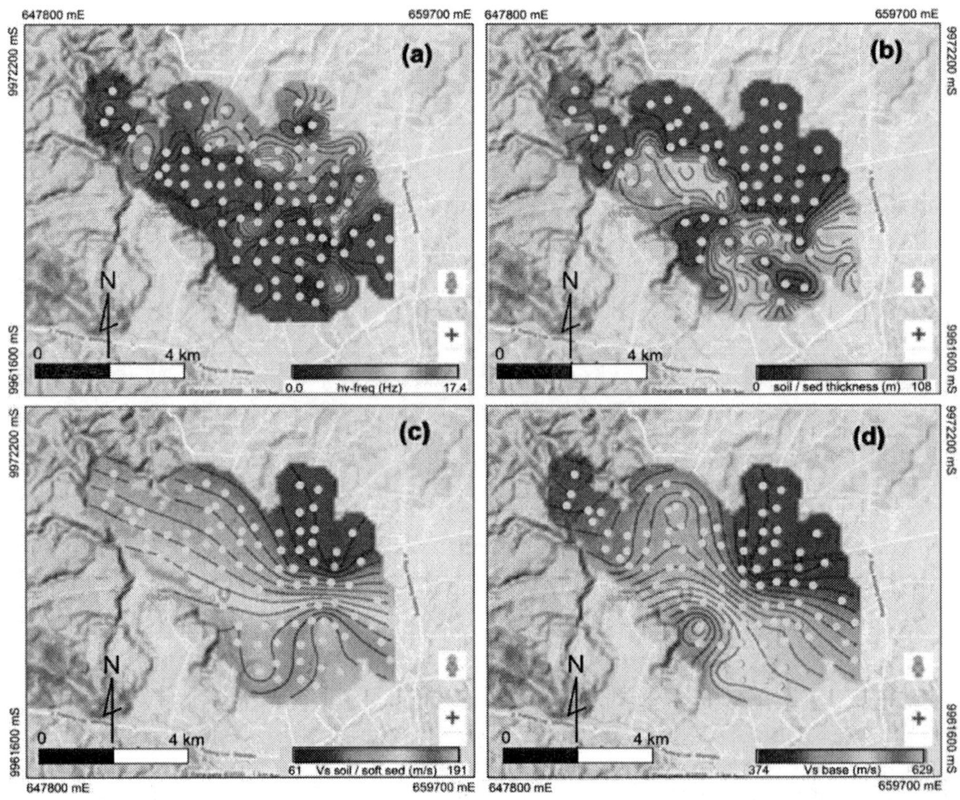

Figure 2. The parameters for site amplification calculation of Bukittinggi area (a) resonant frequency, (b) soil thickness, (c) soil shear wave velocity and (d) base shear wave velocity. The yellow dot represents a single microtremor measurement station.

3. Result and Discussion

There is no doubt that the surface geology as one of dominant factor that effect to ground shacking intensity during an earthquake occurred. Therefore, characterization of surface sediment / soil is become a necessary requirement in seismic hazard reduction.

Since the Bukittinggi region is located near the equator, the weathering intensity is high, and the weathered layer on the ground surface is thick enough. The shear wave velocity profiles in the eight sites indicate that the soil layer is relatively soft, with shear wave velocity values in the range of 61-191 m/s, and the base, which is estimated to be the fresh volcanic clastics layer, has shear wave velocity values in the range of 374-629 m/s (Figure 3). According to equation (2), the site amplification value is primarily a contribution of the impedance contrast between the base and the soil layer.

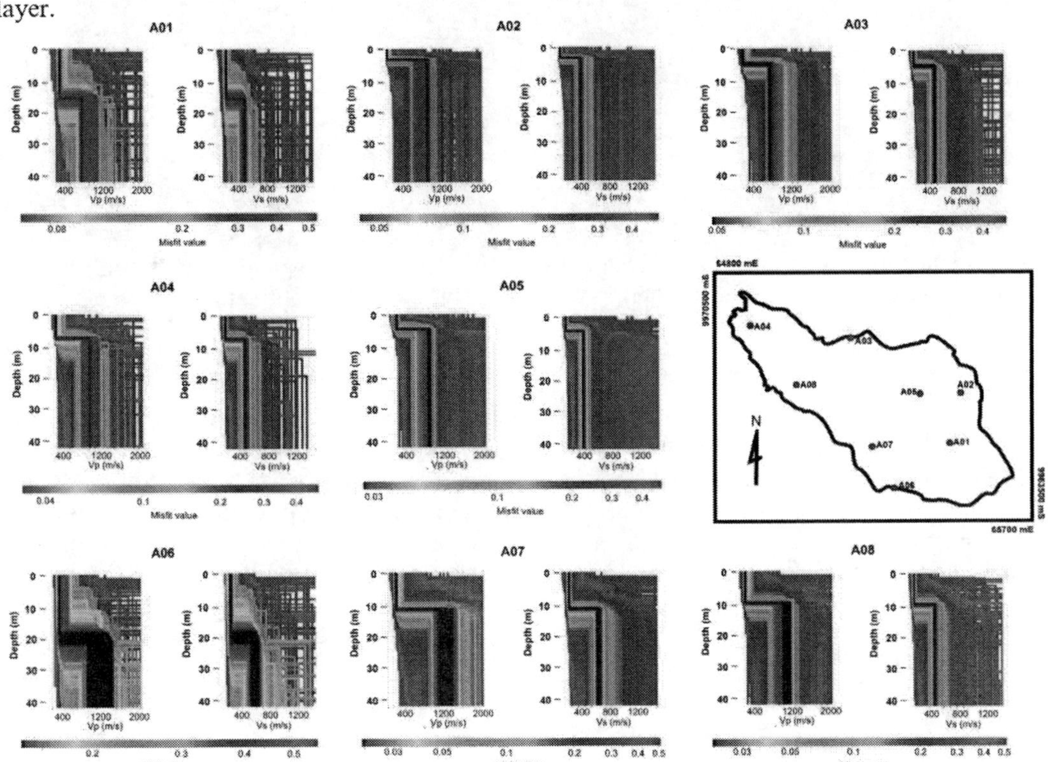

Figure 3. Ground profiles at the Bukittinggi area's eight sites. Soil is seen as the top layer.

The amplification value as a function of frequency in the site amplification spectrum. The maximum amplification value (peak) is correspond to resonant frequency. However, we will not use maximum amplification in this paper since maximum amplification is only valid in a narrow frequency range around the resonant frequency. Instead, we use mean amplification which is valid in the broader frequency range. The result of site amplification calculation shows that the mean amplification in Bukittinggi area is in the range of 1.0-1.91 (Figure 4).

Generally, the high amplification zone is found in the northernmost part of Bukittinggi region, as well as in some areas in the southern part. This importance of site amplification indicates the level of vulnerability due to the possibility of ground shaking during an earthquake.The high amplification zone is not recommended for further development.

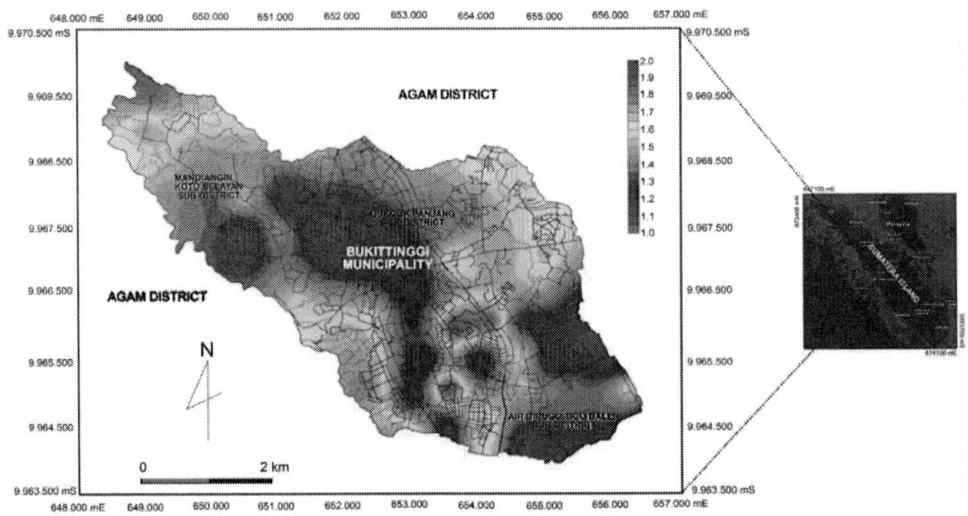

Figure 4. The distribution of mean amplification by soil layer of Bukittinggi area, the strong amplification zones are found in the northernmost with a few spot in the south area.

4. Conclusion

Bukittinggi area is covered by volcanic clastics, which is predicted as the product of Maninjau eruption. The result of shear wave velocity sounding by microtremor array measurements in eight sites in Bukittinggi area, its inferred that the surface ground layer are the weathered soil of volcanic clastics with shear wave velocity in the range of 61-191 m/s. The fresh volcanic clastics at the soil's base are expected to have shear wave velocity in the range of 374-629 m/s. The result of analysis shows that mean amplification value in Bukittinggi area is in the range of 1.0-1.91. The high amplification zones is in the northernmost and locally in some area in the southtern part of Bukittinggi city. This site amplification data is expected used as consideration fo further urban development and also for reevaluation of spatial planning.

References

[1] Tarigan Y A, and Alfindo O 2020 *Bukittinggi Municipality in Figures 2020* BPS-Statistic of Bukittinggi Municipality

[2] Kastowo, Gerhard W L, Gafoer S and Amin T C 1996 *Geological Map of Padang Quadrangle, Sumatera, Scale 1:250.000* Geological Research and Development Centre, Bandung Indonesia

[3] Kokusho T and Sato K 2008 *Site Amplification Formula for Seismic Zonation Based on Downhole Array Records During Strong Earthquake* Proceedings of the 14thWorld Conference on Earthquake Engineering October 12-17 2008 Beijing China.

[4] Kudo K, Kanno T, Okada H, Ozel O, Erdik M, Sasatani T, Higashi S And Yoshida K 2002 Site Specific Issues for Strong Ground Motions during the Kocaeli, Turkey, Earthquake of 17 August 1999 as Inferred from Array Observations of Microtremor and Aftershocks Bulletin of the Seismological Society of America Vol. 92 (1) p 448–465

[5] Luo S, Luo Y, Zhu L and Xu Y 2016 *On the Reliability and Limitations of the SPAC Method with a Directional Wavefield* Journal of Applied Geophysics Vol.126 p 172-182

[6] Ibs-von Seht M and Wohlenberg J 1999 *Microtremor Measurements used to Map Thickness of Soft Sediments* Bulletin of Seismological Society of America Vol. 89 (1) p 250–259

IOP Conf. Series: Earth and Environmental Science **873** (2021) 012009 doi:10.1088/1755-1315/873/1/012009

[7] Field E H and Jacob K H 1995 *A Comparison and Test of Various Site-Response Estimation Techniques* Bulletin of Seismological Society of America Vol. 85 p 1127-1143

The Spatio-temporal Analysis of b-value in the Banda Arc, Indonesia

Muhammad Bani Al-Rasyid[1], Mira Nailufar Rusman[1*], Daniel Hamonangan[1], Pepen Supendi[2], Kartika Hajar Kirana[1]

[1]Department of Geophysics, Universitas Padjadjaran, Sumedang, Indonesia
[2]Agency for Meteorology and Climatology (BMKG), Jakarta 10610, Indonesia

mira17001@mail.unpad.ac.id

Abstract. Banda arc is a complex tectonic structure manifests by high seismicity due to the collision of a continent and an intra-oceanic island arc. Using the relocated earthquakes data from ISC-EHB and BMKG catalogues from the time period of 1960 to 2018, we have conducted a spatial and temporal variation of b-value using the Guttenberg-Richter formula in the area. Our results show that the spatial distribution of low b-values located in the south of Ambon Island and southeast of Buru Island. On the other hand, the temporal variation of b-value shows a decrease in the northern part of the Banda sea probably high potential to produce large earthquakes in the future. Therefore, further mitigation is needed to minimize the impact of earthquakes in the area.

1. Introduction

Banda arc is located in the triple junction of Pacific, Eurasia, and Indo-Australian plates. This region is formed from the convergence of three main plates, the Eurasian continent which is relatively stable, the Australian plate which is relatively moving towards the north, and the Pacific plate which is relatively moving towards the west [1].

Banda Arc is located in a complex tectonic structure and this area is still controversies for many scientists: how it acquired its shape, the origin, and the direction of subduction. During the formation process of the Banda Arc, there was a change in the proto-Banda oceanic crust with the Banda oceanic crust. During the formation of the Banda arc, the Banda Sea was expanded due to the rollback of the Keno-Tethys oceanic crust which subsequently triggered delamination and collision along the edges of the Banda continental crust, forming Timor trough, the Tanimbar trough, and the Seram trough [1]. The Banda Arc region has a frequency of large shallow and deep earthquakes [2,3].

Earthquakes are rapid shaking of the earth caused by the release of energy that occurs suddenly due to rock cracks that stretch beyond their elasticity limit [4]. Historical earthquakes of the Banda Sea show that this area has high seismicity, as shown in Figure 1. The earthquake and tsunami in Seram and Ambon on February 17, 1674, which killed 2322 victims [5]. Another tsunami event was caused by the earthquake that occurred on October 8, 1950, which struck Ambon (Hutumuri, Hative Kecil, and Galala) [5]. Therefore, one of the pieces of information that can be used to analyze the potential for earthquakes is the b-value from the Guttenberg-Richter formula [6]. The b-value will vary depending on the stress distribution [10] and material heterogeneity.

The previous study of spatial variations of b-value has been carried out in several areas. Ngadmanto [8] examined the b-value using the NEIC catalogue from 1973-2009 and stated that the

Banda Arc area where an earthquake with a magnitude 6.5 was around Flores, Alor, and in the west of Seram. The purpose of this study is to an analysis of spatial and temporal of b-value in the Banda arc and its surrounding area.

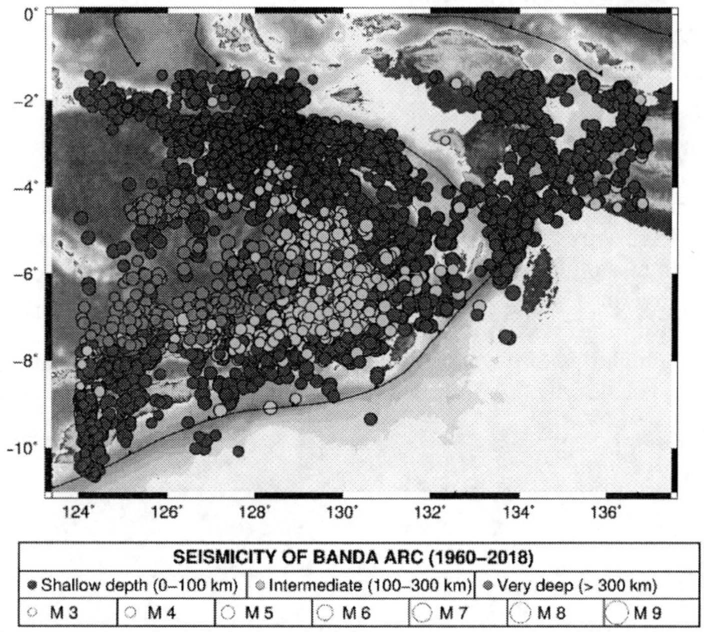

Figure 1. Seismicity of the Banda Arc region from the ISC-EHB and Indonesian Agency for Meteorology, Climatology, and Geophysics catalogue showing all hypocenters with magnitudes greater than 3 Mw between 1960 and 2018. The map shows a pattern of earthquakes that form a spoon-like shape with hypocenters that gets deeper at the inside of a curved Banda Arc.

2. Data and Method

In this study, we used catalogue data from International Seismological Center-Engdahl, van der Hilst, Buland (ISC-EHB) [2] and Indonesian Agency for Meteorology, Climatology, and Geophysics (BMKG) [3] for a time period from 1960 to 2018, with a boundary area of 1.46° S–10.64° S and 124.13° E – 136.87° E, 0-680 km depth. The total number of earthquakes data is 6.779. We used the conversion magnitude formula [8] to converted of initial magnitude from the catalogues to Moment magnitude (Mw). We have 6.488 earthquakes after the declustering process using the Reasenberg method [9]. We used ZMAP [10] to calculate a-value, b-value, and magnitude of completeness (Mc) by plotting the frequency-magnitude distribution.

The seismicity parameters of a region can be calculated using the Gutenberg-Richter formula [6]:

$$log\, n(M) = a - bM \tag{1}$$

n = number of earthquakes
a = a contstant which informed level of seismicity
b = the gradient of the cumulative number relative to the magnitude

b-value can be calculated using the Maximum Likelihood estimation [11] which use the equation:

$$b = \frac{0.434}{M - M\, min} \tag{2}$$

M = average magnitude
Mmin = minimum magnitude

We used grids size of 0.2° x 0.2° with 150 km of a constant radius parameter and 50 minimum number of earthquakes for spatial analysis. Then, we calculated the variation of temporal b-value using the sliding time-window method. The time sliding-window method works by calculating the b-value of the number of earthquakes from the catalogue, then the window is shifted with a fixed number of earthquakes and the b-value for that number of earthquakes, the process is repeated until the last earthquake [8].

3. Result and Discussion

Distribution of b-values for all data from 0.36 to1.61 (Fig. 2a). In the southern part of the Banda arc, shows a void of b-value, this is due to the low of seismicity in the area. Low b-values are seen in the northern and western parts of the study area (the North Banda Arc segment and the Banda Sea part). Further analysis is performed by dividing the event based on depth into three types, which are shallow (less than 60 km), medium (60-300 km), and deep (more than 300 km).

Distribution of b-value for depth less than 60 km range from 0.37 to 1.45 (Fig. 2b). This map shows that the b-value varies greatly, especially in the northern segment, meanwhile the southern segment tends to have a high b-value, especially around Banda Island. Besides that, there are also several earthquake events with Mw > 6.5 located in the southeast of Buru Island.

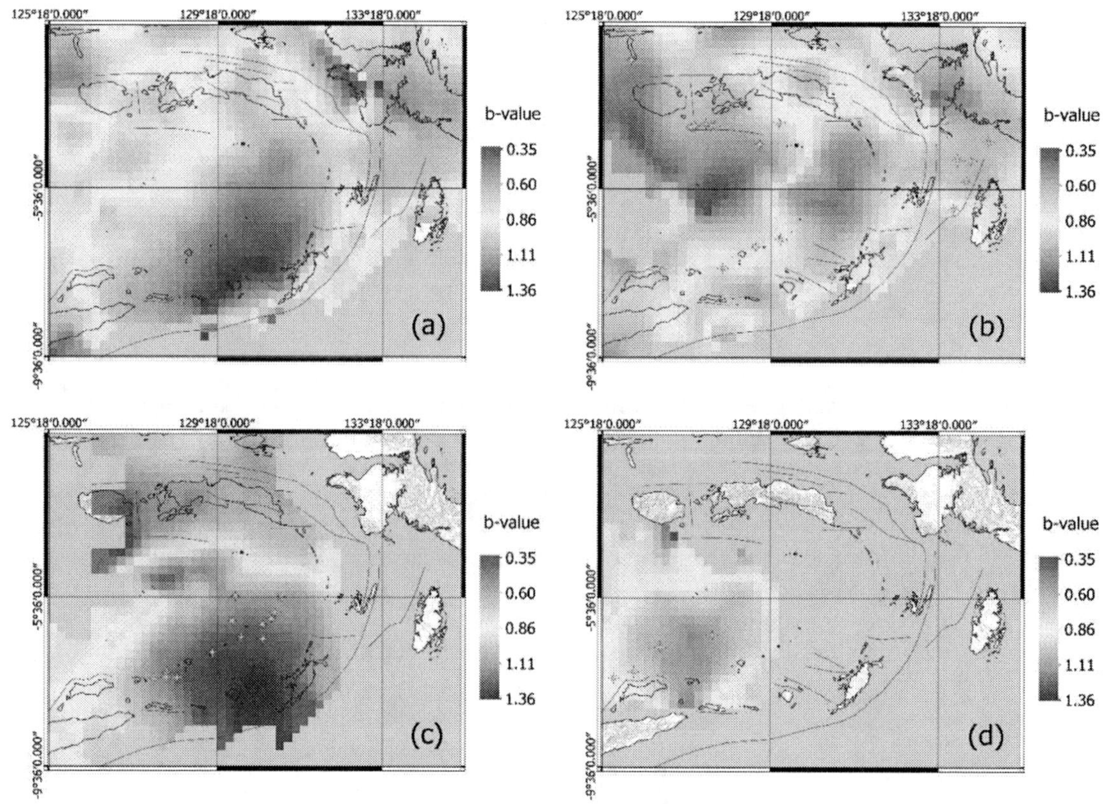

Figure 2. Map of b-value in the Banda Arc region based on the depth of hypocenter: a) Map of all dataset, b) for hypocenter less than 60 km, c) for hypocenter from 60 to 300 km, and d) for hypocenter more than 300 km. Red colour represents high b-value whereas blue colour represents low b-value. Red lines are fault [13]. The yellow stars depict earthquakes $M_w > 6.5$.

The variation of b-value for intermediate-depth (60 - 300 km) is shown in Figure 2c. The number of events at this depth is less than the shallow earthquakes, so the distribution of the b-value tends to be less and local. The b-values range from 0.43 to 1.56. b-value tends to decrease for deeper earthquakes. The distribution of b-value tends to be distributed in the center of the Banda Arc. The difference in b-value is clear between the northern and southern parts of Banda Arc, the distribution of low b-values is located at 2° – 4° S and 126° - 132° E, while high b-values are spread over 4° - 9° E and 126° - 132° E. There are earthquake events with Mw > 6.5 which are located in areas with a high b-value distribution. Figure 1d. shows the distribution of b-values for depths more than 300 km. It shows that the distribution of b-values is centered on the center of the Banda Arc with a variation of the value that has a smaller range that is 0.49 - 1.34. There were also several earthquake events with Mw > 6.5 at depths more than 300 km.

3.1. *Spatial Analysis*

The distribution map of the b-value (Fig. 2) indirectly shows the seismic distribution of the Banda Arc. This map shows that as the depth of earthquakes increases, the seismicity starts to cover a smaller area and more concentrated in the inside of Banda Arc. Generally, the distribution of b-value concentration can describe the distribution of seismic activity that corresponds with the tectonic setting of the Banda arc. The slab reaches a depth of > 300 km in the western part of the study area and shallow to the east in the shape of a spoon [1].

The distribution of b-values at depths > 300 km are quite varied. A low b-value usually correlates with a high level of stress and its release can occur any time which causes a large earthquake [10]. However, at a depth of > 300 km, the b-value distribution shows a value > 1, it means that the area has relatively low stress. Based on the data in this study, large earthquakes at this depth only occurred in 1969 (Mw 7.2) and 1982 (Mw 7.5).

At a depth of 60-300 km, the b-value shows a pattern that forms a segmentation in the research area into northern and southern parts. The north shows a relatively low b-value, while the south shows the opposite. The contrast of b-value (between the northern and southern parts) tends to decrease towards the east, this phenomenon is presumed to be caused by the shape of the slab which resembles a spoon, and by the stress distribution which tends to be more uniform. However, this requires further research. At a depth of 60-300 km, several large earthquakes were recorded, mostly scattered in the Inner-Banda Arc area such as the 2009 event of Mw 7.2, and the 2012 event of Mw 7.5.

The potential for a large earthquake to occur at a depth of 60 - 300 km, especially in the northern part, is greater than that in the southern part because the stress level tends to be low as indicated by the high b-value. Based on the historical earthquakes, most of the major earthquakes occurred in the southern part. Thus, the stress in the south may have been accommodated by previous earthquake events. In contrast to the northern part, which has a relatively low b-value, no earthquakes with Mw > 6.5 are found.

Based on the distance of hypocenter from the surface, large earthquakes that occur at a depth of more than 60 km tend to experience energy absorption by the mantle and the earth's crust so this research focuses on depths less than 60 km. The distribution of b-values for shallow depths (< 60 km) shows varying values, especially in the north. Meanwhile, the southern part shows relatively low value. This variation of b-value in the northern part is thought to be related to the distribution of stress due to the presence of Seram Strike-slip and Fault Weight as an earthquake generator.

In the northern part of the Banda Arc, the distribution of low b-values is visible in the south of Ambon or southeast of Buru Island and Banda Island. It appears that there were large earthquakes concentrated in the Southeast of Buru Island. The low stress at this location is thought to be related to the existence of the South Buru Fault. Based on the catalogue data used, no earthquakes Mw> 7 were found, but there were several earthquakes above Mw> 6.5, including Mw 6.8 occurred in 2015 in the south of Seram Island, and Mw 6.5 occurred in 2001 in the Southeast of Buru Island.

3.2. *Temporal Analysis*

The temporal b-value analysis is only carried out for earthquakes with a depth of fewer than 60 km. Figure 3a. shows the distribution of b-value for the Northern Banda Arc, the highest value is 1.59 that occurred in 1973 and the lowest is 0.6341 in 2012. Earthquakes with a magnitude greater than or equal to Mw 6.5 in the northern part of the Banda Arc usually occurred when the b-value decreases. Although an anomaly occurred in 1995 where an earthquake with Mw 6.8 occurred at the same time with a high b-value. Something similar happened in the southern part of the Banda Arc (Fig. 3b), it is shown that two of the three earthquakes that occurred with a magnitude greater than or equal to Mw 6.5 occurred simultaneously with decreasing b-value. The highest b-value for the southern part occurred in 1979 at 1.521 and the lowest b-value occurred in 2006 at 0.99.

Analysis of the b-value concerning time, generally shows that the occurrence of a large earthquake is associated with a low b-value, but it should be considered that there are generalizations of the b-value in a fairly wide area. This is what creates anomalies in the north (1995) and south (1991). The trend of changes in b-value over time shows that in 2018 there are differences between the northern and southern segments. The northern segment shows a decrease in the b-value, while the southern segment is relatively constant.

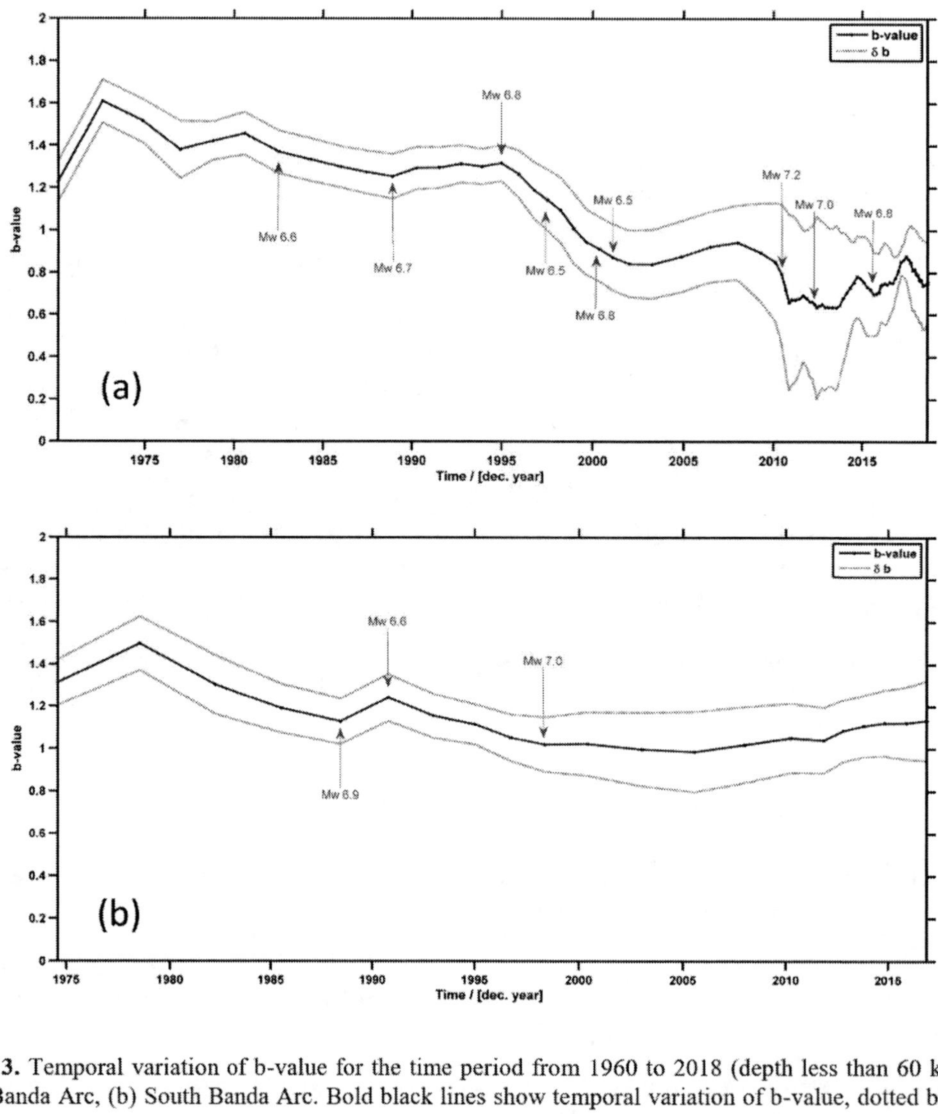

Figure 3. Temporal variation of b-value for the time period from 1960 to 2018 (depth less than 60 km) in (a) North Banda Arc, (b) South Banda Arc. Bold black lines show temporal variation of b-value, dotted black lines

represent the standard deviation of the b-value, the red arrow indicates earthquakes with a magnitude above Mw 6.5.

4. Conclusion

The spatial distribution of b-values using the ISC-EHB and BMKG earthquake catalogues from the time period from 1960 to 2018 at a depth of less than 60 km shows b-value variations with a range from 0.36 to 1.45. Low b-values are spread in the south of Ambon and Southeast of Buru Island. The temporal variation of the b-value shows a decrease in the North Banda Arc. Our results indicate that these areas have the potential to produce large earthquakes. The results of this study can be used as additional information in earthquake mitigation efforts.

Acknowledgements

We would like to thank ESC-EHB and BMKG for earthquake catalogue, QGIS for creating b-value map, and GMT for displaying the seismicity of Banda Arc.

References

[1] Spakman W and Hall R 2010 Surface deformation and slab-mantle interaction during Banda arc subduction rollback *Nat. Geosci.* **3** 562–6
[2] International Seismological Centre (2020), ISC-EHB Dataset
[3] Supendi P, Nugraha A D, Widiyantoro S, Pesicek J D, Thurber C H, Abdullah C I, Daryono D, Wiyono S H, Shiddiqi H A and Rosalia S 2020 Relocated aftershocks and background seismicity in eastern Indonesia shed light on the 2018 Lombok and Palu earthquake sequences *Geophys. J. Int.* **221** 1845–55
[4] Sapiie B, Magetsari N A, Harsolumakso A H and Abdullah C I 2010 *Geologi Fisik* (Bandung: ITB)
[5] Hamzah Latief, Ardito M. Kodijat, Dominic Oki Ismoyo, Bustamam, Dini Adyasari, Navisa Nurbandika H P R 2016 Air Turun Naik di Tiga Negeri Mengingat Tsunami di Ambon 1950
[6] Gutenberg B and Richter C F 1944 Frequency of earthquakes in California *Bull. Seismol. Soc. Am.* **34** 185–8
[7] Scholz C H 1968 The frequency-magnitude relation of microfracturing in rock and its relation to earthquakes *Bull. Seismol. Soc. Am.* **58** 399–415
[8] Ngadmanto D 2010 Penentuan Potensi Gempa Bumi Merusak Berdasarkan Parameter Kegempaan di Wilayah Busur Banda *Widyariset* **13** 125–32
[9] Reasenberg P 1985 Second-order moment of central california seismicity, 1969–1982 *J. Geophys. Res. Solid Earth* **90** 5479–5495
[10] Wiemer S 2001 A Software Package to Analyze Seismicity: ZMAP *Seismol. Res. Lett.* **72** 373–82
[11] Utsu T 1965 A method for determining the value of b in a formula of log N=a-bM showing the magnitude frequency relation for earthquakes *Geophys. Bull. Hokkaido Univ* **13** 99–103
[12] Irsyam M, Widiyantoro S, Natawidjaya D H, Meilano I, Rudyanto A, Hidayati S, Triyoso W, Hanifa N R, Djarwadi D, Faizal L and Sunardjito 2017 Peta Sumber dan Bahaya Gempa Indonesia Tahun 2017.
[13] Wessel P and Smith W H F 1996 A global, self-consistent, hierarchical, high-resolution shoreline database *J. Geophys. Res. B Solid Earth* **101** 8741–3
[14] Wessel P, Luis J F, Uieda L, Scharroo R, Wobbe F, Smith W H F and Tian D The Generic Mapping Tools version 6 *Geochemistry, Geophys. Geosystems* **20** 5556–5564

Preliminary study of stress changes evolution on central part of Sumatran Fault influenced by large interplate earthquakes and tectonic stress rates

Muhammad Taufiq Rafie[1,*], Phil R Cummins[2,3], David P Sahara[2], Sri Widiyantoro[2,4], Wahyu Triyoso[2], Andri Dian Nugraha[2]

[1]Graduate Program of Geophysical Engineering Program, Faculty of Mining and Petroleum Engineering, Institut Teknologi Bandung, Bandung 40132, Indonesia
[2]Global Geophysics Research Group, Faculty of Mining and Petroleum Engineering, Institut Teknologi Bandung, Bandung 40132, Indonesia
[3]Research School of Earth Sciences, Australian National University, Canberra ACT 2601, Australia
[4]Faculty of Engineering, Maranatha Christian University, Bandung 40164, Indonesia

[*]Corresponding author: taufiqrafie@students.itb.ac.id

Abstract. The inland seismic activity in Great Sumatran Fault (GSF) has significantly increased over the past several decades after the occurrence of historical large interplate earthquakes along the plate boundary. This condition led to some occurrences of historical intraplate earthquakes along Sumatran fault. To quantitatively examine the physical mechanisms between intraplate earthquakes and interplate earthquakes, we estimated the static coseismic stress changes of Coulomb failure function (ΔCFF) using receiver fault approach from large historical-recorded interplate earthquakes and the increase in tectonic stress rates. We examined this research in the central part of GSF since this zone is assumed to have the most heterogeneous stress field and thus became our focus study area. The cumulative ΔCFF models showed almost all segments in the central part of GSF suffered negative changes (<-0.1 MPa) which assumed to be unlikely to rupture in short time. However, the preliminary analysis of the increase in tectonic stress rate indicated that large intraplate earthquakes occurred on Angkola and Siulak segments were dominantly influenced by the increase in interseismic stress rate just after the series of large subduction earthquake occurrences, apart from the decreased stress changes from those major interplate earthquakes.

Keywords: Interplate earthquakes, Sumatran Fault, Tectonic stress rates

1. Introduction

The Sunda subduction margin is an active tectonic convergence zone between the subducting Indian-Australian plate and the over-riding Sundaland plate. Along this plate boundary, the accumulated strain energy during earthquake cycle have resulted in numerous large interplate-subduction earthquakes occurrences such as the great 2004 Sumatra-Andaman and 2005 Nias-Simeulue earthquakes (see Fig. 1a and Table 1). Sumatran fault system or known as Great Sumatran Fault (GSF) which is close to these large subduction earthquake ruptures was formed by where the accumulated strain is partitioned into right-lateral/dextral strike-slip component due to oblique plate motion which almost parallel to the subduction trench [1, 2]. One interest GSF segment is the largest fault bends in the central part of GSF

IOP Conf. Series: Earth and Environmental Science **873** (2021) 012011 doi:10.1088/1755-1315/873/1/012011

which formed strike-slip duplex structure and named as 'equatorial bifurcation zone' [2]. For our focus study area, we treated the central part of GSF from Angkola segment to Siulak segment which we have found interesting since it is assumed to have the most heterogeneous stress field along GSF [3]. As shown in Fig. 1b, many historical inland earthquakes have occurred repeatedly for over the past 40 years after many historical large interplate-subduction earthquakes occurrence. To explain the activity of inland earthquakes, we estimated the effect of static coseismic stress changes from historical large subduction earthquake and the tectonic stress rate increase on GSF. This is a crucial thing to consider that if the tectonic stress rate is quite steep, then the stress drop on a fault caused by the earthquakes will vastly recover by the interseismic stress rate increase. Using the calculated coseismic stress changes and the tectonic stress rate increase, we try to analyse the possible linkage that can explain the occurrence of large inland earthquakes in the central part of the Great Sumatran Fault (GSF).

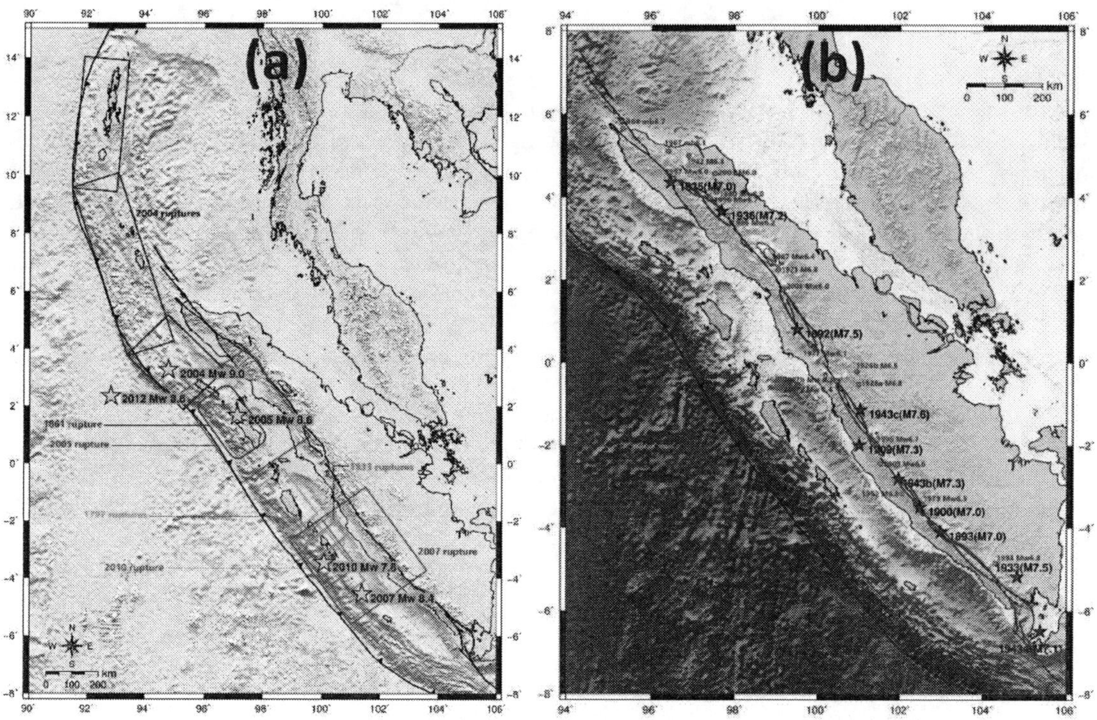

Figure 1. Map of rupture area of each large interplate earthquake in Sumatra subduction zone (a) and map of epicenter location of historical large intraplate earthquakes along GSF (b). The rupture area of large interplate earthquakes based on slip model listed in Table 1 while large intraplate earthquake along GSF based on [4].

Table 1. Lists of historical earthquakes recorded in Sumatra subduction zone.

List historical earthquake	Mw	Slip model used for ΔCFF models
1797	~8.0	[5]
1833	~9.0	[5]
1861	~8.5	[6]
2004 Sumatra-Andaman	~9.0	[7]
2005 Nias-Simeulue	~8.6	[8]
2007 Bengkulu	~8.4	[8]
2009 Padang	~7.5	[8]
2010 Mentawai	~7.8	[8]

2. Methodology

2.1. Estimation of static Coulomb Failure Function (ΔCFF) *by large interplate earthquakes*
ΔCFF is calculated using Coulomb failure assumption which defined as follows [9]:

$$\Delta CFF = \Delta\tau + \mu(\Delta\sigma + \Delta P) \quad (1)$$

Where $\Delta\tau$, $\Delta\sigma$ and ΔP are the changes in shear stress on a fault plane (positive in fault slip direction), normal stress on a fault plane (positive in extension/clamping fault) and changes in pore pressure, respectively. The changes in pore pressure are often related to the mean stress changes and assumed to be proportional to compressional stress changes (normal stress changes), thus the equation (1) is redefined as follows:

$$\Delta CFF = \Delta\tau + \mu'\Delta\sigma \quad (2)$$

Where μ' is the apparent (effective) of friction which derived from the [10]'s coefficient effect and homogeneous medium [11]. Its value can be considered in the range of 0.2-1.0, and in this study, we used the value of 0.2 because we consider the Sumatran fault as major transform fault system which assumed to have low friction coefficient [12]. Here we used the receiver fault of specific orientation approach in resolving the ΔCFF with fixed geometries (strike, dip, and rake) of a well-known GSF orientation. To calculate ΔCFF, we need the source and fault parameter of large interplate earthquake. The fault parameters of historical records of large interplate earthquakes are estimated from previous studies while fault of orientations of Sumatran segment are based on geological measurements by [2]. Then, we estimated the static coseismic Coulomb stress changes under the assumption of elastic dislocation approach [13] from a given fault slip model by using COULOMB3.4 code [14,15].

2.2. Calculation the increase in tectonic stress rate
We estimated the tectonic stress rate increase under the assumption of Time Predictable (TP) model by [16] which stated that stress drop of interplate-subduction earthquake will recover by tectonic stress rate increase prior to the next one, so that we can infer the tectonic stress rate increase as $\Delta\tau/T$, where $\Delta\tau$ is the static coseismic stress drop by a large earthquake occurred on a fault and T is the fault average recurrence time interval. [17] calculated right-lateral slip rates along GSF and used those slip rates to estimate the average event recurrence intervals of each segment from their expected maximum magnitude of earthquake. They roughly found that average recurrence intervals are ranging from about $4x10^2$ year in Southern Sumatra to $2x10^2$ year in Northern Sumatra. In this case, we took the value of 200 year of mean recurrence interval time in central part of GSF. For simplicity, we assumed the maximum earthquake to be occurred along these segments. From the given assumption, we estimated the uniform stress drop to be 4 MPa. Thus, we expected the uniform tectonic stress rate increase to be 0.02 MPa/year.

3. Results and discussion

Coseismic stress changes were calculated using coseismic slip models. These models were obtained from various authors included [5] for 1797 and 1833 earthquake and [8] for 2005, 2009, 2007 and 2010 earthquakes which collected through SCRMOD [18]. The ΔCFFs were resolved at 10 km depth. For simplicity, we also assumed that ΔCFF of 1861 earthquake has the same effect as 2005 Nias-Simeulue earthquake since 1861 rupture area was in the rupture patch of 2005 earthquake [19]. Fig. 2 showed stress changes on each receiver fault for each large interplate earthquakes. For 1797 earthquake, increase stress zone was located from Sianok segment to Siulak segment with > 1 bar while Angkola to Sumpur segments were decreased (< -1 bar). Positive stress changes also recorded in Siulak segment due to 1833 earthquake while the rest had negative stress changes (< -1 bar in Sianok, Sumani and Suliti segments). The ΔCFF model caused by 2004 Sumatra-Andaman earthquake showed any no significant effect, and this is happened since its stress changes distribution is far away and did not reach into the central part of Sumatran segment. When 2005 Nias-Simeulue earthquake occurred, part of Angkola, Barumun, Sumpur and Sianok segments recorded high stress up to > 1 bar. Almost all segments in central part of

Sumatra suffered stress shadow zone caused by 2007 Bengkulu earthquake except Sumpur segment which recorded a slight increase of stress. ΔCFF models of 2009 Padang and 2010 Mentawai earthquakes showed a small effect of stress changes on each segment and it was expected since those two recent earthquakes had moderate magnitude (< 8) compared to other interplate earthquakes.

Fig. 3a showed a cumulative ΔCFF of all large interplate earthquakes, we can see that almost all segments recorded negative ΔCFF (<~ -0.1 MPa) which initially might assume that central part of Sumatra segment is unlikely to rupture in short time. Some of those stress shadow zones correlated well with zone of seismic gap inferred by [4] especially on Barumun, Sumpur, and Sianok segments. Stress changes evolution after 1861 earthquakes (Fig. 3b) showed that Siulak segment is likely to have been advanced towards failure since it recorded increase stress changes and later ruptured as 1909 earthquake. However, Angkola and Suliti segments which suffered decrease stress changes ruptured during 1892 and 1943 earthquakes, respectively. In this case, we still need more analysis on the assumption that the central part of Sumatra is retreat from failure because stress shadow zone can be triggered if the increase in tectonic stress rates is high enough to recover the fault so that the fault could meet its critical point [20 21 22]. To initially investigate this, we have calculated the uniform tectonic stress rate increase which is around 0.02 MPa/year. To make the calculation simple, we used 4 MPa as uniform stress decrease in Sumatra fault system/GSF where this value is actually almost similar to the coseismic stress drop around the rupture zone of large subduction earthquake [23]. We analysed the effect of tectonic stress rate increase in triggering the intraplate earthquakes that occurred in Angkola and Siulak segments. Fig. 4 illustrated the ΔCFF time evolution of Angkola and Siulak segment since 1771 as time reference where the time was taken only after a year of 1770 event [24]. From Fig. 4a, we can see that after the series of large interplate earthquakes occurrence, Angkola segment ruptured around 3 decades (1892 M7.5) and 8 decades (1977 M6.1) afterwards while Siulak segment (Fig. 4b) ruptured for about 5 (1909 M7.3) decades and 8 decades (1995 M6.7) afterwards. Generally, these results are in agreement with study by [25] where the next large inland earthquake can only be occurred during several decades before the subsequent large subduction earthquake. From this preliminary analysis, we initially hypothesize that the increase of tectonic stress rate around central part of GSF is dominantly influence the faults to be closer failure, triggering a major inland earthquake, and less affected by the ΔCFF loading on the fault.

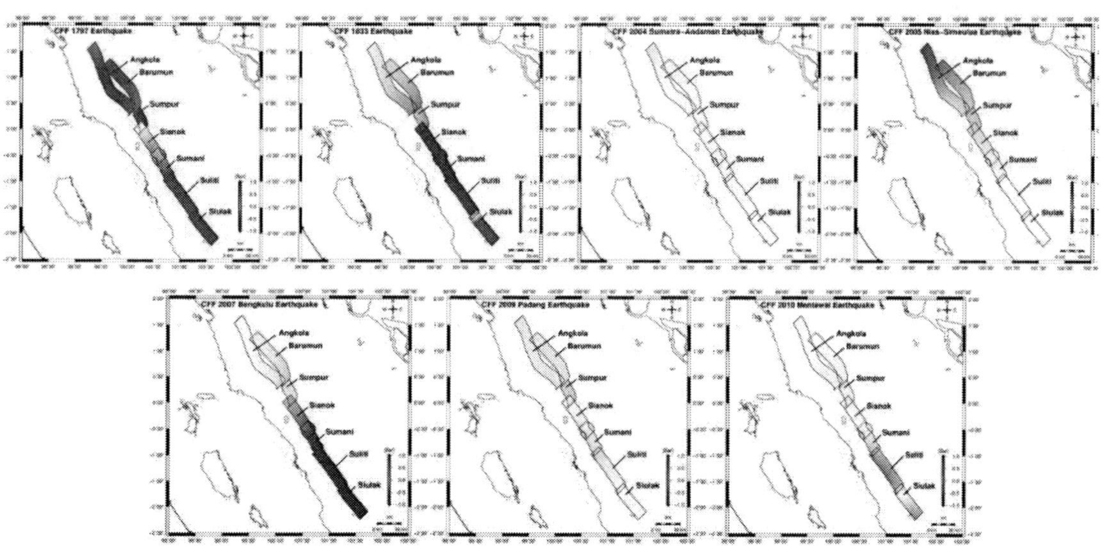

Figure 2. Coulomb stress models resolved on receiver faults of central part of GSF from coseismic slip model of each large interplate earthquakes. The color represents the maximum stress changes at 10 km depth with a scale saturated at 1 bar.

IOP Conf. Series: Earth and Environmental Science **873** (2021) 012011 doi:10.1088/1755-1315/873/1/012011

4. Conclusions

We investigated the effect of stress changes in central part of GSF using static ΔCFF by the large interplate earthquakes along Sunda subduction margin. Based on the estimated stress changes on inland faults, we infer that the stress changes accumulation is low enough around the central part of GSF to be ruptured. However, the initial analysis of the uniform increase in tectonic stress rates on Angkola and Siulak segments showed that the tectonic stress rates is high enough to recover the segments to be critical to failure and become stress trigger zones. Since it is a preliminary analysis, we still need more investigation to better explain the triggering mechanisms, need more historical interplate earthquakes to give well-details ΔCFF models, and detailed tectonic stress rate increase on each segment. Furthermore, an advanced approach is also necessary to explain the postseismic stress loading (i.e., viscoelastic response function).

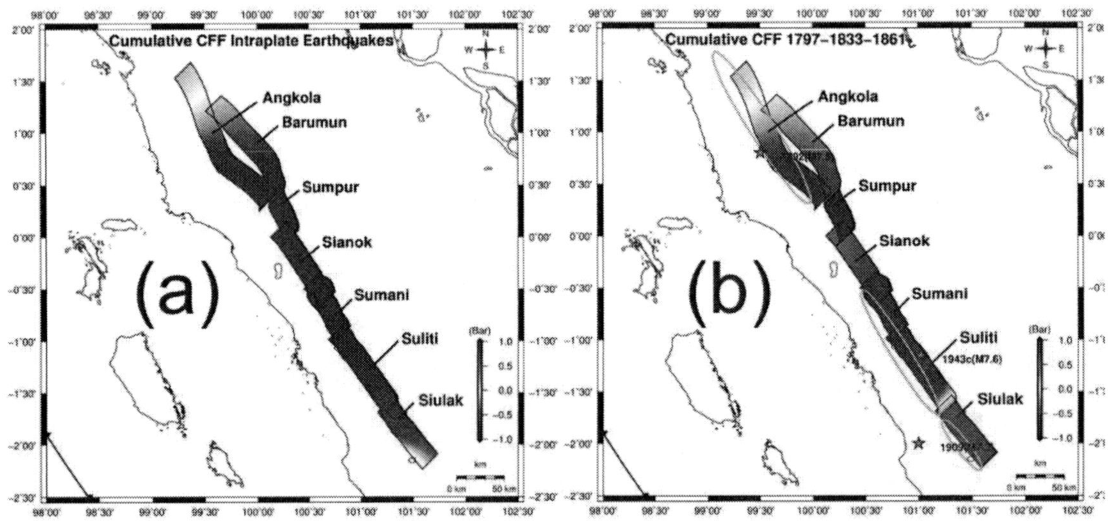

Figure 3. Cumulative ΔCFF of each earthquake listed in Table 1 (a) and cumulative ΔCFF of 1797, 1833, and 1861 earthquakes (b). The cyan ellipses are the damage area of large intraplate earthquakes marked as green star. The ΔCFF is calculated at 10 km depth with a scale saturated at 1 bar.

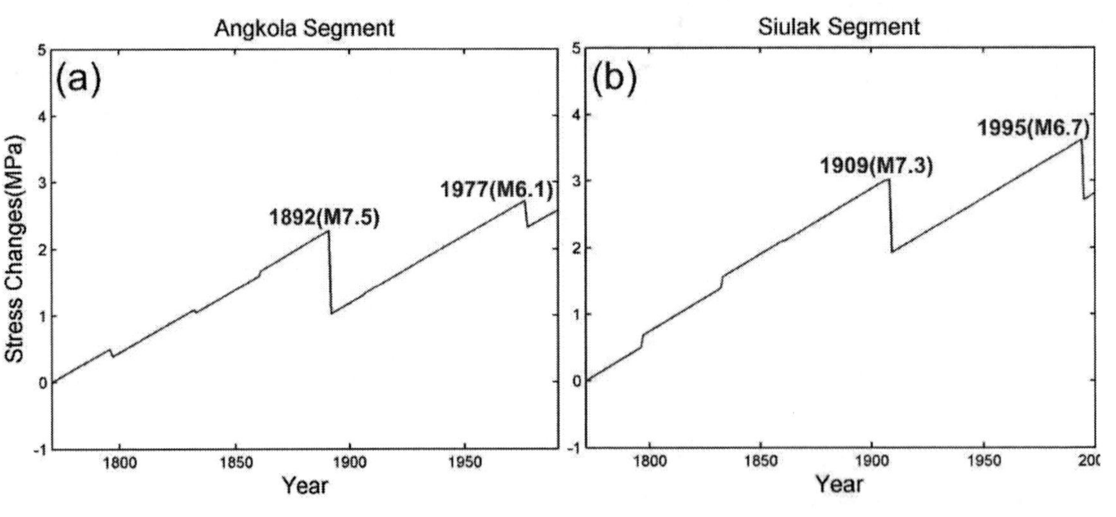

Figure 4. ΔCFF time evolution which included the simple estimation of tectonic stress rates on Angkola segment (A) and Siulak segment (B) with historically recorded large intraplate earthquakes.

Acknowledgments
This research was supported by the Directorate General of Resources for Science, Technology, and Higher Education of KEMRISTEKDIKTI/BRIN through Program Magister menuju Doktor untuk Sarjana Unggul (PMDSU) scheme which has given to MTR.

References

[1] Fitch T J 1972 *Journal of Geophysical Research* **77** 4432-4460
[2] Sieh K, Natawidjaja D 2000 *Journal of Geophysical Research* **105** 28295-28326
[3] Sahara D P, Widiyantoro S, Irsyam M 2018 *Journal of Asian Earth Sciences* **164** 1-8
[4] Hurukawa N, Wulandari B R, Kasahara M 2014 *Bulletin of Seismological Society of America* **104** (4) 1750-1762
[5] Natawidjaja D H, Sieh K, Chlieh M, Galetzka J, Suwargadi B W, Cheng H, Edwards R L, Avouac J P, Ward S N 2006 *Journal of Geophysical Research* **111** (B06403)
[6] Rivera L, Sieh K, Helmberger D, Natawidjaja D 2002 *Bulletin of Seismological Society of America* **92** (5) 1721-1736
[7] Ammon C J, Ji C, Thio H, Robinson D, Ni S, Hjorleifsdottir V, Kanamori H, Lay T, Das S, Helmberger D, Ichinose G, Polet J, Wald D 2005 *Science* **308** 1133-1139
[8] Hayes G P 2017 *Earth and Planetary Science Letters* **468** 94-100
[9] Oppenheimer D H, Reasenberg P A, Simpson R W 1988 *Journal of Geophysical Research* **93** 9007-9026
[10] Skempton A W 1954 *Geotechnique* **4** 143-147
[11] Harris R A 1998 *Journal of Geophysical Research* **103** (B10) 24347-24358
[12] Parson T, Stein R S, Simpson R W, Reasenberg P A 1999 *Journal of Geophysical Research: Solid Earth* **104** 20183-20202
[13] Okada Y 1992 *Bulletin of Seismological Society of America* **82** 1018-1040
[14] Lin J, Stein R S 2004 *Journal of Geophysical Research: Solid Earth* **109** (B2)
[15] Toda S, Stein R S, Richards-Dinger K, Bozkurt S B 2005 *Journal of Geophysical Research: Solid Earth* **110** (B5)
[16] Shimazaki K, Nakata T 1980 *Geophysical Research Letters* **7** 279-282
[17] Bellier O, Sebrier M, Pramumijoyo S, Beaudoiun T, Harjono H, Bahar I, Forni O 1997 *Journal of Geodynamics* **24** 169-183
[18] Mai P M, Thingbaijam K K S 2014 *Seismological Research Letters* **85** (6) 1348-1357
[19] Meltzner A J, Sieh K, Chiang H -W, Wu C C, Tsang L L, Shen C -C, Hill E M, Suwargadi B W, Natawidjaja D H, Philibosian B E, Briggs R W 2015 *Quartenary Science Reviews* **122** 258-281
[20] Shikakura Y, Fukahata Y, Hirahara K 2014 *Journal of Geophysical Research: Solid Earth* **119** 502-518
[21] Hori T, Oike K 1999 *Tectonophysics* **308** 83-98
[22] Mildon Z K, Roberts G P, Walker J P F, Toda S 2019 *Nature communications* **10** (1) 1-9
[23] Yoshioka S, Hashimoto M 1989 *Physics of the earth and planetary interiors* **56** (3-4) 349-370
[24] Newcomb K R, McCann W R 1987 *Journal of Geophysical Research* **92** 421-439
[25] Shimazaki K 1978 *Bulletin of the Seismological Society of America* **68** 181-192

Integrated Pore Pressure Model Estimation – Case Study of Jambi Sub-basin, South Sumatera, Indonesia

Ahmad Farhan Farabi[1]*, Ignatius Sonny Winardhie[2], Noor Cahyo Wibowo[3]

[1]*Geophysical Engineering Study Program, Faculty of Mining and Petroleum Engineering, Bandung Institute of Technology, Bandung, Indonesia*
[2]*Geophysical Engineering Study Program, Faculty of Mining and Petroleum Engineering, Bandung Institute of Technology, Bandung, Indonesia*
[3]*Geophysics Division, Pertamina Upstream Technical Center, Jakarta, Indonesia.*

**Corresponding author's email: farhan.farabi@students.itb.ac.id*

Abstract. Pore pressure estimation is crucial in drilling wells for safety purposes also a very effective method for dealing with drilling accidents. Determination of overpressure is the main foundation in the evaluation to minimize the non-productive time (NPT). Here we present several models to generate pore pressure analysis of well from Jambi Sub-basin, South Sumatera, Indonesia. The model for estimation pore pressure is carried out by 3 methods: Eaton, Yan & Han, and Kan & Swan. Those methods will be compared to gain a more accurate model estimation within the study area. Kan and Swan's model show the best fit for estimation because this method is suitable for the formation of tester like MDT/DST on higher frequency with parameter of C1 = 0.001 and C2 = 0.0003 for Jambi Sub-basin. The velocity data to construct the 3D pore pressure model was also validated with well data using multi-attribute analysis. The multi-attribute analysis used 2 algorithms, namely step-wise regression and probabilistic neural network (PNN). The analysis show that PNN has a better correlation compared to step-wise regression. The analysis shows the overpressure zone depth is ranges from 1700 – 2000m on Gumai Formation with maximum pressure around 6500 psi. The peak of overpressure dominated by Gumai and Talang Akar formation is caused by the loading mechanism because the rate of sedimentation on thick shale sequence is higher than the rate of dewatering on those formations.

Keywords: Pore Pressure, Kan & Swan Model, Eaton Model, Yan & Han Model.

1. Introduction

As E&P activities focus more on complex environments and reduction of drilling costs, pore pressure analysis becomes part of the well planning and drilling process to generate safety drilling accidents. Overpressure, one of the drilling accidents seen globally is the main foundation in the evaluation to minimize the non-productive time (NPT). An accurate model for estimation pore pressure is required to generate suitable information future production. We already did analysis 1D modeling pore pressure of onshore well from South Sumatera Basin especially Jambi Sub-basin with integrated data of formation pressure, well log and drilling event. We applied Eaton, Yan & Han, and Kan & Swan method to compute pore pressure from the sonic log and compared each other to generate an accurate model within the study area. Integration of multiattribute analysis to generate velocity is used to construct 3D pore pressure model. 2 methods step wise regression and probabilistic neural network are compared to generate a better velocity model.

1.1. Geological Setting

Jambi sub-basin is located in the South Sumatera Basin and was on the east of Barisan Mountain. South Sumatera Basin is divided into two sub-basin, Palembang, and Jambi, which is separated each other [1]. The basin was formed due to extensional tectonic during rifting from Eosen-early Oligosen which generates half-graben[3]. It also generates two different basins between Barisan Mountain that is fore-arc basin and back-arc basin. This study area is categorized as back-arc basin. Several large oil and gas fields have been discovered in this basin, which is on Talang Akar Sandstone formation and Batu Raja Limestone formation have been established in the multiple pay zone of various reservoirs.

1.2.Pore Pressure Estimation Method

Pore pressure estimation is well defined by the Terzaghi equation which means pore pressure on formation is vertical stress minus effective stress from the rock. Physically, effective stress is the stress that is borne by grain-to-grain contacts.

$$p = (\sigma_v - \sigma_v') \tag{1}$$

Where p is pore pressure, σ_v overburden pressure, σ_v' effective stress

1.2.1 Eaton Method

Eaton [2] proposed an empirical equation for calculating pore pressure from the sonic log and resistivity. This method using the assumption of Terzaghi equation where overburden pressure formed by pore pressure and effective stress itself. This method was applicated by Eaton in 1975 for well that have overpressure zone on the Gulf of Mexico. An empirical equation for the sonic log is

$$p = \sigma_v - (\sigma_v - P_n)\left(\frac{\Delta t_n}{\Delta t}\right)^n \tag{2}$$

1.2.2 Yan & Han Method

This method shows that velocity model (Figure1.a) is best property to implicated stress effect on velocity where the value of differential pressure is not high or less than 60 MPa[6]. It also using three parameters fitting from formation tester which can be generated from this equation

$$V_p = V_{pa}(1 - C.\exp\left(-\frac{P_d}{b}\right) \tag{3}$$

Where, V_p is the velocity of P-wave, P_d is differential pressure and V_{pa} ,C and b is constant.

Differential pressure of this basin can be obtained from formation tester (MDT/DST) data within 3 well (Figure1.b)

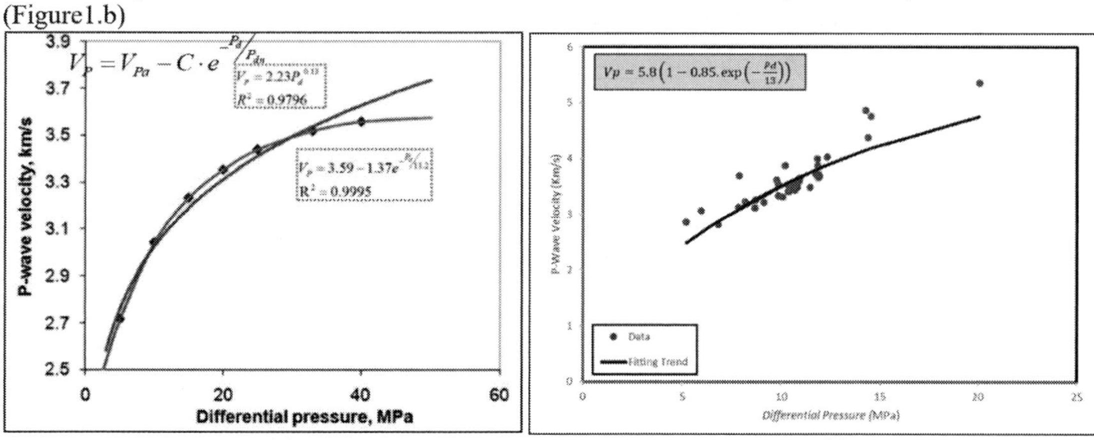

(a) (b)

Figure 1. (a) Pore pressure effect on P-wave velocity from laboratory measurement on Gulf of Mexico. (b) Pore pressure effect on P-wave velocity from analysis on Jambi Sub-basin.

1.2.3 Kan & Swan Method

In 2001 Swan proposed a new scheme for calculated pore pressure using velocity data. This method using a polynomial empirical equation between gradient pore pressure and transit time deviation. This deviation is between normal compaction trend and sonic transit time[4]. We can conclude that a higher deviation generates a higher pore pressure gradient (Figure2). This also needs formation tester to be fitting precisely to generate constant. Gradient relationship can be modeled empirically which can be written

$$\frac{P(z)}{z} = R_w + C_1 \Delta t + C_2 (\Delta t)^2 \quad (4)$$

$$\Delta t = t(z) - t_c(z) \quad (5)$$

Where R_w is hydrostatic gradient, $t(z)$ is sonic transit time, $t_c(z)$ is normal compaction trend and C_1 C_2 is constant.

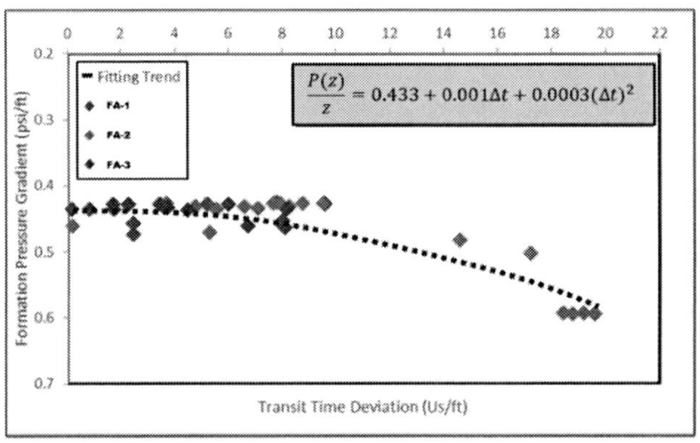

Figure 2. fitting trend between transit time deviation and pressure gradient within 3 well on Jambi sub-basin.

2. Data and Methodology

This study area located on onshore of South Sumatera Basin (Figure3.a). Data of this study include 5 well with MDT data (modular tester), DST data (Drill stem test), marker formation and wireline log which can be shown on (Figure3.b).

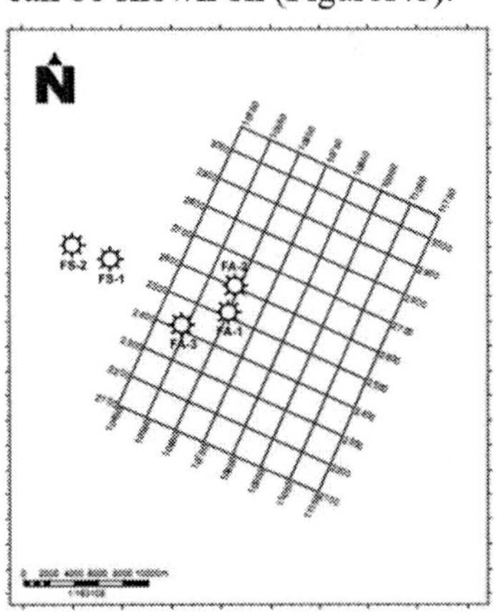

Information	Well				
	FA-1	FA-2	FA-3	FS-1	FS-2
Gamma Ray	√	√	√	√	√
Density	√	√	√	√	√
Sonic	√	√	√	√	√
Checkshot	√	√	√	√	√
MDT/DST	√	√	√	√	√
Marker	√	√	√	√	√

(a) (b)

Figure 3. (a) Basemap study area. (b) availability of data.

Methodology to generate pore pressure could be breakdown into several workflows which can be elaborate below. First, Normal compaction trend (NCT) is determined before calculated pore pressure because of the rock compaction that causes pressure. Determining the volume of shale is necessary because shale rock has the consistency of compaction rather than sandstone or limestone. So we determine pore pressure in shale sequence. (Figure 4.a) Shows that data already cut-off on 0.29 Vshale and 85 gamma-ray value which means below that value cannot be used on further analysis. (Figure 4.b) is sonic log data that already cut-off by shale and from that we can be implied there was deviation between sonic log and normal compaction trend which can generate overpressure zone. A pattern of sonic log on FA-1 shows that the value from 1900m is constant that indicates overpressure which caused by loading mechanism[5]. On formation pressure, we could mapping gradient of hydrostatic and overburden to simplify the calculation and should be fitting due to evaluating a characteristic of fluid density and rock density within the study area (Figure 4.c).

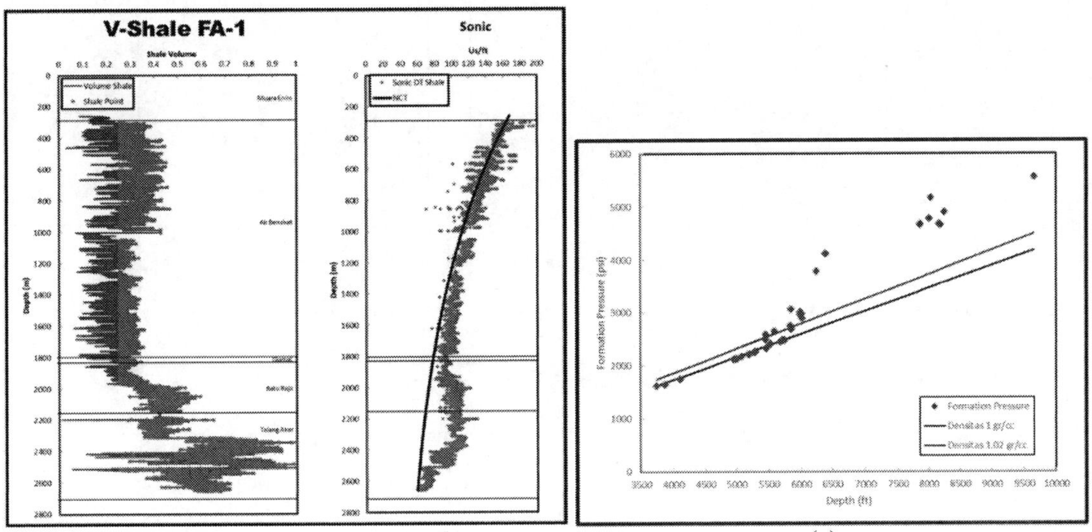

(a) (b) (c)

Figure 4. (a) V-shale FA-1 (b) Sonic data and NCT well FA-1 and (c) Hydrostatic Pressure curve

3. Results and Discussion

1D model using 3 methods are applied to all well within the study area and indicated penetrate through the overpressure zone. On (Figure 5) shows FA-2 well's top overpressure is from 1900m on Gumai formation. Based on those models, we can conclude that Kan & Swan model is the best fit between pore pressure calculation and formation tester rather than two other methods on higher frequency or adjacent sample with parameter of C1 = 0.001 and C2 = 0.0003 for Jambi Sub-basin particularly. The main reason Yan & Han model is not good enough is because in this model using an exponential empirical equation and did not present of hydrostatic gradient/pressure on that model so the result could be good at overpressure but, can not be good enough to fit hydrostatic line. Eaton also not fit enough than model Kan & Swan because parameter Eaton constant could be different every basin and could be "cheating" with Eaton. We can conclude that model Kan & Swan is the best fit and use on well correlation. Pressure excess shows that overpressure has approximately 6500 psi on maximum overpressure zone on Gumai or Talang Akar formation. Well correlation from 5 well (Figure 6) show that top overpressure is between 1700 – 2000m on Gumai formation, it is not too much different between well caused by a post-rift mega sequence that basin height is assumed to equal along the basin. This matches the geological condition

IOP Conf. Series: Earth and Environmental Science **873** (2021) 012012 doi:10.1088/1755-1315/873/1/012012

on Gumai formation which on the transgression phase so the rate of sedimentation is higher rather than the rate of dewatering.

On (Figure 7) we used 10 attribute to compute the algoritm with 3 itteration of each calculation to reduce error every attribute. PNN algoritm show much higher frequency rather than step wise regression or interval velocity for instead and also have the best correlation 0.966 and then PNN velocity would be our initial velocity for 3D pore pressure model.

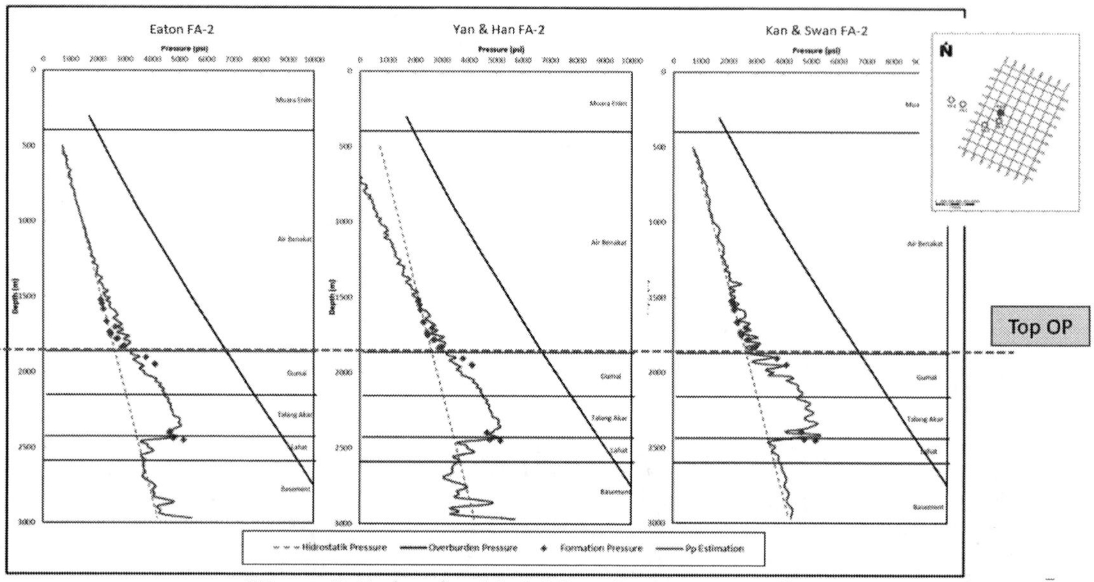

Figure 5. Estimation Model Pore Pressure on well FA-2

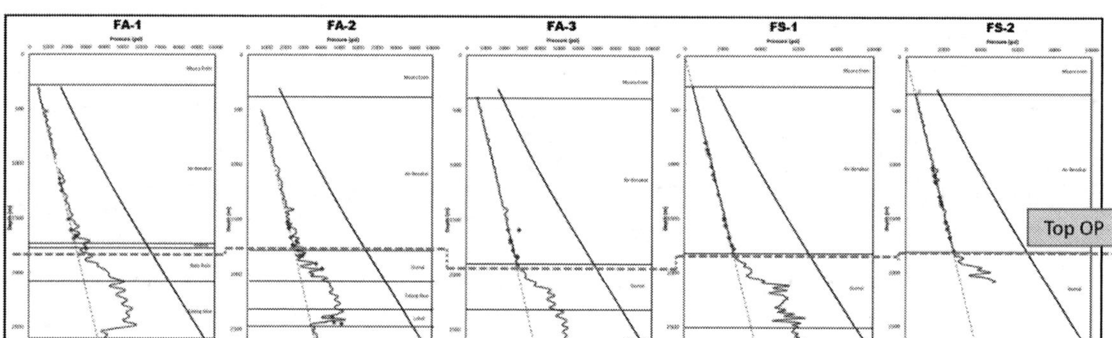

Figure 6. Well correlation model estimation of pore pressure Jambi Sub-basin, South Sumatra.

IOP Conf. Series: Earth and Environmental Science **873** (2021) 012012 doi:10.1088/1755-1315/873/1/012012

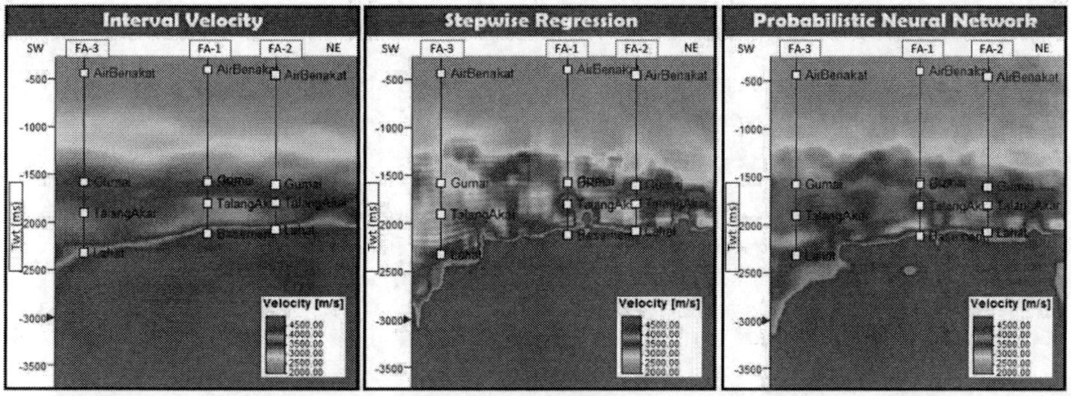

Figure 7. Velocity ouput of multiattribute analysis.

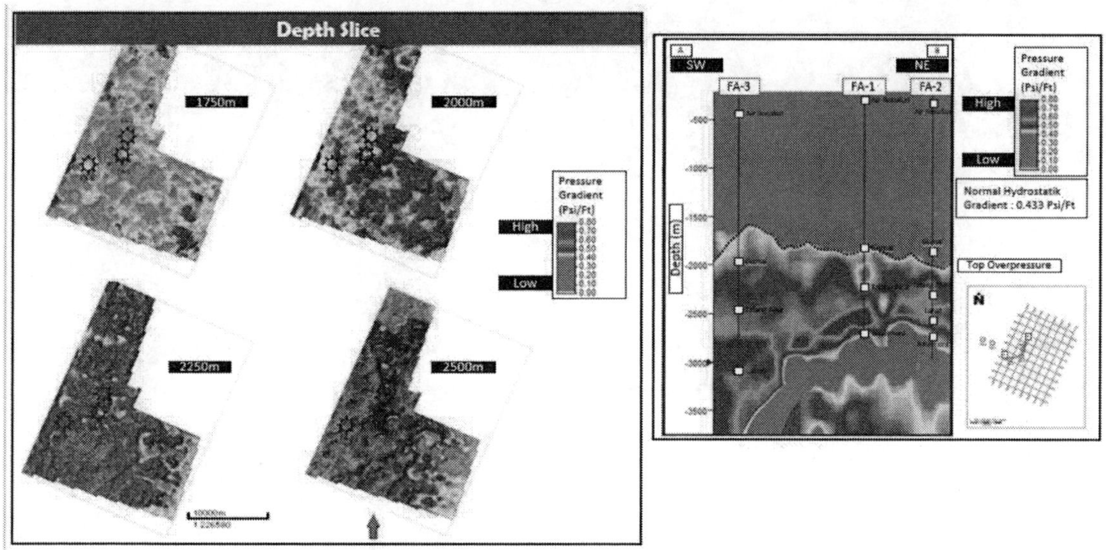

Figure 8. Depth slice and section of 3D pore pressure.

From 3D pore pressure model and slicing (Figure 8) overpressure zone is present on 1700m and keep increasing until 2250m. Talang akar and Lahat formation dominated by highly overpressure equivalent to 0.8 psi/ft. That issue could happen on shale sequence which overburden by thick shale above it for this case is Gumai formation.

4. Conclusions

Based on our study, we can summarize that model Kan & Swan has the best fit on higher frequency with parameter of C1 = 0.001 and C2 = 0.0003 for Jambi Sub-basin. Velocity from multiattribute analysis show PNN has the best correlation rather than step wise regression or interval velocity. From well analysis, we can conclude that the top overpressure zone is present between 1700-2000m on Gumai formation that can consist of 6500psi on the overpressure zone. 3D pore pressure model indicate top overpressure zone along the seismic area was present on 1700m and increasing until 2250m. The pattern on the sonic log and pressure model shows that it has constant velocity on the overpressure zone that can be categorized by loading mechanism or disequilibrium compaction. It is matched with the geological condition on Gumai formation which on the transgression phase which dominated by shale sequence so the rate of sedimentation is higher rather than the rate of dewatering of fluid.

References

[1] Barber, A.J., Crow, M.J., dan Milsom, J.S. (2005): Sumatra: Geology, Resources and Tectonic Evolution, *Geological Society*, London.
[2] Eaton, B. A., 1975. *The equation for geopressure from well logs.* Society Petroleum Engineers of AIME, Paper SPE 5544
[3] Ginger, D. dan Fielding, K. (2005): Petroleum Systems and Future Potential of South Sumatra Basin, *Proceedings 30st Annual Convention Indonesian Petroleum Association*, Jakarta.
[4] Kan, T.K., and Swan, Herbert, W., 2001. *Geopressure prediction from automatically-derived seismic velocities*: Geophysics, Vol. **66**, No.6: Page. 1937-1946.
[5] Ramdhan, A.M. dan Goulty, N.R. (2010): Overpressure Generating Mechanism in The Peciko Field, Lower Kutai Basin, Indonesia. *Petroleum Geoscience*, **16**, page.363-367.
[6] Yan, F., and Han, D. H., 2012. *A new model for pore pressure prediction*: SEG Las Vegas Annual Meeting.

IOP Conf. Series: Earth and Environmental Science **873** (2021) 012013 doi:10.1088/1755-1315/873/1/012013

The M 6.5 Ambon earthquake 26 September 2019: the source mechanism and the aftershock sequence characteristics

S. Rohadi [1], Y.H. Perdana[1], N. Herayndoko[1], B. Sunardi[1], T. A. Prakoso[1], Suliyanti, Sunardi[2] , N.Florida[1] , Robert Edison[2]and D.Karnawati[1]

Agency for Meteorogy Climatology and Geophysics[1]
BMKG Geophysics Station Ambon[2]

Abstract. The area of Ambon, Maluku is located in the subduction zone in bands where the Australian plate meets the Eurasian plate, thus causing tectonic activities. The Ambon earthquake on 26th September 2019 with 6.5 Magnitude, while the Epicentral coordinates of the earthquake were determined as 3,53° S and 128,39° E and a focal depth of 10 km, according to the Agency for Meteorology Climatology and Geophysics, Indonesia. This earthquake was strongly felt at the biggest shock was felt with intensity VI-VII as unified in Ambon City, while several other areas are reported to have experienced small shaking, such as Intensity V in Masohi, and Intensity IV in Namlea and Namrole. We used a dataset of 24 waveforms of seven sensors, we determine a tabular solution, which have a large moment of 0.4573 x 1019 N-m, the depth is 6 km by minimizing the inversion residual. The method resulting strike and rake fault, with strike: 341.8°; dip; 81.5°; rake: 158.4°, and second nodal plane strike: 75.1°; dip; 68.6°; rake: 9.14°. The mechanisms were compared with those from other agency in agreement. The time decay intervals between mainshocks and significant aftershocks follow Mogi and Utsu's Law but with a relatively faster rate of decay than that of aftershocks in general.

1. Introduction

The Ambon earthquake, 26th September 2019 with epicenter coordinates 3,53° S and 128,39° E at a depth of 10 km, 42 km northeast of Ambon City due to the movement of local faults, Ambon Island has a lot of potential for earthquakes and tsunamis, Ambon Island is composed of a complex tectonic arrangement [2]. Tectonically, this island has many faults which are the source of earthquakes. According to data of the National Disaster Management Agency (BNPB), the impact of the Ambon earthquake on September 26, 2019, 23 people died, injured victims occurred in Central Maluku. More than 100 people were injured in Liang Village. In Ambon City, 5 people were injured. Meanwhile, West Seram District, 1 person was injured in Waisama Village. About 15,000 residents are still displaced after the earthquake because their houses were damaged and in anticipation of aftershocks that could endanger them. Meanwhile, infrastructure damage did not only occur in the housing sector, but also educational facilities, places of worship, offices, and public facilities. Damage to houses in the affected areas reached 171 units, with details of 59 being heavily damaged, 45 moderately damaged and 67 slightly damaged. Five units of educational buildings were damaged, including several buildings at Pattimura University [6]. This paper discusses the mechanism and the decay of aftershocks of the Ambon earthquake.

Seram and Ambon are part of the Banda Arc system. The Banda Arc, located in the eastern part of Indonesia between Australia, and Sulawesi. Banda and Ambon are a region of complex and active tectonics. Convergence between Australia, Eurasia and the Pacific Sea plates ([3], [5]) results in both contraction and extension from subduction, lithospheric delamination and slab break-off [3]. Convergence of Indo-Australia, Eurasia, and Pacific plates creates the Banda Arc which is a very complex tectonics system including a number of Microplates involved. Convergence between Australia, Eurasia, and the Pacific Sea plates ([3], [5]) results in both contraction and extension from subduction, lithospheric fracture and slab detached [3]. In the outer arc islands of Seram, Tanimbar, Babar, Timor and Sumba, the geological features have been created as a result of subduction processes. The North Seram Basin is located in the Australian plate, in the south adjacent to the Banda micro plate and the arc-continent collision belt in the margin. The northern part of the basin is bounded by the Seram trough in the north, strike-rake of the Ruru fault belt in the west, Kawa strike-rake fault belts and in the south is metamorphic thrust belts, and in the east is the terra-Aiduna strike-rake fault belt. The North Seram, and it has features of a Foreland basin. According to its structural characteristics, the basin can be divided into two second-order tectonic units: Foreland fold thrust belt and foredeep belt [9].

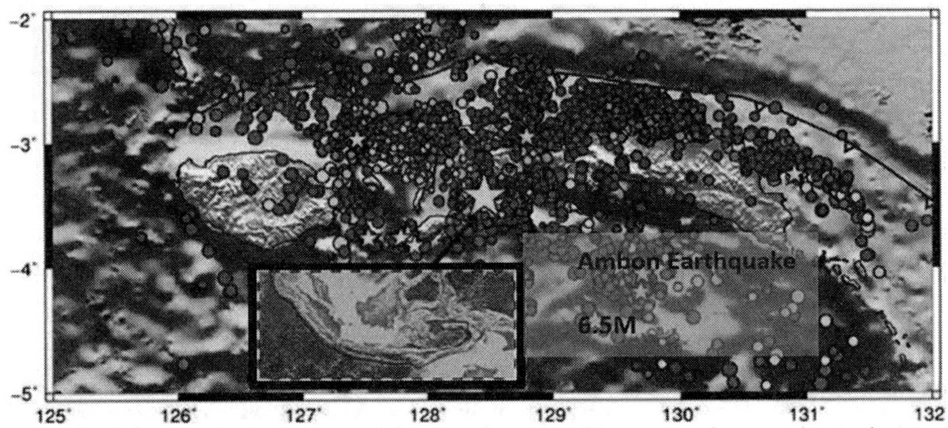

Figure 1. Seismicity in Ambon-Seram island and surounding area, the star is Ambon earthquake Sepetember 26, 2019.

2. Data and Method

We use waveform data from global stations (IRIS) to obtain a reliable and accurate focal mechanism, figure 2.a shows the station distribution which used for waveform inversion. Earthquake source modeling used teleseismic waves at 24 stations 30° to 90° from the epicenter. The data for aftershocks analysis originating from the Agency for Meteorology Climatology and Geophysics (BMKG) catalogue is used to determine the patterns and estimates of the end of aftershocks decay (Fig. 2b**).**

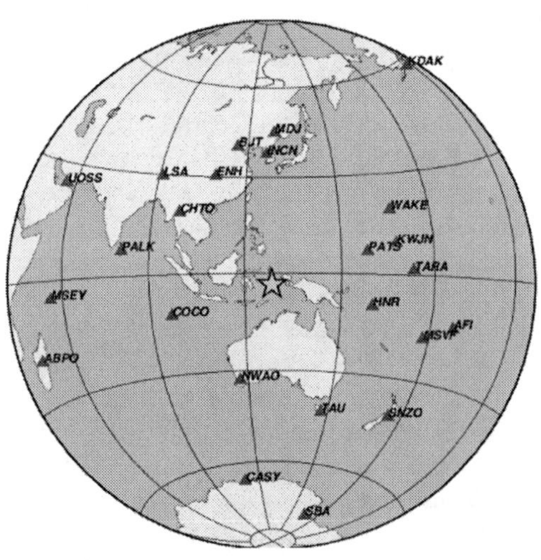

a)

b)

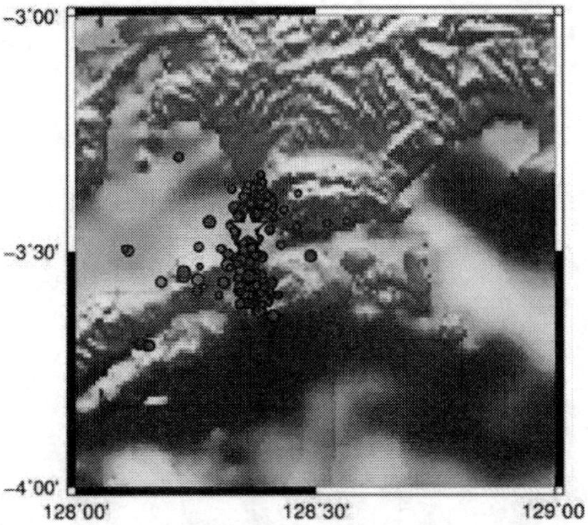

Figure 2. a) Seismic station used for focal mechanism and a) aftershock epicenter distribution method

The fault's during an earthquake are mathematically and geometrically represented with focal mechanism. Focal consists of the motion of two earth fault. Focal mechanisms can consist of the double-couple and symmetric tensor

$$M_{ij} = M_o(\hat{d}_i\hat{n}_j + \hat{n}_i\hat{d}_j) \qquad (1)$$

where $\hat{d}$ and $\hat{n}$ are unit vector perpendicular to the fault along the direction of rake [1].

Decay of the aftershock sequence is determined using the decay formula, according to Mogi [6] and Utsu [8] that shows the activity level of aftershocks (t ≤100 days), in the relationship between earthquake frequency and time is
:

$$n(t) = a.e^{bt} \qquad (2)$$

where n, t, a, and b are number of event, time, and constant, respectively. This relation has been successfully applied to many aftershock sequences of the main shock. The process to determine the decay is fitting the equation to observational data of the aftershock.

3. Result and Discussion

This modeling is done by inverting body waves at low frequencies (0.018 - 0.091 Hz). Modeling results get a magnitude Mw6.4 (BMKG (update) M6.5, GFZ Mw6.4, IRIS Mww6.5) and an earthquake depth of 6 km (BMKG 10 km, GFZ 10 km, IRIS 18.16 km). If the epicenter of the earthquake is viewed as a point source, the modeling results show that the earthquake mechanism is a strike rake type. The focal mechanisms were compared with those from other studies. The results of the inversion of the seismic wave form are N1: Strike, dip, rake = (341.8°, 81.5°, 158.4°) and N2; (Strike, Dip, Rake) = (75.1°E, 68.6°, 9.14°).

The earthquake process can be explained by focal mechanism and energy release of the intraplate earthquake, which indicate the release of energy accumulated inside the plate tectonics. The interplate earthquakes are controlled by relative moving as a rake vector of fault plane. Verification of focal mechanisms of the fundamental concept derived using the spatial distribution of aftershock [1].

IOP Conf. Series: Earth and Environmental Science **873** (2021) 012013 doi:10.1088/1755-1315/873/1/012013

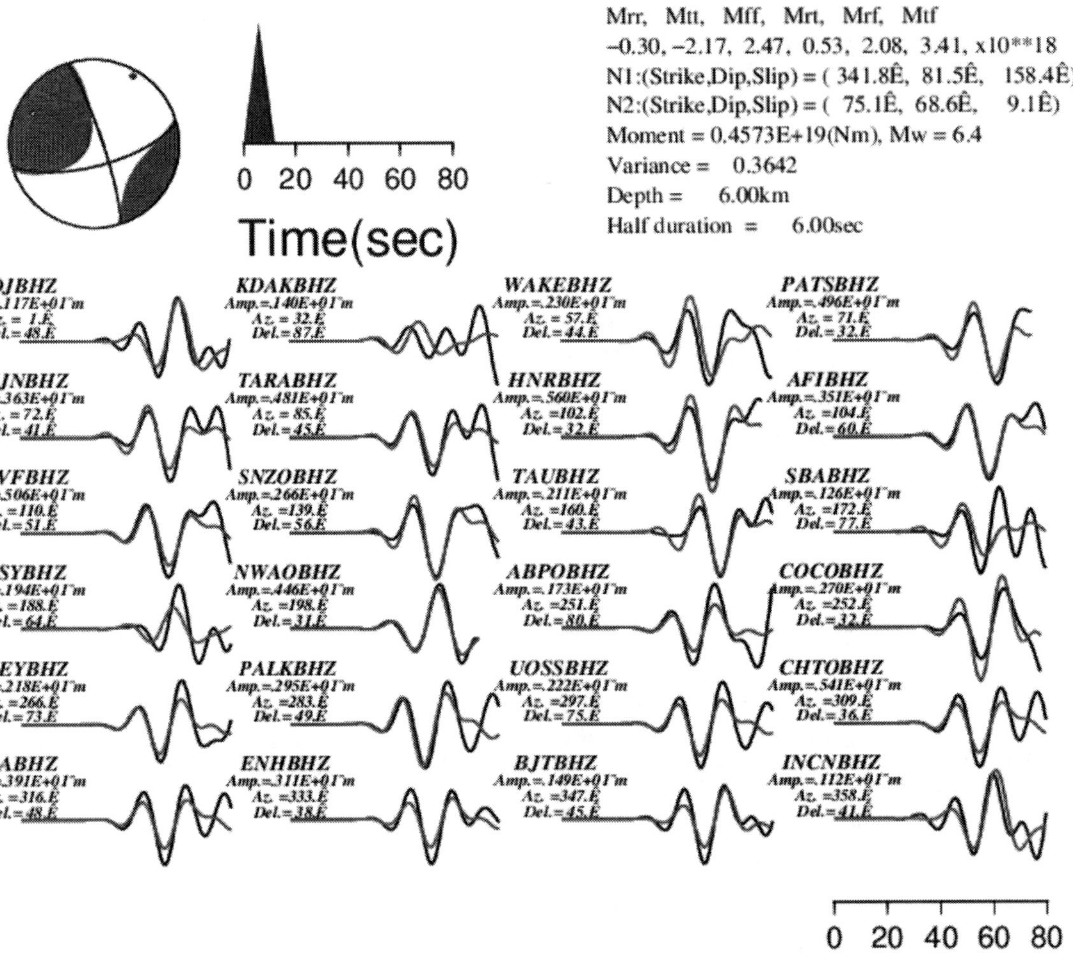

Figure 3. Focal mechanism and the waveform fitting of Ambon earthquake September 26, 2019.

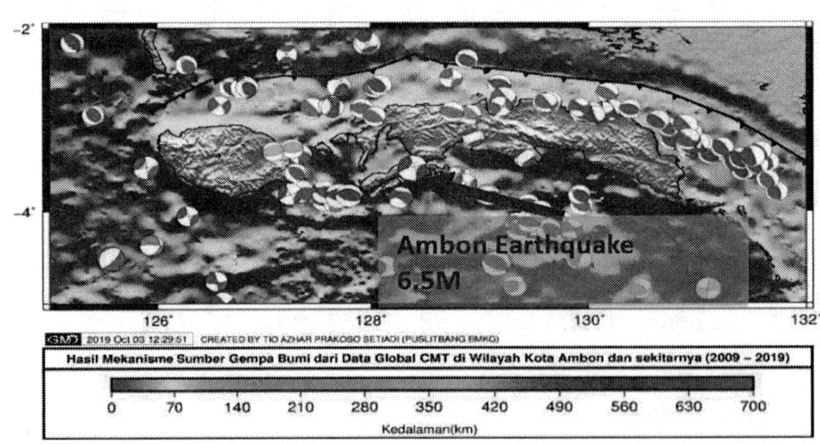

Figure 4. Distribution of some earthquake focal mechanism in Ambon and surounding area.

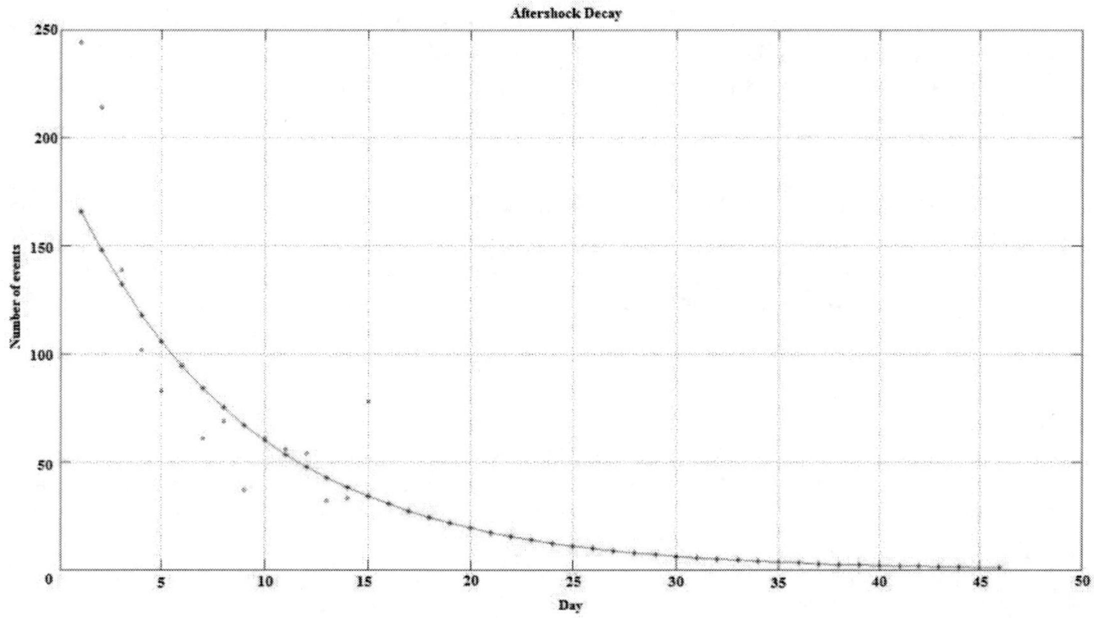

Figure 5. Decay of the aftershock activity of the Ambon earthquake September 26, 2019 (Source: BMKG).

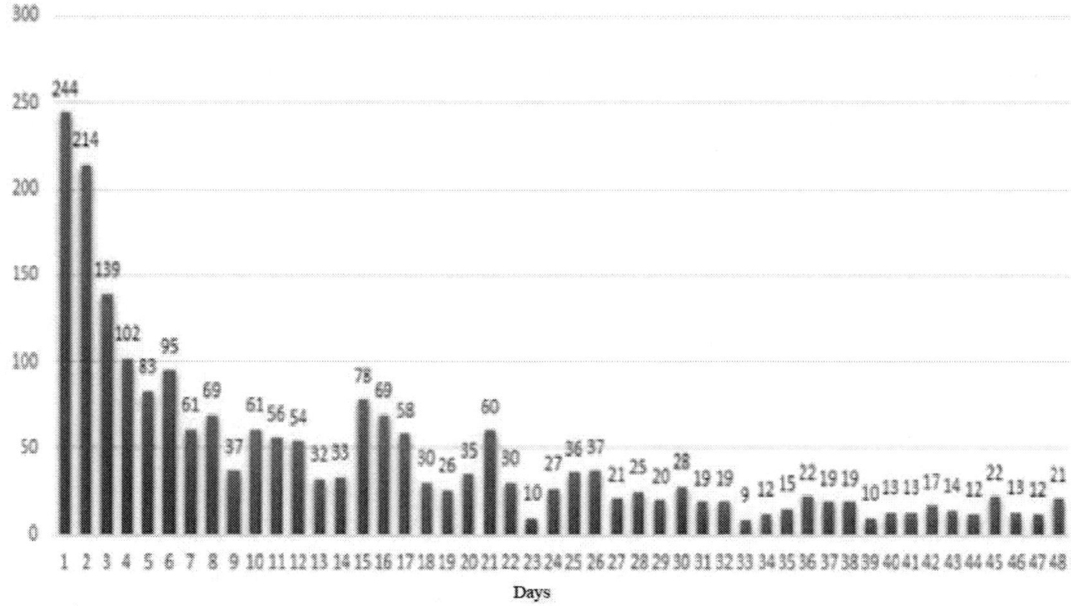

Figure 6. Histogram of aftershock event of Ambon earthquake September 26, 2019 (Source: BMKG).

Based on decay relation, aftershock of the main earthquake will be ending after 46 days (Fig.5), compute using Mogi and Utsu model ([6], [8]). The aftershocks may continue after this time but difficult to separate from the background seismicity. Aftershock decay defines as the duration of aftershock activity as a function of the time to the level of background seismicity. The aftershock activities of the large earthquakes continue in some months or even years.

The background seismicity of the aftershock zone can be estimated using several approaches, mainly the background seismicity similar to the average seismicity before the main shock. The aftershock decay is dependent on background seismicity of the region. The decay of the aftershock defined the duration as the time when the rate decreased to some fixed level (example: 1 event/day).

4. Conclusion

Based on data processing and analysis of the results, it can be determined that the Ambon earthquake was strike rake fault. This mechanism is in accordance with the regional tectonic patterns in the region. The decay pattern of aftershocks shows an exponential decay pattern according to the formula from Mogi and Utsu. The end of the aftershocks is predicted 46 days after the main earthquake. Reliable focal mechanism and accurate estimation of the aftershock decay are very important in earthquake mitigation efforts.

Acknowledgment

We would like to acknowledge financial support from Agency for Meteorology Climatology and Geophysics (BMKG). BMKG staff in Ambon conducted survey and supporting earthquake data. We acknowledge with gratitude the contributions of the many individuals who assisted following the earthquake.

References

[1] Aki K and Richard P G 1980 *Quantitative Seismology* New York: W.H. Freeman p. 932.
[2] Hall R and Wilson M E J 2000 Neogene suture in eastern Indonesia *Journal of Asian Earth Sciences* **18** 781-802.
[3] Hall R 1996 Reconstructing Cenozoic SE Asia. In: Hall, R. & Blundell, D.J. (eds) Tectonic Evolution of SE Asia. Geological Society, London, Special Publications, **106**, 153–184, https://doi.org/10.1144/ GSL.SP.1996.106.01.11
[4] Hall R 2012 Late Jurassic–Cenozoic reconstructions of the Indonesian region and the Indian Ocean *Tectonophysics* **570** 1–41.
[5] Hamilton W 1979 Tectonics of the Indonesian Region. US Geological Survey, Professional Papers, 1078.
[6] Mogi K 1962 On the time distribution of aftershocks accompanying the recent major earthquakes in and near Japan, Bull. Earthquake Res. Inst. Univ. Tokyo **40** 107 – 124.
[7] Sutopo P N 2018 (https://www.bnpb.go.id/kerugian-dan-kerusakan-dampak-bencana-di-sulawesi-tengah-mencapai-1382-trilyun-rupiah), di akses 10 Februari 2019.
[8] Utsu T Magnitudes of earthquakes and occurrence of their aftershocks, Zisin, Ser. 2, 10, 35-45, 1957 (in Japanese).
[9] Xi Zhugang, Hu Xiaolin, Fang Yong, Yin Xinyi,and Du Hongyu, Tectonic evolution of North Seram Basin, Indonesia, and its control over hydrocarbon accumulation conditions, China Petrolium Exploration, Volume 21, Issue 6, November 2016.

The Study of Automatic Picking of P and S Wave Arrival and Identification of Earthquake Sequence Pattern using Scalogram in Obspy (Python)

Sri Kiswanti[1], Indriati Retno Palupi[1], Wiji Raharjo[1], Faricha Yuna Arwa[1], and Nela Elisa Dwiyanti[1]

[1] Department of Geophysical Engineering, University of Pembangunan Nasional "Veteran" Yogyakarta, Yogyakarta, Indonesia

E-mail: *srikiswanti98@gmail.com*

Abstract. Initial identification on an earthquake record (seismogram) is something that needs to be done precisely and accurately. Moreover, the discovery of a series of unexpected successive earthquake events has caused unpreparedness for the community and related agencies in tackling these events. Determining the arrival time of the P and S waves becomes an important parameter to finding the location of the earthquake source (hypocenter) as well as further information related to the earthquake event. However, manual steps that are currently often used are considered to be less effective, because it requires a lot of time in the process. Continuous Wavelet Transform (CWT) analysis can be a solution for this problem. With further CWT analysis in the form of a scalogram, can help to determine the arrival time of P and S waves automatically (automatic picking) becomes simpler. In addition, further CWT analysis can also be utilized to help identify the sequence of earthquake events (foreshock, mainshock, aftershock) through the resulting scalogram pattern.

Keywords: aftershock, CWT, foreshock, mainshock, scalogram, seismogram.

1. Introduction

Geographically, Indonesia is locating between three major plates of the world, including the Eurasian Plate, Pacific Plate, and Australian Plate (**Figure 1**). Each tectonic plate interacted with each other and resulted in earthquake sources, such as faults and subduction. Therefore, it was not surprising that Indonesia was called one of the most vulnerable countries to natural disasters, especially earthquakes. Constraints that must be faced until now, namely not yet predict when the earthquake occurred, still found difficulties in determining the location of the earthquake source (hypocenter) and the difficulty of distinguishing the pattern of earthquake events. At present, to find out information about earthquakes, specifically regarding the time of an earthquake, it is solved using statistical methods. Manual steps were still often used to determine the arrival time of P and S waves as important parameters to determine earthquake hypocenter [8].

However, with various approaches that have been made to be able to speed up the process of determining the arrival time of P and S waves is still considered ineffective and still requires a long time in its completion. Therefore, to speed up the process can be done using the automatic picking method. Automatic picking analysis using the Continuous Wavelet Transform (CWT) method can be used as a solution to accelerate and simplify the process of determining the phase of the wave [3] [10]. Further CWT analysis in the form of scalogram can also be used to help identify the sequence of earthquake events (foreshock, mainshock, aftershock) through the resulting scalogram pattern.

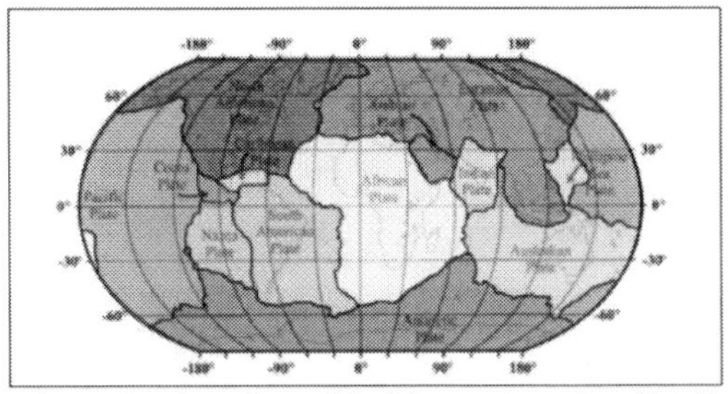

Figure 1. Tectonic plates (top) and worldwide earthquake distribution (bottom) [4].

2. Data and Methodology

Seismic waves are waves that arise due to the release of energy that comes from artificial and natural seismic sources [2] [5]. Seismic waves will be recorded by the seismometer as a seismogram and basically carry information about the seismic source and the medium through which it passes in a function of time or speed with a certain time [1].

A scalogram is a time-scale picture of a signal that is represented in a certain color. The scalogram does not allow for direct interpretation of the frequency, because the scale on the scalogram represents the frequency range and is not a single frequency. To get a scalogram done using the CWT method. CWT is defined as the sum of the signals *f(t)* where the multiplier parameter is a scale which is a shifting position and wavelet function (ω) [7].

$$C(\text{scale, position}) = \int_{-\infty}^{\infty} f(t)\psi(scale, position, t)dt \quad (1)$$

The result of CWT is wavelet C coefficient which is a function of scale and position. Multiplication of each coefficient with the right scale and position will produce a constituent wavelet from the original signal. Wavelets are defined as functions $\psi(t) \epsilon L^2(R)$ with zero-mean, localized in time and frequency. Through dilation and translation wavelet ψ(t) produces the wavelet family as follows:

$$\psi a.b^{(t)} = \frac{1}{\sqrt{a}}\psi\left(\frac{t-b}{a}\right) \quad (2)$$

Where a, b ϵ R and a ≠ 0. a is a scale parameter and b is a shift (position) parameter. CWT is defined as the origin product of the wavelet family ψ ab (t) with the signal f (t) given by the equation:

$$Fw\ (a, b) = (f(t), \psi a, b^t = \int_{-\infty}^{\infty} f(t)\frac{1}{\sqrt{a}}\psi^*\left(\frac{t-b}{a}\right)dt \quad (3)$$

Where ψ * is the conjugate complex of ψ. *Fw* (a, b) is a time-scale map (scalogram). The flow chart in this study can be shown in the figure below (**Figure 2**):

IOP Conf. Series: Earth and Environmental Science **873** (2021) 012014 doi:10.1088/1755-1315/873/1/012014

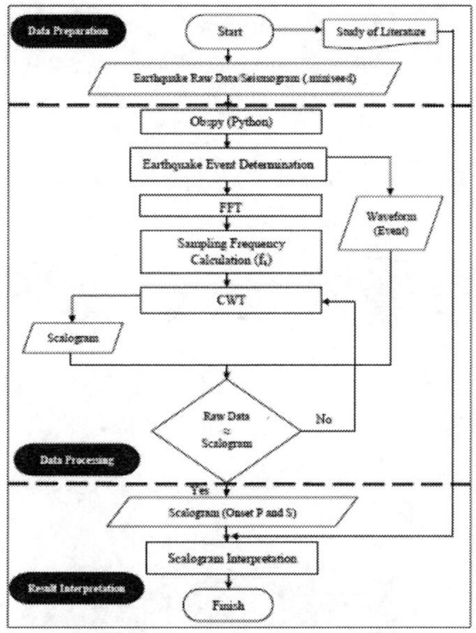

Figure 2. Research flow chart

The data used in this study is in the form of raw earthquake record data (seismogram) that occurred in Indonesia with varying magnitude. Obtaining the data by downloading through the IRIS [6]. Then the data is analyzed from the initial earthquake event, to the CWT analysis to get a scalogram profile. This process is entirely carried out in Python Software by utilizing the Obspy library/application/module. The use of Python was chosen because it is able to execute large data quickly, its programming language is easy to understand, has a complete module to support research in the field of science, opensource so that it can be used by anyone, and also effective because it can be connected directly to the internet [9].

3. Results and Discussion

This study takes several examples of raw data on earthquake events in Indonesia, which have been downloaded on the IRIS [6]. The series of earthquakes included foreshock, mainshock, and aftershock earthquakes that occurred in Lombok in 2018, North Sumatra in 2017, and Palu in 2018. The CWT analysis is performed to obtain the scalogram results. Foreshock earthquake is an earthquake that occurred before the main earthquake occurred. Foreshock earthquake data used in this study are Lombok earthquake on 28 July 2018, North Sumatra earthquake on 6 January 2017, and Palu earthquake on 28 September 2018. Based on **Figure 3** (a) (b) (c) it can be seen that the time of P wave arrival and S is characterized by a color enhancement on the scalogram (red line). P waves always arrive the first time compared to other waves because the speed possessed by the P wave is greater, so the P wave requires a shorter time to reach the recording station than the other waves. The time of arrival of P and S waves in a further foreshock earthquake can be seen in **Table 1**.

Table 1. The P and S waves of foreshock earthquake arrived.

Earthquake Occurrence	Tp (second)	Ts (second)
Lombok, 28 July 2018 (M6.4)	67.18	551.11
North Sumatra, 6 January 2017 (M5.4)	15.59	112.12
Palu, 28 September 2018 (M6.1)	87.72	592.16

Identification of earthquake patterns can be seen through the pattern on the scalogram. **Figure 3** (a) (b) (c) are successively a foreshock earthquake from the Lombok earthquake on July 28, 2018, North Sumatra earthquake on January 6, 2017, and Palu earthquake on September 28, 2018, scalogram patterns are relatively more tenuous (wide) and also collect (black circle). In addition, the color on the scalogram for the foreshock earthquake is not too strong and the surface wave duration is not too long.

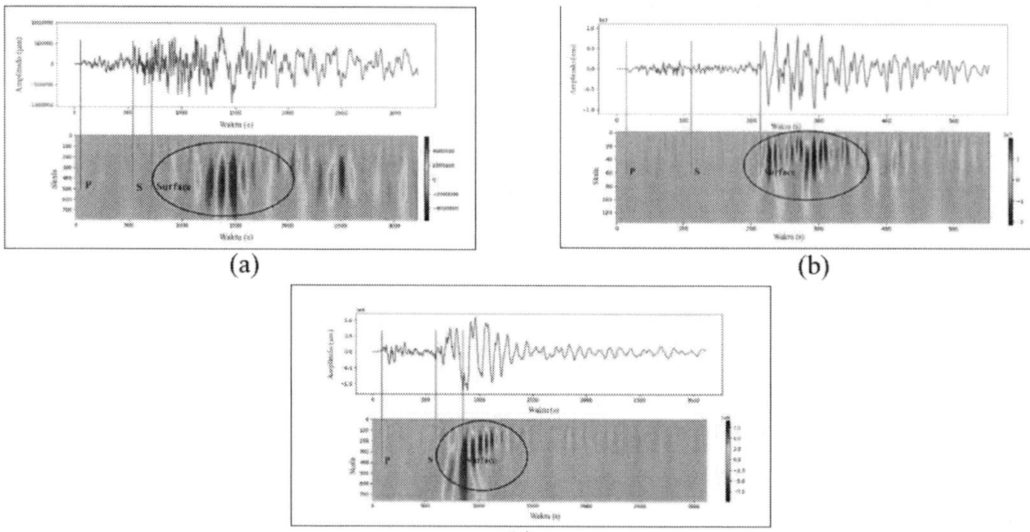

(a) (b) (c)

Figure 3. Lombok earthquake recording and scalogram on July 28, 2018 (a), North Sumatra on January 6, 2017 (b), and Palu on September 28, 2018 (c).

The mainshock earthquake is an earthquake that was present after the foreshock earthquake occurred. Foreshock earthquake data used in this study are Lombok earthquake data on 5 August 2018, North Sumatra earthquake on 16 January 2017, and Palu earthquake on 28 September 2018. Based on **Figure 4** (a) (b) (c) it can be seen that the time of P wave arrival and S is characterized by a color enhancement on the scalogram (red line). The time of arrival of P and S waves in the mainshock earthquake can be seen in **Table 2**.

Table 2. When the P and S waves arrived at the mainshock earthquake.

Earthquake Occurrence	Tp (second)	Ts (second)
Lombok, 5 August 2018 (M6.9)	82.16	587.07
North Sumatra, January 16, 2017 (M5.6)	21.55	192.12
Palu, 28 September 2018 (M7.5)	61.83	672.98

Identification of earthquake patterns can be seen through the pattern on the scalogram. **Figure 4** (a) (b) (c) are respectively the mainshock earthquake from the Lombok earthquake on August 5, 2018, the North Sumatra earthquake on January 16, 2017, and the Palu earthquake on September 28, 2018, a scalogram pattern shows a much stronger and wider color (black circle), then followed by a weakening and reinforcement that ensues (arrow). In addition, it can be observed that the duration at the surface wave is longer.

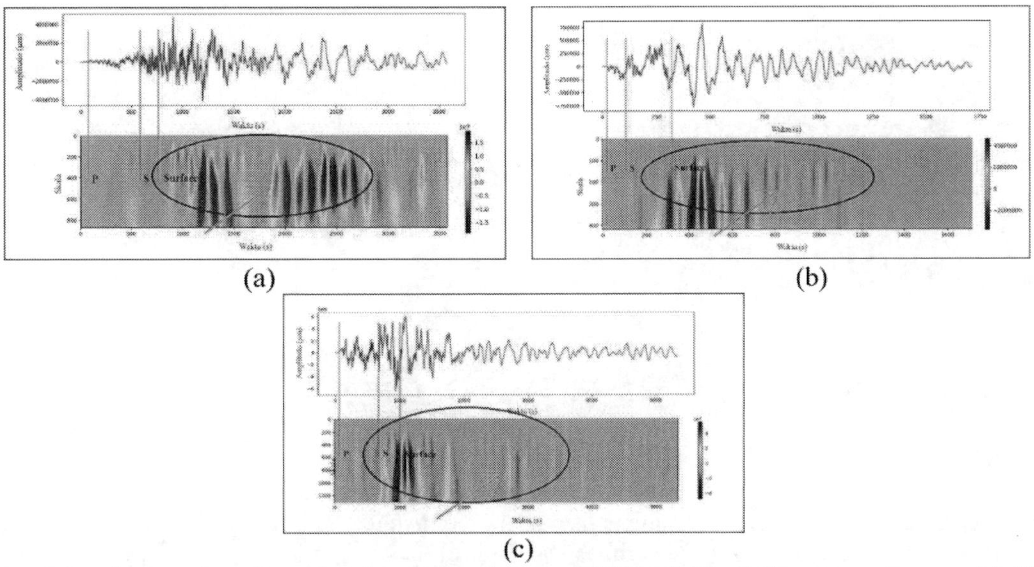

Figure 4. Lombok earthquake and scalogram on August 5, 2018 (a), North Sumatra on January 16, 2017 (b), and Palu on September 28, 2018 (c).

Aftershock earthquake is small earthquakes that are present after the main earthquake occurred. Aftershock earthquake data used in this study are Lombok earthquake data on 5 August 2018, North Sumatra earthquake on 23 January 2017, and Palu earthquake on 28 September 2018. Based on **Figure 5** (a) (b) (c) it can be seen that the time of P and S waves arrives marked by the strengthening of the color on the scalogram (red line). The time of arrival of P and S waves in the aftershock earthquake can be seen in **Table 3**.

Table 3. When the aftershock earthquake P and S waves arrive.

Earthquake Occurrence	Tp (second)	Ts (second)
Lombok, 5 August 2018 (M4.6)	34.99	262.55
North Sumatra, January 16, 2017 (M4.9)	122.06	676.03
Palu, 28 September 2018 (M5.0)	16.41	632.18

Identification of earthquake patterns can be seen through the pattern on the scalogram. **Figure 5** (a) (b) (c) is an aftershock earthquake from the Lombok earthquake on August 5, 2018, North Sumatra earthquake on January 23, 2017, and Palu earthquake on September 28, 2018, the scalogram color on the surface wave is not too strong and tends to be denser (narrower) compared to foreshock and mainshock (black circle) events.

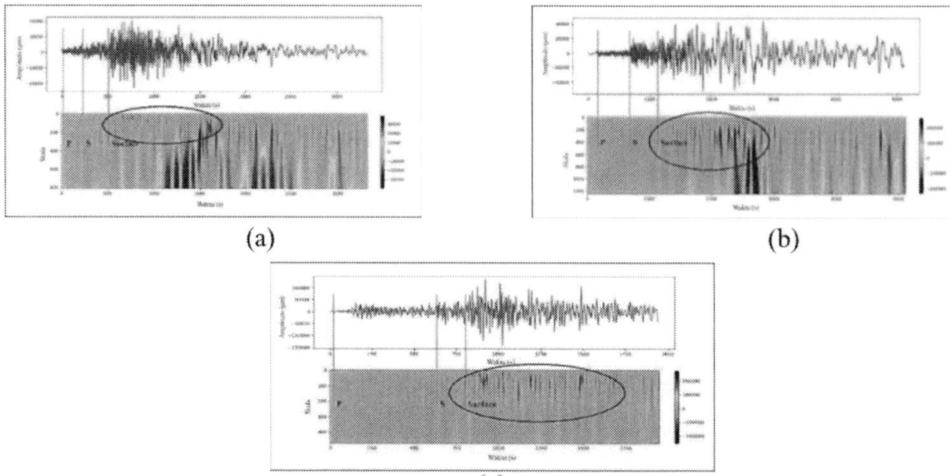

(a) (b)

(c)

Figure 5. Lombok earthquake recording and scalogram on August 5, 2018 (a), North Sumatra on January 23, 2017 (b), and Palu on September 28, 2018 (c).

4. Conclusions

Based on the results of the scalogram profile in several cases of earthquake data, it is known that CWT analysis through a scalogram can be used to map the arrival time of P and S waves (wave phase) on seismograms in a simpler, clearer, and faster manner (for recording earthquake events recorded with good). In addition, this technique can also be used to help differentiate earthquake sequence patterns by looking at patterns on the resulting scalogram.

Acknowledgments

The authors wish to thank to all those who have supported, helped in the collection of data and information required in this study.

References

[1] Afnimar 2009 *Seismologi Edisi Pertama* (Bandung : ITB) Chapter 2 p 7

[2] Bath, M 1979 *Introduction to Seismologi* 2nd edition (Basel : Birkhauser Verlaag) Chapter 6 p 428

[3] Bogiatzis P and Ishii M 2015 *Contionuous Wavelet Decomposition Algorithms for Automatic Detection of Compressional and Shear-Wave Arrival Times* Bulletin of the Seismological Society of America **105** 3 1628-1641

[4] Elnashai SA and Sarno DL 2008 *Fundamental of Earthquake Engineering* (Hongkong : Wiley)

[5] Ibrahim G and Subarjo 2004 *Buku Seismologi* (Jakarta: BMG)

[6] IRIS 2018 Accessed From https://ds.iris.edu/wilber3/find_event1#

[7] Pranoto S C 2016 *Pengolah Sinyal Digital: Transformasi Wavelet* Buletin Cuaca Antariksa, Pusat Sains Antariksa – (Bandung : LAPAN) pp 6-8

[8] Sunarjo, Gunawan MT and Pribadi S 2012 *Gempabumi Edition Populer* (Jakarta: BMKG) Chapter 5 pp 71-83

[9] ObsPy Development Team 2019 *ObsPy Tutorial Release 1.1.1*

[10] Zhang H, Thurber C and Rowe C 2003 *Automatic P-Wave Arrival Detection and Picking with Multiscale Wavelet Analysis for Single-Component Recordings* Bulletin of the Seismological Society of America **93** 5 1904-1912

IOP Conf. Series: Earth and Environmental Science **873** (2021) 012015 doi:10.1088/1755-1315/873/1/012015

The Sector Development at Tanjung Sakti, Mount Dempo, Bengkulu by Using Integration of Remote Sensing Data and TOPEX Gravity Satellite

Zahrah Athirah[1] Muhammad Dhery Mahendra[1]
Program of Study Geophysics Universitas Indonesia[1]

**zahrah.athirah@ui.ac.id*

Abstract Mount Dempo is the highest volcano in South Sumatra, which lies between the Bukit Barisan mountains and Gumai. The mountain located in Dempo Makmur Village, Sub-district of Pagar Alam, Lahat Regency, South Sumatra is located at an altitude of 3173 meters above sea level with coordinates of 4.03 ° S 103.13 °E. Mount Dempo's morphology is formed by pyroclastic deposits consisting of Tuff and Sand rocks. Mount Dempo's vegetation is dominated by Cassia sp. and Camellia sinensis for upper vegetation, while Strobilanthes hamiltoniana and Strophanthus membranifolium dominate the undergrowth. The purpose of this study is to identify geological structures to predict geothermal prospect areas by integrating remote sensing data and TOPEX Gravity Satellite Data. The remote sensing data used in this study is Landsat 8. This data is used to analyze Land Surface Temperature (LST) from a single thermal infrared band, surface emissivity based on Normalization Difference Vegetation Index (NDVI) from the study area and determine structure delineation. Gravity Satellite Data is used to map gravity anomalies in the volcanic complex of Mount Dempo. Gravity data processing produces a high anomaly zone in the northern part of the study area and is predicted as a prospect area because it is assumed to be related to the plutonic body. High density contrast indicates that there is an error in that area. In line with the error, there are several hot springs because the error serves as a pathway for geothermal fluid to rise to the surface. The study believes that with all the facts stated above, the spots which are located in Tanjung Sakti, Mount Dempo district are very prospective to be developed as a geotourism complex, in which could also increase the welfare of the local citizens.

Keywords: *Landsat 8, Topex, Mount Dempo, Sector Development*

1. Introduction

Indonesia has many mountains, one of them is located in South Sumatra with an altitude of 3173 meters above sea level called Gunung Dempo as can be seen on Figure 1 with coordinates of 4.03 ° S 103.13 °E. Mount Dempo is the highest mountain in South Sumatra which has seven craters at its peak. The rocks at Mount Dempo consist of Tuff and Sand rocks which are tertiary deposits. Tertiary deposits are derived from andesite lava – basaltic and pyroclastic flows that are deposited. The geological structure that can be found on Mount Dempo are shear fault (dextral and cynical), normal fault, joint around its peak, and topographical damage (VSI ESDM, 2014). Mount Dempo has a type of volcano strato. The last eruption of the mountain occurred in 2009. When the mountain, which is located in Dempo Makmur Village, Pagar Alam, South Sumatra erupted, it emitted ash rain with a distance ± 10 Km. The character of the eruption from the volcano was a freatik eruption. The status of Mount Dempo is at level 1 (normal). Around Mount Dempo also has geotourism potential. There are four natural tourism, namely Pucak Gunung Dempo Crater, Dempo Tea Plantation Area, Singga Kudai Waterfall, Lawang Agung Waterfall and megalithic sites history sriwijaya.

Figure 1. Tanjung Sakti, Mount Dempo, Bengkulu (vsi.esdm.go.id)

2. Data and Methodology

- Landsat 8

This study uses Landsat 8 imagery data with the path in the southern hemisphere 48 UTM coordinate system, which can be obtained from the download of USGS EarthExplorer. Use ArcGIS 10.8 software to process Landsat 8 images to obtain vegetation index and land surface temperature maps.

1. Vegetation Map

NDVI is a method of measuring vegetation velocity by comparing the spectrum between NIR waves and Red waves. In Landsat 8's OLI sensor, NIR is level 5, while Red is level 4 (Ardiansyah, 2015).NDVI is calculated by the formula below:

$$NDVI = (NIR\text{-}Red)/(NIR+Red) \quad (1)$$

2. Land Surface Temperature

To identify land surface temperature, band used in Landsat 8 as the thermal wave is band 10 and band 11

- Gravity Satellite (TOPEX)

TOPEX or Poseidon is a satellite launch by NASA and CNES on 10th August 1992 to map the topography of ocean. The satellite was operated from 1992 until 2005. This is a geodetic satellite whose purpose is to provide high-precision global sea level (ocean height) measurements in coordinates relative to the center of the earth (Space News, 1992).

The satellite's orbit is a circular, non-sun-synchronous, nearly circular frozen orbit with an altitude of 1,336 kilometers, a period of 112.4 minutes, an inclination of 66.039o, and a repetition period of about 10 days (127 revolutions, equivalent to 9.9156 days) (P. Gaspar, 1994).

The total gravity data acquired from the satellite for this study is 64 points which mostly covered the Dempo Caldera area. Moreover, it is arranged in gridding with spacing for each point about 1km in TOPEX measurement as can be seen on Figure 2.

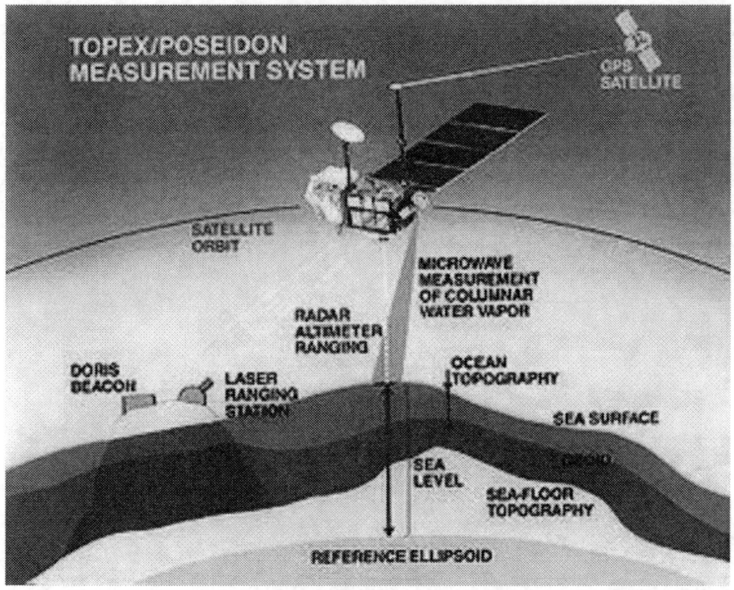

Figure 2. TOPEX measurement

3. Results and Discussion

- Landsat 8 Data

1. NDVI

According to the results of data processing, the minimum NDVI value is -0.307692, and the maximum NDVI value is 0.664967. On the vegetation map (Figure 3), the green area has a higher NDVI value, while the blue area has a lower NDVI value.

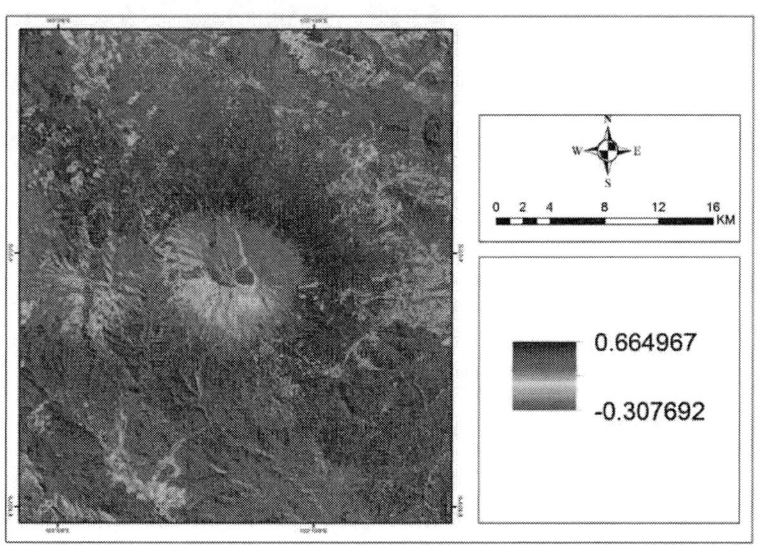

Figure 3. NDVI Map of Tanjung Sakti, Mount Dempo

The NDVI value is low because the surface of clouds and water reflects more visible wave energy than near-infrared light, while a high NDVI value (green area) indicates higher vegetation because the surface of vegetation reflects more radiation than infrared wavelengths many visible light.

This NDVI cannot be used as a reference for distinguishing vegetation objects from non-vegetation objects. There is no clear and consistent threshold for the NDVI value, because the NDVI value always changes with the seasons (Ardiansyah, 2015).

2. Land Surface Temperature

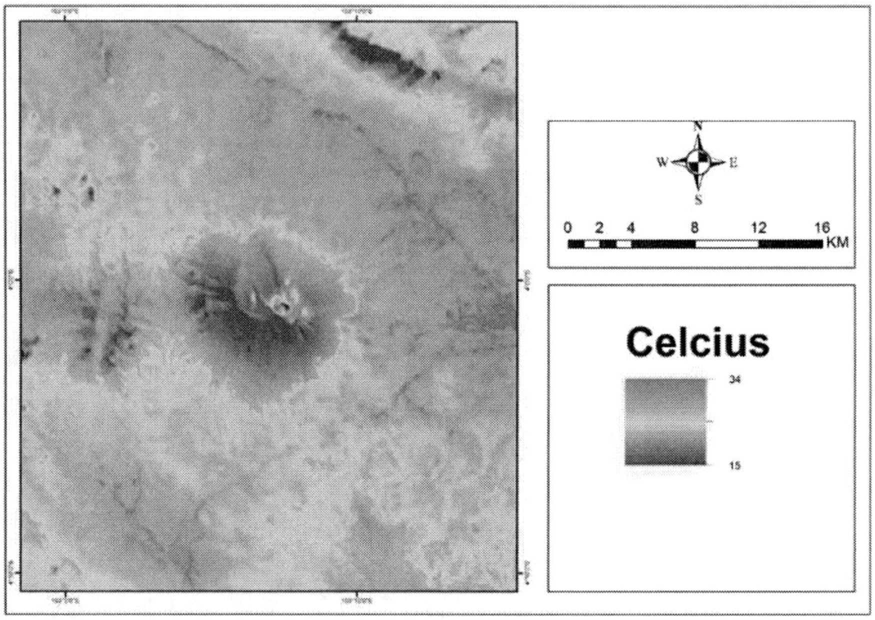

Figure 4. LST Map of Tanjung Sakti, Mount Dempo

The maximum surface temperature is in the range of 34°C, which is indicated in red, while the minimum temperature is 15°C, which is indicated in blue (Figure 4). The highest surface temperature in the northeast of the region. The area is suspected to be the Tanjung Sakti exploration area in Tempo.

- **Gravity Data**

1. Complete Bouguer Anomaly

The complete bouguer anomaly map can be seen on Figure 4. The value ranges from 121.8 to 184.8 mGal and the average density derived from parasnis method is 1.1359 g/cm3. As can be seen on Figure 5, high anomaly gravity zone spotted at southwest direction extend to northeast direction.

Based on the geological map, the high anomaly zone covered almost the ancient caldera which assumed it might be related with high density value of volcanic rock. While on the east and northeast direction low anomaly zones are spotted which represented by green and blue color. The low anomaly zone on the northeast direction interpreted might be associated with the presents of the remain caldera wall.

Further study is needed to predict the geothermal prospect area due to the variative interpretation which can be conclude from the gravity anomaly only. Thus, additional data is needed to get better interpretation.

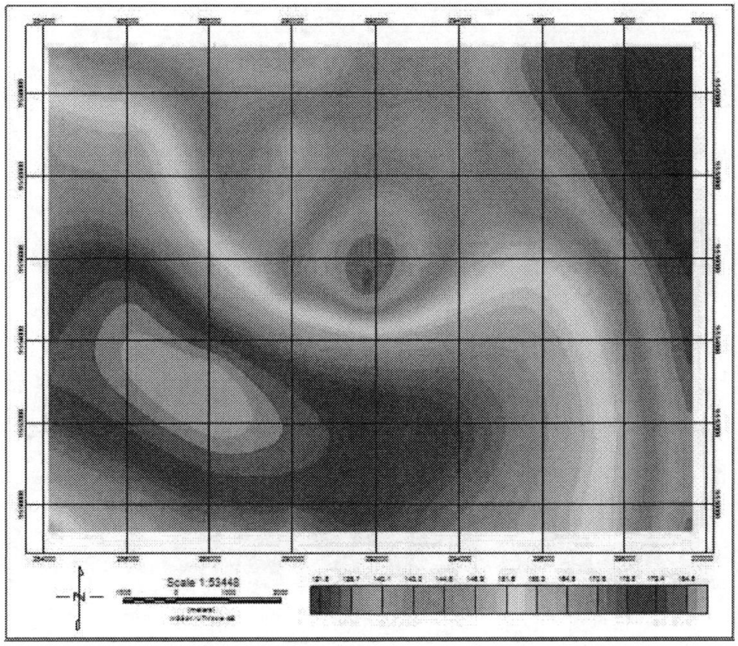

Figure 5. CBA Map of Tanjung Sakti, Mount Dempo

2. Regional Anomaly

Regional anomaly and residual anomaly was separate by using spectrum analysis using Oasis Montaj software. The result of regional anomaly can be seen on Figure 6. The anomaly is lookalike complete bouguer anomaly shown in Figure 5. The tendency of high anomaly zone extends from southwest to south east direction. Considering the study location which located in the volcanic complex, probably the high anomaly zone related with volcanoes which might be intrusive body.

Moreover, consider that regional anomaly images deeper anomaly at the study area, can be interpret that this high anomaly zone extends as it goes deeper. Thus, to get better image of high anomaly zone extension 2-D modelling and upward continuation should be done.

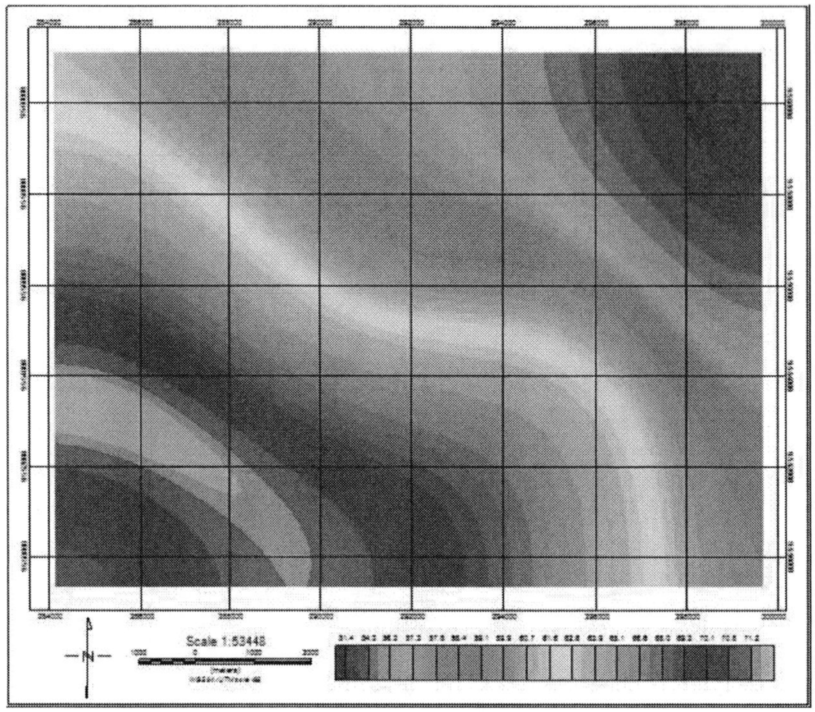

Figure 6. Regional Anomaly Map of Tanjung Sakti, Mount Dempo

3. Residual Anomaly

By separating the regional and residual anomaly, residual anomaly was conducted as can be seen on Figure 7. The pattern of residual anomaly does not lookalike either the CBA or Regional map. As can be seen on the picture, the residual anomaly tends to show shallower anomaly which is more variative and tend to be affected by noise. Thus, several small closures show there.

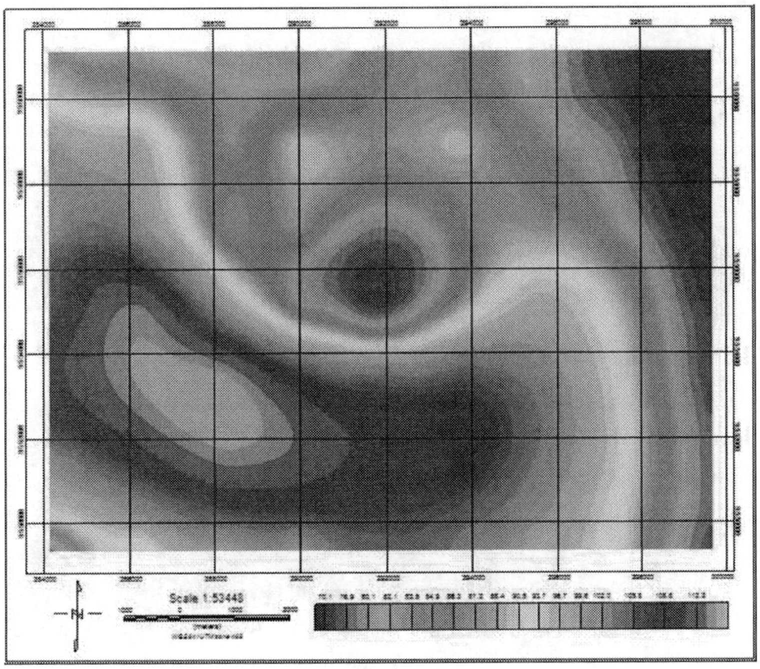

Figure 7. Residual Anomaly Map of Tanjung Sakti, Mount Dempo

- **Sector Development Area**

Pagaralam City, one of the cities in South Sumatra located on the slopes of Mount Dempo holds historical relics. Megalithic relics are scattered at several points of the city. The relics are menhir, dolmen, stone coffins, dimpls, and statues of stones in a static and dynamic style (Kherti, 1953).

In Basemah, menhirs were found stand alone or in group and form formations, and often side by side with other buildings. On the deep reef, a plain menhir found 1.6 meters high, standing on a rock. In the hamlet of Tegurwangi, there are also many plain menhirs 1.5 meters high near dolmens, statues and stone coffins. In addition to menhir, dolmens are also found in the city. The best dolmen is located in Batucawang. Dolmen stone board measuring 3 x 3 m with a thickness of 7 cm, located on top of 4 supporting stones. Another megalithic that has been found is a stone statue depicting the shape of an animal.

The most interesting discovery is a megalithic called "Batugajah". Batugajah is a piece of egg-shaped stone, measuring 2.17 m long and carved all over its surface as can be seen on Figure 8.

Figure 8. Batugajah

4. Conclusions

Based on remote sensing data analysis, the sector development zone is located in the center of this study area, because it has high surface temperature in NDVI. This area also has high gravity anomaly in Gravity Satelite. The high anomaly value in the center area might be associated with ancient intrusive body owned by Dempo caldera or possible heat source of geothermal system. Further study in this boundary should be done to give advance identification of geothermal prospect zone at Dempo geothermal prospect area. And also the site history study believes that with all the facts stated above, the spots which are located in Tanjung Sakti, Mount Dempo district are very prospective to be developed as a geotourism or tourism, in which could also increase the welfare of the local citizens. By integrating the local culture diversities and geodiversities, we believe that this paper could give a new insight for the local people to develop their geotourism potential where their priceless natural diversities are located.

Acknowledgments
The authors would like to thank the Geophysics, Department of Geoscience, Universitas Indonesia, for supporting this research.

References

[1] Ardiansyah. 2015. *Remote Sensing Image Processing Using ENVI 5.1 and ENVI LiDAR*. PT Labsig Inderaja Islim, Jakarta.
[2] Official website vsi ESDM. 2014. *G. Dempo*. Retrieved October 8, 2020, via vsi.esdm.go.id
[3] Official website of Magma Indonesia. 2020. *Dempo Volcano Activity Report*. Retrieved October 8, 2020, via magma.esdm.go.id
[4] P. Gaspar, F. Ogor , P. Y. Le Traon, O. Z. Zanife.1994. *Joint estimation of the TOPEX and Poseidon sea-state biases. Journal of Geophysical Research*. Vol. **99** (*Preprint* 24981-24994)
[5] Topex-Poseidon Partners Discuss Sequel', Space News, Aug. 17-23, 1992, p. 3
[6] Kherti. 1953. Official website of Pagar Alam City Government. 2018. Megalit Basemah. Retrieved 10 October 2020, via pagaralamkota.go.id

IOP Conf. Series: Earth and Environmental Science **873** (2021) 012016 doi:10.1088/1755-1315/873/1/012016

Frequency Dependent Magnetic Susceptibility in Topsoil of Bandung City, Indonesia

K H Kirana[1,2], J Apriliawardani[1], D Ariza[1], D Fitriani[1,2], E Agustine[1,2], S Bijaksana[3], S J Fajar[3], M G Nugraha[4]
[1]Faculty of Mathematics and Natural Sciences, Universitas Padjadjaran, Jalan Raya Bandung-Sumedang Km. 21 Jatinangor, 45363
[2]Centre of Citarum Research, Universitas Padjadjaran, Jalan Raya Bandung-Sumedang Km. 21 Jatinangor, 45363
[3]Faculty of Mining and Petroleum Engineering, Institut Teknologi Bandung, Jalan Ganesa 10 Bandung 40132
[4]Faculty of Mathematics and Sciences Education, Universitas Pendidikan Indonesia, Jalan Dr. Setiabudhi 229, 40154

kartika@geophys.unpad.ac.id

Abstract. Soil contains lithogenic components as well as anthropogenic components including combustion residues from traffic activities. The high traffic activities in major cities such as Bandung have caused the air pollution level to increase significantly. These activities might also produce significant combustion residues that accumulate, among others, in the topsoils. Compared with lithogenic components in topsoil, the anthropogenic combustion residues might have different magnetic signatures that could be detected by magnetic measurements. In this study, 38 topsoil samples from 19 roadside sampling points in Bandung City were collected and magnetically analysed to map the magnetic signatures due to traffic activities. The samples were measured for magnetic susceptibility using Bartington MS2B Susceptibility Meter and hysteresis parameter analysed from Vibrating Sample Magnetometer (VSM). The results show that the values of mass-specific magnetic susceptibility (χ_{LF}) vary from 391.20 to 1835.20×10^{-8} m^3/kg with the average value of 1012.16×10^{-8} m^3/kg while the values of frequency dependent susceptibility ($\chi_{FD}\%$) vary from 0.54% to 4.48% with the average value of 1.9%. The relatively high value of magnetic susceptibility indicates higher concentration of magnetic minerals compared to that of pristine topsoil around Bandung. This is in agreement with similar studies on roadside topsoil elsewhere. The poor correlation between mass-specific magnetic susceptibility and frequency dependent magnetic susceptibility infers that the magnetic minerals are predominantly non superparamagnetic. This finding is supported by magnetic hysteresis parameters showing that the predominant grains are likely to be pseudo-single domain (PSD) if magnetite is assumed to be the predominant magnetic mineral. Similar studies in German and China reported that the predominant magnetic mineral is mixture of single domain to multi domain magnetite.

1. Introduction

The increasing number of vehicles will produce abundant particulates, both as particulate matter that is present as air pollutants as well as those identified from the surface soil. Beckwith et al., (1986) showed a relationship between the high abundance of magnetic minerals in the soil and particulate matter in air pollutants [1]. For decades, identification of magnetic mineral characterization of urban topsoil have been widely used for environmental monitoring.

Rocks, sediments and soil at least contains of natural ferri(o)magnetic minerals. However, anthropogenic sources containing ferri(o)magnetic minerals also but in different phase, that could be deposited and accumulated in soil and sediment [2]. Therefore, by relying on magnetic properties of soil, this study identifies the characterization of magnetic minerals in urban topsoil in Bandung city. Considering that Bandung is one of the big cities in Indonesia with a dense population and a large number of vehicles, which often causes traffic congestion, resulting in an increased abundance of particulates. Since most of the air quality monitoring system in Bandung City not working properly, then the information about particulate derive from anthropogenic activity need to assess. The use of magnetic properties on soil sample is one of the common methods.

Many studies that correlate the anthropogenic sources with the abundance of magnetic minerals from particulates derived from vehicle in urban topsoil have been conducted in German [2], China [3], Argentina [4], and other countries. Most of the studies show that the more polluted an area, the more anthropogenic contains on the soil will produce higher magnetic susceptibility value. Higher magnetic susceptibility detected in the urban topsoil in industrial and roadside area [3]. Moreover, the higher magnetic concentration-magnetic enhancement found also emitted by vehicle [4].

The magnetic mineral derived from vehicle usually has dominantly by coarse multidomain grains, its identified the multidomain particle on the urban dust aerosol [5]. The abundance of multidomain grain also showed from vehicle emission in Abuja, Nigeria [6]. Therefore, by identifying the characterization of magnetic minerals in urban topsoil, it is hoped that it can identify the presence of anthropogenic minerals caused by the presence of motorized vehicle particulates.

2. Material and Methods

This research uses urban topsoil from Bandung City. In total, there are 38 samples of topsoil that collected from 19 locations around Bandung City (Figure 1). Furthermore, all samples were conducted on preparation to do magnetic susceptibility and hysteresis parameter measurement. For magnetic susceptibility measurement, the sample is inserted into a 10 cm^3 volume holder. Then, the sample is measured using a Bartington MS2B Magnetic Susceptibility meter. The results of these measurements will be obtained values of magnetic low-frequency susceptibility (χ_{LF}) and high-frequency magnetic susceptibility (χ_{HF}) which can then be calculated frequency-dependent susceptibility value using the equation ($\chi_{FD}\%$) = $[(\chi_{LF} - \chi_{HF})/ \chi_{LF}] \times 100\%$.

Furthermore, the hysteresis parameter was measured using the Oxford type 1.2H Vibrating Sample Magnetometer (VSM). The plot result of this measurement will obtain a hysteresis curve, which is the plot between the magnetization value M (Am^2/kg) to the applied field H (T). Before measuring hysteresis parameters, preparation was carried out by sieving the sample using a 325-mesh size in order to obtain a more uniform and smoother grain [7]. Through this measurement, a Day plot can also be obtained to determine the type of magnetic domain in the sample. The Day plot is a plot between the ratio of remanent saturation magnetization and saturation magnetization (Mrs/Ms) to the comparison of coercivity remanent field and coercivity field (Hcr/Hc).

3. Results and Discussions

The results of χ_{LF} mapping are shown in Figure 1. The value of low-frequency magnetic susceptibility (χ_{LF}) of topsoil samples in Bandung City shows that the samples have a range of 391.20 – 1835.20 ($\times 10^{-8}$ m^3/kg) with an average of 1012.16×10^{-8} m^3/kg. Based on the mapping in Figure 1, the North, East and South part of Bandung regions appear to have a larger and abundance distribution of magnetic susceptibility values, compared to the West and North Bandung areas which have small to large magnetic susceptibility values.

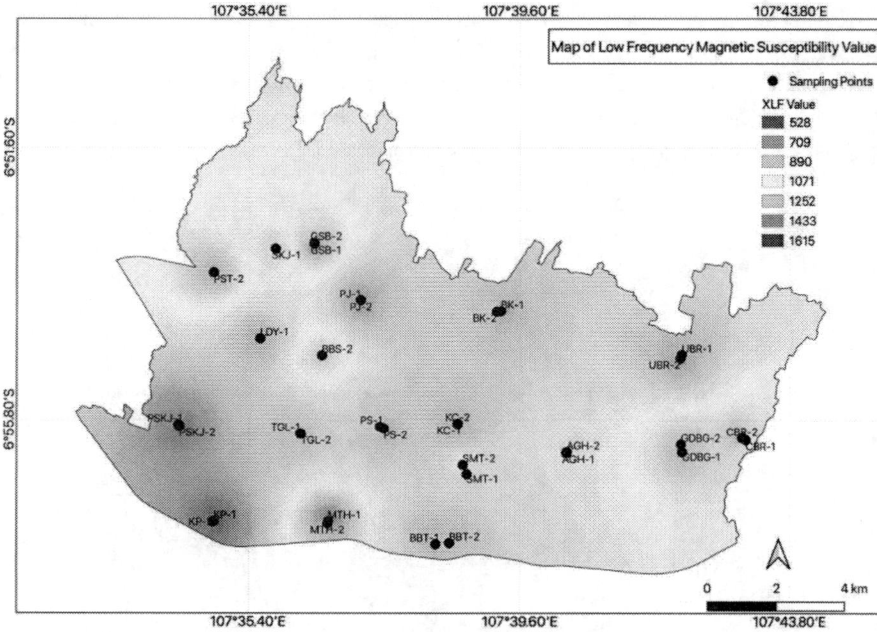

Figure 1. Map of low frequency magnetic susceptibility value of topsoil samples in Bandung City

The magnetic susceptibility value with mean of 1012.16 × 10^{-8} m^3/kg indicates that the urban topsoil sample is highly magnetic and dominated by ferrimagnetic minerals. This high value of magnetic susceptibility indicates higher concentration of magnetic minerals [8]. Soil that has a magnetic susceptibility value above 100 × 10^{-8} m^3/kg is a very magnetic soil and may have magnetic enrichment [9]. This result is supported by the hysteresis parameter measurement where the hysteresis curve shows the shape of the curve for magnetite minerals, which is experiencing magnetization saturation before 0.3 T. Figure 2 and Figure 3 are the hysteresis curves of two representative samples, namely the BBS1 and MTH1 samples.

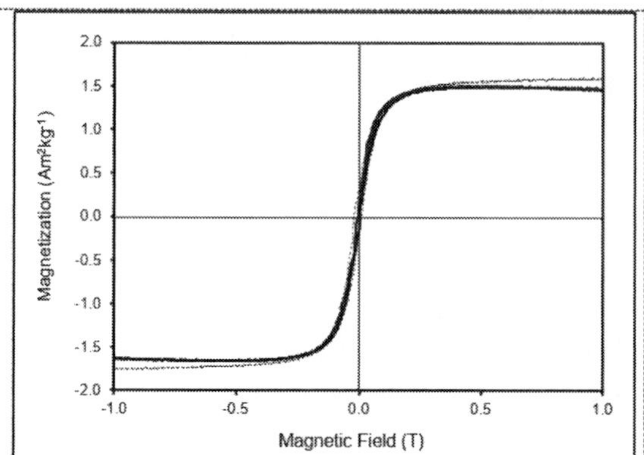

Figure 2. Hysteresis parameter of sample BBS1. Red line and black line is hysteresis curves before and after correction

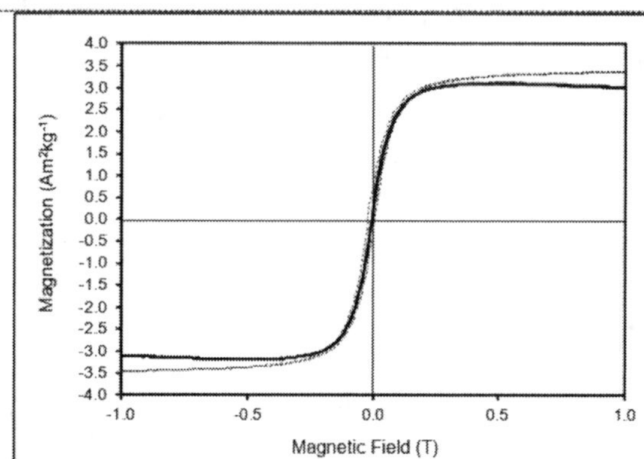

Figure 3. Hysteresis parameter of sample MTH1. Red line and black line is hysteresis curves before and after correction

The value of frequency-dependent susceptibility mapping represents in Figure 4. The figure shows that the North Bandung area is an area that has the highest value of (χ_{FD}%) compared to other areas in

Bandung City. The results of measuring (χ_{FD}%) of urban topsoil samples in Bandung City show that the samples have a range of (χ_{FD}%) 0.38% - 4.52% with an average of 1.89%. The (χ_{FD}%) value below 2% indicates a small amount of the superparamagnetic grain content in the sample (<10%) or almost no superparamagnetic grains, while the (χ_{FD}%) value in the range 2% - 10% indicates a mixture of superparamagnetic and non-superparamagnetic coarse grains [10].

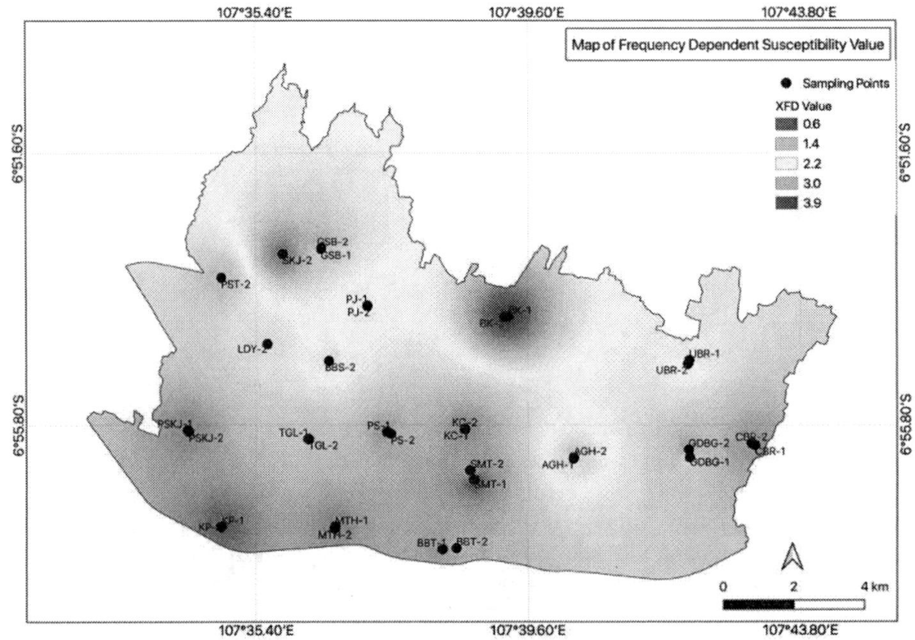

Figure 4. Map of frequency dependent susceptibility value of topsoil samples in Bandung City

Based on the mapping results of χ_{LF} and (χ_{FD}%), it can be said that the areas of East and South Bandung have a high χ_{LF} and an average (χ_{FD}%) was less than 2%, the North Bandung area has a value of χ_{LF} and an average (χ_{FD}%) was more than 2%, while the West Bandung area had χ_{LF} and the average (χ_{FD}%) was less than 2%. Several studies suggest that the high value of urban topsoil magnetic susceptibility is derived from fuel combustion pollutants and vehicle emissions [2] However, the value of χ_{LF} topsoil in Bandung City has a greater value than some areas, for example in Tubingen Germany has a magnetic susceptibility value range of 200 - 300 $\times 10^{-5}$ SI [2], in Lanzhou China with the average magnetic susceptibility value is 449.88 $\times 10^{-8}$ m^3/kg [3], and Luoyang China with an average magnetic susceptibility value less than 1000 $\times 10^{-8}$ m^3/kg [5]. The high value of χ_{LF} topsoil in Bandung City is due to the type of soil in Bandung City is volcanic. Natural volcanic soil (away from roadside vehicle activity) has a susceptibility value range of 339.8 – 806.7 ($\times 10^{-8}$ m^3/kg) [11].

In addition, high magnetic susceptibility values can also be caused by the presence of fine particles (superparamagnetic grains) that come from anthropogenic materials. The samples that has (χ_{FD}%) values ranging from 1 - 4% indicate the presence of anthropogenic minerals [12]. Most of the minerals that contain anthropogenic sources have multi-domain (MD) coarse grain types [5]. This is in line with the results of the Day plot (Mrs/Ms to Hcr/Hc) in Figure 5 which shows that the urban topsoil BBS1 sample has a mixed SD-MD type with a larger MD grain composition, while the MTH1 sample is dominated by SP grains. Magnetic minerals that have a tendency to have a mixture of more MD grains are spherule-shaped magnetite minerals [13]. The magnetic mineral that has spherule magnetite derived from anthropogenic sources [14].

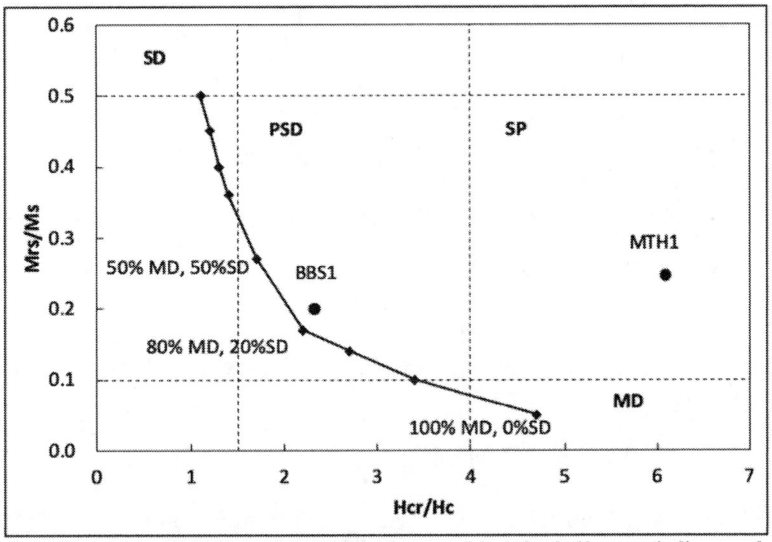

Figure 5. Day plot of samples BBS1 and MTH1 (dot mark). Black line and diamond marks represent percentage of SD-MD abundance in samples adopted from Dunlop (2002) [12].

4. Conclusion

The urban topsoil samples have a high magnetic mineral content indicated by a high χ_{LF} value dominated by magnetite. This high magnetic mineral content is due to the presence of magnetic enrichment by anthropogenic sources as evidenced by the small value (χ_{FD}%) and the abundance of coarse magnetic grains which have SD-MD domains.

Acknowledgments

Financial support for this research was provided by the Ministry of Education and Culture of the Republic of Indonesia through PDUPT Grant (No. 1827/UN6.3.1/LT/2020) to KHK.

References

[1] Beckwith P R, Ellis J B, Revitt D M and Oldfield F 1986 Heavy metal and magnetic relationships for urban source sediments *PEPI* **42** 1 pp. 67-75.

[2] Hoffmann V, Knab M, and Appel E 1999 Magnetic susceptibility mapping of roadside pollution *Journal of Geochemical Exploration* **66** 1-2, pp.313-326.

[3] Wang G, Oldfield F, Xia D, Chen F, Liu X and Zhang W 2012 Magnetic properties and correlation with heavy metals in urban street dust: a case study from the city of Lanzhou, China *Atmospheric Environment* **46** pp.289-298.

[4] Marié D C, Chaparro M A, Gogorza C S, Navas A and Sinito A M 2010 Vehicle-derived emissions and pollution on the road autovia 2 investigated by rock-magnetic parameters: A case study from Argentina *Studia Geophysica et Geodaetica* **54** 1 pp.135-152

[5] Liu H, Yan Y, Chang H, Chen H, Liang L, Liu X, Qiang X and Sun Y 2019 Magnetic signatures of natural and anthropogenic sources of urban dust aerosol *Atmospheric Chemistry and Physics* **19** 2 pp.731-745

[6] Solomon J S, Ahmed A L, Adamu I H and Dimu O O 2017 Identifying Anthropogenic Metallic Pollutants Using Frequency Dependent Magnetic Susceptibility Measurements In Abuja Metropolis *Currents Trends In Natural Sciences* **6** 11 pp.13-22

[7] Novala G C, Kirana K H, Fajar S J and Bijaksana S 2019 Testing the effectiveness of mechanical magnetic extraction in riverine and lacustrine sediments *Journal of Physics: Conference Series* **1204** 1 p. 012085

[8] Kirana K, Aufa N, Huliselan E and Bijaksana S 2011 Magnetic and electrical properties of leachate *ITB Journal of Science* **43** 3 pp.165-178

[9] Gautam P, Blaha U, Appel E, Ghanashyam N 2004 Environmental magnetic approach towards the quantification of pollution in Kathmandu urban area, Nepal *Phys Chem Earth*

[10] Dearing J A 1999 *Using the Bartington MS2 System. Environmental Magnetic Susceptibility* Kenilwort, U K: Chi Publising

[11] Faridlah M, Tohari A and Iryanti M 2016 Hubungan Parameter Sifat Magnetik Dan Sifat Keteknikan Tanah Pada Tanah Residual Vulkanik *Wahana Fisika* **1** 1 pp.54-76

[12] Bijaksana S and Huliselan E K 2010 Magnetic properties and heavy metal content of sanitary leachate sludge in two landfill sites near Bandung, Indonesia *Environmental Earth Sciences* **60** 2 pp.409-419

[13] Dunlop D J 2002 Theory and application of the Day plot (Mrs/Ms versus Hcr/Hc) 2. Application to data for rocks, sediments, and soils *Journal of Geophysical Research: Solid Earth* **107** B3 pp.EPM-5

[14] Kirana K H, Apriliawardani J, Ariza D, Fitriani D, Agustine E and Nugraha MG 2021 Magnetic morphology and mineralogy in urban soil of Bandung City, Indonesia *Journal of Physics: Conference Series* **1876** 1 p. 012006

IOP Conf. Series: Earth and Environmental Science **873** (2021) 012017 doi:10.1088/1755-1315/873/1/012017

The Role of 2D Fast Fourier Transform and High Pass Filter in Regional and Residual Anomaly Separation in Gravity

I R Palupi, W Raharjo, S Kiswanti

Geophysical Engineering, UPN "Veteran" Yogyakarta, Indonesia

email : inder13101986@gmail.com

Abstract. Regional and residual Separation anomaly is one thing that must do in gravity processing data. It is important before calculating the depth of anomaly by power spectrum. There are several ways to do this, one of them is using 2D Fast Fourier Transform (FFT). 2D FFT will calculate the two-dimensional power of the gravity map (Bouger anomaly) to change the spatial domain into the wavenumber domain. 2D FFT result has no unit because it works in the wavenumber domain. Power spectrum do in wavenumber domain map. Besides that, to make the wavenumber map in the frequency domain, it should be convolved with some filter (high–pass filter) and then inverse to separate the regional and the residual map. The design of the filter matrix depends on the number of the data and the location of anomalies will be enhanced. It will influence the separation result. The best result gets from the trial and error process. 2D FFT is act like Upward Continuation or Polynomial Fitting in the gravity method with the simple process. In this paper, the process fully done in Python. Python is an effective and simple language programming because it has many modules to support the processing and covering the big data. It also gives the flexibility to the researcher to determine the specific location that will be enhanced

Keywords: Gravity, Regional and residual separation, 2D FFT, high pass filter, Python

1. Introduction

Regional and residual separation is an important thing in Gravity processing data to get a good interpretation in geology like determine the depth of the anomaly. To do this, upward continuation is chosen as a mathematical method based on Jacobsen in [1]. The concept of upward continuation is to transform a point assumed Q located in mean sea level with the distance R to the higher position, assumed P, with the distance –h (negative sign interpreted as an up direction).

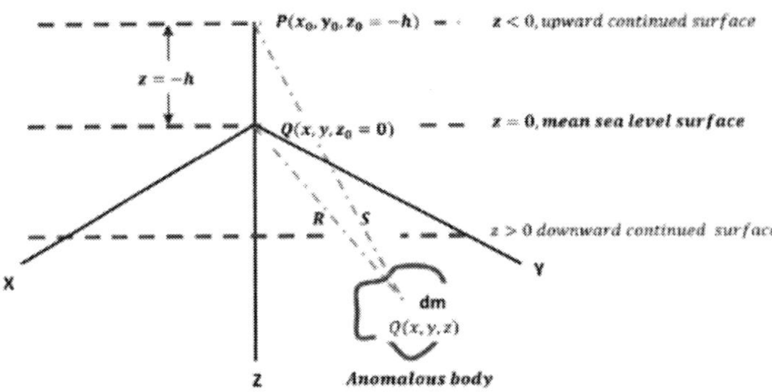

Figure 1. Illustration of upward continuation technique [1]

The other way to separate regional and residual anomaly of gravity data can use the trend surface by Menke in [1]. This is a filtering procedure that uses the polynomial surface fitting concept by applied least square techniques. In other words, the trend surface tries to solve the nonlinear problem (gravity) by using a linear approximation. Besides the two methods mentioned before, Zahra and Oweis applied the 2D Fast Fourier Transform (FFT) to define the residual source by changing the spatial domain into the wavenumber first [2]. After that, a high pass filter is applied by convolving it with the 2D FFT map result to enhance data characteristics and separating the regional and the residual map anomalies as the final model.

This research's purpose to explain the role of 2D Fast Fourier Transform and High Pass Filter in the separation of the regional and residual anomalies in Gravity. The Data that used in this research is gotten from topex and assumed as Bouger Anomaly. We do the process by making the program in Python. It gives us the flexibility to express our ideas unusually in Gravity data processing. Python is chosen because it is easy to understand and has a lot of modules supported in processing and covering big data.

2. Methods

Fast Fourier Transform (FFT) that usually used in seismic to convert or change the data from time to frequency domain, can also be applied to gravity data. It calculates the spectrum energy and changed the spatial data in the gravity map into the wavenumber domain before estimates residual and regional anomalies and to be used in calculating the depth of anomaly. In gravity, 2D FFT will calculate the two-dimensional power spectrum before operated with a filter [3]. The example of the power spectrum to calculate the depth of the anomaly is in **Figure 2**

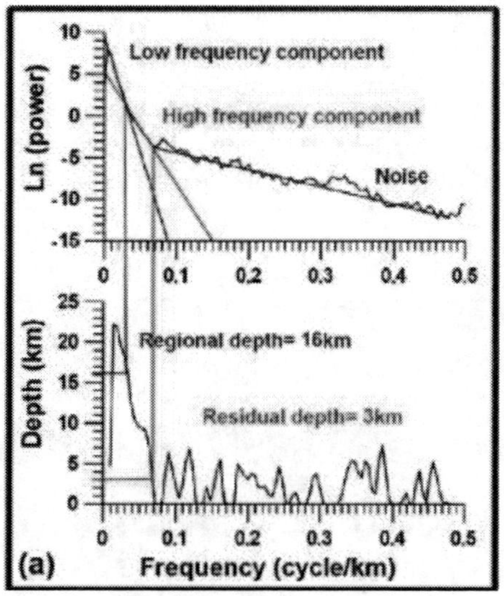

Figure 2. Illustration of the power spectrum to calculate the depth of anomaly [3]

Generally, FFT is the development product of Discrete Fourier Transform (DFT). DFT considered is not effective to solve some problem data related to periodicity, symmetries, and orthogonality. The basic FFT formulation is defined from the DFT equation in equation (1) [4],

$$C(k) = \sum_{n=0}^{N-1} x(n) W_N^{nk} \quad (1)$$

$$W_N^{nk} = e^{-j2\pi/N} \quad (2)$$

where n and k are integers, j is the imaginer sign and N is the numbers of data. However, when FFT is applied to the continued problem, equation (2) will be modified like in equation (3)

$$F(k) = \int_{n=0}^{N-1} f(n) e^{\frac{-2\pi jkn}{N}} dn \quad (3)$$

FFT uses complex number parameters in its formula. It includes real and imaginary parts. When both of them calculate together, magnitude or power spectrum is produced. FFT in equation (3) is 1D FFT and works in one dimension only (n). 2D FFT works in two dimensions spatial like in equation (4)

$$F(u,v) = \iint_{-\infty}^{\infty} f(x,y) e^{-2\pi j(xu+yv)} dxdy \quad (4)$$

where u and v assumed works in x and y.

2D FFT in gravity works in the spatial domain. When it is applied, the result of 2D FFT has no unit. To make it to the frequency domain, the result of 2D FFT should be convolved with some filters (low-pass, band-pass, high-pass, or Butterworth) [5]. After that, inverse 2D FFT should be applied to restore the map and separate the regional and residual anomaly, with equation (5).

$$f(x,y) = \iint_{-\infty}^{\infty} F(u,v) e^{2\pi j(xu+yv)} dudv \quad (5)$$

3. Result and Discussion

Data used in this research is downloaded from topex and assumed as Bouger Anomaly. The focus of this research is about the role of 2D FFT and high pass filter and how to processed them in Python. It just two modules needed to support the process, they are NumPy and Matplotlib. NumPy is used to load and calculate the FFT of the gravity map. While Matplotlib is used to display the map. The Anomaly Bouger can be seen in **Figure 3**.

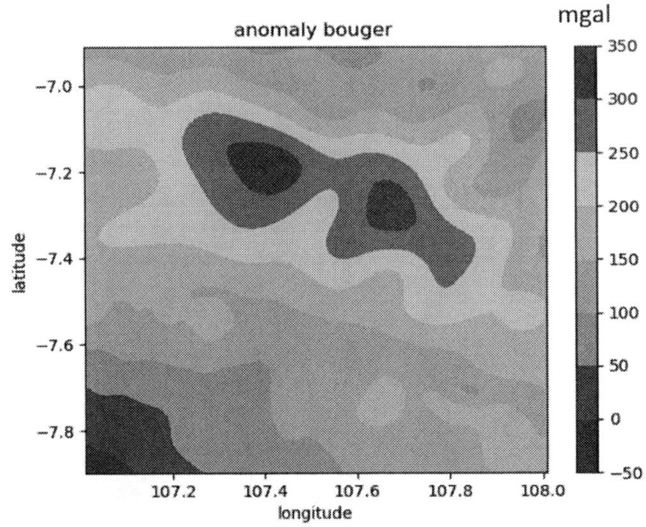

Figure 3. The Example of Anomaly Bouger data

When 2D FFT is applied on the Anomaly Bouger map, Fourier Transform works in each pair of x and y data in each grid of map like in equation (5). The result will have real and imaginary parts of each grid, to get the magnitude of them, it is needed to calculate the root square like in **Figure 4**. In **Figure 4** also there is an illustration digitizing step before calculating the depth of anomaly by power spectrum method.

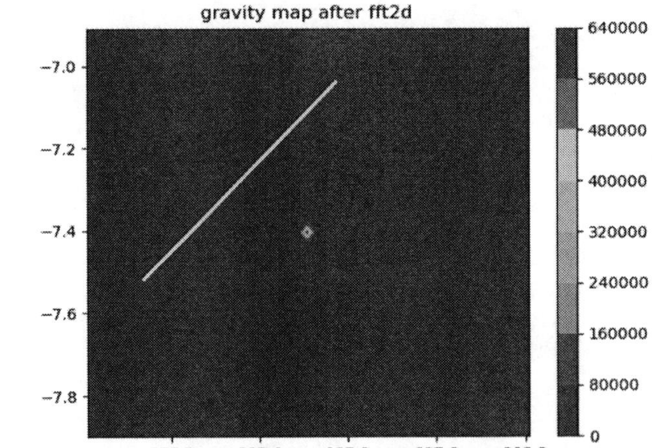

Figure 4. Illustration of anomaly Bouger anomaly map after 2D FFT with the section (yellow line)

Power spectrum analysis is done to get the depth of anomaly can be applied in **Figure 4**. It is done by digitizing the map in **Figure 4** based on the research area. Because the map has no unit, the y axis information is assumed as the amplitude. Then, the 2D FFT map is convolved with the high pass filter and inverse to get regional and residual separation. The example of the high-pass filter can be seen in **Figure 5**.

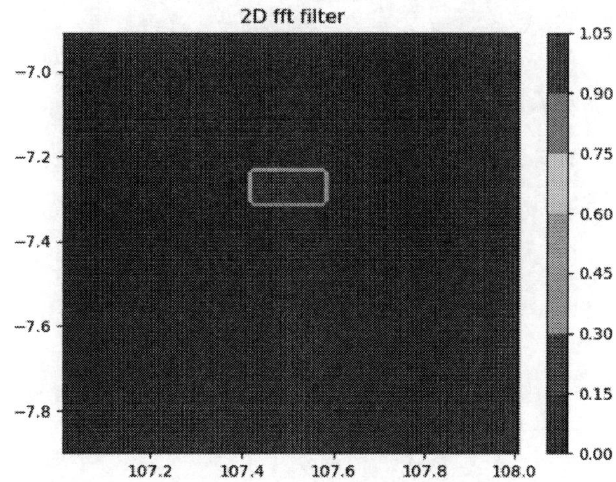

Figure 5. Illustration of high-pass filter

A high-pass filter is a matrix arrangement of some number. In this research, the numbers are 0 (blue color) and 1 (red color). Location and width of red color are gotten from the trial and error process. If both of them are not precise, the result will not maximal. As a comparison, in **Figure 6** there is a display of other illustrations of high-pass filters and the result can be seen in **Figure 7** and **Figure 8**

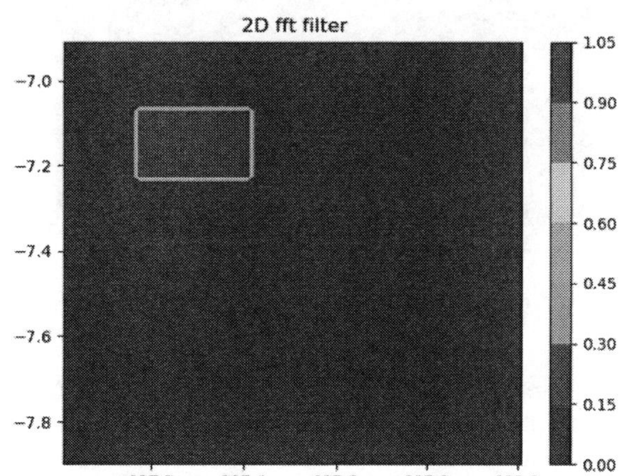

Figure 6. The other illustration of high pass filter

The next process is convolving the result of the 2D FFT map in **Figure 4** with a high-pass filter in **Figure 5** and **Figure 6**. In this process, the convolve will calculate each pair of x and y in a 2D FFT map with a high-pass filter map. So, the size of the matrix in 2D FFT and high-pass filter maps should be the same. Then, after convolve, the map should be inverse of 2D FFT to get the separation of the regional and residual map. Then, inverse the map to separate regional and residual anomalies.

IOP Conf. Series: Earth and Environmental Science **873** (2021) 012017 doi:10.1088/1755-1315/873/1/012017

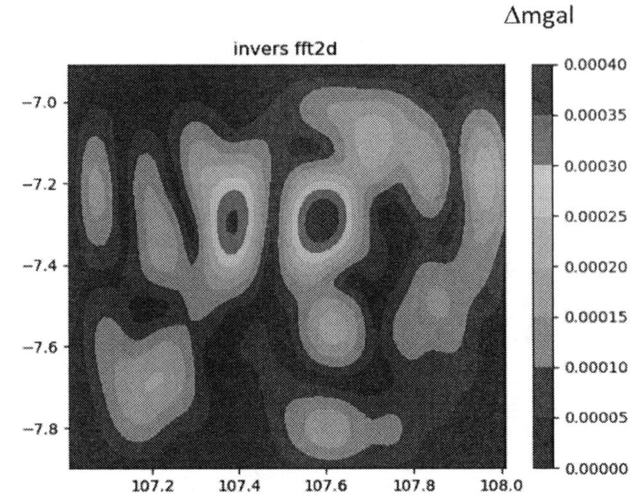

Figure 7. Regional and residual map as a result of inverse result of the convolve between 2D FFT map and high pass filter in **Figure 5**

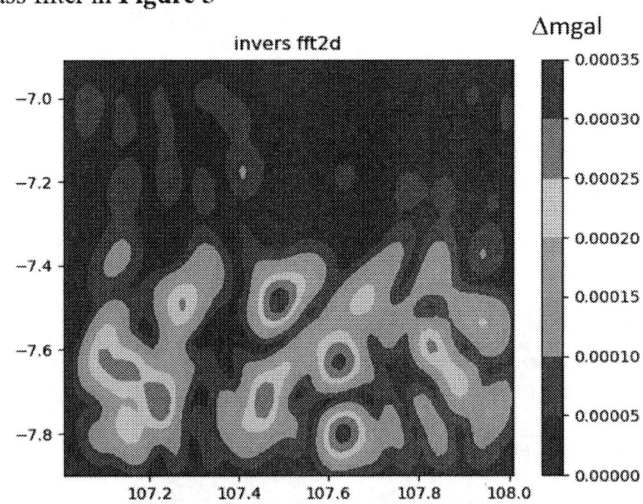

Figure 8. Regional and residual map as a result of inverse result of the convolve between 2D FFT map and high pass filter in **Figure 6**

The result in **Figure 7** is more appropriate than in **Figure 8**. Figure 7 is more relatable with the original data or Anomaly Bouger map. The high anomaly of the Bouger Anomaly map is suitable with the result in **Figure 7**. It means that placement and determination are two parameters that should be considered. In **Figure 7**, the regional and residual anomaly is separated better than in **Figure 8**. In **Figure 7,** two high anomalies of gravity data are good at corresponding with Bouger anomaly, but in **Figure 8,** the arrangement of high anomaly scattered in the south part of the map and not corresponding with Bouger anomaly. 2D FFT and the filtering process give a chance to the geoscientist to determine the specific location will be enhanced.

Conclusion

There are several ways to separate regional and residual anomalies in the gravity method. One of them is 2D FFT that will be convolved with some filters (high-pass filter) and finally inverse to separate the

IOP Conf. Series: Earth and Environmental Science **873** (2021) 012017 doi:10.1088/1755-1315/873/1/012017

regional and residual map. 2D FFT calculates the power spectrum of each pair x and y in a map to define the source of the residual data by changing spatial domain into wavenumber. Besides acting like input data before it convolves with the high pass filter matrix, the 2D FFT result also is basic data to do the power spectrum method to get the depth of anomaly. However, it means that 2D FFT and high pass filter are the alternative way to gravity data processing besides the usual method like the upward continuation. It can be easily done by making the program of ourselves (with Python or others) so the researcher has the flexibility to determine the location will be processed.

References

[1] Kebede H, Alemu A and Fisseha S 2020 Upward continuation and polynomial trend analysis as a gravity data decomposition, case study at Ziway-Shala basin, central Main Ethiopian rift *Heliyon* **6** e03292

[2] Zahra H S and Oweis H T 2016 Application of high-pass filtering techniques on gravity and magnetic data of the eastern Qattara Depression area, Western Desert, Egypt *NRIAG J. Astron. Geophys.* **5** 106–23

[3] Saada S . and El-Khadragy A A 2015 An Integrated Study of Gravity and Magnetic Data on West El-Minya Area, Western Desert, Egypt *J. Am. Sci.* **11** 169–84

[4] Burrus C S 2012 Fast Fourier Transforms *OpenStax CNX*

[5] Handyarso A and Kadir W G A 2017 Gravity Data Decomposition Based on Spectral Analysis and Halo Wavelet Transform, Case Study at Bird's Head Peninsula, West Papua *J. Eng. Technol. Sci.* **49** 423–37

IOP Conf. Series: Earth and Environmental Science **873** (2021) 012018 doi:10.1088/1755-1315/873/1/012018

Flower Pollination Algorithm for the Inversion of Schlumberger Sounding Curve

F Raflesia[1] and W Widodo[2]

[1] Master Program of Geophysical Engineering, Faculty of Mining and Petroleum Engineering, Institut Teknologi Bandung, Bandung 40132, Indonesia
[2] Applied Geophysics and Exploration Group, Faculty of Mining and Petroleum Engineering, Institut Teknologi Bandung, Bandung 40132, Indonesia

farkhan.raflesia@gmail.com

Abstract. Inversion of schlumberger sounding curve is non-linear, and multi-minimum. All linear inversion strategies can produce local optimum, and depend on the initial model. Meanwhile, the non-linear bionic method for inversion problems does not require an initial model, simple, flexible, derivation-free mechanism and can avoid local optimum. One of the new algorithm of the non-linear bionic method for geophysical inversion problem is the Flower Pollination Algorithm (FPA). The FPA is used for the inversion of schlumberger sounding curve. This algorithm was stimulated by the pollination process for blooming plants. The applicability of the present algorithm was tested on synthetic models A-type and KH-type curve. Numerical tests in MATLAB R2013a for the synthetic data and the observed data show that FPA can find the global minimum. For further study, inverted results using the FPA are contrasted with the damped least-square (DLSQR) inversion program, Particle Swarm Optimization (PSO), and Grey Wolf Optimizer (GWO). The outcomes of the comparison reveal that FPA performs better than the DLSQR inversion program, PSO, and GWO.

1. Introduction

Many inversion algorithms that have been applied in geophysical inverse problems; such as the Occam algorithm, Genetic Algorithm (GA) and PSO [1], and Lavenberq-Marquard using Singular Value Decomposition [2]. In the last decade, much of the work carried out on electrical resistivity has mainly concentrated on 2D, 3D and 4D techniques. Therefore, 1D inversion results are useful in constructing initial models for multidimensional interpretation. Therefore, a uniqueness study of 1D inversion is important. [3] used partial curve matching for the interpretation of Schlumberger sounding curve data. This technique is done manually, so it takes a long time. [4] creates MATLAB code to automate this work so that it is not only fast but also produces more consistent results. However, this result is only suitable for use as an initial model for the inversion process. Recently, the global approach has also been used by some researchers, such as the hybrid monte Carlo-based neural network and the Improved Gray Wolf Optimizer (IGWO).

From the simulations conducted by [5], it shows that FPA is more efficient than GA and PSO. The use of FPA in geophysical inversion problems has been carried out by [6] for self-potential (SP) data. The result is that FPA is robust for SP data interpretation. We demonstrate a MATLAB based inversion program that uses FPA for the analysis of schlumberger sounding curve. The applicability of the demonstrate algorithm was proved on synthetic models A-type (three-layered model with $\rho_1 < \rho_2 < \rho_3$) and KH-type (four-layered model with $\rho_1 < \rho_2 > \rho_3 < \rho_4$) curve. Furthermore, a field survey was still operated in Padalarang, Indonesia to analyze the effectiveness of the program on actual data.

2. Methods

2.1. Forward Modeling Formulation

The relationship between apparent resistivity (ρa) and layer parameters (layer thickness and actual resistivity) can be expressed by integral equations that take into account the isotropic and homogeneous earth model. [7] wrote the following equation:

$$\rho_a(s)=s^2\int_0^\infty T(\lambda)J_1(\lambda s)\lambda d\lambda \tag{1}$$

Here s is the half-spaced current electrode $(AB/2)$, J_1 shows the first-order Bessel function of the first form and λ shows the integral variable. The resistivity transform function, $T(\lambda)$ is written as follows:

$$T_i(\lambda) = \frac{T_{i+1}(\lambda)+\rho_i tanh(\lambda h_i)}{[1+T_{i+1}(\lambda)tanh(\lambda h_i)/\rho_i]}, i = n-1, \ldots\ldots, 1 \tag{2}$$

Where n is the amount of layers, ρ_i and h_i are the actual resistivity and thickness of the ith layer. Forward modelling in this study used code from [8].

2.2. FPA optimization method

The pseudo-code of the FPA is provided in figure 1. FPA was influenced by the pollination technique of flower, which is transfer of pollen grains to the ovule. The pollination technique should be global or local, the global pollination can be written as follows:

$$s_i^{t+1} = s_i^t + \alpha L(\lambda)(s_i^t - gbest) \tag{3}$$

Define a switch probability $p \in [0,1]$;
Initialize the randomly generated population within its upper and lower bounds;
Calculate objective function of pollen using Eq. (6);
Find the best solution *gbest* in the initial population;
while $abs(min_{RMS}-max_{RMS}) > 10^{-10}$ && $min_{RMS} > 10^{-10}$;
    for i=1 : *n* (*n* indicates number of pollen);
        if rand <*p;*
            global pollination via Eq. (3);
        else
            do local pollination via Eq. (4);
        end if
        Evaluate new solutions;
        Use greedy selection Eq. (5);
    end for
    Find the current best solution *gbest;*
end while

Figure 1. Pseudo-code of the FPA.

Local pollination can be represented mathematically as:

$$s_i^{t+1} = s_i^t + \in (s_j^t - s_k^t) \tag{4}$$

Where, s_i^t, s_j^t, s_k^t is the pollen of different flowers in the same plant species. This means index $i \neq j \neq k$. $\in$ denotes a uniform distribution between 0 and 1. Probability p describes the change in the two types of pollination. All pollen obtained for each generation is assess adopting greedy selection to decide which pollen could continue to live until the offspring. The equation used for the greedy selection is as follows:

$$s_i^{t+1} = \begin{cases} s_i^{t+1} \; if \; f(s_i^{t+1}) \leq f(s_i^t) \\ s_i^t \; if \; f(s_i^{t+1}) \quad > f(s_i^t) \end{cases} \tag{5}$$

2.3. Objective function

The suitability in the model response and the observed data is stated by a minimum objective function, whose search process is associated with the search for the optimum model. The model is modified in such a way that its response fits the data. In the process, it is clear that inverse modeling can only be performed if the dependence in the data and model parameters (future modeling functions) is known.

The objective function is explained as the total of squares of the data prediction error, which represents the difference in the observed data and the calculated data:

$$E=\sum_{i=1}^{N}\left(rho_i^{cal} - rho_i\right)^2 \tag{6}$$

For the objective function to have a clearer meaning in the same unit as the data (Ωm), it can be represent as the root mean square error (ERMS), with the following equation:

$$E_{RMS}=\sqrt{\frac{1}{N}\sum_{i=1}^{N}\left(rho_i^{cal} - rho_i\right)^2} \tag{7}$$

3. Results

3.1. Synthetic data

DC data sets over three-layered (A-type model) and four-layered models (KH-type model) are inverted to test the performances of the FPA. Furthermore, wide searching area is designed in the test below to prove the robustness of the FPA. The population *npop=50* and the iteration stops for A-type model and KH-type model when *RMS* smaller than 10^{-10} or has converged.

The results of the FPA based on models A-type and KH-types are contrasted with the DLSQR inversion program, PSO, and GWO for verification of the accuracy and efficiency of the FPA. We set inverted parameters (search space and maximum iteration) in the FPA to be the same as those in the DLSQR inversion program, PSO, and GWO. The search space and the experimental results are provided in table 1. Comparisons of RMS errors behaviours between the DLSQR inversion program, PSO, GWO, and FPA are exhibited in figure 2. We can clearly see that the FPA outperform the DLSQR inversion program, PSO and GWO in terms of the accuracy of the final inverted results and their convergence speeds (based on number of iterations).

For the FPA and DLSQR inversion program, the RMS error values are all smaller than the GWO and PSO. For the DLSQR inversion program, the RMS error values always sharply decay at the beginnings of the iterations, but always preceded by FPA in the middle iteration. Figure 2 reveal that

FPA has better performances than the DLSQR inversion program, PSO and GWO in terms of those improved RMS error situation.

Table 1. Results of the inversions based on the DLSQR inversion program, PSO, GWO, and FPA.

Model	Parameters	True value	DLSQR Initial value	DLSQR Estimated value	Search space	PSO	GWO	FPA
A-type	ρ_1 (Ωm)	30		30		30	30.13	30
	ρ_2 (Ωm)	100	300	100.2	0.5-1500	123.4	25.82	100
	ρ_3 (Ωm)	500		500.01		500.46	493.75	500
	h_1 (m)	8	10	8	1-30	8.19	7.32	8
	h_2 (m)	4		4		4.23	1.55	4
KH-type	ρ_1 (Ωm)	500		500		576.23	500.31	500
	ρ_2 (Ωm)	1000	300	1000	0.5-1500	37.17	1021.36	1000
	ρ_3 (Ωm)	211		211		1500	153.43	211
	ρ_4 (Ωm)	1375		1375		1500	1341.97	1375
	h_1 (m)	4		4		30	4.07	4
	h_2 (m)	8	10	8	1-30	1.22	8.13	8
	h_3 (m)	`12		`12		1	8.42	`12

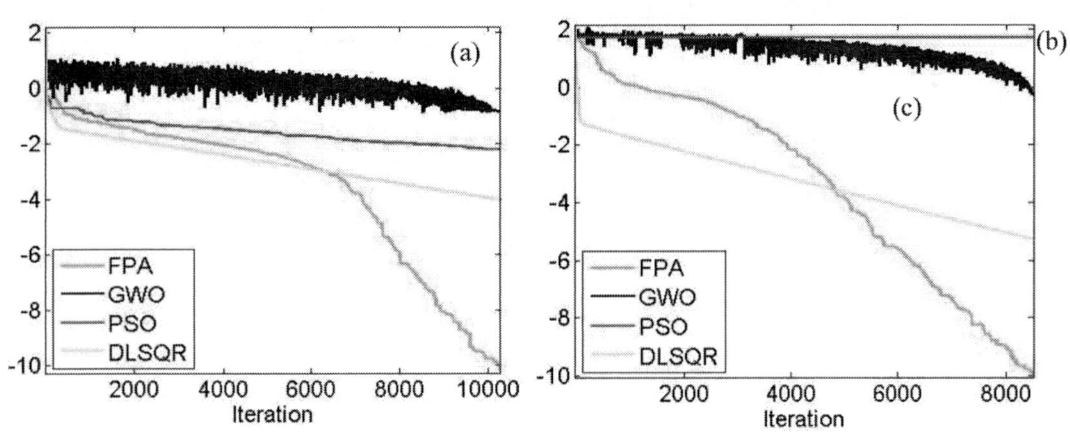

Figure 2. Comparisons of RMS error behaviours between the DLSQR inversion program, PSO, GWO, and FPA. (a) RMS error in the iterative process based on A-type model. (b) RMS error in the iterative process based on KH-type model.

Table 2. Elapsed time needed in the DLSQR inversion program, PSO, GWO, and FPA inversions.

Model	Number of iterations	Elapsed time (s) DLSQR	PSO	GWO	FPA
A-type	10264	404.53	242.17	270.71	3173.83
KH-type	8497	399.21	203.09	216,31	2708.82

The elapsed time (a Core i5-5200U CPU 2.2GHz) needed for all iteration and the number of iterations to obtain a solution in the DLSQR inversion program, PSO, GWO, and FPA inversions are shown in table 2. Table 2 indicates that the FPA is more time consuming than the DLSQR inversion program,

PSO and GWO. Generally, the FPA requires fewer iterations than the DLSQR inversion program, PSO, and GWO to obtain an acceptable solution.

3.2. Field data

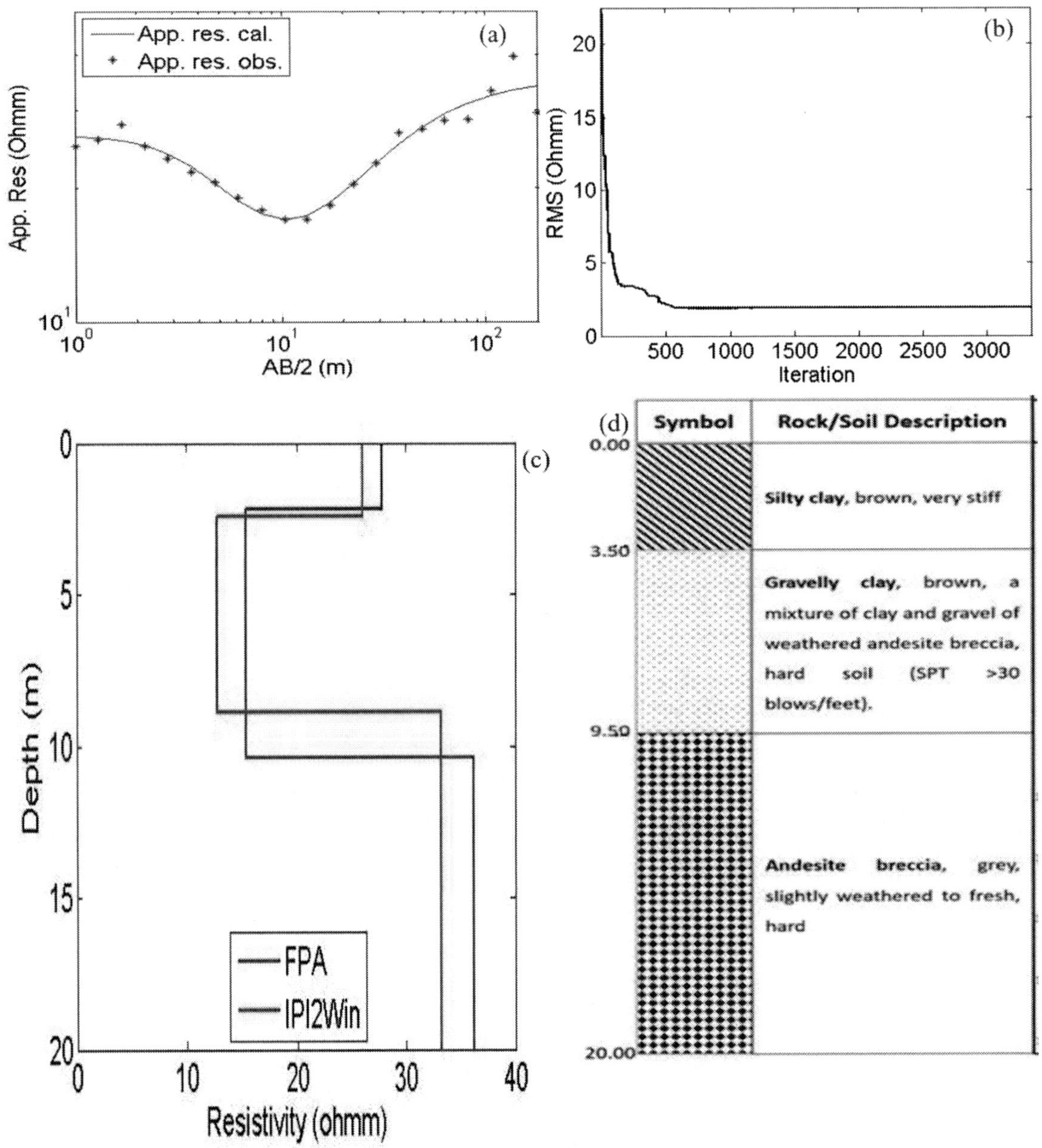

Figure 3. A field data example. (a) The calculated and observed apparent resistivity of the FPA. (b) RMS error behaviour of the FPA. (c) Inversion results using the FPA and IPI2Win. (d) Also shown is the borehole data.

The inverted result of the FPA for the observed DC data are illustrated in figure 3. The RMS error between the observed data and the calculated data is 1.9 Ωm (FPA) and 2.56 Ωm (IPI2Win). Figure 3a illustrates that the calculated apparent resistivity fit the measured observed apparent resistivity quite well. Figure 3b shows the convergence behaviour in the iterative process of the FPA. The RMS error values in figure 3b sharply decay during the first 500 iterations. Figure 3c depicts the inverted solutions from the FPA and IPI2Win. The inverted solution generally conforms to the results obtained using IPI2Win and quite according to the borehole data. Generally, first layer has resistivity value of approximately 27.81 Ωm at depths of 0-2.18 m. The second layer is a low resistivity layer (15.37 Ωm) at depths 2.18 - 10.4 m. The third layer has a higher apparent resistivity 36.21 Ωm in accordance with andesite breccia (igneous rock has higher resistivity than clay at first and second layer).

4. Conclusions

We have shown that the FPA able to inverse problems of Schlumberger sounding curve. In contrast to the DLSQR inversion program, PSO and GWO, the FPA can acquire more accurate results and provides more improved RMS errors. The only downside of FPA is the length of elapsed time it takes to complete the entire iteration. The high performance, efficiency, accuracy, and satisfactory convergence characteristics of the FPA for inverse problems of Schlumberger sounding curve are sufficiently illustrated in this paper.

Acknowledgments

This research received financial support by LPDP (Indonesia Endowment Found for Education), LPPM ITB, and DIKTI (Directorate General of Higher Education of Indonesia).

References

[1] Yogi I B S, Widodo 2017 Time domain electromagnetic 1D inversion using genetic algorithm and particle swarm optimization Available from: http://dx.doi.org/10.1063/1.4990901

[2] Widodo W, Saputera D H 2016 Improving Levenberg-Marquardt Algorithm Inversion Result Using Singular Value Decomposition *Earth Science Research* **5 (2)** 20

[3] Orellana E and Mooney H M 1966 Master tables and curves for vertical electrical sounding over layered structures *Interciencia* Madrid

[4] Ogunbo N J 2018 MATLAB code for data-driven initial model of 1D Schlumberger sounding curve *GEOPHYSICS* **83 (2)** F21–F28

[5] Yang X-S 2012 Flower Pollination Algorithm for Global Optimization. *Lecture Notes in Computer Science* Springer Berlin 240–249

[6] Sungkono 2020 Robust interpretation of single and multiple self-potential anomalies via flower pollination algorithm *Arabian Journal of Geoscience* **13** 100

[7] Koefoed O 1979 *Geosounding Principle, I: Resistivity Sounding Measurements*. Elseveir Scientific Publishing Co

[8] Ekinci Y L and Demirci A 2008 A damped least-square inversion program for the interpretation of schlumberger sounding curves *Journal of Applied Science* **8 (22)** 4070-4078

IOP Conf. Series: Earth and Environmental Science **873** (2021) 012019 doi:10.1088/1755-1315/873/1/012019

Subsurface Sediment Deposits of Tanjung Berikat Coast, Bangka, Determined from GPR Interpretation

L Gustiantini[1,*], U Kamiludin[1], M Zulfikar[1], Y Noviadi[2], U Hernawan[2], A F Ismayanto[3]

[1] Marine Geological Institute, Ministry of Energy and Mineral Resources, Jl. Dr. Junjunan 236 Bandung, Indonesia

[2] Centre for Geological Survey, Ministry of Energy and Mineral Resources Jl Diponegoro 57 Bandung, Indonesia

[3] Research Centre for Geotechnology, Indonesian Institute of Sciences – LIPI, Komplek LIPI, JL Sangkuriang Bandung 40135 Indonesia

gustiantini@yahoo.com

Abstract. Tanjung Berikat Coast in Central Bangka, is a part of the Southeast Asian tin belt. We conducted four Ground-Penetrating Radar (GPR) survey lines and 13 hand auger coring to understand sediment deposition and composition. Two similar units were determined from GPR lines BLG 01–BLG 03: Unit A at the top part, reflected by parallel and continuous reflector configuration, weak–strong electromagnetic wave. Underneath Unit A is Unit B, characterized by subparallel configuration, not continuous–chaotic, weak–medium electromagnetic wave. Unit B is absent in BLG 04. We identify another two units from BLG 04 and BLG 03, Unit C, characterized by subparallel reflector configuration, not continuous–chaotic, weak–strong electromagnetic wave. It exhibits distinctive modulating contact with Unit D. Unit D is characterized by chaotic reflector configuration, relatively stronger electromagnetic wave that might be correlated to the granite intrusion Tanjung Klabat. Sediment deposit is composed of fine–coarse sand, consisting mostly of clastic plutonic and clastic biogenic (coral and mollusk fragments), which increase downward. This indicates marine-fluvial influence, which suggests that sea-level changes strongly influence sedimentation process. Unit A from GPR is correlated to these sediment deposits, the other three units might be correlated to weathering of older insitu deposit.

1. Introduction

Bangka area is a part of the main tin belt granite of Southeast Asia. This tin belt lies from Myanmar, Thailand, Malaysia, Riau Archipelago, Bangka Belitung Singkep to Karimata Island in the Western Kalimantan [1, 2]. Previous study also found indication of uranium, thorium, and Rare Earth Element (REE) at Bangka area contained at monazite mineral as an associated mineral in a tin [3]. Therefore, Bangka coasts are interesting in understanding the occurrence of the placer mineral. According to a previous study, at the water nearby Tanjung Berikat and surrounding found sediment that comparable to Ranggam Formation, which indicates channel filling that allows for placer mineral and accessories mineral deposition derived from granitoid source rocks [3].

In addition to that, the Bangka area has many beaches that become to be a tourist destination, leading to infrastructure development. Unfortunately, the beaches are endangered from erosion. It is known that the coastline in Batu Beriga Beach is moving landward, particularly in the last ten years [2]. Therefore, we conducted this study to understand the subsurface Quaternary deposit related to the sedimentation process. According to the coastal characteristic mapping, which has been conducted in 2017 [3], the beach at Tanjung Berikat and surrounding are composed of 3 different beach types, are (Figure 1):

1. Sand beach, as the dominant type. This type lies along Lubuk Laut beach to the north of Tanjung Berikat beach, and southwest of Tanjung Berikat to the northeast of Batu Beriga beach. It is composed of Holocene alluvium, dominated by marine sedimentation process hence influenced by the fluvial process. This beach is relatively stable, and the process of sedimentation and abrasion is maintained balance.
2. Rocky beach (Rocks), particularly at Tanjung Berikat and Batu Beriga beaches. It is composed of Granit Klabat outcrops in 3 – 25 m in size, from Early Triassic – Early Jurassic [4], forming a medium – steep hill. Undisturbed coral characterizes the beach face. This beach type is dominated by the marine process. Although considered stable, a trace of erosion due to the oceanic current and wave is distinguished.
3. Mud flat and mangrove, formed at the southwest of Batu Beriga to the western part of the study area. This area is composed of Holocene alluvium, sandstone and claystone of Tanjung Genting Formation from Early Triassic [4]. This beach is influenced by the tide and fluviatile process. It is covered by mangrove vegetation.

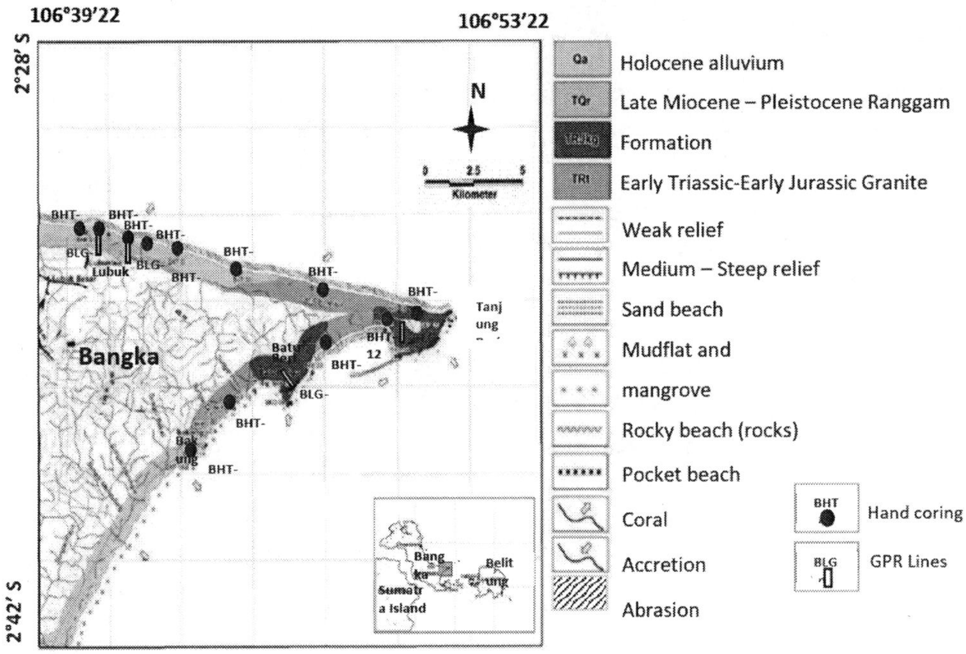

Figure 1. Coastal Characteristic Map and lithology of the Tanjung Berikat and surrounding [3], black circles are the hand coring location for this study, rectangles are GPR lines.

The lithology of the waters surrounding Tanjung Berikat revealed from seismic profile [3] indicated the oldest sequence that resembles Tanjung Genting Formation. This formation is composed of intercalating between sandstone and claystone. It was deposited during Early Triassic in shallow marine environment [4]. The formation was infiltrated by Granite Klabat intrusion during Early Triassic – Early Jurassic. Then sequence above was correlated to Ranggam Formation, intercalating between sandstone, claystone and conglomerate. This formation was deposited during Late Miocene – Early Pliocene, in fluvial environment. The youngest sequence was correlated to Holocene Alluvium, composed of a loose grain of clay, mud, sand, and gravel, as fluvial, swamp, and beach deposit (Table 1).

Table 1. Seismic stratigraphy [3] correlated to the regional stratigraphy [4] of Tanjung Berikat waters

Stratigraphy of Southern Batam [4]						Tanjung Berikat and Surrounding [3]
ERA	PERIODS	EPOCH	Surface Deposits and Sedimentary rocks	Metamorphic Rocks	Intrusive Rocks	Sequence Stratigraphy
CENOZOIC	QUATERNARY	Holocene	Alluvium (Qa)			Sequence A
		Pleistocene				
	TERTIARY	Pliocene	Ranggam Formation (TQr)			Sequence B
		Miocene				
		Oligocene				
		Eocene				
		Paleocene				
MESOZOIC	Cretaceous					Intrusion Sequence C
	Jurassic				Granite Klabat Intrusion (TRJkg)	
	Triassic		Tanjung Genting Formation			
PALEOZOIC	Permian			Pemali Metamorf (CPp)		
	Carboniferous					

Ground Penetrating Radar (GPR) is a non-invasive method for a subsurface geological investigation that relatively shallow and detail, according to the changes of its geological material properties and physical characteristic composition [5], [6], [7] and [8]. The principle of GPR application is similar to the reflection seismic as facies identification and subsurface layer sequence [6]. The GPR signal can be influenced by dielectric constant material properties [9]. Interpretation of sediment type and facies are based on the concept of seismic stratigraphy, notably configuration identification and reflection termination, a configuration of internal reflection and external geometry [6], [10]. The result of GPR image will exhibit geometrical basic form as lithology, freshwater, and subsurface utilization [11].

2. Methods

We conducted Ground Penetrating Radar (GPR) by using SIRVeyor 20 model of GSSI product. It has 40 MHz MLF (Multi-Level Frequency) antenna as separated transmitter, and receiver antenna. Depth penetration optimum of the equipment unit approximately down to 30 m. Supporting equipment, including accumulator survey wheel, GPS, and measurement type, were also used.

GPR method is based on the emission of electromagnetic wave to the ground and the capture of electromagnetic waves that are transmitted, reflected and scattered by subsurface structure and anomalies beneath the earth surface. Reflected and scattered electromagnetic wave is then received by receiving antenna on the earth surface [12]. An electromagnetic pulse will be transmitted through the ground, and the return time of the reflected pulse will be recorded. The velocity of electromagnetic wave transmission and backscattering is swift and stated in the nanosecond time unit [13].

The resolution of penetration depth of radar signal relies on the transmitted pulse and antenna choices, mostly between 10 and 1000 MHz. Higher frequencies will result in higher resolution but shallower penetration depth. However, the depth of penetration will also according to the dielectric and the properties of conductive ground material [7].

In general, georadar data processing consist of several steps: data conversion to digital format, eliminating or reducing direct wave from air to the surface, setting and adjustment of amplitude and gain, static data adjustment for removing elevation distinction, data filtering, and velocity analysis [10]. To process the data, we used Radar 5 software of GSSI product. The first step of processing data was time zero correction. This step removes direct wave between antenna and land surface.

IOP Conf. Series: Earth and Environmental Science **873** (2021) 012019 doi:10.1088/1755-1315/873/1/012019

Afterwards, were spatial filter process, deconvolution, migration, and adjustment of amplitude and gain. The last step was interpretation of GPR data.

GPR data interpretation was based on interpretation of radar facies. The concept is similar to the methodology applied to seismic stratigraphy surveys [14]. The characteristics of radar facies were interpreted according to the internal configuration, reflection continuity, and the pattern of reflection termination [15]. It is generally characterized by the basic of shape, amplitude, continuity, and internal reflection configuration and external form [16], [17].

GPR survey was conducted on four traverse lines, with seaward direction, perpendicular to the beach line. Two lines were conducted at Pantai Lubuk Laut (BLG-01 and BLG-02) at S-N direction, 50 m each in length. One line (BLG-03) is located at Pantai Gusung Tanjung Berikat (S-N) along 175m. The last one is located at Batu Beriga (BLG-4) aligned from NW-SE along 200 m (Figure 2).
In order to understand the texture and characteristic of sediment in horizontal distribution, 13 Hand coring (shallow coring) was also performed on the sand beach by using Eijkelkamp hand auger. Due to sand texture and shallow seawater, it is known that coring on sand beach barely reaches 100% recovery. Part of the sediment will escape once the barrel core is lifted to the surface.

3. Result and Discussion

From shallow hand coring, we obtained sediment core varied between 20 to 260 cm in length (Table 2). Core sediment recovery indicates relatively thicker sediment along the beach of Lubuk Laut to Tanjung Berikat sediment profile (NW – SE) compared to that of Tanjung Berikat to Batu Beriga (SW – NE) (Figure 2 and 3). The sediment length from the first sediment profile indicates more than 1 m lengths, relatively longer at the NW side (> 1.5 m). The size of core sediment at the second profile exhibits shorter (mostly < 1 m). The most extended sediment core is BHT 11 at the northeast of Batu Beriga (1.6 m), and BHT 13 at Batu Beriga (1 m length). The other sediment cores are less than 1 m. seawater contents might be the significant factors that lead to various sediment core recovery, as mentioned above. Particularly along the beach Batu Beriga to the southwest part is characterized by mudflat and mangrove, thus, its sediment is composed of relatively higher water content. The NW beach, at Lubuk Laut is marked by accretion due to more elevated sedimentation, lead to thicker sand deposits, particularly from the marine side transported by a wave that forming berms as beach ridge [3]. While nearby Tanjung Berikat and at Batu Beriga are more characterized by abrasion process lead to lower sediment deposit.

Table 2. Sediment core location and depth

Sample	Coordinate		lengths
	Longitude	Latitude	(cm)
BHT-01	106°40'40.98"	2°32'21.624"	160
BHT-02	106°41'18.816"	2°32'26.052"	220
BHT-03	106°41'52.26"	2°32'32.244"	170
BHT-04	106°42'24.264"	2°32'42"	210
BHT-05	106°43'17.508"	2°32'55.32"	260
BHT-06	106°45'5.688"	2°33'20.916"	160
BHT-07	106°47'3.768"	2°33'42.12"	130
BHT-08	106°49'41.268"	2°34'9.948"	120
BHT-09	106°43'58.728"	2°37'23.916"	60
BHT-10	106°44'50.892"	2°36'28.44"	20
BHT-11	106°47'27.096"	2°35'7.116"	160
BHT-12	106°49'1.02"	2°34'41.088"	20
BHT-13	106°49'1.02"	2°34'41.088"	100

IOP Conf. Series: Earth and Environmental Science **873** (2021) 012019 doi:10.1088/1755-1315/873/1/012019

Furthermore, from 13 coring, the sediment composition are rather similar, characterized by sand deposit, brownish yellow wet color, whitesh brown for dry color, fine – coarse in size, rounded to subangular, medium - bad shorting, composed mainly by clastic plutonic and partly by carbonaceous biogenic clastic. Plutonic material is particularly composed of quartz, feldspar, mafic mineral, and rock fragment. Bioclastic material is composed of rock fragments and biogenic material, which increase downward.

GPR record from BLG-01 and BLG-02 at Pantai Lubuk Laut indicate similar characteristics, and each line penetrates down to ± 25 m, exhibit 2 different unit characteristics. Unit A at the top part down to 14 m, indicated by parallel reflector configuration, continuous with a weak – strong electromagnetic wave. Below Unit A, separated by a parallel unconformity, is Unit B, characterized by subparallel reflector configuration, not continuous – chaotic, with a weak – medium electromagnetic wave (Figure 4 and 5). This finding indicates that these areas composed of at least two different facies that might be resulted from 2 different sedimentation processes.

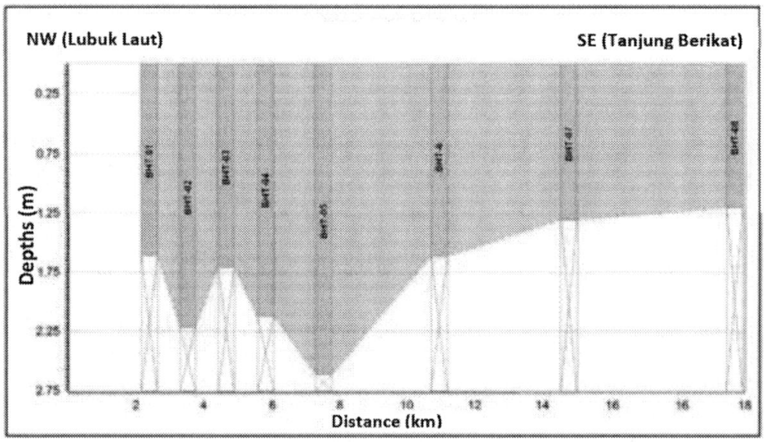

Figure 2. Sediment profile of the hand coring along the beach of Lubuk Laut – Tanjung Berikat (northern part of the study area, NW – SE)

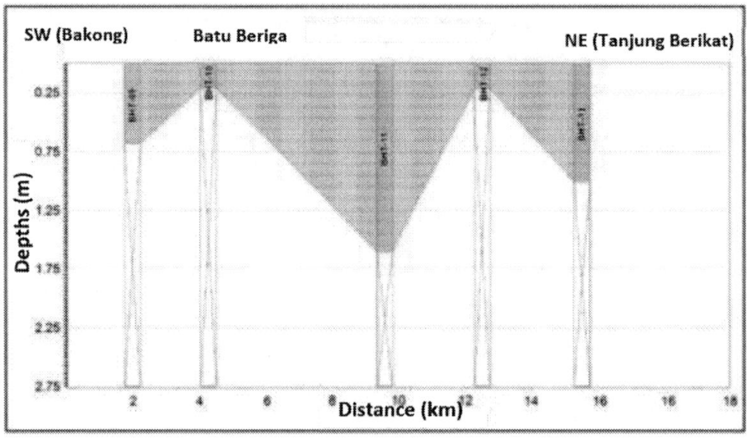

Figure 3. Sediment profile of the hand coring along beach of Bakung – Batu Beriga - Tanjung Berikat (SW – NE)

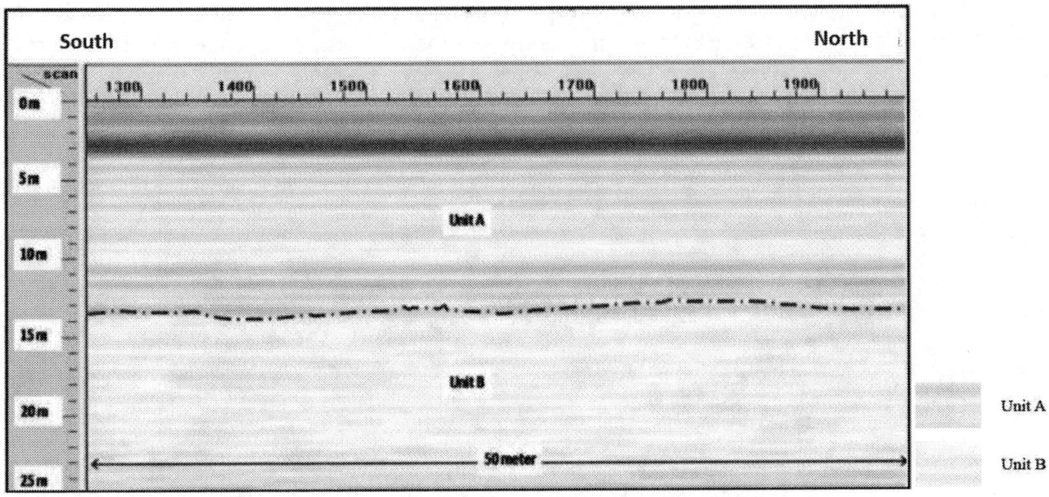

Figure 4. GPR Record of BLG-01 (Pantai Lubuk Laut)

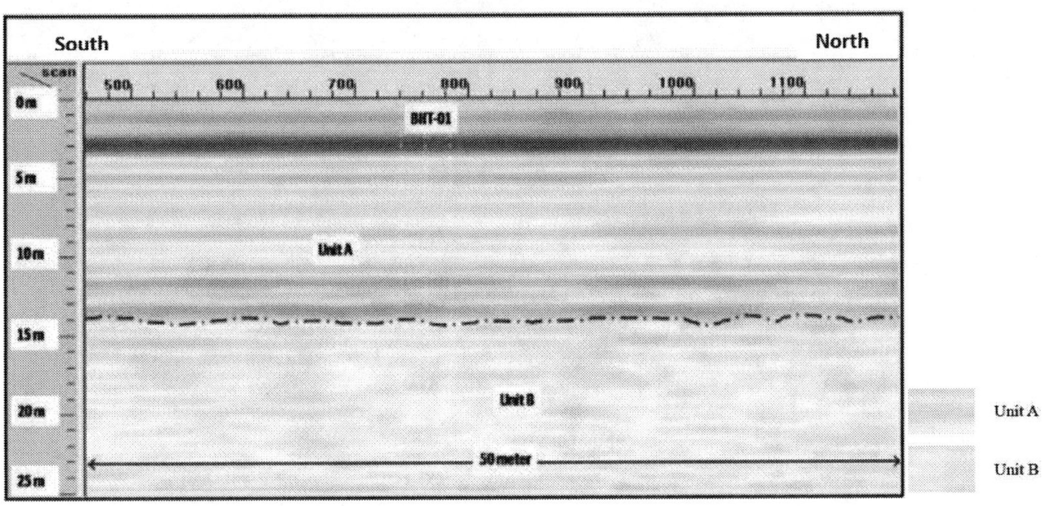

Figure 5. GPR Record of BLG-02 (Pantai Lubuk Laut)

BLG-03 and BLG-04 are located at Pantai Gusung Tanjung Berikat, and Batu Beriga also indicates similar characteristics, each line penetrates down to ± 40 m. Subsurface geological condition reveals three different characters. The top part of these two lines is identical to that of BLG-01 and BLG-02, characterized by parallel reflector configuration, continuous with a weak – strong electromagnetic wave, hence marked as Unit A. The thickness of this configuration is ± 16 m. Below Unit A from BLG-03 is Unit B, characterized by subparallel reflector configuration, not continuous – chaotic, with a weak – medium electromagnetic wave, comprised for approximately 2 m. While at BLG-04, Unit B is absent. Afterwards, confined with parallel unconformity below, is Unit C, determined by subparallel reflector configuration, not continuous – chaotic, with a weak – strong electromagnetic wave. This unit's bottom part is undulated, resulting in various thickness between 3 and 20 m from BLG-03. The bottom part of BLG-03 and BLG-04 is Unit D, characterized by chaotic reflector configuration with stronger electromagnetic waves (Figure 6 and 7). At BLG-03, Unit D is distinguished from 20 m to 36 m, while at BLG-04, the configuration is found from 20 – 25 m depth.

Correlating GPR record to the shallow core profile, the first horizon of radar (Unit A) indicating parallel reflector. It can be correlated to the sandy sediment deposit, which is characterized by a parallel layer with fine – coarse grain size, deposited by marine-fluviatile process related to the sea level changes, or the Unit Holocene Alluvium of regional stratigraphy [4] (Table 3).

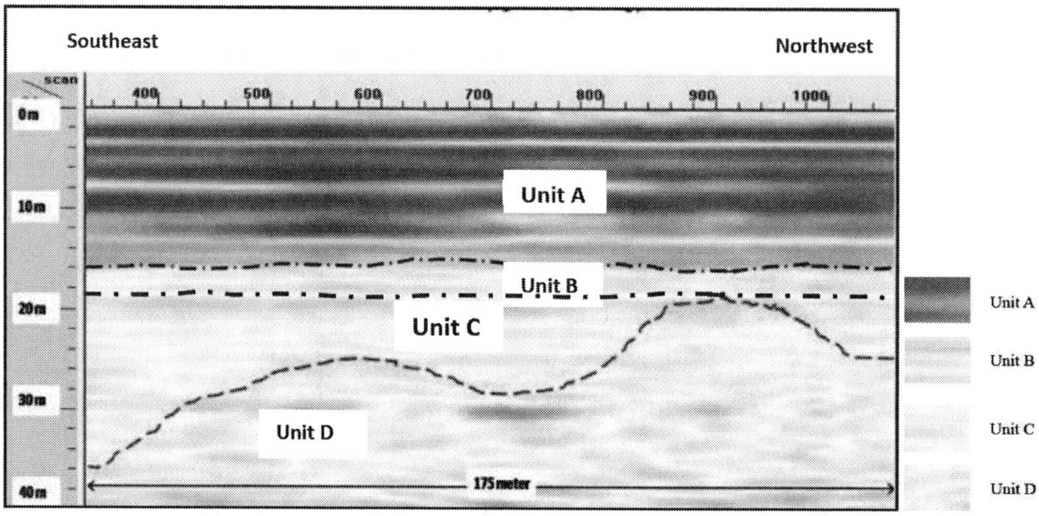

Figure 6. GPR Record of BLG-03 (Pantai Tanjung Berikat)

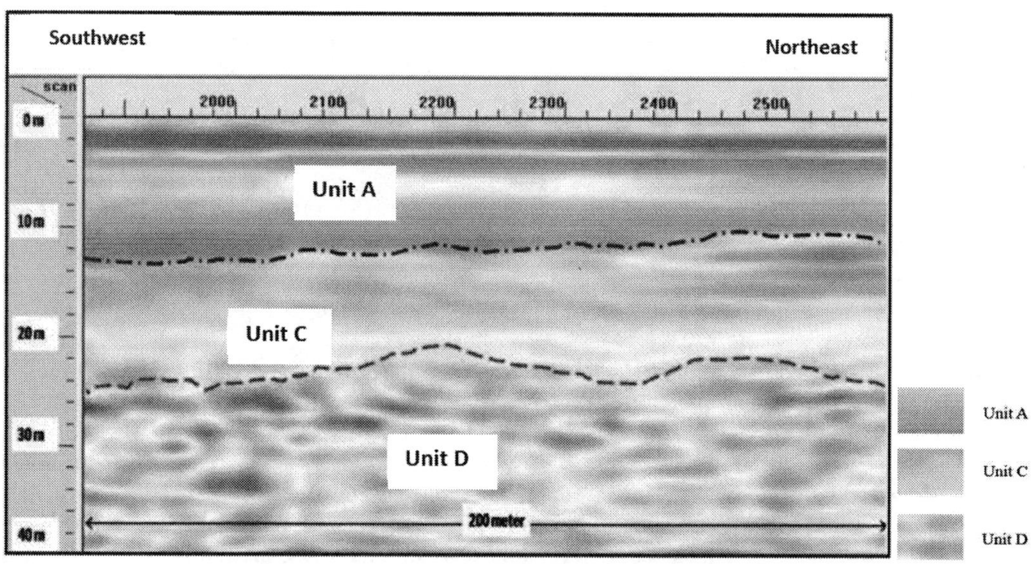

Figure 7. GPR Record of BLG-04 (Pantai Batu Beriga)

Unit B that characterized by subparallel reflector configuration, not continuous – chaotic, with a weak – medium electromagnetic wave, clearly represents different facies than Unit A, we assumed that this profile is comparable to stratigraphy below the alluvium, which is Ranggam Formation. And we correlated Unit C to Tanjung Genting Formation, below the Ranggam Formation. The shape of chaotic reflector configuration with stronger electromagnetic wave (D) that infiltrate Unit C is comparable to granite intrusion. The intrusion is also distinguished from the seismic profile from waters surrounding the area [2]. This granite is also found at the beach outcrop, particularly at Tanjung Berikat and Batu Beriga, where the Line BLG-03 and BLG-04 of GPR were performed. Ranggam Formation, according to [2], indicates channel filling that allows for placer mineral and accessories mineral deposition derived from granitoid source rocks. Therefore, the presence of this formation revealed from the GPR profile might evidence the occurrence of placer minerals in the study area.

Table 3. Stratigraphy of Tanjung Berikat and surrounding from GPR data compared to the regional stratigraphy

Stratigraphy of Southern Batam [4]						Tanjung Berikat and Surrounding [3]
ERA	PERIODS	EPOCH	Surface Deposits and Sedimentary rocks	Metamorphic Rocks	Intrusive Rocks	GPR
CENOZOIC	QUATERNARY	Holocene	Alluvium (Qa)			Unit A
		Pleistocene	Ranggam Formation (TQr)			Unit B
	TERTIARY	Pliocene				
		Miocene				
		Oligocene				
		Eocene				
		Paleocene				
MESOZOIC	Cretaceous					
	Jurassic				Granite Klabat Intrusion (TRJkg)	Unit D
	Triassic		Tanjung Genting Formation			Unit C
PALEOZOIC	Permian			Pemali Metamorf (CPp)		
	Carboniferous					

4. Conclusion

The GPR interpretation indicates that the lithology of the area surrounding Tanjung Berikat is composed of 4 deposit types: Holocene Alluvium, Ranggam Formation, Tanjung Genting Formation, and the granite intrusion of Klabat. The beach sediment is composed of particularly plutonic and clastic biogenic, which indicates marine-fluvial influence, suggesting that sea-level changes strongly influence the sedimentation process. Ranggam Formation's presence indicates a possibility of the placer mineral deposit occurrence, although the form of channel filling was not found.

Acknowledgement

We aknowledge the Head of Marine Geological Institute of the Ministry of Energy and Mineral Resources for all supports, including funding and all facility for field survey and laboratory analysis. Gratitude also to all the Team member of the Tanjung Berikat, Bangka field survey. We acknowledge the District Government of Tanjung Berikat, Bangka, and to all colleagues for discussion and suggestion.

References

[1] Cobbing E J, Mallick D I J, Pitfield P E J and Teoh L H 1986 The granites of the Southeast Asia tin belt, *J. Geol. Soc. London.*, pp. 537–550
[2] Kamiludin U, Aryanto N C D, Pertala A W and Zulfikar M 2018 Paleo channel indication as place of placer minerals and rare earth elements in Tanjung Berikat waters and surrounding, Central Bangka, Bangka Belitung (in bahasa Indonesia), *Jurnal Geologi Kelautan*, **16(2)**: 115-123
[3] Kamiludin U, Noviadi Y, Setyanto A, Pertala A W, Zulfikar M, Astawa I N, Mustofa M A, Surachman M, Piranti S A, Sampurno P, Saputro E, Hutabarat Y, Aryanto N C D, Setiady D, Widiatmoko H C, 2017 Penelitian Mineral Plaser dan Unsur Tanah Jarang di Perairan Tanjung Berikat dan Sekitarnya, Project Report, Pusat Penelitian dan Pengembangan Geologi Kelautan, unpublished

[4] Margono U, Supandjono R J B and Partoyo E 1995 Peta Geologi Lembar Bangka Selatan, Sumatera, skala 1 : 250.000. *Pusat Penelitian dan Pengembangan Geologi, Bandung*

[5] Moysey S, Rosemary J K and Hary M 2006 Texture – based on classification of ground-penetrating radar images. *Geophysics* **71**: k111-k118

[6] van Heteren S, Fitzgerald D M, Mckinlay P A and Buynevich I V 1998 Radar fasies of paraglacial barrier systems: coastal New England, USA. *Sedimentology* **45**:181-200

[7] Sjöberg Y, Marklund P, Pettersson R and Lyon S W 2015 Geophysical Mapping of Palsa Peatland Permafrost. *The Cryosphere* **9**: 465-478

[8] Carrivick J L 2007 GPR-Derived Sedimentary Architecture and Stratigraphy of Outburst Flood Sedimentation Within a Bedrock Valley System, Hraundalur, Iceland. *Journal of Environmental and Engineering Geophysics* **12 (1)**: 127-143

[9] van Dam R L and Schlager W 2000 Identifying Causes of Ground Penetrating Radar Reflection Using Time-Domain Reflectometry and Sedimentological Analyses. *Sedimentology* **47**: 435-449

[10] Beres Jr M and Haeni F P 1991 Application of Ground Penetrating Radar Methods in Hydrogeologic Studies, *Ground Water* **29 (3)**:375-386

[11] Budiono K, Noviadi Y, Latuputty G and Hernawan U 2012 Investigation of Ground Penetrating Radar for Detection od Road Subsidence Northcoast of Jakarta, Indonesia. *Bulletin of the Marine Geologi* **27 (2)**:87-97

[12] Busby J P, Cuss R J, Raines M G and Beamish D 2004 Application of ground penetrating radar to geological investigations, *British Geological Survey Internal report*, **IR/04/21**, 33pp

[13] Allen R L 1979 Studies In Fluviatile Sedimentation: Anelementary Geometric Model For The Connectedness Of Avulsion-Related Channel Sand Bodies. *Sedimentary Geology* **24**:253-267

[14] Tamura L N, de Almeida R P, Taioli F, Marconato A and Janikian L 2016 Ground penetrating radar investigation of depositional architecture: the São Sebastião and Marizal Formations in the Cretaceous Tucano Basin (Northeeastern Brazil), *Brazilian Journal of Geology* **46 (1)**: 15-27

[15] Shan X, Yu X, Clift P D, Tan C, Jin L, Li M and Li W 2015 The Ground Penetrating Radar Facies and Architecture of a Paleo-spit from Huangqihae Lake, North China: Implication for genesis and evolution, *Sedimentary Geology* **323**: 1-14

[16] Chowksey V, Joshi P, Maurya D M and Chamyal L S 2011 Ground penetrating radar characterization of fault-generated Quaternary colluvio-fluvial deposits along the seismicity active Kachchh Mainland Faults,Western India, Research communication, *Current Science* **100 (6)**: 915-921

[17] Ekes C and Hickin E J 2001 Ground penetrating radar facies of the paraglacial Cheekye Fan, southwestern British Columbia, Canada, *Sedimentary Geology* **143**: 199-217

Integration of Post-stack Inversion and Rock Physics to Determine Sandstone Reservoir Quality: Barakan Sub-basin case

T B Nainggolan[1*], M P Adhar[2] and I Setiadi[1]

[1]Marine Geological Institute, Jl. Dr. Djundjunan No. 236, Bandung, 40174
[2]University of Jambi – Muara Bulian Km. 15, Mendalo Indah, Jambi, 36361

[*]tumpal.nainggolan@esdm.go.id

Abstract. Barakan Sub-basin is assessed as potential basin for hydrocarbon reserves in the eastern region of Indonesia because it is adjacent to Masela block giant gas field. Reservoir rocks in this sub-basin are sandstones from Middle Jurassic (Lower Flamengo Formation) until Oligocene (Adi member Formation). Main sandstone reservoir rocks are knowingly studied to have good porosity in Upper Flamengo, Kopae, Ekmai and Adi member Formations. But, there is no significant study to determine sandstone reservoir distribution that have good porosity quality. Therefore, an integrated method of inversion and rock physics study are needed to determine sandstone reservoir quality. This study uses 2D marine seismic post-stack time migration and 2 wells namely Barakan-1 and Koba-1 wells. Sensitivity analysis with cross-plot of gamma ray log versus acoustic impedance values range of 20-60 API and 9000-42000 (ft/s)*(g/cc) shows a strong correlation of good porosity sandstone to low impedance in Ekmai Formation of both wells. Model based of post-stack inversion reveals sandstone distribution in Ekmai Formation of both wells. Time structure maps of top and bottom horizons in Ekmai Formation indicates Barakan-1 well within anticline height structure and Koba-1 well are deposited in a middle of sub-littoral environment.

Keywords: sandstone reservoir, rock physics, post-stack inversion, Ekmai Formation, Barakan Sub-basin

1. Introduction

Barakan Sub-basin is located on Western end of the Arafura Region and is part of the Northwestern Australian Continent. To the north and west it is bounded by the Aru Trench and Tanimbar Island, while to the east it is bounded by the Aru Islands. This basin is filled with series of Paleozoic, Mesozoic and Cenozoic sediments which are deposited inland into deep sea environment [1]. The direction that formed its structures are N-S and NE-SW, relatively. The N-S trending directional structures are formed during the Paleozoic due to tectonic extension occurring in this area. The tectonic event series are responsible to form Barakan Sub-basin. Furthermore, in the extension phase, sedimentary series filling occurs in the basin area and compression is occurred to lift the area. The last tectonic event is compression in the Late Tertiary associated with the subduction of plates in Banda arc. The tectonic above causes the erroneous of substances that have been formed to reactivate and become an upward and downward fault. The study area situated on Southwest Aru Waters in between 132^{0}E – 135^{0}E and 6^{0}S – 9^{0}S (**Figure 1**).

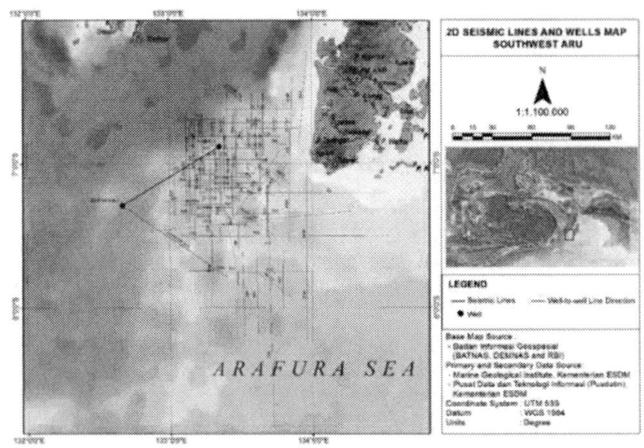

Figure 1. Wells and seismic lines location map

2. Data and Methodology

2.1. Well Data
Well log data and information such as marker formation, corrected check-shot of two wells, namely Barakan-1 and Koba-1 are acquired in this study. The types of well logging used are gamma ray, density neutron, resistivity and sonic logs.

2.2. Seismic Data
The seismic of total 85 2D post-stack seismic lines are acquired in this study. There are also 2 selected seismic lines, namely UT91KR and 83ARAFU which are nearest to Barakan-1 and Koba-1, and used for model based acoustic impedance inversion.

2.3. Petrophysical Analysis
Qualitative interpretation in this study uses gamma ray logs to identify permeable or impermeable zones due to the presence of natural radioactive elements which indicate rocks lithology. A quick look at the cross-over NPHI or Neutron Porosity Hydrogen Index curve against RHOB or bulk density curve which is overlaid each other with a range of opposite curves to indicate hydrocarbon reservoir [2]. Then, quantitative petrophysical analysis is applied to obtain the shale content of rocks, effective porosity, water saturation and permeability [3].

2.4. Acoustic Impedance
Acoustic Impedance (AI) is defined as the ability of rocks to pass seismic waves. The harder a rock is, the greater its acoustic impedance value [4]. Mathematically, acoustic impedance is a product of the multiplication of compression wave velocity and rock density, as described in following equation:

$$AI = V\,.\,\rho\ (ms^{-1}kgm^{-3}) \quad (1)$$

2.5. Seismic Inversion
Seismic inversion can be defined as a subsurface modeling technique using seismic data as input and well log data as control. This definition explains that the inversion method is the opposite forward modeling which is associated with making synthetic seismogram based on earth model. Based on the data source, the seismic inversion method into two groups, namely pre-stack inversion and post-stack inversion [5]. Inversion is a geophysical modeling that is carried out to predict information on the physical properties of the earth based on information obtained from seismic data.

3. Results and Discussion

Well correlation between Barakan-1 and Koba-1 shows significant differences where the two wells have complex formations in different vertical depths. Barakan-1 well has deeper and ticker formations than the Koba-1 well, so the target zones in both wells are fill in different depths (**Figure 2**). Identification of target zone aims to determine the lithology type of reservoir rock composition in the study area that has its own characteristics in the form of porous rocks and have good permeability. In this study, the reservoir rock is in sandstone lithology in both wells lies in Ekmai Formation at depth of 5360-6240 ft in Barakan-1 well and 3640-4200 ft in Koba-1 well.

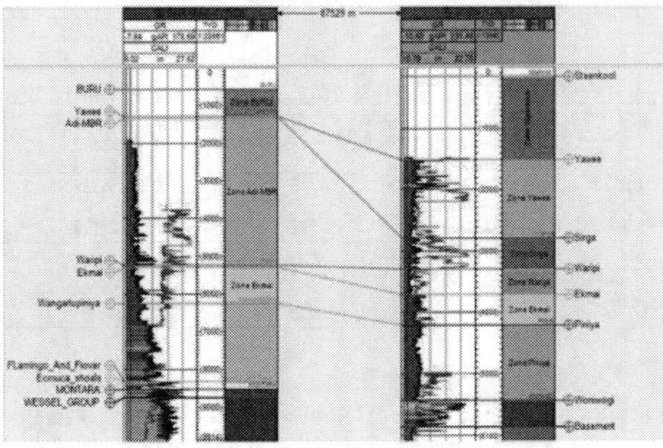

Figure 2. Barakan-1 and Koba-1 wells correlation

In Barakan-1 well, reservoirs in Ekmai Formation are identified in two interval zones, at depths of 5630-5660 ft and 5990-6245 ft. These interval zones are limestone and sandstone lithology with gamma ray values range of 15.4-36.8 API. There are four interval zones identified as sandstone lithology in Ekmai Formation of Koba-1 well, at depths of 3703-3723 ft, 3877-3890 ft, 3933-3947 ft and 3957-4171 ft. Through quantitative calculations, these interval zones produce petrophysical parameters, namely shale volume (V_{sh}), effective porosity (PHIE), water saturation (S_w) and permeability (**k**) (**Table 1**).

Table 1. Petrophysical parameters for Barakan-1 and Koba-1 wells

Formation	Well	Zone	Depth (ft)	Petrophysical parameters (Mean)				Litho
				V_{sh} (v/v)	PHIE (%)	S_w (%)	**k** (mD)	
Ekmai	Barakan-1	1	5630-5660	0.32	0.04	0.94	0.03	LS
		2	5990-6245	0.46	0.15	0.97	1.9	SS
	Koba-1	1	3703-3723	0.34	0.18	0.95	0.9	SS
		2	3877-3890	0.45	0.13	0.99	1.3	SS
		3	3933-3947	0.33	0.18	0.99	1.4	SS
		4	3957-4171	0.34	0.20	1	1.3	SS

Shale volume in Ekmai Formation of both wells show a relatively small shale content which values range of 32-46%. These results signify the interval zones as reservoirs with less impurity content. Furthermore, shale volume will be accounted into water saturation (S_w) calculation. The porosity calculation in this study is performed to obtain effective porosity (PHIE) which is not influenced by the clay/shale content. These interval zones show fair to good porosity results except poor porosity (4%) in interval zone at depth of 5630-5660 ft. The classification of effective porosity value itself are based on ability to hold more hydrocarbon fluids [6]. High values of water saturation (S_w) are shown in all interval zones. It indicates that Ekmai Formation of both wells contain only small amount of hydrocarbon fluids and are dominated by water. These results also conclude that Ekmai Formation of Barakan-1 and Koba-1 are not qualified as hydrocarbon reservoir. Permeability values in these interval zones are less than 5 mD which are categorized as tight limestone and sandstone reservoirs.

Tuning thickness is influenced by the wavelength that passes through the rock, mathematically the value is $\lambda/4$ of the seismic wave that passes through the layer. In determining the value of tuning thickness, the average velocity (*Vp*) and dominant frequency of the target zone are needed [5]. The thickness of the target zone is greater than the thickness of the tuning, it causes the lack of accuracy of seismic cross section in the target zone (**Table 2**). This can be overcome by using acoustic impedance inversion that include low frequency information from the log data, whereas seismic data only has only a certain frequency ranges.

Table 2. Comparison between tuning thickness and target zone thickness

Well	*Vp* average (m/s)	Dominant frequency (Hz)	$\lambda = Vp/f$ (m)	Tuning thickness $\lambda/4$ (m)	Target zone thickness (m)
Barakan-1	3791	22	172.32	43.08	268.22
Koba-1	3207	21	152.71	38.18	170.69

Pre-inversion analysis is done by comparing the curve value AI inversion, initial model and impedance in the quantitative calculation. The inversion method chosen in this analysis is model based which shows the best quantitative results from other two methods, namely bandlimited and sparse spike. Model based method show the correlation results of 0.991 with an error calculation of 0.136 in the Barakan-1 well and 0.992 with an error calculation of 0.127 in the Koba-1 well (**Figure 3**). The control parameters used in inversion analysis using this model based method are the constraints, the average block size, and the many iterations. High correlation and low error calculation results will increase the suitability of inversion model with the earth geological structures [7]. Based on the correlation and error values in the two wells, the results are good enough for inversion.

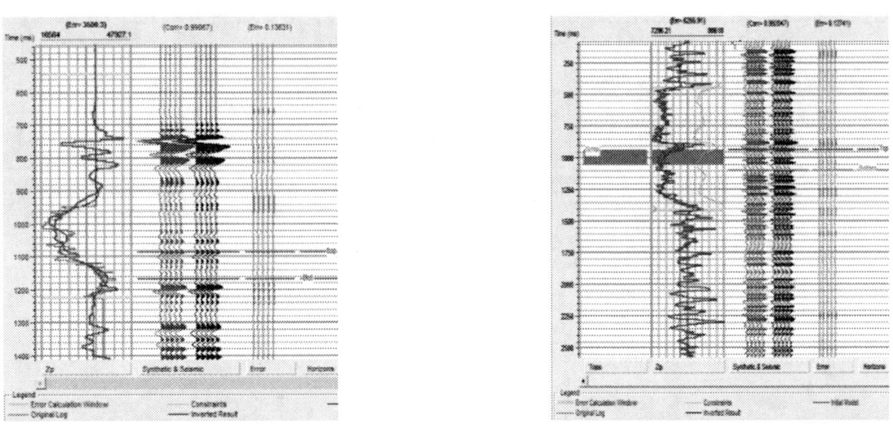

(a) (b)

Figure 3. Correlation results and error calculations of (a) Barakan-1 and (b) Koba-1 wells

Model based of acoustic impedance inversion of Barakan-1 well in sandstone Ekmai Formation target zone shows that the acoustic impedance value range of 36,000-38,500 ((ft/s)*(g/cc). The result of the AI inversion indicates the distribution of AI values spread across the target horizon which some exceptional areas of less representative the acoustic impedance (**Figure 4**). The result of the inversion of acoustic impedance values tend to be homogeneous and relatively high. It is suspected that the high content of clay materials at these depths can affect the sensitivity of the sonic log reading. High acoustic impedance values can be caused by cementation process in the rock layers so that the rocks will become denser or more compact [8].

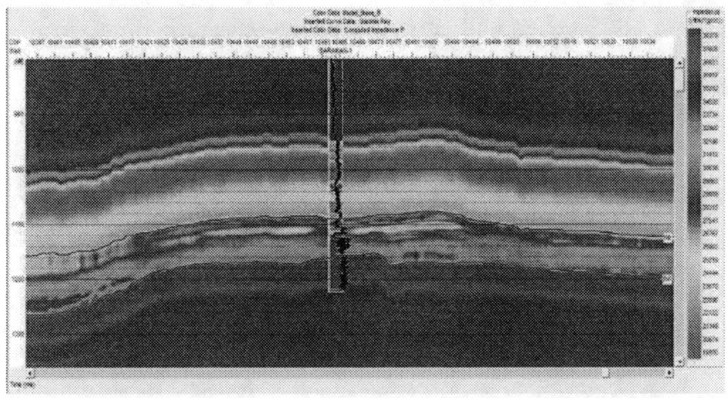

Figure 4. Ekmai Formation acoustic impedance inversion of Barakan-1 well

Model based of acoustic impedance inversion of Koba-1 well in sandstone Ekmai Formation target zone shows that the acoustic impedance value range of 9000-42000 ((ft/s)*(g/cc), with green color indicates AI results at low values and purple color indicates high values. The distribution of low AI values are started at top horizon of Ekmai Formation and increased to high AI values to bottom horizon instantaneously. The inversion results show a match with the AI values in both wells which are indicated by similar color within formation from top to bottom horizons (**Figure 5**). The areas of formation with high AI values indicate denser sandstone reservoir, while areas with low AI values indicate more porous sandstone reservoir. That is because the AI values are the result of P-wave (*Vp*) and density that describes characters of the rocks within formation, where P-waves are propagating inside the matrix of the rock in its path, so that the denser the matrix of the rocks and smaller pores resulting in higher acoustic velocity. Whereas density values in the rocks are also influenced by the matrix density of them. So, the AI values are able to describe the character of the rocks properly and AI values are largely controlled by P-wave velocity values.

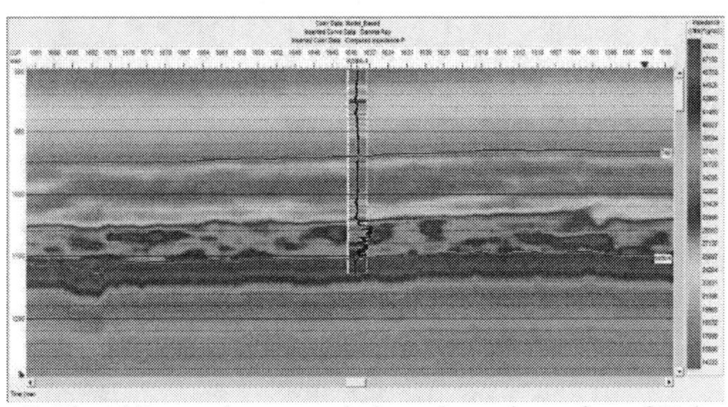

Figure 5. Ekmai Formation acoustic impedance inversion of Koba-1 well

Gridding method is applied in making top and bottom horizons of Ekmai Formation time structure maps to get the results of interpolated and extrapolated interval velocity values due to the limited amount of data. These two maps indicate the delineation of Ekmai Formation geological structures. On the basis of qualitative analysis, the location of Barakan-1 well within anticline height structure that is more towards the north of the well (**Figure 6**). The Ekmai Formation in Koba-1 well are comprised mainly sandstones with minor clays which were deposited in a middle sub-littoral environment.

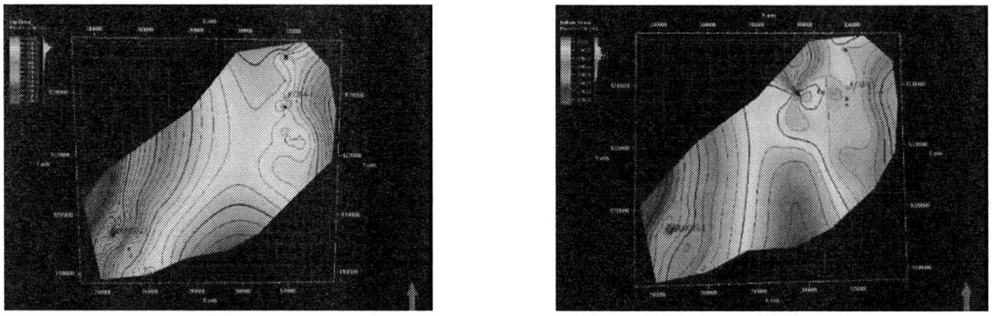

(a) (b)

Figure 6. Time structure map of Ekmai Formation of (a) Top and (b) Bottom horizons

4. Conclusion

Based on the results of the correlation between Barakan-1 and Koba-1 wells, there are complex geological formations and considerable depth differences. Well log data show lithology limestone and sandstone in Ekmai Formation. Petrophysical analysis conclude that Ekmai Formation of Barakan-1 and Koba-1 are not qualified as hydrocarbon reservoir. The AI values in the Ekmai Formation of Barakan-1 well are range of 36000-38500 ((ft/s)*(g/cc) with the type of reservoir being sandstones, while in Koba-1 well Ekmai Formation is 9000–42000 (ft/s)*(g/cc) which is sandstone as well. This result is in accordance with the cutoff value in the sensitivity analysis. Time structure maps describe no show of anticline structure as indication of hydrocarbon trap, so that Barakan-1 and Koba-1 are dry wells, but there is a possibility of height structure in the southeast of Barakan-1 well and northern of Koba-1 well.

Acknowledgment

Sincere appreciation and many thanks to honorable Head of Marine Geological Institute for the trusting and supervising to the authors. Our truly appreciation to Pusdatin - Ministry of Energy and Mineral Resources for providing us the well logs and seismic data.

References

[1] Roberts G, Ramsden C, Christoffersen T, Wagimin N and Muzaffar Y 2011 East Indonesia: Plays and Prospectivity of the West Aru, Kai Besar and Tanimbar Area – Identified from New Long Offset Seismic Data *Search and Discovery Article* 10348 pp 10-13

[2] Nopiyanti T, Nainggolan T B, Dewato O and Haq D A 2020 Well Log Analysis and Geochemical Data to Identify Source Rock and Hydrocarbon Reservoir: Northeast Java Basin Study Case *AIP Conference Proceedings* **2245** 070017 pp 1-6

[3] Dewanto O, Maulina and Nainggolan T B 2020 Identification of biogenic gas reservoir zone using log, petrophysics and geochemical data in S-1 well of Nias basin, North Sumatera *Journal of Physics: Conference Series* **1572** 012037 pp 1-10

[4] Sukmono S 2000 *Seismik Inversi untuk Karakterisasi Reservoar* Departemen Teknik Geofisika ITB, Bandung

[5] Rifai F Y, Nainggolan T B and Manik H M 2019 Reservoir Characterization Using Impedance Inversion and Multi-Attribute Analysis in Nias Waters, North Sumatra *Bulletin of the Marine Geology* vol 34 pp 51-62

[6] Koesoemadinata R P 1980 *Geologi Minyak dan Gas Bumi Jilid 1 Edisi Kedua* Penerbit ITB p 269

[7] Nainggolan T B and Winardhi I S 2012 Sebaran Reservoir Batupasir pada Formasi Talang Akar di Cekungan Sumatra Selatan dengan Metode Inversi EEI dan Multiatribut *Majalah Geologi Indonesia* vol 27 pp 121-130

[8] Putri I A and Santosa B J 2014 Integrasi Seismik Inversi Akustik Impedance (AI) dan Elastic Impedance (EI) untuk Karakterisasi Reservoir, Studi Kasus: Lapangan Muon *Jurnal Pomits* vol 1 pp 1-5

IOP Conf. Series: Earth and Environmental Science **873** (2021) 012021 doi:10.1088/1755-1315/873/1/012021

Potential Eruption and Current Activity of Anak Krakatau Volcano, Indonesia

Kristianto[1], N Indrastuti[1], A Basuki[1], H D Purnamasari[1], S Adi[1], C Patria[1] and N Haerani[1]

[1] Center for Volcanology and Geological Hazard Mitigation (CVGHM), Jl. Diponegoro 57, Bandung 40122, Indonesia

kris0432@yahoo.com

Abstract. Anak Krakatau Volcano is located in the Sunda Strait known for its paroxysmal eruption in 1883. During the January - November 2019 period, seismicity was dominated by types of quakes which indicated the occurrence of magma supply (VA and VB), near-surface volcanic activity (LF, Hybrid, Harmonic Tremors), and volcanic activity above the volcanic surface (eruptions, emission, and continuous tremors). In the period December 2019 - July 2020, there was an increase in the types of quakes near the surface (LF, Hybrid) and the types of quakes on the surface (emission and continuous tremors). Volcanic deformation monitors changes in tilt over the 2019-2020 period associated with pressure releases before, during and after the eruption. The results of GPS data modeling, the shallow pressure source is at a depth of 0.22 km below sea level. Volcanic activity until July 2020 was dominated by activity near and above the volcanic surface associated with the growth of lava domes. The volcanic system of Anak Krakatau is currently an open system, with the potential for eruptions. Strengthening the early warning system for the eruption of Anak Krakatau is important in mitigating efforts and understanding its eruption potential

1. Introduction

Anak Krakatau volcano is one of the most active volcanoes in Indonesia which is located in the Sunda Strait. The last big eruption of Krakatau in 1883 was one of the well known in the world, a awesome volcanic eruption at the sea and followed by a tsunami that effected the people death in the tens of thousands.

Since its appearance on 11 June 1927 up to 2011, Anak Krakatau volcano has experienced more than 100 eruptions, both explosive and effusive, with repose periods ranging from 1 - 6 years (Data Dasar Gunungapi Indonesia, 2011) (Figure 1). There were several eruptions during the 2012-2013 time period, resulting in a ash column reaching 1000 meters high, followed by a strombolian eruption and ending with lava flows to the southeast and southwest. Since 2016, Anak Krakatau volcano has erupted every year. In 2016 the eruption occurred on 20 June 2016, while in 2017 the eruption occurred on 19 March 2017 in the form of a strombolian eruption (Kristianto, et al., 2019).

In 2018, it erupted again since June 29, 2018 in the form of a strombolian eruption. During the period from July to December 2018 there were eruptions of ash, strombolian, and lava flows. This eruption material fall and flowed around Anak Krakatau volcano. On 22 - 28 December 2018, the activity of G. Anak Krakatau increased with the recording of continuous volcanic tremors associated with continuous eruptions. This eruption resulted in the avalanche of the flank of Anak Krakatau from the west to the southwest, as well as a tsunami that inundate the area around the Sunda Strait. The eruption caused the volcanic monitoring system on Anak Krakatau to be totally damaged, so seismic monitoring can only be done from SRTG Station located on Sertung Island (Kushendratno, 2019).

During the period of 2019, monitoring equipment was installed with 2 seismic stations, deformation (2 tiltmeters, 1 GPS permanent, and 2 GPS temporer), 3 infrasound, and 2 CCTV. The equipment installation aims to replace equipment damaged by the 2018 eruption, continuously monitor volcanic activity, and strengthen the eruption early warning system of Anak Krakatau volcano.

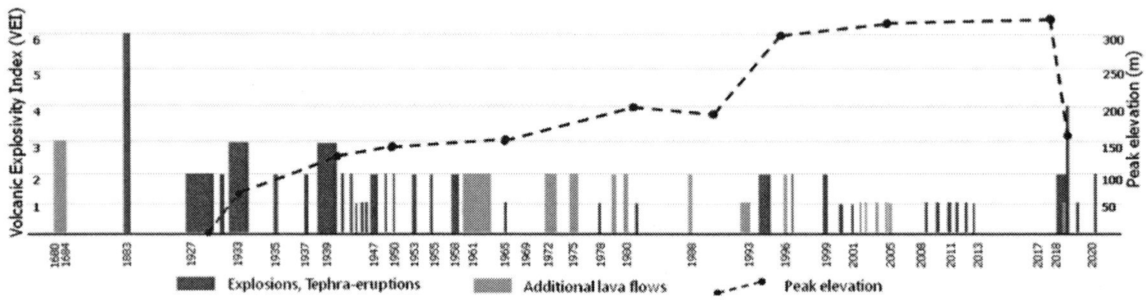

Figure 1. Graph of eruption history, eruption index, and peak elevation of Anak Krakatau. Explosive eruptions and tephra eruptions are drawn in red, lava flows in green, and peak elevations in dashed lines

2. Potential of geological hazard in the Sunda Strait

The tectonic setting of the area around the Sunda Strait is an active area of tectonic activity. The Sunda Strait area is strongly influenced by tectonic activity in the form of collisions between the Eurasian Continental Plate moving southeast at a speed of about 0.4 cm / year and the Indo-Australian Ocean Plate moving north at a speed of about 7 cm / year (Minster and Jordan , 1978 in Yeats, et al., 1997). In the Sunda Strait area, it is interpreted to have an open zone (extension zone) controlled by the Sumatra horizontal fault system and grabens in the western region of Banten (Pramumijoyo and Sabrier, 1991). The Sunda Strait area is also an opening zone with the appearance of a structure in the form of normal faults (Malod, et al., 1995; Susilohadi, et al., 2009. While Handayani and Harjono (2008) state that the zone of openings that occurs from south of the Sunda Strait to southern Java is caused by by the accommodation of the impact force in the subduction zone, the Sumatra Fault and the Mentawai Fault.

The Krakatau complex is located right at the junction of two northeast-southwest and northwest-southeast directions (Verbeek, 1885). The Krakatau volcanic complex is controlled by tectonic movements associated with the South Sumatra Fault System (Effendi, et al., 1986). In the Sunda Strait area, it is interpreted to have an open zone (extension zone) controlled by the Sumatra and graben horizontal fault system in the western region of Banten (Pramumijoyo and Sabrier, 1991).

3. Monitoring system

After the 2018 eruption, the monitoring systems that are still functioning due to the eruption material are one seismic station on Sertung Island and one Pulosari seismic station in the southwestern part of the Pasauran Observatory. Some permanent monitoring equipment was then installed, in February 2019 1 seismometer L4C was installed at the Tanjung Seismic Station in the eastern part of Anak Krakatau (Kushendratno, et al., 2019), on March 2019 1 Biaxial Tiltmeter Jewel type 701-2 sensor and 1 CCTV around the summit of Anak Krakatau were installed (Rosadi, et al., 2019), on March 26-27, 2019 1 Biaxial Tiltmeter Jewel type 701-2 sensor was installed on the eastern slope of Anak Krakatau (Triastuti, et al., 2019), in November 2019 1 3-component broadband digital seismometer (Radian) was installed unit - Guralp), and 1 Chaparal Physics Model 25 infrasonic sensor at a location in the northern part of Anak Krakatau Island (Lava-93 Station), 1 Chaparal Physics Model 25 infrasonic sensor at Kalianda Observatory, and 1 Chaparal Physics Model 25 infrasonic sensor at Pasauran Observatory (Kuswandarto, et al., 2019), as well as the addition of 1 permanent GPS sensor at the Lava-93 station and 2 temporary GPS stations (KRA1 Station and KRA2 Station) (Patria, et al., 2019). In March 2020, 1 CCTV was added to the Lava-93 Station (Patria, et al., 2020) (Figure 2), because the CCTV around the summit was damaged due to bursts of volcanic material.

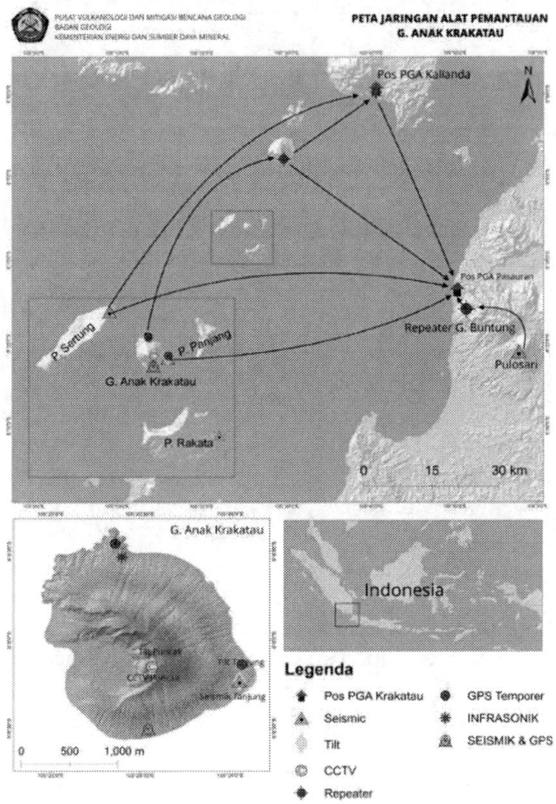

Figure 2. Map of the monitoring network of Anak Krakatau activities using seismic equipment, GPS, Tiltmeter, CCTV, and infrasound (Patria, et al. 2019).

4. Data and methods

The methods used in this paper are: visual observation, seismic observation (seismic classification and analysis of RSAM, SSAM), and deformation observation (GNSS method and tiltmeter method).

The seismic data used in this paper is continuous seismic data from 1 January 2019 to 30 September 2020, to calculate the release of seismic energy Real-time Seismic Amplitude Measurement (RSAM), (Endo and Murray, 1991), Seismic Spectral Amplitude Measurement (SSAM) (Rogers and Stephens, 1995) which is spectral data (frequency content) from earthquake records within a certain time interval, classification, and number of quakes. RSAM data is related to earthquake energy while SSAM data is related to the dominant earthquake mechanism. RSAM and SSAM data are used in evaluating seismic energy and interpreting the phenomena that occur (Basuki, et al., 2019).

The classification of earthquakes records at Anak Krakatau volcano consists of Eruption, Emission, Tremor, Shallow Volcanic (VB), Deep Volcanic (VA), Low Frequency (LF), Local Tectonic (TL), and Far Tectonic (TJ). The Eruption quake has a low dominant frequency content, which is around 1.3 s.d. 2.4 Hz, strong earthquake onset and duration of 15 - 255 seconds. The Emission quake is also characterized by a low dominant frequency content, which is around 1.3 s.d. 3.0 Hz, quake onset is generally emergent, lasting about 9 - 180 seconds. Low Frequency quakes have a bandlimited frequency spectrum between 1 and 8 Hz. Hybrid quakes contain a dominant frequency of 2 - 3 Hz. Tremor quakes generally contain a frequency of about 1 - 2.4 Hz with a duration of about 14 - 101 seconds. VB quakes and VA quakes are Volcano Tectonic / VTs quakes characterized by a broader band and have a higher frequency content, VTs band quakes ranging from 1 to 20 Hz, generally having a dominant frequency peak of 8 - 15 Hz. Local Tectonic quakes are high frequency quakes and have an S-P greater than 4 seconds.

The 10-minute RSAM data and SSAM data are calculated automatically using the earthworm data acquisition system (http://www.earthwormcentral.org/), from 10 minutes of data a daily average is calculated to be compared with the total daily quakes. RSAM provides information on the relative total of seismic energy release. Seismic classification is carried out based on direct analysis of the frequency content and waveforms.

Deformation data used in this paper is GNSS data with dual frequency L1 (1575.42 MHz) and L2 (1277.60 MHz) for the period of November and December 2019. Monitoring the ground deformation of Anak Krakatau uses the GNSS (Global Navigation Satellite System) method. conducted in the period November - December 2019. GNSS data is processed automatically in the Webobs application using the GipsyX / JPL software (Beauducel, et al., 2020). The results of GNSS data processing yield the speed and depth of the source of deformation.

The tiltmeter data is used from the recordings at Tanjung Station for the period January - April 2020. Deformation monitoring with the tiltmeter uses the tilting method on each of its axes (X and Y axes), the X (tangential) axis is located 90 ° from the Y (radial) axis. With this Tiltmeter device, it can be used to determine the leverage pattern caused by changes in pressure on the volcanic ground and can calculate the volume of magma injection (Nishi, et al., 2007).

5. Visual monitoring

Visual monitoring of Anak Krakatau's activities is carried out using CCTV (Closed Circuit Television) cameras and direct monitoring in the field. The equipment used to support visual monitoring consists of 2 CCTV cameras installed at the G. Anak Krakatau Observation Post in Pasauran and on Anak Krakatau Island. The CCTV equipment located on Anak Krakatau Island (Lava-93 Station) has been operating since March 11, 2020 until it was damaged at the end of April 2020 due to being hit by the eruption material, it can resume operation after being repaired on September 7, 2020. Visual monitoring via CCTV can also be shared via the web: www.vsi.esdm.go.id, so that it can be accessed by anyone who needs a visual image (Photo 1b).

Visual monitoring was also carried out by conducting direct checks to locations around Anak Krakatau (Photo 1a). A visit to this location is very important to find out the current condition of the Anak Krakatau Crater after the 2018 eruption. In December 2019, the crater visuals often observed surtseyan eruptions in the central and northern parts of the crater lake (Patria, et al., 2019) (Photo 2a). The last monitoring towards the crater was carried out on March 11, 2020 (Kuswandarto, et al., 2020) (Photo 2b), and photo donation from Samsul as a tour guide who had passed Anak Krakatau volcano in the southwestern part of April 17, 2020 (Photo 3a). When compared to the results of monitoring March 2020 with September 2020, there are very significant changes in the conditions around the crater. On March 11, 2020, it was observed that the crater had begun to fill with volcanic material in the middle, and the crater lake was divided into two parts. In the photo on April 17, 2020, the condition of the crater is covered in volcanic material and lava flows, the crater lake water is no longer visible. On monitoring on September 8, 2020, the crater conditions were increasingly changing, especially at the height of the lava flow which was getting higher and evenly distributed to the west-southwest. The height of the lava dome is still below the current highest peak (157 m), estimated to be about 100 meters above sea level (Photo 3b).

(a)

(b)

Photo 1. Visual Crater of Anak Krakatau from the north-eastern crater rim on September 7, 202, observed activity of gas emission from the center and southeast side of the lava dome. While the lava flow leads to the west - southwest (a), and from the CCTV at Lava-93 Station on September 8, 2020 (b), it is observed gas emission with a height of 25 - 50 meters from the summit.

Continuously gas emission (Constant degassing) was observed predominantly from the central location of the lava dome and around the crater, with a height of up to 500 meters. On September 7, it was observed that the dominant gas emission came from the center of the lava dome and the eastern part of the crater, with a height of 25 - 50 meters from the peak of the lava dome and occasionally observed exceeding the peak of Anak Krakatau as high as 50 meters (Photo 1b).

(a)

(b)

Photo 2. Visual crater in December 2019, often observed surtseyan eruptions in the central and northern parts of the crater lake (a). Visually around the crater on March 11, 2020, the crater lake has started to fill with volcanic material in the middle, and the crater lake is divided into two parts, Constant degassing is observed in the eastern crater (photo :Jumono, March 11, 2020) (b).

(a)

(b)

Photo 3. Visual Crater of G. Anak Krakatau from the northwest using a speedboat, observed activity of gas emission from the center and around the lava dome. While the lava flow leads to the west - southwest part (Photo: Samsul, 17 April 2020) (a). Visually on September 8, 2020, activity of gas emission was observed from the center of the lava dome of Anak Krakatau as high as 25 - 50 meters from the top of the lava dome (b).

(a)

(b)

Photo 4. Volcanic rock falls resulting from the eruption of 10-11 April 2020, around Lava-93 Station with a diameter of about 10-30 cm (a), around the crater rim generally measuring 30-100 cm in diameter (b).

The distribution of the eruption material from 10-11 April 2020 in the form of gravel was observed to be scattered to the northeast - north - west - south from the active crater. Volcanic material falls with a diameter of 30-100 cm are found around the crater rim and a small portion of about 10-30 cm in diameter can be found around the Lava-93 station (Photo 4).

6. Seismic monitoring (2019-2020)

Seismicity records during 2019 were marked by continuous surtseyan eruption quakes and there was no clear precursor before the eruption occurred in the period February to October and December 2019. The increase in VA quakes in January - February was followed by an increase in VB, LF, Eruption, emission, and continuous tremors quakes starting March 2019. The eruption on February 14, 2019 was preceded by the recording of the VA quake swarm the day before. The VA Earthquake shows a new intrusion, building up pressure and causing surface quakes (gas events) to be recorded, that are VB quake, emission quake, and Harmonic Tremor quake. Continuous tremors correlate with constant degassing or constant magma in contact with water (which produces steam) - heat (Kristianto, et al. 2020).

During the period from January to September 2020, the seismicity was characterized by non-continuous eruption quakes until April 17, and increased Emission, Continuous Tremors, LF, Hybrid, VB, TL, and TJ quakes (Figure 3). Changes in the release of seismic energy calculated using RSAM

(Figure 4) are associated with an increase in the amplitude of continuous tremors, corresponding to the SSAM pattern (Figure 5) dominated by a dominant frequency of 0.5 - 4 Hz related to the mechanism of Eruptions, Emission, and Continuous Tremors quakes. The dominant amplitude occurs with a maximum between 3 - 40 mm.

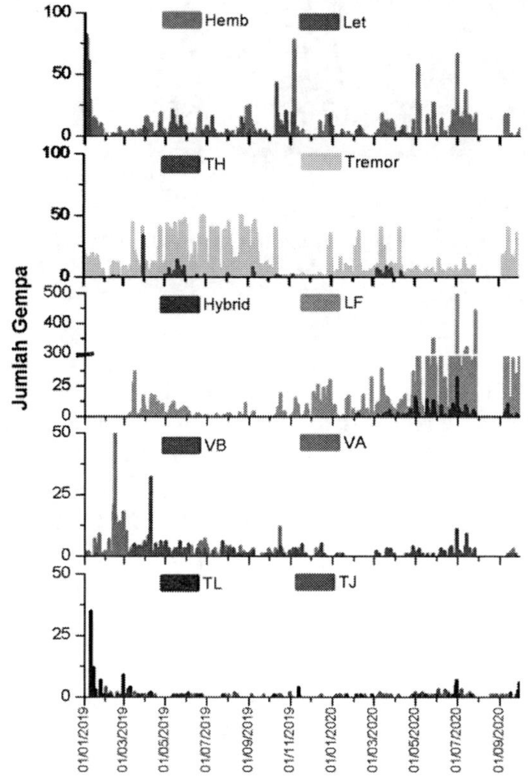

Figure 3. Daily quakes at Tanjung Station for the period January 2019 to September 2020, from top to bottom Eruption (Let), Emission (Hemb), Harmonic Tremor (TH), Continuous Tremors, Hybrid, Low Frequency (LF), Shallow Volcanic (VB), Deep Volcanic (VA), Local Tectonic (TL), and Far Tectonic (TJ).

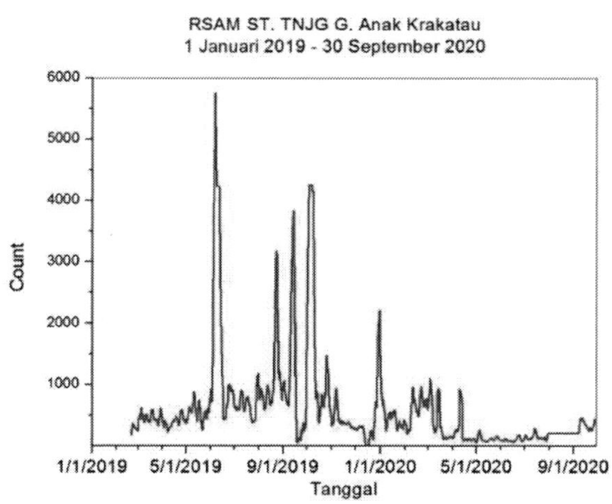

Figure 4. Graph of the value of RSAM Tanjung Station for the period February 2019 to September 2020, during the period 28 July - 6 September 2020 there was no data because the equipment was damaged.

IOP Conf. Series: Earth and Environmental Science **873** (2021) 012021 doi:10.1088/1755-1315/873/1/012021

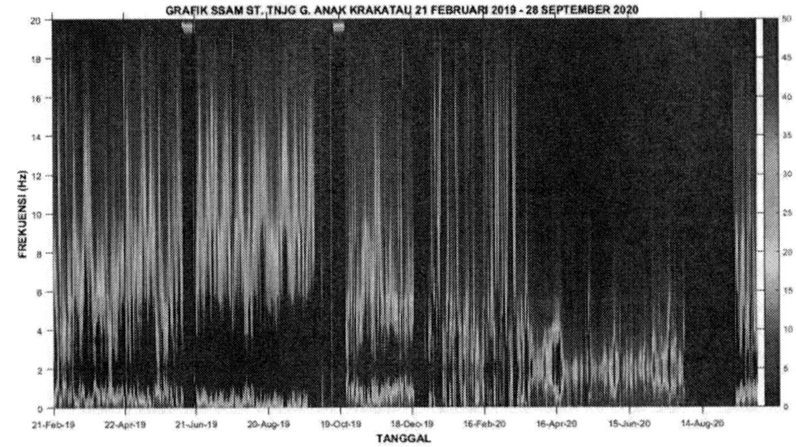

Figure 5. Graph of the SSAM value of Tanjung Station for the period January 2019 to September 2020, during the period 29 July - 6 September 2020 there was no data because the equipment was damaged.

7. Deformation monitoring (2019-2020)

Deformation monitoring of Anak Krakatau using GNSS data at 3 GNSS stations, taht are KRA1, KRA2 and KRA3 produces velocity and depth of the source of deformation. Webobs uses a depth-varying point source solution (Williams & Wadge, 1998) for a topographic correction approach in deformation, with isotropic point model type, 3-D probability map, volume variation and probability deflation - inflation model (Beauducel, et al., 2014) shown in Figure 6.

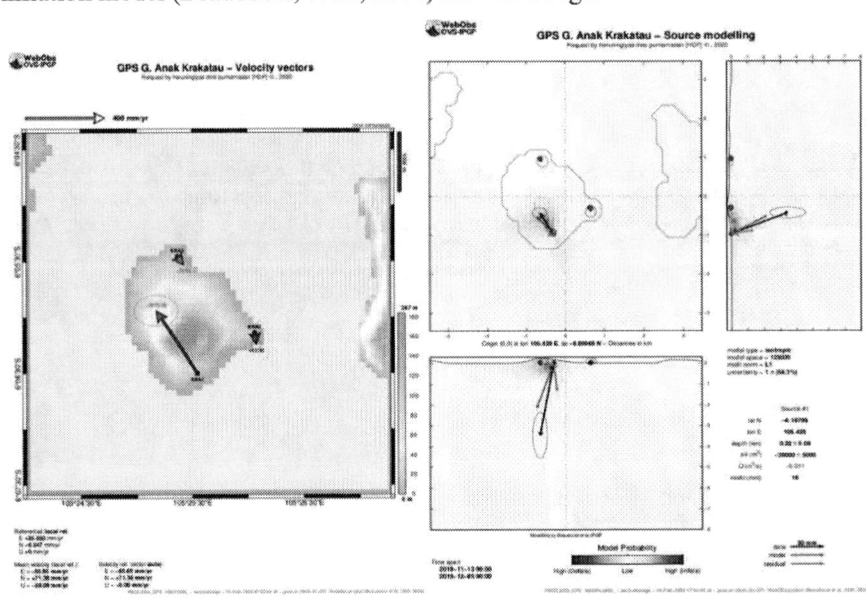

(a) (b)

Figure 6. Velocity Vector of GPS (a) and SourceModeling GPS (b) for the period November 13 - December 5, 2019 at KRA1, KRA2 and KRA3 stations.

The tiltmeter data of Tanjung Station for the period January 1 - April 30, 2020 which is correlated with the eruption quake that occurred at Anak Krakatau is shown in Figure 7, and the tilting pattern after the eruption on April 10 and before the eruption on April 11 2020 is shown in Figure 8.Changes in the average magma injection (Nishi, et al., 2007) that occur at the tiltmeter after calculating the average magma injection of Anak Krakatau is 37-54 m3 / minute at a depth of 4-5 km below the surface of the activity center. This is greater than the magma injection for the period October - December 2019 of 15 - 34 m3 / minute (Kristianto, et al., 2020).

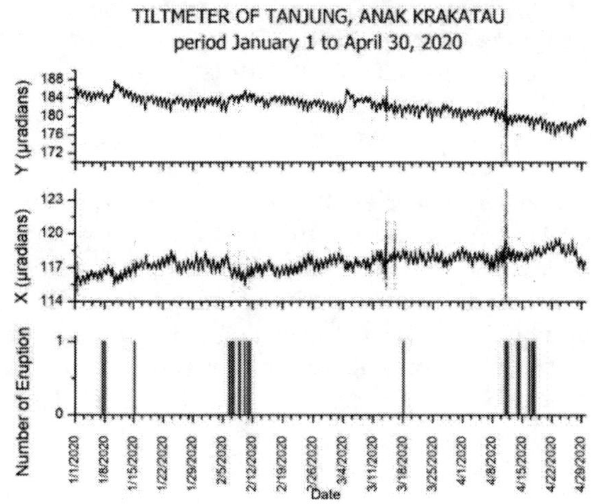

Figure 7. Tanjung station tiltmeter (X axis and Y axis) correlated with the eruptive quake that occurred during the period January 1 - April 30, 2020.

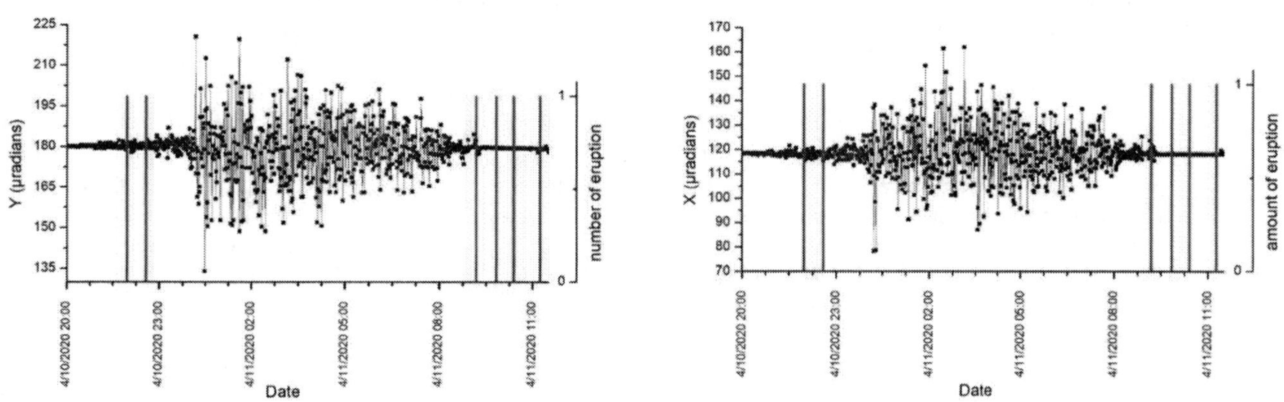

(a) (b)

Figure 8. Y axis (a) and X (b axis on the Tiltmeter showing the tilting pattern at Tanjung Station after the eruption on April 10, 2020 and before the eruption on April 11, 2020.

8. Discussion

The potential for flank landslides that can produce tsunami waves in the near future is still very small, because at this time a large enough cone has not been formed. Cones that are built through eruptive activity (explosive or effusive). Explosive eruption material is deposited around the crater and lava flows out to the west - southwest become an effective system in building new cones, currently reaching a height of around 100 meters above sea level. By comparing the visuals of the crater on March 11, April 17, and September 7, 2020, a significant change occurred in the March - April 2020 period marked by changes in the height of the lava dome and lava flows towards the west-southwest. This can be seen also in the seismic recordings in that period which are dominated by types of emission quakes are directly associated with the occurrence of passive gas emission, LF quakes associated with increased pressure in shallow areas under volcanic craters, Harmonic Tremors and Continuous Tremors associated with volcanic or strombolian eruption activity on 10-11 April 2020 which produces ash plume, steam, and ballistic trajectory rocks around the crater and the northern part of the crater.

In the period January - March, mid-April, and September 2020 there is a corresponding pattern of increasing data for RSAM, SSAM, eruption quakes, emission quakes, and continuous tremors. RSAM data shows the relative value of the total high seismic energy release, SSAM data is dominated by a dominant frequency of 0.5 - 4 Hz related to the mechanism of eruptions quakes, emission quakes, associated with the release of volcanic material to the surface in the form of eruptions (volcanic and strombolian), emission gas, and lava flows.

Seismicity during the period January 2020 to mid-September 2020 is indicated by an increase in LF, Emission, Hybrid, and continuous tremors quakes. The seismicity of the VA and the VB quakes was insignificant, on average less than 1 event per day, but there were several times the increase in VB quakes in April, June and July 2020. This shows that magma intrusion still occurred in this period. The seismicity of LF, Emission, and Hybrid quakes was quite significant, indicating the occurrence of gas movements, as well as the release of volcanic material in the form of lava and gas flows to the surface. Continuous tremor quakes were recorded significantly in this period, which correlates with the constant degassing process.

The strombolian or surtseyan eruptions that occurred on February 6 -11, 2020 were not preceded by changes in the seismicity that could be identified significantly (Kristianto, et al., 2020), this event was similar as the eruption that occurred on 10-11 April 2020 not preceded by changes in the VA and VB quakes significantly, some quakes that increase in the 5 days before the eruption are the Emission, LF, and continuous tremor quakes. The volcanic system of Anak Krakatau is currently open, so volcanic or strombolian eruptions can occur without a significant change in seismicity that can be identified. During the eruption period the seismicity was dominated by Continuous Tremors, Harmonic Tremors, Emission, LF, and Eruptions. Continuous tremors correlate with constant degassing or constant magma in contact with water (producing steam) - heat and gas (McNutt and Roman, 2015; White and McCausland, 2019).

The tectonic activity around Anak Krakatau volcano has an effect on its volcanic activity, the Krakatau volcanic complex is controlled by tectonic movements associated with the South Sumatra Fault System (Effendi, et al., 1986). Faults in the Sunda Strait region are controlled by a horizontal fault system in Sumatra and grabens in the western region of Banten (Pramumijoyo and Sabrier, 1991), seen in an increase in Local and Far Tectonic quakes from 29-30 June 2020, which triggered an increase in Emission, LF, and Hybrid quakes. This shows that tectonic activity can affect the stability of magma that is already at a shallow depth to near the surface.

The results of GNSS data observations show that the velocity vector that points to the source of the crater activity is interpreted as deflation. In the monitoring period from November to December 2019, seismic activity was dominated by emission, so that observations of deformation that showed deflation had an effect on the release of energy that occurred in the body of Anak Krakatau. The tilting pattern showed stability before and after the eruption on April 10, 2020, but after that the tilting on the X axis and Y axis increased for about 8 hours before finally stabilizing and one hour 45 minutes later the eruption returned to be precise at 09.45 WIB , this indicates the occurrence of magma migration from shallow pockets to the surface.

9. Conclusion

The results of visual, seismicity and deformation data analysis of Anak Krakatau volcano show that the potential hazard from the current activity is vulcanian and strombolian eruptions which can be accompanied by ballistic trajectory rocks, lava flows and heavy ash fall around the crater within a radius of 2 km from the active crater. The volcanic system of Anak Krakatau volcano is currently open, so that volcanic or strombolian eruptions can occur without significant changes in volcanic phenomenon.

The current volcanic activity of Anak Krakatau is in the form of the release of volcanic material to the surface in the form of lava flows and constant degassing or constant magma in contact with water (which produces steam) - heat and gas. In order to face the potential hazard of Anak Krakatau eruption, strengthening the early warning system is important in mitigating efforts and understanding its eruption potential.

Acknowledgements
The author would like to thank the Head of the Center for Volcanology and Geological Hazard Mitigation, Kasbani, and the Head of Volcanic Mitigation, Hendra Gunawan, who gave the team the opportunity to repair the monitoring equipment of Anak Krakatau volcano and study its volcanic activity. We would like to thank Samsul from the tour guide who has donated his photos for use in this paper.

References

[1] Basuki A, Kristianto, Mathovanie A 2019 Sistem pemantauan gunung Anak Krakatau *Dinamika Geologi Selat Sunda Dalam Pembangunan Berkelanjutan*, Badan Geologi, Kesdm

[2] Beauducel F, Nurnaning A, Iguchi M, Fahmi A, Nandaka M, Sumarti S, et al. 2014 Real-time source deformation modeling through GNSS permanent stations at Merapi volcano (Indonesia) in *AGU Fall Meeting Abstracts* (San Francisco, CA)

[3] Data Dasar Gunungapi Indonesia, 2011. Badan Geologi Kementerian Energi Dan Sumber Daya Mineral.

[4] Effendi AC, Bronto S and Sukhyar R, 1986 Geological map of Krakatau Volcanic Complex. Volcanological Survey of Indonesia.

[5] Endo ET and Murray T 1991 Real-time seismic amplitude measurement (RSAM): a volcano monitoring and prediction tool *Bulletin of Volcanology*, **53(7)**, pp.533-545.

[6] Handayani, L., dan Harjono, H., 2008, Perkembangan tektonik daerah busur muka Selat Sunda dan hubungannya dengan zona sesar Sumatera, *Jurnal Riset Geologi dan Pertambangan Jilid 18* **No.2**, pp. 31-40.

[7] Kristianto, Triastuty, H, Mulyana I, Rosadi U, Basuki A, 2019, Revitalization of early warning systems of the Anak Krakatau volcano eruption post-eruption 2018. *The 6 th Annual Scientific Meeting on Disaster Research 2019 International Conference on Disaster Management Proceeding Book* **Vol. 4**, Indonesia Defense University, Bogor 18 – 19 June 2019.

[8] Kristianto, Patria C, Indrastuti N, Basuki A, Heruningtyas, Adi A, Mardiono D, Kuswandarto, H, Yoga A and Haerani N, 2020 Kajian seismik gunung Anak Krakatau pasca erupsi tahun 2018, *Prosiding Kolokium Kajian Kegunungapian 2019 Pusat Vulkanologi Dan Mitigasi Bencana Geologi* Hal. 35-44.

[9] Kushendratno 2019 Laporan tanggap darurat letusan gunungapi Anak Krakatau, Banten 14 – 23 Februari 2019. Arsip Pusat Vulkanologi dan Mitigasi Bencana Geologi, Badan Geologi KESDM.

[10] Kuswandarto H, Mulyana I, Setiono S and Rahmanto, 2019. Laporan instalasi peralatan pemantauan G. Anak Krakatau, Lampung 11-24 November 2019 *Arsip Pusat Vulkanologi dan Mitigasi Bencana Geologi Badan Geologi KESDM.*

[11] Malod JA, Karta K, Beslier MO and Zen MT 1995 From normal to oblique subduction :Tectonic relationships between Java and Sumatera, , *Journal of Southeast Asian Earth Sciences, Vol. 12. No.1 2*, pp. 85 – 93.

[12] McNutt SR and Roman DC 2015 Volcanic seismicity. In *The Encyclopedia of Volcanoes* (pp. 1011-1034). Academic Press.

[13] Nishi K, Hendrasto M, MulyanaI, Rosadi U and Purbawinata MA 2007 Micro-tiltchanges preceding summit explosions at Semeru volcano, Indonesia *Earth Planets Space 59, 151-156*
[14] Patria C, Kristianto, Adi S and Basuki A 2019 Laporan kajian Anak Krakatau tahap pertama 8– 14 November 2019 Arsip Pusat Vulkanologi dan Mitigasi Bencana Geologi Badan Geologi KESDM.
[15] Patria C, Mulyana I, Setiono S, Hidayat D A and Yosa Putra D E 2020 Laporan tanggap darurat letusan gunungapi Anak Krakatau, Banten Maret 2020. Arsip Pusat Vulkanologi dan Mitigasi Bencana Geologi Badan Geologi KESDM.
[16] Pramumijoyo S and Sebrier M 1991 Neogene and Quaternary fault kinematics around the Sunda Strait area, Indonesia, *Journal of Southeast Asian Earth Sciences,* **Vol. 6**. *No. 2*, pp. 137 – 145.
[17] Rogers JA dan Stephens CD 1995 SSAM: real-time seismic spectral amplitude measurement on a PC and its application to volcano Monitoring *Bulletin of the Seismological Society of America***Vol. 85** No.2, pp.632 - 639.
[18] Rosadi U 2019 Laporan singkat survey gunung Anak Krakatau tanggal 7-8 Maret 2019 Arsip Pusat Vulkanologi dan Mitigasi Bencana Geologi, Badan Geologi KESDM.
[19] Susilohadi, Gaedicke C dan Djajadihardja Y 2009 Structures and sedimentary deposition in the Sunda Strait, Indonesia *Journal of Tectonophysics 467*, pp. 55-71.
[20] Triastuty H 2019 Laporan tanggap darurat letusan gunungapi Anak Krakatau Banten 26 – 27 Maret 2019 Arsip Pusat Vulkanologi dan Mitigasi Bencana Geologi Badan Geologi KESDM.
[21] Verbeek RDM, 1885 Krakatau. In: Simkin T and Fiske R S (Editors), Krakatau 1883 : the volcanic eruption and its effects. *Smithsonian Inst. Press,* Washington, D.C.
[22] White, R.A. and McCausland, W.A 2019. A process-based model of pre-eruption seismicity patterns and its use for eruption forecasting at dormant stratovolcanoes. *Journal of Volcanology and Geothermal Research*, **382**, pp.267-297.
[23] Williams C and Wedge G 1998 The effects of topography on magma chamber deformation models: Application to Mt. Etna and radar interferometry, *Geophys*. Res. Lett., 25, **No.10**, 1549-1552.
[24] Yeats RS, Sieh K and Allen CR 1997 The geology of earthquakes *Oxford university press*: 567.

IOP Conf. Series: Earth and Environmental Science **873** (2021) 012022 doi:10.1088/1755-1315/873/1/012022

Bayesian Inference of Centroid Moment Tensors of the 2019 Ambon (Mw 6.5) Aftershock Earthquake Sequence, Indonesia: A Preliminary Result

A W Baskara[1*], D P Sahara[2], A D Nugraha[2], A Muhari[3], A A Rusdin[4], S Rosalia[2], A Priyono[2], Z Zulfakriza[2], S Widiantoro[2,5], N T Puspito[2], A Lesmana[2], D Kusumawati[2], A Ardianto[1], Y Halauwet[4]

[1]Geophysical Engineering Study Program, Faculty of Mining and Petroleum Engineering, Institut Teknologi Bandung, Indonesia

[2]Global Geophysics Research Group, Faculty of Mining and Petroleum Engineering, Institut Teknologi Bandung, Indonesia

[3]National Disaster Management Authority of Indonesia, Jakarta, Indonesia

[4]Agency of Meteorology, Climatology, and Geophysics of Indonesia, Jakarta, Indonesia

[5]Faculty of Engineering, Maranatha Christian University, Jalam Surya Sumantri No 65, Bandung 40164, Indonesia

*widhiaria077@gmail.com

Abstract. *The Ambon Mw 6.5 earthquake on September 26th, 2019, had contributed to give severe damages and significantly increased seismicity around Ambon Island and surrounding areas. Mainshock was followed by aftershocks with spatial distribution added to the impact of destructions in this region. We investigated aftershocks sequences to reveal the effect of mainshock toward the change in the in-situ stress field, including the possibility of the existing faults reactivation and the generation of aftershocks. We inferred centroid moment tensor (CMT) for significant aftershock events with Mw more than 4.0 using waveform data recorded from October 18th to December 15th, 2019. The aftershock focal mechanism was determined using the Bayesian full-waveform inversion code ISOLA-Obspy. This approach provides the uncertainty of the CMT model parameters. From ten CMT solution we had inferred in three seismic clusters, we found that majority of events have a strike-slip mechanism. Four events located on the south of the N-S trendings have a dextral strike-slip fault type, reflected the rupture of the mainshocks fault plane. Three events in the cluster of Ambon Island are dextral strike-slip, confirming the presence of the fault reactivation. Meanwhile, three CMT solutions in the north show the dextral strike-slip faulting and may belong to the mainshock main fault, connected with the cluster in the south.*

Keywords: Aftershocks, Ambon Mw 6,5 earthquake, Bayesian full-waveform inversion, fault reactivation

1. Introduction

The Ambon earthquake on September 26, 2019, with a magnitude of 6.5 Mw had increased the seismicity level in the surrounding area, where aftershocks with a significant amount consistently occurred over several months after the mainshock. The earthquake exhibited a strike-slip fault type of focal mechanism with N-S strike trends [1]. Most of the aftershock events are parallel with the north-south trendings from the SW of Seram to the area between Ambon and Haruku island. We noticed that two cluster events appeared to be outside the supposed main fault area where most of the aftershock occurred. One cluster is located on Ambon Island with a North-South to NE-SW trend parallel to the surface faults in the region. On November 12, 2019, an aftershock event with a magnitude of 5.1 Mw occurred in this area. another suspected cluster is in the north that shows a branching of the aftershock distribution. These two findings raise our consideration that the Ambon earthquake aftershock occurred in more than one sequence, and may involve the presence of fault reactivation.

The analysis of the earthquake aftershock sequence has become an interesting issue to be analyzed. Mainshock is probably followed by direct aftershock that triggered by a given primary shock and also secondary aftershock that are not related to the original mainshock [2]. Understanding the sequences of aftershock is quite essential to reveal the rupture process of an earthquake sequence. In other words, we need to clarify how the aftershock generation occurs. Many seismologists reveal the geometry of the mainshock fault plane based on the distribution of aftershocks [3]. Meanwhile, a study of the focal mechanism could give us a better understanding of the physical process of an individual seismic event as it provides information about the fault plane parameter. In the present study, we combined accurate determination of hypocenter and estimated focal mechanism of aftershocks to reveal these issues. However, This study focuses more on the CMT solution in which the focal mechanism will be analyzed further to interpret the generation of aftershock events.

2. Methodology

2.1. Waveform Data
We used earthquake waveform data of aftershock events on the period October 18 to December 15, 2019, which were recorded by 11 temporary local seismic networks deployed by ITB, BMKG, and BPNP and four regional BMKG seismic stations.

2.2. Estimation of focal mechanism
The Bayesian full-waveform CMT inversion approach was applied using python-based ISOLA-Obspy programming [4]. This method uses a combination of assesing the full moment tensor by linear least-squares approach and the estimation of centroid location and time by grid search [5]. The full-wavefield Green Functions are constructed by the discrete wavenumber method [6] in a 1D velocity model with homogenous layers. Then, the inverse problem is solved within the Bayesian framework [7]. The events were selected based on the seismic cluster of the hypocenter and then we inferred CMT of aftershock earthquakes with Mw magnitude more than 4.0. we also included events with magnitude below the criteria to accomodate the CMT solution in all clusters.

Waveform velocity data are corrected from the instrument responses and then filtered by a bandpass filter and integrated into displacement waveforms. We selected signals by executing Mouse Trap code [8], where station components with the high noises and disturbances are rejected. The bandpass filters cut-off were determined manually thorough evaluation based on the best waveform fitting. Filtered and integrated waveforms were downsampled to the sampling frequency of about 1.0-1.2 Hz to ease the computational process.

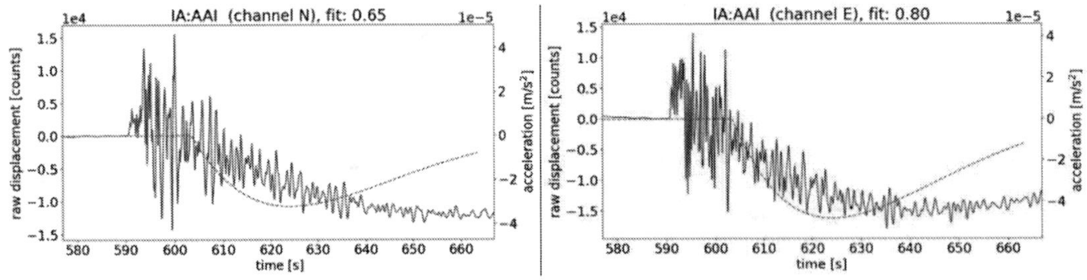

Figure 1. Example of selected signals by executing Mouse Trap code. The output visualize steplike disturbance in the integrated (raw displacement) from AAI station on N and E component

We set a dense grid in our grid search parameter where the step of the grid point was on the range of 0.1-0.4 km in all three coordinates within the distance of 2-5 km from the CMT centroid location. With standard time grid steps of 0.6-0.8 s, the time grid covers 0-3 s after the hypocenter time. 400 possible solutions was generated from ten-dimensional posterior PDF. We decompose the MTs solution into DC and CLVD component.

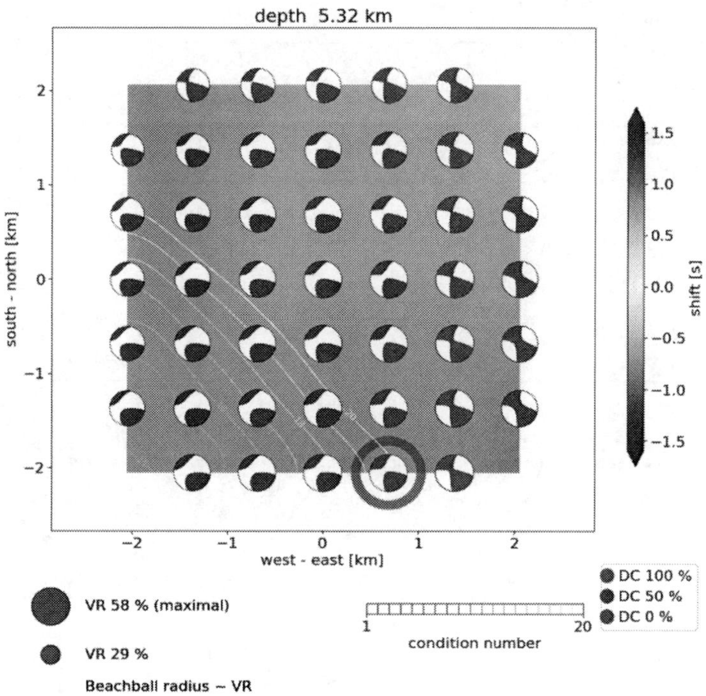

Figure 2. Example of selected CMT solution on agrid point by spasial grid search from an aftershock event occured on Oktober 18, 2019 11:36:37 UTC. The beachball size reflected the value of variance reduction (VR), where the best solution is marked by red circle. The background color represents the inverted centroid time at a specific grid point, while the contour lines represent the condition number.

3. Result and Discussion

The CMT solution parameters for all nine events are shown in Table 2, consisting of CMT location, magnitude, Variance reduction (VR), fault parameters (strike, dip, and rake) for the two nodal planes, DC percentage, and DC uncertainty. We checked the reliability of our results based on VR percentage, DC precentage, and DC uncertainty. The CMTs solutions are acceptable if VR value> 40%, DC percentage> 50%, and uncertainty <30%. we defined these criteria based on [4] where we simplify and downgrade the criteria for VR from 50% to 40%. Based on our results, 8 out of 9 events had VR values> 40%. Then there are five events with a DC component> 50% where events 1 and 7 have a significant percentage of DC (74 and 88%). Most of the events have low uncertainty, except events numbers 4 and 7 with an uncertainty of more than 20%. We noticed that these two events have a small magnitude (below 3.2 Mw) for which the inversion process in our data processing didn't give robust solutions.

Figure 3 below shows the focal mechanism distribution derived from The Bayesian CMT full-waveform inversion. The majority of CMT solution showed strike-slip fault type. Four solutions in the southern cluster were dextral strike-slip mechanism with N-S strike orientation. Similar to the solutions in the south, The northern cluster was also characterized by dextral strike-slip faulting with N-S strike trends but the dip was not so steep as the south. We considered that The CMT solution in the southern and northern part of the aftershocks distribution was still correlated with the rupture of the mainshock fault plane. However, the CMT solution in Ambon was strike-slip with N-S strike trends. These events may belong to the fault reactivation triggered by stress change during the aftershock sequences which impacted to the surrounding areas.

Table 1. Parameters of the inferred CMT solutions

Event	CMT Location					DC Component			DC
ID	Lat.	Lon.	Depth (km)	Mw	VR(%)	S/D/R(1)	S/D/R (2)	DC (%)	Unc. (%)
1	-3.642	128.388	5.3	4.2	58	281/81/20	187/70/171	88	1
2	-3.61	128.230	7.0	4.2	47	357/73/-134	250/46/-24	11	nearly 0
3	-3.636	128.348	7.4	4.0	40	92/49/-28	201/70/-136	40	1
4	-3.399	128.345	4.0	3.2	50	273/69/-32	16/69/-155	42	22
5	-3.635	128.345	6.5	3.8	53	90/87/1	0/89/177	61	2
6	-3.609	128.377	5.0	3.8	44	182/89/-176	92/86/-1	61	2
7	-3.396	128.313	9.0	3	9	1/66/165	97/76/24	74	25
8	-3.579	128.243	10.7	3.6	44	264/69/-5	355/85/-159	48	3
9	-3.564	128.255	6.6	4.2	68	238/74/-21	334/69/-163	54	nearly 0

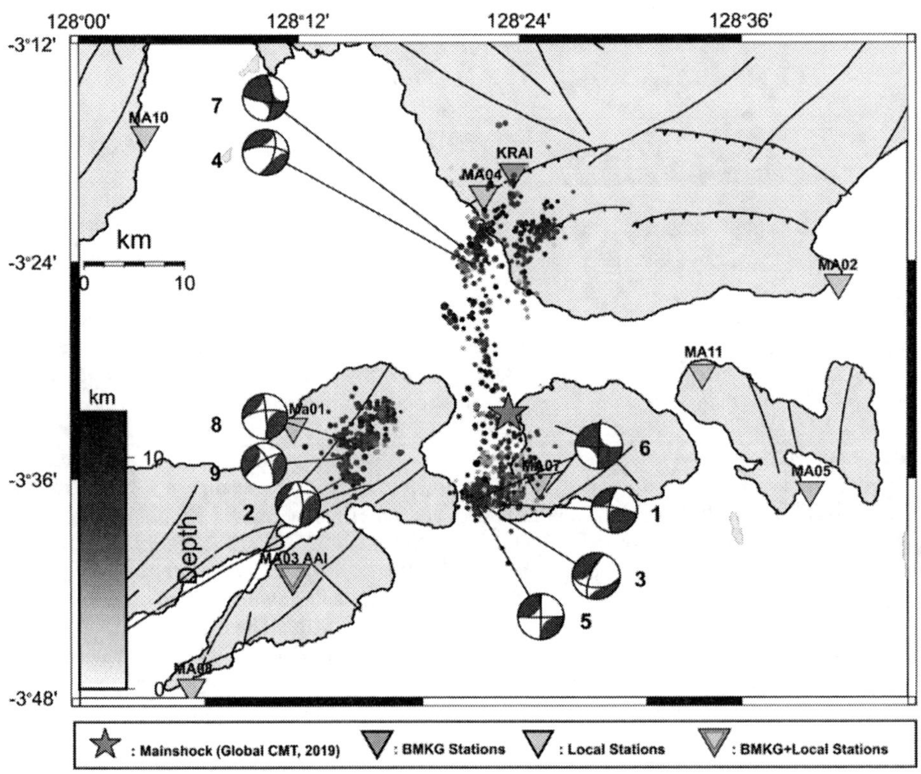

Figure 3. Map of Focal mechanism distribution derived from Bayesian CMT full-waveform inversion for nine aftershock events. The solution is shown by blue beachball with event numbers correspond to data in Tabel 1. A green star symbol depicts the mainshock event. The cyan and yellow reverse triangles depict the BMKG stations and local stations respectively.

4. Conclusions

CMT solutions of nine aftershocks had been inferred using CMT inversion with a Bayesian framework. In general, the parameter of the CMT solution followed the criteria we had set before in which 90% of data give VR values more than 0.40. Most CMT solutions give a high percentage of DC components with low uncertainties for events with a magnitude greater than Mw 4.0 and show a higher value of uncertainties for events with a small magnitude with the order below 3.5 Mw. From the focal mechanism solution, we concluded that the event cluster located on the north and the south which have a dextral strike-slip mechanism reflect the rerupture of the mainshock fault plane. While three solutions in Ambon Island with dextral strike-slip faulting proved the presence of fault reactivation.

Acknowledgments
The Indonesian Agency for Meteorology, Climatology, and Geophysics (BMKG) provided us with access to 2019 Ambon aftershock data from their four regional stations, which we used in this study. The deployment of local seismographic stations was funded by the BNPB.

References

[1] Ekström G, Nettles M and Dziewoński A M 2012 The global CMT project 2004–2010: Centroid-moment tensors for 13,017 earthquakes *Physics of the Earth and Planetary Interiors* **200** 1–9.
[2] Felzer K R, Thorsten W, Becker T W, Abercrombie R E, Ekstrom G and Rice J R 2002 Triggering of the 1999 Mw 7.1 Hector Mine Earthquake by Aftershocks of the 1992 Mw 7.3 Landers Earthquake *Journal of Geophysical Research* **107** 2190.
[3] Shibutani T, Iio Y, Matsumoto S, Katao H, Matsushima T, Ohmi S, Takeuchi F, Uehira K, Kin'ya N and Enescu B 2005 Aftershock distribution of the 2004 Mid Niigata Prefecture Earthquake derived from a combined analysis of temporary online observations and permanent observations. *Earth Planets Space* **57** 545–549.
[4] Vackář J, Gallovič F, Burjánek J, Zahradník J and Clinton J 2017 Bayesian ISOLA: new tool for automated centroid moment tensor inversion *Geophysical Journal International* **210** 693–705.
[5] Zahradník J and Sokos E 2018 ISOLA code for multiple-point source modeling: review. In: D'Amico S (ed) Moment tensor solutions: a useful tool for seismotectonics. Springer, Berlin
[6] Bouchon M 1981 A simple method to calculate Green's functions for elastic layered media *Bulletin of the Seismological Society of America* **71** 959–971
[7] Tarantola A 2005 Inverse problem theory and methods for model parameter estimation. Society for Industrial and Applied Mathematics, Philadelphia.ISBN 0-89871-572-5
[8] Vackář J, Gallovič F, Burjánek J and Zahradník J 2015 Automated detection of longperiod disturbances in seismic records; MouseTrap code *Seismological Research Letters* **86** 442–450.

Ambient Seismic Noise Cross - correlation of Ambon Island and Surrounding Area, Eastern Indonesia: Preliminary Result

Muhammad Fachrul Rozi Kurniawan[1*], Shindy Rosalia[2,6], Andri Dian Nugraha[2,6], Zulfakriza[2,6], David P Sahara[2], Abdul Muhari[3], Andi Azhar Nurdin[4], Sri Widiyantoro[2,6], Nanang T Puspito[2], Awali Priyono[2], Aditya Lesmana[1], Ardianto[1], M Zain Tuakia[5], Erfin Elly[5]

[1]Geophysical Engineering Study Program, Faculty of Mining and Petroleum Engineering Bandung Institute of Technology, Bandung, Indonesia
[2]Global Geophysics Research Group, Faculty Mining and Petroleum Engineering, Bandung Institute of Technology, Bandung, Indonesia
[3]National Disaster Management Authority of Indonesia, Jakarta, Indonesia
[4]Agency for Meteorology, Climatology and Geophysics of Indonesia, Jakarta, Indonesia
[5]Geological Engineering Study Program, Pattimura University, Ambon, Indonesia
[6]Center for Earthquake Science and Technology (CEST), Research Center for Disaster Mitigation, Bandung Institute of Technology, Bandung 40132, Indonesia

fachrulrozi23@gmail.com

Abstract. The island of Ambon lies on complex tectonics, part of Banda Arc which is driven by the Australia – Eurasia collision. Historical earthquake data show that an earthquake resulting the greatest tsunami in Indonesia had occurred at Ambon Island. On 26 September 2019, Ambon was shaken by an M 6.5 earthquake at a depth of 10 km (BMKG). In this study, we use ambient noise data from 11 temporary stations deployed by ITB and 4 permanent stations owned BMKG which are recorded from October until December 2019. Here, we purely use the vertical component of seismogram to retrieve the Empirical Green's Function of Rayleigh waves. Cross-correlations were obtained from the daily data series and stacked the day-by-day cross-correlation data into one inter-station cross-correlation. The Empirical Green's Function is seen at the band period 1-15 s. As a part of our study, we analyze the Green's Function with frequency-time analysis (FTAN) to get Rayleigh wave group velocity. The group velocity of Rayleigh waves varies from 1.04 km/s – 3.75 km/s. Low group velocity might be indicated the presence of sediment or volcanic deposits and high group velocity might be indicated metamorphic rocks. The result of this study might give a finer velocity model of the shallow crustal beneath Ambon Island and the surrounding area.

Keywords: *Ambon, tomography, ANT*

1. Introduction

Ambon Island, part of Banda Arc lies on a complex and unique convergence plates zone which is producing a high heterogeneity structure and seismically active area that has been experiencing great earthquakes and tsunamis [1,2,3]. Based on the historical earthquake, Ambon Island has been hit by large earthquakes and several of them were followed by tsunamis. One of the biggest tsunamis in Indonesia was documented at Ambon Island that was triggered by an earthquake [4]. In recent years, another large earthquake, the M 6.5 earthquake hit Ambon Island on 26 September 2019, causing many injuries and buildings damaged. The mainshock located between Ambon and Haruku Islands at the depth of 10 km and could be classified into a shallow earthquake with a strike-slip source mechanism (BMKG). The mainshock M 6.5 followed by a dozen aftershock for more than 2 months with the aftershocks pattern trending almost in the N-S direction. The biggest aftershock occurred in Ambon Island with a magnitude of 5.1 located 10 km from the mainshock [5]. However, according to the magnitude and number of events, the aftershock decreased with time period. Although the aftershock dominated with small magnitude, it could be felt by civilization due to the hypocenter is in shallow depth.

The interesting phenomena lead to ITB, BMKG, and BNPB deployed temporary stations in Ambon Island and surrounding area to learn about the phenomena for future disaster mitigation. Recently, the application of ambient noise has emerged to inspect closely subsurface of earth. This method utilizes the ambient noise for estimating the Green's function (EGF) by calculating the cross-correlation between two seismic stations as if one station acts as a virtual source and another station acts as a receiver [6].

Ambient noise to obtain shallow subsurface structure, that based on cross-correlation, has been successfully applied in several places in Indonesia. There are in the volcanic region of Toba Lake [7] and Agung – Batur volcanoes [8], complex tectonic of continent collision in Banda Arc [9], and Java Island: Western part of Java [10, 11,12], Central Java [13], Eastern part of Java [14].

In this paper, we describe our preliminary results of cross-correlation process of ambient noise data and frequency-time analysis (FTAN) process to procure the dispersion curve corelated of the group velocity and tried to correlate with geological subsurface. We hope this study could be used to understand the earth structure by analyzing the group velocity of Ambon Island and surrounding island.

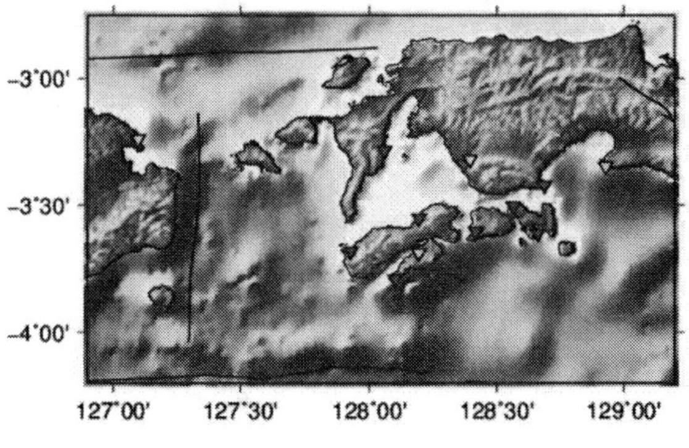

Figure 1. Distribution of temporary seismic station from the ITB temporary network (red inverted triangles) and permanent seismic station from the BMKG network (yellow inverted triangles).

2. Data and Method

In this study, we use 11 temporary stations from the ITB network and 4 permanent stations from the BMKG network supported by BNPB. The temporary stations were installed from October to December 2019, but not all stations have the same recording time because the battery was running out. Figure 1 shows the deployment of seismic stations at Ambon Island and its surrounding. In this study, we focused on the vertical component to infer the EGFs of Rayleigh waves to obtain the group velocity.

Data processing, we follow common procedures to get empirical green functions (EGF) [15]. The process contains single station preparation and preprocessing, cross-correlation and stacking, and measurement of the group velocity. We merged data from 10 minutes to a daily data waveform and converted the data from MSEED to SAC. In this study, we used different type of seismometers, so we have to apply instrument correction to the data. The preprocessing data contains de-trending, de-meaning, and bandpass filtering to accentuate the ambient noise signal. The data filtered in the period band 1 – 15 s and resampled to 20 Hz. We performed temporal normalization and spectral whitening in 1-hour segment data prior to the cross-correlation process to suppress the effect of earthquake signals. Second, the cross-correlation process performed in the frequency domain with lag time -100 s (acausal part) to 100 s (causal part) and set lag time – 60 s to 60 s to arrange all of EGF. To calculate cross-correlation, we used a package from [16], and this process yields $n\ (n\text{-}1)\ /\ 2$ stations pair, where n is the stations that we used. The number of cross-correlation in this study is 105 from the availability of the data. After measuring daily cross-correlation, we stacked the daily cross-correlation by using the linear stacking method for all-day recorded if the data in both stations are available. The stacking process will increase the coherency of the EGF signal and increase the signal to noise ratio (SNR) (shown in figure 2). Third, the measurement of the group velocity of Rayleigh waves was obtained based on frequency-time analysis (FTAN). We manually picked the EGFs and set the SNR greater than 3 to get good dispersion curves. SNR is calculated using the ratio of maximum amplitude of envelope around the frequency signal length and the mean amplitude envelope after the signal window. From all the 105 data we have only extracted 52 good quality dispersion curves.

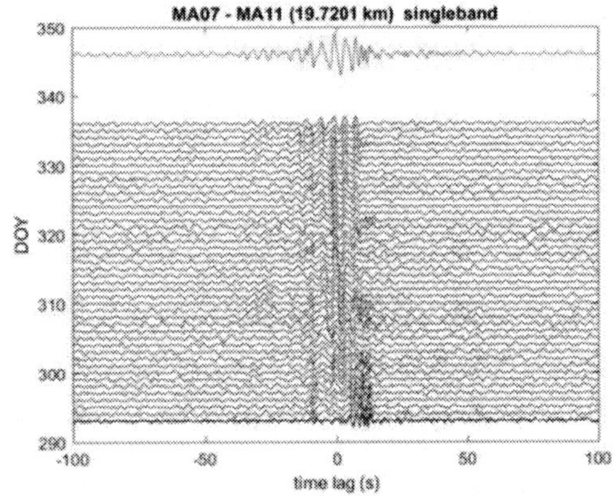

Figure 2. The daily EGF vertical component from station MA07 – MA11 (black lines) and the stacked EGF from the daily cross-correlation result (red line).

3. Result and Discussion

We arrange all of the EGFs from cross correlation as interstation distant function. Here, we assume that EGF is a representation from medium passed by seismic waves between both of stations. We can see that EGF cross correlation in figure 3 (left) has a reflection pattern in both causal and acausal sides. Anyway, the "V" pattern not clearly seen, but it has a little pattern shaped "V". The result was probably affected by the sea waves around the stations deployment.

The group velocity of fundamental mode of Rayleigh waves was obtained from cross-correlations and analysis the time-frequency of ambient noise data. Fundamental mode of Rayleigh waves at small periods (high frequencies) are sensitive to the physical properties at the shallow structures and high periods (low frequencies) at the deep structures. The dispersion curves were obtained from all possible EGF that have SNR greater than 3 shown in figure 3. We use a method from Yao [17] to determine group velocity of fundamental mode Rayleigh waves. Figure 3 (right) shows the dispersion curved that has been extracted from the EGFs result. The group velocity in the study area varies from 1.33 km/s – 3.75 km/s at the band period of 0.1 s – 6 s (most energetic Rayleigh waves). The low group velocity might be related to the sediment layer that can amplify the seismic wave propagating from the earthquake. Then, the high velocity might be correlated with metamorphic rocks that dispersed in the study area.

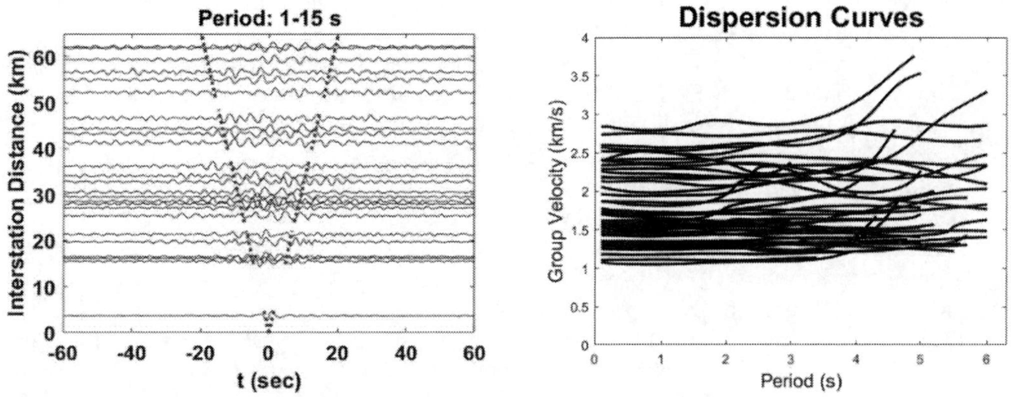

Figure 3. Empirical Green Function (EFG) cross correlation filtered bandpass in 1 – 15 s with SNR greater than 3 (left). The dispersion curves of Rayleigh waves that can be extracted from EFG with SNR greater than 3 (right).

4. Conclusions

In this preliminary study, we have conducted 2 months ambient noise data from 15 stations that supported ITB, BNPB, and BMKG to obtain group velocity. We obtained the group velocity of the Rayleigh wave using cross-correlation of the vertical component of ambient noise data in Ambon Island and surrounding. The EGF is seen at period band 1 – 15 s and from the EGF's arranged as interstation distant function ("V" shape pattern) show that the distribution of ambient noise source might be distributed randomly from the surrounding stations (close to sea). The dispersion curves, that have SNR greater than 3, is ~49 percent.

The group velocity varies from 1.04 km/s – 3.75 km/s at band period 0.1 s – 6 s (most energetic Rayleigh waves). Low group velocity indicated the existence of a sediment layer and high group velocity indicated metamorphic rocks that needed further study to know better about the distribution of the rock or layer types in the study area.

Acknowledgements

We thank the Indonesia Endowment Fund of Education (LPDP) for the magister fully funding program given to the author, the Meteorological, Climatological, and Geophysical Agency of Indonesia (BMKG) and BNPB for the seismic data.

References

[1] J M Pownall, R Hall, and I M Watkinson. 2013. *Solid Earh.* **4** 277 – 314
[2] J M Pownall, R Hall, R A Armstrong, and M A Forster. 2014. *Geology.* **42** 279 – 282
[3] A Spicak, R Matrjkova, and I Vanek. 2013. *J. Asian Earth Sci.* **64** 1 – 13
[4] D Sahara, A D Nugraha, A Muhari, A A Rusdin, S Rosali, A Priyono, Z Zulfakriza, S Widiyantoro, N T Puspito, A Rietbrock, A Lesmana, D Kusumawati, A Ardianto, A W Baskara, YHalauwet, H A Shiddiqi, M T Rafie, R Pradisti, P W Mozefg, M Z Tuakia, E Elly. 2021. *Techtonopycics.* **799**
[5] I R Pranantyo and P R Cummins. 2019. *Pure Appl. Geophysics.* DOI:10.1007/s00024-019-02390
[6] N M Shapiro, M Campillo, L Stehly, and M H Ritzwoller. 2005. *Science.* **307** 1615 – 1618
[7] J Stankiewicz, T Ryberg, C Haberland, Fauzi, and D Natawijaya. 2010. *Geophysical Research Letters.* **37** 1-5

[8] Zulfakriza, A D Nugraha, S Widiyantoro, P R Cummins, D P Sahara, S Rosalia, A Priyono, Kasbani, D K Syahbana, I C Priambodo, Martanto, Ardianto, Y M Husni, A Lesmana, D kusumawati, and B S Prabowo. 2020. *Front. Earth Sci.* **8** 1 - 11
[9] R W Porrit, M S Miller, L J O'Driscoll, C W Harris, N Roosmawati, LT da Costa. 2016. *Earth and Planetary Science Latters.* **449** 246 – 258
[10] E Saygin, P R Cummins, A Cipta, R Hawkins, R Pandhu, J Murjaya, Masturyono, M Irsyam, S Widiyantoro, and B L N Kennet. 2016. *Geophysical Journal International.* **204** 918 – 931
[11] B Pranata, T Yudistira, S Widiyantoro, B Brahmantyo, P R Cummins, and E Saygin. 2019. *Geophysical Journal International.* **220** 1045 – 1054
[12] S Rosalia, P R Cummins, S Widiyantoro, T Yudistira, A D Nugraha, and R Hawkins. 2020. *Geophysical Journal International.* **220** 1260 - 1274
[13] Zulfakriza, E Saygin, P R Cummins, S Widiyantoro, and A D Nugraha. 2014. *Geophysical Journal International.* **197** 630 – 635
[14] A A Martha, P R Cummins, E Saygin, amd S Widiyantoro. 2017. *Geosci. Lett.* **4** 1 – 12
[15] G D Bensen, M H Ritzwoller, M P Barmin, A L Levshin, F Lin, M P Moschetti, N M Shapiro, and Y Yang. 2007. *Geophysical Journal International.* **169** 1239 – 1260
[16] H Yao, C Beghein, and V D Hilst. 2008. *Geophysical Journal International.* **173** 205 – 219
[17] H Yao, P Gouedard, J A Collins, J J McGuire, and V D Hilts. 2011. *Comptes Rendus Geoscience.* **343** 571 – 583

IOP Conf. Series: Earth and Environmental Science **873** (2021) 012024 doi:10.1088/1755-1315/873/1/012024

Landslide Monitoring using Inclinometer with Micro Electromechanical System (MEMS)

Tedy Setiawan[1*], Fatkhan[1], Ramadhani Yasyfi Cysela[1]

[1]Department of Geophysical Engineering, Faculty of Mining & Petroleum Engineering, Insititute of Technology Bandung (ITB), Indonesia

*Corresponding author tedyset@gmail.com

Abstract. Being located on the Pacific Ring of Fire and a tectonically active country, Indonesia has to cope with the constant risk of volcanic eruptions, earthquakes, floods and landslides. Landslides and other mass movements are serious geo-environmental hazards in Indonesia. Following report from the Indonesia National Disaster Management Authority (BNPB), landslides are among high disaster death toll throughout the archipelago. It claimed 248 lives last year alone. The number of landslides in Indonesia increase steadily to 376 in 2014 from 291 in 2012. Hence, landslide monitoring system is required to determine style of landslide movement, for risk and even emergency risk management assessments and to assist with the design of mitigation works. A landslide instrumentation program including an inclinometer is designed for landslide monitoring. The inclinometer, or tilt sensor, is an instrument used for measuring slope, tilt, or inclination. In this paper we use Micro Electromechanical System (MEMS) as a sensor to measure changes in an angle. Then information is transferred to a central server soon after real-time accelerations are monitored. A data logger also used as a data recording. With low-cost MEMS accelerometers, the results show this instrument is able to provide reliable ground-motion data in network-scale deployments.

1. Introduction

Indonesia is located at the joining point of four major tectonic plates. Most areas in Indonesia are highly prone to various hazards. According to Meteorology, Climatology and Geophysics Agency (BMKG) chairwoman Dwikorita Karnawati said that the nation lacked programs to raise disaster awareness, despite the country sitting on the Pacific Ring of Fire [1]. Indonesia is prone to earthquakes and volcanic eruptions. Landslides and other mass movements are serious geo-environmental hazards in Indonesia. Landslides are among the highest disaster death toll throughout the archipelago, according to the BNPB,

Landslide monitoring is required to determine style of landslide movement, for risk and even emergency risk management assessments. This is used to assist design of mitigation works. The monitoring of surface displacement is crucial for the prevention and forecast of landslides. There are several systems and techniques for landslide monitoring, one of them is ground-based techniques using inclinometer to measure slope, tilt or inclination.

2. MEMS

In order to obtain high quality of micro earthquake data, micro earthquake sensors deployed in the field have several main characteristics, such as high sampling rate response, sensitivity, wide frequency bandwidth with low frequency coverage, high temperature resistant, and high pressure and humidity resistant. There are numerous applications of MEMS. The usage of MEMS in electronics such as disk drive heads and inkjet printer heads. MEMS sensor also used in different domains which include medical, automotive, communications and defense. The following parts are used to build MEMS sensor. Each part has its own role and specific characteristics.

2.1. The Core

Conventional seismometers use coil as the main sensor. Coil is relatively cheap, but in order to cover low frequency range, it becomes frail and has inappropriate sampling rate. Moreover, coil has harmonic frequencies which add some noise to the recorded data. There is often a second resonance at much higher frequency, which is produced by the combination of the inductance of the sensor coil combined with capacitance [2]. This millennium breakthrough technology has the answer for overcome the limitation of conventional seismometers, namely a MEMS sensor. The MEMS is made from glass and silicon substrate which has no harmonic frequency characterization. It enables covering wide bandwidth with low frequency cover and high sampling rate response with high sensitivity. It has been used three MEMSes Colibrys 1500SN.A for developing a 3-C borehole seismometer. Each MEMS represents one axis direction. The specification of this type of MEMS can be seen in Table 1.

Table 1. Specification of MEMS 1500SN.A

Parameter	Value	Unit
Linear Output Range	± 3	g peak
Sensitivity	1.2	V/g
Frequency Response	DC to 1500	Hz
Dynamic Range	120	dB
Noise (10 to 1000 Hz)	300 to 500	ngrms/√Hz
Cross-axis rejection	> 40	dB
Shock Limit (0.5 ms 1/2 sine)	500	g peak
Vibration (20 Hz - 2000 Hz)	60	g pk-pk
Operating Temperature	- 40 to + 125	°C
Sensitivity Temperature Coef.	75	ppm/°C
Offset Thermal Coefficient	± 100	µg/°C
Linearity Error	± 0.1 %	Full Scale
Input Voltage	± 6 to ± 15	Volts DC
Quiescent Current	< 11.6	mA

Instrumentation system of MEMS based seismometer borehole consists of 3 units 1500SN.A MEMS, amplifier, low pass filter, analog to digital circuit, and microcontroller. MEMS based seismometer borehole block diagram is shown in Figure 1.

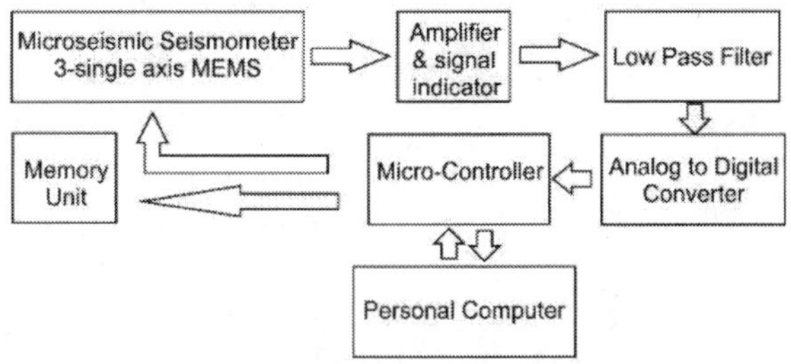

Figure 1. Block diagram of planned borehole seismometer

Micro earthquakes are sensed by MEMS as input signals. Then the MEMS converts the land vibration into electronic signals, having low level of voltages. Thus, the signals need to be amplified. The captured signal by MEMS is not only micro earthquake signal but they consist also many land vibration sources such as man activities, sea waves, vehicles vibrations, etc. The various input signals are separated by filter. Afterward, the filtered signals are converted into digital signal by ADC. The results are sent to memory unit by microcontroller.

Colibrys 1500SN.A is an MEMS accelerometer having several requirements compared to a low frequency geophone, e.g. it needs a DC power supply. As can be seen in Figure 2, MEMS accelerometer has a good stability character compared to low frequency geophone. Monitoring micro earthquake events for geothermal engineering takes months even years for real time monitoring, so a proper seismometer should have a good stability response [3].

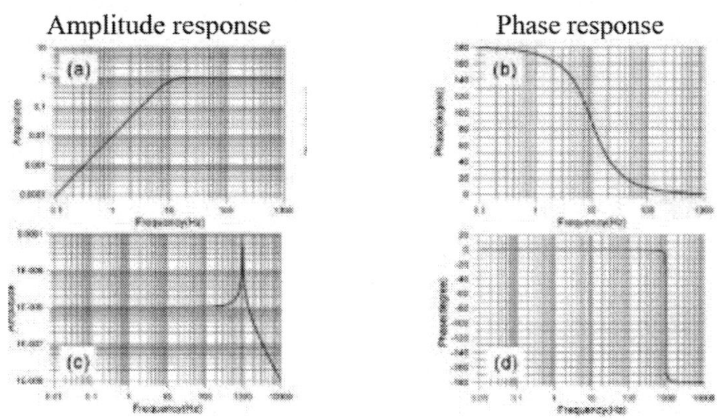

Figure 2. Amplitude and phase responds comparison between MEMS accelerometer and 10 Hz geophone [4] a and b for 10 Hz geophone; c and d for MEMS accelerometer.

2.2. *Amplifier*

Since microearthquake events have low amplitudes producing very low output voltage, thus an amplifier is needed. The MEMS 1500SN.A's sensitivity is 1,2mV/gm. It means that the output voltage for weak motion is in the order of 10-3 volt. This very low output voltage needs to be amplified in order to be read by the ADC. Figure 3 shows the circuits of amplifier.

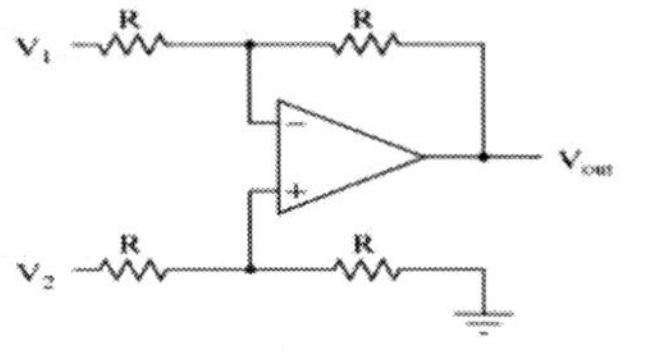

Figure 3. Basic schematic of differential amplifier

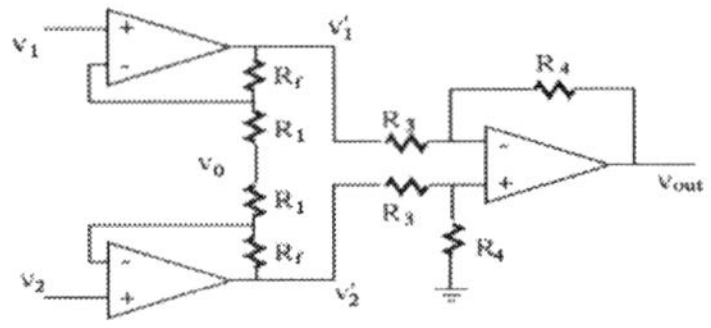

Figure 4. Basic schematic of amplifier instrumentation.

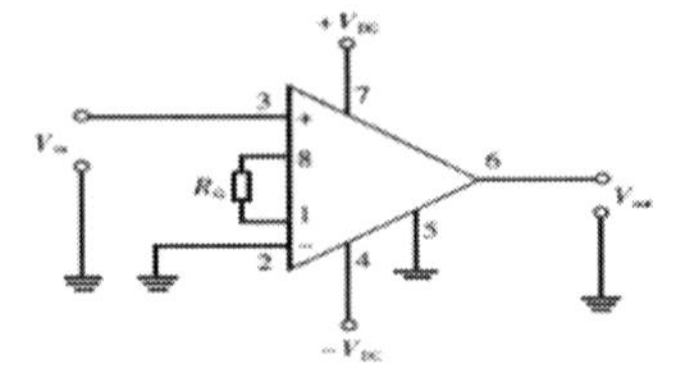

Figure 5. Simplified schematic of AD620AN instrumentation amplifier

Differential amplifier is an op-amp based active amplifier. It has low precision because there are two resistors which has to be adjusted precisely. Thus, this amplifier must be modified becoming a better one. Basic circuit (known as instrumentation amplifier) can be seen in Figure 4. It has been used an AD620AN device as main amplifier component. AD620AN has low offset voltage drift, low noise performance, high common mode rejection ratio, high input impedance, and low power consumption. Thus, there is one adjustable resistor (R1 or Rg) as amplification factor, in which its circuit can be seen in Figure 5.

2.3. Filter

The MEMS 1500SN.A detect any motion within its range of measurement. It sends all signals in every frequencies to ADC and CPU. Desired signal is only signal with frequency below 500 Hz. Therefore, a filter is needed for separating signals based on frequency. Active LPF with 500 Hz as a cut-off frequency is used to reject signal above 500Hz. MEMS does not have harmonic frequency, thus band pass filter is not needed. Only a low pass filter (LPF) is needed to separate desired frequencies from the others. Figure 6 shows basic schematic of one order LPF.

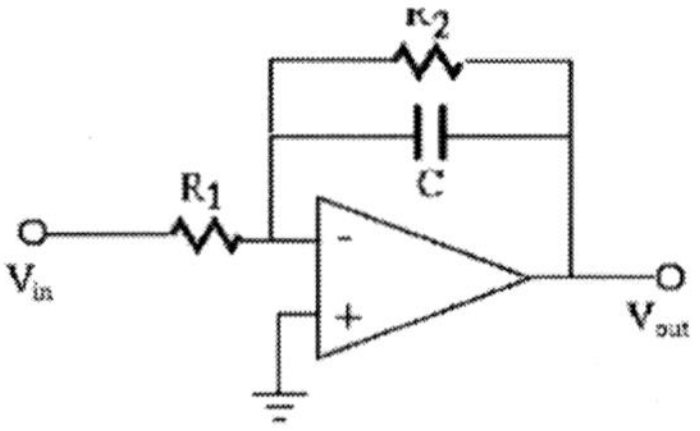

Figure 6. Basic schematic of a one-order LPF

2.4. ADC and Microcontroller

Analog signal that come out from LPF is already in voltage level that possible to be processed by ADC. A 12-bit ADC is used to convert analog signal into digital signal. The conversion result is sent to PC/ memory unit from ADC through microcontroller. Microcontroller has a special task to control all the data acquisition system. It plays also an important role as a connector between the main controller units (PC) and the MEMS.

2.5. Casing

The casing For the MEMS is designed to protect the core from high temperature, high pressure, high humidity and shock. Current prototype MEMS casing is made of ordinary stainless steel to support laboratory test and small scale field test in geotechnical , which will be conducted in a near future. The next stage of MEMS development includes the research of the best type of casing materials and its structures.

IOP Conf. Series: Earth and Environmental Science **873** (2021) 012024 doi:10.1088/1755-1315/873/1/012024

2.6. Clamping
Current prototype MEMS- has not attached to any clamping system. The function of the clamping system will be temporary replaced by special acquisition installation technique. This technique is aimed to achieve similar environment as if it is attached to a clamp in a borehole, in most of rock formation types.

3. Laboratory Test
Physical simulation as laboratory scale test is performed to obtain result seismometer core part performance. The test is done by putting the seismometer core (already inside a cilindrycal casing) on a spring. The idea is to test whether the seismometer can detect the small disturbance (represented with shaking spring in three axis) as illustrated in Figure 7. The response is recorded by a laptop computer. The MEMS-based seismometer borehole shows the performance of its recorded small signal.

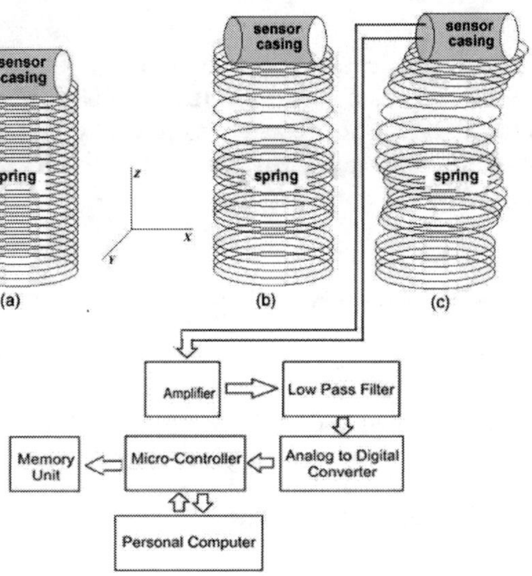

Figure 7. LPF Spring test (a) stable position. (b) disturbance in z axe. (c) disturbance in 3-axis.

Figure 8 describes laboratory test equipment for testing core's signal to noise ratio performance. The signal generator is used to calibrate the signal indicator, amplifier, low pass filter and analog to digital circuits both independently and dependently. Afterward, the prototype of 3-C MEMS borehole seismometer is connected and the signal-to-noise ratio performance test is started and recorded. While the test is conducted, the oscilloscope shows the output signals and it can be compared to the displayed signal on laptop.

The prototype of MEMS-based seismometer borehole is able to record desired signal and stable to measure and record continuous data. The noises, recorded and displayed, are caused by the fan of power supply. This means that the seismometer's signal to noise (S/N) ratio is high.

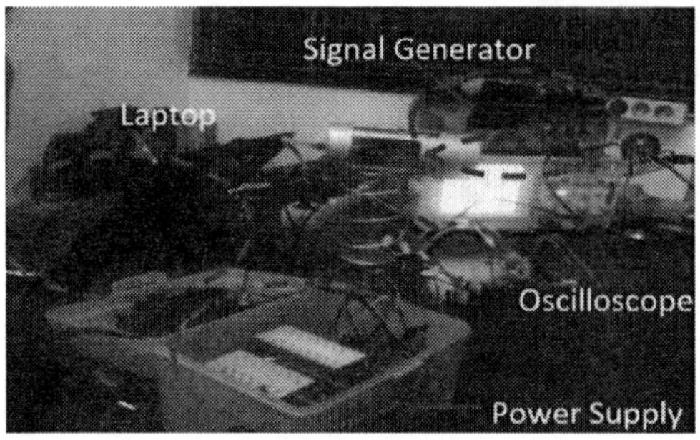

Figure 8. Laboratory scale test – core part

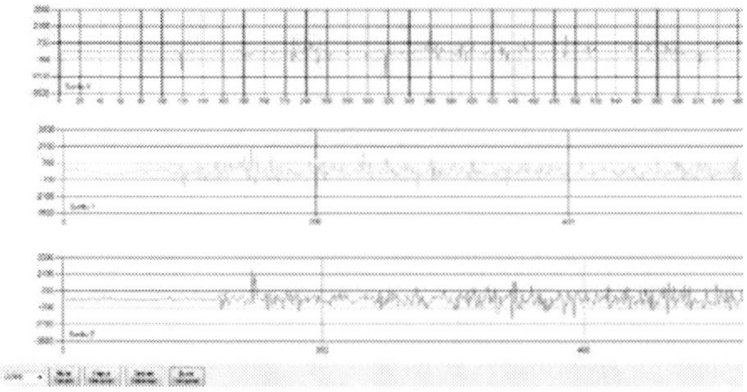

Figure 9. Result of signal to noise ratio test

4. Conclusions

Micro earthquake signals are low level in term of frequency and amplitude. However the signal amplitudes can still be amplified by using instrumentation amplifier. In the other hand, active LPF is the solution for separating signals based on frequency. From physical simulation as laboratory test, the MEMS works properly as their functions. The next project is to finalize the casing, to attach the core to be inside the casing, and to conduct small scale the field test.

Acknowledgement

We are grateful for receiving funding from the P3MI program at Institut Teknologi Bandung.

References

[1] F Van Kann and J Winterflood 2005 *Rev. Sci. Instrum.* **76** 034501
[2] F Van Kann and J Winterflood 2005 *Rev. Sci. Instrum.* **76** 034501
[3] Aizawa T et al 2008 Application of MEMS accelerometer to geophysics. *Int. J. JCRM.* **4**.
[4] Speller K. and Yu D 2004 A low-noise MEMS accelerometer for unattended ground sensor applications *Proc. SPIE Int. Opt. Eng.* **5417** 63

Site characterizations of BMKG seismic network, Indonesia, based on HVSR analysis

Muzli Muzli[1,*], R A P Kambali[1], J Nugraha[1], S Sulastri[1], A R Hakim[1], S Rohadi[1], R Triyono[1], N F Riama[1] and D Karnawati[1]

[1]Agency for Meteorology, Climatology, and Geophysics (BMKG), Jakarta 10610, Indonesia

*Email: muzli@bmkg.go.id

Abstract. In the last few years BMKG (Agency for Meteorology Climatology and Geophysics) has increased the number of seismic stations significantly. Until mid-2020, when this study was conducted, the number of BMKG stations has reached 339 units. Development of the network is aimed to improve the data quality, speed of earthquake data processing and information dissemination to the public, accuracy of the hypocenter, as well as the magnitude. The objective of this study is to identify the site characteristics of the network. We analysed the noise recorded at BMKG stations throughout Indonesia using the HVSR (Horizontal to Vertical Spectral Ratio) method. The results of HVSR analysis were used to classify the site conditions of each station. We got the number of stations with the classification of hard rock, rock, hard soil, medium soil, and soft soil, 47, 57, 52, 30, and 84, respectively. This site condition represents the stations characteristics and affects the quality of seismic waveform data.
Keywords: BMKG, seismic network, earthquake, HVSR

1. Introduction

Since the Sumatra earthquake in 2004 [1,2] which caused enormous damage and casualties, the Indonesian government has paid great attention to disaster management caused by the earthquake and tsunami. Since then, the government through BMKG (Agency for Meteorology Climatology and Geophysics) has increased the number of seismic stations for earthquake monitoring throughout Indonesia. In the last few years BMKG has added a significant number of new stations. Until mid-2020, at the time of this study was conducted, the number of BMKG seismic stations has reached 339 units which are distributed throughout Indonesia (Figure 1).

With the addition of the number of seismic stations, it is expected that the determination of hypocenter location, magnitude and mechanism of the earthquake source can be done more quickly with better accuracy. Moreover, by increasing the density of the number of stations, earthquakes with relatively small magnitudes can be recorded well by the national seismic network. The completeness of earthquake data up to a small magnitude can help research related to earthquakes and the identification of active faults which until now have not been well mapped [e.g. 3].

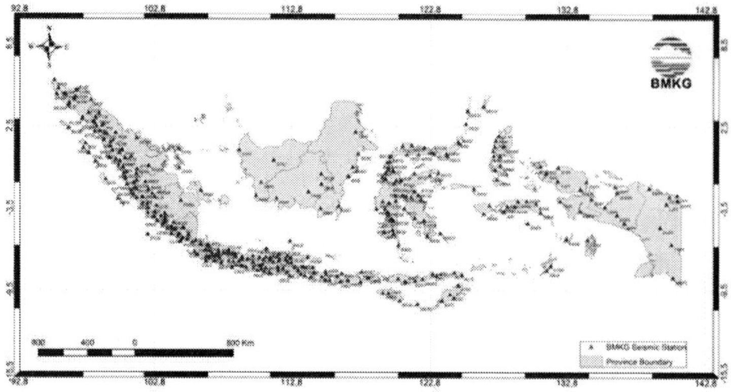

Figure 1. Indonesian permanent seismic stations in April 2020 operated by BMKG (Agency for Meteorology Climatology and Geophysics).

Rapid and accurate earthquake data processing of the hypocenter location, magnitude, and source mechanism can help the community to reduce the impact of earthquake disasters. Rapid and accurate information received by the community becomes an early warning for evacuation or decision making for emergency response. This issue is the main objective of BMKG through the project of InaTEWS (Indonesian Tsunami Early Warning System).

The significant development of BMKG national seismic network increases the dissemination speed and accuracy of earthquake parameters. However, there are a number of rules that must be considered in the installation of seismic stations in order to obtain good data quality. Several factors that must be considered are the type of sensor used, the station density according to the tectonic structure or identified faults, azimuthal coverage of the station distribution, installation procedures or quality, orientation of sensor installation, quality of station location, security, availability of communication networks, and accessible [4].

The objective of this study is to investigate the quality of BMKG national seismic network based on the site characteristics of the station location. The quality of station location is usually associated with the level of noise and surface soil structure [4,5]. This study focuses on discussions related to surface soil structure, identifying and classifying surface soil types from all BMKG stations throughout Indonesia to obtain the characteristics of each station.

2. Data and Method

In general, it is well known that the soft surface soil structure causes amplification of the seismic waves [6,7]. The stations located on the soft soil causing relatively higher noise [8]. If an earthquake occurs, especially with a relatively small magnitude, the signal to noise ratio is small, making it difficult to identify earthquake signal, even though the wave arrival time is not affected. In addition, seismic wave amplification also affects the accuracy of determining the magnitude of the earthquake and also the source mechanism using the waveform inversion method.

One method of identifying surface soil structures can be done by using HVSR (Horizontal to Vertical Spectral Ratio) of the ambient noise at the station location [9,10]. The HVSR method is simple and effective. This method has been widely used and successful for the interpretation of surface soil structures [11]. The basis of this method is that in the case of soft soil the noise spectrum of horizontal component has relatively larger amplitude than the vertical component. Whereas in the case of hard rock, the spectrum of noise for both components, horizontal and vertical, have more or less the same amplitude [12].

The interpretation of the results of HVSR curve is that if the dominant frequency is in the low frequency range it indicates the soft surface soil structure, and vice versa if the dominant frequency is high it indicates a typical rock or hard surface soil structure [12]. The development of ambient noise analysis using this method shows that the dominant frequency indicated from HVSR can be used to estimate sediment thickness [e.g. 11].

The weakness of this method is that the large amplitude at the dominant frequency is inconsistent for the interpretation [e.g. 13]. This is possible due to the heterogeneity of the surface soil structure based on the layer structure, there are systematic, multi layers with a horizontal arrangement and another case are not systematic. In the case of many horizontal layers that are systematic allows the emergence of the optimal amplitude, and vice versa. Another possible cause of the difference in the HVSR amplitude is the difference in the level of impedance or contrast between the hard rock layer and the sediment above it. However, interpretation of the dominant frequency alone can be used to indicate the classification of surface soil types at the measurement location [14].

The HVSR analysis in this study was conducted using Geopsy software (e.g. Figure 2). Ambient noise data was collected from 339 BMKG seismic network entire Indonesia. At the time of the data was collected, there was 69 stations have no data. The time span used for each station with duration of 30 minutes is 00:00-00:30 WIB (local time). Waveform was downloaded from BMKG Webdc application (accessed on 18 April 2020).

Table 1. The reference of site classes based on natural period and the approximately corresponding NEHRP site classes [14]

Site Class	Description	Natural Period (sec)	V_{30} Calculated from Site Period (m/s)	NEHRP Site Classes
Hard Rock			$V_{30} > 1100$	A
SC I	Rock	$T < 0.2$	$V_{30} > 600$	A + B
SC II	Hard soil	$0.2 = T < 0.4$	$300 < V_{30} = 600$	C
SC III	Medium soil	$0.4 = T < 0.6$	$200 < V_{30} = 300$	D
SC IV	Soft oil	$T = 0.6$	$V_{30} = 200$	E + F

The classification of surface soil types is based on the reference table [14] and NEHRP (National Earthquake Hazard Reduction Program). Each station is adjusted to the results of the dominant frequencies and reference table (see Table 1).

3. Results and Discussion

In general, the dominant frequency based on the results of HVSR analysis on the ambient noise at each station can be seen clearly. This value is indicated by a significant spike or peak on the HVSR curve. However, at some station locations, the HVSR curve pattern shows more than one spike. In that case, the spike determination is based on additional information such as elevation and environmental conditions around the station location.

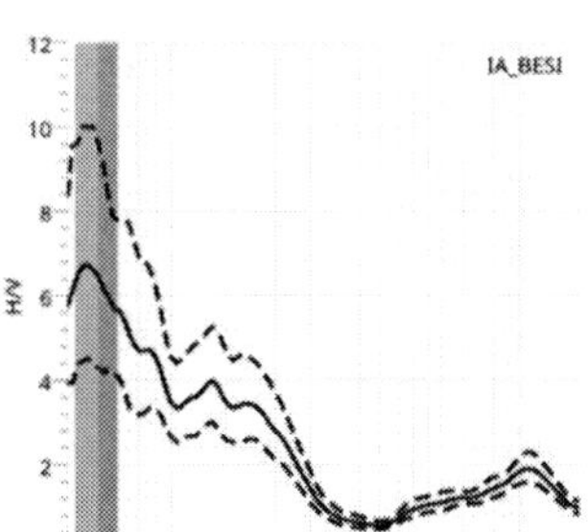

Figure 2. The result of HVSR analysis for BESI station located at 4.72 N and 96.87 E with the lowest dominant frequency of 0.57Hz.

IOP Conf. Series: Earth and Environmental Science **873** (2021) 012025 doi:10.1088/1755-1315/873/1/012025

We limited the HVSR interpretation of dominant frequency between 0.2 to 20Hz based on the optimal reference value of the dominant frequency at Mount Batu Lembang [15] and the low dominant frequency value at the measurement location with sediment that is about 200m thick [11]. Figure 2 shows the lowest dominant frequency, 0.57 Hz, precisely measured at station BESI located at 4.72 N and 96.87 E. On the other hand, the highest dominant frequency 19.32 Hz was measured at station KMPI located at 3.66 S and 133.70 E (Figure 3). Interpretation of the classification of surface soil types based on the dominant frequency shows that the stations BESI and KMPI have soft soil and hard rock structures, respectively, with classifications of IV and hard rock based on the reference table [14] and NEHRP.

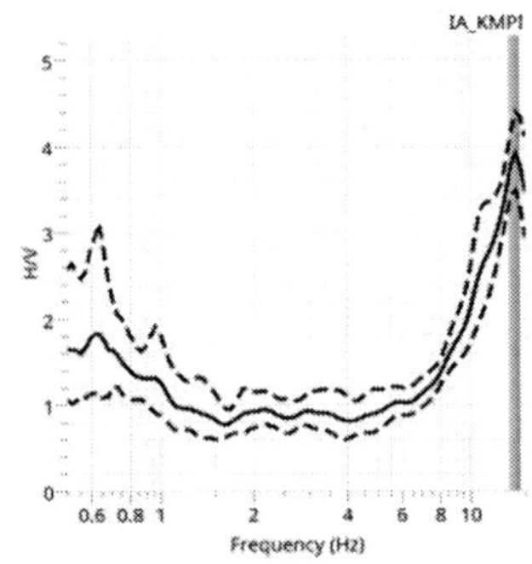

Figure 3. HVSR of KMPI station representing the typical hard rock station with the dominant frequency of 19.32Hz. The station located at 3.66 S and 133.70 E.

In some locations the result of HVSR analysis shows that there is no dominant frequency or the curve fluctuates at certain amplitude of HVSR value (e.g. Figure 4). Determination of the dominant frequency is then selected manually by observing the HVSR curve, elevation, and conditions around the measurement location. Moreover, some other measurement results exhibit the significant spike with high amplitudes at high frequencies greater than 20Hz, even greater than 30Hz. Based on the reference of the HVSR value on Mount Batu Lembang which shows the dominant frequency around 18Hz [15], the interpretation was considered within the dominant frequency range smaller than 20Hz. Likewise, the results of HVSR method, which show the dominant frequency is smaller than 0.2Hz, were not interpreted. The selection of the dominant frequency values was considered within 0.2-20Hz frequency range.

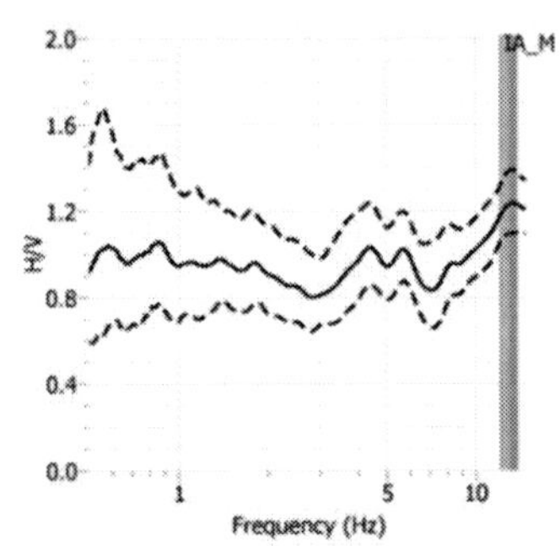

Figure 4. Typical HVSR result for the hard rock station. The HVSR amplitude fluctuates at approximately one.

Figure 5 shows that from a total of 339 BMKG seismic stations, the number of stations with the typical hard rock, rock, hard soil, medium soil, and soft soil is 47, 57, 52, 30, and 84 stations, respectively. This classification of surface soil types represents the characteristics and quality of the data recorded at the station. A good station is placed on a foundation with hard rock and a quiet environment. The station will produce data with good quality. However, other factors in selecting a station location must also be considered, such as accessible and security of sensors.

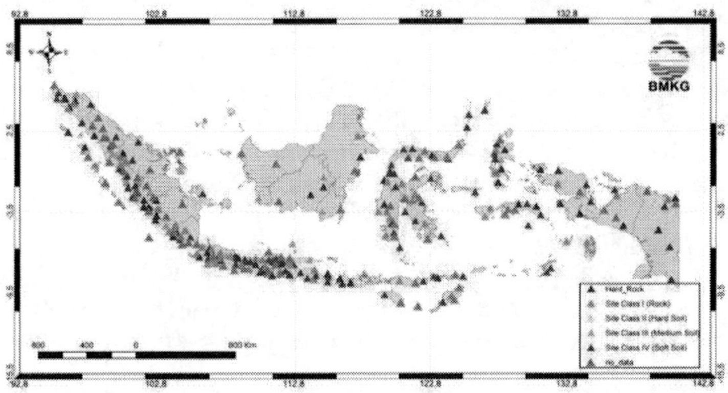

Figure 5. Site classification of BMKG national seismic network based on HVSR of the ambient noise.

4. Conclusions

The characteristics of BMKG seismic station are indicated through the classification of surface soil types which are estimated based on the dominant frequency of the HVSR curve. The quality of the station location represented by the surface soil structure varies from the typical hard rock, rock, hard soil, medium soil, and soft soil. The highest number of typical structure was soft soil with 84 stations. The locations of these stations are distributed throughout Indonesia. Meanwhile, the stations with tolerable quality with the criteria of at least having a surface soil structure with medium soil classification are 186 stations. This number of stations is 69 percent of the total number of BMKG national seismic stations. The use of BMKG data for the purpose of arrival time data can be used well for all stations, regardless of amplification factors. However, the use of waveforms for inversion or amplitude, as well as the magnitude needs to be more careful, especially for the waveform data from soft soil stations.

Acknowledgments

We would like to thank the Deputy of Geophysics BMKG for their support. Thanks to all colleagues in the Division of Geophysical Research and Development, RDC (Puslitbang) BMKG. Thanks also to the colleagues from the Deputy of Instrumentation, Engineering, and Calibration BMKG for sharing information.

References

[1] Lay, T., Kanamori, H., Ammon, C. J., Nettles, M., Ward, S. N., Aster, R. C., ... & DeShon, H. R. (2005). The great Sumatra-Andaman earthquake of 26 december 2004. Science, **308**(5725), 1127-1133.

[2] Stein, S., & Okal, E. A. (2005). Speed and size of the Sumatra earthquake. Nature, **434**(7033), 581-582.

[3] Nasional, P. S. G. (2017). Peta sumber dan bahaya gempa Indonesia tahun 2017. Pusat Penelitian dan Pengembangan Perumahan dan Permukiman, Badan Penelitian dan Pengembangan, Kementerian Pekerjaan Umum.

[4] Bormann, P. (2012). New manual of seismological observatory practice (NMSOP-2), IASPEI, GFZ German Research Centre for Geosciences, Potsdam.

[5] Picozzi, M., Strollo, A., Parolai, S., Durukal, E., Özel, O., Karabulut, S., ... & Erdik, M. (2009). Site characterization by seismic noise in Istanbul, Turkey. Soil Dynamics and Earthquake Engineering, **29**(3), 469-482.

[6] Seed, H. B., Romo, M. P., Sun, J. I., Jaime, A., & Lysmer, J. (1988). The Mexico earthquake of September 19, 1985—Relationships between soil conditions and earthquake ground motions. Earthquake spectra, **4**(4), 687-729.

[7] Bard, P. Y., Campillo, M., Chavez-Garcia, F. J., & Sanchez-Sesma, F. (1988). The Mexico earthquake of September 19, 1985—A theoretical investigation of large-and small-scale amplification effects in the Mexico City Valley. Earthquake spectra, **4**(3), 609-633

[8] Smith, K., & Tape, C. (2019). Seismic noise in central Alaska and influences from rivers, wind, and sedimentary basins. Journal of Geophysical Research: Solid Earth, **124**(11), 11678-11704.

[9] Nakamura, Y., & Saito, A. (1983). Estimations of amplification characteristics of surface ground and PGA using strong motion records. Proc. 17th JSCE.

[10] Nakamura, Y. (1989). A method for dynamic characteristics estimation of subsurface using microtremor on the ground surface. Railway Technical Research Institute, Quarterly Reports, **30**(1).

[11] Parolai, S., Bormann, P., & Milkereit, C. (2002). New relationships between Vs, thickness of sediments, and resonance frequency calculated by the H/V ratio of seismic noise for the Cologne area (Germany). Bulletin of the seismological society of America, **92**(6), 2521-2527.

[12] Nakamura, Y. (2008). On The H/V Spectrum. The 14th World Conference on 619Earthquake Engineering, October 12-17, 2008, Beijing, China.

[13] Sylvette, B. C., Cécile, C., Pierre-Yves, B., Fabrice, C., Peter, M., Jozef, K., & Fäh, D. (2006). H/V ratio: a tool for site effects evaluation. Results from 1-D noise simulations. Geophysical Journal International, **167**(2), 827-837.

[14] Zhao, J. X., Zhang, J., Asano, A., Ohno, Y., Oouchi, T., Takahashi, T., ... & Fukushima, Y. (2006). Attenuation relations of strong ground motion in Japan using site classification based on predominant period. Bulletin of the Seismological Society of America, **96**(3), 898-913.

[15] Tim Survey BMKG (2014). Laporan survei lokasi stasiun bedrock di wilayah Jawa Barat, Badan Meteorologi Klimatologi dan Geofisika (BMKG). Jakarta.

Preliminary Results: Probabilistic Non-Linear Method to Determine the Hypocenter Location in the Molucca Sea Collision Zone from BMKG Networks

Gazali Rachman[1,a], Bagus Jaya Santosa[2], Supriyanto Rohadi[3], Andri Dian Nugraha[4,5], Shindy Rosalia[4,5]

[1]Student of Physics Department, Institut Teknologi Sepuluh Nopember (ITS), Surabaya 60111, Indonesia
[a] Physics Education-FKIP, Pattimura University of Ambon-Molucca 97233, Indonesia
[2] Physics Department, Institut Teknologi Sepuluh Nopember (ITS), Surabaya 60111, Indonesia
[3]Agency for Meteorology, Climatology and Geophysics (BMKG), Jakarta, Indonesia
[4]Global Geophysics Research Group, Faculty of Mining and Petroleum Engineering, Institut Teknologi Bandung, Bandung, Indonesia
[5]Center for Earthquake Science and Technology (CEST), Research Center for Disaster Mitigation, Institut Teknologi Bandung, Bandung 40132, Indonesia

Email : gazali.rachman@gmail.com

Abstract. Molucca Sea collision zone is a region which has very complex geology and tectonic setting, producing high seismicity and volcanoes activities. In this study, we have determined hypocenter location around the region using local & regional network of Agency of Meteorology, Climatology, and Geophysics, Indonesia (BMKG). We used 1,647 events that recorded by 32 seismic stations. We repicked the P-and S-phase manually and have been succesfully determined ~17,628 P and ~17,628 S arrival times. The P- and S-arrival times are used to determine the hypocenter location by applying NonLinLoc method which estimating the probability density function (PDF) using the oct-tree importance sampling algorithm. Our preliminary results show that the seismicity beneath the Molucca Sea collision zone forming a double subduction pattern which is dipping westward under the Sangihe Arc, reaching a depth of ~ 600 km and eastward under the Halmahera Arc, reaching a depth of ~ 250 km. The seismicity pattern under the Sangihe Arc deepens to the north and the deep earthquake events increase in number. The seismicity is related to the Molucca Sea Plate which is dipping into west and east direction beneath Sangihe-Halmahera Arc. To have a further understanding of the complex tectonic activity in this area, our future work will focus on conducting a seismic tomographic inversion to determine the 3D seismic velocities structure around the Molucca Sea collision zone.

1. Introduction

The Molucca Sea plate, a microplate flanked by three main plates: the Pacific plate, the Eurasian plate, and the Australia-Asian plate, is one of the regions with the highest seismic activity on earth [1,2,3].

The Molucca collision zone is a very complex region due to the interaction of the Halmahera Plate, the Molucca Sea Plate, and the Sangihe Plate. The interaction of the plates creates the Halmahera volcanic arc, the Sangihe volcanic arc, the double subduction, active faults, ridges, and earthquakes [1,4–7].

The Halmahera arc which is located in the eastern part and the Sangihe arc which is located in the western part converging each other at the Molucca Sea to form a collision zone [5]. This tectonic activity makes the earthquake distribution occurred in two Benioff zones, dipping away from each other. Beneath the Sangihe arc in the western part, the earthquakes distribution is deeper than beneath the Halmahera arc in the eastern part [6–8]. It is also shown in the latest research which presents a vertical cross-section of an inverted "V" seismicity pattern beneath the Molucca Sea area [9–12].

In this study, we reidentify the P- and S-phases from BMKG waveform data to determine the hypocenter location. We use a probabilistic non-linear location method with an oct-tree importance sampling approach to determine the hypocenter location [13]. We aimed to get a more accurate earthquake hypocenter location to better identify the patterns and the geometric structures beneath the Molucca Sea.

2. Data and Methods

The waveform data used is from BMKG network from period of January 1, 2010, to December 31, 2017. We used earthquake with a magnitude≥ 4. The total earthquakes recorded are 1,647 events which distributed in an area of 1.5° S - 4.25° N and 121.5° E - 129.5° E. The number of stations used is 32 stations which distributed in an area of 4.8° S - 4.25° N and 120.5° E - 134.5° E (Figure 2). The re-picking process of P- and S- waves is carried out manually and to help with quality control of the re-picking process we used Wadati diagrams.

The hypocenter determination was carried out by applying the probabilistic non-linear location method [13]. This method determine the "optimal" hypocenter by estimating the probability density function (PDF) using an oct-tree importance sampling approach. The algorithm used was developed by Moser et al., Tarantola and Valette, Wittlinger et al., [14–16], summarized in the NLLoc program [13,17]. We used a 3D seismic velocity model from high-resolution tomographic imaging results beneath Indonesia [18] which is a regional-global seismic velocity model.

3. Results and Discussions

From the re-picking process of 1,647 events, we can identified ~ 17,628 P- and ~ 17,628 S-arrival times with the Wadati diagram for all data shown in Figure 1. The Wadati diagram is a diagram that shows a linear relationship of the P-phase arrival time with the difference between the P- and S-phases arrival times which could be used to determine the origin time of the earthquake and the Vp/Vs value [19].

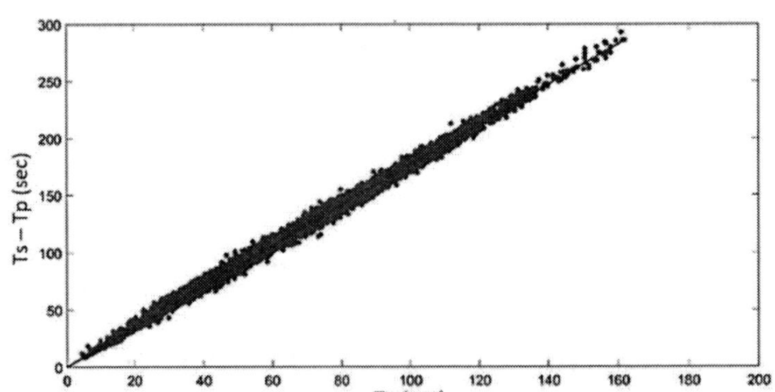

Figure 1. Wadati diagram from all of the re-picked P- and S- arrival times with the Vp/Vs ratio is 1.76.

The horizontal results of hypocenter location from the probabilistic non-linear method using the NLLoc program are shown in Figure 2, and vertical cross-section results in Figures 3 and 4. Based on Figure 3 (profiles A-A ', B-B', C-C '), seismicity in the Molucca Sea Collision Zone is mostly occurred in the subduction area. In the profile A-A' of Figure 3, seismicity clusters around the slab of Sulawesi (blue line) are at a depth of ≤ 100 km and are above the Molucca Sea Plate which is subducting to the west. Seismicity clusters around the Molucca Sea Plate (green line), forms a double subduction pattern. In the western part, beneath the Sangihe arc, the earthquakes reach depths of ~600 km and in the eastern part, beneath the Halmahera arc reach depths of ~ 250 km, this result is in good agreement with the previous research [6,7,21]. We also can see that from south to the north, the earthquakes are getting deeper which shown as green and blue circles in Figure 3. Seismicity cluster around the slab of the Philippine (brown line) dips to the west at a depth of ~100 km and is above the Molucca Sea Plate which is subducting to the east.

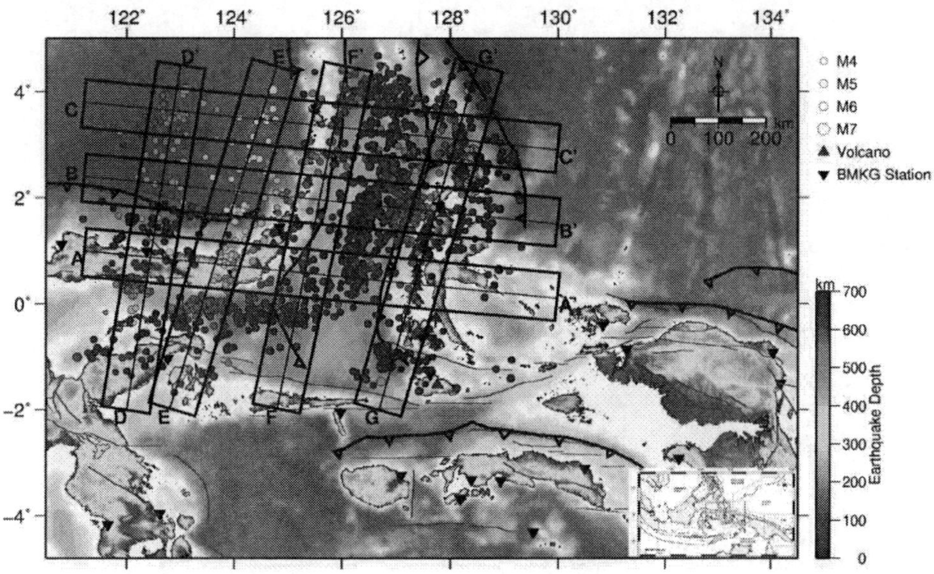

Figure 2. Seismicity maps from the NLLoc results around the Molucca Sea collision zone. Colored dots indicate the earthquakes. Inverted black triangles are BMKG station. The black boxes A-A', B-B', C-C', D-D', E-E ', F-F' are the areas of the cross-section in Figure 3.

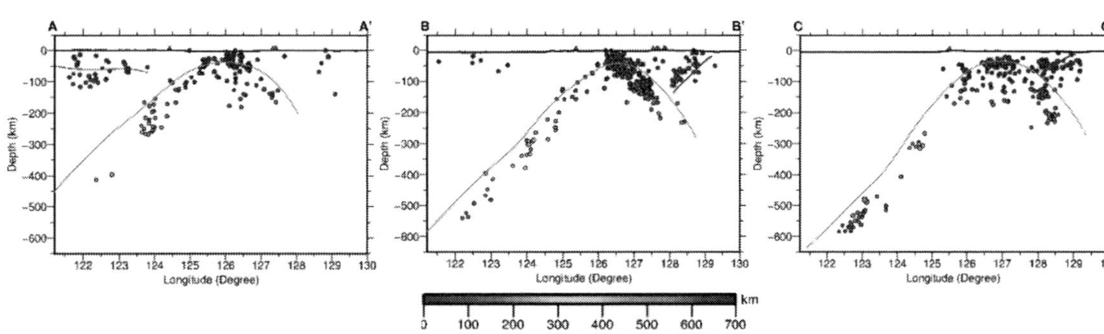

Figure 3. The West-East vertical cross-section of A-A', B-B', C-C' that show the seismicity pattern beneath Molucca Sea and forms a double subduction pattern. Red triangles are the volcanoes, colored dots are earthquakes, the green line describes the Molucca Sea Plate, the blue line describes the slab2 model of Sulawesi, and the brown line describes the slab2 model of Philippine [22].

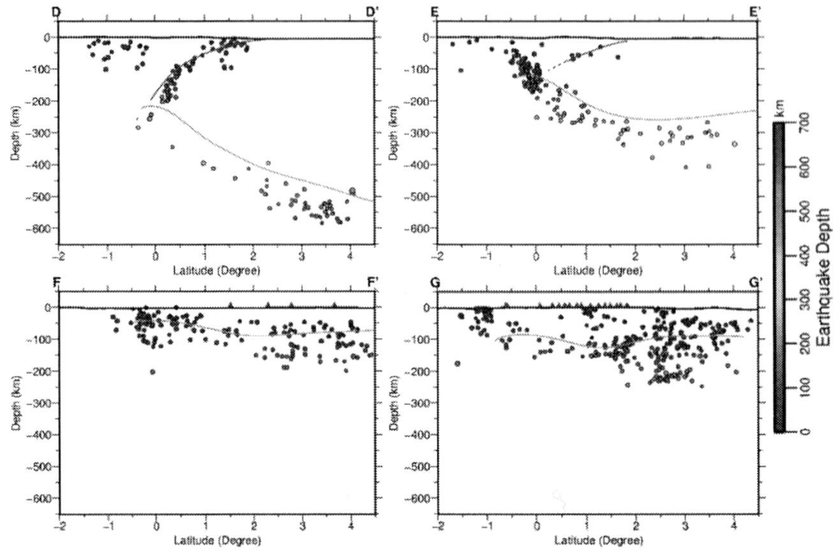

Figure 4. The North-South vertical cross-section of D-D ', E-E', and F-F ' (beneath the Sangihe arc) and G-G' (beneath Halmahera arc). Red triangles are volcanoes, colored dots are earthquakes, the green and blue lines describe the Molucca Sea Plate and slab2 model of Sulawesi, respectively [22].

Based on Figure 4, Profile D-D' shows a deep earthquakes around 300-600 km which indicate the Molucca Sea Plate (green line), meanwhile a south dipping shallower earthquake from ~0-200 km indicate the subduction zone in the north of Sulawesi Island (blue line). The E-E' profile shows the N-S earthquakes pattern in the subduction zone of Molucca Sea Plate, and the slab of Sulawesi shows the same pattern as Profile D-D' only the depth is shallower. Profile F-F' beneath the Sangihe volcanic arc shows the shallow and medium earthquake distribution which shows the earthquake pattern near the top of western flank of double subduction of Molucca Sea Plate. Profile G-G' beneath the Halmahera volcanic arc shows the earthquake distribution in the eastern flank of Molucca Sea Plate.

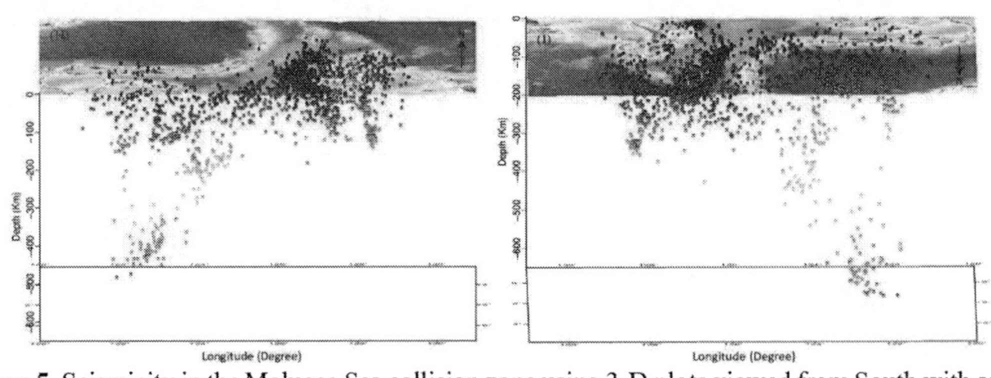

Figure 5. Seismicity in the Molucca Sea collision zone using 3-D plots viewed from South with azimuth 180° (H) and from North with azimuth 0° (I).

Seismicity in the Molucca Sea collision zone using 3-D plots are shown in Figure 5. Based on Figure 5, the seismicity pattern beneath the Sangihe arc forms westward dipping plane associated with Molucca Sea Plate.

4. Conclusions

We have determined the hypocenter location in the Molucca Sea collision zone using a probabilistic non-linear method. We observed that the seismicity follows the pattern of the double subduction of the Molucca Sea Plate, to westward beneath the Sangihe arc which the earthquake could reach a depth of ~600 km and to eastward beneath the Halmahera arc which the earthquake reach a depth of ~250 km, from south to north, the number of deep earthquake beneath the Sangihe Arc is increasing. We also observed that the seismicity following the zone pattern of subduction of Sulawesi and Philippine. This preliminary result will be used for further research of hypocenter relocation and travel time tomographic inversion beneath the Molucca Sea collision zone.

Acknowledgments

We would like to thank the collaboration of BUDI-DN and the Indonesian Endowment Fund for Education (LPDP) which provided scholarship funds to GR. We also would like to thank the Agency for Meteorology, Climatology and Geophysics (BMKG) for providing the waveform data used in this study, the Volcanology and Geothermal Laboratory of ITB, Prof. Anthony Lomax for the NLLoc software [13] and Seisgram2Kv.7.0 [23] used in this research. All figures in this paper were plotted by using the GMT tool [24].

References

[1] Hamilton W B 1979 *Tectonics of the Indonesian region*
[2] Moore G F and Silver E A 1983 Collision processes in the northern Molucca Sea *The Tectonic and Geologic Evolution of Southeast Asian Seas and Islands: Part 2. (1983), Geophys. Monogr. Ser., vol. 27, edited by D. E. Hayes, pp. 360-372, AGU, Washington, D. C.* vol 27 pp 360–72
[3] McCaffrey R 1983 Seismic wave propagation beneath the Molucca Sea arc-arc collision zone, Indonesia *Tectonophysics* **96** 45–57
[4] McCaffrey R, Silver E A and Raitt R W 1980 Crustal structure of the Molucca Sea collision zone, Indonesia (American Geophysical Union (AGU)) pp 161–77
[5] Silver E A and Moore J C 1978 The Molucca Sea Collision Zone, Indonesia *J. Geophys. Res. Solid Earth* **83** 1681–91
[6] Cardwell R K, Isaacks B L and Karig D E 1980 The spatial distribution of earthquakes, focal mechanism solutions, and subducted lithosphere in the Philippine and Northeastern Indonesian Islands *The Tectonic and Geologic Evolution of Southeast Asian Seas and Islands. (1980), Geophys. Monogr. Ser., vol. 23, edited by D. E. Hayes, pp. 1-35, AGU, Washington, D. C.* vol 23 pp 1–35
[7] Hatherton T and Dickinson W R 1969 The relationship between andesitic volcanism and seismicity in Indonesia, the Lesser Antilles, and other island arcs *J. Geophys. Res.* **74** 5301–10
[8] McCaffrey R 1982 Lithospheric deformation within the Molucca Sea arc-arc collision: Evidence from shallow and intermediate earthquake activity *J. Geophys. Res.* **87** 3663
[9] Nugraha A 2014 Teleseismic Double-Difference Earthquake Hypocenter Relocation in the Indonesian Region
[10] Nugraha A, Shiddiqi H, Widiyantoro S, Ramdhan M and Wandono W 2015 Seismicity analysis in Indonesia region from high precision hypocenter location *EGU Gen. Assem. 2015, held 12-17 April. 2015 Vienna, Austria. id.8769* **17**
[11] Setiadi T A P, Rohadi S and Heryandoko N 2017 Earthquake relocation in Mollucas Sea using teleseismic double difference method for tectonic setting analysis *AIP Conference Proceedings* vol 1857 (AIP Publishing LLC) p 020007
[12] Yulianto Y, Nugraha A D and Wiyono W S 2017 Hypocenters relocation using double-difference method around Molucca Collision Zone *AIP Conference Proceedings* vol 1857 (AIP Publishing LLC) p 020002
[13] Lomax A, Virieux J, Volant P and Berge-Thierry C 2000 Probabilistic Earthquake Location in

3D and Layered Models (Springer, Dordrecht) pp 101–34
[14] Moser T J, van Eck T and Nolet G 1992 Hypocenter determination in strongly heterogeneous Earth models using the shortest path method *J. Geophys. Res.* **97** 6563
[15] Tarantola A and Valette B 1982 Inverse Problems = Quest for Information *J. Geophys.* **50** 159–70
[16] Wittlinger G, Herquel G and Nakache T 1993 Earthquake location in strongly heterogeneous media *Geophys. J. Int.* **115** 759–77
[17] Lomax A 2005 A Reanalysis of the Hypocentral Location and Related Observations for the Great 1906 California Earthquake *Bull. Seismol. Soc. Am.* **95** 861–77
[18] Widiyantoro S and Hilst R 1997 Mantle structure beneath Indonesia inferred from high-resolution tomographic imaging *Geophys. J. Int.* **130** 167–82
[19] Havskov J and Ottemöller L (Lars) 2010 *Routine data processing in earthquake seismology : with sample data, exercises and software* (Springer)
[20] Hayes G Slab2 - A Comprehensive Subduction Zone Geometry Model - ScienceBase-Catalog
[21] Nugraha A, Shiddiqi H A, Widiyantoro S, Ramdhan M, Wandono, Sutiyono, Handayani T and Nugroho H 2015 Preliminary results of teleseismic double-difference relocation of earthquakes around Indonesian archipelago region *AIP Conference Proceedings* vol 1658 (AIP Publishing LLC) p 030002
[22] Hayes G P, Moore G L, Portner D E, Hearne M, Flamme H, Furtney M and Smoczyk G M 2018 Slab2, a comprehensive subduction zone geometry model *Science (80-.).* **362** 58–61
[23] Lomax A and Michelini A 2009 Mwpd: A duration-amplitude procedure for rapid determination of earthquake magnitude and tsunamigenic potential from P waveforms *Geophys. J. Int.* **176** 200–14
[24] Wessel P and Smith W H F 1998 New, improved version of generic mapping tools released *Eos, Trans. Am. Geophys. Union* **79** 579–579

IOP Conf. Series: Earth and Environmental Science **873** (2021) 012027 doi:10.1088/1755-1315/873/1/012027

Geomagnetic inverse modeling to estimate the volume of magnetization in areas with topographic variations

Eko J Wahyudi

Institut Teknologi Bandung

ekojw@gf.itb.ac.id

Abstract. A three-dimensional geomagnetic modeling is proposed to obtain information about a magnetization over topographic area from magnetic total force from land data survey. The model is divided into three-dimensional mesh grid of rectangular prisms that cover a finite depth below topographic elevation. The model parameter is assigned to each prism describing magnetization. A set of linear equations was formulated in terms of volume magnetization for each prism. The inverse calculation is obtained using the solution of over-determined problems. In this inverse calculation over three-dimensional mesh grid, the weighted matrix will be used to determine distribution of volume magnetization in the subsurface and keep the zero value of volume magnetization above the surface. A computer program has been tested with artificial topographic data and applied to digital elevation model of Bandung. The result of the inversion shows the comparison between synthetic model and inverse model to provide several analysis of algorithm performance. The high magnetization value can be characterize using inverse model up to 77% of volume area with average success recovery of model parameter value up to 56%.

1. Introduction

Qualitative interpretation of geomagnetic data through inversion modeling is generally used to obtain a general distribution of the magnetic susceptibility of the study area. Several developments for computational inversion modeling with synthetic data simulation as shown by several publications [1, 2, 3, 4] are able to provide a good reconstruction of subsurface conditions. In the early stages of work, this research is part of laboratory work related to geomagnetic data modeling which aims to obtain an estimate of the volume of magnetization distributed in the discretization of rectangular prisms below the surface. Inversion modeling calculations were developed to perform topographic evaluations of the geomagnetic data observed near the surface (ground survey). Furthermore, the focus of this work is to analyze the performance of the algorithm based on the comparison of the synthetic model and the inversion calculation model

2. Modeling Computation

The discretization beneath the surface can be approximated by a fairly simple set of elements in a form that represents the model of a three-dimensional magnetic body [5, 6, 7, 8]. For practical purposes, the subsurface discretization of a body can be divided into an array of small volume elements in the form of a rectangular prism. Each prism is arranged in a direction parallel to the x-axis, y-axis, and z-axis. Each volume element is a dipole having a uniform magnetization value (M) in terms of its prism volume

IOP Conf. Series: Earth and Environmental Science **873** (2021) 012027 doi:10.1088/1755-1315/873/1/012027

$(dV = dxdydz)$. Each volume element can then add up its effects cumulatively to a set of coordinates at the geomagnetic observation point.

The geomagnetic anomaly of a prism also involves a regional field whose direction is assumed parallel to $\hat{\boldsymbol{F}} = (\hat{F_x}, \hat{F_y}, \hat{F_z})$. As described in the textbook [9] magnetization inside a boundary of volume elements $x_1 \leq x \leq x_2, y_1 \leq y \leq y_2, z_1 \leq z \leq \infty$ is as follows:

$$\boldsymbol{M} = M(\hat{\imath}\widehat{M_x} + \hat{\jmath}\widehat{M_y} + \hat{\boldsymbol{k}}\widehat{M_z}). \tag{1}$$

The mbox subroutines [9] provides an explanation related to mathematical equations that can be used in the numerical calculation of the total-field anomaly of a prism with an upper limit at z_1 and a lower limit at ∞. The observed total-field anomaly (ΔT) at the station coordinates (0, 0, 0) is as follows:

$$\begin{aligned}\Delta T = C_m M \Bigg[& \frac{(\widehat{M_y}\widehat{F_z} + \widehat{M_z}\widehat{F_y})}{2} \log\left(\frac{r-x}{r+x}\right) + \frac{(\widehat{M_x}\widehat{F_z} + \widehat{M_z}\widehat{F_x})}{2} \log\left(\frac{r-y}{r+y}\right) \\ & - \frac{(\widehat{M_x}\widehat{F_y} + \widehat{M_y}\widehat{F_x})}{2} \log(r + z_1) - \widehat{M_x}\widehat{F_x} \tan^{-1}\left(\frac{xy}{x^2 + rz_1 + {z_1}^2}\right) \\ & - \widehat{M_y}\widehat{F_y} \tan^{-1}\left(\frac{xy}{r^2 + rz_1 - x^2}\right) + \widehat{M_z}\widehat{F_z} \tan^{-1}\left(\frac{xy}{rz_1}\right) \Bigg] \Bigg|_{x=x_1}^{x=x_2} \Bigg|_{y=y_1}^{y=y_2}, \end{aligned} \tag{2}$$

where $r^2 = x^2 + y^2 + z^2$ and C_m is proportionality constant (in SI units $C_m = \frac{\mu_0}{4\pi} = 10^{-7}$ H/m). The equation (2) can be evaluated two times with:
(a) value of $z_1 = z_a$ as upper limit of the prism and $M = Mo$,
(b) value of $z_1 = z_a + dz = z_b$ as lower limit of the prism and $M = -Mo$.
Based on the principle of superposition (illustration is shown in Figure 1) will produce a dipole field of a prism with magnetization Mo inside the boundary of the volume element $x_1 \leq x \leq x_2, y_1 \leq y \leq y_2, z_a \leq z \leq z_b$.

The form of linear matrix notation from the prism by accommodating Mo and $-Mo$ for a set number of data $(\boldsymbol{d} = [d_i] = [d_1, d_2, ... d_P] = [(\Delta T_{z_a} - \Delta T_{z_b})_i])$ over a set of discretized number of model parameters $(\boldsymbol{m} = [m_j] = [m_1, m_2, ... m_Q] = [(Mo)_j])$ then the system of linear equations from the forward calculation can be arranged in more detail as follows:

$$\begin{pmatrix} (\Delta T_{z_a} - \Delta T_{z_b})_1 \\ (\Delta T_{z_a} - \Delta T_{z_b})_2 \\ \vdots \\ (\Delta T_{z_a} - \Delta T_{z_b})_N \end{pmatrix} = \begin{bmatrix} G_{11} & G_{12} & \cdots & G_{1Q} \\ G_{21} & G_{22} & \cdots & G_{2Q} \\ \vdots & \vdots & \ddots & \vdots \\ G_{P1} & G_{P2} & \cdots & G_{PQ} \end{bmatrix} \begin{pmatrix} (Mo)_1 \\ (Mo)_2 \\ \vdots \\ (Mo)_Q \end{pmatrix}. \tag{3}$$

The model parameters in equation (3) show the corresponding physical properties of the geomagnetic method, namely magnetization (Mo) with uniform value for each element of the prism volume. Based on equations (2) and (3) it shows that the dipole field depends linearly on the value of its physical property (magnetization). A more commonly known physical property (namely magnetic susceptibility) can be derived from the conversion of the magnetization value (Mo) at each discretization position of the model parameter ($j = 1, 2, ... Q$). The more general and simpler form of matrix notation of equation (3) is as follows:

$$\boldsymbol{d} = \boldsymbol{Gm}, \tag{4}$$

where $\boldsymbol{G} = G_{ij}$ is the Kernel matrix with dimension $P \times Q$. Figure 2 shows a plot illustration of the horizontal slice of a synthetic model at a certain depth with a representation of colored pixels to represent variations in the model parameter values.

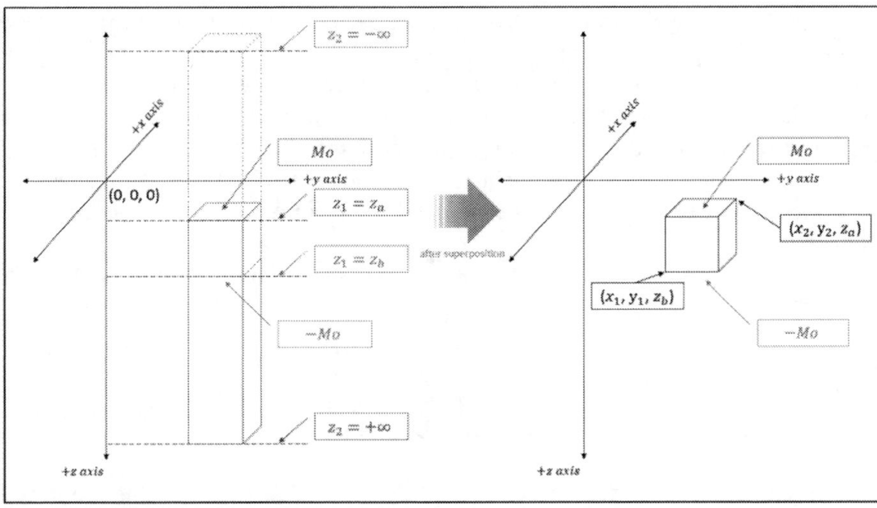

Figure 1. An illustration of the calculation with equation (2) which is evaluated at $z_1 = z_a$ and $z_1 = z_b$ to produce a rectangular prism shaped volume element.

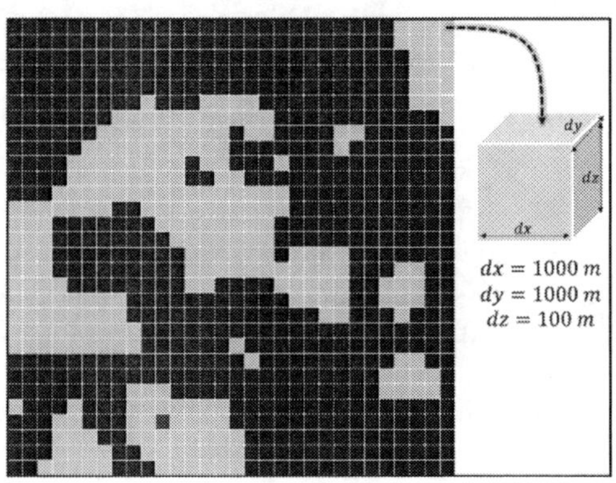

Figure 2. Illustration of the plot (view on the X-Y plane) of the discretization model showing single square (pixel) as a rectangular prism representation of uniform dimensions $(dx \times dy \times dz)$ for the volume of magnetization shown by the color scale.

Generally, the linear inverse problem for equation (4) is solved to get the estimated value of a model parameter ($\widehat{\boldsymbol{m}} = \left[(Mo)_j\right]$) which is divided into several rectangular prisms. Observational data input ($\boldsymbol{d}$) used in the computation for the solution of inversion problem (a minimum-norm solution). The inverse problem is under-determined with the dimensions $P < Q$ or in other words the number of observation data matrix ($\boldsymbol{d}^{obs}$) are smaller than the number of model parameter matrix ($\widehat{\boldsymbol{m}} = \left[(Mo)_j\right]$). The computational scheme in this study also involves the initial information as the initial model in the default form of an uniform model parameter matrix value ($\widehat{\boldsymbol{m}}^{k=0} = 0$). The miss-fit data ($\phi^k$) for k-iteration calculated using *root-mean-square-error* (RMSE) as follows:

$$\phi^k = \left(\frac{\sum_{i=1}^{i=P}\left(d_i^{obs} - G_{ij}\widehat{m}_j^k\right)^2}{P}\right)^{1/2} \quad (5)$$

For observational data measured on a rounded topography, the mesh grid is designed to involve sufficient size with at least the upper boundary of the mesh grid covering the maximum height of the topography in the study area. Topographic evaluation in a mesh grid for discretization of model parameters needs to be involved in the iterative scheme. Each mesh grid coordinate as the center point of the rectangular prism volume in the vertical direction is compared to the coordinates of the observation station which is located on its vertical component. For each mesh grid coordinate whose position is above the coordinates of the observation station on its vertical component, the change in the model solution in the k-iteration ($\Delta \boldsymbol{m}^k$) is given a zero multiplier factor, while for other mesh grid coordinates is given one multiplier. The model parameter values ($\widehat{\boldsymbol{m}}^{k+1}$) which are updated on every k-iteration are evaluated based on the bounds model for the minimum and maximum limits of the magnetization value.

The minimum-norm solution which is updated iteratively in the k-iteration is calculated with the following equations:

$$\widehat{\boldsymbol{m}}^{k+1} = \widehat{\boldsymbol{m}}^k + \Delta \boldsymbol{m}^k, \tag{6}$$

$$\Delta \boldsymbol{m}^k = \boldsymbol{G}^T \boldsymbol{D}[\boldsymbol{D}\boldsymbol{G}\boldsymbol{G}^T \boldsymbol{D} + \varepsilon \boldsymbol{I}]^{-1} \boldsymbol{D}\left(\boldsymbol{d}^{obs} - \boldsymbol{G}\widehat{\boldsymbol{m}}^k\right). \tag{7}$$

Equations (6) and (7) are use superscripts k and T, which describe the k-iteration and transpose matrix respectively. In equation (7) the symbol ε is a damping factor and $\boldsymbol{I}$ is the identity matrix. Mendonca and Silva [10] used a normalized matrix $\boldsymbol{D}$ with diagonal matrix elements as follows:

$$D_{ii} = \left(\sum_{j=1}^{Q} G_{ij}^2\right)^{-1/2}. \tag{8}$$

3. Modeling Results

Inversion algorithm testing is carried out with a synthetic model with a mesh grid size which is a set of rectangular prisms totaling (30 × 30 × 25) with topographic variations using DEM data in the Bandung Basin (www.tides.big.go.id/DEMNAS downloaded at 24 April 2019). Topographic variations that range from 500 to 2000 meters above sea level, are modeled inside the mesh grid in a three-dimensional direction with the upper boundary of the mesh grid at an elevation of 1450 meters above sea level to the lower boundary of the mesh grid at an elevation of 1050 meters below sea level (Figure 3). In this study, calculations with forward modeling were carried out for two simulation schemes of the distribution of the observation points from the ground survey and airborne survey. The position for the plane, which is a simulation of a synthetic airborne path, is carried out at an elevation of 1600 m above sea level (the illustration is shown in Figure 4A). The position of the observation point for the ground survey simulation is close to the topographic elevation (Figure 4B). The synthetic data from the ground survey and airborne survey simulations are shown in Figures 4C and 4D, respectively. Progress of iterative calculation shown Figure 5 with the changes of RMSE.

IOP Conf. Series: Earth and Environmental Science **873** (2021) 012027 doi:10.1088/1755-1315/873/1/012027

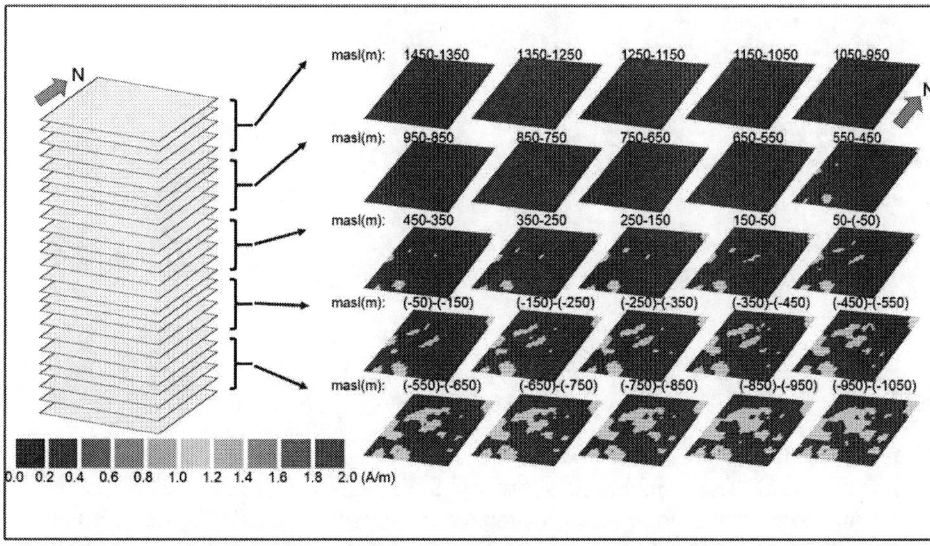

Figure 3. Synthetic model with mesh grid that cover DEM in the Bandung Basin.

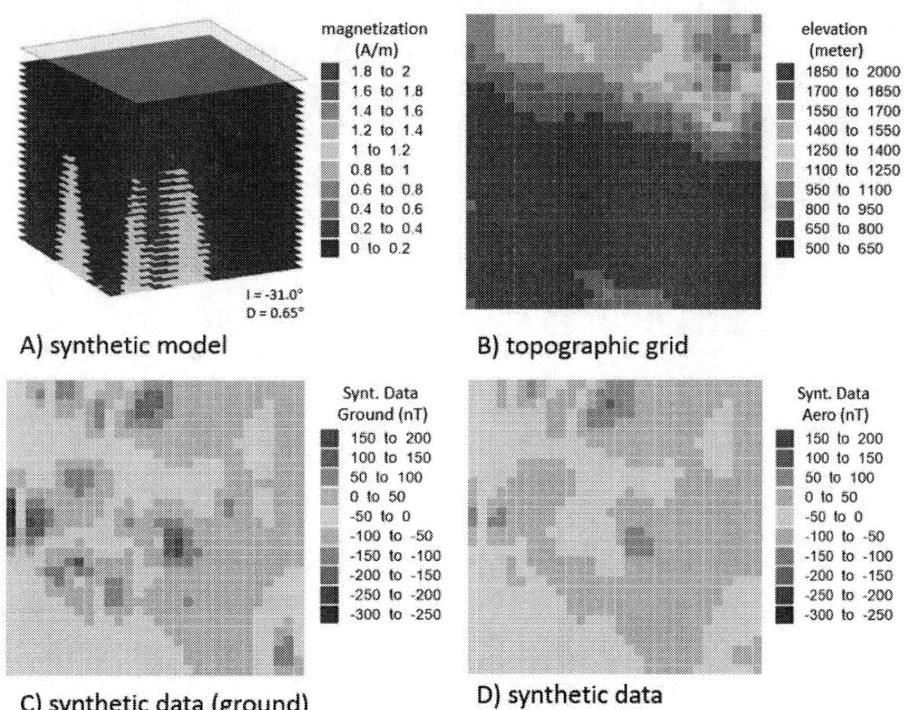

Figure 4. Illustration of design and synthetic data at the forward calculation stage is shown by (A) synthetic model mesh grid illustration, (B) topographic discretization, (C) synthetic data for ground survey simulation, and (D) synthetic data for airborne survey simulation.

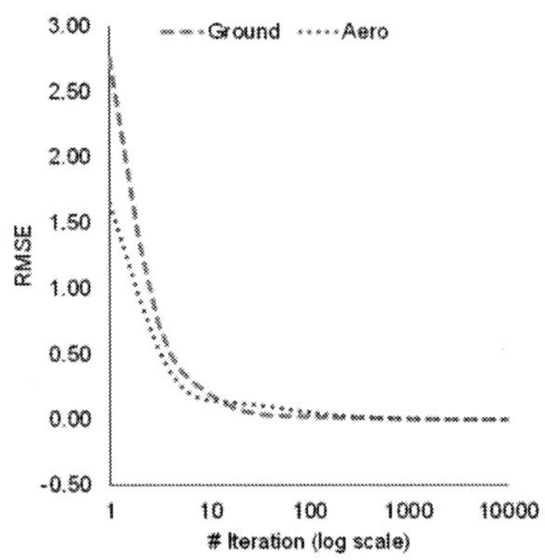

Figure 5. RMSE vs iteration number from two synthetic data (ground and airborne survey).

Comparison model shown in Figure 6 with the example of vertical slice of the mesh grid in X=788000 (South to North section). The topographic shown with black dash-line and the parameter model value of magnetization shown using color scale. Comparison synthetic model (Figure 6A) with the inverse model (Figure 6B, 6C, and 6D) showing an agreement the image of high magnetization body location. The test inverse model (Figure 6B) is compute using free-value of model parameter (bound-less), while the inverse model (Figure 6C and 6D) are compute using boundary of parameter model 0 to 5 A/m. Qualitatively, the capability of the iterative calculation can be used to characterize high magnetization in the subsurface. From this study (using ground and airborne survey simulation) the value of parameter model from the inverse model successfully characterize 77.01% and 63.63% synthetic model (of high magnetization value). Average success of recovery synthetic value of parameter model are 42.18% and 56.01% for ground and airborne survey simulation, respectively.

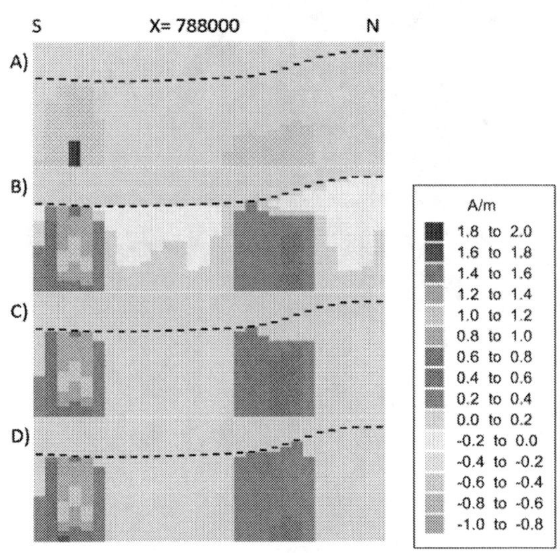

Figure 6. Example of vertical slice (X=788000): A) synthetic model, B) test inverse model, C) inverse model using ground data simulation , and D) inverse model using airborne data simulation.

4. Conclusion

The algorithm of inverse modeling using topographic variation can be used to estimate the solution of three dimensional distribution volume magnetization in the subsurface. From this synthetic study, the numerical example of high magnetization value can be characterize using inverse model up to 77% of volume area with average success recovery of model parameter value up to 56%.

References

[1] Li Y and Oldenburg D W 1996 3-D inversion of magnetic data *Geophysics* **61** 394-408
[2] Lelievre P, Oldenburg D and Williams N 2008 Constraining geophysical inversions with geologic information, *SEG Las Vegas Annual Meeting* 1223-1227
[3] Lelievre P G and Oldenburg D W 2009 A 3D total magnetization inversion applicable when significant complicated remanence is present *Geophysics* **74** L21-L30
[4] Li Y, Shearer S E, Haney M W and Dannemiller N 2010 Comprehensive approaches to 3D inversion of magnetic data affected by remanent magnetization *Geophysics* **75** L1-L11
[5] Wahyudi E J, Santoso D, Kadir W G A and Alawiyah S 2014 Designing a genetic algorithm for efficient calculation in time-lapse gravity inversion *Journol of Engineering and Technological Sciences* **46** 59-79
[6] Wahyudi E J and Kadir W G A 2018 Inverse modeling of gravity data with two layers density in sedimentary basin structure *IOP Conf. Series: Journal of Physics: Conf. Series* **1090** 012048
[7] Wahyudi E J, Santoso D, Kadir W G A, and Aji W 2019 Bouguer anomaly comparison using elevation data measured with hand-held device and geodetic GPS in Blora and Pandan Mountain *IOP Conf. Series: Earth and Environmental Science* **318** 012031
[8] Gunawan I, Wahyudi E J, Alawiyah S, Kadir W G A, and Fauzi U 2020 Annotation of using borehole time-lapse gravity by genetic algorithm inversion for subsurface modeling *Journol of Engineering and Technological Sciences* **52** 153-165
[9] Blakely R J 1996 *Potential Theory In Gravity And Magnetic Applications* (Cambridge: Cambridge University Press)
[10] Mendonca C A and Silva J B C 1995 Interpolation of Potential-Field Data by Equivalent Layer and Minimum Curvature: A Comparative Analysis *Geophysics* **60** 399-407

IOP Conf. Series: Earth and Environmental Science **873** (2021) 012028 doi:10.1088/1755-1315/873/1/012028

Elastic Property Modeling, Extended Elastic Impedance (EEI) and Curve-Pseudo Elastic Impedance (CPEI) Inversion for Pore Type Analysis and Hydrocarbon Distribution in Carbonate Reservoir, Kujung I Formation, "Humaira" Field, North East Java Basin

I N Kumalasari[1*] and I S Winardhi[2]

[1]Sumatera Institute of Technology, Lampung, Indonesia
[2]Exploration and Engineering Seismology Research Group, Faculty of Mining and Petroleum Engineering, Bandung Institute of Technology, Bandung, Indonesia

*Corresponding author's email: istinurkumalasari@gmial.com

Abstract. The complexity of the pore shape in carbonate rocks causes the need for a special strategy to characterize carbonate reservoir. The more information used, the more accurate the reservoir characterization will be. Pore type analysis is the important study because it relates to the fluid flow properties. The elastic property modeling show a good match to the actual data. The results of the well log and petrophysical data analysis show that the gas zone is located at the upper side of Kujung I Formation. Based on rock physics modeling result, the possible pore type developing in the Kujung I Formation is reference pore with the dominance of the aspect ratio value of about 0.17-0.19. The carbonate layer containing hydrocarbons is characterized by low Lamda-Rho, Lamda/Mu values and a low Poisson ratio. Porous carbonate layer, characterized by a low Mu-Rho value. The slice results show that the gaseous area is located on the anticline. The zone that has good porosity indicated by low Mu-Rho. In the IN-3 well there are no hydrocarbons, this analysis is in accordance with the geological condition of the IN-3 well which is in a low area on the time structure map. The inversion results show a good match between CPEI against water saturation log and CPEI against porosity log.

1. Introduction

Apart from sandstone reservoirs, oil and gas production in the world is also largely produced from carbonate reservoirs. The complexity of the pore shape in carbonate rocks causes the need for a special strategy to characterize this reservoir. The accuracy in characterizing the reservoir depends on how much information or data we use. The integration of available data will increase the accuracy of obtaining reservoir character information [1]. The diaganesis process causes heterogeneity in the pores of carbonate rocks. Based on laboratory measurements, variations in pore shape in carbonate rocks cause the relationship between velocity and porosity to become scatter [2]. Pore type analysis is important for well production studies (fluid flow). The aspect ratio is a parameter that represents the pore shape of the rock. The intensive cementation and diagenesis processes cause carbonate rocks to have more complex pore shape variations. In rock physics, the pore types of carbonate rocks are divided into three, namely soft pore (i.e., microporosity and crack), reference pore (i.e., interparticle), and stiff pore (i.e., vuggy and moldic porosity). The rounder the pore shape in the rock, the greater the Vp value in the rock when compared to the flattened pore shape or soft pore [3]. In this study, a modification of the [4] method was used to predict the aspect ratio and the Kuster-Toksöz equation [5] to calculate the effective modulus of rocks. Analysis of rock physics is a step that needs to be continued before carrying out the seismic inversion process. The Extended Elastic Impedance (EEI) method is the development of the concept of Elastic Impedance (EI) proposed by Connolly with a wider angle range (-90 ° to 90 °) [6]. To further improve the results of reservoir character analysis, the Curved Pseudo Elastic Impedance (CPEI) can be used. The CPEI method is the development of the EEI method based on the Rock Physics Template (RPT) with the concept of connecting rock elastic parameter variations with reservoir parameters so that the CPEI approach can improve the results of reservoir characterization in describing fluid distribution [7]. Avseth and Veggeland found that the CPEI attribute correlates well with saturation and resistivity. In this study area. CPEI inversion is necessary, considering this method has never been applied in the research area. The purpose of this study was to determine the pore shape in the study area and to determine the distribution of gas hydrocarbons. By integrating the results of log data analysis, petrophysical data, pore type analysis, EEI and CPEI inversion analysis, will be able to maximize reservoir characterization (pore type and gas distribution) in the research area that reflects the true geological state for further field development analysis.

2. Data and Methodology

2.1. Data
The data used in this study are marker data, check shot data, 3 well log data (IN-1, IN-2, IN-3), horizon data, petrophysical data and partial angle stack seismic data (near, mid, and far).

2.2. Method
In this study, there are 3 main steps, namely elastic property modeling, extended elastic impedance (EEI) inversion, and curve pseudo elastic impedance (CPEI) inversion.

2.2.1. Elastic property modeling
First, data with Vcl more than 0.2 need to be eliminated before data processing is performed. We need to minimize the effect of claystone, so that the modeling focus on carbonate rocks only. In this study, there was no XRD (X-Ray Diffraction Data). So, the mineral modulus for each depth predicted by using Dvorkin and Krief equations. A modified Kumar and Han method, the aspect ratio of soft pore, reference pore, and stiff pore are 0.09, 0.177, and 0.506, respectively. Modulus effective of inclusion rock calculate by using Kuster-Toksöz (K_{sat} and μ_{sat}). The Pmi and Qmi used in this study is penny crack pore geometry. The comparison between the original log data and the predicted result (P-wave and S-wave velocity) need to be done to test the accuracy of the predicted elastic property values used in the modeling are correct. Furthermore, rock physics processing is continued to obtain rock physics template (RPT) to obtain the relative CPEI (wet saturated line in figure 2) equation from the poisson ratio and Lamda-Rho curves. In addition, the trigonometric CPEI equation is used to find the optimum angle which has high correlation with porosity (figure 1).

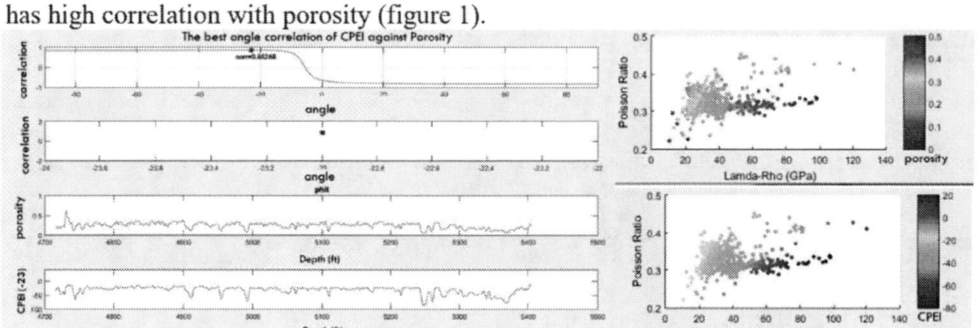

Figure 1. Comparison between CPEI (-23°) against porosity

2.2.2. Hydrocarbon prospect zone and sensitivity analysis

Hydrocarbon prospect zone analysis is done by integrating well and perophysical data. The most sensitive parameter for separating reservoir zones containing hydrocarbons is Lamda-Rho (8-26 GPa) and Poisson Ratio less than 0.3 (unitless). The Mu-Rho parameter is good for separating the lithology of shale, porous carbonate, and tight carbonate.

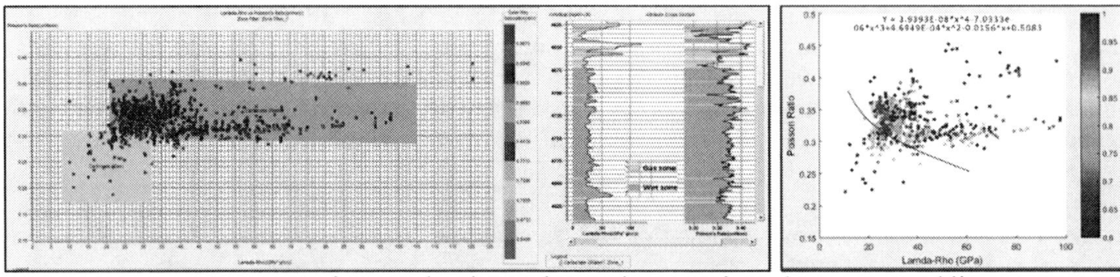

Figure 2. Cross plot Lamda-Rho against Poisson Ratio and wet saturated line.

2.2.3. S-wave velocity prediction

The prediction of the S-wave velocity is done using the neutron porosity log. In the IN-2 well, a cross plot was carried out between the neutron porosity as the x-axis and Mu as the y-axis in the wet zone (because the presence of hydrocarbon makes the NPHI log smaller and does not reflect the actual state of porosity). From the cross plot, the Mu equation is obtained as an NPHI function to predict Mu parameters in IN-1 wells to calculate S-wave.

2.2.4. EEI and CPEI inversion

Based on the cross plot analysis, it was found that the elastic parameters needed in this study were Lamda-Rho and Mu-Rho. The best correlation angle for EEI with Mu-Rho, Lamda-Rho, and Poisson Ratio are 77°, -24°, -45° and -46° respectively. Initial model was carried out using IN-2 and IN-3 wells because both wells had P-wave and S-wave data. Meanwhile, other well data with S-wave obtained from the prediction results is used as a blind test inversion results. The next step is pre-inversion analysis as QC parameters that will be used in the inversion process. After obtaining the Lamda-Rho, Mu-Rho, Lamda /Mu, and Poisson Ratio Inversion Volume from the EEI inversion using model based and sparse spike method and combine the result using frequency domain merge. The next process is to obtain the relative CPEI volume using the wet saturated line and CPEI equations which correlate with porosity. The results of this CPEI volume need to be scaled so that a range of values is equivalent to the value range with saturation and porosity.

3. Result and Discussion

3.1. Well log and petrophysical data analysis

The results of the well log and petrophysical data analysis show that the gas zone is located at the upper side of Kujung I Formation (Figure 3). This result is in accordance with the well report data that there is a gas show of IN-1 and IN-2 wells at the top of the Kujung I Formation. Meanwhile, in the IN-1 well there is no gas zone found. This is supported by the location of the IN-3 well which is in the low area on the time structure map. Table I is the result of gas zone analysis from log and petrophysical data.

Table 1. Target zone analysis based on well log and petrophysical data

Well no	Gas zone depth	GR	*Vclay*	NPHI and RHOB curves	Mean total porosity	Water saturation
IN-1	4635-4714 (ft)	Low	0.22	*Crossover*	0.27	0.52-0.90
IN-2	4617-4684 (ft)	Low	0.25	*Crossover*	0.28	0.24-0.86

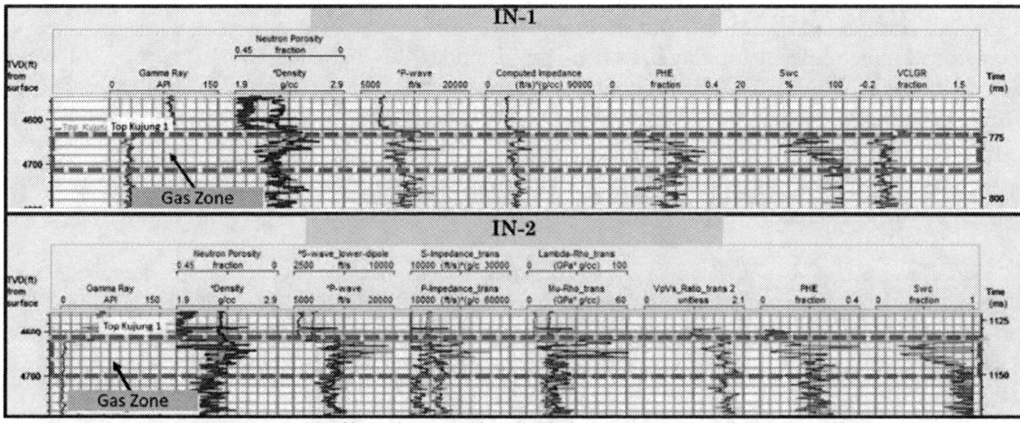

Figure 3. Well and petrophysical data at IN-1 and IN-2 well

3.2. Pore type analysis

Figure 5 shows the result of elastic property modeling. Modeling results show a good match against the original data. The elastic property modeling results indicate that the possible pore type developing in the Kujung I Formation is reference pore with the dominance of the aspect ratio value of about 0.17-0.19 (figure 4). Carbonate rocks with reference and soft pore types have effective porosity and good permeability [8]. However, if it has a lot of clay content, the effective porosity decreases because the clay blocks the pores. Meanwhile, carbonate rocks with stiff and reference pore types have lower effective and permeability, because the vuggy or moldic porosity types are not connected to each other.

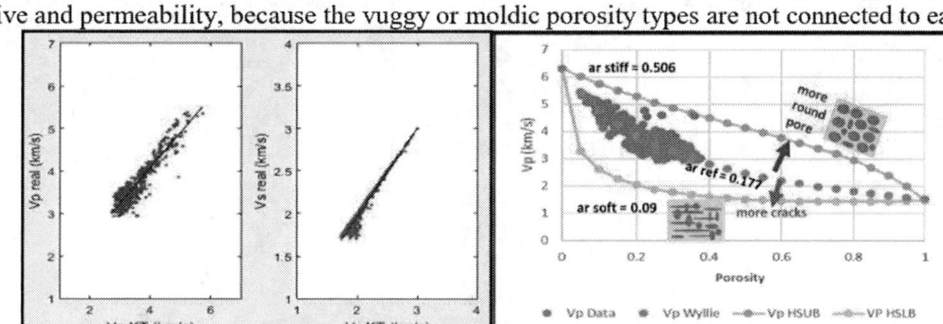

		Correlation	RMSE
Kujung I Formation	Vp	0.9269	0.0719
	Vs	0.9803	0.0342

Figure 4. The result of property elastic modelling (Kumar and Han (2005) for predicting aspect ratio and Kuster-Toksöz for effective modulus calculation).

Figure 5 shows that the gas-saturated prospect zone is located at a depth of 4617-4684 ft has a higher aspect ratio (in some location has a moderate aspect ratio). The bigger the aspect ratio, the rounder the pore shape. This means that, perhaps at the upper side of formation has a more rounded pore shape or stiff pore such as vuggy or moldic porosity.

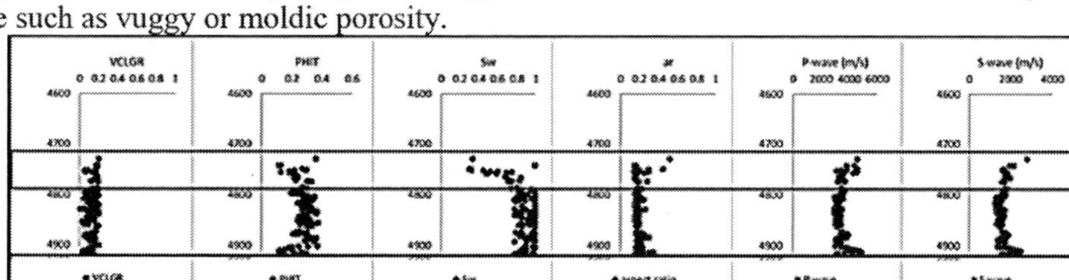

Figure 5. Display of well log data, petrophysical data, and aspect ratio modeling (prospect zone).

3.3. Porous carbonate and gas distribution analysis

The best correlation angle EEI log against Mu-Rho, Lamda-Rho, and Lamda/Mu are 77°, -24°, and -45° respectively. The results of the EEI inversion show a match to the well log data. In Figure 6, it can be seen that the hydrocarbon zone is located at the top of Top Kujung 1 which is marked by low Lamda-Rho inversion results (20-28 GPa), low Lamda/Mu (1.19-2.5) and low poisson ratio (less than 0.3) . The carbonate layer contains gas with good porosity located in the high area on the time structure map (Figure 7).

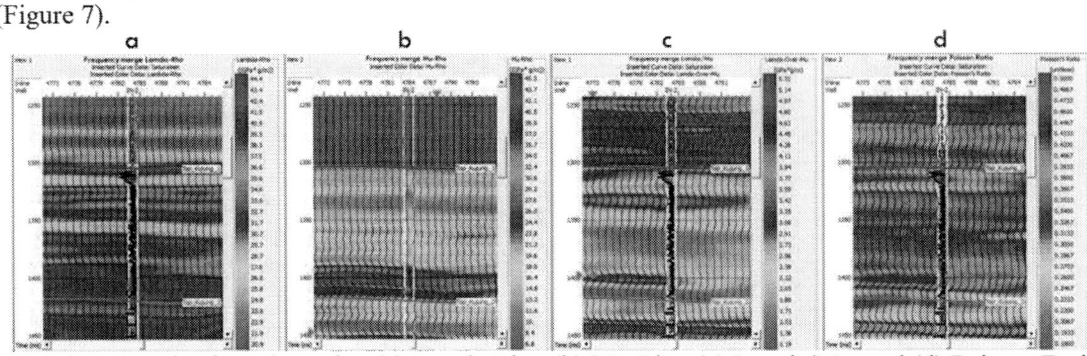

Figure 6. Result of EEI inversion (a) Lamda,Rho, (b) Mu-Rho, (c) Lamda/Mu, and (d) Poisson Ratio

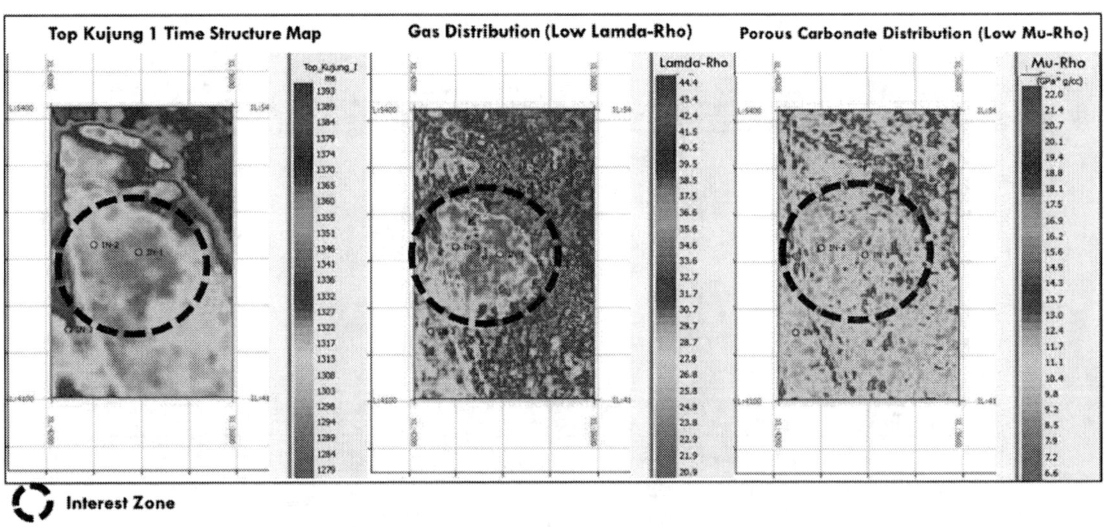

Figure 7. Distribution of gas and porous carbonate from EEI inversion result

Based on the trigonometric CPEI equation, the best correlation angle to porosity is -23° with a correlation value of 0.85. Meanwhile, the gas availability is searched by the CPEI relative equation (wet saturated line). The results of the CPEI scaling process can be seen in Figure 8 which shows a good match to the saturation and porosity log data. The gaseous layer is located below the Top Kujung 1 horizon with a thickness of 10-15 ms at the IN-2 well. In the IN-3 well there are no hydrocarbons with high CPEI results, this analysis is in accordance with the geological condition of the IN-3 well which is in a low area on the time structure map.

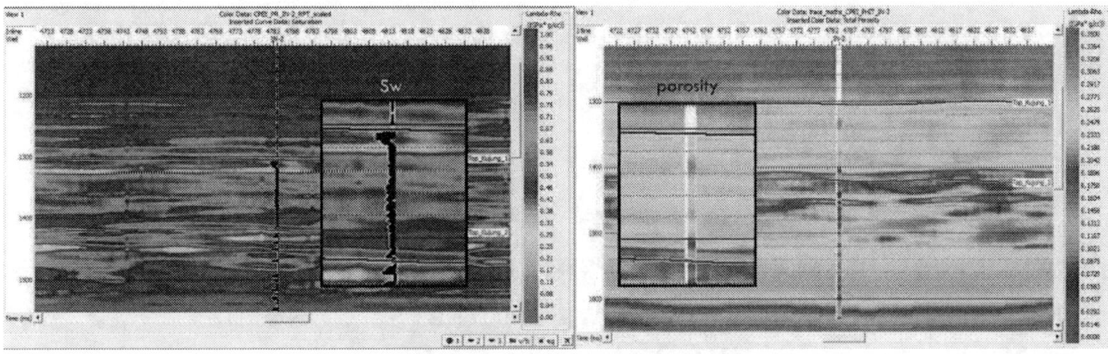

Figure 8. CPEI Inversion Result

4. Conclusion

Fluid flow properties control by pore type in carbonate, so it is important to do a study on pore type analysis. In this study, the elastic properties modeling show a good match to the actual data. The possible pore type developing in the Kujung I Formation is reference pore. The gas-saturated prospect zone has a higher aspect ratio (in some points it has a moderate aspect ratio). The bigger the aspect ratio, the rounder the pore shape. This means that, perhaps at the upper side of formation has a more rounded pore shape or stiff pore such as vuggy or moldic porosity. The carbonate layer containing hydrocarbons is characterized by low Lamda-Rho, Lamda / Mu values and a low Poisson ratio. Porous carbonate layer, characterized by a low Mu-Rho value. The slice results show that the gaseous area is located on the anticline (on the time structure map). The zone has good porosity, which is indicated by low Mu-Rho. The inversion results show a good match between CPEI against Sw and CPEI on porosity.

Acknowledgments
The author would like to thank all those who have helped towards the entire process of drafting and publishing this paper.

References

[1] Fanchi J R 2002 Shared Earth Modelling Butterworth-Heinemann Oxford, England 170-181
[2] Eberli G P Baechle G T Anselmetti F S, and Incze M L 2003 Factors controlling elastic properties in carbonate sediments and rocks The Leading Edge **22** 654-660
[3] Xu S and Payne MA 2009 Modelling Elastic Properties in Carbonate Rocks The Leading Edge **28** 66-74
[4] Kumar M and Han D 2005 Pore shape effect on elastic properties of carbonate rock Society of Exploration Geophysicists Expanded Abstracts 1477-1480
[5] Kuster G T and Toksoz M N 1974 Velocity and attenuation of seismic waves in two-phase media Geophysics **39** 587–618
[6] Whitecombe D A Connolly P A Reagan R and Redshaw T C 2002 Extended elastic impedance for fluid and lithology prediction Geophysics **67** 63–67
[7] Avseth P Veggeland T and Horn F 2014 Seismic screening for hydrocarbon prospects using rock-physics attributes The Leading Edge **33** 266–274
[8] Prananda A Rosid M S and Widodo R W 2020 Pore type characterization of Kais Carbonate Reservoir using differential effective medium *IOP Conference Series Earth and Environtmental Science* **538** 1-8

IOP Conf. Series: Earth and Environmental Science **873** (2021) 012029 doi:10.1088/1755-1315/873/1/012029

Analysis of Source Mechanism and Coulomb Stress Change (Case Study: Molucca Sea Earthquakes November 15th, 2014 and November 14th, 2019)

Indra Josua Purba[1], Iman Suardi[1], Gatut Daniarsyad[2], Defni Lasmita[1]

[1]School of Meteorology Climatology and Geophysics, Tangerang Selatan, Indonesia
[2]Meteorology Climatology and Geophysical Agency of Indonesia (BMKG), Jakarta, Indonesia

indra.purba@stmkg.ac.id

Abstract. On November 15, 2014, and November 14, 2019, two major earthquakes occurred in the Molucca Sea with a moment magnitude of Mw 7.0 and Mw 7.1, respectively. These earthquakes were caused by the convergence activity between the Sunda Plate and the Philippine Sea Plate which form a double subduction zone in the Molucca Sea. We carried out the moment tensor inversion using Kiwi Tools to analyze the source mechanism for both of the earthquakes. The results show a thrust fault mechanism with the strike, dip, and rake of the ruptured fault planes are 187°, 63°, 85° and 196°, 43°, 83°, for the first and second events, respectively. We refine the location of the two mainshocks and their aftershocks by performing hypocenter relocation using the double difference method. This resulted in NE-SW aftershocks distribution for both events which occured close to the Molucca Sea Plate boundaries with the mainshocks location are relatively close to each other (± 50.32 km). Finally, we calculate the Coulomb stress changes to analyze the triggering effect between the two major events and between the mainshock and its aftershocks for each event. The results show that the hypocenter of the November 14, 2019 earthquake is in the increased zone of Coulomb stress changes produced by the November 15, 2014 earthquake with the value of 1.2 bar. The aftershocks of both events also occurred in the increased Coulomb stress changes with the range value of 0.5 - 1.8 bar for the first event and 0.2 - 0.8 bar for the second event.

1. Introduction

Molucca Sea is an area with active seismicity due to the historical complex collision that include the Eurasian Plate, Indo-Australian Plate, Pacific Plate and Philippine Sea Plate [1]. The collision formed the Sangihe Arc on the west and Halmahera Arc on the east, and enforce the Molucca Sea Plate subduct beneath the Sangihe and Halmahera arc which form a double subduction. The Meteorology Climatology and Geophysical Agency of Indonesia (BMKG) has recorded 13 earthquakes with magnitude of M>6 in the Molucca Sea from 2014 to 2019. From this series of earthquakes, there was two M>7 earthquakes that occurred in a relatively close distance between each other. Based on BMKG catalog the earthquakes was occurred on November 15, 2014 02:31:41 UTC with moment magnitude of Mw 7.0 and on November 14, 2019 16:17:40 UTC with moment magnitude of Mw 7.1.

The purpose of this study is to enhance our understanding of the seismotectonic characteristics of the Molucca Sea region by identifying the source mechanism of these two large earthquakes and analyzing its Coulomb stress changes. A detailed analysis is carried out by relocating the mainshocks and aftershocks hypocenter using double difference method, obtaining the focal mechanism using moment tensor inversion on regional broadband waveform, and calculating the static Coulomb stress changes for both of Mw>7 earthquakes.

2. Data and method

It is necessary to an accurate parameter of the earthquakes to provide a reliable information for better analysis. Therefore, we first relocate the hypocenter of the mainshock and the aftershocks using the double difference method [2]. The method assumes that two earthquakes which has hypocentral distance smaller than its distance to the station will have a same ray path. The relocation process is done by minimizing the time residual between the observation and calculation data.

$$dr_{ck}^{ij} = (t_k^i - t_k^j)^{obs} - (t_k^i - t_k^j)^{cal} \quad (1)$$

where *i* and *j* show the index of two adjacent hypocenter and k is the index of the station. We used the arrival time data of P and S waves obtained from BMKG catalog for mainshocks and aftershocks of the November 15, 2014 event and the November 14, 2019 event. We use the aftershocks for both events in the period of 30 days after the mainshock. The range of the study area is 1.14° - 2.26°N and 126.14° - 126.75°W (Figure 1).

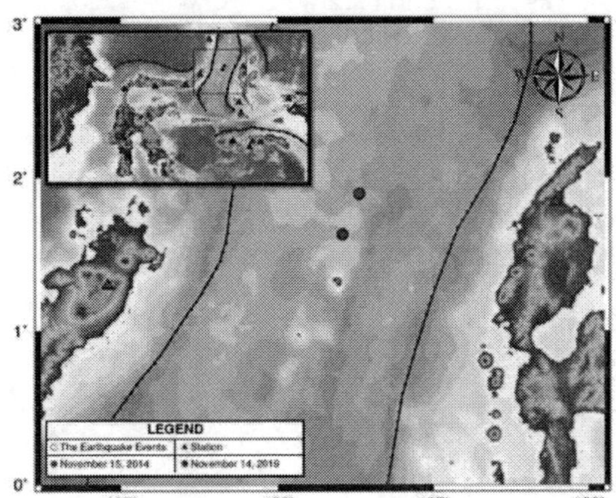

Figure 1. Map view of study area

Focal mechanism is estimated by performing the moment tensor inversion to the regional broadband waveform. The focal mechanism is used to determine the fault plane parameters, those are *strike* (Φ_f), *dip* (δ), *slip* along the fault plane, and rake (λ). Moment tensor inversion solves the equation of the moment tensor which can be represented as follows:

$$u_n(x,t) = M_{pq}(\xi,t) * \frac{\partial}{\partial \xi} G_{np}(x,\xi,t)|_{\xi 0} = M_{pq}(\xi,t) * G_{np,q}(x,\xi,t) \quad (2)$$

where u_n is the component of soil displacement at a point *x* (station). M is the moment tensor of point source at location ξ (source), and G_{np} is the spatial derivative to ξ_k is the component of Green's function (Green's tensor) [3]. We use the Kiwi Tools that resolve not only the focal mechanism but also a simplified finite source model which is called as eikonal source model. The inversion is following the procedure in [4] which consist of 3 steps; the first step uses the amplitude spectra inversion to obtain the focal mechanism parameters (without polarity), the second step uses the waveform inversion in time domain to provide the polarity and determine the compression and dilatation quadrant, and the final step uses the amplitude spectra inversion to obtain the eikonal source model that provide the rupture area, rupture duration, and the rupture propagation direction.

The focal mechanism solution obtained from the last step is then used to calculate the Coulomb stress changes. Characterization using the Coulomb failure criteria [5] has been used to explain the occurrence of failure at the rocks. When earthquake is generated by a fault, this fracture will encourage the changes in stress on the fracture around the fault. In the Coulomb criteria, a fault occurs in a plane when the Coulomb stress exceeds a certain value.

$$\Delta CFF = \tau_\beta - \mu(\sigma_\beta - p) \quad (3)$$

where τ_β is the shear stress in the fault plane, σ_β is the normal stress, p is the pore pressure of the fluid, and μ is coefficient of friction. The value of τ_β in this equation must be positive, but the process of calculating the voltages in an area can be positive or negative depending on the potential slip. By assuming simple Coulomb friction models for the earthquake, slip potential will increase or decrease in Coulomb failure stress [6].

3. Result and discussion

3.1. Hypocenter relocation

We have successfully relocated 166 and 284 aftershocks for the first and second event, respectively. The relocation results for two earthquakes are shown in Figure 2. The mainshock of two earthquakes are located in relatively close distance (± 50.32 km). The aftershocks are mostly distributed at the shallow depth (< 35 km). The aftershocks distribution shows an SSW-NNE oriented trend which indicate that the strike orientation of the fault. In the vertical distribution also indicate a thrust fault trend with the dip is facing to the west direction for both events. This result corresponds well with previous study [7] about the existence of a splay fault that generated the coseismic of November 15, 2014 earthquake.

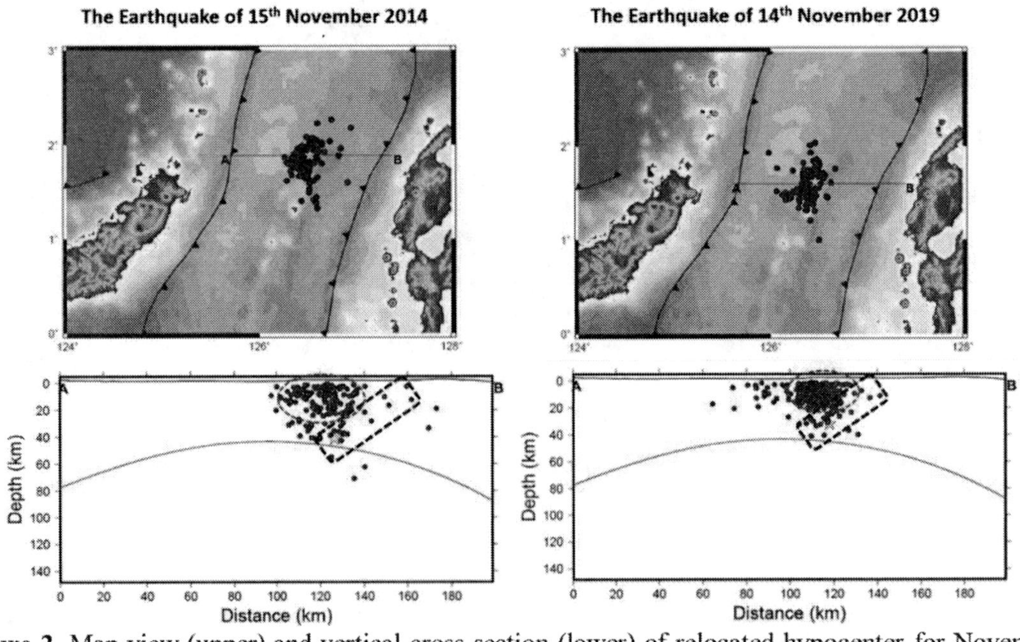

Figure 2. Map view (upper) and vertical cross section (lower) of relocated hypocenter for November 15, 2014 (left) and November 14, 2019 (right). Yellow stars represent the mainshock, red circles show the aftershocks. Curved red lines show the Molucca Sea plate that subducted beneath Sangihe arc (west) and Halmahera arc (east). Black dashed rectangles indicate the suggested rupture plane.

3.2. Focal mechanism

We use unrotated data of the three-component seismogram from the broadband seismic stations with a distances 10 degrees within the epicenter. The inversion is conducted in the frequency range of 0.01-0.04 Hz with a time window of 60 s for the first and second step, refer to [8] for focal mechanism inversion. The amplitude spectra and displacement waveform fit are shown in Figure 3.

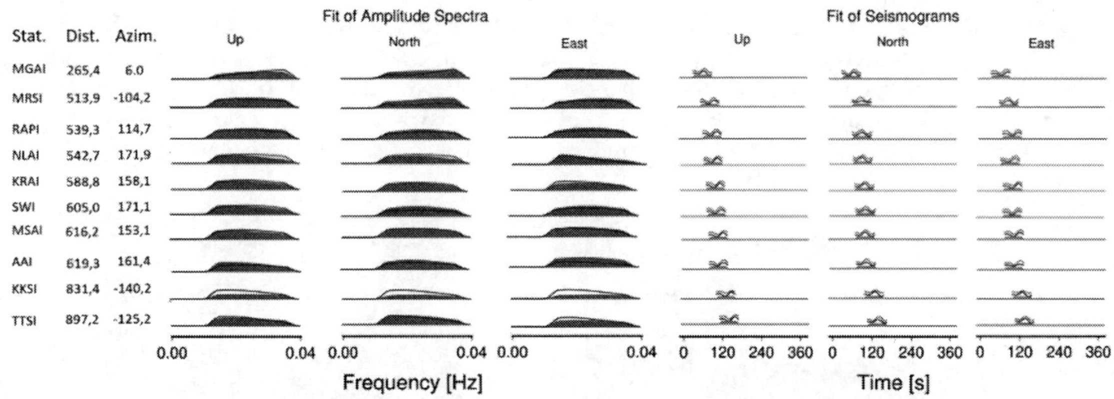

Figure 3. Example of the fitting on amplitude spectra (left) and displacement waveform (right) for the November 14, 2019 event. Red area and lines represent the observed while black area and lines represent the synthetics. Station name with each distance and azimuth respect to the epicenter are on the left side.

The final step is conducted in the frequency domain with a higher frequency range (0.01 – 0.09 Hz) to obtain the earthquake rupture parameters. The inversion result for both Mw>7 mainshocks are presented in Figure 4. The results show that the rupture for both earthquakes were occurred in the west-dipping fault plane with the strike, dip, rake are 187°, 63°, 85° and 196°, 43°, 83° for the first and second event, respectively. Both events also have similar rupture area (30 × 30 km) with the centroid depth is relatively close (39.9 km and 37.4 km). A slight difference for both events is on the rupture propagation. The rupture for the first event propagates predominantly downward along the fault plane or to the west direction, while the second event has an oblique rupture propagation toward the southwest direction. Based on the source mechanism solution and the relocated hypocenter distribution, the November 15, 2014 and the November 14, 2019 events were originated from the same fault with a similar thrust fault mechanism.

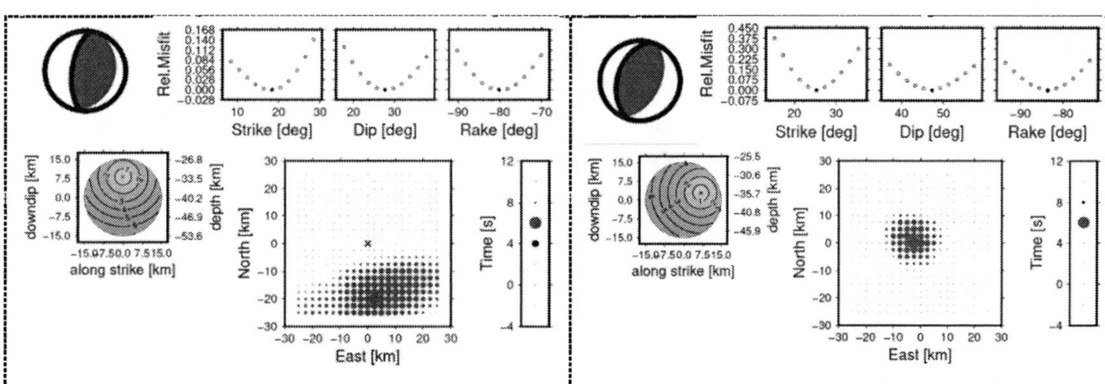

Figure 4. Results of inversion on step 3 in the event on November 15, 2014 (left) and November 14, 2019(right) that shows actual fault plane (strike,dip,rake), rupture propagation, focus mechanism with polaritas, rupture time, rupture area, mean of slip.

3.3. Coulomb stress changes

The Coulomb stress change for the November 15, 2014 and November 14, 2019 events was calculated by using Coulomb3.3 program. In its processing, we used the equation by input magnitude and orientation of the fault (strike, dip, and rake) [9], so the result shows the fault area is 48.37 × 22.32 Km. Then, the coefficient of friction area of subduction (thrust faults) has high coefficient, that is 0.8 [10]. From the results showing that change of stress in aftershocks that occurred on November 15, 2014, we know that the range of Coulomb stress is from 0.5 to 1.8 bar and on November 14, 2019 the range of Coulomb stress is from 0.2 to 0.8 bar (Figure 5). The result of calculation, we also know the Coulomb stress change during coseismic on November 15, 2014 affected the location of the hypocenter on November 14, 2019, which Coulomb stress is 1.2 bar (Figure 6). That means we can conclude that the earthquake on November 15, 2014 gave the increase in stress at the hypocenter location of the earthquake on November 14, 2019.

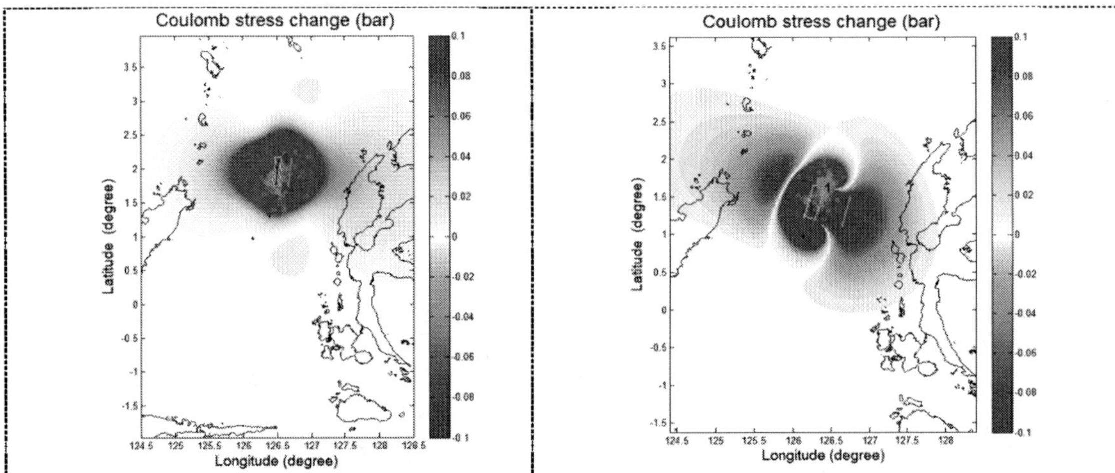

Figure 5. The distribution of Coulomb stress changes on November 15, 2014 (left) and on November 14, 2019 (right) which is seen horizontally at the depth of the fault plane. The red dots are the distribution of the aftershocks, the green line is the projection of the fault trail horizontally on the surface, and the red box in the middle of the blue zone is the area of fault rupture.

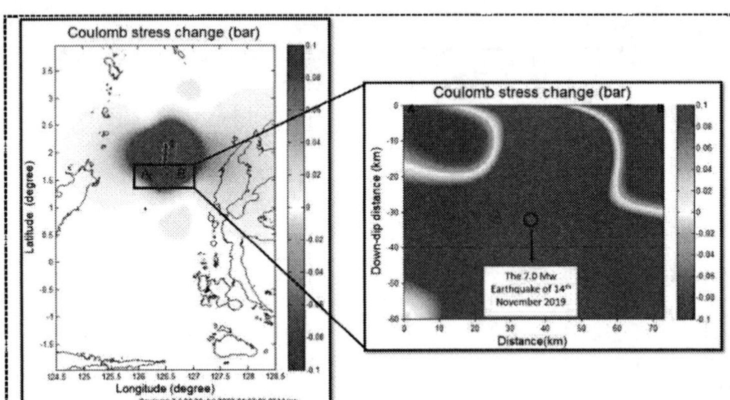

Figure 6. The figure on right side is result of slice from line AB in the figure on left. East-West slice Coulomb stress changes from the event on November 15, 2014 for the event on November 14, 2019. Red area is positive stress and blue area is negative stress.

4. Conclusions

Based on the analysis that we have discussed, aftershocks from the two mainshocks have a similar pattern which shows a SW-NE orientation. The two mainshocks hypocenter also obtained in relatively close distance between each other (± 50.32 km). The November 15, 2014 and November 14, 2019 earthquakes were originated from the same fault and with a similar thrust fault source mechanism. The rupture area for both events are 30 × 30 km with a predominantly downward rupture propagation for the first event and an oblique downward rupture propagation for the second event. The Coulomb failure stress (CFS) changes analysis shows that the aftershocks of both events were occurred in the positive CFS area and the November 14, 2019 event has been triggered by the November 15, 2014 event with a Coulomb stress changes value about 1.2 bar

Acknowledgments

Authors are so thankful to Meteorology Climatology and Geophysical Agency of Indonesia (BMKG) for providing the events catalog data. We also thank to Mr. Puji Ariyanto and Mr. Wandono for all suggestions and discussions in this paper.

References

[1] Moore G F, Silver E A 1983 *The Tectonic and Geologic Evolution of Southeast Asian Seas and Islands* **27(Part 2)** 360-372.
[2] Waldhauser F, Ellsworth W L 2000 *Bulletin of the Seismological Society of America* **90(6)** 1353-1368.
[3] Lay T, Wallace T C 1999 Elsevier.
[4] Cesca S, Heimann S, Stammler K, and Dahm, T 2010 Journal of Geophysical Research 115.
[5] King G C, Stein R S, and Lin J 1994 *Bulletin of the Seismological Society of America* **84(3)** 935-953.
[6] Toda S, Stein R S, Richard-Dinger K, and Bozkurt S B 2005 *Journal of Geophysical Research: Solid Earth* **110(B5)**.
[7] Gunawan E, Kholil M, and Meilano I 2016 *Physics of the Earth and Planetary Interiors* **259** 29-33.
[8] Suardi I, Afnimar, Widiyantoro S, Yagi Y 2014 International Journal of Tomography and Simulation **25** 63-74.
[9] Toda S, Stein R S, Sevilgen V, and Lin J 2011 *US Geological Survey open-file report* **1060** 63.
[10] Wells D L, Coppersmith K J 1994 *Bulletin of the seismological Society of America* **84(4)** 974-1002.4.

Indonesian Earth's Lithospheric Magnetic Field modelling using Spherical Cap Harmonic Analysis Method

Ilham,[1,a)] M Syirojudin,[2] R Margiono,[1,2] A Marsono,[1,2] N Ardiana,[1]

[1]College of *Meteorological Climatology and Geophysical*
[2]*Meteorology Climatology and Geophysical* Agency

[a)]ilhammaret1998@gmail.com

Abstract. The earth's lithospheric magnetic field is part of the main earth's magnetic field. The lithospheric field has a very small value compared to the Earth's main magnetic field, approximately less than 1%, and this field is generated at the earth's crust and upper mantle. Modelling of lithospheric field is useful mainly for predicting the distribution of the value of lithospheric fields and to determine the magnetic anomaly. In this research, modelling the Earth's lithospheric magnetic field uses Spherical Cap Harmonic Analysis (SCHA) method and this method can do modelling using regional magnetic data. The data used for the modelling are magnetic repeat station data in Indonesia region (BMKG's Epoch) and SWARM satellite data. The results of the modelling using integrated SWARM satellite and repeat station data produce RMSE values of 64.0834 nT and the expansion of index K is 70. In addition, the results of the modelling resolution is 1.5^0. The value's range of modelling's result are -987.192 – 998.239 nT for X component, -968.189 – 949.438 nT for Y component, -981.266 – 608.676 nT for Z component, and -904.151 – 997.389 nT for total intensity are.

Keywords: Lithosphere, SCHA, Modelling

1. Introduction

The lithospheric field is a part of the geomagnetic field that has small value comparing to the main field, about 1% of the main field [1]. This field are generated at the earth's crust and the upper mantle. In order to know the value of the lithospheric field over a region, some modelling method are used. One of the modelling method used for modelling lithospheric field in a region is Spherical Cap Harmonic Analysis (SCHA) [2]. There are two types of data that used as input for this model. The first one is repeat station data measured by BMKG (Meteorological Climatological and Geophysics Agency of Indonesia) and this type of data is used before for mapping the main field [3]. The second one is SWARM satellite data [4].In order to knowing the local anomaly magnetic field cover some regions, the lithospheric field can be modelled. The local anomaly is caused by the rock and mineral that available there. Furthermore, the result of modelling can interpret the geological feature in the subject region [4]. By knowing the local anomaly in the subject place, the result can be used for some purposes (e.g. prospecting, determination of the kiblah direction).

2. Data and methods

This study uses data from the magnetic repeat station measurement which has been used for mapping the Indonesian main magnetic field for the year of 2015 [3], and SWARM satellite data in the same year. Based on [3], repeat stations are measured at the place that relatively free from geomagnetic disturbance. There are 68 stations all over Indonesian region that can be seen in Figure 1.

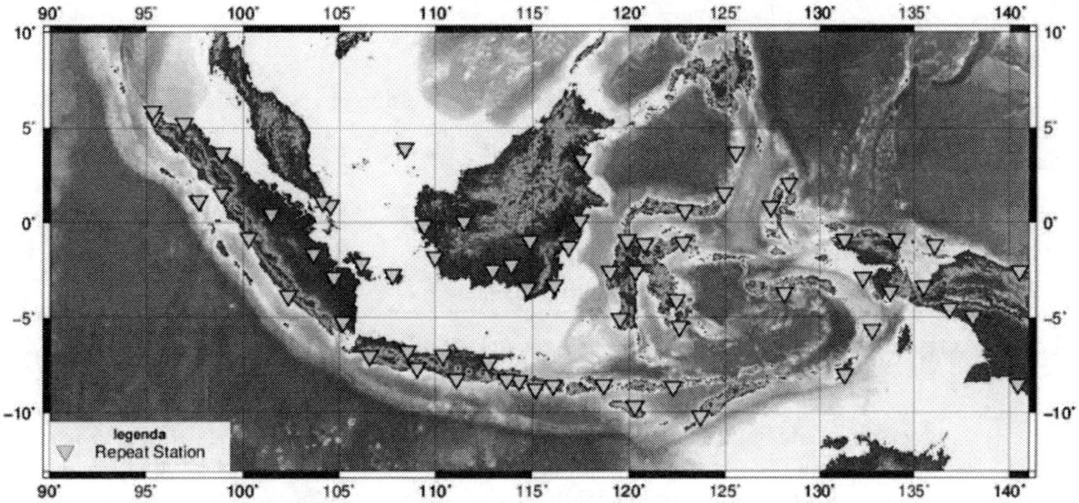

Figure 1. Location of repeat station in Indonesia

The repeat station's data are relatively free from magnetic field disturbance, while the satellite data are not. The satellite data are easily affected by the external source of geomagnetic field such as solar wind [5]. In order to minimize the effect of external disturbance, the data are selected based on Dst index and Kp index [4]. Based on [4] and [6], data selected with respect to $K_p \leq 2$, $|D_{st}| \leq 50$ nT, and $|dD_{st}/dt| = 3$ nT.

The lithospheric field of each repeat stations must be known by reducing the measurement result by the main field. In this study, BMKG's magnetic calculator data [3] are used as the main field. This process mathematically described by

$$\partial \boldsymbol{B} = \boldsymbol{B}_m - \boldsymbol{B}_c \tag{1}$$

Where $\partial \boldsymbol{B}$ is the residual value or lithospheric field for each stations, $\boldsymbol{B}_m$ is the measurement result, and $\boldsymbol{B}_c$ is the main field at the same position and time as the measurement point.

Spherical Cap Harmonic Analysis (SCHA) is method for modelling earth's magnetic field in regional scale. This method introduced by Haines [2] by solving Laplace equation in spherical coordinate. The Equations stated in [1], described by

$$\Delta X = \sum_{k=1}^{Kmax} \sum_{m=0}^{k} \left(\frac{a}{r}\right)^{n_k+2} (g_k^m \cos m\Lambda + h_k^m \sin m\Lambda) \frac{dP_{n_k}^m(\cos\Theta)}{d\Theta} \tag{2}$$

$$\Delta Y = \sum_{k=1}^{Kmax} \sum_{m=1}^{k} \frac{m}{\sin\Theta} \left(\frac{a}{r}\right)^{n_k+2} (g_k^m \cos m\Lambda - h_k^m \sin m\Lambda)\, P_{n_k}^m(\cos\Theta) \tag{3}$$

$$\Delta Z = \sum_{k=0}^{Kmax} \sum_{m=0}^{k} (n_k + 1)\left(\frac{a}{r}\right)^{n_k+2} (g_k^m \cos m\Lambda + h_k^m \sin m\Lambda)\, P_{n_k}^m(\cos\Theta) \tag{4}$$

$$n_k = \left(2k + \frac{1}{2}\right)\frac{\pi}{2\theta_0} - \frac{1}{2} \quad (5)$$

Where a is the earth mean radius (6371.2 Km), Θ is colatitude, Λ is longitude, θ_0 is the regional radius, r is the radius of station to the earth's centre, g_k^m and h_k^m are model's parameter determinated through inversion. Then, the resolution of model's result is determinated by

$$l = \frac{360^0}{\left(\frac{(K_{max} + 0.5)90^0}{\theta_0}\right) - 0.5} \quad (6)$$

After the model's parameter and the coordinate based on resolution are calculated, Equation (2), (3), and (4) are applied again with the model's parameter and the new coordinate as input in order to know the model's result of lithospheric field.

3. Results and discussion

In this study, we do the inversion process two times with the expansion of SCHA's order (K_{max}) reach 30. Firstly, the inversion process use repeat station data only. Secondly, the inversion process is applied by integrated repeat station and satellite data. The results are shown in Table 1. In the first process, the RMS value is still high (285.10367 nT). However, in the second process, the RMS value is 64.03337 nT and this value is reasonable enough according having similarity with the research were done by [1] and [7]. Both processes show huge difference of RMS value caused by amount of data being used. We repeat the second inversion process with expanding $Kmax$ to 70. The results show that RMS values decrease significantly until K_{max} = 30 and they are mostly stable beyond those point. However, if K_{max} = 30 is being used, the resolution of the map becomes wider, as can see in Table 2. Consequently, K_{max} = 70 is elected for better resolution.

Table 1. The difference of RMS value based of amount data used

Data	Number of data	RMSE (nT)
Repeat station	68	285.10367
with satellite data	1806	64.03337

Table 2. Resolution and RMS of K_{max} = 30 and K_{max} = 70

$Kmax$	RMSE (nT)	Resolution
30	64.03337	3.66^0
70	64.0834	1.58^0

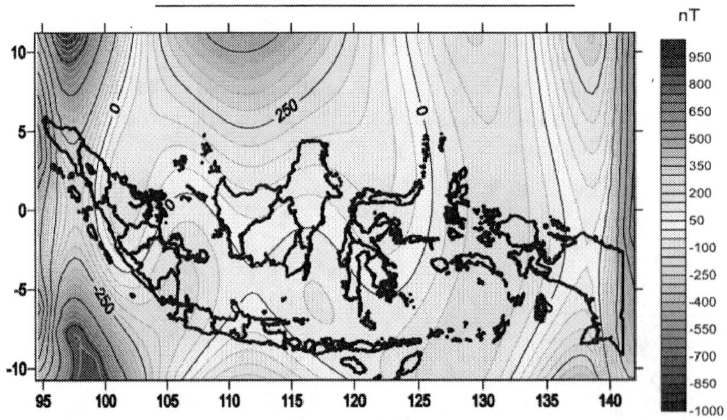

Figure 2. The X component of lithospheric magnetic field

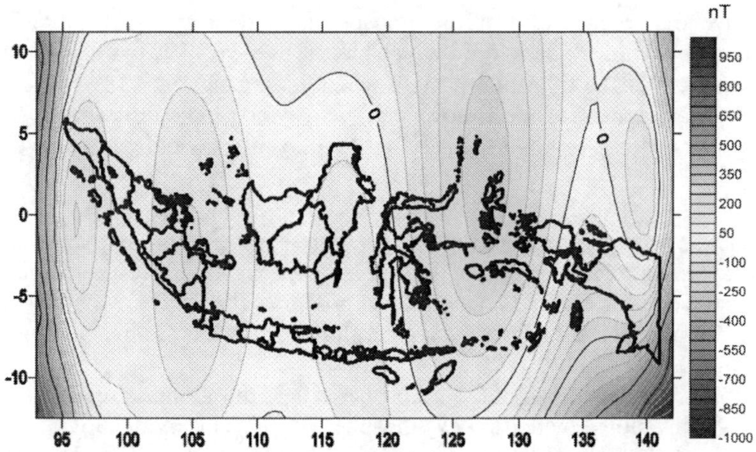

Figure 3. The Y component of lithospheric magnetic field

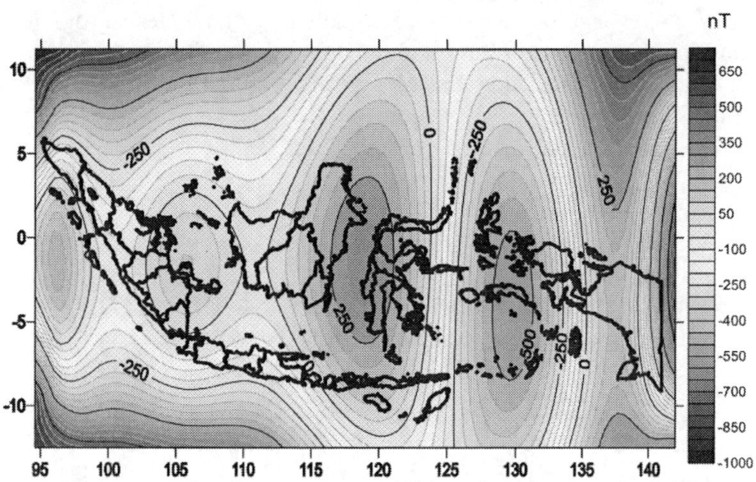

Figure 4. The Z component of lithospheric magnetic field

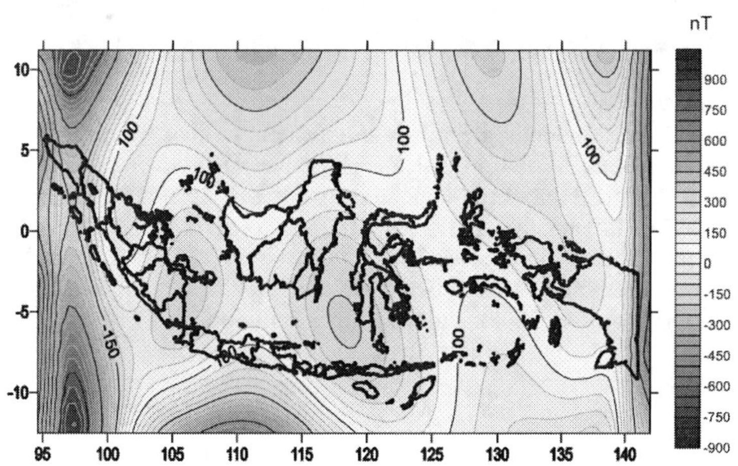

Figure 5. Total intensity of lithospheric magnetic field

Those pictures above are the map result of the lithospheric modelling in this study. Figure 2 shows the map of X component of the lithospheric field which have range -987.192 to 998.239 nT. Figure 3 shows the map of Y component of the lithospheric field which have range 968.189 to 949.438 nT. Figure 4 shows the map of Z component of the lithospheric field which have range -981.266 to 608.676 nT. Lastly, Figure 5 shows the map of total intensity of the lithospheric field which have range -904.151 to 997.389 nT.

The lithospheric field contour in oceanic region has shorter wavelength, higher amplitude, and higher frequency than in continent region [8]. The lithospheric contour in the west of Sumatera and the south of Java which have the oceanic contour characteristic is depicted in Figure 5. In the continent region has contour pattern of continental crust. Based on the result this study has similar finding who stated by [8].

Figure 6 shows the total intensity map of the lithospheric field in Indonesia region from EMAG2 model [9]. EMAG2 model is the global modelling of lithospheric field. This model do not use data from repeat station only, but they use both the shipborne data and aeromagnetic data. They have more complex data and smaller total intensity value's range than this study. EMAG2 model has resolution about 2 minutes degree, this caused by the expand of Legendre equation's order (K-index in regional modelling) reach 150, meanwhile this study only has resolution about 1.5 degree and the expand of K-index just reach 70. However, both EMAG2 and this study's result agree that Indonesia region mostly has negative total intensity of lithospheric field.

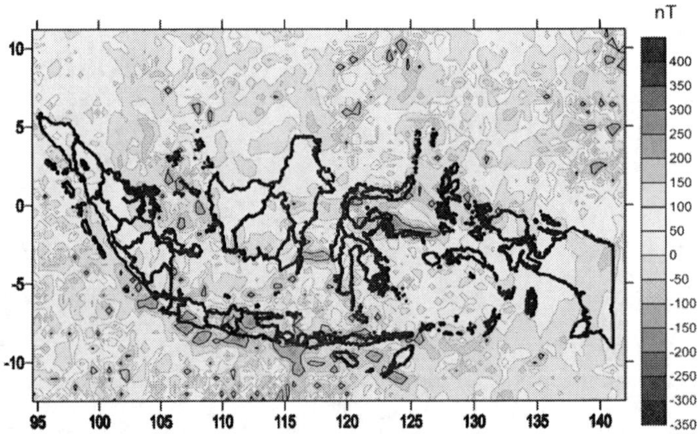

Figure 6. EMAG2 model (global modelling of lithospheric field)

4. Concluding remark

This study shows the regional modelling result of lithospheric field as we can see in Figure 2 to Figure 5. The results of the modeling resolution is 1.5^0. The range of lithospheric magnetic field values of component X, Y, Z, and the total magnetic field intensity are -987.192 – 998.239 nT, -968.189 – 949.438 nT, -981.266 – 608.676 nT, and -904.151 – 997.389 nT respectively. These results are considered reasonable by comparing them with the previous result from [1], [7], and the statement of [8]. These results still need improvement and it would be possible by including shipborne and aeromagnetic data, and expanding the number of K-index to achieve better resolution.

References

[1] Gu Z 2006 Earth Planets Space **58** 741
[2] Haines G V 1985 Journal of Geophysical Research **90(B3)** 2583
[3] Syirodjudin M et al 2018 *IOP Conf. Ser.: Earth Environ. Sci.* **132** 012003
[4] The´bault E, Mandea M, Schott J J 2006 *Journal of Geophisical Research* **111**
[5] Maus S et al 2008 Geochemistry Geophysics Geosystems **9**
[6] Qiu Y et al 2017 *Sensor* **17** 238
[7] Duzgit Z, Malin, S R C 2000 *Geophys. J. Int.* **141** 829
[8] Sharma, P. V., 1976. Geophysical Methods in Geology, 1976, Methods in Geochemistry and Geophysics, 428 pages
[9] Maus S et al 2009 Geochemistry Geophysics Geosystems **10**

IOP Conf. Series: Earth and Environmental Science **873** (2021) 012031 doi:10.1088/1755-1315/873/1/012031

Characteristic of the 2010-2020 Earthquakes in North Arm of Sulawesi based on Focal Mechanism and b-value

A P Astuti[1], E M Elsera[2], M F I Massinai[1], and M A Akbar[1]

[1]Department of Geophysics, Hasanuddin University, Makassar, Indonesia
[2]The Indonesian Agency of Meteorology, Climatology and, Geophysics (BMKG) Region IV, Makassar, Indonesia

*Email: astutiap17h@student.unhas.ac.id or auliapuji24@gmail.com

Abstract. The north arm of Sulawesi has a fairly high level of seismicity. The North Sulawesi arm is bounded in the south by the Palu-Koro Fault, the northern part is bounded by the North Sulawesi Trench and the Molluca Sea Thrust in the east. Therefore, this study aims to analyze the characteristic of the 2010-2020 earthquakes in the north arm of Sulawesi by analyzing the earthquake's focal mechanism and mapping the b-value using the maximum likelihood method. From this study, we obtained the focal mechanism consist of thrust and strike-slip, this is due to the activity of faults and subduction zones in the North arm of Sulawesi such as the Palu-koro fault, the Gorontalo Fault, North Sulawesi Trench, Molucca Sea Collision, and several other faults that affect the seismicity of this region. The variation of the b-value ranging from 0.5-1.1 These studies indicate that thrust fault regions have lower b-values, while strike-slip fault regions have intermediate b-values. Meanwhile, areas with active volcanoes tend to have high b-values. The results of this research can be used as a basis for decision making related to earthquake mitigation in this area in the future.

1. Introduction

Sulawesi Island and its surroundings, especially northern Sulawesi, are one of the most complex active margins in terms of geology, structure, and tectonics. This region is the center of the meeting of three convergent plates due to the interaction of the three main earth plates in the Neogene era. This convergence has led to the development of all types of structures at all scales, including subduction and collision, fault, and thrust zones. Currently, most of the Neogene structures and some praneogene structures are still active or reactive [1].

The geological structures identified in Sulawesi are still actively moving at different shifting speeds and often produce earthquakes [2]. According to [3], the North Sulawesi Arm is bounded to the south by Palu-Koro Fault, to the north it is bordered by North Sulawesi Trench and Molluca Sea Thrust in the east. This is indicated by the presence of a crust that has a high shear strain due to the movement of the Sula Block around the Palu-Koro Fault. In this zone, there are also some strike-slip fault activities, including the Poso fault and the Gorontalo fault. On the western side of the Molluca Sea there is a Sangihe plate with a west-dipping direction. And there is also a plate under the North Sulawesi Arm called the Celebes plate. This plate subducts the Sulawesi Sea lithosphere to the south from the North Sulawesi trench [4].

One of the parameters that can be used in studying the characteristics of an earthquake that occurs is the focal mechanism. Seismograms record various distances and azimuths which are used to study fault geometry due to earthquakes [5]. The focal mechanism is a term used to describe the propagation properties of the hypocenter of earthquake energy. When an earthquake occurs, the seismic waves caused by the earthquake will be emitted in all directions in the form of wave phases. The initial phase recorded first is the P wave because it has the greatest speed compared to other waves. The direction of the first P wave motion recorded by the seismogram is then used to study the focal mechanism. This is because the P waves are the clearest to read and easy to spot. Besides, the polarity and amplitude of the S wave and the moment tensor inversion are also used in determining the focal mechanism [6]. The figure below is an example of the focal mechanism of a beach ball earthquake and its fault geometry.

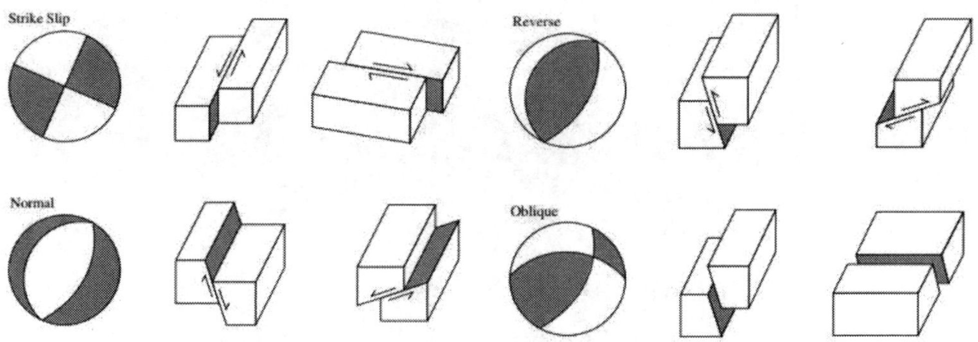

Figure 1 Some examples of focal planes and their fault geometries [7].

Physically, the value of b can be said to be the seismotectonic parameter of a region. Meanwhile, regionally, a change in the b-value is believed to have an inverse relationship with changes in stress levels. The method commonly used in determining the b-value is the classical frequency distribution (FMD) [8]. From FMD, information about the number of earthquakes is obtained as a function of magnitude (M) that has occurred in an area, with the following formula:

$$log\, N = a - bM \tag{1}$$

where N is the total number of earthquakes that have occurred, M is the magnitude of the earthquake, constant a characterizes the general seismic activity in an area, the higher the number, the higher the seismic level. Constant b is believed to depend on the stress level and tectonic character of the area. This study seeks to analyze the fault characteristics that caused earthquakes in the North arm of Sulawesi based on the focal mechanisms and possible implications for future earthquakes using b-value. The results of this research can also be used as a reference for mitigating natural disasters.

2. Data and Method

The research location focused on the North Arm of Sulawesi. The data used in this research is the earthquake data online catalog belonging to the International Seismological Center (ISC) [9]. The data used in this research include information about origin time, magnitude, hypocenter, and fault parameters such as strike, dip, rake, etc. The total earthquakes data used are 10,094 events that occurred from January 2010 to August 2020 with a variation of the magnitude ranging from 1.1 to 6.9. The research is focused on the coordinate $119^0 - 127^0$ E dan -1^0 - 3^0 N. The research began with hypocenter mapping to see the distribution of earthquake depths and the focal mechanisms using the Generic Mapping Tools (GMT) [10]. To study variations of the b-value, we use the maximum likelihood method and it was done using Zmap code [11].

IOP Conf. Series: Earth and Environmental Science **873** (2021) 012031 doi:10.1088/1755-1315/873/1/012031

3. Results and Discussion

3.1 Analysis of the focal mechanism

The distribution of the hypocenter (figure 2) in the North Arm of Sulawesi, both on land and coast, shows high seismicity in this decade. We analyze the earthquake focal mechanisms by dividing it into 3 regions, namely western North Arm, central North Arm, and Eastern North Arm. This is because there are significant differences in earthquake focal mechanisms in these three areas, so that analysis is needed to determine the characteristics of the earthquake causes.

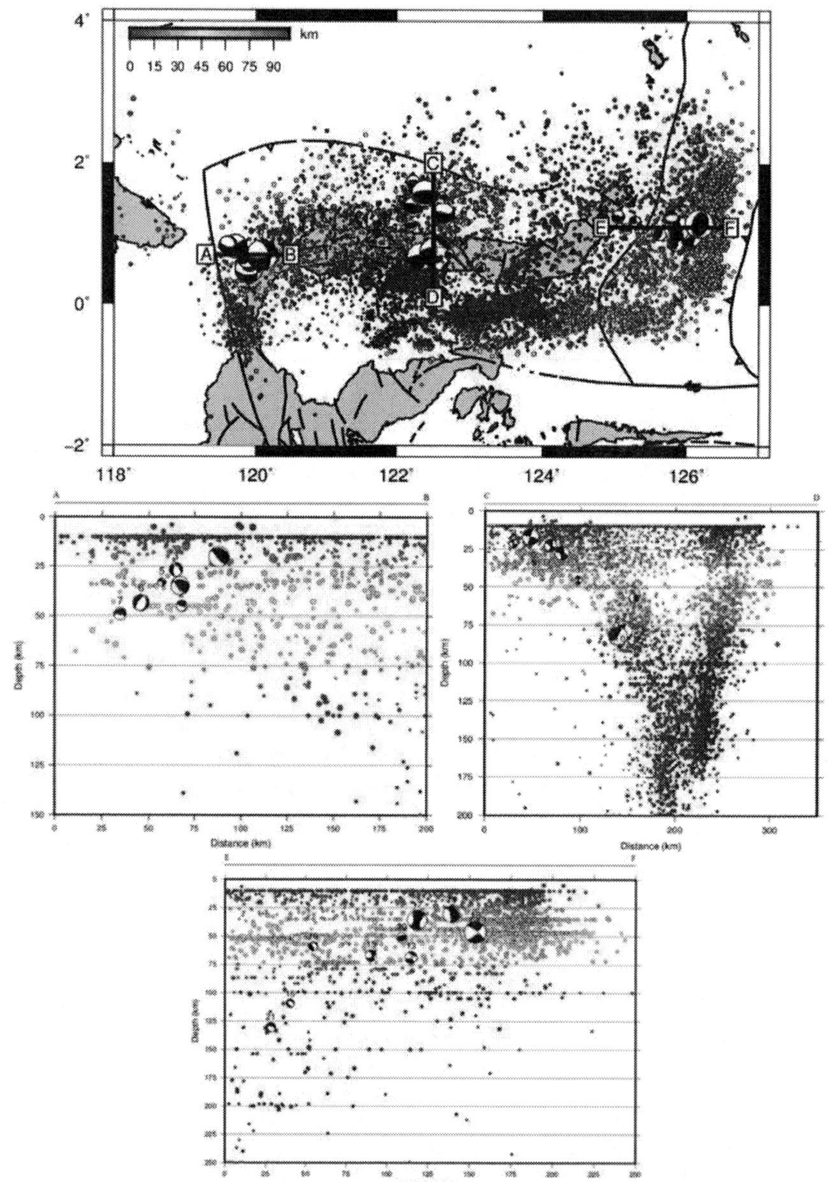

Figure 2 Cross-section results of the hypocenter and earthquake focal mechanisms.

Figure 2 shows the seismic activity that occurred in the western, center (especially land and coast) and in the eastern part of the north arm of Sulawesi. In the area traversed by the A-B line, the results obtained show that the earthquake that occurred around the area was dominated by strike-slip, so it is interpreted that the earthquake that occurred were influenced by the Palu-Koro Fault. The Palu-Koro fault is one of the longest and active faults in Sulawesi. According to [3], several segments of the Palu-Koro Fault cover an area with a width of 50 km around Palu. Research by [12] states that the focal mechanism of the Palu-Koro fault is the left lateral strike-slip. The Palu-Koro fault is connected to the Matano fault and the Lawanopo fault, while the northern end of the fault passes through the Makassar Strait and is connected to the North Sulawesi subduction zone at the western end [13].

In the area traversed by the C-D line, the analysis result of the focal mechanism show the northern part of Sulawesi North Arm is dominated by thrust faults, on land by several strike-slip faults. We interpreted that the earthquakes that occurred were caused by North Sulawesi Trench on the coast and on the land caused by Gorontalo Fault. North Sulawesi Trench or better known as Minahasa Trench is the longest and largest thrust in Sulawesi. North Sulawesi Trench is a subduction zone between the Philippine Plate and the North Sulawesi microblock [2]. The North Sulawesi Trench belongs to a relatively old subduction zone whose rips develop eastward along the northern edge of Sulawesi. North Sulawesi subduction infiltrates with a slope angle of about 14^0 and the Benioff zone dips to a depth of 170-180 km, with an angle of about 45^0 [14]. Then, [15] in their analysis state that the Gorontalo Fault has a displacement speed of 11 mm/year, with a focal mechanism of right-lateral strike-slip [2], while [16] state that this fault is a type of oblique fault.

In the area traversed by the E-F line, we found that the focal mechanism of the earthquakes show thrust fault, so we interpreted that the earthquakes that occurred were caused by the Molucca Sea Collision. This zone consists of West Sangihe Thrust and East Halmahera Thrust. According to [17], the Sangihe plate with a movement towards the west has carried out deeper subduction than the Halmahera plate which points to the east. Analysis by [4,17] states that the Sangihe plate has undergone subduction of up to 660 km, while the Halmahera plate has reached a depth of 350–400 km.

3.2 Analysis of b-value

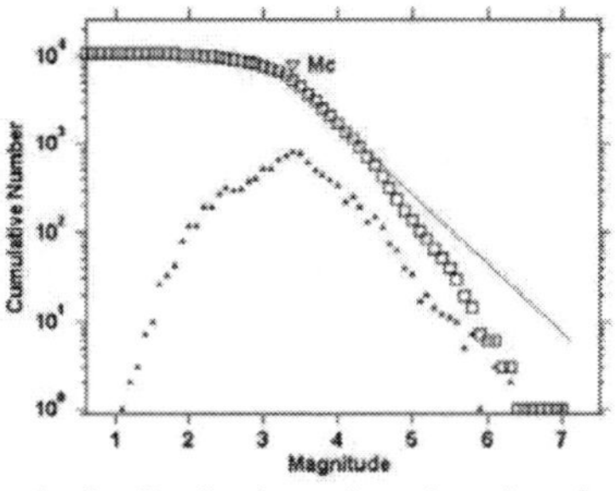

Figure 3 Frequency–magnitude distribution of earthquakes from January 2010-August 2020. The slope of the line represents the Gutenberg-Richter relationship.

This Mc value is the lower limit of the magnitude or magnitude that most often occurs when an earthquake occurs. According to [18], the prediction of the Mc value is based on the magnitude of the cumulative frequency. In earthquake studies, one of the parameters often used in studying the seismotectonic parameters of a region is the b-value, which varies from 0.2 to 2.0. Generally, very low b-values are found in the case of direct aftershocks and higher values are found in the case of earthquake swarms [19]. The decrease in the value of b can be affected by the increase in shear stress or the effective stress that occurs. A smaller b-value can indicate that the stress in the area is high, and vice versa.

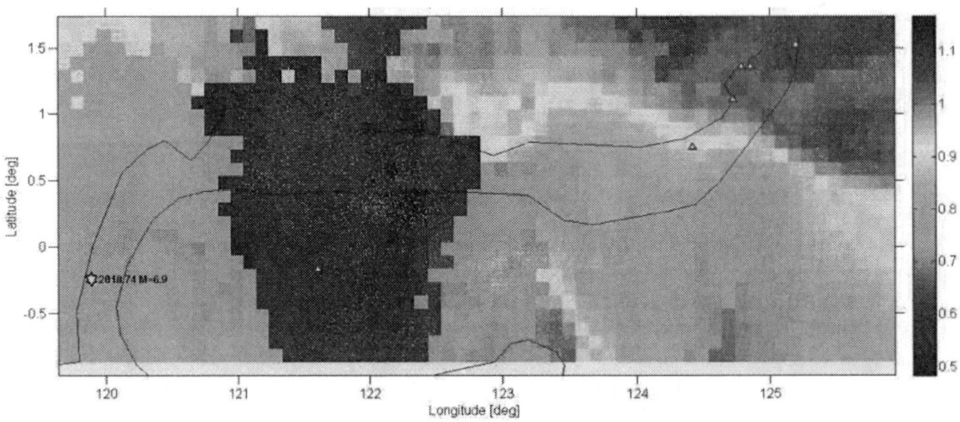

Figure 4 Distribution of b-value in the North Arm of Sulawesi. Stars mark the large earthquakes M 6.9. Higher b-value is indicated in red and lower b-value is indicated in blue.

Based on figure 4, it is obtained that the b-value in the North Arm of Sulawesi ranges from 0.5-1.1. A high b-value means an earthquake is dominated by a small earthquake. Conversely, a low b-value indicates that earthquakes of greater magnitude predominate in the region [20]. In the area around the Palu-Koro fault, the b-value ranges from 0.7-0.8. Research by [18] state that the area around the strike-slip fault has a b-value of 0.5-1.0, while [21] mention that the temporal b-value for Palu region was in the range of 0.55-0.961. According to [22], this b-value depends on stress conditions, such as spatial and temporal changes in applied shear stress. In the central part of the north Sulawesi arm affected by the North Sulawesi Trench, there is a contrasting difference in b-values, which is around 0.5-0.6, this low b-value indicates that the stress level in the region is high. According to [3], lower b-values indicate higher seismic activity while high b-values indicate low and medium seismic activity. In the area around the Gorontalo Fault, the b-value ranges from 0.7-0.8. Meanwhile, at the eastern end of the North Sulawesi Arm, the b-value tends to be high, ranging from 0.9-1.1. From the volcanic side, this value can be influenced by magmatic activity. This is supported by the number of active volcanoes in this region. Research by [22] mentioned that high b-values are found in volcanic areas. Meanwhile, in the area around the Molucca Sea Collision, the b-value ranges from 0.8-1.1. From a tectonic perspective, areas such as the Molucca Sea Collision will have a high b-value. Research by [3] state that the b-value in West Molucca Sea and East Molucca Sea is 0.95.

4. Conclusions

Based on the analysis of the focal mechanism, the earthquakes that occurred around the North arm of Sulawesi consisted of thrust fault and strike-slip mechanisms. This was caused by the activity of faults and subduction zones existing in the North arm of Sulawesi such as the Palu-Koro fault, Gorontalo Fault, North Sulawesi Trench, Molucca Sea Collision, and several other faults that affect the seismicity of this area. These studies indicate that thrust fault regions have lower b-values, while strike-slip fault regions have intermediate b-values. Meanwhile, areas with active volcanoes tend to have high b-values. The b-value in the North Arm of Sulawesi ranges from 0.5-1.1, this indicates that the North Arm of Sulawesi region is affected by tectonic activity and volcanism.

Acknowledgment
We would like to thank the International Seismological Centre (ISC) [9] for providing the earthquakes event catalog and focal mechanisms that were used in this research. Indonesian Meteorology, Climatology, and Geophysics Agency (BMKG Region IV, Makassar) and Department of Geophysics Hasanuddin University which has provided support to the authors is also gratefully acknowledged. And also to all participants who have helped and provided criticisms and suggestions to the authors.

References

[1] G Pasau and A Tanauma 2011 Pemodelan Sumber Gempa di Wilayah Sulawesi Utara Sebagai Upaya Mitigasi Bencana Gempa Bumi *Jurnal Ilmiah Sains* **11**

[2] PuSGen 2017 *Peta Sumber dan Bahaya Gempa Indonesia Tahun 2017*

[3] Cipta A, Robiana R, Griffin J D, Horspool N, Hidayati S and Cummins P 2016 A Probabilistic seismic hazard assessment for Sulawesi, Indonesia *The Geological Society of London* **441**

[4] Hall R and Spakman W 2015 Mantle structure and tectonic history of SE Asia *Tectonophysics* **658** 14–45

[5] Stein S and Wysession M 2015 *An Introduction to Seismology, Earthquakes, and Earth Structure* (UK: *Blackwell Publishing)*

[6] Harmadhoni D 2011 Analisis Mekanisme Fokus Gempa di Blitar Jawa Timur 17 Mei 2011 *Thesis* Universitas Islam Negeri Syarif Hidayatullah

[7] Shearer P M 2009 *Introduction To Seismology Second Edition* (New York: Cambridge University Press)

[8] Nuannin P 2003 The Potential of b-value Variations as Earthquake Precursors for Small and Large Events *Dissertations* Uppsala University

[9] International Seismological Centre 2020, On-line Bulletin, https://doi.org/10.31905/D808B830

[10] Wessel P, Smith W H F, Scharroo R, Luis J, and Wobbe F, 2013 Generic Mapping Tools: Improved Version Released *EOS Trans. AGU* **94** 409–410.

[11] Wiemer S 2001 A Software Package to Analyze Seismicity: ZMAP *Seismological Research Letters* **72** 374-383.

[12] Massinai M F I, Astuti A P, Kiraman M R B, Massinai M A and Ramdhan M 2019 Hypocenter determination and focal mechanism solution of May 29 2017 earthquake around Poso, Central Sulawesi, Indonesia *Journal of Physics: Conference Series* **1341**

[13] Pakpahan, Ngadmanto D, Masturyono, Rohadi S, Rasmid, Sulistyo H S and Susilanto P 2015 Analisis Kegempaan di Zona Sesar Palu Koro, Sulawesi Tengah *Jurnal Lingkungan dan Bencana Geologi* **6** 253-264

[14] Kertapati E 2006 *Aktivitas Gempa Bumi di Indonesia Perspektif Regional Pada Karakteristik Gempa Bumi Merusak* (Bandung: Pusat Survey Geologi)

[15] Rangin C, Pichon X L, Mazzotti S, Pubellier M, Chamot-Rooke N, Aurelio M., Walpersdorf A, and Quebral R 1999 Plate convergence measured by GPS across the Sundaland/Philippine Sea Plate deformed boundary: The Philippines and eastern Indonesia *Geophysical Journal International* **139** 296-316

[16] Nurfitriani, Pasau G and Raharjo S 2014 Identifikasi Sesar di Wilayah Gorontalo dengan Analisis Mekanisme Bola Fokus *Jurnal Mipa Unsrat Online* **3** 40-43

[17] Zhang Q, Guo F, Zhao L and Wu Y 2017 Geodynamics of divergent double subduction: 3-D numerical modeling of a Cenozoic example in the Molucca Sea region, Indonesia Journal of Geophysical Research: Solid Earth **122** 3977-3998

[18] Massinai M A, Harimei B, Agustiawati A and Massinai M F I 2019 Seismicity analysis Sulawesi North Arm based on B-Values *Journal of Physics: Conference Series* **1341**

[19] Adagunodo T A, Lüning S, Adeleke A M, Omidiora J O, Aizebeokhai A P, Oyeyemi K D and Hammed O S 2018 Evaluationof $0 \leq M \leq 8$ earthquake data sets in African – Asian region during 1966–2015 *Data in Brief* **17** 588-603

[20] Wu Y M, Chen S K , Huang T C, Huang H H, Chao W A and Koulakov I 2017 Relationship Between Earthquake b-Values and Crustal Stresses in a Young Orogenic Belt *Geophysical Research Letters* **45** 1832–1837

[21] Simangunsong A V, Priadi R, Dwilyantari A A I and Marsono A 2019 Determination of temporal value of a-value and b-value to identify the level brittle of rock and seismic activity in the Palu region *Journal of Physics: Theories and Applications* **3** 1-8

[22] Farrell J, Husen S and Smith R B 2009 Earthquake swarm and b-value characterization of the Yellowstone volcano-tectonic system *J. Volcanol. Geotherm. Res* **188** 260-276

IOP Conf. Series: Earth and Environmental Science **873** (2021) 012032 doi:10.1088/1755-1315/873/1/012032

Focal Mechanism Analysis of the September 25th, 2019 Mw 6.5 Ambon Earthquake and Its Implication for Seismotectonics

Gatut Daniarsyad[1,2*], Aprilia Nur Vita[1,2], Shengji Wei[3]

[1]Agency for Meteorology Climatology and Geophysics (BMKG), Jakarta, Indonesia
[2]Master Program of Earth Sciences, Faculty of Earth Sciences and Technology, Bandung Institute of Technology, Bandung, Indonesia
[3]Earth Observatory of Singapore, Nanyang Technological University, Singapore

lp.daniarsyad@gmail.com

Abstract. On September 25th, 2019, an Mw 6.5 earthquake occurred in Ambon, Maluku Province, Indonesia, and caused casualties and infrastructures damages. The epicenter located in a tectonically active region with the potential strike-slip and thrust faulting earthquake sources, yet the responsible fault is still not well understood. Based on focal mechanism solutions from available seismological agencies, i.e. USGS, GFZ, GCMT, and BMKG, the earthquake has a similar strike-slip focal mechanism, although there are discrepancies on detailed source parameters. To provide a better understanding of the earthquake mechanism and seismotectonic, we apply the Cut-and-Paste (CAP) focal mechanism inversion method to broadband seismic waveforms from regional and teleseismic distances. The CAP inversion results on the regional data grouped in different distance ranges show a robust strike-slip solution. We then refine the earthquake focal depth by performing the CAPtele inversion and resulted in a depth of 12 km with similar fault plane solution as the regionals. The ruptured fault plane is resolved by a directivity analysis using azimuthal pattern of the apparent source durations, which indicates an obvious unilateral rupture propagation toward SSE direction. Our result suggests the NNW-SSE orientated fault is the ruptured fault plane, which is also consistent with the near N-S distributed aftershocks. This fault is located in a narrow sea between Seram, Ambon and Haruku island and was not reported yet in previous studies. The Coulomb failure stress (CFS) changes analysis of the mainshock shows that the Ambon earthquake has promoted the off-fault aftershocks which occurred to the west of the ruptured fault.

1. Introduction

A moderate-shallow earthquake hit Ambon, Indonesia with magnitude Mw 6.5 on September 25, 23:46 UTC. The earthquake caused strong shaking throughout Seram Island with the maximum intensity of VI MMI, as shown by felt reports in the Provincial Capital Ambon. Thousands of structures were damaged and dozens of fatalities were also reported in the aftermath of the earthquake. The mainshock was then followed by hundreds of aftershocks within a week. According to the Indonesian Meteorology Climatology and Geophysical Agency (BMKG) the earthquake has a strike-slip focal mechanism and the fault planes have sub-vertical dips, showing either a right-lateral N-S orientation fault or a left-lateral E-W orientation fault. The moment tensor solution from other seismological agencies (e.g. GFZ, USGS, GCMT) also present a strike-slip focal mechanism although there are detailed differences between the source parameters (Table 1).

The inconsistency of the source parameters in a tectonically active region with the potential strike-slip and thrust faulting earthquake sources requires further investigation of the earthquake source parameters, including the ruptured fault plane. The Banda Arc region including Seram Island is a product of complex ongoing collision and rotation interaction between the Eurasia and Australia plates along with Philippine Sea microplate. Due to this complex interaction, region of Seram Island and Ambon have at least three major earthquake sources (North Banda subduction, Seram Fault Thrust Belt, Kawa Fault) and numerous small faults [1]. This earthquake could occur on the east-west fault crosses Kairatu that corresponds to lineament of topography in northern and southern part of Seram, or an unknown north-south trending fault. Left-lateral Kawa strike-slip fault was first presumed to be the responsible fault because the epicenter of the earthquake is located close to its lineament. Seemingly after further analysis, despite numerous fracture zone and faults, the Ambon earthquake is more likely to have occurred on an unknown or unmapped fault. Therefore, it is important to provide more robust source mechanism solution in order to reveal the unmapped faults in the region because such faults near densely populated city could be the hosts for unexpectedly seismic events.

Here we perform the Cut-and-Paste (CAP) inversion method using regional and teleseismic data to obtain the focal mechanism and depth of the Ambon earthquake. We then conduct a directivity analysis to define the actual fault plane as well as the rupture direction propagation. We also relocate the aftershocks and calculate the Coulomb failure stress (CFS) changes of the mainshock to verify the effect on the seismicity of the surrounding area.

2. Data and method

In this study we carried out the Cut-and-Paste (CAP) method [2, 3] to determine both depth and source mechanism of the Mw 6.5 Ambon earthquake. The idea of CAP method is to divide the seismogram into five segments including two segments of Pnl wave and three segments of surface wave and invert them independently. This approach provides the advantage due to the different sensitivity of the body and surface wave to varied parts of crustal structure. The CAP method also allows time shifting on each segment which make the inversion is less dependent to the assumed velocity model and earthquake location inaccuracy. The dominance of surface wave amplitude then can be avoided by applying different weight to the body and surface wave, while the dominance of near station records is balanced by the distance scaling factor. The best solution of the focal mechanism, source depth, and moment magnitude are obtained using grid search procedure by minimizing the misfit between the observed and synthetic seismogram. We applied the CAP inversion to both regional and teleseismic data to get better constrain for the solution. The regional data were collected from the BMKG broadband seismic network within 20 degrees from the epicenter (Figure 1b), while the teleseismic data were retrieved from the stations available at IRIS server (https://www.iris.edu) within the range of 30-90 degrees (Figure 1c).

The full waveform recorded at regional stations were used in the inversion with the intention of using the body and surface wave. We first corrected the data by removing the instrument response and rotating the two horizontal components to radial and tangential components. The Green's functions were computed using frequency-wave number (F-K) method [4]. The modified PREM 1-D velocity model [5] with the addition of a detailed crustal model from Crust1.0 [6] was used to calculate the synthetics seismogram for both regional and teleseismic distances. We then applied the inversion on 50 s window length of Pnl wave and 180s window length of surface wave. The synthetic and observed signals were filtered using 4th order Butterworth bandpass in the frequency band of 0.01-0.07 and 0.005-0.02 for Pnl wave and surface wave, respectively. The example of waveform fit for regional distance is shown in Figure 2a.

We used only P and SH wave records for the CAPtele inversion considering the teleseismic surface waves are more distorted due to the lateral heterogeneities over long propagating distances. The tel3 program was used to compute the synthetics waveform for teleseismic distance. The teleseismic waveform inversion was conducted at the frequency band of 0.01-0.05 Hz and 0.01-0.04 Hz for P and SH wave, respectively, which fit of the waveform is shown in Figure 2b.

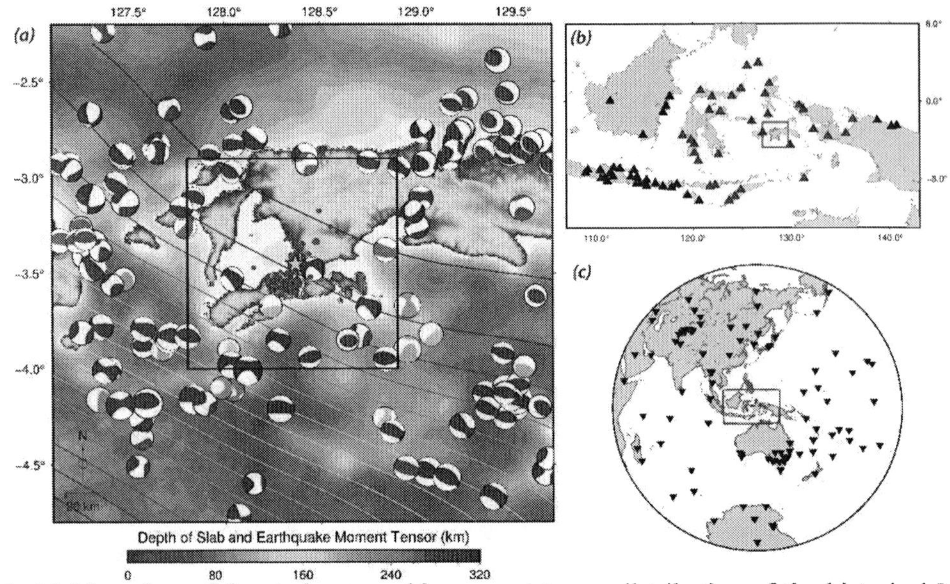

Figure 1. (a) Map view of the study area with moment tensor distribution of the historical Mw≥5.0 earthquakes from GCMT catalog around the 2019 Ambon earthquake (yellow star) and its aftershocks (purple circles). Colored slab contours obtained from slab2.0 [7]. Black box shows the map in Figure 4. (b) Regional stations used in the CAP inversion, different colors indicate different distance from the epicenter (blue for 5°, red for 10°, black for 20°). Red box shows the map in panel (a). (c) Teleseismic stations used in the CAPtele inversion (black inverted triangles). Red box shows the map in panel (b).

3. Result and discussion

The CAP inversion was performed to the regional data for three different station distance ranges and resulted in a strike-slip mechanism for all of the models (Table 1). We excluded three closest stations to the epicenter because the near-source ground shaking of the earthquake was much affecting the waveform. The best solution is the outcome from the inversion on all stations within 20 degrees from the epicenter (Figure 2a). The focal mechanism indicated a near-vertical dip for both nodal planes which had NNE-SSW and WSW-ENE orientation with the source parameters were 252/82/-8 and 343/82/-173 for strike, dip, and rake, respectively. The earthquake focal depth from the three inversions were obtained at 11 km and 16 km (Figure 2c), although there was local minimum at 12 km depth for 5- and 20-degrees inversions.

Table 1. Summary of the double couple point source solutions for the 2019 Ambon earthquake from this study and its comparison with the solutions from available seismological agencies.

Model	Fault Plane 1	Fault Plane 2	Depth (km)	Magnitude
GlobalCMT	253/84/-12	345/78/-174	12.7	6.5
USGS	71/75/-12	164/78/-165	11.5	6.46
GFZ	76/76/9	343/81/165	14	6.4
BMKG	256/89/9	346/81/179	10	6.6
CAP (0-5)	252/76/-15	346/75/-165	16	6.5
CAP (0-10)	251/74/-17	346/74/-163	11	6.46
CAP (0-20)	252/82/-8	343/82/-173	16	6.48
CAPtele	253/74/-7	345/83/-164	12	6.59

We carried out the CAPtele inversion at a relatively high frequency to include the depth phases to better constrain the focal depth. The inversion yielded the focal depth of the Ambon earthquake in a depth of 12 km (Figure 2d). The obtained magnitude was 6.59, slightly larger than the regional result. A strike-slip mechanism was also acquired with the parameters almost similar compared to the regional result (Figure 2b). The good agreement came from the strike and dip orientation with insignificant differences. However, an obvious difference appeared from the rake angle with the CAPtele result provided a bigger dip-slip component compared to the CAP solution. This was probably caused by the incident angle which is more evident for the teleseismic stations than the regionals.

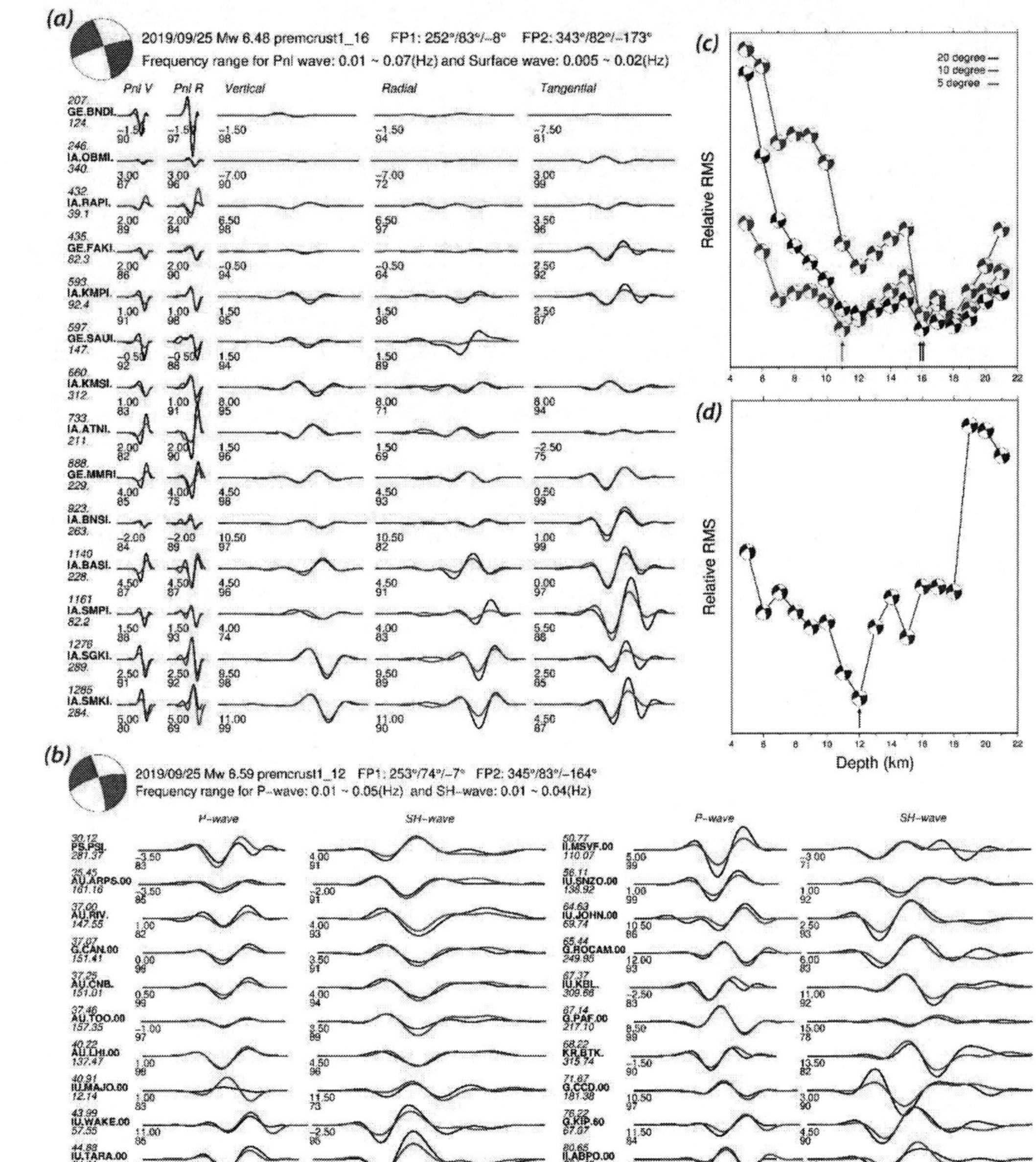

Figure 2. Best solution and representative waveform fit for (a) CAP inversion on regional data and (b) CAPtele on teleseismic data (black lines are observed, red lines are synthetic). Depth misfit curve for (c) CAP and (d) CAPtele inversion.

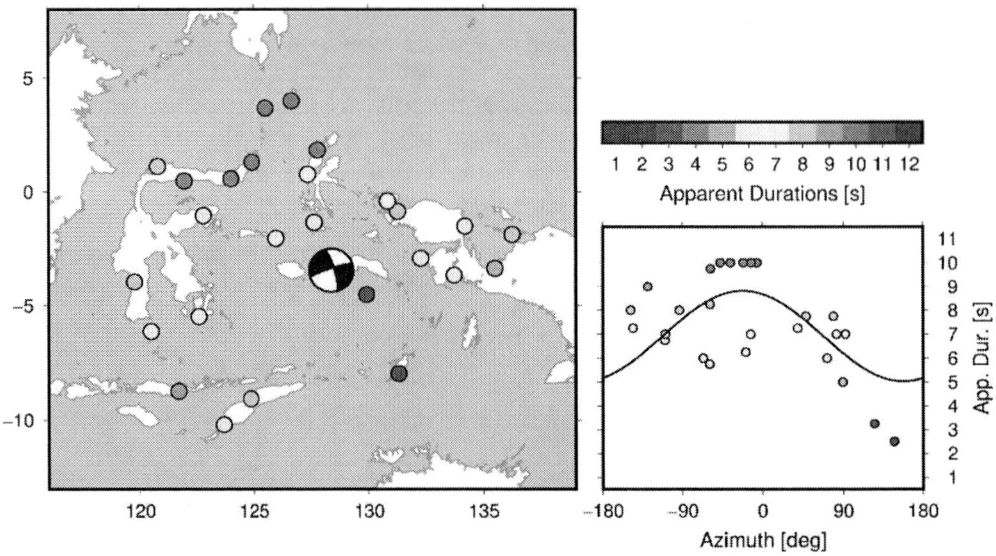

Figure 3. Azimuthal distribution of apparent durations obtained from fast directivity inversion, including the map view (left) and azimuthal distribution (right). Colored circles show the apparent durations according to the color scale, black line shows the best matching pattern for unilateral rupture.

We further clarify the source of the Ambon earthquake by performing a directivity analysis to determine the actual fault plane. The directivity analysis was carried out using the amplitude spectra inversion to derive the apparent duration at each station [8], which can reveal the direction of the rupture propagation. The idea is the stations which located toward the rupture propagation will have shorter duration, and longer duration for stations in the opposite direction [9]. We used only the P-wave recorded at seismic stations within 10 degrees from the epicenter and applied a bandpass filter in the range of 0.01-0.1 Hz to get the amplitude spectra for each station. The inversion was conducted by fitting the observed and synthetic spectra from several possible source duration using grid search procedure. The apparent durations were well distributed and resulted in a unilateral rupture propagating to southeast direction (Figure 3). This indicates that the Ambon earthquake was generated by a fault with NNW-SSE orientation. This also implies that the earthquake occurred in a northeast-dipping plane and the rupture propagated opposite to the strike direction.

Based on the focal mechanism and the focal depth, the Ambon earthquake is estimated did not take place by the subduction occurring in the region. The slab beneath the epicenter location is around 30-40 km for a subduction event with the strike orientation is NW-SE (Figure 1a). Furthermore, a shallow subduction event is commonly indicated by a thrust fault. Therefore, it is expected to be a local fault that was responsible for the earthquake. However, we cannot determine the possible existence of local faults based on the surface morphology since the epicenter and mostly of the aftershocks were distributed at the sea between the Seram, Ambon and Haruku Island. Historical seismicity with similar mechanism at the same or adjacent locations was also unavailable.

We did the aftershock relocation using the double-difference algorithm [10] and resulted in a predominantly N-S distribution with mostly occur at the depth of 2 to 10 km. This result corresponds well to the NNW-SSE orientation of the Ambon earthquake. However, a detailed study using local seismic network shows that the aftershock distribution deviates from the main rupture plane with a diffuse pattern at both ends and is therefore thought to be related to reactivation of minor faults around the fault plane [11]. The aftershocks also occurred to the west of the earthquake fault in the Ambon Island. We explain this off-fault occurrence using Coulomb failure stress analysis. Figure 4 shows the Coulomb stress changes distribution caused by the 2019 Ambon earthquake using the focal mechanism from the regional CAP inversion result with the focal depth of 12 km.

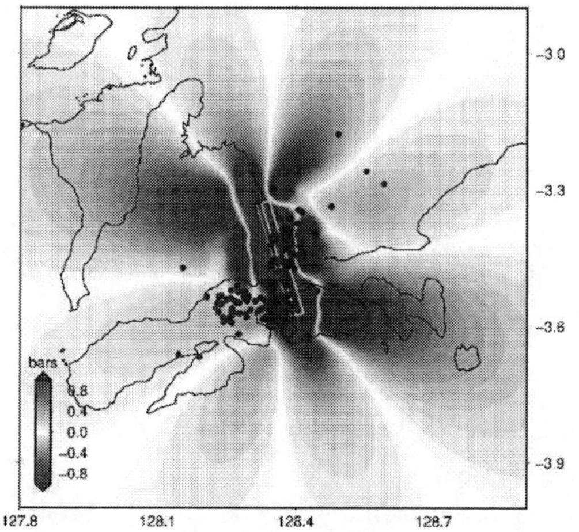

Figure 4. Coulomb stress changes of the 2019 Ambon earthquake at depth of 5 km. Dark red circles show the M>3 relocated aftershocks, yellow rectangle and green line show the map and surface projection of the fault, respectively.

The off-fault aftershocks on the western side occurred in the positive Coulomb stress changes which imply that this aftershock cluster has been triggered by the mainshock. This occurrence also indicates that there could be a potential earthquake source in the region. The results of this study are expected to be useful for increasing public awareness and assisting the earthquake disasters mitigation efforts.

4. Conclusions

We have successfully investigated the focal mechanism of the 2019 Ambon earthquake using broadband waveform inversion. The results show that the earthquake had a right-lateral strike-slip mechanism with the focal depth of 12 km. The directivity analysis provided a clear azimuthal pattern of the apparent durations which conclude that the earthquake was occurred in a northeast-dipping plane with a unilateral rupture propagate to SSE direction, as opposed to the strike direction. These results correspond well with the N-S aftershock distribution. The Ambon earthquake is expected to be generated by an N-S orientated unmapped fault at a narrow sea between Seram, Ambon, and Haruku Island. Based on the Coulomb failure stress analysis, the mainshock has generated an off-fault aftershocks cluster that occurred to the west of the ruptured fault.

Acknowledgments

The authors are very thankful to the Indonesian Agency for Meteorology Climatology and Geophysics (BMKG) for providing the seismic waveform used in this study. Figures are created by using the Generic Mapping Tools [12].

References

[1] Pownall J M and Hall R 2014 *Proc. Thirty-Eighth Annual Convention & Exhibition* **vol 1** (Jakarta: Indonesia Petroleum Association) IPA14-G-305
[2] Zhu L and Helmberger D V 1994 *Bull. Seismol. Soc. Am.* **86(5)** 1634-1641
[3] Zhao L and Helmberger D V 1996 *Bull. Seismol. Soc. Am.* **84(1)** 91-104
[4] Zhu L and Rivera L A 2002 *Geophys. J. Int.* **148(3)** 619-627
[5] Dziewonski A M and Anderson D L 1981 *Phys. Earth Planet Int.* **25** 297–356
[6] Laske G, Masters G, Ma Z and Pasyanos M E 2012 *Geophys. Res. Abs.* **14** EGU2012-3743-1
[7] Hayes G 2018 Slab2 – A Comprehensive Subduction Zone Geometry Model: *USGS data release*
[8] Cesca S, Heimann S and Dahm T 2011 *J. Seismol.* **15** 147–164
[9] Haskell N A 1964 *Bull. Seismol. Soc. Am.* **54(6A)** 1811-1841
[10] Waldhauser F and Ellsworth W L 2000 *Bull. Seismol. Soc. Am.* **90 (6)** 1353–1368
[11] Sahara D P *et al* 2021 *Tectonophysics* **799** 228709
[12] Wessel P, Smith W H, Scharroo R, Luis J and Wobbe F 2013 *Eos, Transactions AGU* **94(45)** 409–410

IOP Conf. Series: Earth and Environmental Science **873** (2021) 012033 doi:10.1088/1755-1315/873/1/012033

Slip distribution effect in spatial coulomb stress analysis (Case study: Palu earthquake on September 28, 2018)

Kevin Hanyu Clinton Wulur[1,*], Iman Suardi[1], Sesar Prabu Dwi Sriyanto[2], Yusuf Hadi Perdana[3]

[1] Sekolah Tinggi Meteorologi Klimatologi dan Geofisika (STMKG), Jl. Perhubungan 1 No.5, South Tangerang, Banten, Indonesia
[2]Stasiun Geofisika Manado, Badan Meteorologi Klimatologi dan Geofisika (BMKG), Jl. Harapan No.42, Manado, North Sulawesi, Indonesia
[3]Pusat Penelitian dan Pengembangan, Badan Meteorologi Klimatologi dan Geofisika (BMKG), Jl. Angkasa 1 No.2, Kemayoran, Jakarta, Indonesia

Email: kwulur@gmail.com

Abstract. On September 28, 2018, the Palu-Koro fault released the accumulated stress that caused the earthquake. An earthquake with magnitude 7.5 caused large and massive damage around Palu. There were many aftershocks along the Palu-Koro fault. This research aims to calculate a model of spatial Coulomb stress based on this event to find a correlation between mainshock and the aftershocks. The slip distribution was used as an input of the spatial stress Coulomb modeling to increase the accuracy. We use the Teleseismic Body-Wave Inversion method to calculate slip distribution along the fault plane. As a result, this earthquake was generated by the Palu-Koro fault movement with Mw 7.48, strike 350°, dip angle 67°, and rake -9°. There are three asperity zones along the fault plane located in the north and southern parts of the fault plane. The location of the most energy discharge is in the south asperity zone of the fault plane model with a maximum slip value of 1.65 meters. The spatial Coulomb stress change of this event shows that aftershocks concentration are in areas experiencing increased stress after the earthquake.

1.Introduction

The source mechanism is a study to improve a better understanding of earthquake sources and detailed knowledge of the rupture process, which can be used to analyze earthquake sources [7]. Determination of the source mechanism using a waveform taken from a local station (near-source) still has a high level of noise due to the heterogenity of the upper mantle structure. Kikuchi and Kanamori [1] developed a Teleseismic Body-Wave Inversion method to obtain an interpretation of the slip distribution, asperity, and orientation of the fault plane.

Seismic energy radiation caused by the source mechanism process can be interpreted using Coulomb stress. In Coulomb stress method, an earthquake can be modeled as a slip distribution in a uniform elastic half-space medium [3]. The direction of seismic energy radiation can be modeled properly in Coulomb stress analysis. Therefore, it is necessary to analyze the slip distribution using the inversion method of teleseismic waves and Coulomb spatial stress from the earthquake on September 28, 2018 in Palu. This research conduct to explain the source mechanism, slip distribution, and Coulomb spatial stress, as well

IOP Conf. Series: Earth and Environmental Science **873** (2021) 012033 doi:10.1088/1755-1315/873/1/012033

as validate the distribution of aftershocks that occurred after the release of energy in the Palu earthquake on September 28, 2018.

Previously, the research of Coulomb Stress changes due to the Palu earthquake on September 28, 2018 has been conducted by Utaminingtyas [12]. The parameters of the hypocenter location, strike, dip, slip, rake, and magnitude was used by Utaminingtyas [12] to model the fault plane based on the Wells-Coppersmith equation. From the calculation of Coulomb spatial stress changes with a fault area of 134.13 x 26.83 km, it was found that about 78.2% of the aftershocks that occurred were in the stress zone. In contrast to the Utaminingtyas [12], this study uses a fault plane model with slip distribution resulting from the Teleseismic Body-Wave Inversion method to improve the accuracy of Coulomb spatial stress change modeling.

2. Data and Methods

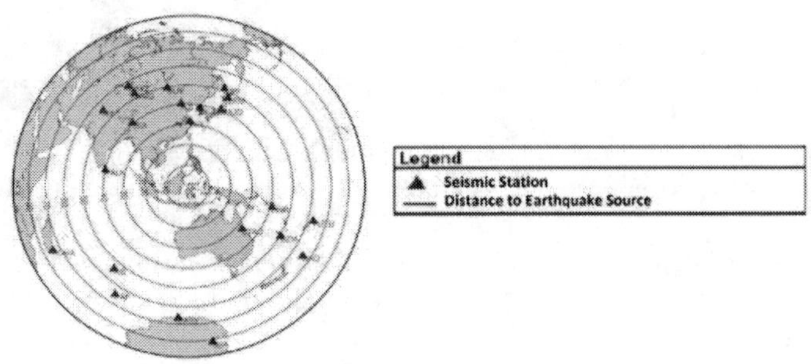

Figure 1. Distribution of seismic station locations within 30-90 ° from the earthquake source whose waveform data is used in slip distribution modeling.

This research is a case study that discusses and analyzes the source mechanism, slip distribution, and Coulomb stress from the Palu earthquake of September 28, 2018. This research consists of 2 stages, namely modeling the slip distribution and calculating Coulomb stress. The method used in slip distribution modeling is the teleseismic body wave inversion method from Kikuchi and Kanamori [2]. Then, the results of the slip distribution are used to calculate Coulomb spatial stress change. This process has done using the Coulomb 3.3 program from USGS (United States Geological Survey) [8].

The 2018 Palu earthquake waveforms used in this study were acquired from the database of IRIS (Incorporated Research Institutions for Seismology). The seismogram component used in this study is the Z component of the broadband seismometer. We collected the waveforms from the seismic stations with a distance of 30 ° - 90 ° from the earthquake source to get the teleseismic waveforms data. The seismic station distribution shows in figure 1.

3. Result and Discussion

3.1. Slip Distribution Model

The results of the inversion processing of teleseismic waveform data in figure 2 show the moment rate, focus mechanism, and slip distribution of the Palu earthquake source on September 28, 2018. The seismic velocity structure model used is the default program Jeffreys-Bullen velocity model, while the rupture velocity used is 5 km/hour. The source time function model used is the triangle model, with the assumption that the rock will immediately break after the stress energy has accumulated to a certain level.

The results of the inversion process obtained a focus mechanism that is similar to the initial input. From these results, it can be seen that the type of fault from the September 28, 2018 Palu earthquake was strike-slip with strike parameters of 350 °, dip 67 °, and rake -9 °. Besides, the seismic moment value (M_0) is 0.210E^{21} Nm or equivalent to the moment magnitude (Mw) of 7.48. The length of the moment rate or source duration which is the result of the convolution of the slip velocity and the expansion of the rupture area, is 60 seconds from 15th to 75th seconds. Within 60 seconds of the rupture duration, there are two peak energy moments in the 40th and 50th seconds.

IOP Conf. Series: Earth and Environmental Science **873** (2021) 012033 doi:10.1088/1755-1315/873/1/012033

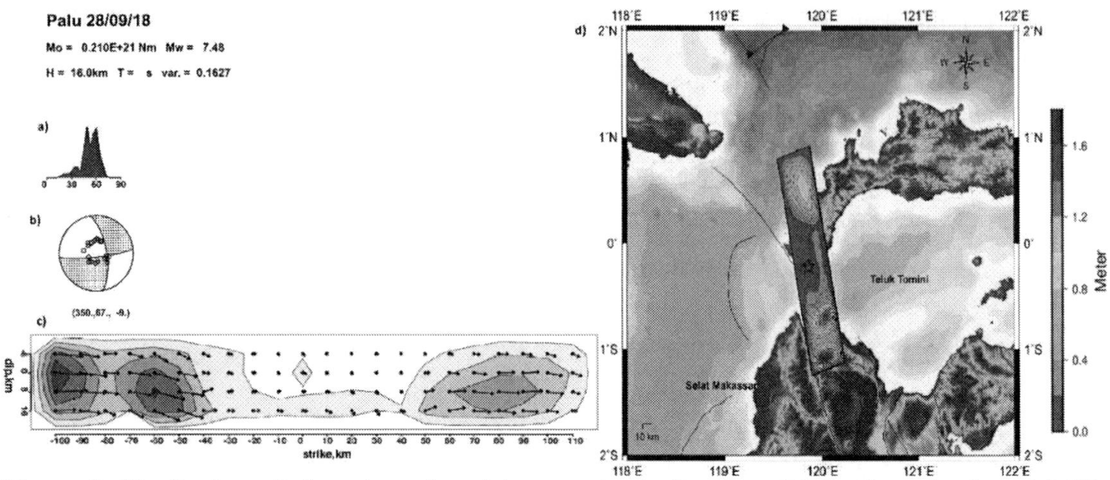

Figure 2. The final result from inversion. (a) moment rate function. (b) focal mechanism. (c) Slip distribution. (d) Slip distribution projected on the map. Asterisk indicated the initial breakpoint.

The results of the earthquake source mechanism obtained from this study are similar to the results issued by several worldwide institutions, such as the GFZ (Geo Forschungs Zentrum), USGS, BMKG (Badan Meteorologi, Klimatologi, and Geofisika), and Global CMT (Centroid Moment Tensor). The beachball produced by GFZ, USGS, BMKG, Global CMT, and this study show that the 2018 Palu earthquake source was a horizontal displacement with a left-lateral direction. Based on the epicenter location and the focal mechanism, this earthquake is associated with the Palu-Koro fault which divides Palu City in a north-south direction with a relative strike-slip type of sinistral (left-lateral) movement [13]. It is also supported by the fact that the 2018 Palu earthquake was a shallow depth earthquake [4].

The estimation of the area of the fault area takes references from several previous studies [5] [6] [10]. The results of the slip distribution show that the dominant earthquake slip is pointing to the left of the fault plane model with a maximum slip of 1.65 m. In figure 2c, it can be seen that 3 asperity zones that broke during the Palu earthquake on September 28, 2018. Based on this figure, it can also be seen that the location of the asperity zone is to the north and south of the hypocenter point with two asperity zones in the south and one asperity zone located to the north of the hypocenter. Figure 2d shows the projection of the slip distribution model in figure 2c onto the map. Based on the modeling results, the rupture process in the fault plane is preceded by an initial break or hypocenter indicated by an asterisk. Then move south to break the two zones of asperities and to the north solve one zone of asperity. The maximum slip in the Palu earthquake is in the southern part of the fault plane model. The asperity zone located in the southern part of the fault plane correlates with the location of the strongest damage in Palu City.

These results are also consistent with several studies that have been carried out, including the previous results [5] [6] [10] [11]. The comparison of the slip distribution model from those study showed in the table 1. The similarity of the slip distribution model of this study with these studies is the pattern of the aspherical zone and the location of the highest slip. Although using different data and methods, the results of the slip distribution pattern show that the aspherical zone pattern is mostly in the southern part of the fault model, while the location of the highest slip is around Palu Bay. However, these studies have differences in the maximum slip value. This difference is estimated due to differences in the data and input parameters used. The maximum slip value obtained in this study is not much different from the results of USGS [11] which obtained a maximum slip value of 1.8 meters. Meanwhile, the results of other studies show that the maximum slip varies from 2 to 8 meters at most. When compared with USGS [11], this study uses a single fault model and does not take into account the surface rupture in the earthquake source modeling. However, UGSG [11] does not only use teleseismic waveform data but also combined with SH waveform data and long-period surface wave data.

IOP Conf. Series: Earth and Environmental Science **873** (2021) 012033 doi:10.1088/1755-1315/873/1/012033

Meanwhile, Okuwaki [5] also uses teleseismic waveform data, single fault models, and a maximum rupture velocity of 5 km/s. However, the number of teleseismic waveform data and calculating the surface rupture is likely to be the cause of the maximum slip value reaching 4.25 meters. The surface rupture that appears in several locations in Palu City is estimated to cause the maximum slip calculation to be high. It also can be seen in the results of study by Socquet [6] and Ulrich [10] which make surface rupture data from satellite images for the main data. The use of this satellite image data results in the maximum slip estimates by Socquet [6] and Ulrich [10] of 5 meters and 8 meters, respectively.

Table 1. Comparison of the slip distribution model from various study

Parameter	Okuwaki (2018)	USGS (2018)	Socquet (2019)	Ulrich (2019)	This study
Data	46 Z tele-seismic P waveform	• 40 teleseismic broadband P waveforms • 12 broadband SH waveforms • 57 long period surface waves	Insar: Landsat Sentinel-2 ALOS-2	Combine: Insar + Physic based model + teleseismic	22 Z tele-seismic P waveform
Velocity Model	Crust 1.0	(1D) Crust 2.0	-	T. Yudistira Crustal Velocity Model of Sulawesi + PREM	(1D) Jeffreys-Bullen
Geometry Model	Single fault	Single fault	Multiple fault	Multiple fault	Single fault
Max Vr	5 km/s	4 km/s	4 km/s	5 km/s	5 km/s
Plane Model	240 km x 30 km (5 x 5 Subfault)	200 km x 30 km	42 x 7 Subfault	200 km strike	220 km x 32 km (10 x 8) subfault
Mo	3.2×10^{20} Nm	2.5×10^{17} Nm	3.4×10^{20} Nm		2.1×10^{20} Nm
Surface Rupture Input	Yes	No	Yes	Yes	No
Max slip	4.25 meter	1.8 meter	Strike slip 5 meter Dip slip 2 meter	8 meter	1.65 meter

3.2. Spatial Coulomb Stress Model

Coulomb spatial stress modeling due to the Palu earthquake on September 28 2018 was processed using the results of the slip distribution model. Parameters used in Coulomb stress modeling include strike, dip, rake, slip, depth, epicenter coordinates, magnitude, length and width of the fault plane model, net slip, and rake per net slip. Another input used is the friction coefficient, with a value of 0.4 for the strike-slip fault type [8]. Initial break modeling in the Coulomb stress model uses a fault center with the hypocenter point in the middle of the fault plane. The calculation of the area of the fault plane is assumed to be an isotropic homogeneous rectangular area. The fault plane model is made along 220 km × 32 km which is divided into 22 grid boxes with the strike direction and 4 grid boxes with the dip direction. The size per grid is 10 km × 8 km in the direction of strike and dip, respectively. Furthermore, in each grid box, the net slip and rake values of the grid points in the fault plane model are entered in the previous process.

IOP Conf. Series: Earth and Environmental Science **873** (2021) 012033 doi:10.1088/1755-1315/873/1/012033

Figure 3 shows the changes in the spatial Coulomb stress caused by the Palu earthquake on September 28, 2018. Areas in red indicate an increase in Coulomb stress while the blue ones indicate a Coulomb stress decrease. The width of the stress change area is the result of the influence of magnitude. The greater the magnitude of an earthquake, the larger the area of Coulomb stress changes. The pattern of increasing Coulomb stress value which is dominated at each end of the fault is caused by the relatively sharp dip angle and the slip direction which tends to be parallel to the strike direction. The results of the Coulomb stress change plot resulted in 12 areas that experienced an increase and decrease in Coulomb stress. In addition to being at the ends of the fault model, the increase in Coulomb stress is also spread over several areas above the fault plane model, this is probably because the input used in this study is a fault plane model with slip distribution.

Different from previous research by Utaminingtyas [12], changes in Coulomb stress results from this study are more dominated by stress reduction areas both in the east and west of the fault area. Meanwhile, Utaminingtyas [12], which does not use slip distribution inputs, shows that the eastern area of the fault area has increased stress while the west part is dominated by a stress reduction zone. Differences also exist in the zones of increased stress with a small area around the fault plane due to variations in the length and direction of the slip along the fault plane. The use of uniform length and slip direction values in Utaminingtyas [12] causes the absence of stress-increasing zones with this small area.

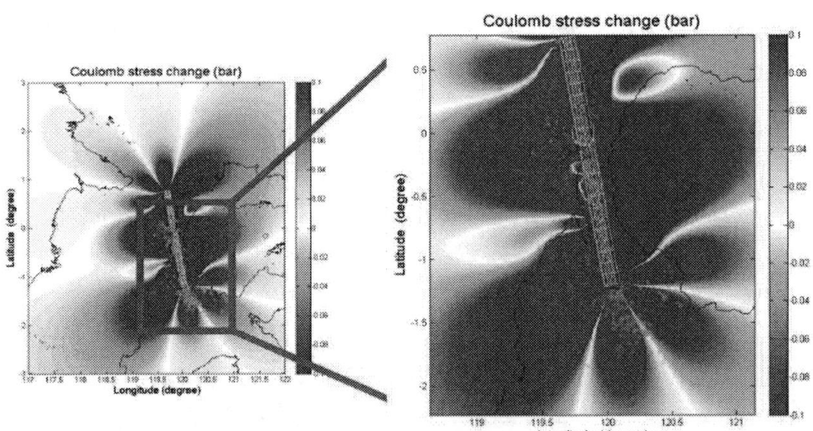

Figure 3. Coulomb stress changes due to the major earthquake. The orange circle indicates aftershocks distribution.

When an area experiences increased stress, the area has the potential to experience an earthquake triggered by a previous earthquake. Seismic activity triggering is a process where changes in stress activity caused by an earthquake can cause or slow down seismic activity in the surrounding area. In the 2018 Palu earthquake, there were a total of 573 aftershocks that occurred between September 28, to November 21, 2018 [12]. To compare the relationship between the spatial distribution of changes in Coulomb stress and aftershocks, the aftershocks parameter that was relocated by Utaminingtyas [12] used hypoDD method.

In figure 3, which shows an overlay of the epicenter location of aftershocks with the distribution of Coulomb stress changes, it can be seen that the concentration of aftershocks tends to be in the northwest hypocenter area and the southern part of the fault plane model. The concentration of aftershocks in the southern part of the fault plane model correlates with an increase in Coulomb stress, which is consistent with previous studies [9] [14]. Although these aftershocks are outside the main fault plane, they are also potentially triggered by changes in stress caused by major earthquakes [14]. Meanwhile, the aftershocks in the northwest of the hypocenter were seen in the stress reduction zone. This discrepancy is thought to be due to the simplification of the fault plane model using only a single fault. This study has not yet accommodated a more complex multi-fault model as in the research of Socquet [6]. Based on Socquet

[6] who analyzed surface deformation using satellite data, the rupture of the Palu September 28, 2018 earthquake had a turn to the northwest after the hypocenter so that it is estimated that the area of concentration of aftershocks is still in the fault plane.

4. Conclusions

The Palu earthquake of 28 September 2018 which had a magnitude of Mw 7.48 was a left-lateral strike-slip fault with a strike parameter of 350 °, dip 67 °, and rake -9 °. Based on the results of the slip distribution, three asperity zones occurred during the Palu earthquake on September 28, 2018. One asperity zone is in the north of the fault plane model and two are in the southern part of the fault plane model. The greatest energy release zone is in the southern asperity zone with a maximum slip value of 1.65 meters. Compare to the other studies, this slip distribution model result is similar to various approaches although using minimum data. This study also confirms that the calculation of the maximum slip requires a surface rupture input. Some studies that use surface rupture data generate higher maximum slip than those that only use seismic data.

The results of the Coulomb spatial stress of the Palu 28 September 2018 earthquake showed an increase in the dominant stress in the northern and southern ends of the fault plane model. Also visible are several areas above the fault plane model that experience increased stress. The concentration of aftershocks in the area of increased stress in the outer part of the fault plane model is estimated to be due to the triggering of an earthquake by the main Palu earthquake on September 28, 2018. Meanwhile, for aftershocks that are still in part of the fault plane model, it is estimated that they are still a series of releases of main earthquake energy.

References

[1] Kikuchi M and Kanamori H 1982 Inversion of complex body waves *Bull. Seismol. Soc. Am.* **72** 491-506.
[2] Kikuchi M and Kanamori H 2003 *Manual of Teleseismic Body-Wave Inversion Program.* Retrieved from http://wwweic.eri.u-tokyo.ac.jp/ETAL/KIKUCHI/.
[3] King G C, Stein R S and Lin J 1994 Static stress changes and the triggering of earthquakes. *Bull. Seismol. Soc. Am.* **84** 935 - 953.
[4] Massiani M F, Nugraha A D, Ramdhan M and Wandono 2017 Precise hypocenter determination around Palu Koro fault: a preliminary result *IOP Conf. Ser. Earth Environ. Sci.* **62**.
[5] Okuwaki R, Hirano S, Yagi Y and Shimizu K 2020 Inchworm-like source evolution through a geometrically complex fault fueled persistent supershear rupture during the 2018 Palu Indonesia earthquake *Earth Planet. Sci. Lett.* doi:10.1016/j.epsl.2020.116449
[6] Socquet A, Hollingsworth J, Pathier E and Bouchon M 2019 Evidence of supershear during the 2018 magnitude 7.5 Palu earthquake from space geodesy. *Nat. Geosci.* **12** 307.
[7] Suardi I, Yagi Y, Widyantoro S, Afnimar and Meilano I 2013 Analysis of source rupture process of the September 2, 2009 Tasikmalaya earthquake by using the joint inversion method of near field and teleseismic data *Int. J. Tomogr. Simul.* **27**.
[8] Toda S, Stein R S, Sevilgen V and Lin J 2011 *Coulomb 3.3 Graphic-Rich Deformation and Stress-Change Software for Earthquake, Tectonic,and Volcano Research and Teaching—User Guide* (Virginia: USGS) p 63.
[9] Toda S and Stein R S 2015 2014 Mw 6.0 South Napa earthquake triggered exotic seismic clusters near several major faults *Bull. Seismol. Soc. Am.* **86**.
[10] Ulrich T, Vater S, Madden E, et al. 2019 Coupled, physics-based modeling reveals earthquake displacements are critical to the 2018. *Pure Appl. Geophys.* **176** 4069–4109.
[11] USGS 2018 M 7.5 - 70km N of Palu, Indonesia. Retrieved from https://earthquake.usgs.gov/earthquakes/eventpage/us1000h3p4/finite-fault.
[12] Utaminingtyas B 2019 Analisis perubahan static coulomb stress dan pemicuan aktivitas gempa (Studi kasus: gempa Palu-Donggala 28 September 2018) *Bsc Thesis* Sekolah Tinggi Meteorologi Klimatologi dan Geofisika.

IOP Conf. Series: Earth and Environmental Science **873** (2021) 012033 doi:10.1088/1755-1315/873/1/012033

[13] Watkinson I M and Hall R 2017 Fault systems of the eastern Indonesian triple junction: evaluation of Quaternary activity and implications for seismic hazards *Geohazards in Indonesia: Earth science for disaster risk reduction* ed Cummins P R and Meilano I (London: Geological Society of London) pp 71-120.

[14] Wu J, Cai Y, Li W and Feng Q 2017 Strong aftershock study based on coulomb stress triggering—A case study on the 2016 Ecuador Mw 7.8 earthquake *Appl. Sci.* **7**(1):88

Results of land subsidence measurement using GPS method in the Jakarta groundwater basin in 2015-2019

F M Abdullah, H Andriyanto, J R Nababan, F Abdillah and R I H Sulistyawan

Groundwater Conservation Unit, Geological Agency, Ministry of Energy and Mineral Resource, Indonesia

E-mail : firmanmaliki@gmail.com, firman.abdullah@esdm.go.id

Abstract. The development of population and development activities in big cities in Indonesia, especially in the city of Jakarta and surrounding areas is very rapid. From several land subsidence studies, several factors have been identified that cause land subsidence, namely: excessive groundwater extraction, reduction due to building/infrastructure loads, subsidence due to natural consolidation of soft soil layers, and subsidence due to tectonic forces. At present the exploitation of ground water for industrial and residential needs is at a level that needs attention. Excessive pumping of groundwater will cause a decrease in the quantity of ground water, entry of seawater into the land (sea water intrusion) and land subsidence. Symptoms of the negative impact of land subsidence have been felt in several areas, especially in industrial areas located in the northern part of Jakarta. This land subsidence can be measured by GPS or satellite geodetic method, which have begun to develop in Indonesia in the past two decades. Measurements were made using the radial method at 53 GPS points in 2015 up to 100 measurement points in 2019 in Jakarta Groundwater Basin. The result of these campaign GPS surveys that is northern part of Jakarta relatively had higher subsidence rate than the southern. The largest subsidence almost reached 6.2 cm/year in Muara Baru in northern area which is southern area only suffered an average rate of 1.16 cm/year.

Key words : land subsidence, GPS survey method, Jakarta Groundwater Basin

1. Introduction

Jakarta is the capital city of Indonesia and considered to be the largest and densest population in the South East Asia. Jakarta is located at Jakarta Groundwater Basin which is consist of mainly young volcanic and sediment deposits. The Jakarta area is relatively flat in the northen part and slightly hilly in the southern part. Administratively, the Jakarta Groundwater Basin consists of DKI Jakarta itself, eastern part of Tangerang City and Regencies, Tangerang Selatan City, western part of Bekasi City and Bekasi Regencies, and northern part of Depok City and Bogor Regencies (Figure 1).

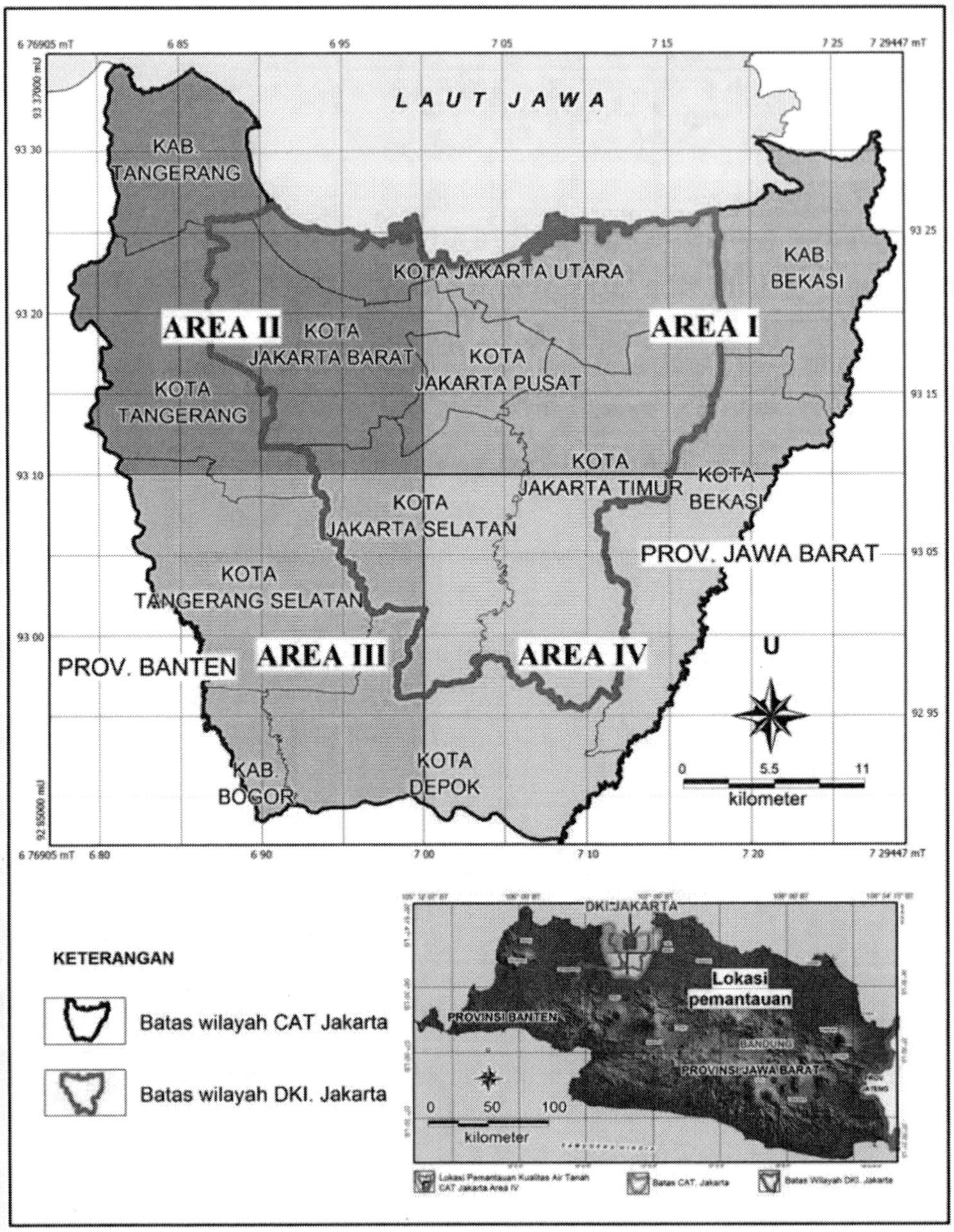

Figure 1. The location of Jakarta Groundwater Basin

Hydrogeologically, the base of aquifer system is formed by impermeable miocene sediments which also cropped at southern boundary of the basin. The basin mainly consist of quarternary alluvial volcanic fan deposit that formed from tuffaceous and conglomerate sandstone, young river deposit, swamp deposit, sand bar deposit, Banten Tuff, Serpong Formation, and Bojongmanik Formation. Fachri et all called quarternary alluvial volcanic fan deposit as Citalang Formation which is the main aquifer of Jakarta [16].

We used the term of Jakarta Groundwater Basin rather than Jakarta itself because of our main task and duty not only considering Jakarta but our main responsibility covered the whole of Jakarta Groundwater Basin. The objective of this research is to estimate the rate of land subsidence that has occurred from 2015 to 2019 by using GPS- satellite based positioning system- in the Jakarta Groundwater Basin, and therefore the results can be used as basic data or information in carrying out further planning and policy by the government or interested party.

2. Data and Methods

Land subsidence in Jakarta and its neighboring area had been reported in several scientific investigation reports. There are some scientific methods usually used for investigating or studying land subsidence phenomena such as GPS campaign survey, analysis of In Sar data, extentiometer analysis method, and so on. In this study we used GPS survey method which is relatively flexible and could cover wide area.

The method of GPS survey is re-measuring GPS points every year in the Jakarta Groundwater Basin. The GPS points that are measured annualy must be the same points. After recording the GPS data from satellites for 6-8 hours, data will be processed later (post processing). The GPS surveys used dual frequency geodetic type GPS receivers, data was taken with interval 15-30 seconds, elevation mask 10-15°.

The GPS receiver used in these surveys was Trimble R7 (2015-2017) and Trimble R9 (2018-2019) and the processing software was Trimble Business Centre v 2.5 (2015-2017) and v 4.0. (2018-2019). The GPS data processed with the nearest reference station to the GPS points. Those reference stations which will be used as benchmark of measurement are managed by Geospatial Information Agency. Every baselines must be resolved to get the best results. Satellites with large standard deviations or have too many cycle slips when data was recorded must be edited or eliminated.

The results are ellipsoid heigths in those points. More experience in processing data will generate better results. These results will be compared every year, and therefore can describe the ellipsoid heigths differences from time to time at the same points. Every year we got ellipsoidal heights in all GPS points, then we got height differences on these points. Average results are the average changes per year in the year 2015 to 2019. The positive results indicated the ground surface level are up, and the negative results indicated ground surface level are down or subside. The location of the initial GPS points in 2015 can be seen in figure 2.

IOP Conf. Series: Earth and Environmental Science **873** (2021) 012034 doi:10.1088/1755-1315/873/1/012034

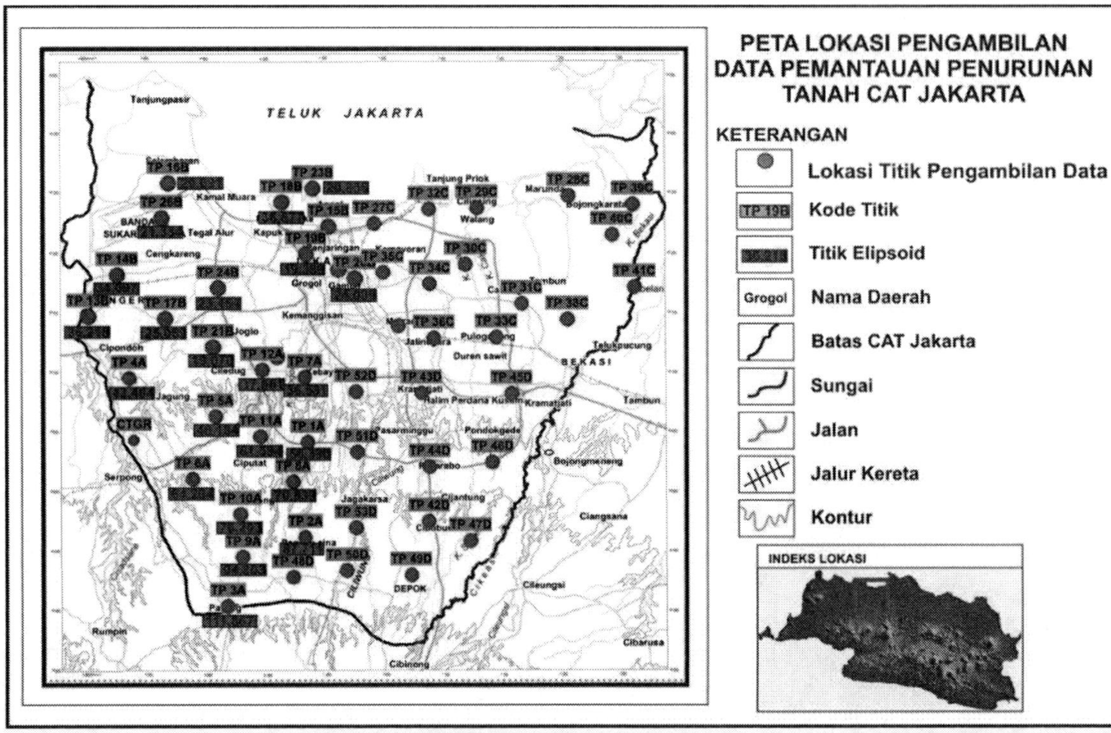

Figure 2.. The initial GPS Points in 2015

GPS survey campaign first started in 2015 with 53 GPS points. In 2019 we measured up to 100 GPS points. The selected results will be showed in Table 1. which are only contain the results of GPS measurement (ellipsoid height) in northern area of Jakarta Groundwater Basin and Table 2 which had the results in southern basin[1-12,14-15,17-20]. The positive average changes mean the ground surface was up whereas negative average changes mean the land surface was subsided. There are no GPS survey conducted in 2017.We only selected the results with relative good acccuracy.

Table 1. Ellipsoid height results in northern area

GPS points	2015	2016	2017	2018	2019	Average changes (cm/year)
13B, Cikokol	36.161	36.157	36.168	36.102	36.117	-1.467
15B, Joglo	32.969	32.914	32.902	32.877	-	-3.067
17B, Duri Kosambi	25.046	25.054	25.024	24.898	24.807	-5.975
19B, Tongkol	19.389	19.371	19.327	19.273	19.249	-3.500
21B, Muara Baru	19.130	19.031	19.043	18.944	18.883	-6.175
23B, Latumenten	20.839	20.834	20.846	20.794	20.762	-1.925
24B, Gambir	23.422	23.364	23.409	23.372	23.399	-0.575
26B, Dadap	21.333	21.262	21.233	21.156	21.092	-6.025
28B, Kamal Muara			20.507	20.426	20.443	-1.600
30B, Kali Deres 2			21.473	21.432	21.362	-2.775
31B, PIK			19.231	19.094	19.063	-4.200
35B, Muara Angke			18.756	18.724	18.638	-2.950
27C, Kayu Manis	28.898	28.884	28.877	28.915	-	0.567
28C, Klender		29.116	29.167	28.977	28.986	-4.333
29C, Medan Satria	27.424	27.424	27.340	27.299	27.304	-3.000
31C, Babelan	22.080	22.071	22.067	21.977	21.940	-3.500
32C, Buni Bhakti	21.539	21.537	-	21.525	21.525	-0.267
33C, Penggilingan	30.166	30.135	30.261	30.171	30.189	0.575
35C, Kemayoran		20.578	20.555	20.468	20.561	-0.567
36C, Tj. Priok	20.271	20.263	20.266	20.240	20.235	-0.900
38C, Cakung	26.932	26.939	27.030	26.972	-	1.333
40C, Marunda	22.142	22.070	21.978	22.032	22.033	-3.633
41C, Tarumajaya	21.455	21.349	21.332	21.320	21.284	-5.700
44C, Pulomas 2			21.973	21.903	21.844	-4.300

Table 2. Ellipsoid height results in southern area

GPS points	2015	2017	2018	2019	Average changes (cm/year)
01A, Fatmawati	56.316	36.168	36.102	56.266	-1.150
04A, Alam Sutera	43.494	43.477	43.423	43.499	0.125
08A, Cinere	70.961	70.963	70.931		-1.000
09A, Cinangka	95.088	95.082	95.100	95.094	-0.150
10A, Pamulang	74.515	74.457	74.458	74.444	-1.775
11A, Rempoa	59.773	59.744	59.652	59.712	-1.525
42D, Cibubur	83.599	83.559	83.445	83.437	-4.050
44D, Halim	37.204	37.210	37.158	37.161	-1.075
45D, Jati Sampurna	90.317	-	90.287	90.247	-1.750
46D, Pondok Melati	59.895	59.877	59.888	59.887	-0.200
50D, Margonda	104.155	-	104.102	104.064	-3.033
51D, Kebagusan	54.286	-	54.309	54.294	0.200
53D, Pondok Cina		85.743	85.735	85.752	0.300

3. Conclusion and Discussions

It appears from the results of the GPS measurement.survey the northern part is relatively have higher rate subsidence than the southern part, and then can be said that the subsidence is heteregenous in time and place. There is no land subsidence more than 10 cm / year in the period of 2015-2019 using GPS surveys method. The largest subsidence almost reached 6.2 cm/year in Muara Baru in northern area which is southern area only suffered an average rate of 1.16 cm/year.

What is interesting is, there are some GPS points in the southern part of the Jakarta Groundwater Basin which have decreased up to several cm, namely the measurement points in Cibubur (42D) and Margonda (50D), but still need further research to confirm this condition. Implementing a continuously GPS network may be usefull because of possibly large temporal variations in the rates of subsidence which is difficult to observed by episodic GPS measurements[15]. However the GPS survey method is not the only way to study or investigate land subsidence phenomena. Other researchers can use extensiometer method and/or InSAR data to estimates the rate of land subsidence.

References

[1] Abdillah F 2019 Pemantauan Penurunan Tanah Cekungan Air Tanah Jakarta Area III 2019, Balai Konservasi Air Tanah, Pusat Air Tanah dan Geologi Tata Lingkungan, Badan Geologi, 2019
[2] Abdullah F M 2015 Pengukuran / Topografi Geodetik Perubahan Muka Tanah dan Amblesan Cekungan Air Tanah Jakarta Area I 2015, Balai Konservasi Air Tanah, Pusat Sumber Daya Air Tanah dan Geologi Lingkungan, Badan Geologi, 2015
[3] Abdullah F M 2015 Pengukuran / Topografi Geodetik Perubahan Muka Tanah dan Amblesan Cekungan Air Tanah Jakarta Area II 2015, Balai Konservasi Air Tanah, Pusat Sumber Daya Air Tanah dan Geologi Lingkungan, Badan Geologi, 2015
[4] Abdullah F M 2015 Pengukuran / Topografi Geodetik Perubahan Muka Tanah dan Amblesan Cekungan Air Tanah Jakarta Area III 2015, Balai Konservasi Air Tanah, Pusat Sumber Daya Air Tanah dan Geologi Lingkungan, Badan Geologi, 2015
[5] Abdullah F M 2015 Pengukuran / Topografi Geodetik Perubahan Muka Tanah dan Amblesan Cekungan Air Tanah Jakarta Area III 2015, Balai Konservasi Air Tanah, Pusat Sumber Daya Air Tanah dan Geologi Lingkungan, Badan Geologi, 2015
[6] Abdullah F M 2016 Pemantauan Penurunan Tanah Cekungan Air Tanah Jakarta Area I 2016, Balai Konservasi Air Tanah, Pusat Air Tanah dan Geologi Tata Lingkungan, Badan Geologi, 2016
[7] Abdullah F M 2016 Pemantauan Penurunan Tanah Cekungan Air Tanah Jakarta Area II 2016, Balai Konservasi Air Tanah, Pusat Air Tanah dan Geologi Tata Lingkungan, Badan Geologi, 2016
[8] Abdullah F M 2017 Pemantauan Penurunan Tanah Cekungan Air Tanah Jakarta Area II 2017, Balai Konservasi Air Tanah, Pusat Air Tanah dan Geologi Tata Lingkungan, Badan Geologi, 2017
[9] Abdullah F M 2017 Pemantauan Penurunan Tanah Cekungan Air Tanah Jakarta Area III 2017, Balai Konservasi Air Tanah, Pusat Air Tanah dan Geologi Tata Lingkungan, Badan Geologi, 2017
[10] Abdullah F M 2018 Pemantauan Penurunan Tanah Cekungan Air Tanah Jakarta Area II 2018, Balai Konservasi Air Tanah, Pusat Air Tanah dan Geologi Tata Lingkungan, Badan Geologi, 2018

[11] Abdullah F M 2018 Pemantauan Penurunan Tanah Cekungan Air Tanah Jakarta Area III 2018, Balai Konservasi Air Tanah, Pusat Air Tanah dan Geologi Tata Lingkungan, Badan Geologi, 2018

[12] Abdullah F M 2018 Pemantauan Penurunan Tanah Cekungan Air Tanah Jakarta Area I 2019, Balai Konservasi Air Tanah, Pusat Air Tanah dan Geologi Tata Lingkungan, Badan Geologi, 2019

[13] Abidin H Z Andreas H., Djaja R., Darmawan D., Gamal M., Land Subsidence Characteristics of Jakarta between 1997 and 2005, as estimated using GPS Surveys, Springer Verlag, 2007

[14] Andriyanto H 2018 Pemantauan Penurunan Tanah Cekungan Air Tanah Jakarta Area IV 2018, Balai Konservasi Air Tanah, Pusat Air Tanah dan Geologi Tata Lingkungan, Badan Geologi, 2018

[15] Andriyanto H 2019 Pemantauan Penurunan Tanah Cekungan Air Tanah Jakarta Area IV 2019, Balai Konservasi Air Tanah, Pusat Air Tanah dan Geologi Tata Lingkungan, Badan Geologi, 2019

[16] Fachri M, Hutasoit L M, Djuhaeni and Ramdhan A M 2002 Stratigrafi dan Hidrostratigrafi Cekungan Airtanah Jakarta, Buletin Geologi, **34**, No 3 Departemen Teknik Geologi ITB, 2002

[17] Nababan, J R S 2017 Pemantauan Penurunan Tanah Cekungan Air Tanah Jakarta Area I 2017, Balai Konservasi Air Tanah, Pusat Air Tanah dan Geologi Tata Lingkungan, Badan Geologi, 2017

[18] Nababan, J R S 2017 Pemantauan Penurunan Tanah Cekungan Air Tanah Jakarta Area IV 2017, Balai Konservasi Air Tanah, Pusat Air Tanah dan Geologi Tata Lingkungan, Badan Geologi, 2017

[19] Nababan, J R S 2018 Pemantauan Penurunan Tanah Cekungan Air Tanah Jakarta Area I 2018, Balai Konservasi Air Tanah, Pusat Air Tanah dan Geologi Tata Lingkungan, Badan Geologi, 2018

[20] Nababan, J R S 2018 Pemantauan Penurunan Tanah Cekungan Air Tanah Jakarta Area II 2019, Balai Konservasi Air Tanah, Pusat Air Tanah dan Geologi Tata Lingkungan, Badan Geologi, 2018

Development Prototype of Indonesia Seismic Microzonation Information System (Inasmis)

Thomas Hardy*, Arif Rachman Hakim, Supriyanto Rohadi, Angga Setiyo Prayogo, Drajat Ngadmanto, Pupung Susilanto, Rasmid, Bambang Sunardi, Suyanti Pakpahan, Wiko Setyonegoro, Telly Kurniawan, Jimmi Nugraha, Yusuf Hadi Perdana, Tio Azhar P. Setiadi, Sulastri, Aditya Rahman, Aprilia Nur Vita, Priyobudi, Agustya Adi Martha, Asep Nur Rachman, Jaya Murjaya, Muzli, Rahmat Setyo Yuliatmoko

Meteorological, Climatological and Geophysical Agency of Indonesia (BMKG), Jl Angkasa 1 No. 2 Kemayoran Jakarta Pusat 10720

thomas.hardy@bmkg.go.id

Abstract. Indonesia has a high level of seismic vulnerability because located in the junction area of four large plates, namely the Eurasian plate, Indo-Australian plate, Pacific plate, and Philippine Plate, and also the number of active faults that stretch throughout its regions. One of the efforts to mitigate earthquake disasters is to conduct seismic hazard microzonation efforts. The Research and Development Center of BMKG has carried out ongoing research in the field of seismic microzonation to obtain information on the level of seismic hazard in some regions based on microtremor parameter values, including in the Sukabumi (2009), Bantul (2010), Padang (2011), Cilacap (2012), Kulonprogo (2014), Tasikmalaya (2017), Garut (2018) and Pangandaran (2019). This study tries to prototype the seismic microzonation information system in a web-based spatial information system called InaSMIS (Indonesia Seismic Microzonation Information System). InaSIMS contains information about the analysis of microtremor survey results, including the value of the dominant period (T0), Ground Shear Strain (GSS), and USGS Vs30, which states the level of seismic hazard in an area. InaSMIS is still being developed in the research stage and is not yet an operational service for BMKG. In the future, InaSMIS expected to be a source of public information to determine seismic hazards in Indonesian regions.

Keywords: *InaSMIS, microzonation, seismic hazards*

1. Introduction

Indonesia has a high seismic vulnerability level because located in the junction area of four large plates, namely the Eurasian plate, Indo-Australian plate, Pacific plate, and Philippine Plate. The sources of earthquakes on land also come from active fault lines that stretch almost all parts of Indonesia. The occurrences of the earthquake and tsunami that have occurred have certainly caused casualties, damage to buildings, and significant economic losses. BNPB recorded losses due to natural disasters, including earthquakes in 2019, reaching 80 trillion [1].

Until now, earthquakes has not been able to accurately predict the location, time, and how much strength [2]. Therefore, earthquake hazard mitigation efforts are very important to do to build preparedness so that casualties and the risk of socio-economic loss can be minimized. One of the many earthquake mitigation efforts that have been carried out is sesimic microzonation. Seismic microzonation is defined as the process of subdividing an area into zones with respect to geological characteristics of the sites, so that seismic hazards at different locations within a city can correctly be identified [3]

The Research and Development Center of BMKG has carried out ongoing research in the field of seismic microzonation to obtain information on the level of seismic hazard in some regions, including in the Sukabumi (2009), Bantul (2010), Padang (2011), Cilacap (2012), Kulonprogo (2014), Tasikmalaya (2017), Garut (2018), and Pangandaran(2019) [4]. Several methods have applied in seismic microzonation research, such as the Horizontal-to-Vertical Spectral Ratio (HVSR) method on microtremor data to determine the local site effect [5, 6, 7, 8, 9], the Multichannel Analysis of Surface Waves (MASW) method to determine the value of Vs30 so that the soil type of an area can be determined [10, 11, 12, 13] and Probabilistic Seismic Hazard Analysis (PSHA) method for knowing the Peak Ground Acceleration (PGA) of a region [14, 15, 16, 17]

Unfortunately, the results of seismic microzonation measurements have not been collected in an integrated and easily accessible database (web-based). The data from microzonation research are still scattered and have not yet become a single unit of data. This research tries to collect all the data from the microzonation measurements that have been carried out by the Research and Development Center of BMKG from 2009-2018 and present the results in a web-based prototype spatial database. We expect in the future, the scientific community and the general can get information about the vulnerability of earthquakes in a particular region.

2. Data and method

This research tries to make a web-based spatial database prototype called InaSMIS (Indonesia Seismic Microzonation Information System) to display the results of microzonation research that has been carried out by the BMKG Research and Development Center. InaSMIS designed using the concept of an interactive web-based Geographical Information System (GIS) where users can display earthquake hazard level data in an area without any prior GIS knowledge. The maps displayed include the predominant period, ground shear strain (GSS) Map, USGS Vs30 Model, and Earthquake Catalog. Figure 1 shows the InaSMIS application design concept.

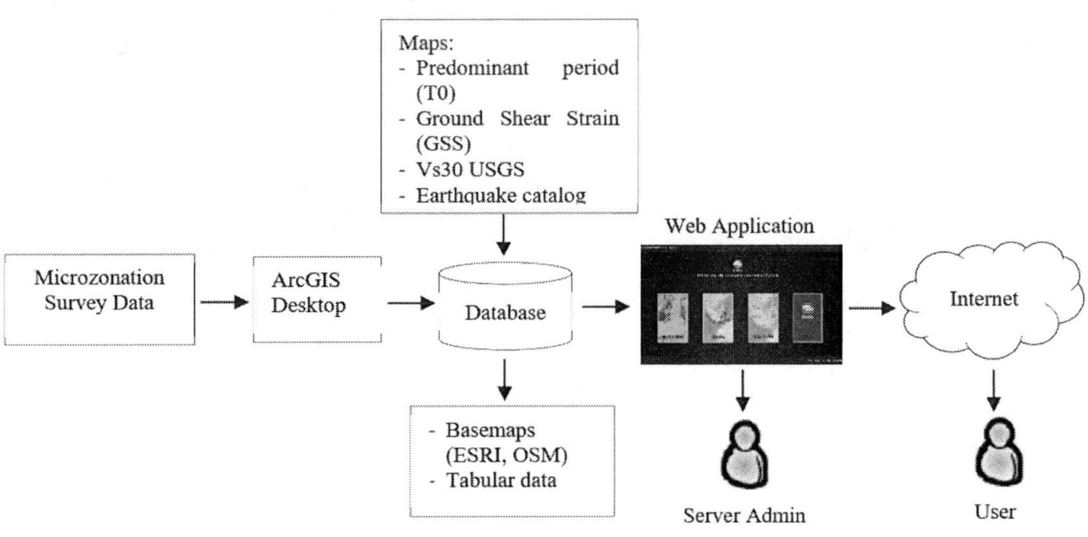

Figure 1. InaSMIS application design concept

The predominant period is the time required for microtremor waves to propagate in the surface sediment layer. Predominant period value can evaluate the subsurface layer's site class to characterize soil properties in a particular area. We use site classification proposed by [18]. Ground Shear Strain (GSS) is a soil layer material's ability to stretch or shift during an earthquake. The relationship between strain and dynamic properties of soil uses the classification introduced by Nakamura [19]. Areas with a high GSS value means having an increased risk of ground motion due to earthquakes, leading to landslides, land subsidence, and liquefaction. The Vs30 USGS Model, based on high-resolution topographic data, is the time-averaged shear-wave velocity (VS) in the upper 30 meters and adopted by the earthquake engineering community to account for seismic site conditions [20]. For catalog data, it obtained from the BMKG earthquake repository [21]

3. Results and Discussion

The InaSMIS prototype has been developed and can access at http://puslitbang.bmkg.go.id/inasmis. This application consists of several menus, including Spatial Data, Earthquake Catalog, Slab Model, and About. Figure 2 shows the interface of InaSMIS. The map view of the results of seismic microzonation for several regions displayed in the Spatial data menu, which consists of the USGS Vs30 Model Map for all of Indonesia and the Dominant Period and Ground Shear Strain (GSS) Maps for each study area. For the basemap, we can use three options to display, namely: Esri Topographic Map, Light Gray Canvas, and Open Street Map (OSM).

The Vs30 map of the USGS model for all Indonesia regions shown in Figure 3, where the Vs30 values obtained are correlated with SNI 1726-2012 to get rock classes. Information on seismic microzonation research results for several areas that have been carried out shown in Figure 4. Examples of maps of the predominant period and ground shear strain for the Garut City area shown in Figure 5 and Figure 6.

An earthquake catalog map from BMKG is also displayed, with a magnitude range of 1-9. We can show all magnitudes' earthquake events or select them based on the magnitude range: 1 - 4.5, 4.5 - 5.5, and 5.5 - 9.5. The earthquake catalog data from BMKG shown in Figure 7. The Slab Model menu is still under development, while the About menu contains an explanation of the InaSMIS application.

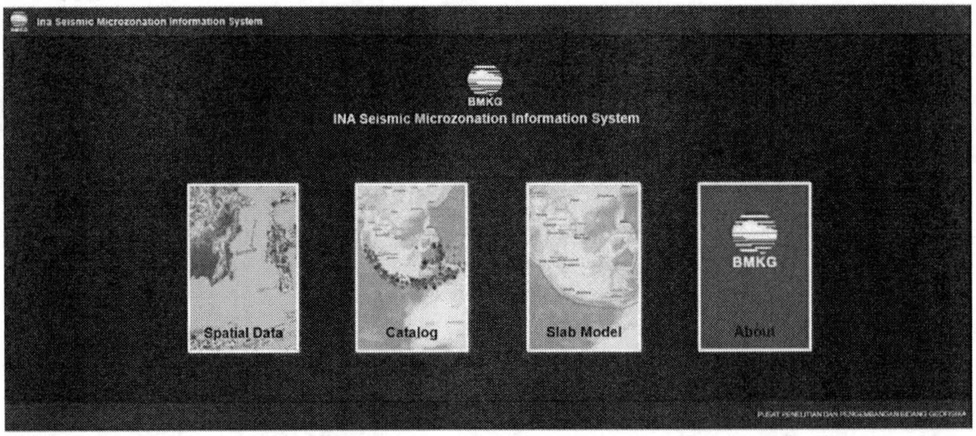

Figure 2. InaSMIS interface, consisting of 4 main menus: Spatial data, Catalog, Slab Model, and About

IOP Conf. Series: Earth and Environmental Science **873** (2021) 012035 doi:10.1088/1755-1315/873/1/012035

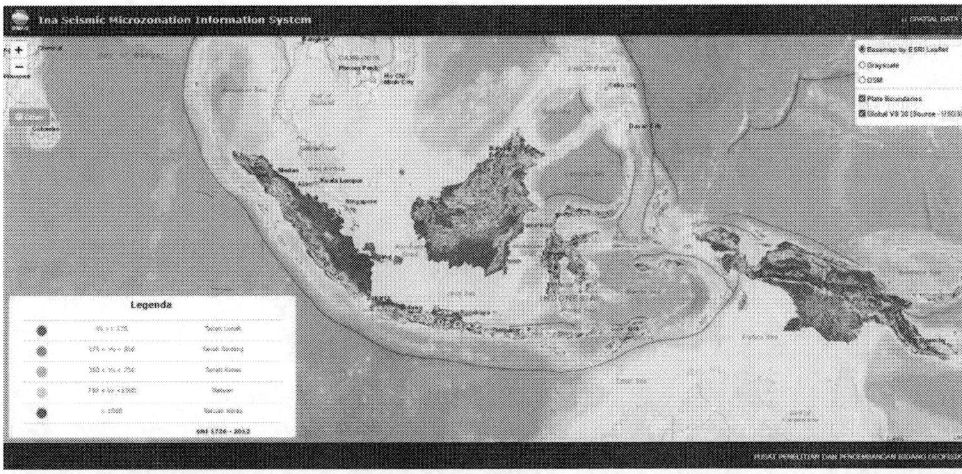

Figure 3. The USGS Vs30 model for all regions of Indonesia, correlated with SNI 1726-2012 for site classification

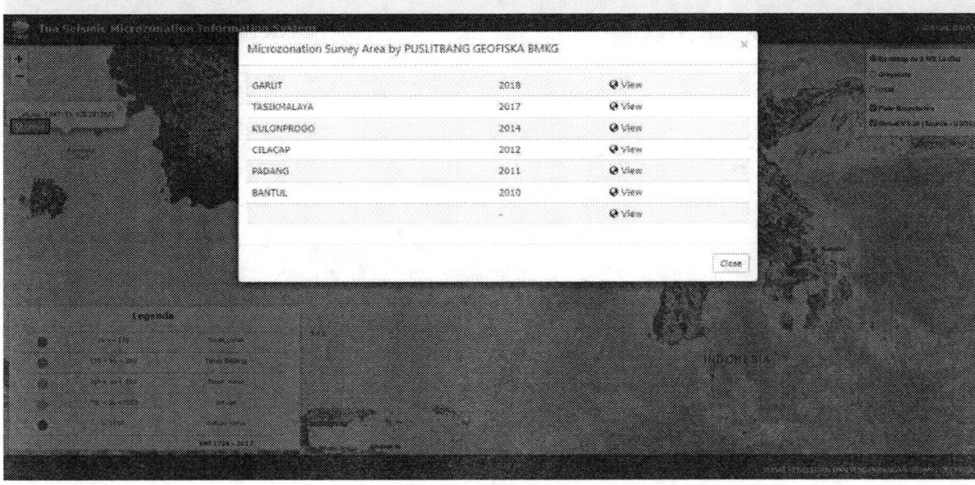

Figure 4. Selected display of seismic microzonation research results in several areas that have been carried out.

IOP Conf. Series: Earth and Environmental Science **873** (2021) 012035 doi:10.1088/1755-1315/873/1/012035

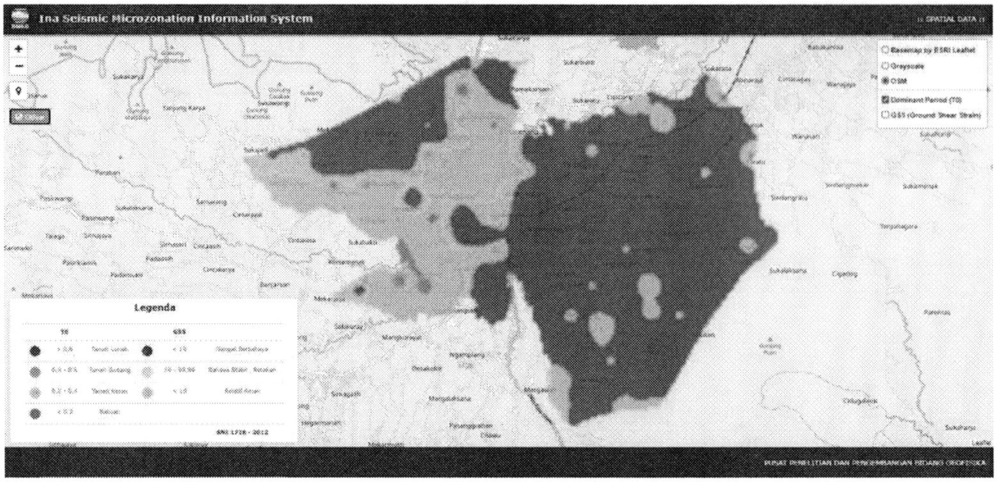

Figure 5. Map of the value of the predominant period for the Garut City area

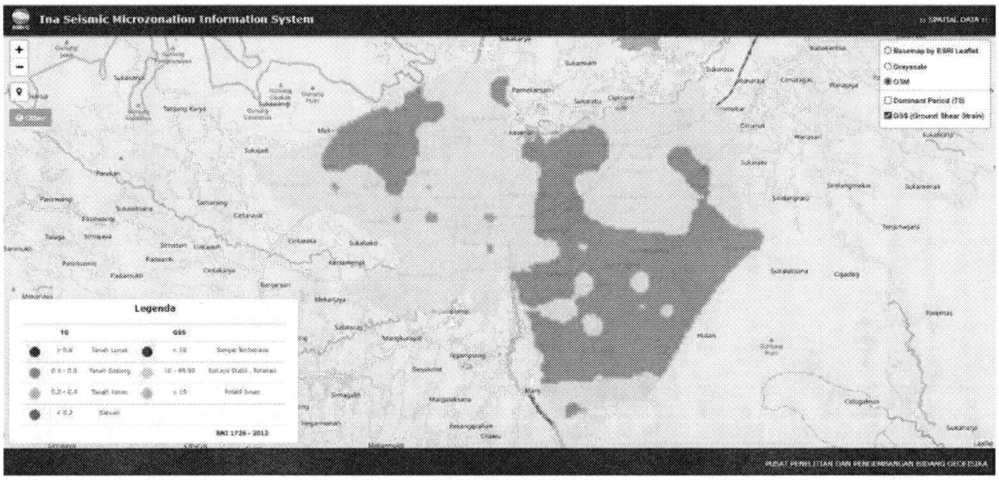

Figure 6. Map of Ground Shear Strain (GSS) for the Garut City area

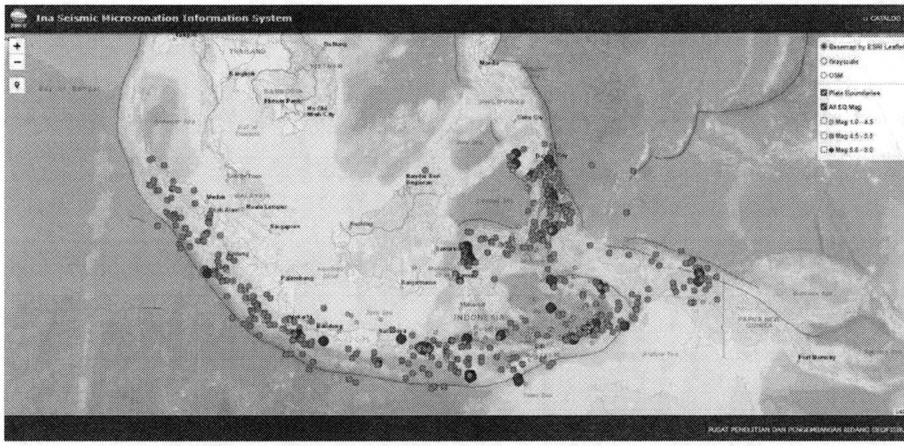

Figure 7. Display of BMKG's earthquakes catalog

InaSMIS application is still being developed in the prototype stage because there are still a lot of seismic microzonation data that has not been included, including Vs30 measurement results using the Multichannel Analysis Surface Waves (MASW) method, Vs30 from spatial autocorrelation technique (SPAC), Probability Seismic Hazard Assessment (PSHA), Seismic Analysis Risk, and the collection of seismic microzonation measurement results that have been carried out by the other division of BMKG.

4. Conclusion

InaSMIS has been developed to provide information on the level of potential seismic hazards in several regions in Indonesia, measured by the BMKG Research and Development Center. InaSMIS is still being developed in the research stage and is not yet an operational service for BMKG. In the future, InaSMIS is expected to be a source of public information to determine seismic hazard level in Indonesian regions.

5. Acknowledgments

This study was funded by the Research and Development Centre of BMKG. The support is gratefully acknowledged.

References

[1] https://nasional.kontan.co.id/news/bnpb-kerugian-yang-diderita-sepanjang-bencana-2019-capai-rp-80 triliun#:~:text=Sepanjang%20tahun%202019%2C%20Badan%20Nasional,bulan%20Juli%20hingga%20November%20lalu., accesed 23rd October 2020
[2] Geller R J 1997 *Geophysical Journal International* **131** 425-450
[3] Tuladhar R, Yamazaki F, Warnitchai P and Saita J 2004 *Earthquake Engineering and Structural Dynamics* **33** 211-225
[4] Puslitbang BMKG 2019 Laporan Penelitian Pasca Gempabumi: Pengembangan Sistem Informasi Seismik Microzonasi dan Crowd Model (in Indonesian)
[5] Sokolov V Y, Loh C H, and Jean W Y 2007 *Soil Dynamics and Earthquake Engineering* **27** 314-323
[6] Ghofrani H, and Atkinson G M 2014 *Soil Dynamics and Earthquake Engineering* **67** 30–43
[7] Ji K, Refen Y, and Wen R 2017 *Journal of Asian Earth Sciences* **147** 79-94
[8] Stanko D, Markušić S, Strelec S, and Gazdek M 2017 *Soil Dynamics and Earthquake Engineering* **92** 666–677
[9] Setiawan B, Jaksa M, Griffith M, and Love D 2018 *Soil Dynamics and Earthquake Engineering* **110** 244-261
[10] Parolai S, Richwalski A M, Milkereit C, and Fah D 2006 *Bulletin of Earthquake Engineering* **6**

65–94
[11] Karubulut S 2018 *Journal of Applied Geophysics* **150** 254-266
[12] Nortey G, Armah T K, and Amponsah P 2018 *Journal of African Earth Sciences* **142** 158-169
[13] Mohammed M A, Abudeif A M, and Abd el-aal A K 2020 *Journal of African Earth Sciences* **162** 103721
[14] Silacheva N V, Kulbayeva U K, and Kravchenko N A 2018 *Geodesy and Geodynamics* **9** 131-141
[15] Barania S, Ferrettia G, and Ferrari R D 2020 *Engineering Geology* **267** 105479
[16] Roshan A D, and Basu P C 2010 *Nuclear Engineering and Design* **240** 3443–3454
[17] Mostafa S I, Abdelhafiez H E, and Abd el-aal A K 2019 *Journal of African Earth Sciences* **160** 103655
[18] Zhao J X, Irikura K, Zhang J, Fukushima Y, Somerville P G, Asano A, Ohno Y, Oouchi T, Takahashi T, and Ogawa H 2006 Bulletin of the Seismological Society of America **96** 914-925
[19] Nakamura Y 1997 *Seismic Vulnerability Indices for Ground and Structures Using Microtremor.* Florence: World Congres on Research Institute.
[20] Allen T I. and Wald, D J 2009 *Bulletin of the Seismological Society of America* **99** 935-943.
[21] BMKG's earthquakes repository, http://repogempa.bmkg.go.id/repo_new/, accesed 23rd October 2020

The determination of prospective reservoirs in the slope environment of the Tarakan Basin (Indonesia) by using sequence stratigraphy, rock physics and AI inversion

D Aminulloh[1], B T Tampubolon[1], S Sukmono[1]

[1]Geophysical Engineering Program Study, Bandung Institute of Technology, Jl. Ganesha 10 Bandung, Indonesia 40132.

E-mail : dzikru09xiu1@gmail.com

Abstract. Prospective reservoirs in the Tarakan Basin are mostly deposited in the Middle Miocene to Pliocene and related to the change of depositional environment from transgressive to regressive sedimentary environment. Log sequence stratigraphic analysis in well AST 1 shows that transgressive and lowstand system tracts were deposited during the study area. Hashin-Shtrikman method is used in the rock physics analysis to identify the relative rock hardness and integrated with gamma ray log, NPHI, and resistivity data analysis. The result show that potential reservoir exist at the depth range of 7650-7725 feets and associates with low velocity, low gamma ray, low NPHI, high resistivity values. The potential reservoir interval was deposited under low stand system tract of slope depositional environment. The acoustic impedance (AI) map shows that the low AI's are mostly located in the northwestern part of the study area.

1. Introduction

The Tarakan Basin is geographically located in the north of the Mahakam Basin (Figure 1). The regional stratigraphy of the Tarakan Basin is shown in Figure 2. In the Middle to Late Miocene, the Meliat, Kabul and Santul formations were deposited. Meliat Formation is composed mainly of quartz sandstones interspersed with conglomerate shale and sandstones. Tabul Formation is dominated by shales with alternating sandstone and silt. Santul Formation comprises thicker sandstone intercalation with claystone and streak of lignites. In the Late Miocene to Pliocene, the delta system in the Tarakan Formation was deposited and composed mainly of sandstones with a cross-sectional structure and conglomerate sandstones, alternating shale, and coal. In the Pliocene to Recent, there is a reactivation of transform movements along the horizontal faults through the Makassar Strait which began in the Upper Pliocene period and continues until now.

Prospective reservoirs in the Tarakan Basin are mostly deposited in the Middle Miocene to Pliocene and related to the change of depositional environment from transgressive to regressive sedimentary environment [1]. This study integrates sequence stratigraphy, rock physics and bandlimited acoustic impedance (AI) inversion to identify Middle Miocene to Pliocene prospective reservoirs in the slope environment of the Tarakan Basin.

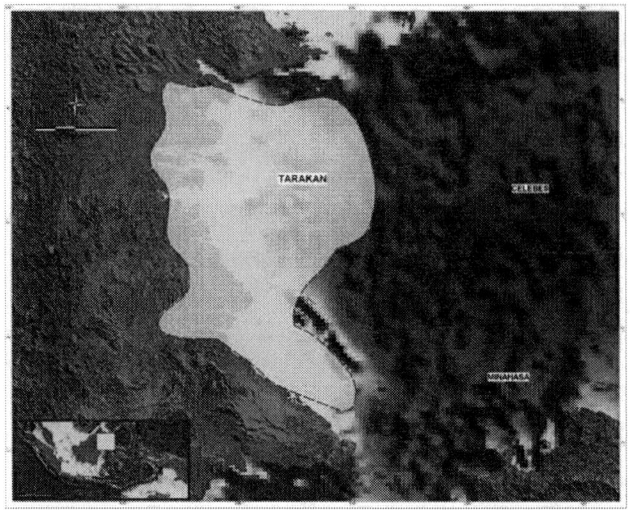

Figure 1. Location of the Tarakan Basin [1]

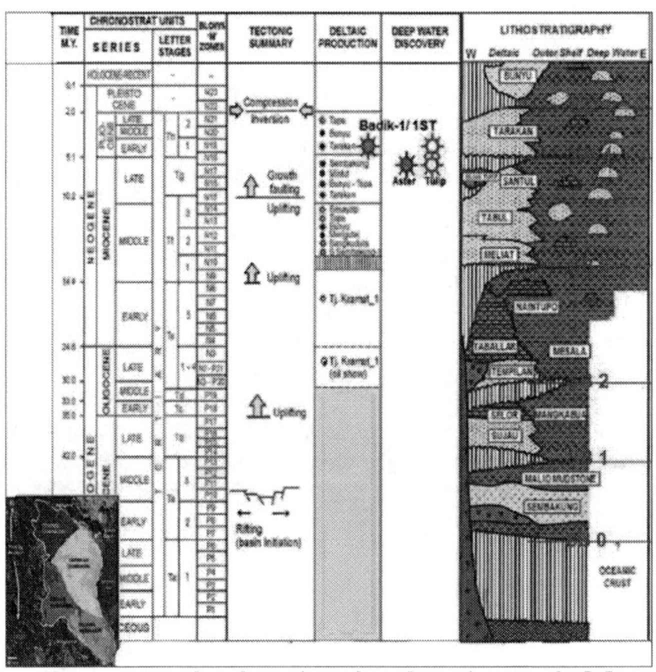

Figure 2. Regional stratigraphy of Tarakan Basin [1]

2. Data and Method

The AST-1 well is used as the main data for the sequence stratigraphy and rock physics analysis. The well penetrates the Pleistocene to Late Miocene rock interval deposited under stable regressive cycle in slope depositional environment.

Sequence stratigraphy analysis in the Well AST-1 found that there are 4 depositional cycles in the studied interval. From the oldest to the youngest, they are the slope fan of the low-stand system tract (LST), the low stand wedge of the LST, the transgressive system tract (TST), and finally another LSW LST. The LST slope fan is deposited when the relative sea level rises slowly and sediment deposition shifting to the land. In the study area, the LST slope fan is equivalent to Tabul Formation and characterized by fining upward pattern in the gamma-ray log (Figure 3).

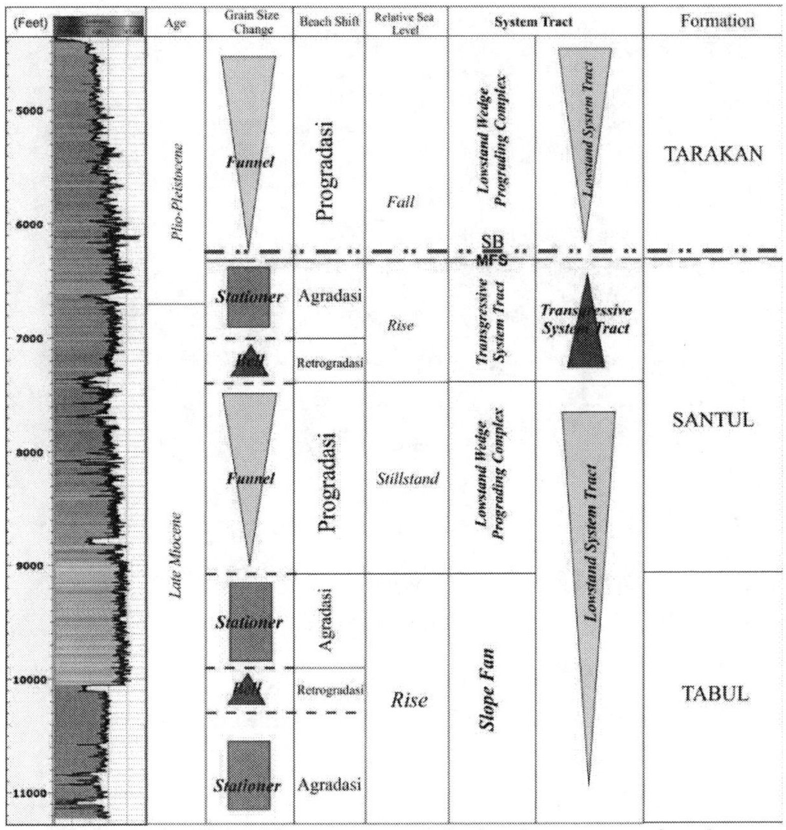

Figure 3. Stratigraphic sequence analysis of gamma ray log data

The low stand wedge (LSW) prograding complex is formed when sea level is relatively stillstand and the sediments were deposited rapidly. In the study area, the LSW is equivalent to the lower-middle Santul Formation. In the gamma-ray log, this system tract is indicated by upward pattern of grain coarsening related to the decrease in the gamma ray value (Figure 3). The TST cycle is deposited when there is a rapid rise of the relative sea and sedimentation that cannot catch-up the sea level rises. The cycle usually ends at the maximum flooding surface (MFS), so that the sediment grains will show fining upward pattern. In the study area, the TST cycle is represented by the upper Santul Formation (Figure 3).The Santul formation is upper bounded by a sequence boundary indicating sedimentation hiatus after the cycle reaches the MFS. Another LSW prograding complex, equivalent to the Tarakan Formation with Plio-Pleistocene in age, was deposited above the sequence boundary.

Rock physics analysis result by using the Hashin-Shtrikman boundary method is shown in Figure 4. The Hashin-Shtrikman method is one of the best approaches to predict the upper and lower limits of the effective elastic modulus of a medium [7]. In the cross plot between Vp data versus porosity, when the data points clustered to the lower bound line, then they area interpreted as soft and deformable lithology and vice versa. If the data points are distributed to the right of the critical porosity line and attached to the lower bound line, the data is interpreted as suspension lithology. By using this principle, the prospective reservoir zone is interpreted as the interval with Vp< 2300 m/s and porosity >20% (marked by yellow zone in Figure 4).

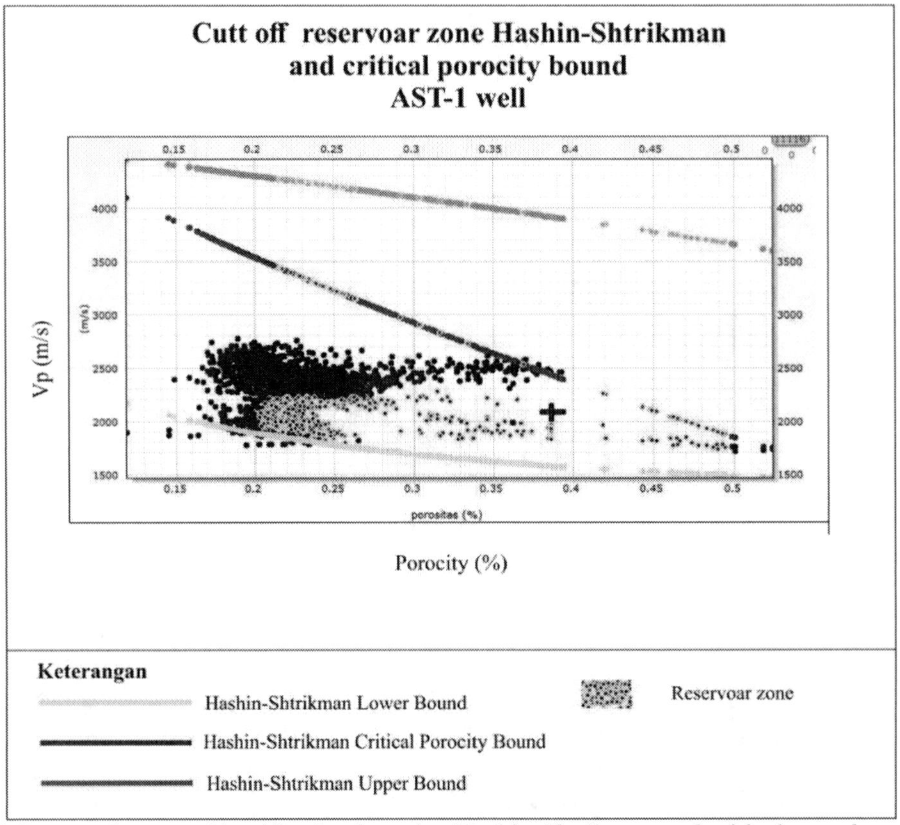

Figure 4. Prospective zone identification by using Hashin-Shtrikman and critical porosity cross-plot

The log data analysis (Figure 5) is used to narrow down the hydrocarbon reservoir prospect zone and to increase the validity confidence on the prospective zone determination. The yellow zone in the DT (Vp) and the density log (RHOZ) in Figure 5 is equivalent to the prospective reservoir interpreted in Figure 4. The prospective zone of this reservoir is lied in the LSW LST prograding complex cycle and marked by yellow interval bounded by dashed red. The log-based sequence stratigraphy analysis is ideally combined with the seismic-based and anisotropy analysis as discussed in [2], [3], [4], [5] and [6]. However, due to the data limitation, in this study the analysis is done by using log data only.

IOP Conf. Series: Earth and Environmental Science **873** (2021) 012036 doi:10.1088/1755-1315/873/1/012036

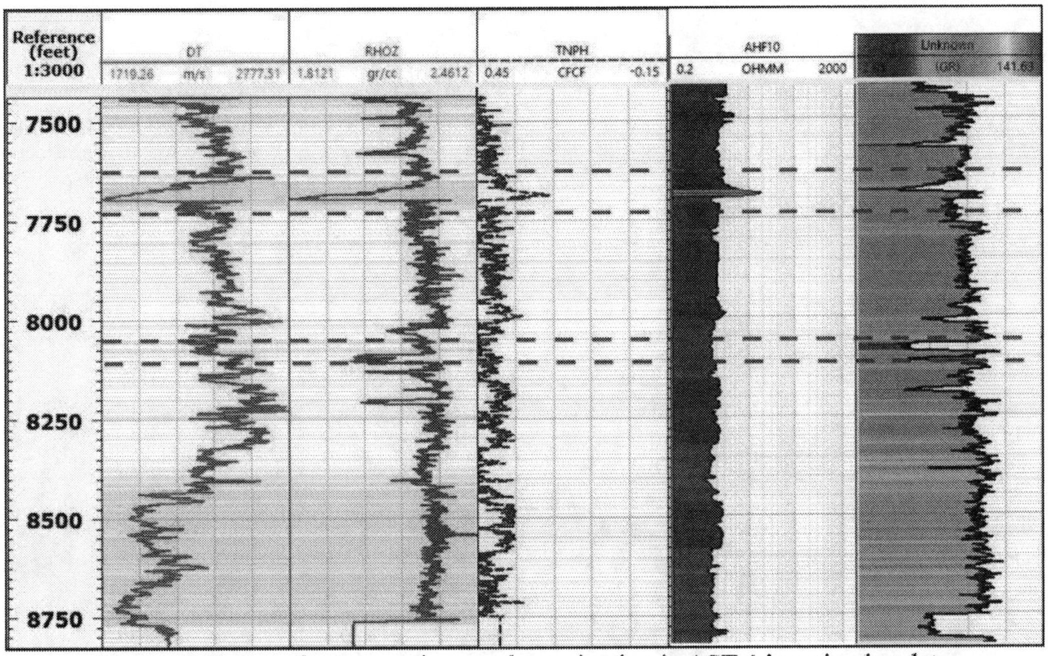

Figure 5. Prospective reservoir zone determination in AST-1 by using log data

Figure 6 is a bandlimited AI section passing thru the AST-1. In this section, the prospective reservoir is marked by green area inside blue line. It has acoustic impedance values ranging from 14000 ((ft/s) * (gr/cc)) to about 10000 ((ft/s) * (gr/cc)). On the related AI map in Figure 7, the prospective reservoir around the AST-1 well is shown low AI area bounded by pink line in the northwestern part of the map. Another prospective zone can be identified in the southwest part of the map and marked as low AI area bounded by blue.

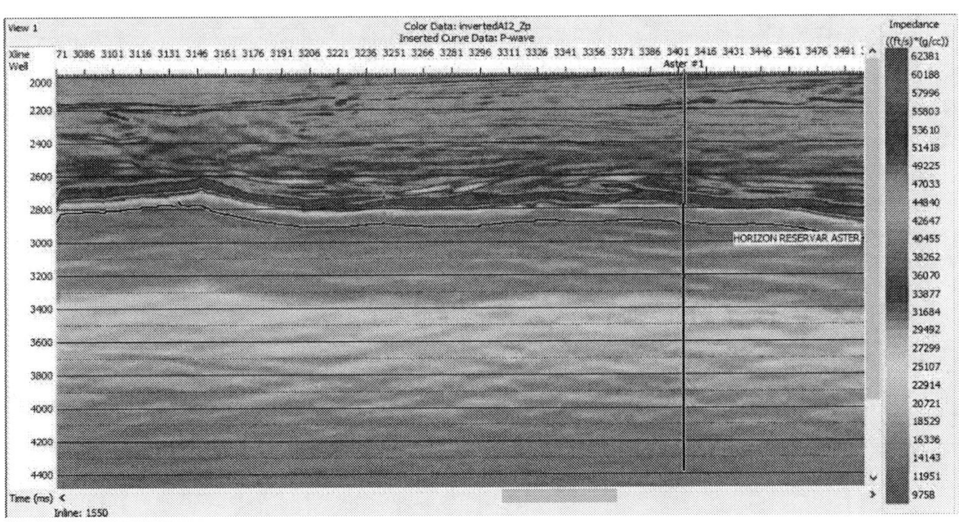

Figure 6. AI Seismic Section *Inline* 1550 passing thru AST-1

3. Conclusion

The integration of sequence stratigraphy, rock physics and log data analysis show that the prospective reservoir interval in AST-1 well is located at a depth of 7650–7725 feet. It is characterized by low velocity, low gamma ray, low NPHI, low density and high resistivity. The prospective interval was deposited in the Late Miocene slope setting as the LSW prograding complex equivalent with bottom to middle interval of Santul Formation. The AI map shows that the distribution of the prospective zone is around the AST-1 and in the northwestern part of the study area.

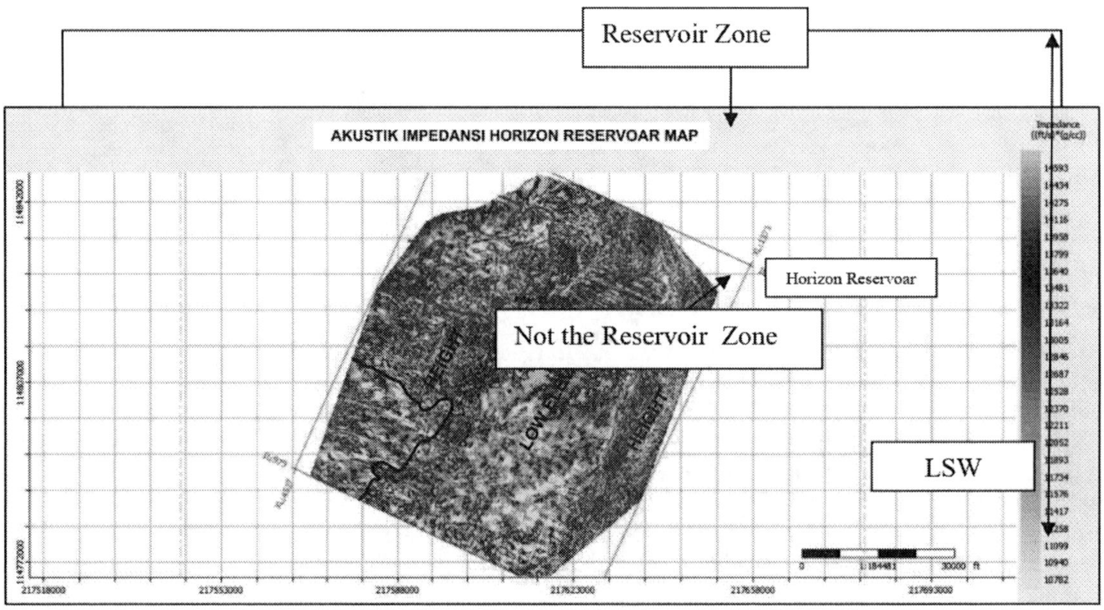

Figure 7. AI horizon slice map showing prospective areas

References

[1] SKK Migas., 2019. Final Report Studi Play Identification Petroleum System and Resource Assesment Cekungan Tarakan. Bandung: Institut Teknologi Bandung.
[2] Sukmono S, Samodra A, Sardjito and Waluyo W 2006 *The Leading Edge* **25** 532
[3] Sukmono S 2007 *The Leading Edge* **26** 126
[4] Alamsyah M N, Sukmono S, Marwosuwito S, Sutjiningsih, W and Marpaung L 2008 *The Leading Edge* **27** 260
[5] Sari, V M, Sukmono S, Sanny, T A and Sapiie, B 2018 *First Break* **36** 61
[6] Sukmono and Ambarsari 2019 *Practical seismic interpretation for petroleum exploration* (Bandung: ITB Press)
[7] Mavko G, Mukerji T, Dvorkin, J 2009 *The Rock Physics Handbook* (USA: Cambridge University Press)

IOP Conf. Series: Earth and Environmental Science **873** (2021) 012037 doi:10.1088/1755-1315/873/1/012037

Integration of sequence stratigraphy, rock physics and AI inversion to determine the prospective reservoir in the shelf environment of the Tarakan Basin, Indonesia

M Irsyad[1], B T Tampubolon[1], S Sukmono[1]

[1]Geophysical Engineering Program Study, Bandung Institute of Technology, Jl. Ganesha 10 Bandung, Indonesia 40132.

E-mail:contact.muhammadirsyad@gmail.com

Abstract. The Tarakan Basin is one hydrocarbon basin in Indonesia with approximately 40 discoveries and more than 1000 MBOE reserves. This study discusses an approach to integrate the sequence stratigraphy, rock physics and acoustic impedance (AI) inversion analysis to determine the prospective reservoirs in the basin. PRG-1 well data is used in the sequence stratigraphy and rock physic analysis. The sequence stratigraphy analysis of PRG-1 shows that there are three system tracts: transgressive, low stand tract and high stand system tracts. The integration of sequence stratigraphy, rock physics and log data analysis show that the prospective reservoir interval in PRG-1 well is located at a depth of 4730-4780 feet. It is characterized by low gamma ray, low NPHI, low density and high resistivity. The prospective interval was deposited in early Pliocene as Tarakan Formation in the low stand system tract of shelf depositional environment. The AI map shows that the distribution of the prospective is around the PRG-1 and in the eastern part of the area.

1. Introduction

The Tarakan Basin is in East Kalimantan, Indonesia (Figure 1). It is a productive basin with approximately 40 discoveries and more than 1000 MBOE reserves [1]. The hydrocarbon in the Tarakan Basin is mostly discovered in the Middle to Late Miocene interval (Figure 2). During this era, the depositional facies vary from the proximal delta to deep sea. Shallow and continental marine facies also develop at local high in the south, east of the carbonate shelf. Super tidal found in the western and northern basins. Facies delta front of the Meliat Formation is composed of quartz sandstones interspersed with conglomerate shale and sandstones in several places. In the Bunyu and Tarakan Islands, the facies change to the delta front and prodelta. South of the Tarakan Basin, the entire clastic sequence was terminated by shallow marine limestone in the Late Miocene. The middle to late Miocene Tabul Formation is a delta complex that is graded to the east and is dominated by shale with alternating sandstone and silt. The Menumbar Formation, which is laterally equivalent, is composed of calcareous claystone, marl and limestone. In the northeastern part of the Tarakan sub-basin, Menumbar Bawah extends as a thick limestone layer which is equivalent to the Tabul Formation in the west of the Tarakan and Tidung Sub-basins. Delta front deposition during Middle to Late Miocene shifted southward in the Late Miocene to Early Pliocene period. The Sejau Formation delta deposit in the Tarakan is composed of sandstones with a cross-sectional structure and conglomerate sandstones, alternating shale and coal.

This study discusses an approach to integrate the sequence stratigraphy, rock physics and AI inversion to determine the Middle to Late Miocene prospective reservoirs deposited in the shelf environment of the Tarakan Basin.

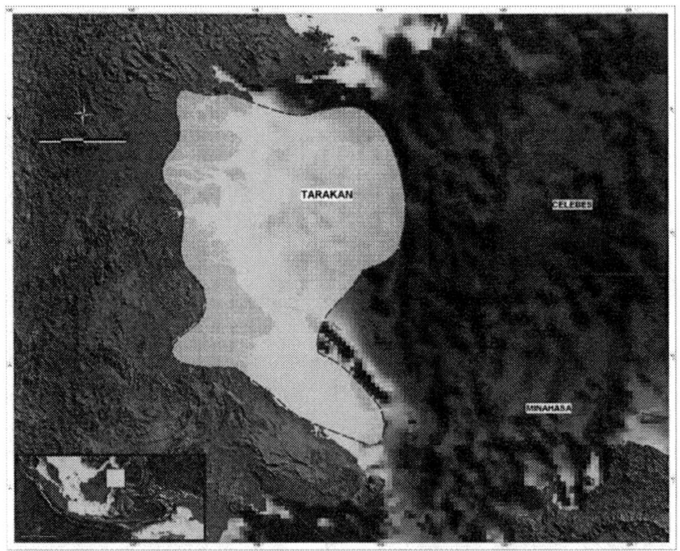

Figure 1. Location of the Tarakan Basin [1]

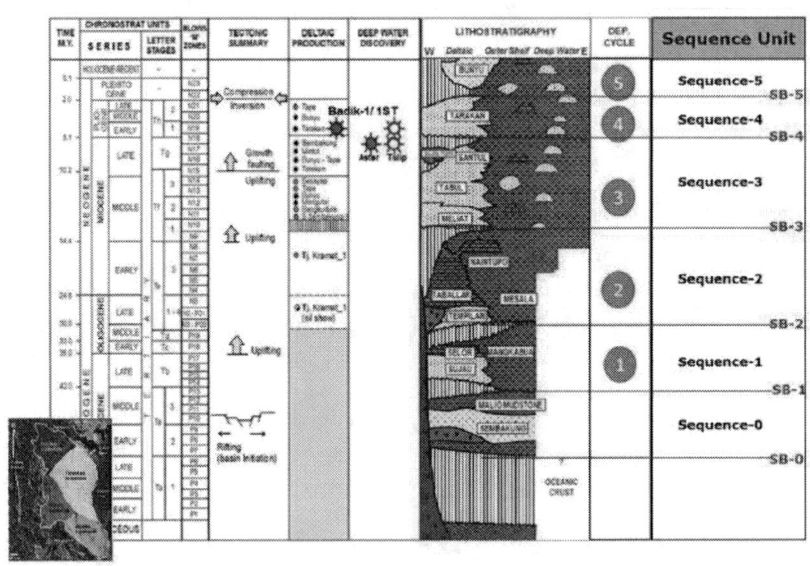

Figure 2. Stratigraphy of the Tarakan Basin [1]

2. Data and Methods

For sequence stratigraphy and rock physics analysis, the PRG-1 well is used. PRG-1 is located on the Middle to Late Miocene shelf environment and contains Bunyu, Tarakan, Santul and Tabul formations. Based on the system tract analysis, PRG-1 well has 4 cycles of deposition, from the bottom to the top: the 1st transgressive system tract (TST), low stand system tract (LST), the 2nd TST and high stand system tract (HST) (Figure 3). Tabul and Santul formations were deposited in the 1st TST cycle. Tarakan formation was deposited on the LST overlying the 1st TST. Above the LST, the 2nd TST cycle containing Bunyu formation was deposited.

The transgressive system tract occurs when the sea level rises, and the sediment deposition rate is smaller than the sea level rise. It is characterized by fining upward trend and upward increase of gamma ray values. The LST, iss deposited when the sea level relatively falls, and the sedimentation rate is high. It is characterized by coarsening upward pattern where the gamma ray values decrease upwards. The HST is formed when the sea level is still standing, and the rate of the sediment is high. It is characterized by coarsening upward sequence and upward decrease of gamma ray value.

The log-based sequence stratigraphy analysis is ideally combined with the seismic-based and anisotropy analysis as discussed in [2], [3], [4], [5] and [6]. However, due to the data limitation, in this study the analysis is done by using log data only.

Rock physics analysis was done by using the Voigt-Reuss bounding method [6] and the result is displayed in Figure 4. When the data distribution is closer to the Reuss lower bound line, the data point is interpreted as a softer and more deformable lithology and vice versa. The reservoir prospect zone is interpreted to have a low P wave velocity below 2.5 km/s and porosity above 20%. It is indicated by yellow zone in the Vp vs Porosity cross-plot in Figure 4. The cross-plot also suggests that PRG-1 well data distribution tends to be closer to the lower bound line.

3. Results

The rock physics result is used to guide the determination of reservoir interval in the log data analysis (Figure 5). In the log data, the reservoir zone is marked as an interval bounded by dashed red lines. It is characterized by a low gamma ray, low P wave velocity, high porosity, low density, low neutron porosity and high resistivity values. On the contrary, the interval bounded by dashed blue lines is not a reservoir zone because it has high neutron porosity value. The correlation with the system tract and stratigraphic sequence analysis shows that the reservoir interval is in the low stand wedge prograding complex of the LST cycle.

Figure 6 is an AI section crossing the PRG-1 well. It is derived from the model-based AI inversion. The reservoir interval is marked with a layer in green with the AI value of 17,000- 20,000 (ft /s) * (gr/cc). The related AI map of the reservoir level is shown in Figure 7. As the AI value is proportional to the density value, then by using the log analysis result, the distribution of the reservoir zone in the AI map can be identified. It is marked by areas with red line around the PRG-1. Using the same principle, another prospective zone with low AI is identified in the eastern part of the map.

IOP Conf. Series: Earth and Environmental Science **873** (2021) 012037 doi:10.1088/1755-1315/873/1/012037

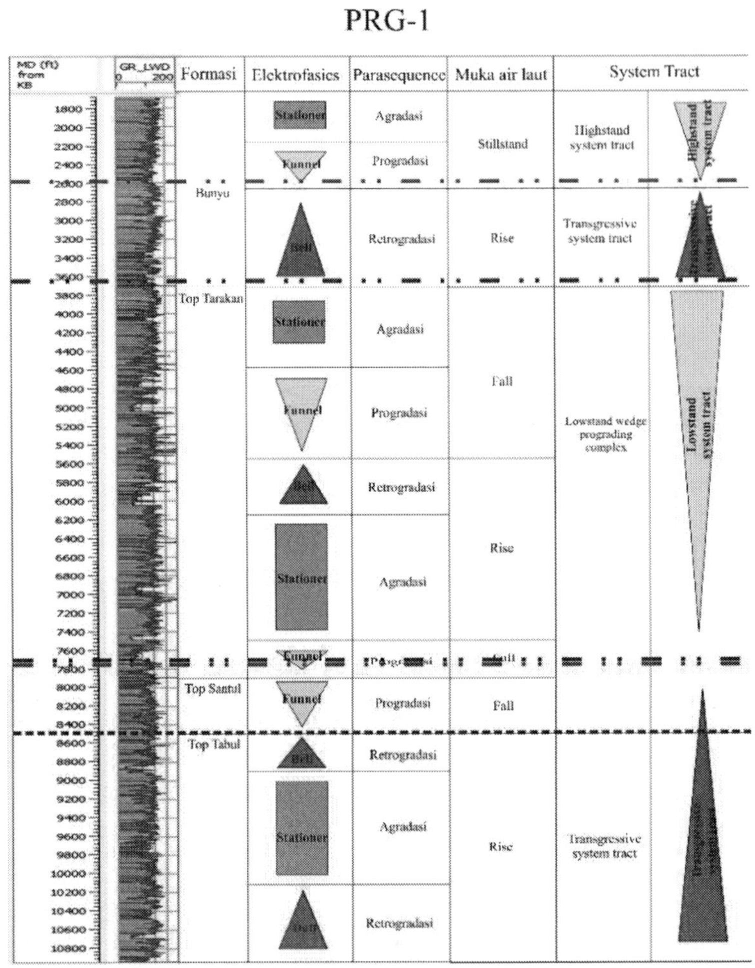

Figure 3. Sequence stratigraphy analysis of PRG-1well

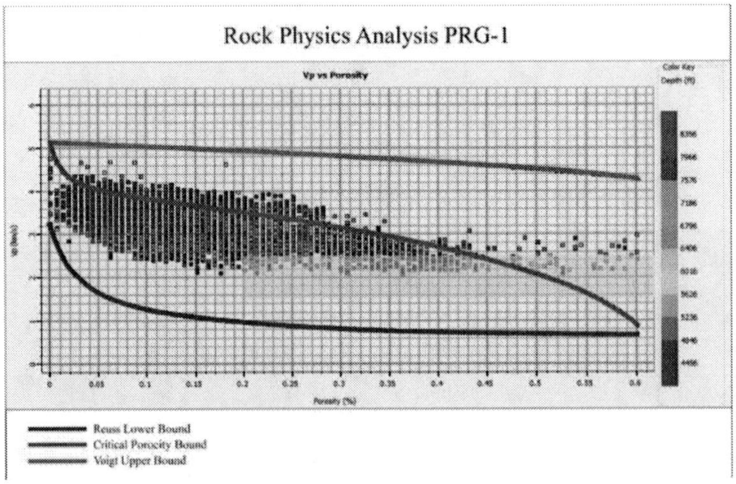

Figure 4. Rock physics analysis to determine the prospective reservoir zone

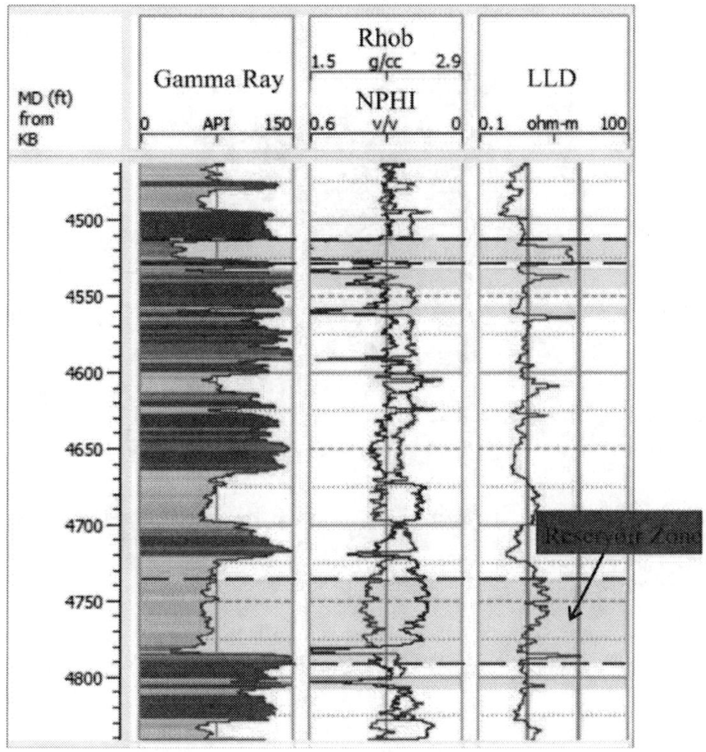

Figure 5. PRG-1 Well log data analysis

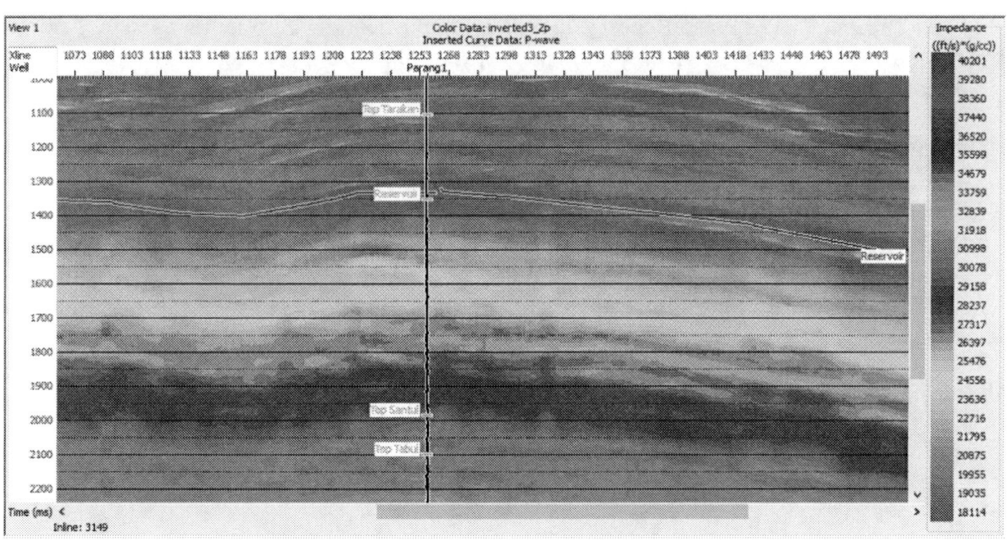

Figure 6. AI section crossing the PRG-1 well.

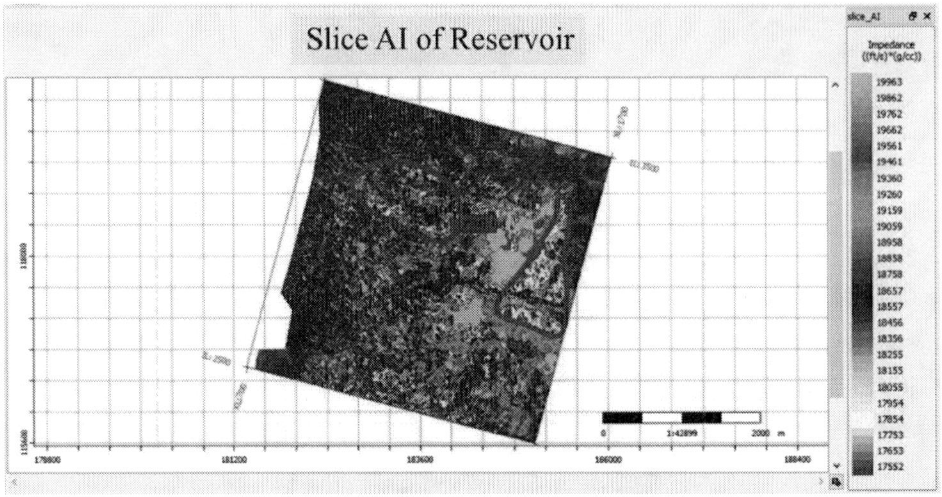

Figure 7. AI slice map showing the distribution of prospective zones.

4. Conclusions

The integration of sequence stratigraphy, rock physics and log data analysis show that the prospective reservoir interval in PRG-1 well is located at a depth of 4730-4780 feet. It is characterized by low velocity, low gamma ray, low NPHI, low density and high resistivity. The prospective interval is equivalent to the early Pliocene Tarakan Formation which deposited as LST in the shelf depositional environment. The AI map shows that the distribution of the prospective is around the PRG-1 and in the eastern part of the area.

References

[1] SKK Migas., 2019. Final Report Studi Play Identification Petroleum System And Resource Assesment Cekungan Tarakan. Bandung: Institut Teknologi Bandung.

[2] Sukmono S, Samodra A, Sardjito and Waluyo W 2006 *The Leading Edge* **25** 532

[3] Sukmono S 2007 *The Leading Edge* **26** 126

[4] Alamsyah M N, Sukmono S, Marwosuwito S, Sutjiningsih, W and Marpaung L 2008 *The Leading Edge* **27** 260

[5] Sari, V M, Sukmono S, Sanny, T A and Sapiie, B 2018 *First Break* **36** 61

[6] Sukmono and Ambarsari 2019 *Practical seismic interpretation for petroleum exploration* (Bandung: ITB Press)

[7] Mavko G, Mukerji T, Dvorkin, J 2009 *The Rock Physics Handbook* (USA: Cambridge University Press)

The Use of Core & Log Anisotropy Parameters into Seismic Data Processing: A Case Study of Deep-Water Reservoir

Madaniya Oktariena[1], Wahyu Triyoso[2], Dona Sita Ambarsari[2], Sigit Sukmono[2], Erlangga Septama[3], Pongga D Wardaya[3], Befriko S Murdianto[4], Rusalida Raguwanti[3], Richie R Pratama[3]

[1] Postgraduate Program of Geophysical Engineering, Faculty of Mining and Petroleum Engineering, Institut Teknologi Bandung, Jalan Ganesha 10, Bandung 40132, Indonesia
[2] Faculty of Mining and Petroleum Engineering, Institut Teknologi Bandung, Jalan Ganesha 10, Bandung 40132, Indonesia
[3] PT Pertamina (Persero) Upstream Research & Technology Innovation, SopoDel Office Tower 50th Floor, Jalan Mega Kuningan Barat III Jakarta 12950, Indonesia
[4] Pertamina Hulu Kalimantan Timur, CIBIS NINE Tower 9th Floor, Jalan TB Simatupang 2, Jakarta Selatan 12560, Indonesia

madaniyaoktariena@outlook.com

Abstract. The seismic far-offset data plays important role in seismic subsurface imaging and reservoir parameters derivation, however, it is often distorted by the hockey stick effect due to improper correction of the Vertical Transverse Isotropy (VTI) during the seismic velocity analysis. The anisotropy parameter η is needed to properly correct the VTI effect. The anisotropy parameters of ε and δ obtained from log and core measurements, can be used to estimate the η values, however, the upscaling effects due to the different frequencies of the wave sources used in the measurements must be carefully taken into account. The objective is to get better understanding on the proper uses of anisotropy parameters in the the velocity analysis of deepwater seismic gather data. To achieve the objective, the anisotropy parameters from ultrasonic core measurements and dipole sonic log were used to model the seismic CDP gathers. The upscaling effects is reflected by the big difference of measured anisotropy values, in which the core measurement value is about 40 times higher than the log measurement value. The CDP gathers modelling results show that, due to the upscaling effect, the log and core-based models show significant differences of far-offset amplitude and hockey sticks responses. The differences can be minimized by scaling-down the log anisotropy values to core anisotropy values by using equations established from core – log anisotropy values cross-plot. The study emphasizes the importances of integrating anisotropy parameters from core and log data to minimize the upscaling effect to get the best η for the VTI correction in seismic velocity analysis.

1. Introduction

Physical property variation for various measurement angles within a medium defines as anisotropy [1]. The variation could be in the form of velocity, density, porosity, and other means of physical properties representing rock heterogeneity conditions. Concerning seismic, velocity is the main parameter that defines anisotropy recorded in the rock formation. [2] describes three unitless parameters that correlate

IOP Conf. Series: Earth and Environmental Science **873** (2021) 012038 doi:10.1088/1755-1315/873/1/012038

anisotropy as seismic wave velocity variation within the direction of wave propagation. These parameters are better known as Thomsen's Anisotropy Parameter ε, γ, and δ.

Value of V_0, V_{45}, and V_{90}, ε, γ, and δ could be estimated based on the rock formation's velocity data. The measurement is specifically possible only for core plug and dipole logging configuration due to its flexibility in measuring the rock from various directions. Core & log data have a different resolution that leads to a possibility of varying anisotropy magnitude. The more significant the anisotropy magnitude, the bigger the physical properties variation within a rock formation. This logic could lead to a different response in the reflectivity due to the presence of anisotropy.

A deepwater environment has a known special treat regarding the anisotropy condition. The reservoir's depth demands a more extended offset to analyze the anisotropy information stored in the far offset. In the Vertical Transverse Isotropy (VTI) case, the anisotropy affects the long offset of seismic data, forming an artifact known as the hockey stick. Deepwater reservoir further amplifies this effect, requiring a decent length offset to create relevant seismic imaging amidst the anisotropy. The anisotropy correction in VTI media is constructed by evaluating the η based on offset during the velocity analysis. η is a combination of Thomsen's Anisotropy Parameter ε and δ heavily associated with long seismic offset. Incorporating anisotropy measurement from core & log as a guide for long offset seismic correction needs to mind the resolution difference brought by the measurement technique.

Delving more into the resolution difference between core & log, we propose to analyze the effect of these differences in the resulted seismic record. This paper intends to showcase the impact of anisotropy value calculated from Ultrasonic Core Measurement & DSI Dipole Sonic Log on long offset seismic in a deepwater environment.

2. Methods

[3] introduced three anisotropic parameters in case of weak anisotropy condition. In the VTI case, Thomsen's Anisotropy Parameter ε, γ, and δ are related to velocity variation with incident angle, precisely at 0^0, 45^0, and 90^0. Epsilon is the fractional difference of vertical and horizontal P-wave, which is associated with the long offset effect. Whereas Gamma is related to horizontal & vertical S-Wave, while Delta is influenced by the move-out velocity versus the zero-offset velocity. In a way, as explained by [4], Seismic Anisotropy Eta could be described by Thomsen's Anisotropy Parameters, mainly Epsilon and Delta. These anisotropy values could be calculated from velocity through several means. This research will compare the anisotropy value from Ultrasonic Core Measurement and Dipole Shear Sonic Log. Both of the methods have different hard data as the input and also different resolution as the output. This study aims to better understand the effect of anisotropy value calculated from core & log in long offset seismic reflection through well-based anisotropy modeling.

2.1 Ultrasonic Core Measurement

Ultrasonic Core Measurement application in core data measurement has begun with [5] who used ultrasonic waves to define the relationship between P-Wave, S-Wave, and Poisson's Ratio in dry sedimentary rocks. Since then, the ultrasonic wave is widely known as one method to obtain an elastic and anisotropy parameter from core data. Attempts to compare the result of ultrasonic measurement and sonic log have been conducted to analyze heterogeneity and anisotropy of rocks [6][7]. To estimate the *Eta* based on seismic data, Thomsen's Anisotropy Parameter is being calculated from the core plug using ultrasonic wave and compared the anisotropy value from seismic processing [8].

The core plug data with ultrasonic waves in several directions or angles are used in this study. They are Vp_0, Vp_{45}, Vp_{90} as well as Vs_0 and Vs_{90}. Each angle represents seismic wave velocity propagation at adjacent, 45 degrees, and perpendicular to the possible anisotropy symmetrical axis. In Vertical Transverse Isotropy, the symmetrical anisotropy axis is perpendicular to the sedimentary layer [9].

2.2 Dipole Shear Sonic Log

Thomsen's Anisotropy Parameter is related to velocity variation within different angles, such as V_0, V_{45}, and V_{90}. Gamma, which is calculated from Shear Sonic, cannot be obtained from seismic data with a compressional source and one component geophone. The different angles' velocity could be measured through dipole logging configuration in well-logging acquisition with acoustic log source types. It is applicable for both Sonic Log and Shear Sonic Log.

This research uses DSI Dipole Sonic measurement in cross-dipole configuration to detect the anisotropy and dispersion within the target formation.

2.3 Well-based Modelling with Anisotropy

The degree of anisotropy has various effects on seismic reflection. Well-based synthetic modeling is a preliminary step to analyze further the impact of anisotropy value in long offset seismic. In the Vertical Transverse Isotropy case, anisotropy has a remarkable effect on the far offset of seismic gathering in the form of the hockey stick effect. The reservoir located in a deepwater environment needs a more extended offset to compensate for the more profound depth. The anisotropy analysis in deepwater reservoir should be conducted in seismic data with a long enough offset. Well-based modeling is a way to support further exploration or development decisions in a deepwater environment.

The possible different effect of anisotropy obtained from Ultrasonic Core Measurement and Dipole Sonic Log is to be modeled. The reflectivity calculation of the well-based modeling adopts a Vertical Transverse Isotropy condition. The modeling applies the Zoeppritz equation to ensure a more accurate estimation of the reflection event amplitude from the target interface compared to the linearized one. As for the reflectivity calculation, Schoenberg-Protázio Extension to Anisotropic Zoeppritz Theory is incorporated into the analysis. Schoenberg-Protázio extended the Zoeppritz equations to anisotropy by introducing the (2 x 2) block impedance matrices providing the solution to four components of both reflected and transmitted waves [10]. Azimuthal Plane Wave technique is applied to move out the synthetic gather data for the reflection to appear as NMO corrected seismic gather.

The resulted synthetic gather seismic is sorted into Common Offset Common Azimuth (COCA), Common Angle Common Azimuth (CACA), and Common Offset Vector (COV). The maximum offset to be modeled follows the far offset value of the real seismic data around 6500 meters. The maximum angle of up to 65^0 is adapted from angle information from the available ultra-far angle seismic. Modeling is conducted for the whole total depth of log data where the P-Wave, S-Wave, and density are known. However, the detailed calculation is focused on the level target on which the core measurement available.

3. Results & Discussions

The well of SAD-4 is located at a deepwater Sadewa Field in East Kalimantan. It is drilled to observe a clastic reservoir at around 11000 ft. The target clastic reservoirs are quartz dominant, moderately highly porous, somewhat compacted, and grain size varies from upper the very fine sand to upper fine sand with moderate sorting. Observation is focused on the reservoir zone of Vertical Transverse Isotropy condition. Synthetic well-based modeling is performed to model the anisotropy effect.

This study aims to analyse the effect of different anisotropy values recorded in seismic gather data. The well-based modeling will cover the following issue—first, synthetic seismic well-based modeling anisotropy from Ultrasonic Core Measurement. Second, synthetic seismic well-based modeling with anisotropy from Dipole Sonic Log. Third, the anisotropy value and data sampling effect on synthetic seismic.

3.1 Seismic well-based modeling with anisotropy from Ultrasonic Core & Dipole Sonic Log

Anisotropy value calculated from Ultrasonic Core Measurement and Dipole Sonic Log has an entirely different value range. The anisotropy from the core plug sample in SAD-4 has value ranged from 0.09 to 1.23. Meanwhile, the anisotropy from the log has value ranged from 0.007 to 0.010. Anisotropy value that close to zero has a similar characteristic of the isotropy one. Figure 1 shows the difference in magnitude between Thomsen's Epsilon and Gamma from both datasets. The anisotropy of the log data is weak to very weak, and the anisotropy of the core plug is moderate to strong.

IOP Conf. Series: Earth and Environmental Science **873** (2021) 012038 doi:10.1088/1755-1315/873/1/012038

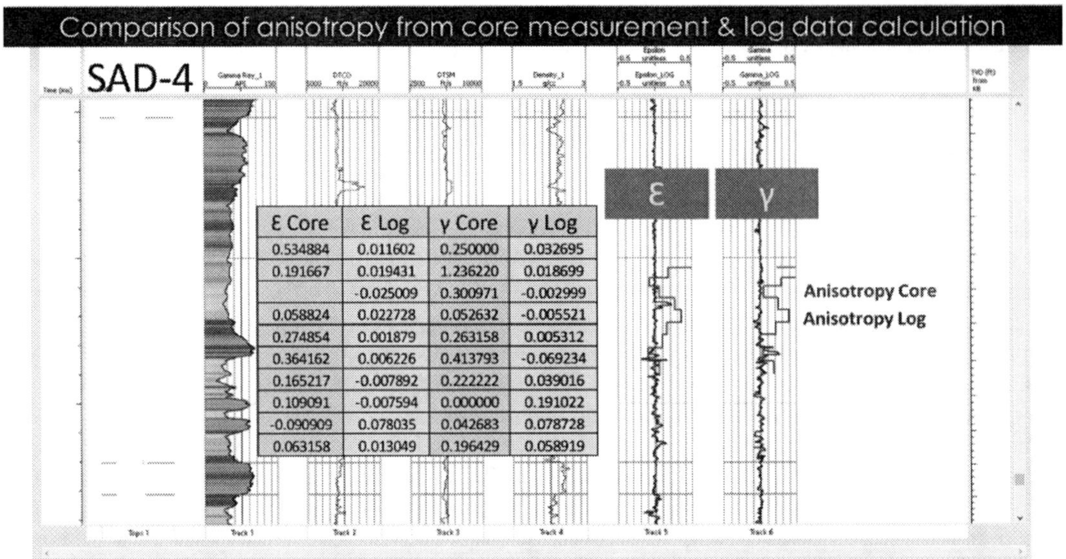

ε Core	ε Log	γ Core	γ Log
0.534884	0.011602	0.250000	0.032695
0.191667	0.019431	1.236220	0.018699
	-0.025009	0.300971	-0.002999
0.058824	0.022728	0.052632	-0.005521
0.274854	0.001879	0.263158	0.005312
0.364162	0.006226	0.413793	-0.069234
0.165217	-0.007892	0.222222	0.039016
0.109091	-0.007594	0.000000	0.191022
-0.090909	0.078035	0.042683	0.078728
0.063158	0.013049	0.196429	0.058919

Figure 1. Anisotropy value calculated from Ultra Sonic Measurement (blue) and Dipole Sonic Log (brown) in SAD-4. The detailed analysis of the anisotropy effect is conducted on the level target where the core plug sample is obtained.

In this study, there are three scenarios to be applied in the well-based modeling: The first scenario is modeling using the original log condition as a baseline to its in-situ state. The second scenario is modeling with anisotropy information from Ultrasonic Core Measurement in VTI media. The third scenario is modeling with anisotropy information from Dipole Sonic Log in VTI media. The synthetic gathers seismic resulted from modeling is sorted into Common Angle Common Azimuth gather to better showcased the kick out of hockey stick effect on the long offset reflection.

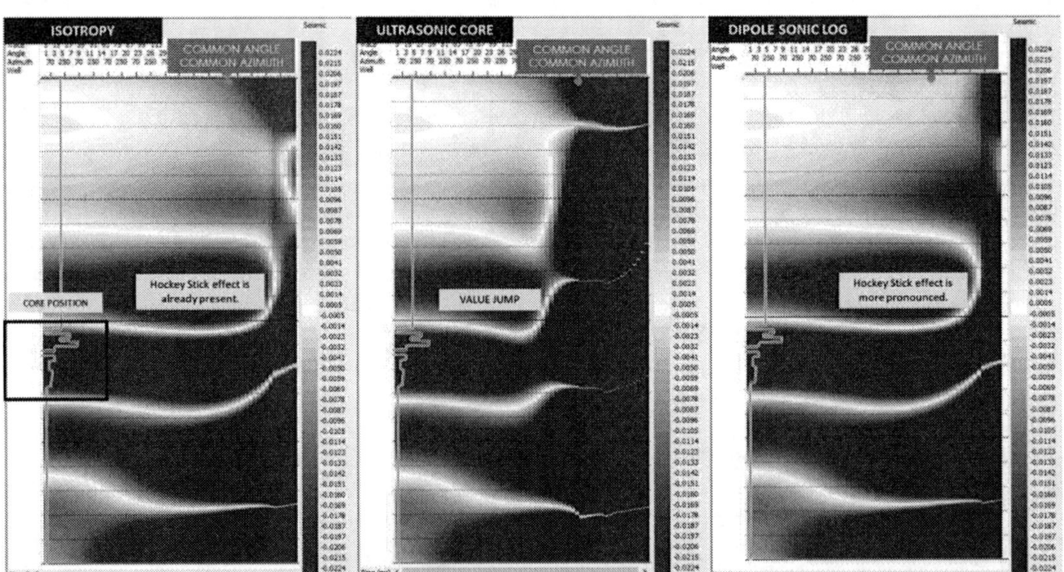

Figure 2. The three scenarios of the synthetic gathers of the well-based modeling result of SAD-4: in-situ log (left), core anisotropy (middle), and log anisotropy (right). The core measurement is conducted around the target, as pointed inside the black box. The synthetic seismic is sorted into Common Angle Common Azimuth gather.

IOP Conf. Series: Earth and Environmental Science **873** (2021) 012038 doi:10.1088/1755-1315/873/1/012038

Synthetic gathers seismic from the in-situ log condition of SAD-4 well showcases the hockey stick even before the anisotropy value incorporated into the model. After the anisotropy value integrated into the model, the synthetic gathers seismic still shows the hockey stick effect but with some enhancement on the kick out curving. The synthetic gathers seismic with log anisotropy has a slight enhancement in the kick-out, starting from angle 55^0 to 65^0. In contrast, the original synthetic gathers seismic forms the curving around the angle of 50^0. A different result is detected in the core anisotropy; the synthetic gathers seismic shows a value jump as early as 40^0 before the kick out, starting at the 50^0 on the long offset (Figure 2). This phenomenon is associated with an anisotropy injection to the model with a reasonably high magnitude level.

Anisotropy magnitude calculated from Dipole Sonic Log is relatively weak compared to the magnitude from Ultrasonic Core Measurement. The implication is that the synthetic gather seismic from log anisotropy tends to be similar to the synthetic gather seismic from in-situ log modeling. The small value of magnitude doesn't have much influence on the seismic reflectivity response.

The synthetic gather model based on the anisotropy that is estimated from Ultrasonic Core Measurement shows a different comparison to the other two in terms of amplitude and reflectivity. The difference in response leads to the possibility of anisotropy magnitude value or irregular data sampling being the factor. To investigate this issue, we come up with another experiment that would be explained later.

3.2 Anisotropy value and data sampling effect on synthetic seismic

There exist two issues that differentiate anisotropy from Ultrasonic Core Measurement and Dipole Sonic Log. They are the range of value and data sampling. The anisotropy from Ultrasonic Core Measurement in SAD-4 has magnitudes 40 times larger than Dipole Sonic Log. The sample is slightly different. Figure 3 shows the illustration of the value difference in Thomsen's Epsilon and Gamma of SAD-4. Furthermore, core plug measurement has a sampling of 9 – 12 ft while the log has a sampling rate of 0.5 ft.

This experiment will explore more about these two factors that influence the seismic response during well-based modeling. The anisotropy value from the log is scaled-up to the anisotropy range value from the core. Likewise, the core's anisotropy value is scaled-down to the anisotropy range value from the log. The scaled-down process in core anisotropy intends to mimic the weaker anisotropy value found in the log. The scaled-up is meant to produce substantial anisotropy with regular sampling. A linear function does the scaling processes derived based on data to data cross plot. Equation (1) represents the scaling function for Epsilon data and Equation (2) represents the scaling function for Gamma.

$$\mathcal{E}_L = -0.0012719\mathcal{E}_C + 0.00182828 \qquad \textit{(Eq. 1)}$$

$$Y_L = 0.0265119Y_C + 0.00479782 \qquad \textit{(Eq. 2)}$$

Figure 4(a) shows the synthetic gather seismic resulting from well-based modeling with scaled-down core anisotropy. The value jump phenomenon found in Figure 2 is no longer detected. Overall, the reflector at the target level has similar behavior to synthetic gather seismic with anisotropy from Ultrasonic Core Measurement. On the contrary, the synthetic gathers seismic due to well-based modeling with scaled-up log anisotropy shows an intensive value jump kick out, as shown in Figure 4(b).

These results confirm the previous guess that value range and data sampling modification influence the seismic response on well-based modeling with anisotropy. The bigger or smaller anisotropy value is implied on the synthetic gathers seismic, more specifically on the long offset. The use of the anisotropy parameter based on core and log data in the long offset seismic data gather, the regularization and value normalization are necessary.

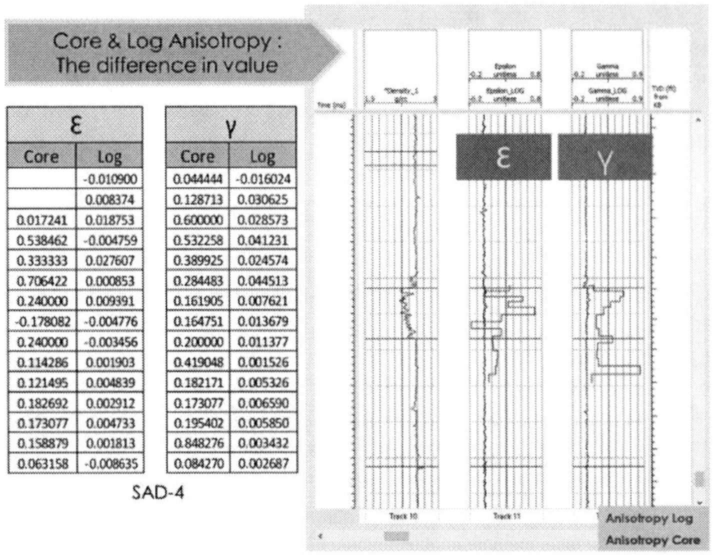

ε		γ	
Core	Log	Core	Log
	-0.010900	0.044444	-0.016024
	0.008374	0.128713	0.030625
0.017241	0.018753	0.600000	0.028573
0.538462	-0.004759	0.532258	0.041231
0.333333	0.027607	0.389925	0.024574
0.706422	0.000853	0.284483	0.044513
0.240000	0.009391	0.161905	0.007621
-0.178082	-0.004776	0.164751	0.013679
0.240000	-0.003456	0.200000	0.011377
0.114286	0.001903	0.419048	0.001526
0.121495	0.004839	0.182171	0.005326
0.182692	0.002912	0.173077	0.006590
0.173077	0.004733	0.195402	0.005850
0.158879	0.001813	0.848276	0.003432
0.063158	-0.008635	0.084270	0.002687

Figure 3. Thomsen's Epsilon and Gamma from Ultrasonic Core Measurement and Dipole Sonic Log in SAD-4.

(a)

(b)

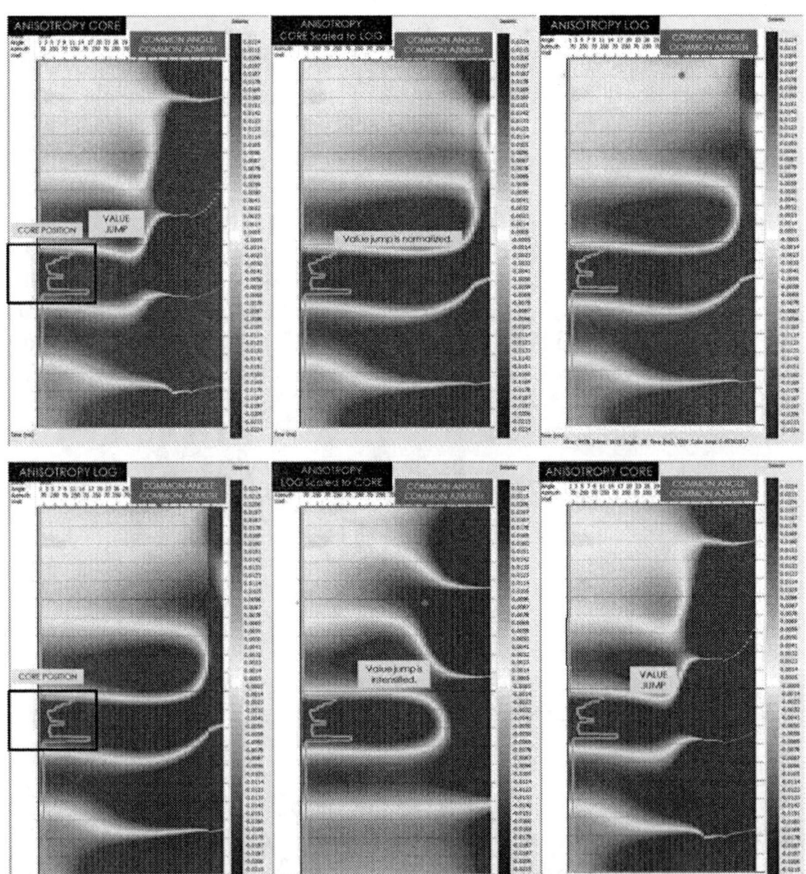

Figure 4. (a) Synthetic gathers seismic result of well-based modeling with core anisotropy scaled into value range of log anisotropy. **(b)** Synthetic gathers seismic result of well-based modeling with log anisotropy scaled into value ranged from core anisotropy.

4. Conclusions

The deepwater reservoir has been posing a challenge in hydrocarbon exploration. Useful seismic imaging is the primary tool to support a thriving and cost-efficient production. Facing the hockey stick effect as a tribute from Vertical Transverse Isotropy condition, seismic processing needs the right anisotropy parameter Eta (η) for the long offset correction. An elasticity parameter η could be derived from Thomsen's Parameter ε and δ. Integrating anisotropy measurement from core & log data has proven to help hand in choosing the best η value. The massive difference in dominant frequency and probably bandwidth of seismic wave source leads to a difference in estimating the anisotropy parameter value between Ultrasonic Core Measurement and Dipole Sonic Log. It could lead to a vague understanding of anisotropy's degree in the same target better. It still needs to find a method for further analysis on transforming the estimation anisotropy parameter value of core and log into the seismic gather data's long offset. In this study, the application of both estimated anisotropy core and log parameters into seismic bandwidth of the long offset data by scaled up or down is made by neglecting the frequency of source or bandwidth dependence effect. The reflection of the weaker anisotropy magnitude shows a similar comparison to the isotropy.

References

[1] Winterstein D F 1990 *Geophysics* **55** 1070-1088
[2] Tsvankin I and Thomsen L 1994 *Geophysics* **59** 1290-1304
[3] Thomsen L 1986 *Geophysics* **51** 1954-1966
[4] Alkhalifah T and Tsvankin I 1995 *Geophysics* **60** 1550-1566
[5] Castagna J P, Batzle M L, and Eastwood R L 1985 *Geophysics* 50 571-581
[6] Lucet N, Rasolofosaon P N, and Zinszner B 1991 *The Journal of the Acoustical Society of America* **89** 980-990
[7] Sondergeld C H and Rai C S 2011 *The Leading Edge* **30** 324-331
[8] Vernik L and Nur A 1992 *Geophysics* **57** 727-735
[9] Rokhlin S I and Wang L 2002 *International Journal of Solids and Structures* **39** 5529-5545
[10] Protázio J S and Schoenberg M 1991 *The 2nd International Congress of the Brazilian Geophysical Society*

IOP Conf. Series: Earth and Environmental Science **873** (2021) 012039 doi:10.1088/1755-1315/873/1/012039

Optimization of 2-D Magnetotelluric (MT) Modes Based on Data Dimensionality to Delineate Potential Area for Shale Gas Prospect in Kutai Basin, East Kalimantan, Indonesia

Hidayat[1], G.M Lucky Junursyah[1], Ahmad Setiawan[1], Andrawan Erlang Pradana[2]

[1] Center for Geological Survey, Geological Agency of Indonesia, Ministry of Energy and Mineral Resources, Indonesia, Jalan Diponegoro No. 57, Bandung 40122, Indonesia
[2] Graduate program of Geophysics, Faculty of Mathematics and Natural Science, University of Indonesia, Jalan Lingkar Kampus Raya, Depok 16424, Indonesia

*email: hidayathidayat01@gmail.com

Abstract. We conducted a study using the magnetotelluric method in the Kutai Basin, which is one of the largest and deepest tertiary sedimentary basin located in the province of East Kalimantan, Indonesia. The Kutai Basin, which is one of the sedimentary basin that is proven to produce hydrocarbons in Indonesia, also has the potential for shale gas with all the complexities of its geological structure. Inversion of 2-D MT can generally be done in three modes with different sensitivity. We perform data processing objectively to obtain the best quality data. We continued our data processing to the inversion process with a range from 80.78% to 97.09% coherency data. We also performed sensitivity skewness calculations to determine the dimensionality of our data. The map of sensitivity skewness is shown for the vertical path A – A' with direction N – S in our study area. Based on the calculation results, the skewness value below 0.3 is obtained around the frequency 320 - 0.002 Hz, and associated with the 2-D structure while value above 0.3 are obtained around the frequency 0.00198 - 0.00034 Hz at KT34 and KT36 stations. Based on dimensionality calculations, it is concluded that the MT data in the Kutai Basin is dominated by 2-D structural responses, so that the TE + TM (invariant) mode is the best measurement mode for inversion modeling. We also performed calculations to obtain the optimum smoothness factor (tau) using a trade-off curve. Based on the results of the inversion with the optimization of these data parameters, we obtained a subsurface geological structure pattern such as fault and fold structure along the vertical path of A – A'. The low resistivity anomaly is interpreted as a response to the presence of black shale which is part of the Pamaluan Formation. The top of the Pamaluan Formation is estimated at the depth that varies from 2000m to 4000m below the surface along the A – A' vertical cross-section.

1. Introduction

Black Shale is thought to be a possible cause of the low resistivity anomaly found in the Earth's upper mantle [1, 2, 3]. The high thermal maturity (> 4000) is the reason for the low resistivity value in black shale. The increase in temperature causes a change in the chemical composition of organic carbon found in black shale which causes a decrease in the resistivity value of the rock [4]. A recent study was conducted to describe the black shale layer in South Africa as a continuous layer with low resistivity values mapped to hundreds of kilometres [5]. Based on this study, measurements in the laboratory were also carried out on black shale samples and confirmed that low resistivity values were found in some rock samples with high thermal maturation [6]. With the unique electrical resistivity properties of black shale, we conducted a study using magnetotelluric method that can depict subsurface geological structures based on electrical properties of rocks from shallow to very deep depths to delineate the potential for shale gas in the Kutai Basin.

2. Geology of Study Area

Based on the geological map of Samarinda sheet [7] shown in Figure 1, there are geological structures with fold pattern in the form of anticline and syncline with a relatively NE - SW oriented axis around this study area. Our A – A' vertical cross section intersecting one axis of the anticline and Mahakam River in station KT36 and KT34 respectively. Based on the order of sedimentary rocks deposition around the study area, the Pamaluan Formation is the oldest rock formation which were deposited during the Oligocene to Early Miocene periods. Furthermore, it was continued with the deposition of the Pulaubalang Formation which was deposited unconformably with the Balikpapan Formation during the Middle Miocene to Late Miocene. The last one is the quaternary sediment deposited in a fluviatile, paludal, deltaic and coastal environment. It is estimated that the black shale in the Pemaluan Formation could be a potential source rock for the Kutai Basin petroleum system due to its high value of both TOC and thermal maturity.

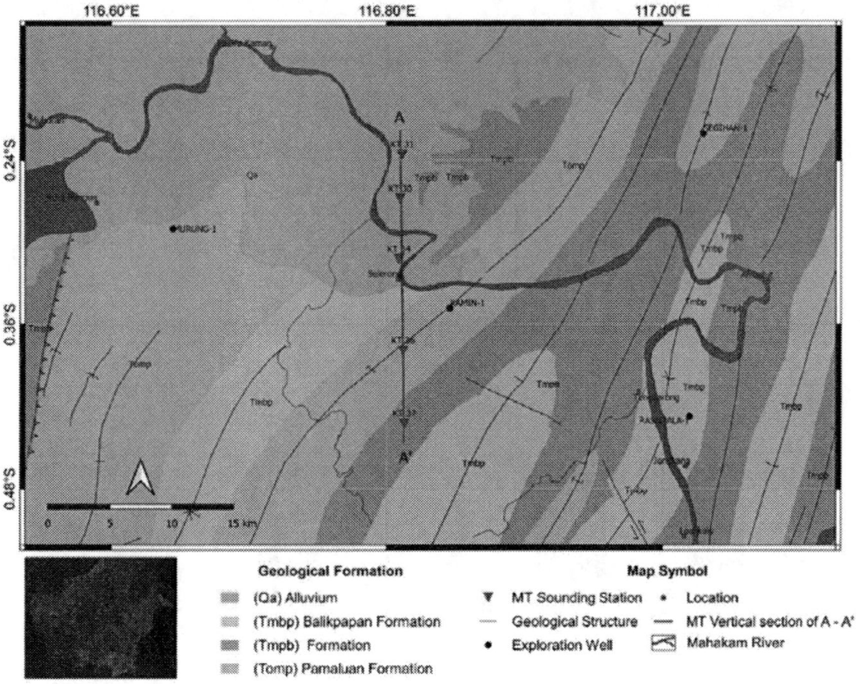

Figure 1. Geological map of the Kutai Basin overlayed with magnetotelluric stations and vertical cross section A – A' [7]

3. Data and Methods

In this study, we used the magnetotelluric data of the Kutai Basin consisting of 5 sounding points. Respectively from North to South are KT31, KT30, KT34, KT36, and KT37 along the section A – A' with a ± 20 km-length section. We performed data processing to increase the quality of MT data based on the coherence value. The term coherence is quite popular to use to determine data quality. For MT data, high coherence value is one of an acceptance criteria and indicates a low contaminated coherent noise in signal [8]. A series of data processing techniques such as robust processing and cross power selection were also conducted in order to obtain an apparent resistivity curve as shown in Figure 2. Histogram comparison of the coherence of raw data and data processing results is shown in Figure 3.

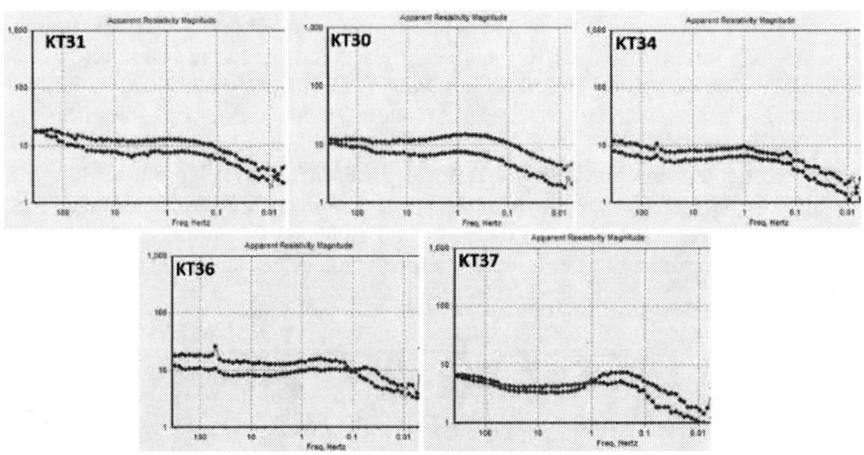

Figure 2. Apparent resistivity curve after performing data processing stages. Smooth curves of our MT data indicate high signal to noise ratio

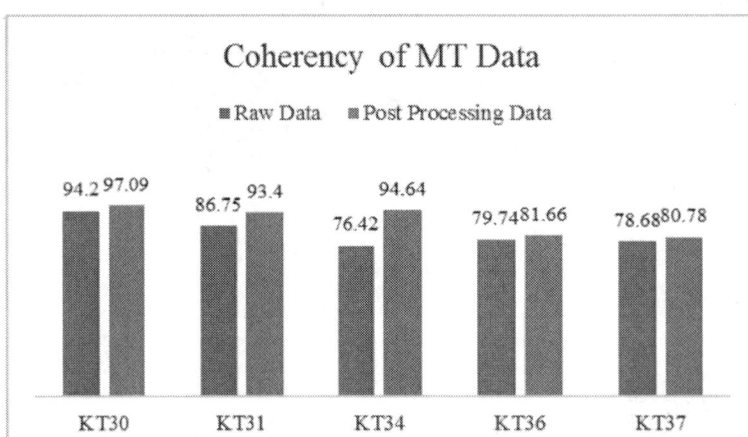

Figure 3. Comparison of the coherence value between pre and post processing data. We managed to improve our quality data from 76.42 – 94.2% to 80.78 – 97.09%

3.1 Dimensionality of MT Data

Dimensionality of MT data for regional electrical structures was originally introduced by [9] based on skewness data (S) which satisfies equation (1).

$$S = \left| \frac{Z_{xx} - Z_{yy}}{Z_{xy} - Z_{yx}} \right| \quad (1)$$

Where Z_{xx}, Z_{xy}, Z_{yx} and Z_{yy} are the impedance tensor elements.

Skewness can be used to evaluate whether the subsurface electrical structure beneath the sounding point of MT qualitatively influence by 1-D, 2-D or 3-D distortion. However, the skewness value [9] in practice still deviates when it is exposed to telluric distortion in the impedance tensor components. [10] proposed a skewness (η) measurement formula that takes into account the effects of distortion on the impedance tensor components. The calculation of skewness (η) by Bahr is shown by equation (2).

$$\eta = \sqrt{2\frac{\left|\mathrm{Re}\,Zxx.\mathrm{Im}\,Zxy - \mathrm{Re}\,Zyy.\mathrm{Im}\,Zxy + \mathrm{Re}\,Zxy.\mathrm{Im}\,Zyy - \mathrm{Re}\,Zyx.\mathrm{Im}\,Zxx\right|}{\left|Zxy - Zyx\right|}} \quad (2)$$

Where *Re* is the real component and *Im* is the imaginary component of a complex number. η values above 0.3 can be categorized as the response of the 3-D electric structure below the surface, η values below 0.3 are categorized as the response of 2-D electric structures, while the $\eta = 0$ indicates the ideal 1-D or 2-D structure conditions [10, 11]. Previous studies of dimensionality and its effect on the sensitivity of MT measurement modes have been carried out by creating a 3-D synthetic model and testing the inversion results of each MT measurement mode which shows that TM is more sensitive to 3-D structures and TE+TM (invariant) is more sensitive to structure 2-D [12].

4. Result and Discussion

We used all frequencies (320 - 0.000034 Hz) in the data set to calculate sensitivity using the impedance tensor data decomposition. The plot of the calculation results is shown in Figure 4.

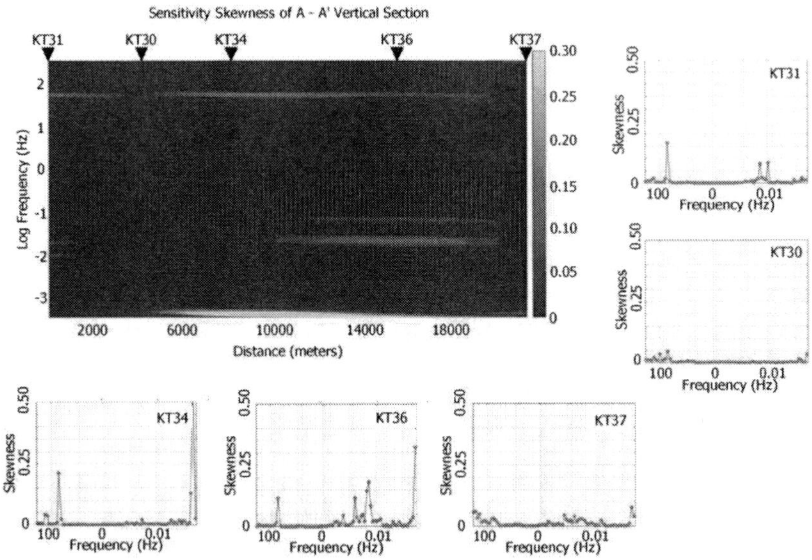

Figure 4. Map of the distribution of sensitivity on the vertical cross section A - A' which shows the distribution of calculated values η is dominated by values in the range 0-0.3

It can be concluded that the 2-D structure is more dominant in the vertical section A – A', so that the TE+TM mode is the most sensitive mode for the 2-D inversion model. We applied a 2-D inversion model based on the NLCG inversion algorithm [13]. For the purposes of making the inversion model, we used MT data with a frequency range of 320-0.01 Hz considering that data below 0.01 Hz is highly contaminated by coherent noise. Our model parametric test was carried out based on the Roughness and RMS plots on the trade-off curve as shown in Figure 5. The comparison between the 2 most optimal model parameterization values (tau) (1 and 5) is shown in Figure 6. Based on these comparisons it can be concluded that tau 1 shows the best results and is more sensitive to anticline in vertical sections A – A'.

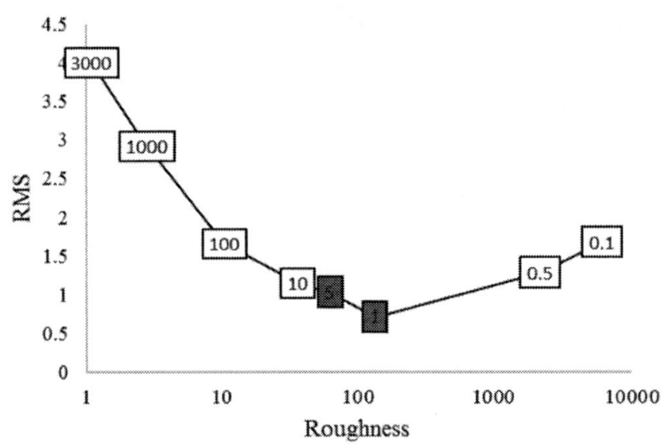

Figure 5. The trade-off curve to find the optimum model parameterization for MT 2-D inversion

IOP Conf. Series: Earth and Environmental Science **873** (2021) 012039 doi:10.1088/1755-1315/873/1/012039

The comparison between the 2 most optimal model parameterization values (tau) (1 and 5) is shown in Figure 6. Based on these comparisons, it can be concluded that tau 1 shows the best results and is more sensitive to anticline in vertical sections A – A'.

The interpretation of the vertical section A – A' (Figure 7) is validated by the geological map of the Samarinda sheet which shows the presence of an anticline and syncline fold structure pattern on the surface and crossed by A - A' vertical section. Subsurface geological feature e.g. fault and fold structure can be well imaged based on the contrast of resistivity. The contrast of resistivity anomaly is interpreted as the top of the black shale bed which is characterized by low resistivity value (<5 ohm m).

The low resistivity anomaly are well resolved below the A – A' vertical section which then interpreted as the response of matured organic black shale bed from the Pamaluan Formation at depths varying from 2000 m to 4000 m below the surface. Constraints on the geological map that show the outcropping of the Pamaluan Formation at the KT36 station can also be characterized well by the results of the 2-D vertical section inversion model A - A '.

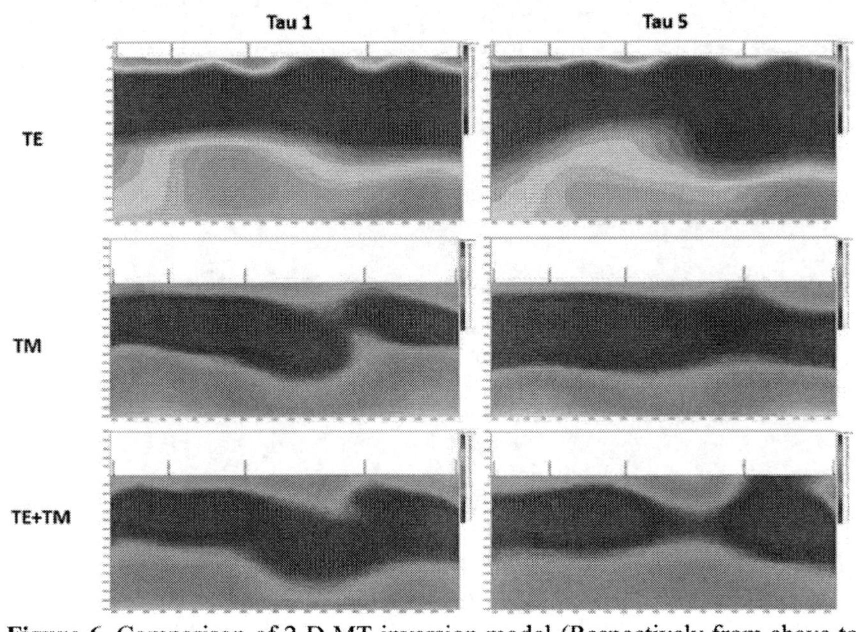

Figure 6. Comparison of 2-D MT inversion model (Respectively from above to below; TE, TM, TE+TM) at value of tau 1 and 5

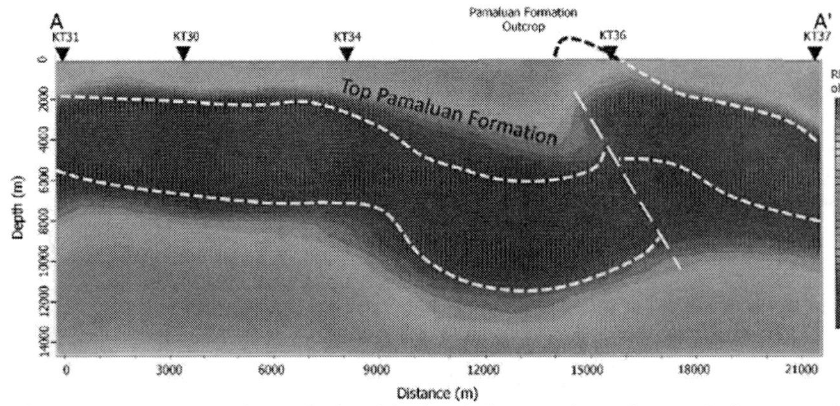

Figure 7. Interpretation of the black shale top boundary of the Pamaluan Formation and the ramp anticline/fault-bend fold along the vertical section A –A '

5. Conclusion

Optimization of parameters in 2-D MT inversion is very influential on 2-D MT inversion model, the wrong parameter selection will cause misinterpretation of subsurface geological information. We managed to determine the optimum parameters based on dimensionality data and optimum model parameterization values to obtain a vertical section that delineates the top blackshale bed from the Pamaluan Formation and subsurface geological features such as the ramp anticline/fault-bend fold.

Acknowledgments

We would like to thank Center for Geological Survey (CGS) Indonesia and all technical staff in MT Kutai data acquisition participating in the fieldwork.

References

[1] Gaucher E 1983 *Estimation of Sulfide Content of a Potential Orebody from Surface Observations and its Role in Optimising Explorations Programmes* A A Fitch (ed.) Developments in Geophysical Exploration Methods 4
[2] Loukola-Ruskeeniemi K 1989 Early Proterozoic Metamorphosed Black Shales in the Kainuu Schist Belt and in the Outokumpu Region *Geological Survey of Finland* **10** (Special Paper) 103-106
[3] Korja T and Koivukoski 1990 *Magnetotelluric Soundings Along the SVEKA Profile in the Central Fennoscandian (Baltic) Shield* Finland *Report n°17*
[4] Jödicke 1992 Water and graphite in the Earth's crust: An approach to interpretation of conductivity model *Surv. Geophys* **13** 381-407
[5] Weckmann U, Jung A, Branch T and Ritter O 2007b Comparison of electrical conductivity structures and 2D magnetic modelling along two profiles crossing the Beattie magnetic anomaly *South African Journal of Geology* **110** 449–464
[6] Branch T, Ritter O, Weckmann U, Sachsenhofer R and Schilling F 2007 The Whitehill formation: a high conductivity marker in the Karoo Basin *South. Afr. J. Geol.* **110** 465–476
[7] Supriatna S and Rustandi E 1995 *Geological Map of the Samarinda sheet, Kalimantan* Geological Research and Development Centre, Bandung
[8] Fontes S L et al. 2002 Processing of noisy magnetotelluric data using digital filters and additional data selection criteria *Physics of the Earth and Planetary Interiors* **52** 30–40

[9] Swift C M 1967 *A magnetotelluric investigation of an electrical conductivity anomaly in the southwestern United States* Ph. D. thesis, Massachusetts Institute of Technology Cambridge MA
[10] Bahr K 1991 Geological noise in magnetotelluric data: a classification of distortion types *Physics of the Earth and Planetary Interiors* **66** 24–38
[11] Lezaeta P 2002 The confidence limit of the magnetotelluric phase sensitive skew *Earth Planets and Space* **54** 451–457
[12] Xiao Q et al. 2011 Application of 2D magnetotelluric methods in a geological complex area, Xinjiang, China *Journal of Applied Geophysics* **75** 19–30
[13] Rodi W and Mackie R L 2001 Nonlinear conjugate gradients algorithm for 2-D magnetotelluric inversion *Geophysics* **66** 174–187

IOP Conf. Series: Earth and Environmental Science **873** (2021) 012040 doi:10.1088/1755-1315/873/1/012040

Simple design to estimate time-lapse microgravity response due to shallow subsurface density redistribution caused by land subsidence

E J Wahyudi, M H Hafidza, and M A Tahta

Institut Teknologi Bandung
ekojw@gf.itb.ac.id

Abstract. A simple design for modeling shallow subsurface density redistribution due to land subsidence is designed to obtain the time-lapse microgravity response. The subsurface model at each point of gravity observation is represented by a rectangular prism. A numerical example of computational modeling is performed to estimate the effect of land subsidence to the data of a time-lapse microgravity. Simple numerical simulations with an initial model that have flat topography, homogeneous density, and homogeneous compaction thickness are carried out in variations of geological and hydrological information that are often found in a study area. Additional algorithms to accommodate information on topographic variations, density variations, and compaction thickness variations in the horizontal direction also shown with illustration. Field data application for this study utilize rough estimation of the geology and the land subsidence rate in Bandung Basin. The estimation results with numerical simulations give time-lapse microgravity anomaly 0.78 to 28.61 μGal/m and field data application give an anomaly up to 10 μGal.

1. Introduction

The change in the value of gravity that occurs at the time-lapse microgravity (TLM) observation point can be caused by several factors. In the research of Tampone et al. [1], changes in gravity values for simulation studies in hydrocarbon production areas consist of: changes in the position of the observation point, deformation below the earth's surface, changes in groundwater level, and changes in the reservoir layer (fluid substitution). For case studies in the research area where there is no hydrocarbon production activity, in general, the target depth of the TLM anomaly is relatively shallower (changes in the water table and changes in the soil surface). Groundwater level changes involve fluid substitution in the rock pore space below the surface (not discussed in this study), while subsidence estimates (which are often monitored by repeated GPS surveys and/or InSAR) provide information on changes in the position of the TLM observation points and redistribution estimates. density of rock near the surface.

In this study, the main focus of research is related to TLM modeling due to changes in the soil surface followed by a redistribution of rock density near the surface. The modeling approach is carried out with a simple volume element in the form of a rectangular prism. The subsidence estimation results provide a change in the dimension of volume elements in the vertical column (z-axis of the Cartesian coordinates) below the surface. The modeling computation is carried out by taking into account two main changes, namely: changes in the position of the observation point and redistribution of density below the surface (illustration in Figure 1). Several researchers have conducted several studies ([1]-[3]) for subsurface deformation modeling. Several approaches and study results for the effect of gravity on changes in the position of the observation point have also been carried out by several researchers ([4]-

[5]) using theoretical approach of the free-air gradient, using a theoretical approach of free-air and Bouguer slab [6], vertical correction of gravity gradient ([7]-[8]), and observational studies [9].

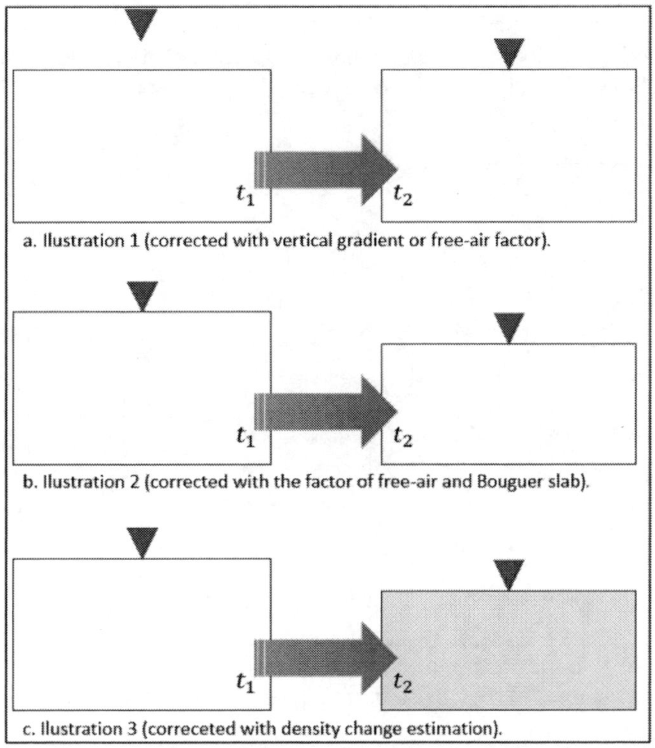

Figure 1. An illustration of a comparison of observation points based on changes in the subsurface from the initial and final conditions in the TLM data acquisition scheme which explains the correction due to subsidence factors using the following approach: a) vertical gradient value or free-air correction, b) free-air coarse and correction slab Bouguer, c) modeling changes in density and observation points due to subsidence.

The illustration in Figure 1 shows a sketch of changes in the position of the observation point and conditions below the surface. The theoretical approach of the free-air gradient or vertical gradient correction as a correction for changes in the earth's surface vertically is shown in Figure 1a. The approach to correction like this is still very rough and can still be improved because based on the illustration (Figure 1a) the position of the observation in the initial conditions is considered to be a certain height above the surface (although at the time of the data acquisition activity, the gravimeters were both placed on the surface of the earth and very much endeavored to do. at the same position in the initial and final conditions). Figure 1b is an illustration for the relatively same observation position from the earth's surface between the initial and final conditions, but also accompanied by subsidence information. In the illustration in Figure 1b, it is shown that the density distribution is still the same in the initial and final conditions but has different volume sizes (one of the near-surface activities that allows this illustration is the activity of taking material from the surface by a hoe or heavy equipment). In general, the free-air and Bouguer slab correction in Figure 1b does not involve redistribution of subsurface density due to subsidence. Figure 1c describes an illustration that is almost the same as Figure 1b, but the mass below the surface does not change so that there is a density redistribution due to subsidence which is shown in gray.

2. Modeling Computation (Simple Design)

The discretization below the surface is approached by a simple set of elements (Figure 2). For practical purposes, discretization of observational data corresponds to discretization of bodies under the surface. Each observational data on a particular topography below is modeled with a prism that is parallel to the x-axis, y-axis, and z-axis. Each volume element has a uniform density change value ($\Delta\rho$) with the prism volume being $dV = dxdydz$. Each volume element can then add up its effect cumulatively at each point of gravity observation on the surface.

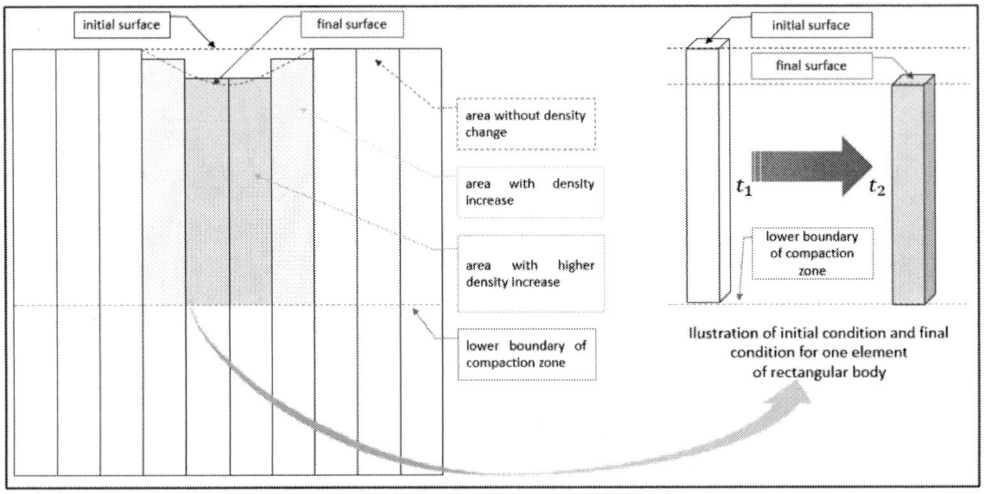

Figure 2. Illustration of a simple design of a subsurface model for the purposes of computation of TLM anomalies due to changes in the position of the observation points on the surface and due to changes in density below the surface.

Based on Figure 2, a simple design illustration of the subsurface model for computation of the TLM anomaly is shown by the initial surface and the final surface at a certain time interval with the change in subsurface density from the initial state (white color) to the final state (gray color). The change in density is a consequence of the mass conservation formulation as shown by Tempone et al. [1] as follows:

$$\rho_2 = \frac{\rho_1 V_1}{V_2}. \tag{1}$$

At a certain time interval, a prism volume element is simply modeled where only the dz value changes, while the dx and dy values are assumed to be constant. Based on Figure 2, the dz value is limited to the lower limit estimate of the compacting. This approach simplifies the lower bound (z_b) which is the same at a certain time interval, so that the calculation formulation of density change is shown in equation (2) as follows:

$$\rho_2 = \frac{\rho_1(dxdydz_1)}{(dxdydz_2)} = \frac{\rho_1 dz_1}{dz_2} = \frac{\rho_1(z_1 - z_b)}{(z_2 - z_b)}. \tag{2}$$

The z_1 and z_2 values are input data from the topographic differences in the initial and final conditions. The change in density ($\Delta\rho$) is then shown in equation (3) as:

$$\Delta\rho = \rho_2 - \rho_1 = \frac{\rho_1(z_1 - z_b) - \rho_1(z_2 - z_b)}{(z_2 - z_b)} = \frac{\rho_1(z_1 - z_2)}{(z_2 - z_b)}. \tag{3}$$

Synthetically, the TLM anomaly value (Δg) from two gravitational observations (g_1 and g_2) at different topographic coordinates (z_1 and z_2) from a certain time interval (t_1 to t_2) is shown by the mathematical expression in equations (4) and (5):

$$\Delta g = g_2(x, y, z_2, t_2) - g_1(x, y, z_1, t_1), \tag{4}$$

$$\Delta g = \gamma \rho_2 \int \frac{z_2}{(r_2)^3} dV_2 - \gamma \rho_1 \int \frac{z_1}{(r_1)^3} dV_1. \tag{5}$$

In equation (5) the values of r_1 and r_2 are the distance between the observation point and the prism body in the initial and final conditions. Equation (5) for TLM involves two integrals, each of which can be solved numerically. Numerical formulations can be used as has been shown by Wahyudi et al [10].

3. Program Testing with Numerical Examples

This section shows program testing with numerical examples based on the simple modeling design descriptions described in the previous section. The simplified visualization for this explanation will be used for the vertical section sections of a simple numerical subsidence model example. Figure 3 shows an illustration of the model change due to subsidence at certain time intervals (zoom in on a depth scale of 0 to 3 meters). At each point of observation (in red) there is an initial model under the surface represented by a discretization of a simple object in the form of a rectangular prism with a density value of 2.5 g/cc (Figure 3a). The illustration for surface change due to subsidence is shown with a vector arrow in the vertical direction (Figure 3b). The final model after the subsidence information is accounted for for the density change in each rectangular prism model is shown in Figure 3c. Equation (3) which is used to obtain the density value estimate for each discretization of the model elements gives density values in the range of 2.5 g/cc to 2.55 g/cc.

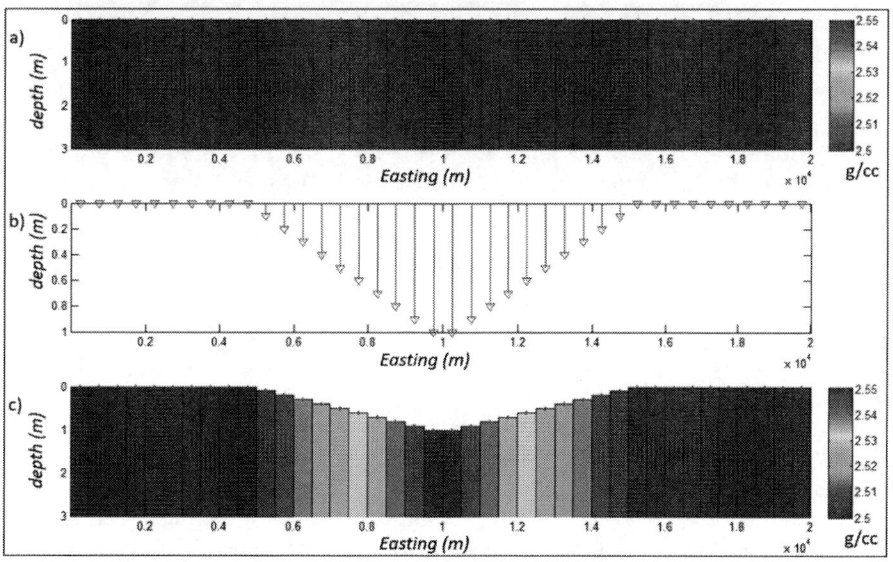

Figure 3. An illustration of model change due to subsidence involves a simple design: (a) initial model with $dy = 500\ m$, (b) vector change in vertical direction (subsidence direction) at certain time intervals, and (c) final model with near surface density change.

Synthetically, the models in Figures 3a and 3c can be used in calculating the gravity anomaly response, where each anomaly is gravitational observations (synthetic data) g_1 and g_2, respectively. An illustration of the synthetic data calculation is shown in Figure 4. If we look at Figures 4a and 4c, it can be seen that the synthetic data on the resulting edge shows a very drastic change as the edge effect of the limited computation model. Regarding the edge effects of the model computation, the main focus regarding the TLM anomaly will be on the center of the model. Figures 4b and 4d illustrate the zoom out of the models in Figures 3a and 3c, to provide more complete information about the models involved in computation. The model in figure 4 shows the thickness of the compacting zone for calculations determined to be 50 meters below ground level.

As a simplification of the illustrations in Figures 3 and 4, the computation of the TLM anomaly and its change model is shown in Figure 5. The TLM anomaly in Figure 5a is generated based on calculations using equation (5). The subsurface model (Figure 5b) is represented by the density change values from the initial and final models. The change in density in Figure 5b is shown in the range of 0 to 0.05 g/cc with an illustration of subsidence at certain intervals represented by a vector in the vertical direction.

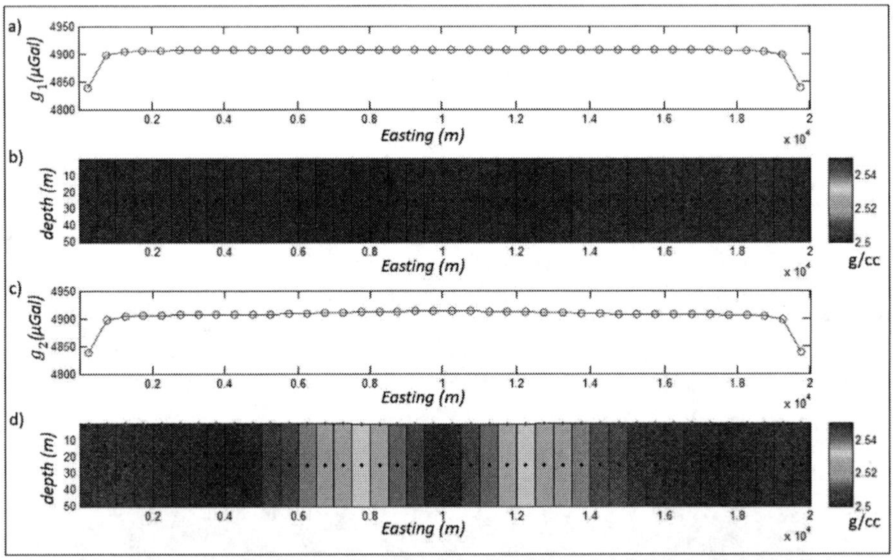

Figure 4. Illustration of synthetic data calculation (g_1 and g_2): (a) initial model gravity response, (b) zoom out initial model, (c) final model gravity response, and (d) zoom out final model.

In the illustration in Figure 5, it is shown that the maximum TLM anomaly value correlates with the position of the maximum subsidence as well, this is related to the maximum compacting process for the compacting zone scenario and the uniform initial model density (2.5 g/cc and 50 meters respectively). As shown in the subsidence arrow vector (Figures 3b and 5b) the calculated value of vertical change is in the range of 0 to 1 meter, this will help provide a rough estimate of the maximum effect of subsidence on a flat topography in units of $\mu Gal/m$. Figure 6 shows some of the estimation results (in $\mu Gal/m$) for the effect of subsidence in the initial model which has a flat topography and uniform rock density. The estimation results in Figure 6 are consecutive simulations for several densities of 1.5 to 3.0 g/cc and successive simulations for several rock layer thicknesses in the computed zone calculation (10 to 300 meters). The curve in Figure 6 provides a larger estimated value for the initial large-density model (3.0 g/cc) and occurs in a thicker compacted zone (300 meters). Estimates of the effect of subsidence and change in density were simple designs with the computational input parameters described in the range 0.78 to 28.61 $\mu Gal/m$.

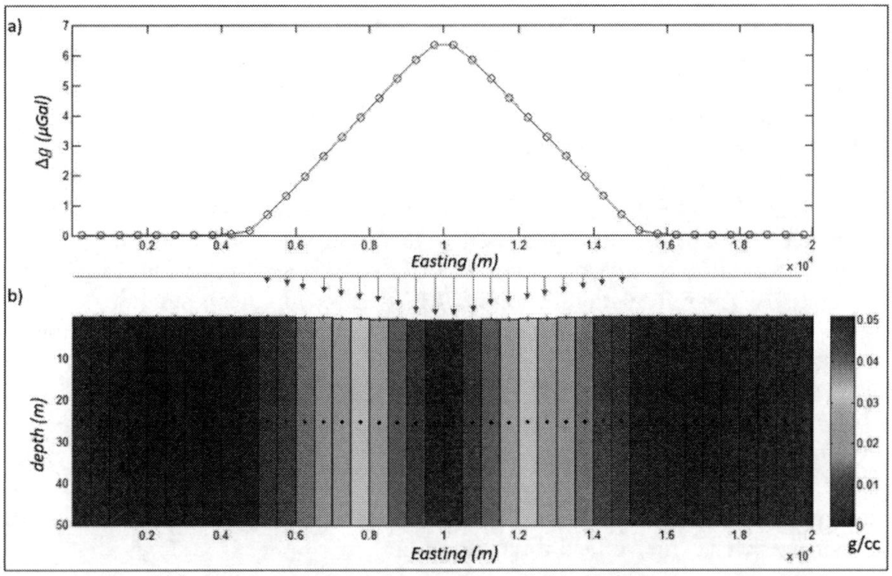

Figure 5. Computational illustration: (a) TLM anomaly and (b) density change model due to subsidence at a certain time interval (subsidence is represented by a vector in the vertical direction).

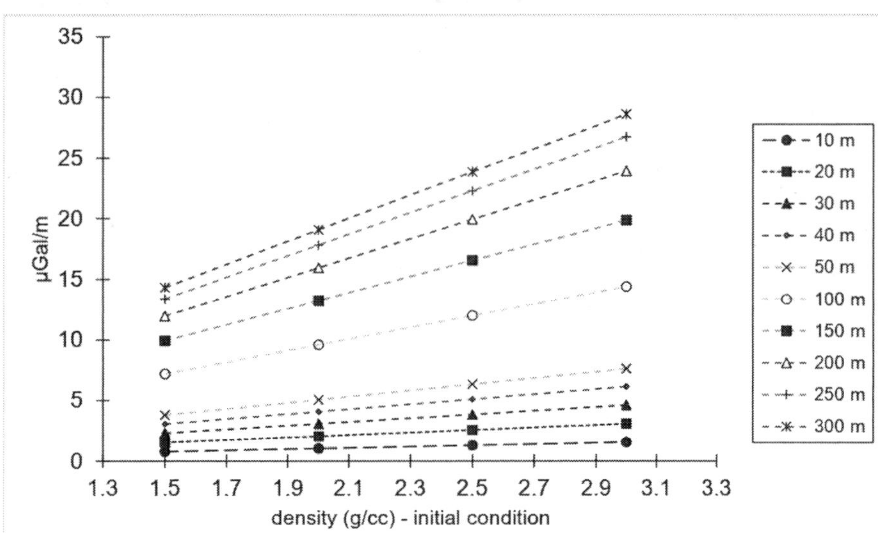

Figure 6. Estimation of the effect of subsidence and near surface density change (in *μGal/m*) for the successive simulations of the initial model densities as well as the successive simulations for several thickness packed layers.

The estimation results of the model in Figure 6 need to be considered for: the initial model with a flat topography, the initial model with a homogeneous density value throughout the discretization of the model under its surface, and the thickness of the homogeneous compacted zone. To carry out an application in an area, the simple computational design that has been described needs to be added with an algorithm to accommodate some of the more real information found in the research area. The topography in the study area in the horizontal direction will tend to vary in topography, as well as the density of the initial model also tends to be not homogeneous (varies according to the geological information in the study area). The compacted zone in a study area also tends to vary in thickness in the horizontal direction. Information on compacted zone thickness can be used based on groundwater utilization well data or conceptual information on hydrological models in a study area. If this information is not available, calculations for a more rough estimate can use a specific, homogeneous value of the thickness of the hydrological layer of a study area.

Figure 7a shows an illustrative example for the initial model with variations in topography, variations in density, and variations in the thickness of the compacted zone in the horizontal direction. Then an illustrative example for subsidence information (which can often be obtained from the results of monitoring with repeated GPS surveys and/or InSAR) is shown with a vector arrow in the vertical direction in Figure 7b. The information in Figure 7 is used for input computation of TLM anomalies. The topographic variation of the initial model is in the elevation range of 700 to 1000 meters above sea level (depth = 0) or has a negative value for the slope of the positive vertical axis downward or depth. The density variation of the initial models is in the range 2.0 to 2.5 g/cc, while the variation for the thickness of the compacted zone is also shown in the range of 100 to 200 meters. At certain time intervals, subsidence information is also synthetically varied in the range of 0 to 1 meter. Figure 8a shows the TLM anomaly calculated by equation (5) in the range 2 to 80 μGal. The density change in Figure 8b is shown in the range of 0 to 0.015 g/cc with an illustration of subsidence at a certain time interval shown by a vector in the vertical direction.

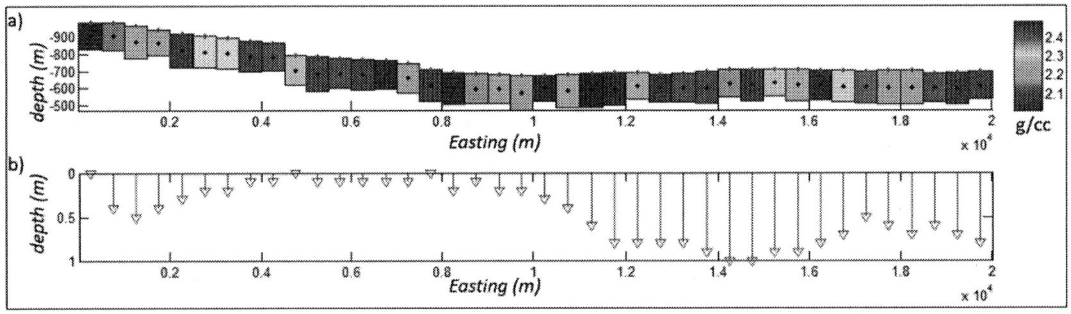

Figure 7. Illustration of the initial model and vector changes in the vertical direction. (a) Initial model with varying topography, density, and thickness of the compacted zone in the horizontal direction. (b) Vector change in the vertical direction (subsidence direction) occurs at certain intervals.

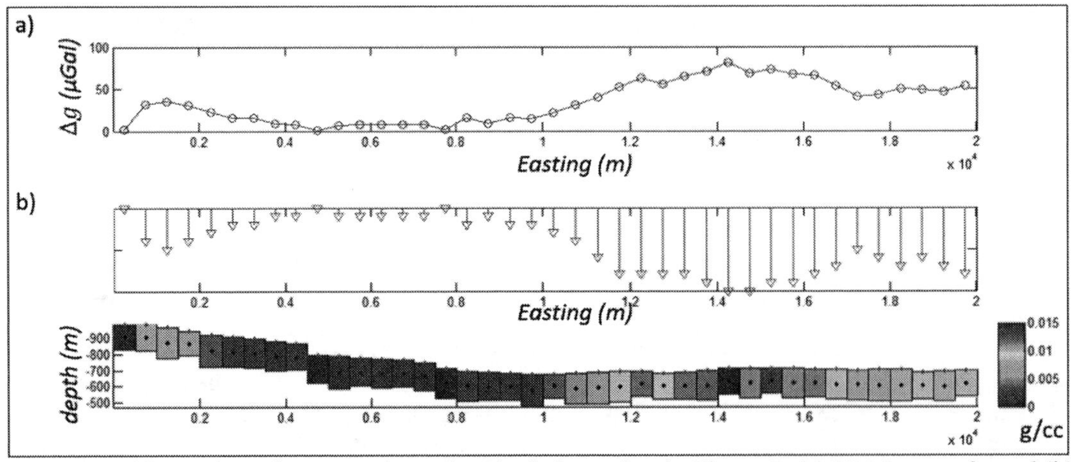

Figure 8. An illustration of computation using the information in Figure 7: (a) TLM anomaly and (b) density change model due to subsidence at a certain time interval (subsidence is shown as a vector in the vertical direction).

4. TLM Anomaly Due to Changes of Topography and Near Surface Density in Bandung

The hydrological information of the research area in Bandung as the initial model to be used in this study is based on the discretization of the conceptual geological model in the publication of Nurliana and Widodo [11]. The discretization of the geological model as shown in Figure 9 provides information on rock formation in the study area. This information will be used as an estimate of the average rock density value of the rectangular vertical prism volume element below the surface. In the discretization of the model, information on topographic variations and variations in the thickness of the zone layers can be extracted in the horizontal direction of the study area. Figure 9 also shows the subsidence estimate in the study area based on the results of the discretization of the average research results of Abidin et al [12]. The subsidence estimate is shown according to the position of the North-South line according to the model trajectory in Figure 9.

Figure 10a shows the initial model with variations in topography, variations in density, and variations in the thickness of the compacted zone in the horizontal direction. Subsidence information is shown with a vector arrow in the vertical direction in Figure 10b. The information in Figure 10 is used for the computational input of the TLM anomaly as shown in Figure 11a. The density change over time as shown in Figure 11b is 15×10^{-4} g/cc. The computational results by accommodating the input data in the study area gave TLM anomalies that reached 10 μGal. The descriptive statistics and histograms of the computational results of the TLM anomaly along the North-South path are shown in Figure 12.

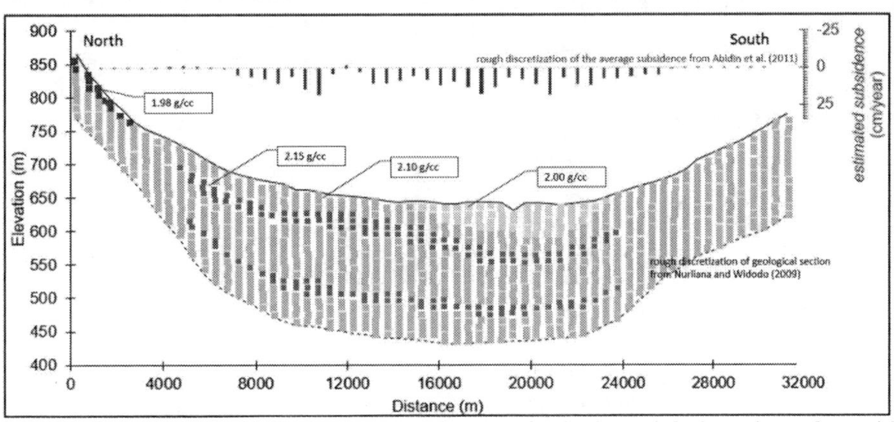

Figure 9. Illustration of discretization of subsurface geological models based on the publication of Nurlina and Widodo [11] and also the estimated average subsidence based on the publication of Abidin et al [12].

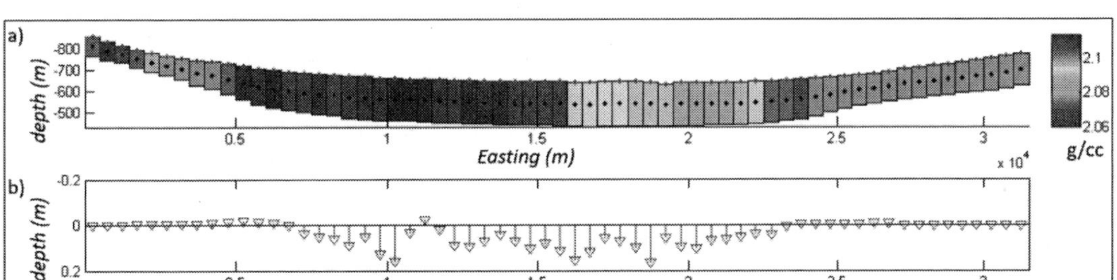

Figure 10. Illustration of the initial model and vector changes in the vertical direction in the study area. (a) Initial model with varying topography, density, and thickness of the compacted zone in the horizontal direction. (b) Vector change in the vertical direction (subsidence direction) occurs at certain intervals.

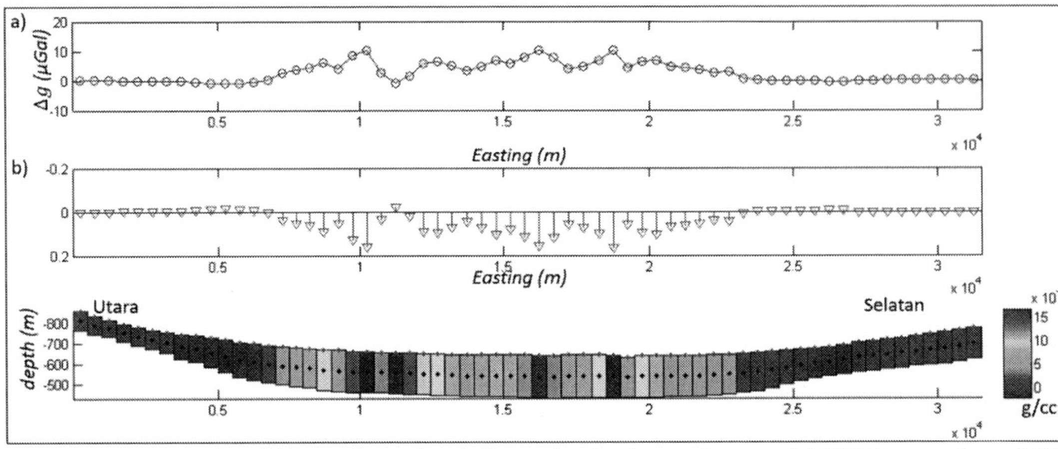

Figure 11. Computational illustration using information in the study area: (a) TLM anomaly and (b) density change model due to subsidence at a certain time interval (subsidence is shown as a vector in the vertical direction).

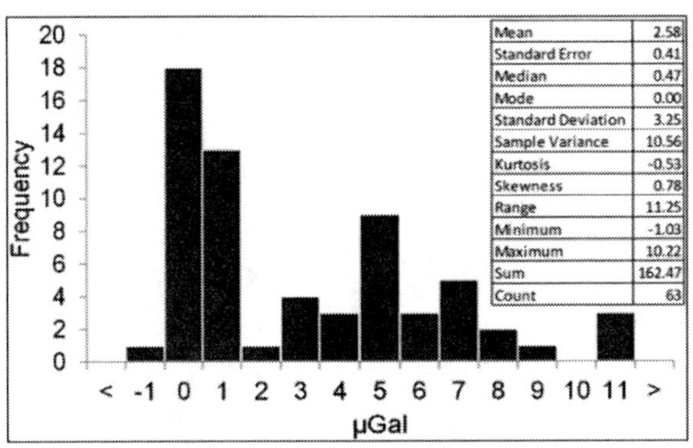

Figure 12. Histogram and descriptive statistics of the TLM anomaly calculated from input data in the study area along the North-South section.

5. Conclusion

In this study, a simple design for TLM modeling of near-surface density redistribution with subsidence estimation results has been shown. A simple numerical simulation on an initial model that has a flat topography, homogeneous density, and homogeneous compressed zone thickness is carried out consecutively for variations in geological and hydrological information that are often found in a study area. The estimation results with numerical simulations of the change in the position of the measurement point and the change in density to the initial model at certain time intervals give TLM anomaly 0.78 to 28.61 μGal/m. Additional algorithms to accommodate topographic variations, initial model density variations, and compacted zone layer thickness variations can be used to obtain computational estimates of the results in the study area. The estimation results in the study area by accommodating the annual trend of subsidence rate gave a time-lapse microgravity anomaly response reaching 10 μGal.

References

[1] Tempone P, Landrø M and Fjær E 2012 4D gravity response of compacting reservoirs: Analytical approach *Geophysics* **77** G45-G54

[2] Bonafede M and Mazzanti M 1998 Modelling gravity variations consistent with ground deformation in the Campi Flegrei caldera (Italy) *Journal of Volcanology and Geothermal Research* **81** 137-157

[3] Battaglia M and Segall P 2004 The Interpretation of Gravity Changes and Crustal Deformation in Active Volcanic Areas *Pure appl. geophys* **161** 1453-1467

[4] Battaglia M, Gottsmann J, Carbone D and Fernández J 2008 4D volcano gravimetry *Geophysics* **73** WA3-WA18

[5] Kobe M, Gabriel G, Weise A and Vogel D 2019 Time-lapse gravity and levelling surveys reveal mass loss and ongoing subsidence in the urban subrosion-prone area of Bad Frankenhausen, Germany *Solid Earth* **10** 599–619

[6] Wahyudi E J, Dahrin D and Alawiyah S 2010 Interpretasi Data Anomali Gayaberat Mikro Antar Waktu untuk Menentukan Pola Pergerakan Air Tanah di Semarang *Buletin Meteorologi Klimatologi dan Geofisika* **6** 287-294

[7] Hunt T M and Sugihara M 2000 Correcting for Effects of Ground Subsidence Microgravity Monitoring *New Zealand Geothermal Workshop* 109-114

[8] Jacob T, Bayer R, Chery J and Le Moigne N 2010 Time-lapse microgravity surveys reveal water storage heterogeneity of a karst aquifer *Journal of Geophysical Research* **115** 1-18

[9] Olsson P–A, Milne G, Scherneck H–G and Ågren J 2015 The relation between gravity rate of change and vertical displacement in previously glaciated areas *Journal of Geodynamics* **83** 76-84

[10] Wahyudi E J, Santoso D, Kadir W G A and Alawiyah S 2014 Designing a Genetic Algorithm for Efficient Calculation in Time-Lapse Gravity Inversion *J. Eng. Technol. Sci.* **46** 59-79

[11] Nurliana L and Widodo L E 2009 Potensi Imbuhan dan Imbuhan Airtanah Cekungan Airtanah Bandung *JTM* **XVI** 261-268

[12] Abidin H Z, Gumilar I, Andreas H, Sidiq T P and Fukuda Y 2011 Study on Causes and Impacts of Land Subsidence in Bandung Basin, Indonesia *FIG Working Week Bridging the Gap between Cultures* 1-18

IOP Conf. Series: Earth and Environmental Science **873** (2021) 012041 doi:10.1088/1755-1315/873/1/012041

River sedimentation modeling using ground-penetrating radar

M A Firdaus*, Widodo and Fatkhan

Geophysical Engineering, Faculty of Mining and Petroleum Engineering, Bandung Institute of Technology, Indonesia.

*Corresponding author: aldimfirdaus2@gmail.com

Abstract. In recent years, siltation has become quite a problem. It has been the main cause of flooding and a rapid decline in water quality. It is usually caused by a high river sedimentation rate and/or uncontrolled waste disposal. The increased rate of erosion also means that river sedimentation occurs faster than normal and could lead to environmental hazards, wildlife deaths, and the disruption of food and drinking water supply among other things. The question is how to monitor the sedimentation process of rivers without damaging the river itself. The suitable geophysical method is GPR. GPR is an active, non-intrusive geophysical method in which electromagnetic radiation and the reflected signals in the form of radar pulses are used for subsurface imaging. The objective is to investigate river sedimentation using GPR, we created the synthetic models based on geological models of rivers with different depths to create their 2-D radargrams to predict the actual model. We set up the first model RSM-I as control which consists of a layer of freshwater with $\rho = 16\ \Omega\text{m}$, $k = 81$ and $\mu_r = 1$ of depth 5 m, two layers of sandstone with $\rho = 850\ \Omega\text{m}$, $k = 2.5$ and $\mu_r = 1$ of total depth 4 m, and a layer of claystone with $\rho = 120\ \Omega\text{m}$, $k = 11$ and $\mu_r = 1$ of depth 1 m. RSM-II and III are added with a buildup of saturated sediment with $\rho = 30\ \Omega\text{m}$, $k = 15$, and $\mu_r = 1$ of depth 2.5 and 4 m, respectively. The radargrams' reflector for each model shows a two-way travel time of 300-350, 150-200, and 60-90 ns in their respective order. GPR models can differentiate between the saturated sediment and freshwater, it shows good results regarding sediment investigation in rivers.

1. Introduction

Siltation is one of the reasons for flooding, wildlife deaths, and the downgrade of water quality. Siltation is caused by erosion and uncontrolled waste disposal that increases the river sedimentation rate. River sediment transport mechanism is divided into two types. Bedload where the sediment is transported along the surface of the bed and suspended load where the sediment is suspended in the fluid [1].

To prevent siltation, there needs to be a method that detects underwater sediment efficiently while providing high-quality results. To overcome this, we use the EM method, specifically ground-penetrating radar (GPR). GPR transmits EM radar waves and when the waves hits a boundary layer, some of the waves are reflected to the receiver and the rest of the waves are either refracted or scattered through the subsurface. The reflected waves are then plotted against its two-way travel time (*TWTT*) creating a radargram, a subsurface image with high resolution that is similar to a seismogram in seismic reflection. GPR will detect the contrast between the water and sediment layer and can determine the depth of the river according to the thickness of the sediment. The advantage of GPR besides the high-resolution image is that the data acquisition requires no direct contact between the instrument and the freshwater layer, making it an effective method in preventing siltation in rivers. We used synthetic GPR data to model the river subsurface condition and to determine the river depth caused by sediment buildup at various sediment thicknesses.

2. Method

We used split-step algorithm [2], a method for modeling and migrating GPR data in a heterogeneous media. The wavefield is extrapolated within the frequency-wavenumber (*f-k*) domain, from the 2-D Maxwell's equations. Using the phase-shift technique, the extrapolation uses constant *K* (relative permittivity) and *Q* (quality factor). Then, the model parameters are characterized using the spatial distribution of the corrected *K* and *Q* factor values that are then integrated into the algorithm by a regular grid configuration. The radar wave attenuation and dispersion are calculated by a linear frequency-dependent *Q* model and shown in the propagation equation.

the total electric current density *J* is a function of the electric field *E*, is presented as

$$J = (\sigma + i\omega\epsilon)E = \in_T \frac{\partial E}{\partial t} \quad (1)$$

$\in_T$ is complex permittivity which can be written as

$$\in_T = \epsilon' - i[\epsilon'' + \sigma(\omega)^{-1}] \quad (2)$$

Q is the quality factor that shows how fast the EM wave dissipates, it is defined as the inverse of the loss tangent δ, the ratio of the imaginary to the real part of $\in_T$[3].

$$Q = tan\,\delta^{-1} = \epsilon'\,[\epsilon'' + \sigma(\omega)^{-1}]^{-1} \quad (3)$$

Bano (4) using K dan Q, determined the phase velocity $V(\omega)$ and attenuation factor $\alpha(\omega)$ are connected to a non-dispersive velocity V_0, and an exponential term n, written as

$$V_0 = \{(\mu K \varepsilon_0)^{1/2} cos[\frac{\pi}{4}(1-n)]\}^{-1} (4)$$

$$n = \frac{2}{\pi}\, tan^{-1}(Q) \quad (5)$$

$$V(\omega) = V_0\,[\omega\,(\omega_c)^{-1}]^{\frac{1-n}{2}} \quad (6)$$

$$\alpha(\omega) = \omega\,[V(\omega)]^{-1} \tan[\frac{\pi}{4}(1-n)] \quad (7)$$

The wavenumber k is then characterized using $V(\omega)$ and $\alpha(\omega)$ as

$$k = \omega\,[V(\omega)]^{-1} + i\alpha(\omega) \quad (8)$$

The radar wave propagation solution for a homogeneous and horizontally layered medium [2] is given by

$$E(k_x, z, \omega) = E(k_x, z = 0, \omega)\exp(-ik''_z z)\exp(-ik'_z z) \quad (9)$$

IOP Conf. Series: Earth and Environmental Science **873** (2021) 012041 doi:10.1088/1755-1315/873/1/012041

$$E(k_x, z, \omega) = E(k_x, z = \Delta z, \omega) \exp(-ik''_z\Delta z) \exp(-ik'_z\Delta z) \quad (10)$$

3. Results

These synthetic models are made considering the stratigraphy of a standard river (assuming the boundary layers are smooth because the sediment deposition is the size of fine sand or less). The resistivity and dielectric constant values are reference values [5-6]. First, a homogeneous model of the sediment is created to check if the grid size and data response fit. The model can be seen in figure 1.

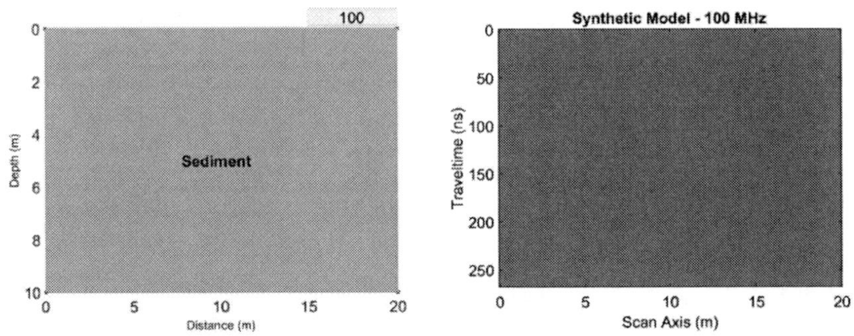

Figure 1. (a) Homogeneous model of saturated sediment with $\rho = 30\ \Omega$m, $\varepsilon_r = 15$ and $\mu_r = 1$. (b) Homogeneous model radargram response.

Three models were made to simulate rivers with varying depths. The models are named RSM-I, RSM-II, and RSM-III and respectively have an ascending sediment thickness. The models' antenna frequency is 100 MHz. The sediment layer is saturated to simulate the river sediment deposit. The models and their radargrams are shown in figures 2-4.

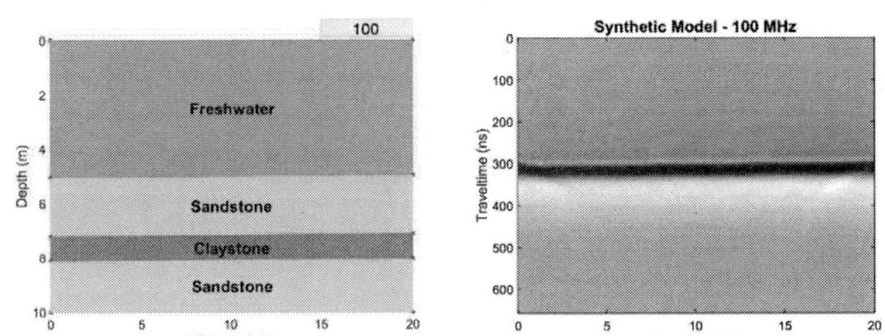

Figure 2. (a) RSM-I models the initial river condition. The stratigraphy consists of a layer of freshwater with $\rho = 16\ \Omega$m, $\varepsilon_r = 81$ and $\mu_r = 1$ of depth 5 m, two layers of sandstone with $\rho = 850\ \Omega$m, $\varepsilon_r = 2.5$ and $\mu_r = 1$ of total depth 4 m, and a layer of claystone with $\rho = 120\ \Omega$m, $\varepsilon_r = 11$ and $\mu_r = 1$ of depth 1 m. (b) RSM-1 radargram response with reflector two-way travel time of 300-350 ns.

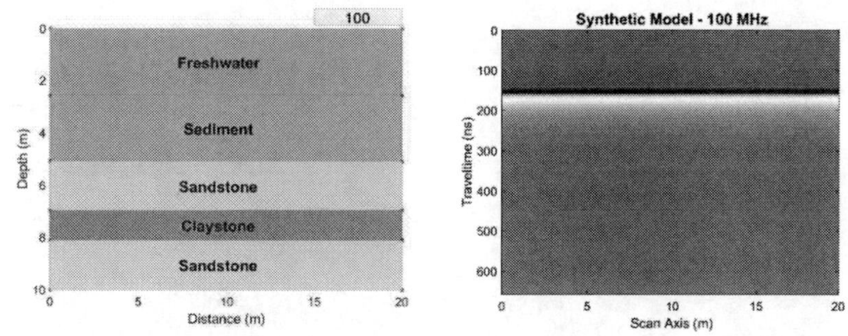

Figure 3. (a) RSM-II. The stratigraphy consists of a layer of freshwater with $\rho = 16\ \Omega m$, $\varepsilon_r = 81$ and $\mu_r = 1$ of depth 2.5 m, a layer of saturated sediment with $\rho = 30\ \Omega m$, $\varepsilon_r = 15$ and $\mu_r = 1$ of depth 2.5 m, two layers of sandstone with $\rho = 850\ \Omega m$, $\varepsilon_r = 2.5$ and $\mu_r = 1$ of total depth 4 m, and a layer of claystone with $\rho = 120\ \Omega m$, $\varepsilon_r = 11$ and $\mu_r = 1$ of depth 1 m. (b) RSM-II radargram response with reflector two-way travel time of 150-200 ns.

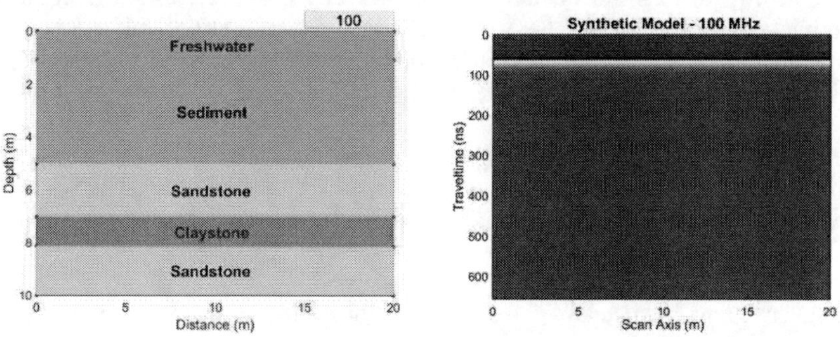

Figure 4. (a) RSM-III. The stratigraphy consists of a layer of freshwater with $\rho = 16\ \Omega m$, $\varepsilon_r = 81$ and $\mu_r = 1$ of depth 1 m, a layer of saturated sediment with $\rho = 30\ \Omega m$, $\varepsilon_r = 15$ and $\mu_r = 1$ of depth 4 m, two layers of sandstone with $\rho = 850\ \Omega m$, $\varepsilon_r = 2.5$ and $\mu_r = 1$ of total depth 4 m, and a layer of claystone with $\rho = 120\ \Omega m$, $\varepsilon_r = 11$ and $\mu_r = 1$ of depth 1 m. (b) RSM-III radargram response with reflector two-way travel time of 60-90 ns.

The radargrams show a clear image of the river's subsurface. The strong reflector indicates a contrast dielectric constant difference between the layers. RSM-I's reflector has the highest travel time range at 300-350 ns, while RSM-III's reflector has the lowest travel time range at 60-90 ns.

To calculate the depth of the river, we use the relationship between the propagating velocity of the electromagnetic wave in a medium and the travel time (*TWTT*) [7]. Using the velocity structure [4], the propagating velocity in the freshwater model is calculated to be 0.033231 m/ns. The thickness of a layer is defined using equation (11).

$$\boldsymbol{Thickness} = \boldsymbol{V_{\mathrm{med}}} \times \frac{(\boldsymbol{TWTT_{\mathrm{B}}} - \boldsymbol{TWTT_{\mathrm{T}}})}{\mathbf{2}} \quad (11)$$

Where V_{med} is the propagating velocity of the electromagnetic wave in a medium, $TWTT_{\mathrm{T}}$ and $TWTT_{\mathrm{B}}$ are the two-way travel time of the top and bottom layers, respectively. Based on equation (11), the thickness of freshwater in RSM-I, II, and III are estimated to be 4.98465-5.815425 m, 2.492325-3.3231 m, and 0.99693-1.495395 m, respectively. The actual and calculated water depths are presented in table 1.

Table 1. River sedimentation model water depths comparison.

River Sedimentation Model	Actual Water Depth	Calculated Water Depth
RSM-I	5 m	4.98465 - 5.815425 m
RSM-II	2.5 m	2.492325 - 3.3231 m
RSM-III	1 m	0.99693 - 1.495395 m

4. Conclusion

GPR synthetic modeling using split-step algorithm successfully simulated the radargram of rivers, the contrast between the freshwater and sediment layer in the radargram is the key in determining the river depth. Each model's river depth can be calculated and sediment buildup can be determined. This shows that GPR can be used as a mitigation for river siltation and further research is required regarding the use of field data in conjunction with synthetic data to produce a higher quality subsurface image of the riverbed.

5. References

[1] Nichols G 2009 *Sedimentology and stratigraphy* (Oxford: Wiley-Blackwell) vol 2 p 46

[2] Bitri A and Grandjean G 1998 Frequency-wavenumber modelling and migration of 2D GPR data in moderately heterogeneous dispersive media *Geophys. Prospect.* **47** 287-301. https://doi.org/10.1046/j.1365-2478.1998.00091.x

[3] Gueguen Y and Palciauskas V 1992 Introduction à la physique des roches ed Hermann (Paris: Editeurs Des Sciences Et Des Arts)

[4] Bano M 1996 Constant dielectric losses in ground-penetrating radar waves *Geophys.J.Int.* **124** 279-288. https://doi.org/10.1111/j.1365-246X.1996.tb06370.x

[5] Davis J and Annan A 1989 Ground-penetrating radar for high-resolution mapping of soil and rock stratigraphy *Geophys. Prospect.* **37** 535. https://doi.org/10.1111/j.1365-2478.1989.tb02221.x

[6] Palacky G 1988 Resistivity characteristics of geologic targets *Electromagnetic Methods in Applied Geophysics* 55. https://doi.org/10.1190/1.9781560802631.ch3

[7] Lachhab A, Booterbaugh A and Beren M 2015 Bathymetry and sediment accumulation of Walter Lake, PA using two GPR antennas in a new integrated method *JEEG* **20** 245-255. http://dx.doi.org/10.2113/JEEG20.3.245

Acknowledgments

This research received financial support from LPPM ITB and DIKTI.

Exploring Reservoir within Hugin Formation in Theta Vest Structure using 4-D Seismic and Machine Learning Approach

Lilik T. Hardanto, Mirzam Abdurrachman, Dwiharso Nugroho

Institute Technology Bandung (ITB), Bandung, West Java, Indonesia

lhardanto19@gmail.com

Abstract. This paper aims to identify the oil distribution using 4-D seismic below a complex 3-D surface in Hugin Formation using machine learning and geobody detection. The exploration well 15/9-19-SR, drilled to the Theta Vest structure, was based on the interpretation of reprocessed ST8215R 3-D seismic survey data from 1991 in the Sleipner area, encountered oil in the Jurassic Hugin Formation. The drills stem test showed outstanding production capacities through time, with low water cut and low GOR. 4-D seismic has all the traditional benefits of 3-D seismic. A significant additional potential benefit is that fluid-flow processes can be directly imaged. The 4-D seismic analysis was conducted in 2012 to repeat the 3-D seismic surveys and analyze images in time-lapse mode to monitor time-varying fluid-flow processes during reservoir production. A comprehensive study of the structure and the discovery has been performed and is reported. The DNN method to predict facies far away from existing production wells by using facies log well to supervise seismic inversion created by the Seismic Color Inversion method. It can detect some oil pockets distribution and risk the well planning and the right candidate for new proposed wells.

1. Introduction

The Theta Vest structure is dome-shaped to trap oil. The structure has a well-defined spill point to the east and is a base case believed to be filled down to this point. However, several faults cutting the structure's crest could be sealing the hydrocarbon accumulation off from spilling eastwards. The hydrocarbon discovery of exploration well 15/9-19-SR has been evaluated using 3-D seismic data taken in 1991 with comprehensive analyzes have been carried out, such as seismic interpretation, structural model, petrophysical evaluation, DST test, and geochemical analysis [1]. Finally, a resource analysis of the discovery and associated prospects has been performed. The exploration well 15/9-19-SR is located on the Theta Vest structure in the Sleipner Øst area (Figure 1). Well planning on the Theta Vest structure was based on the interpretation of reprocessed ST8215R 3-D survey data from 1991. The Top Hugin Formation horizon has been mapped after the well was drilled. With the limited resolution of the seismic data and a weak reflector, the horizon has been challenging to interpret consistently.

IOP Conf. Series: Earth and Environmental Science **873** (2021) 012042 doi:10.1088/1755-1315/873/1/012042

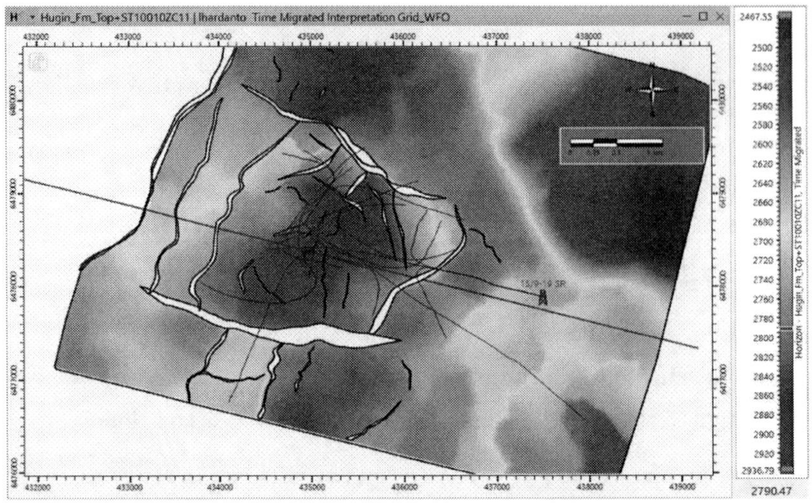

Figure 1. Location of the 15/9-19 SR well and the Theta Vest Structure

In a single 3-D seismic survey, seismic waves are generated by sources (dynamite, airguns, and various sources.) at or near the subsurface. These source waves reflect subsurface seismic impedance contrasts: rock and fluid compressibility, shear modulus, and bulk density. They are recorded as they arrive back at the subsurface. The recorded waves form the classic wiggle traces where high positive amplitude portions are often filled in on a black and white image to enhance visual contrast and show lateral continuity. A wave-equation imaging algorithm is applied to the recorded reflection data to create 3-D seismic images of the reservoir rock and fluid property (seismic impedance) contrasts. 4-D seismic reservoir monitoring is the process of repeating 3-D seismic surveys over a reservoir in time-lapse mode to look for differences caused by the production [2]. The potential exists for dramatic benefits to reservoir management because it is the first technique that may directly image dynamic reservoir processes such as monitoring fluid movement, pressure build-up, and heat flow in a reservoir in a true volumetric sense. In these reservoirs, production causes subsurface deformations and changes in seismic velocity. These changes in the overburden can be monitored and interpreted. Optimal interpretation of this data enables enhancement, well placement, production and injection operations, production history matching, and reservoir simulations by defining reservoir fluid-flow characteristics to improve flow models.

4-D seismic has all the traditional benefits of 3-D seismic, plus a significant additional potential benefit that fluid-flow processes can be directly imaged. It involves repeating the 3-D seismic surveys and analyzing images in time-lapse mode to monitor time-varying fluid-flow processes during reservoir production [3]. In the first order, seismic images are sensitive to spatial contrasts in two distinct types of reservoir properties: 1) non-time-varying static geology properties such as lithology, porosity, shale content, and 2) time-varying dynamic fluid-flow properties such as fluid saturation, pore pressure, and temperature. A 3-D seismic survey is representing a single snapshot in time. It is impossible to distinguish an oil-water contact from a horizontal depositional boundary in a single seismic image. However, with 4-D seismic surveys, examining the difference between time-lapse 3-D seismic images allows the non-time-varying geologic contributions to cancel, resulting in a direct image of the time-varying changes caused by reservoir fluid flow. For example, an oil-water contact may move with time in a series of time-lapse seismic images, whereas a depositional boundary should not.

The 4-D time-lapse seismic technique has the potential to image in a large volume encompassing many wellbores, changes in fluid saturation, pore pressure, and temperature during production. It can provide more robust and reliable volumetric constraints between wells than those developed by interpolating borehole properties [4]. This technique allows observing the dynamic processes in and around the reservoir to help place wells more effectively, increasing development and production success.

IOP Conf. Series: Earth and Environmental Science **873** (2021) 012042 doi:10.1088/1755-1315/873/1/012042

Some successful 4-D case examples have been acquired at reservoirs containing high porous clastic rocks. Examples of such fields are Draugen and Gannet [5], Gullfaks ([6], Troll [7], Schiehallion [8], Foinhaven, Norne ([9] , and Heidrun [10]. Like the Ekofisk [11] and Valhall [12] fields in the Southern North Sea, some significant chalk fields have reported good 4-D results.

The seismic impedance can help compute each reservoir model grid cell through a large but highly parallelizable cell-by-cell inverse problem easily integrated into the reservoir property updating workflow [13], [14]. One of the seismic impedance methods is seismic colored inversion. It is a post-stack inversion technique that transforms seismic data into layered rock properties [15]. Seismic colored inversion performs significantly better than traditional inversion methods like model-based inversion, band-limited inversion methods, fast, easy to use, less expensive, robust, and suitable for application to 3-D datasets [16]. It is a process where the acoustic impedance spectra derived from log data are used to compute the operator's spectrum [17]. Other inversion methods are time-consuming, expensive, and are not performed routinely by the interpreter.

Before its expanded use, the introduction of facies definition is described as a distinctive rock formed under a particular process or environment [18]. Thus, the response of well logs could vary in terms of rock changing through measurement and applied with electrofacies.

Seismic facies automated classification processes are a proven technology. It plays an increasingly vital role in seismic stratigraphy interpretation and reservoir characterization workflows. Due to advances in seismic acquisition, processing, and imaging, geoscientists now have access to high-definition subsurface images; Well characterization was collaborated with seismic attributes and distributed using Democratic Neural Network Association Methods (DNNA). DNNA performs facies inversion based on lithology and seismic data and is a robust alternative to conventional approaches. It comprises different neural networks, having various activation functions but trained with the same labeled dataset. The workflow for DNNA is consists of two training phases. The first training phase performs learning using a set of training samples issued from seismic attributes or amplitudes and lithofacies distribution measured along well boreholes to provides a suite of independently predicted responses. The second training phase is to perform using two kinds of data: soft data and hard data. Soft data are data labeled by the democratic system, which contain lower weights value in training. Hard data are from well data that contain large-amplitude weights value—the DNNA method to reduce the training bias associated with each network. DNNA uses cross-validation and bootstrap error values. They are considerably lowered on synthetic and real datasets to perform facies inversion based on lithology and seismic attributes [20].

2. Data and Methods

The dataset is provided by Equinor, a Norwegian multinational energy company. The data is from the Volve Field in the North Sea. Data used in this research are grouped into two kinds of data: 3-D and 4-D data. The 3-D data and 4-D data are divided into two seismic amplitudes, 3-D and 4-D seismic amplitude data. Four wells (15/9-19-SR, 15/9-F-15-C, 15/9-F-1, and 15/9-F-5) were used to analyze the Hugin Formation is Callovian in age, probably of Late Callovian. The 15/9-19-SR well proved oil in Jurassic sandstones of the Hugin Formation. The 18 meters thick Hugin Formation was filled with hydrocarbons to its base at 2882.0 m MSL TD and reached in Triassic sediments at 3110.3 m MSL. The Hugin Formation's petrophysical analysis in the 15/9-19 SR well results in a mean net/gross of 0.924, the porosity of 22.6%, the permeability of 2913.2 mD, and water saturation of 11.0%. The sandstones are interpreted to be deposited in a high-energy marine environment, possibly a mouth bar setting. The reservoir interval consists of highly variable fine to coarse-grained, well to poorly sorted, subarkosic arenite sandstones with good to excellent reservoir properties. The underlying Skagerrak Formation is entirely tight due to extensive kaolinite and dolomite cementation. Within the Sleipner Øst area, the Jurassic and Triassic reservoir distribution are complicated. Based on lithofacies in 15/9-19-SR well, it can be applied to other wells such as 15/9-F-15-C, 15/9-F-1, and 15/9-F-5, as shown in Figure 2.

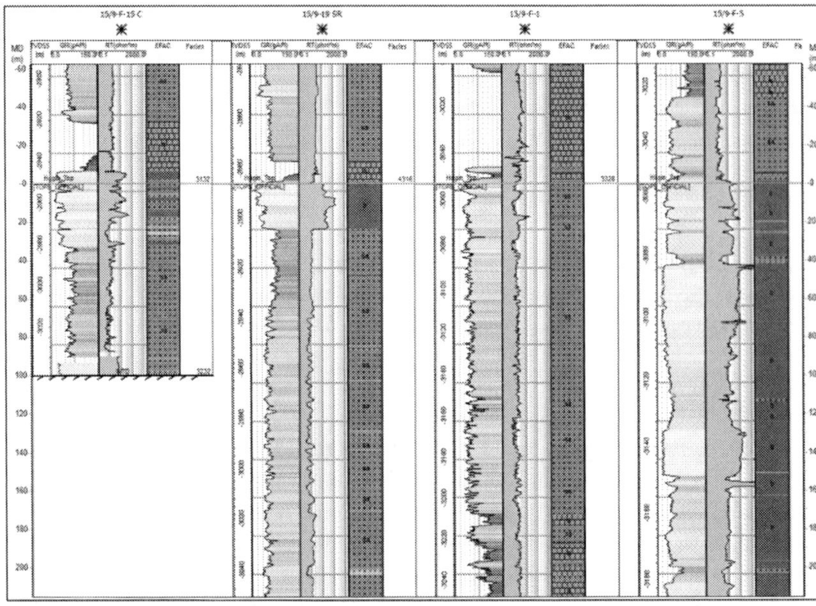

Figure 2. Well 15/9-19 SR in Hugin Formation.

Secondary data resulted from field samples, field geological structures measurements, and tops of geological age markers using paleontology samples from well reports to control geological interpretation and reconstruction. Horizon and fault interpreted on 3-D and 4-D post-stack-final-psdm seismic processed in 26-11-2011 and 30-03-2012, respectively, as shown in Figure 3.

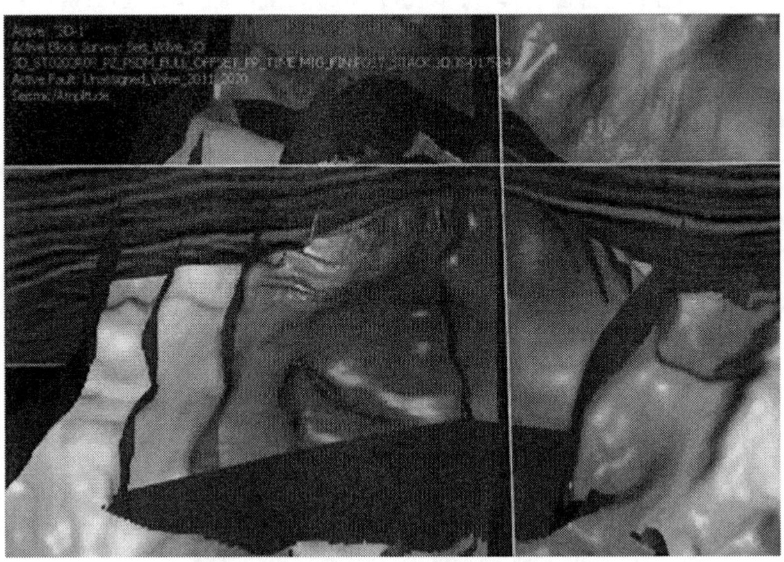

Figure 3. Well 15/9-19 SR Hugin Formation.

Before making any interpretation and processes the seismic, the first well tie needs to be computed based on the seismic section from West to East direction for some wells data. Figure 4 below for Inline 10135 and 15/9-19-SR well data, one of the well tie results.

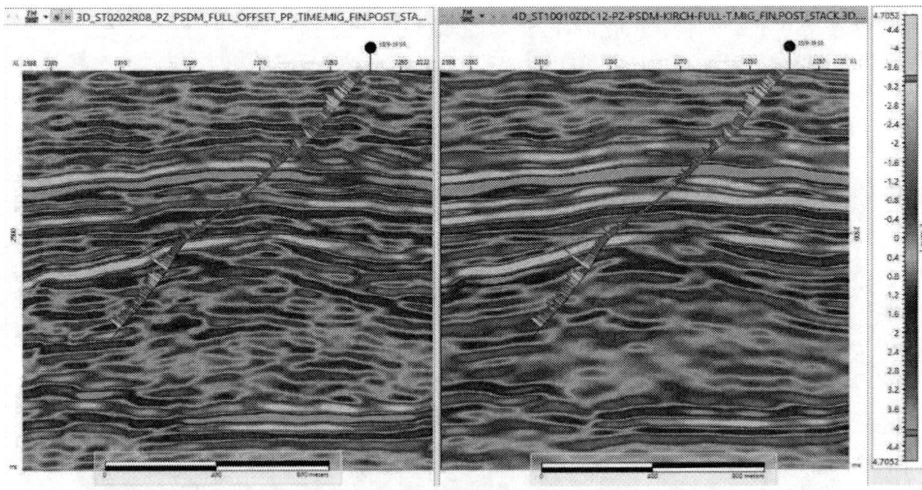

Figure 4. Seismic section of 4-D and 3-D at Well Marker of Top Hugin.

2.1. 3-D and 4-D seismic color inversion

Seismic colored inversion is used because it is easy to use, less expensive, and suitable for a field without any gathers seismic. It is applied to the 3-D and 4-D post-stack seismic data to estimate petrophysical parameters by using the same parameters to process wells and seismic data to generate an impedance seismic, as shown in Figure 5. For the inversion to be as accurate as possible, the 3-D seismic volume ideally would be multiple free, zero-offset, migrated, and (preferably) zero-phased. The seismic wavelet is extracted from the data via a matching filter between the operator derived from the impedance log and the seismic trace at the well location. The operator is derived in the following steps: First, the acoustic impedance is calculated and plotted against frequency for the wells in the seismic section. Second, a curve line is fit to the amplitude spectrum of the acoustic impedance to represent the impedance spectrum in the subsurface in the log scale. Third, the average seismic spectrum is calculated from the seismic traces closed to the wells. These two spectra are used to calculate the operator spectrum, transforming the seismic spectrum into the average impedance spectrum. Fourth, the final spectrum is combined to create the operator in the time domain.

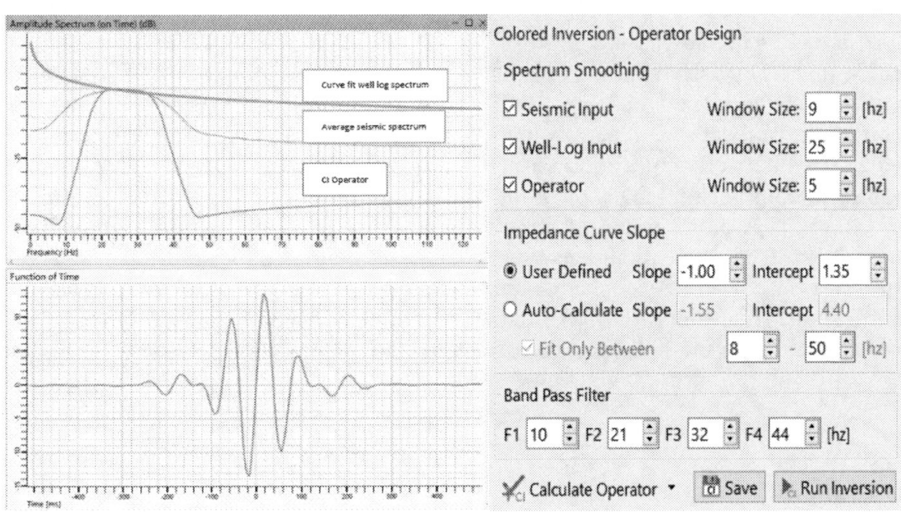

Figure 5. 3-D and 4-D Seismic color inversion

IOP Conf. Series: Earth and Environmental Science **873** (2021) 012042 doi:10.1088/1755-1315/873/1/012042

The entire seismic data is inverted into impedance volume with a correlation factor is estimated at 0.35. Low impedance values are estimated between 2550ms - 2590ms time intervals, which are interpreted as Hugin Formation oil sand, as shown in Figure 6. After seismic color inversion is calculated, we post the result and overlay it with the seismic amplitude shown in Figure 6.

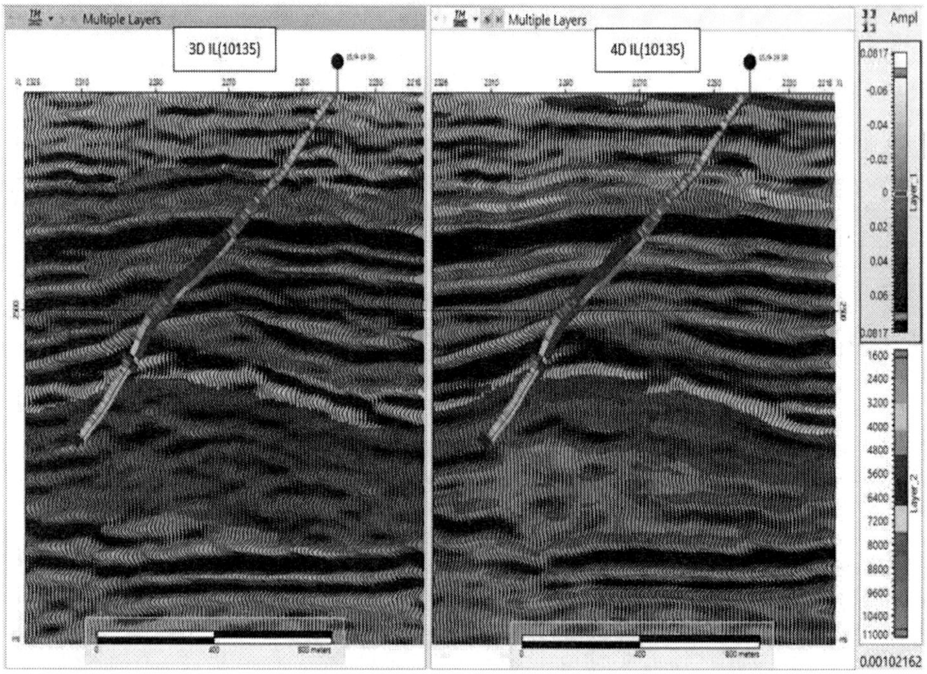

Figure 6. Seismic Section of 3-D and 4-D of Seismic Color Inversion

The purpose of overlay seismic color inversion with seismic amplitude is to see the impedance differences and determine their lithologies contents on the zone of interest. From Figure 6, we can see the yellow color below the blue color is a low impedance value indicated as oil sand or a good reservoir in the formation.

2.2. Seismic facies automated classification

Seismic color inversion derived from post-stack seismic and log data is typically used to predict reservoirs' lithology and fluid content. Well-analysis is considered reliable when describing the reservoir; however, well data are generally too sparse of a data set to describe a prospect adequately. So, We applied a neural network application through the use of DNNA to infer facies defined at wells data. The probabilistic approach of DNNA combines all seismic-related information to build facies probability cubes. This paper proposed a new method for DNNA analysis using Seismic 3-D and Seismic 4-D of their Color Seismic Inversion and has been evaluated to achieve the best prediction. We focus on the reservoir zone between the top and 100 ms below the Hugin formation base to run DNNA facies classification. The result of DNNA is a facies volume between the interval time-lapse of 4-D and 3D seismic, and the facies predicted can be extracted along the borehole, as shown in Figure 7.

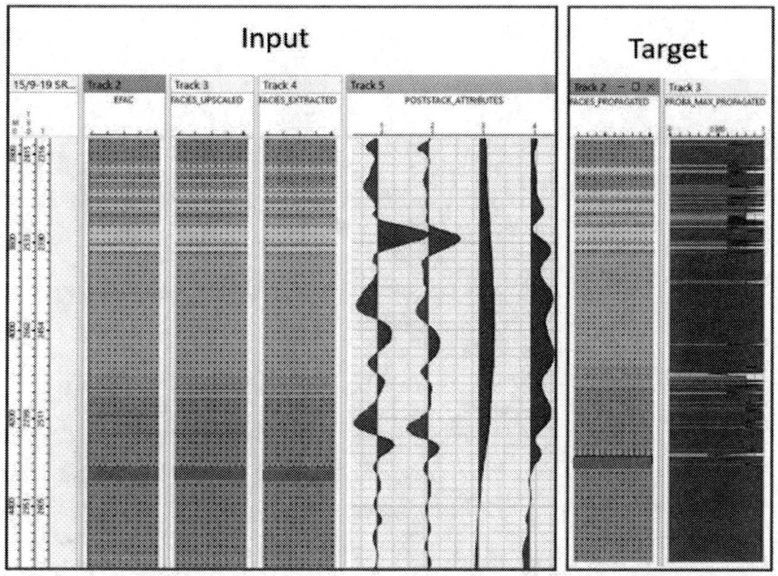

Figure 7. Facies log and Seismic Color inversion data as input and target for DNNA learning

3. Results and discussion

The lithofacies and seismic inversion data at the well location represent the training dataset used as input to the DNNA process (Figure 7). On the data input tracks, there is a lithofacies log and then another that may be extracted or upscaled to eliminate thin layers to give more homogeneity to the vertical distribution of the facies. Track number five corresponds to the seismic traces, each associated with a different attribute, extracted along the well 15/9-19 SR.

Once that process has been run, the intermediate stage is to validate the quality of the operator by visualizing the reconstruction results at the wells (Figure 7). The eighth track is the reconstructed lithofacies log curve. In this context, "reconstructed" is defined as the predicted lithofacies curve that results from applying the neural network model at the well location using only seismic data as input. Therefore, it is a best-case scenario for the reconstruction of lithofacies using this method. Finally, the maximum probability curve for each of the created classes of the reconstructed lithofacies curve is on the far right side.

Machine Learning with DNNA algorithm can produce the facies volume with the prediction number of bootstrap sets: 100, Bootstrap (training) error: 0.165361 and Bootstrap classification rate: 78.8264 % also the Matthews correlation coefficient (all wells): 0.912. The facies classification volume can be extracted along the well path, as shown in Figure 8. below, to blind test the well.

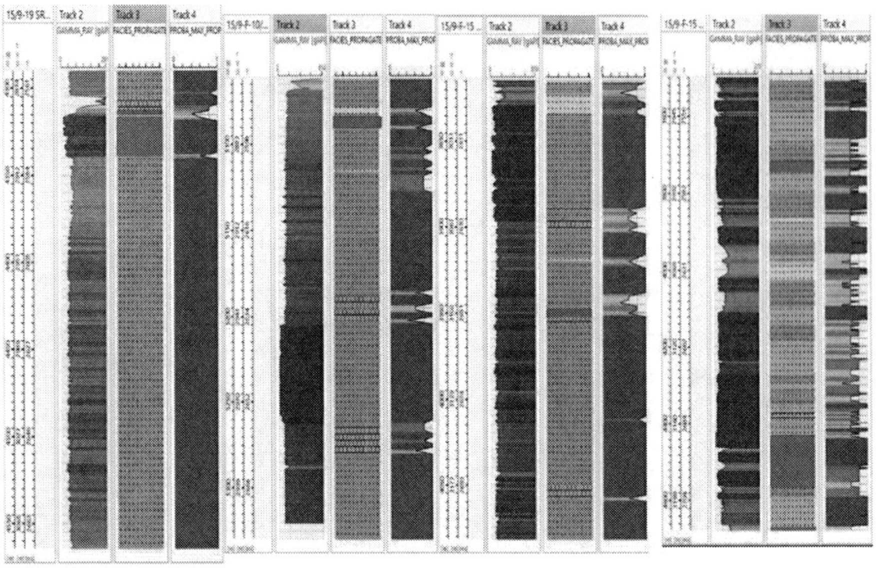

Figure 8. Blind test wells with facies extracted from Facies volume

The blind test well as predicted 0.912, produced high accuracy at well positions. Furthermore, for facies distribution, we can extract the facies volume on top of the Hugin Formation, see in Figure 9, as shown below.

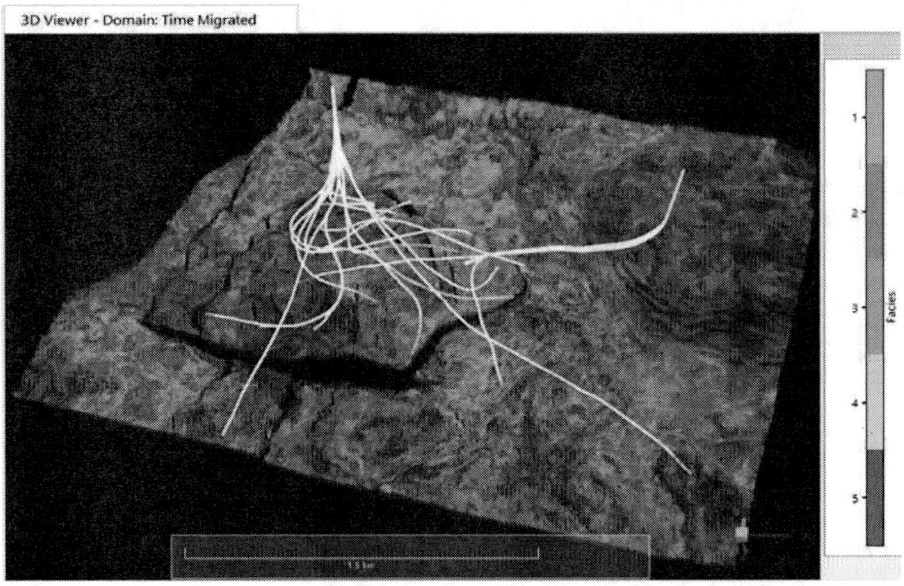

Figure 9. Well 15/9-19 SR Hugin Formation.

The fuchsia color is the Hugin Formation's oil sand, showing some interesting geological-stratigraphy features inside the Hugin formation interval. Geobody detection can detect and create surfaces for bodies of facies five, as shown in Figure 10. Furthermore, by looking at the picture above, some places need to be considered candidates for the following proposed wells.

IOP Conf. Series: Earth and Environmental Science **873** (2021) 012042 doi:10.1088/1755-1315/873/1/012042

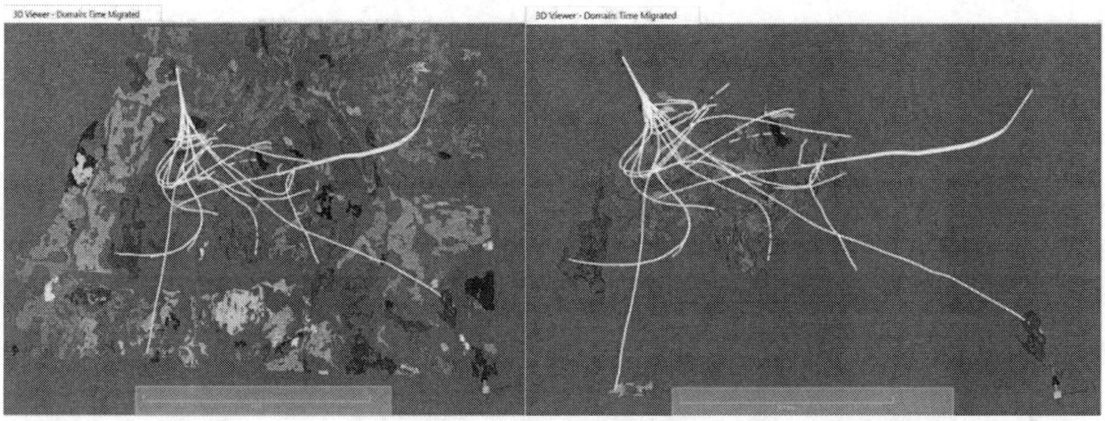

Figure 10. Geobody detection of Hugin Formation (left) and Wells that penetrated geobody (right).

Geobody detection inside the Hugin formation interval can help us to determine the lateral distribution of facies 5. Furthermore, we found some geobodies penetrating existing wells, and the blind test wells confirmed them.

4. Conclusion

DNNA analysis using 3-D and 4-D seismic data using facies log and color seismic inversion is discussed in the above section. The analyses showed that the seismic colored inversion method and 3-D and 4-D seismic supervised by facies logs provide consistent facies volume output. In addition, the methods estimated the reservoir distribution within the reservoir zone with the Mathew correlation 0.912, which confirmed hydrocarbon distribution that needs to be considered to reduce the risk and for the following proposed wells. Moreover, it can be done ineffectively to achieve the best accuracy of predictions compared to the traditional way. Furthermore, it helps to optimize the production in Hugin Formation within the Theta Vest Structure.

Acknowledgments
We would like to thank Equinor for the seismic and well data and Emerson for providing the software.

References

[1] Statoil 1993 *Discovery Evaluation Report Well 15/9-19-SR Theta Vest Structure PL406.*
[2] Rafael S David L and Jeffrey S 2016 Estimation of reservoir fluid saturation from 4-D seismic data: Effects of noise on seismic amplitude and impedance attributes February 2017 *Journal of Geophysics and Engineering* 14(1):51-68. DOI: 10.1088/1742-2132/14/1/51.
[3] Lumley D.E. 1997 Practical Engineering Issues of 4-D Seismic Reservoir Monitoring SPE Annual Technical Conference and Exhibition San Antonio Texas, 449.
[4] Simm R and Bacon M 2014 Seismic Amplitude an interpreter's Handbook 1st edn (Cambridge: Cambridge University Press).
[5] Koster K Gabriels P Hartung M Verbeek J Deinum G. and Staples R 2000, Time-lapse seismic surveys in the North Sea and their business impact: *The Leading Edge*,19, no. 03, 286-293.
[6] Landrø M Solheim O.A Hilde E Ekren B.O. and Strønen, L.K 1999, The Gullfaks 4-D Seismic Study: Petroleum Geoscience 5 213-226.
[7] Elde R Zachariassen E Sandoe I and Haaland A 2000 Troll West—Time lapse processing and analysis 70th Ann Internat Mtg: Soc of Expl. Geophys 1433-1436.
[8] Parr R Marsh J and Griffin T 2000, interpretation and integration of 4-D results intoreservoir management, Schiehallion Field UKCS 70th Ann. Internat. Mtg: Soc. of Expl. Geophys., 1464-1467.
[9] Osdal B and Alsos T 2002 Seismic Modelling of Eclipse Simulations and Comparsion with Real 4-D Data at the Norne Field, 64th Mtg.: Eur. Assn. Geosci. Eng., A029.

[10] Furre A-K Munkvold F.R. and Nord bL. H 2003, Improving Reservoir Understanding Using Time-Lapse Seismic at the Heidrun Field, 65th Mtg.: Eur. Assn. Geosci. Eng., A20.
[11] Guilbot J and Smith B 2002 4-D constrained depth conversion for reservoir compaction estimation: Application to Ekofisk Field: *The Leading Edge*, 21, no. 03, 302- 308.
[12] Barkved O 2004 Continuous seismic monitoring 74th Ann. Internat. Mtg.: Soc. of Expl. Geophys., 2537-2540.
[13] Stephen KD and Macbeth C, 2006, Inverting for the petro-elastic model via seismic history matching *SEG Annual Meeting* (NewOrleans) pp 1688–92.
[14] Sagitov I and Stephen K, 2012, Assisted seismic history matching in different domains: what seismic data should we compare? Proc. of *SPE Europec/EAGE Annual Conf.*
[15] S. P. Maurya and N.P. Singh, 2017 Seismic colored inversion: A fast way to estimate rock properties from the seismic data, Department of Geophysics, Institute of Science, Banaras Hindu University, Varanasi, U.P. -221005 India.
[16] Neep, J. 2007 Time variant coloured inversion and spectral blueing. In: 69th *EAGE Conference & Exhibition.*
[17] Lancaster S Whitcombe D et al 2000 Fast track colored inversion.
[18] Reading HG. Sedimentary environments and facies (2nd Ed). *Oxford: Blackwell Scientific Publishing* 1986.
[19] Hami-Eddine K., P. Klein, and L. Richard 2009 Well based supervised classification, application to a western Africa turbidite reservoir: *SEG Expanded Abstracts*, v. 28, p. 1885-1889.
[20] Hami-Eddine K., Pascal Klein, Loic Richard, Bruno de Ribet, and Maelle Grout. (2015),3(1): A new technique for lithology and fluid content prediction from prestack data: An application to a carbonate reservoir, SC19, http://dx.doi.org/10.1190/INT-2014-0049.1.

IOP Conf. Series: Earth and Environmental Science **873** (2021) 012043 doi:10.1088/1755-1315/873/1/012043

Unexpected Shallow Earthquake of August 1st, 2020 in the North of Indramayu, West Java, Indonesia

Jaya Murjaya[1], Pepen Supendi[1,2], Dwikorita Karnawati[1], Subagyo Pramumijoyo[3]

[1] Agency for Meteorology, Climatology and Geophysics (BMKG), Jakarta 10720, Indonesia
[2] Center for Earthquake Science and Technology (CEST), Research Center for Disaster Mitigation, Institut Teknologi Bandung, Bandung 40132, Indonesia
[3] Gadjah Mada University, Yogyakarta 55281, Indonesia

E-mail: murjaya1@gmail.com

Abstract. During the last one hundred years, there are no shallow seismicity in the north of Java. This area is dominated by intermediate and deep focus earthquakes due to the subducted Indo-Australian slab. An earthquake with magnitude M_L 4.5 struck Indramayu, north of West Java on August 1, 2020. According to the Agency for Meteorology, Climatology, and Geophysics (BMKG), the earthquake was felt III MMI scale in Indramayu and its vicinity. We used waveform data from BMKG seismic station in West Java, then we picked P-and S-waves arrival times from each station and hypocenter location was determined by Geiger method. We have detected Pn before Pg phase on four BMKG seismic stations, indicating a shallow crustal earthquake. Our inversion show that the earthquake occurred in 6.1805° S, 108.2612° E with 5 km focus depth at 16:24:38 GMT+7. Our focal mechanism solution was determined by using moment tensor inversion shows a strike-slip faulting, which corresponds to the active fault in the north of Indramayu.

1. Introduction

One of the high seismic hazards area in Indonesia located in West Java [1]. The complex tectonic in the area caused by the subducted Indo-Australian plate. The relocated earthquakes catalog over 50 years from EHB bulletin [2] combine with Widiyantoro et al. [3] shows this area has high seismicity (Fig. 1). Based on the BMKG data, over this period there have been 23 destructive earthquakes from magnitude 3.3 on land faults to magnitude 7.7 in the subduction zone (see the yellow stars in Figure 1). Meanwhile, there are no-existing shallow seismicity in the north of Indramayu, West Java. Hayes et al. [4] show the Indo-Australian slab beneath this area is about 230 km depth, so that the seismicity in this region is relatively sourced from intermediate to deep earthquakes. Interestingly, on August 1st, 2020, the earthquake M_L 4.5 shook Indramayu area in III Modified Mercalli Intensity (MMI) scale. We have obtained a more acurate hypocenter location and conducted source mechanism analysis of this event.

2. Data and Method

We extracted waveform data form the BMKG seismic stations (the black inverted triangles in Fig. 2). We used Seisgram2K program [5] to picked onset of P-and S-wave carefully. Earthquake hypocenter determination was conducted use Hypoellipse code [6] with Geiger method [7]. This method has succeeded in determining hypocenter locations in previous studies in West Java ([8,9]) and in Madiun, East Java [10]. We used ISOLA program [11] to carry out moment tensor inversion. The displacement data from six BMKG seismic stations (JCJI, JPJI, LEM, PBJI, KPJI, and BTJI) have been inverted. We filtered the observed waveforms of 0.05 Hz to 0.09 Hz. The data processing procedure for focal mechanism analysis is as follows: converting the data from SAC to ASCII format; preparing the input files: the earthquake hypocenter location (longitude, latitude, and depth), origin time, and velocity model; selecting station; removing the instrument response, experimenting trial seismic source depth, and calculating the Green's function. We used a 1-D velocity model from previous study [12, 13] for this research.

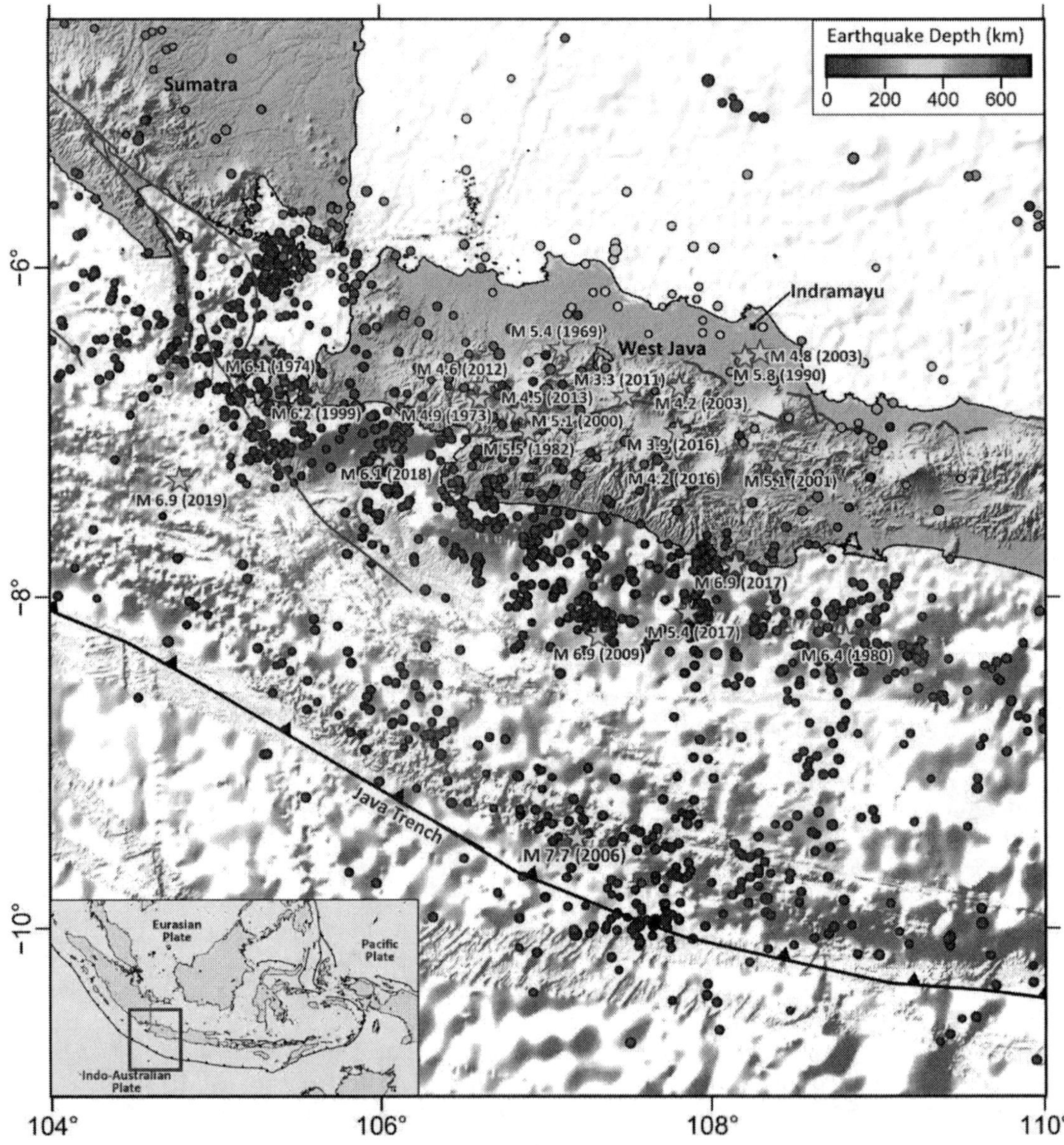

Figure 1. Map of relocated earthquake epicenters [2,3] in the western part of Java in the time period of 1963 to 2018, inset map is the Indonesian region. The yellow stars are the destructive earthquakes from BMKG catalog, they are not colour-coded by depth.

3. Results and Discussion

Our result shows the final hypocenter located in 6.1805° S, 108.2612° E, and 5 km focus depth (Fig. 2), which has 2.5 km and 2.8 km horizontal and depth uncertainties respectively, occurred at 16:24:38 UTC on August 1st, 2020, and RMS 0.7. The focal mechanism solution show strike-slip fault (strike 1: 47°, dip 1: 69° rake 1: -165°; strike 2: 311°, dip 2: 76°, rake 2: -22°) (Fig. 2), the best fitting of waveform inversion shows a fairly good data processing (Fig. 3).

We have determined the hypocenter locations in the north of Indramayu occurred, at 16:24:38 UTC. All of the BMKG seismic stations which recorded this event shows a strong shear wave. We have detected Pn before Pg phase on four BMKG seismic stations (JCJI, KPJI, PKJM, and SEJM) indicating a shallow crustal earthquake and the hypocenter located relatively far away from these stations.

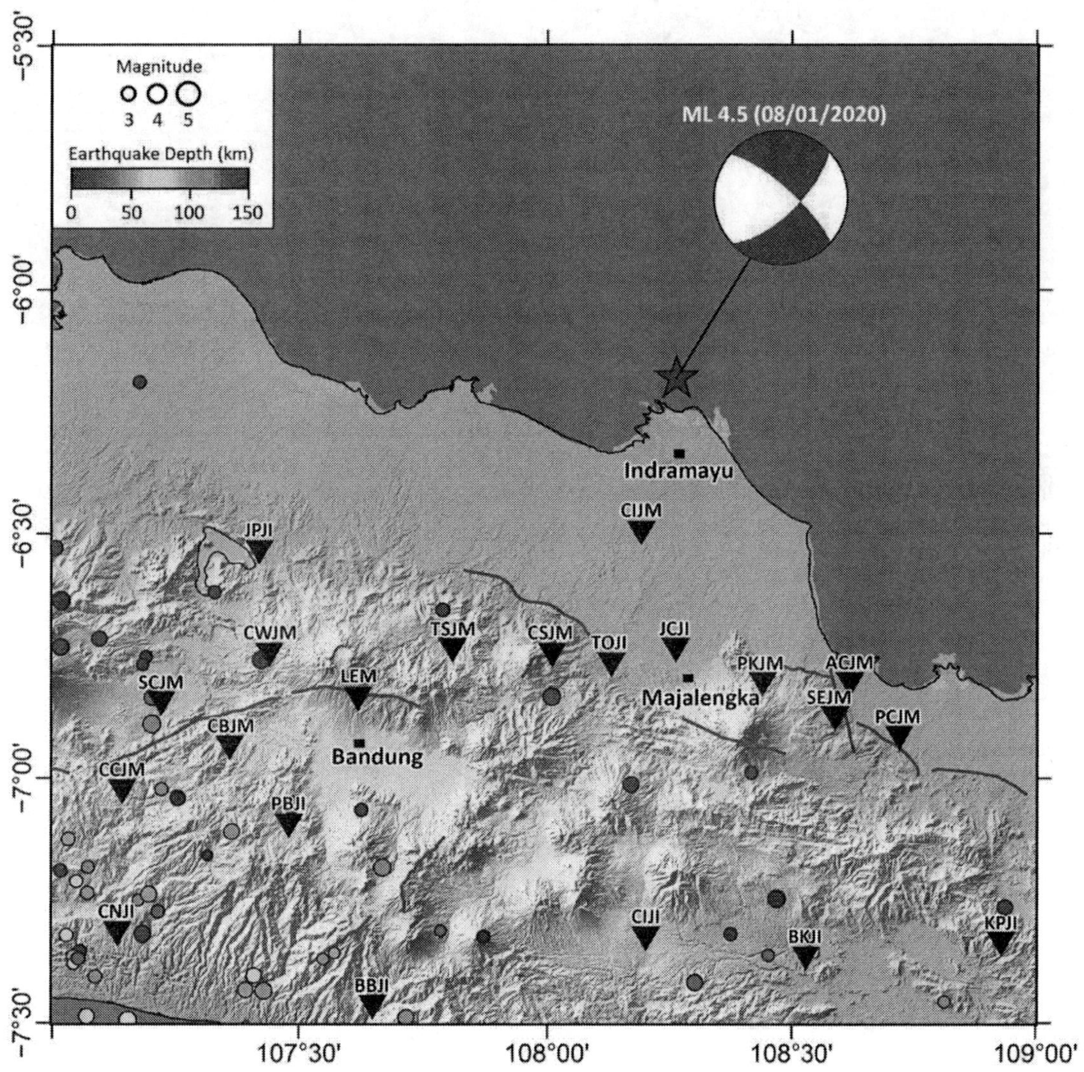

Figure 2. Epicenter location and focal mechanism solution for the earthquake that were determined in this study (August 1, 2020). Relocated background seismicity for Mw≥ 4.0 and focal depth less than 150 km [2,3]. The red lines depict faults in the area are taken from Irsyam et al. [14]. The black inverted triangles are the BMKG seismic stations.

Our interpretation for this event is likely sourced from an active fault in this area. Generally, the felt earthquakes in Indramayu and its vicinity is caused by the shallow earthquake in Baribis fault in the south of this region. Moreover, there are sources from intermediate and deep earthquakes in the north of this area. The unexpected shallow earthquake, which was felt on August 1st, 2020, is located close to Indramayu. Thus, this earthquake probably one of the evidences that proves there has been an active fault in the north of Indramayu.

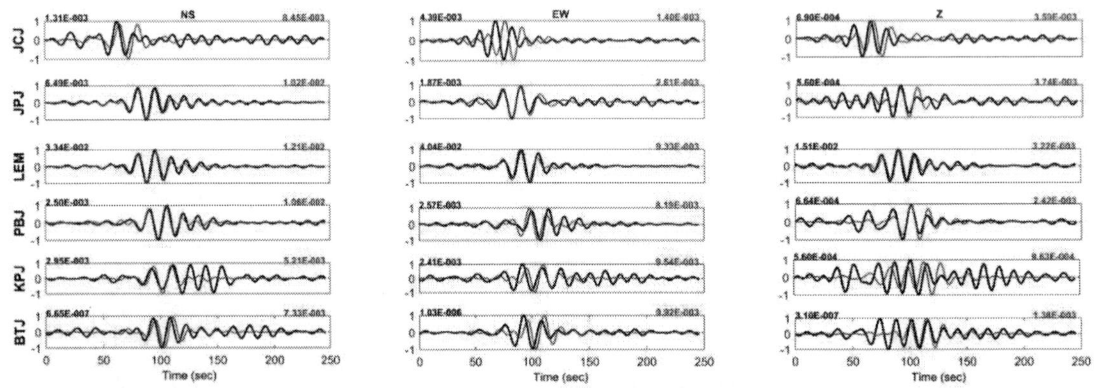

Figure 3. Observed and synthetic seismograms fitting process upon focal mechanism determination for August 1st, 2020, events. Red and black seismograms indicate the synthetic and observed waveforms, respectively.

Then we made the simulation of shakemap. The simulation shows that the maximum impact of the earthquake is about III to IV MMI (Modified Mercalli Intensity) scale (Fig. 4). It means that the Indramayu earthquake was felt weak to light shaking and had no potential for damage. These results are in agreement with the public report–that the earthquake was only caused more people to panic in Indramayu city.

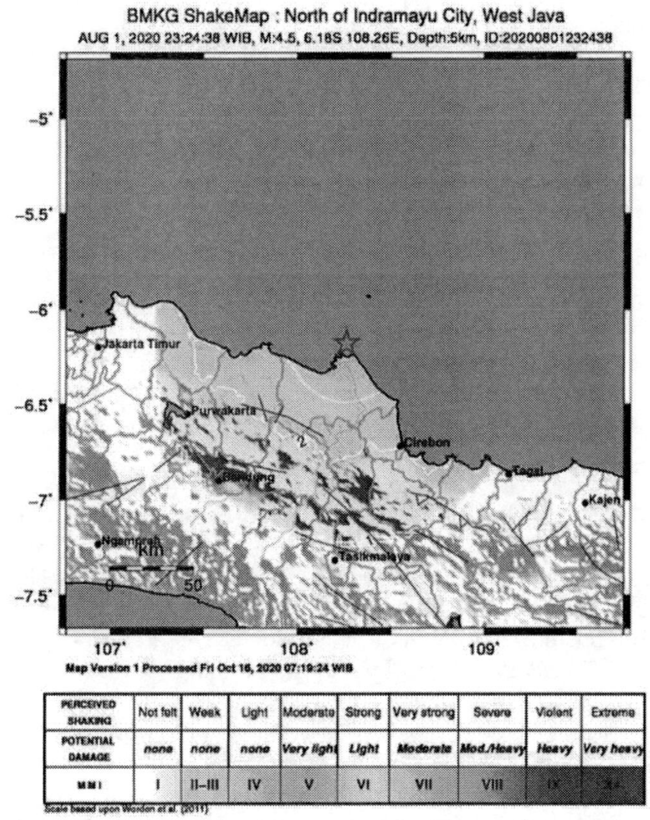

Figure 4. Shakemap simulation of the Indramayu eartquake. We used the epicenter was located in 6.1805° S, 108.2612° E with 5 km focus depth and magnitude moment is about of 4.5.

The modelling results for the Peak Ground Acceleration (PGA) value in the Indramayu area is 2% of g or about of 20 gals and the Peak Ground Velocity (PGV) value is about 5 cm/s (Fig. 5).

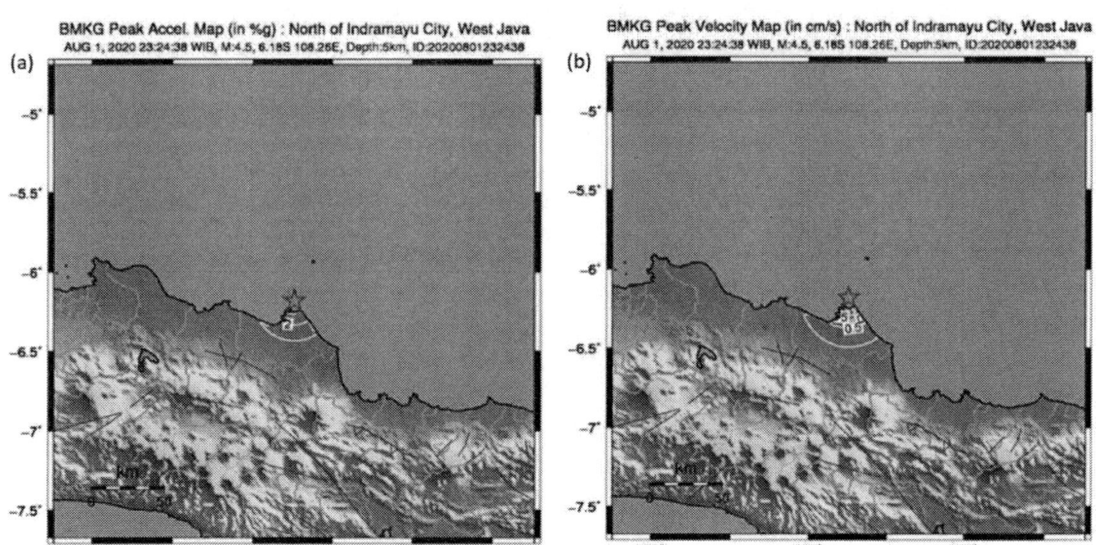

Figure 5. (a) PGA modeling, (b) PGV modeling of the Indramayu earthquake on 1st August 2020.

4. Concluding Remarks

The determination of a hypocenter location of felt earthquake in the Indramayu area shows a shallow depth (about 5 km). Shake map modeled about III to IV MMI scale, while the PGA value in the area 2% of g is too small for ground acceleration and has not destroyed any infrastructures. The source mechanism solutions of this event indicate a strike-slip fault. The occurrence of this event indicates that there is possibly an active fault exist-in the north of Indramayu, West Java.

Acknowledgements
We are grateful to BMKG for support and provide data in this research.

References

[1] Irsyam M, Cummins P R, Asrurifak M, Faizal L, Natawidjaja D H, Widiyantoro S, Meilano I, Triyoso W, Rudiyanto A, Hidayati S, Ridwan M, Hanifa N R and Syahbana A J 2020 Development of the 2017 national seismic hazard maps of Indonesia *Earthquake Spectra* 875529302095120

[2] Engdahl E R, Di Giacomo D, Sakarya B, Gkarlaouni C G, Harris J and Storchak D A 2020 ISC -EHB 1964-2016, an Improved Data Set for Studies of Earth Structure and Global Seismicity *Geophys. J. Int.* **214** 474–84

[3] Widiyantoro S, Gunawan E, Muhari A, Rawlinson N, Mori J, Hanifa N R, Susilo S, Supendi P, Shiddiqi H A, Nugraha A D and Putra H E 2020 Implications for megathrust earthquakes and tsunamis from seismic gaps south of Java Indonesia *Sci Rep* **10** 15274

[4] Hayes G P, Moore G L, Portner D E, Hearne M, Flamme H, Furtney M and Smoczyk G M 2018 Slab2, a comprehensive subduction zone geometry model *Science* **362** 58–61

[5] Lomax A and Michelini A 2009 Mwpd: A duration–amplitude procedure for rapid determination of earthquake magnitude and tsunamigenic potential from P waveforms *Geophysical Journal International* **176** 200–214

[6] Lahr J C 1979 *HYPOELLIPSE: a computer program for determining local earthquake hypocentral parameters, magnitude, and first motion pattern*

[7] Geiger L 1912 Probability method for the determination of earthquake epicenters from the arrival time only *Bulletin St. Louis University* **8** 60–71

[8] Supendi P, Nugraha A D and Widiyantoro S 2018 Recent destructive earthquakes around Garut area, West Java, Indonesia: An unidentified fault? INTERNATIONAL SYMPOSIUM ON EARTH HAZARD AND DISASTER MITIGATION (ISEDM) 2017: The 7th Annual Symposium on Earthquake and Related Geohazard Research for Disaster Risk Reduction p 020077

[9] Supendi P, Nugraha A D, Puspito N T, Widiyantoro S and Daryono D 2018 Identification of active faults in West Java, Indonesia, based on earthquake hypocenter determination, relocation, and focal mechanism analysis *Geoscience Letters* **5**

[10] Nugraha A D, Supendi P, Shiddiqi H A and Widiyantoro S 2016 Unexpected earthquake of June 25th, 2015 in Madiun, East Java *AIP Conference Proceedings* vol 1730 p 020001

[11] Sokos E N and Zahradnik J 2008 ISOLA a Fortran code and a Matlab GUI to perform multiple-point source inversion of seismic data *Computers & Geosciences* **34** 967–977

[12] Koulakov I, Bohm M, Asch G, Lühr B-G, Manzanares A, Brotopuspito K S, Fauzi P, Purbawinata M A, Puspito N T, Ratdomopurbo A, Kopp H, Rabbel W and Shevkunova E 2007 P and S velocity structure of the crust and the upper mantle beneath central Java from local tomography inversion *Journal of Geophysical Research: Solid Earth* **112**

[13] Kennett B L N, Engdahl E R and Buland R 1995 Constraints on seismic velocities in the Earth from traveltimes *Geophysical Journal International* **122** 108–124

[14] Irsyam M, Cummins P R, Asrurifak M, Faizal L, Natawidjaja D H, Widiyantoro S, Meilano I, Triyoso W, Rudiyanto A, Hidayati S, Ridwan M, Hanifa N R and Syahbana A J 2020 Development of the 2017 national seismic hazard maps of Indonesia *Earthquake Spectra* 875529302095120

IOP Conf. Series: Earth and Environmental Science **873** (2021) 012044 doi:10.1088/1755-1315/873/1/012044

Extensive Investigation of the Land Subsidence Impressions on Gedebage District, Bandung, Indonesia

I Gumilar[1*], TP. Sidiq[1], I Meilano[1], B Bramanto[2,1] And G Pambudi[1]

[1]Geodesy Research Group, Institut Teknologi Bandung (ITB), Indonesia
[2]Faculty of Science and Technology, Norwegian University of Life Sciences (NMBU), Norway

E-mail : igumilar@gd.itb.ac.id

Abstract. Gedebage district is presently experiencing rapid and mass infrastructure development and becoming one of the developed districts in Bandung, Indonesia. A football stadium, several luxury housing, the grand mosque of West Java province, and a business center have been built in this district. However, it is well known that the Gedebage district has turned into one of the Bandung districts that suffers from land subsidence phenomena. Since 2000, the Gedebage district has suffered land subsidence at an average rate of 10 cm per year and becoming one of the fastest sinking districts in Bandung. This fast land subsidence phenomenon is suspected of affecting the infrastructure in this district. Therefore, this work aims to capture the current subsidence rate in the Gedebage district using the geodetic approach of the combination of the Global Navigation Satellite System (GNSS) with Interferometric Synthetic Aperture Radar (InSAR) and investigate the impact of land subsidence on infrastructures in Gedebage district. We use GNSS campaign datasets from the years 2016 and 2019. Each GNSS campaign was performed at least 10-12 hours of observations. We also utilize a similar period of 2016 to 2019 for the InSAR datasets. Utilizing both GNSS and InSAR datasets, we can capture the subsidence with the rate reaching 4 -15 cm per year between 2016 and 2019, and it occurs uniformly in this district. The impact of land subsidence occurred in almost all urban areas in the Gedebage district. These impacts include cracks in buildings, bridges and roads, and also tilted buildings.

1. Introduction

Land subsidence disaster is a problem for big cities in Indonesia. This disaster has a long-term adverse effect on the people in the affected area. One of the areas affected by land subsidence is the Bandung Basin (Cekungan Bandung). Land subsidence in Bandung Basin has long been identified using Global Positioning System (GPS) data and Interferometry Synthetic Aperture Radar (InSAR) [1]; [2]; [3]; [4]; [5]. According to the findings of ten GPS surveys conducted between 2000 and 2012, many locations in the Bandung Basin experienced land subsidence, with an average subsidence rate of about –8 cm/year and a maximum subsidence rate of about –16.9 cm/year in a specific location and time span. [5]. Besides, several studies from the geological and geophysical aspects also show that there has been a decline in the Bandung Basin [6]; [7]. Groundwater extraction is believed to be one of the causes of land subsidence in the Bandung Basin

[6]; [2]; [5], alongside other contributing factors such as natural compaction, building loads, tectonic activity [6] agricultural activities [9].

Land subsidence has various effects in the Bandung Basin, including cracks in buildings, infrastructure damage (roads and bridges), tilting and damaged buildings, and increased flooding inundation areas. Flooding occurs regularly in the Bandung basin during the year, with major flooding generally occurring after heavy rainfall. Flooding in the Bandung basin has resulted in significant economic losses [6]. One area that has experienced significant land subsidence is the Gedebage district in Bandung City.

Gedebage is a district in Bandung, where the plan is to create a Technopolis area that includes government centers, housing, recreation, sports, and a business center. In the future, the Gedebage is expected to become the center of information, communication, and technology (ICT) of Bandung.

Several things must be considered in relocating the city center to the Gedebage area. One aspect that must be considered is related to the disaster (hazard) at the Gedebage is the land subsidence phenomenon. Therefore, this work aims to capture the current subsidence rate in the Gedebage district using the geodetic approach of the combination of the Global Navigation Satellite System (GNSS) with Interferometric Synthetic Aperture Radar (InSAR) and investigate the impact of land subsidence on infrastructures in Gedebage district.

2. Method

2.1. Study Site

Gedebage is a district in Bandung City, West Java Province, Indonesia. Gedebage district has four villages: Rancabolang Village, Rancanumpang Village, Cisaranten Kidul Village, and Cimincrang Village. Administratively, the Gedebage district is placed in the eastern part of Bandung City. Topographically, the Gedebage district is positioned in a lowland area with a land elevation of 627 meters above sea level. With an area of 979,930 hectares, Gedebage is currently a district with a large agricultural area compared to other districts in Bandung City. The largest area of the paddy field is located in Cisaranten Kidul Village. However, the paddy field is decreasing significantly by the incessant conversion of land functions that mostly happen in Bandung. On the surface, the formation of Kosambi dominates the Gedebage area (which is primarily affected by land subsidence). Kosambi formation mostly consists of clay, siltstone, and sandstones compact with Holocene age [10]. Aerial map of Gedebage Subdistrict resulting from aerial photography using the Unmanned Aerial Vehicle (UAV) photogrammetry method is displayed in Figure 1.

2.2. Data Accuisition

GPS data used in this study consists of 2016 GPS data and 2019 GPS data. The data is taken from field surveys that have been conducted in 2016 and 2019. The interval from GPS data is 1 second, and the length observation is at least 10 to 12 hours. 18 GPS observation points are located at a number of points in Bandung Basin, including in Gedebage District. The GPS data is tied to CORS ITB which is located on the ITB campus.

InSAR data used in this study is Sentinel-1 data. Sentinel-1 is a series of earth observation satellites which use the SAR (Synthetic Aperture Radar) instrument launched by the European Space Agency (ESA). Sentinel-1 is designed to work in a programmed mode of operation to image all global terrestrial, coastal zones at high resolution.

The sentinel image used in this study consists of four images taken from different years. Interferometric Wide Swath Mode (IW) acquisition mode with Single Look Complex (SLC) format. The details of data collection can be seen in Table 1.

In addition, field surveys were also carried out both directly and using the web scraping method to identify and calculate the damage caused by subsidence.

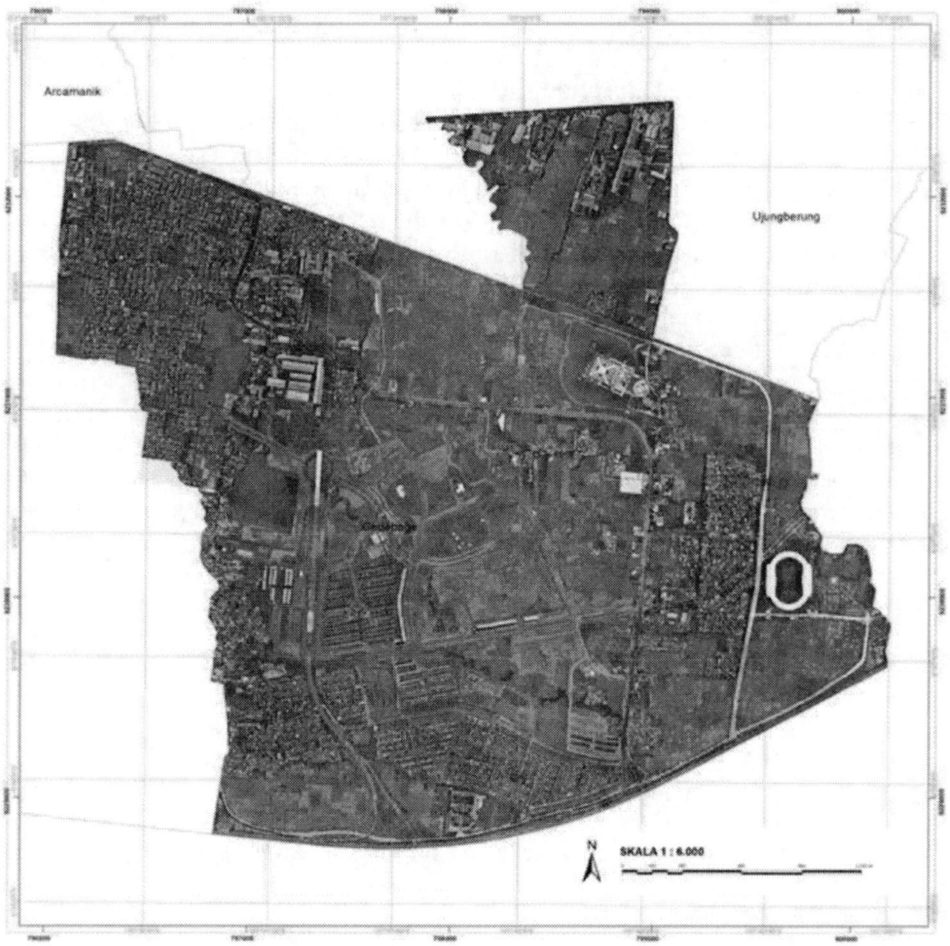

Figure 1. Image map of the Gedebage District, Bandung, West Java, Indonesia.

Table 1. Interferograms master-slave pair selection

Master date	Slave date
12 October 2016	17 October 2017
17 October 2017	18 September 2018
18 September 2018	13 September 2019

IOP Conf. Series: Earth and Environmental Science **873** (2021) 012044 doi:10.1088/1755-1315/873/1/012044

2.3. GPS and InSAR data Processing

Dual-frequency geodetic-type GPS receivers were used to conduct surveys at all GPS stations. ITB1 was used as the reference point as its coordinates were assumed to not differ from time to time. The ITB1 station's coordinate is defined in ITRF 2008 epoch 2012 coordinate system, and the antenna is permanently attached to the roof of a building. Standard pillars or benchmarks were established at the monitoring points, and the monitoring was done periodically using sets of GNSS receivers attached to the tripod.

In GPS data processing, we implemented the radial GPS method and investigated the height variation at the monitoring points. Using this scenario, we obtained sets of coordinates relative to ITB1. Final precise ephemerises (SP3) were used for data processing. The differencing method and the use of dual-frequency measurements limit the influence of atmospheric biases, i.e., troposphere and ionosphere. To further minimize tropospheric impacts, residual tropospheric bias parameters for individual stations were estimated. The majority of the cycle ambiguities of the phase observations were successfully resolved during the processing of GPS baselines.

Sentinel-1 data are processed using the Differential Interferometry method. The acquisition time difference is set to be around one year in order to optimize the error to signal ratio in the unwrapped image. The interferograms are formed by the closest and sequential acquisition date, as listed in **Table 1**.

The topographic effect is eliminated by using 3-arc-second SRTM data. The phase unwrapping is done using SNAPHU routine [11]. We set the coherence value threshold to 0.4. We find that this threshold is suitable for the Bandung area to minimize the processing time but still yield good results confirmed by ground truth checks. The remaining atmospheric error is corrected using Generic Atmospheric Correction Online Service or GACOS [12]. The resulted from unwrapped interferograms are then georeferenced to produced a deformation map in Bandung.

2.4. Accuracy Assessment

The InSAR validation process was done by comparing the InSAR results with the estimated land subsidence from GPS data processing. Since GPS results are more reliable than InSAR and InSAR have a higher spatial resolution than GPS, combining, the two methods will provide more accurate land subsidence information [6]. Thus, the InSAR validation procedure ensures e the consistency of the estimated land subsidence map. Besides, field validation was carried out to see the damage in areas with significant land subsidence.

3. Results and Discussion

Processing of Sentinel master and slave image pairs in 2016 and 2017, 2017 and 2018, as well as 2018 and 2019, produces maps of land subsidence values in the Bandung Basin. The total land subsidence map for three years in the Bandung Basin is displayed in Figure 2 Kopo, Dayeuhkolot, and Gedebage are the three areas in the Bandung Basin suffering the most significant land subsidence. The average land subsidence in the region is 9.2 cm/year, 10.06 cm/year, and 13.51 cm/year, respectively.

IOP Conf. Series: Earth and Environmental Science **873** (2021) 012044 doi:10.1088/1755-1315/873/1/012044

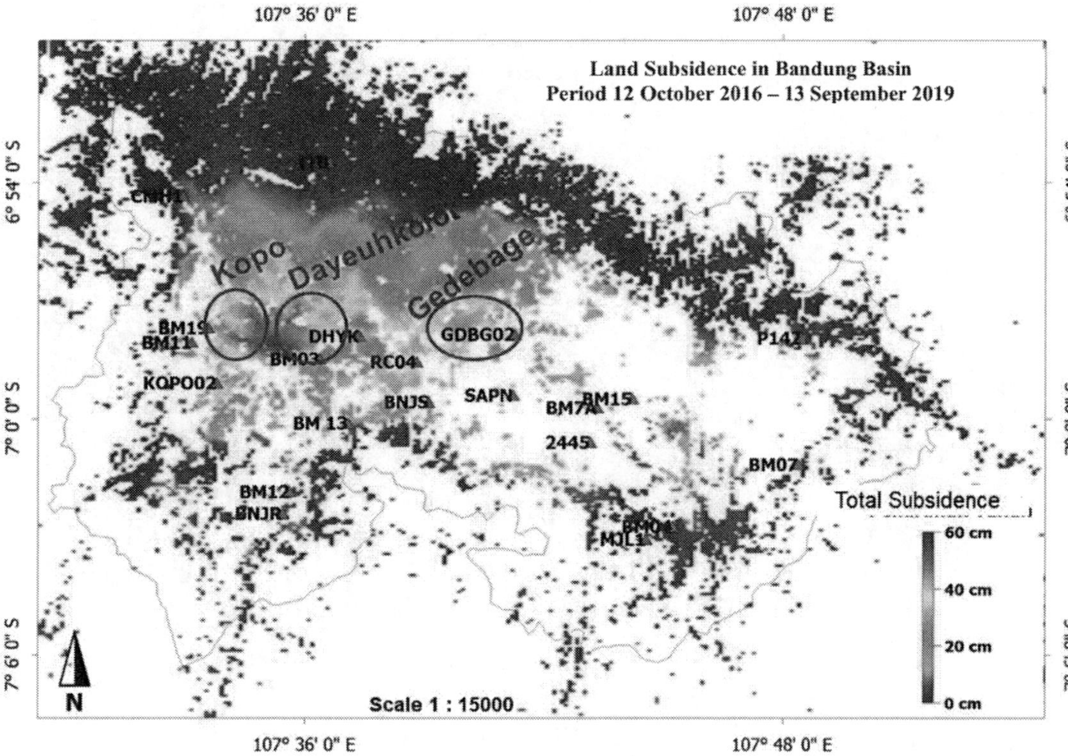

Figure 2. The total land subsidence in the Bandung Basin for the period 2016-2019.

As previously explained, the results of the InSAR method will be validated with GPS observation data. GPS is used as a validator because it has higher accuracy than InSAR. Meanwhile, InSAR results can be influenced by many factors, primarily image resolution, coherence level, and radar waves' ability to penetrate the earth's surface. The validation process is carried out by comparing the GPS results and the InSAR results at the GPS observation point. In this study, it can be ascertained that the ITB1 CORS, which is used as the GPS tie point, does not experience a shift due to land subsidence. Figure 3 shows a comparison of the results of subsidence from InSAR data and GPS data. In general, land subsidence resulted from the InSAR method with GPS gave similar results.

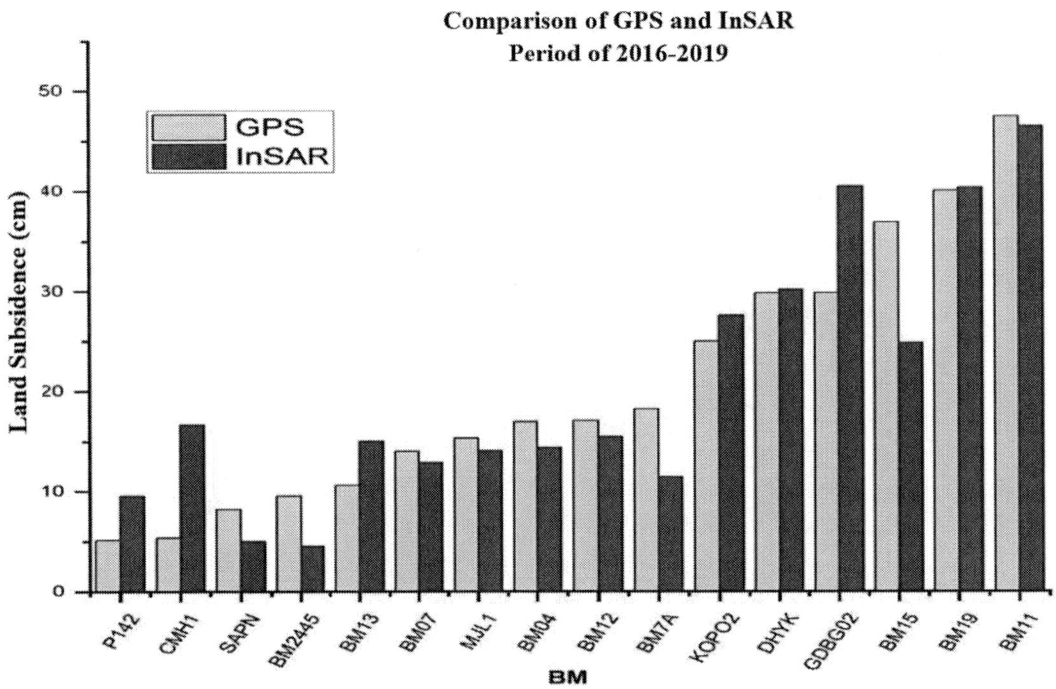

Figure 3. Comparison of the total subsidence of InSAR and GPS for the period 2016-2019.

As explained above, the Gedebage Subdistrict area is one area in the Bandung Basin that has experienced the most considerable land subsidence. Figure 4 shows in more detail the subsidence that occurred in the Gedebage area. Land subsidence occurs in all urban villages with a reduction rate ranging from 4-15 cm per year. Several samples were selected to see the time series of subsidence in this area. The time series of each sample can be seen in Figure 5. Concluding from the land subsidence's pattern, land subsidence that occurs in Gedebage tends to be linear and has occurred to date. Rancabolang Village is the area with the most extensive land subsidence. However, Cisaranten Kidul Village underwent the largest land subsidence. Natural compaction and building loads are deemed to be the causes of land subsidence in the Gedebage Area. Land subsidence occurs in residential areas, roads and bridges, mosques, sports facilities, and business areas.

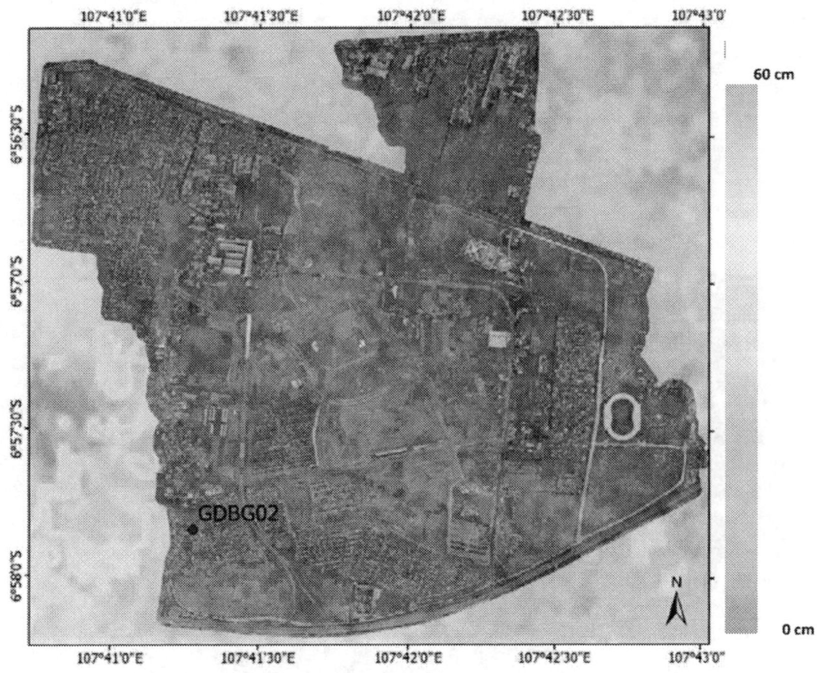

Figure 4. Total land subsidence in the Gedebage District for the period 2016-2019. Dot indicates the GPS Point.

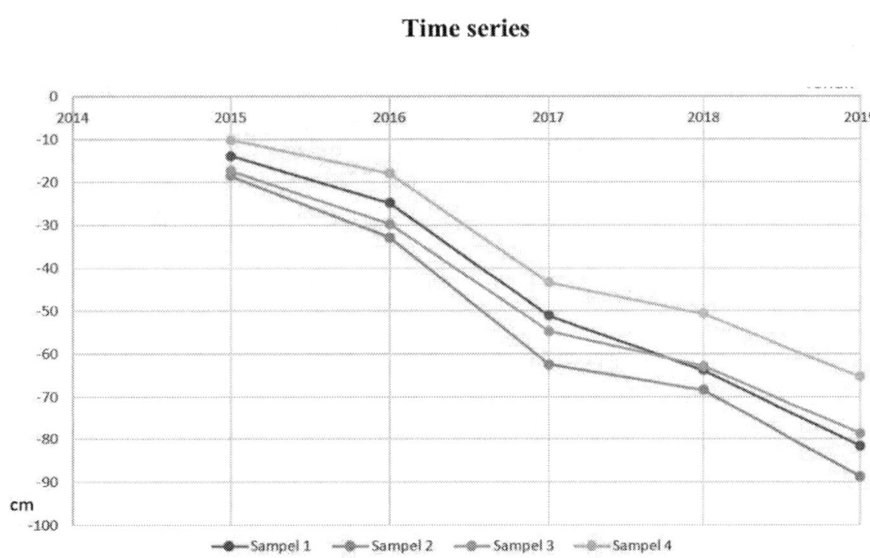

Figure 5. Selected time series of land subsidence in Gedebage District, 2014-2019.

IOP Conf. Series: Earth and Environmental Science **873** (2021) 012044 doi:10.1088/1755-1315/873/1/012044

According to the in-situ observation field survey and web-scraping activities, several pieces of evidence of land subsidence in the affected area were identified. For example, cracks in houses, short houses (because the ground ahead was continuously raised), sloping houses, damage to infrastructure (roads and toll roads), and allegedly expanded the flood inundation area. The entire Cisaranten Kidul area has the most house damage and the longest road damage. Figure 6 and Figure 7 show the damage due to subsidence in Gedebage District.

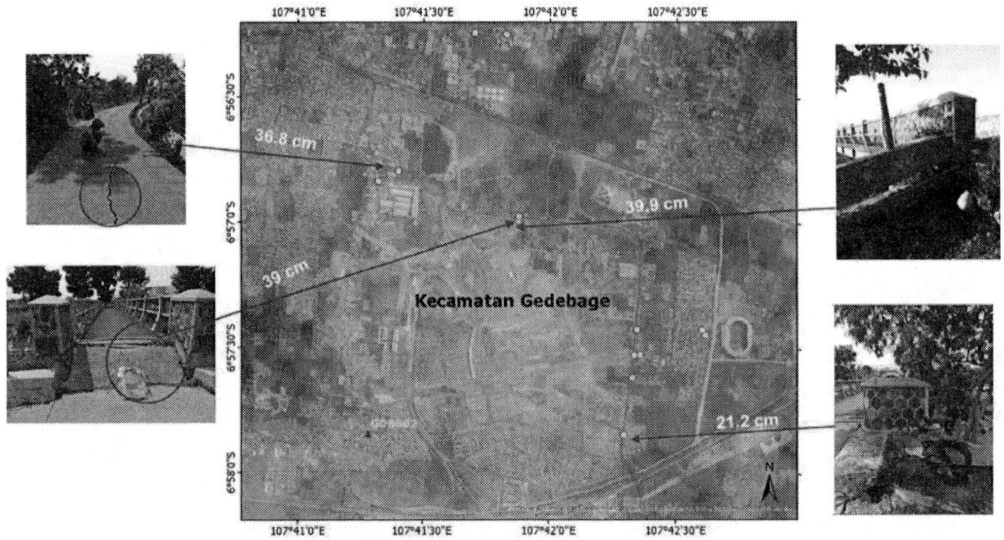

Figure 6. Damage to roads and bridges due to subsidence in Gedebage District

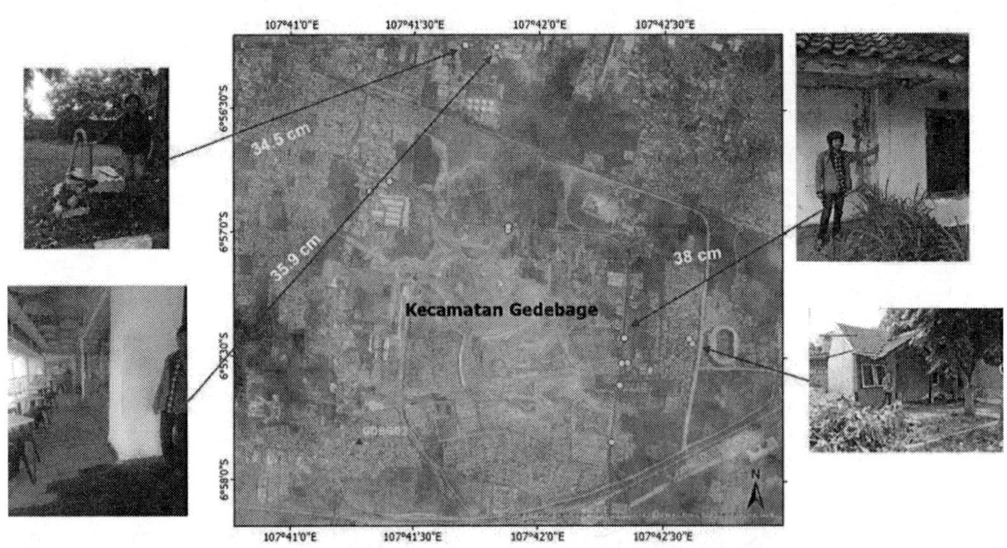

Figure 7. Building damage due to land subsidence in Gedebage District

4. Conclusion

InSAR data processing shows that land subsidence occurs in all districts in Gedebage District, with subsidence rates reaching 4-15 cm per year. The results from GPS data processing show that the land subsidence rate in Gedebage reaches 9.96 cm per year. The pattern of land subsidence in the Gedebage district tends to be linear and is still happening today. Land subsidence occurs in strategic places such as sports facilities, major mosques, housing estates, roads, and bridges. In these locations, many damages can interfere with the function of these objects. In the future, this land subsidence disaster must be given special attention in the development process at Gedebage.

References

[1] Gumilar I, Abidin H Z, Andreas H, Sidiq T P, Gamal M, and Fukuda Y 2013 Land Subsidence, Groundwater Extraction, and Flooding in Bandung Basin (Indonesia) *IAG Symp. 139 Earth on the Edge: Science for a Sustainable Planet Chris Rizos Pascal Willis (Eds)* ISBN 978-3-642-37221-6.

[2] Abidin H Z, Andreas H., Gamal M, Wirakusumah A D, Darmawan D W, Deguchi T, and Maruyama T 2008 Land Subsidence Characteristic of The Bandung Basin, Indonesia, as Estimated From GPS and InSAR *Journal of Applied Geodesy* **2** 167-177.

[3] Abidin H Z, Gumilar I, Andreas H, Murdohardono D, and Fukuda Y 2012 On Causes and Impacts of Land Subsidence in Bandung Basin, Indonesia *Environmental Earth Sciences* DOI 10.1007/s12665-0121848-z.

[4] Ge L , Man Ng A H, Li X, Abidin H Z , Gumilar I 2014 Land subsidence Characteristics of Bandung Basin as Revealed by ENVISAT ASAR and ALOS PALSAR Interferometry *Remote Sensing of Environment* **154** (2014) 46–60.

[5] Estelle C, Amelung F, Abidin H Z, dan Hong S H 2012 Sinking Cities in Indonesia: ALOS PALSAR Detects Rapid Subsidence Due to Groundwater and Gas Extraction. *Remote Sensing Environment* **29** 266 – 278.

[6] Gumilar I, Abidin, H Z, Hutasoit L M, Hakim D M, Sidiq T P, Andreas, H 2013 Land subsidence in Bandung Basin and its Possible Caused Factors. *Procedia Earth and Planetary Science* ***12*** (2015) 47-62.

[7] Taufiq A 2010 Land subsidence study for Bandung and surrounding areas (case study area: Dayeuhkolot, Rancaekek and Cimahi) (in Indonesian) *MSc Thesis, Bandung Institute of Technology (ITB)* Bandung Indonesia

[8] Setyawan A, Fukuda Y, Nishijima J, and Kazama T 2015 Detecting Land Subsidence Using Gravity Method in Jakarta and Bandung Area, Indonesia. *Procedia Environmental Sciences* ***23*** (2015)

[9] Sidiq T P, Gumilar I, Abidin, H Z, Gamal M 2019 Land subsidence induced by agriculture activity in Bandung, West Java Indonesia *IOP Conf. Series: Earth and Environmental Science* ***389*** (2019) 012024. doi:10.1088/1755-1315/389/1/012024

[[10] Hutasoit, L M 2006 Recharge Area and the Origin of Brackish Water in East Bandung: Result of exploration Well, Proceeding of 9th International symposium on Mineral Exploration (ISME IX), Bandung, Indonesia, 19 – 21 September 2006.

[11] Chen C W and Zebker H A 2002 Phase unwrapping for large SAR interferograms: Statistical segmentation and generalized network models *IEEE Transactions on Geoscience and Remote Sensing* **40** pp 1709-1719 (2002).

[12] Yu C, Li Z, Penna N T 2018 Interferometric synthetic aperture radar atmospheric correction using a GPS-based iterative tropospheric decomposition model *Remote Sensing of Environment* **204** pp 109-121 (2018)

IOP Conf. Series: Earth and Environmental Science **873** (2021) 012045 doi:10.1088/1755-1315/873/1/012045

Verification of the Jakarta Geoid Model from the Gravity Data of 2.5 km Resolution with Gravimetric Geoid

D Ramdani, A N Safi'i, P Hartanto, N Oktaviani, M I Hariyono

Indonesia Geospatial Information Agency, Jl. Raya Jakarta-Bogor km. 46, Cibinong, West Java, Indonesia
ayunursafii.10@gmail.com

Abstract. To use the Global Navigation Satellite System (GNSS) correctly, the height information should be transformed into orthometric height by subtracting geoid undulation from it. This orthometric height is commonly used for practical purposes. In 2015 geoid of Jakarta has been produced, and it has an accuracy of 0.076 m. In the year 2019, airborne gravimetry has been done for the entire Java Island. The area of DKI Province cannot be measured because there is inhibition from Airnav. For this reason, terrestrial gravimetric measurements are carried out in this region by adding points outside the previously measured area. To compute the geoid in the Jakarta region is needed the Global Geopotential Model (GGM). In this paper, the GMM used is gif48. The "remove and restore" method will be used in calculating the geoid in this Jakarta region. Besides that in this geoid calculation also uses Stokes kernel and FFT to speed up the calculation. The verification of the resulting geoid is carried with 11 points in DKI Jakarta Province. This verification produces a standard deviation of 0.116 m and a root mean square of 0.411 m.

1. Introduction

Based on Regulation of The Head of Indonesia Geospatial Information Agency No. 15 2013, the Indonesian Geospatial Reference System 2013 (SRGI2013) has been appointed as the only mapping reference system in Indonesia. In SRGI2013, the vertical datum of any Indonesian map, both base, and thematic maps, is geoid. The base map is a product of Basic Geospatial Information (IGD) dan the thematic map is a product of Thematic Geospatial Information (IGT). IGD and IGT are part of the geospatial information Agency. The geoid used in mapping should be modelled or computed based on terrestrial gravity measurements tied to the national gravity control network (JKG). The geoid is formed due to variations in the earth's gravity. An adequate number of gravity measurements with high precission data are needed to obtain an ideal geoid result [1].

In making a large-scale topographic map, the geoid datum should be precise or have a submeter precision number[2] [1]. One good example is the preprocessing step of LiDAR point cloud data which has elevation values measured by the Global Positioning System (GPS). The GPS (or GNSS in a broader context) elevation values are measured relative to the mathematical ellipsoid reference and then called ellipsoidal height (h) [3]. To give these values more scientifically and practically, they should be corrected by geoid undulation (N). These corrected ellipsoidal heights are known as orthometric heights (H). Relation among ellipsoidal heights, geoid undulation, and orthometric height was defined in **Equation 1** and as illustrated in **Figure 1**.

Conversely, from **Equation 1,** it is clear that one can compute geoid undulation (shorten as undulation) by subtracting orthometric height from ellipsoidal height [4]. This geoid determination method is often called geometric geoid or GNSS-Levelling. It is difficult to obtain reliable results to compute the high accuracy geoid from a regional gravimetric with sparse data, which can meet today's geodetic community [5].

$$H = h - N \tag{1}$$

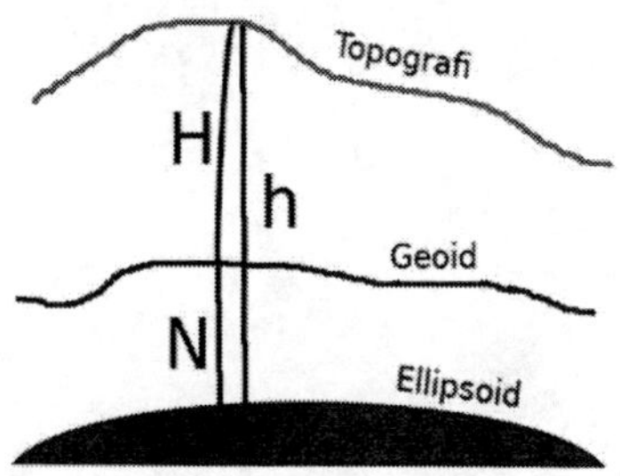

Figure 1. The relation among orthometric height/ plumb line (H), geoid undulation (N), and geoid/ellipsoidal height (h).

Each point on the earth's surface can be expressed as an orthometric height (H) or a geodetic height (h). For mapping and engineering practice, orthometric height is used as a height reference [1]. However, measuring geodetic height is more accessible in the real world than measuring orthometric height by observing from a satellite orbiting above the reference ellipsoid. Therefore, geodetic heights in mapping, geoid undulation values are required, and specific calculations to obtain orthometric height. From **Equation 1,** it is clear that one can compute geoid undulation (shorten as undulation) by subtracting orthometric height from ellipsoidal height [4]. If these undulation values are plentiful and evenly distributed, the national scale geoid model could be calculated. This geoid determination method is also known as geometric geoid or GNSS-Levelling.

Besides the aforementioned geometric geoid determination, the geoid can also be modelled using the gravimetric method. In this method, the geoid was computed based on terrestrial gravity data and the global geopotential model (GGM). One of these GGM is the GIF 48 made by Ries et al. in the year 2011, which is based on GRACE and GOCE gravimetric satellite [6]. Based on previous research by Ramdani et al., GIF 48 is the most optimum GGM for the Jakarta region [7]. It thus will be used as the long-wavelength component of our regional geoid model. The GIF 48 is made by Ries et al. in the year 2011, which is based on GRACE and GOCE gravimetric satellite [8]. In 2015 The geoid of Jakarta had been calculated from terrestrial gravity data measured back in 2012. The data used in this study has a 1.5 km spatial resolution, with a 0.076 m standard deviation and 0.626 m RMS. [9]. The precision value was computed by comparing the gravimetric geoid and the geometric geoid.

In 2019, Geospatial Information Agency conducted an extensive survey of the previous measurement area (Jakarta province). This extension covered area in coordinate 106°-107.2° E and 6°-6.5° S with a spatial resolution of 5 km for each measurement point. Further augmentation was done by the Research Division, BIG, with a spatial resolution of 2.5 km (see **Figure 2**). Those datasets were used as the input for Greater Jakarta Regional Geoid Model.

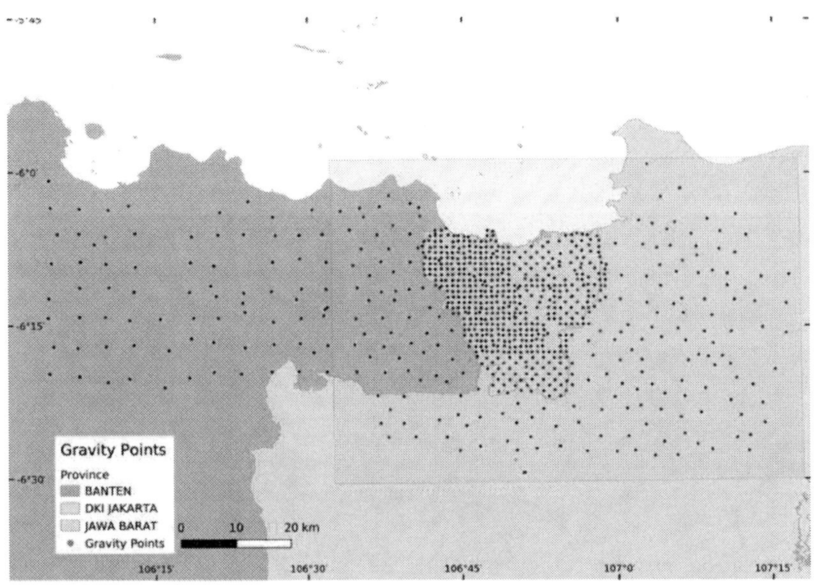

Figure 2. The distribution of 2019 terrestrial gravity measurement points.

2. Methods

The basic equation to compute any geoid model is as written in **Equation 2** [4].

$$N_{GM}(\varphi,\lambda) = \frac{GM}{r\gamma}\sum_{n=2}^{N_{max}}\left(\left(\frac{a}{r}\right)^{n}\sum_{m=0}^{n}(\overline{C_{nm}}\cos m\lambda + \overline{S_{nm}}\sin m\lambda\right)P_{nm}(\sin\phi) \qquad (2)$$

The GGM-derived long-wavelength geoid component has an immense contribution to the whole geoid model, where φ dan λ are spherical coordinates; *GM* is gravitational constant (including earth atmospheric mass); *a* is the semi-major axis of reference ellipsoid; *r* is the distance from the computed point to the center of the earth; γ is normal gravity; N_{max} is the maximum number of *m* degree and *n* order; *C* and *S* are harmonic coefficients, and *P* is the Legendre polynomial.

Meanwhile, to compute the local geoid model, Brun formulae will be applied (see **Equation 3**) [10].

$$N = \frac{T-(W-U)}{\gamma} \qquad (3)$$

with *T* is a potential disturbance outside of the geoidal surface, *W* is gravity potential on the geoid, and *U* is normal potential on the geoid. According to the Laplace equation, this parameter will give zero value outside the earth's surface (see **Equation 4**) [4].

$$\Delta T = \frac{\delta^2 T}{\delta x^2} + \frac{\delta^2 T}{\delta y^2} + \frac{\delta^2 T}{\delta z^2} = 0 \qquad (4)$$

x,y and *z* are the Cartesian coordinates. *T* in **Equation 4** needs a single solution with one boundary condition equation [4] as written in **Equation 5,** which uses the approximation of the sphere with radii of *R*.

$$\Delta g = -\frac{\delta T}{\delta R} - \frac{2T}{R} \qquad (5)$$

with $\Delta g$is free-air gravity anomaly defined as the difference between gravity value on geoid surface (g) and normal gravity (γ) as written in **Equation 6**.

$$\Delta g = g - \gamma \tag{6}$$

The free-air gravity anomaly can be computed from terrestrial gravity measurement data reduced to geoid surface and corresponding normal gravity. This method is called boundary condition solution with Robin's problem. In this method, the geoid is the boundary itself; thus can be computed by using Stokes' integrals as written in **Equation 7** [11].

$$N(\phi,\lambda) = \frac{R}{4\pi\gamma}\int\int_{\sigma} S(\psi)\Delta g(\phi',\lambda')d\sigma \tag{7}$$

$S(\psi)$ is the Stokes' function which can be stated as **Equation 8** or **Equation 9**

$$S(\psi) = \textstyle\sum_{n=0}^{\infty} \frac{2n+1}{n-1} P_n(cos\ \psi) \tag{8}$$

or

$$S(\psi) = \frac{1}{\sin\frac{1}{\psi}} - 6\sin\frac{\psi}{2} + 1 - 5\cos\psi - 3\cos\psi\ \ln(sin\frac{\psi}{2} + \sin^2\frac{\psi}{2}) \tag{9}$$

with ψ is the angular distance on the earth's surface.

Gravity values on the earth's surface should be reduced to geoid surface except in areas of the ocean. There are two ways to calculate the undulation: Equation 7, Fast Fourier Transform (FFT), and least square collocation. The FFT algorithm needs regularized gridded data to optimize the computation's quality [12]

Gridded data in the FFT algorithm were gathered using Trilinear Interpolation embedded in python's matplotlib plugin. The overall geoid model was calculated remove-restore method. The contributive of short and long-wavelength components were removed from the measured terrestrial gravity and restored on the final geoid. The step is written as in **Equation 10** and **Equation 11** [13].

$$N = N_{long} + N_{res} + N_{short} \tag{10}$$

$$\Delta g = \Delta g_{long} + \Delta g_{res} + \Delta g_{short} \tag{11}$$

where:

N = total geoid undulation
N_{long} = long-wavelength geoid
N_{res} = residual geoid
N_{short} = short-wavelength geoid (terrain correction values)
Δg = total free-air gravity anomaly
Δg_{long} = long-wavelength gravity anomaly
Δg_{res} = residual gravity anomaly
Δg_{short} = short-wavelength gravity anomaly (terrain correction values)

IOP Conf. Series: Earth and Environmental Science **873** (2021) 012045 doi:10.1088/1755-1315/873/1/012045

For this research purpose, we made use of about 707 gravity measurement data over the Greater Jakarta Area. Those data consisted of 2010 data which was measured by using Lacoste & Romberg gravimeter (1.5 km spatial resolution), and 2019 data, which was gathered by Scintrex CG-5 gravimeter (2.5 km spatial resolution).

3. Results and Discussion

Gravity measurement data were subtracted by the GGM gravity values as mentioned in the explanation about the remove-restore method and then interpolated into 1.5 x 1.5 km grids. The total interpolated area has the dimension of 50 x 25 km (see **Figure 3**), spanned from 6.4 ° S – 6.06 ° S and 106.47 ° E – 107.15 ° E. This step is known as the removing step of the remove-restore geoid computation method. As shown in **Figure 3**, the residual gravity data has a large value (75.079 mGal) in the dark blue area of North Jakarta and gradually decreases in the yellow area of South Jakarta (31.376 mGal).

This gridded gravity measurement data will be used as the primary input in an FFT processing algorithm embedded in spfour modules on Gravsoft [12]. In this spfour module, the long-wavelength geoid component from GGM will be added back to the aforementioned residual gravity data. This step is called the restore step of the remove-restore method. The final product of the Greater Jakarta local geoid model is depicted in **Figure 4.** As shown in **Figure 4**, the geoid has a low value (17.875 mGal) in the dark blue area of the Tangerang area and gradually increases in the yellow area of the Bekasi area (19.176 m).

A verification procedure will be conducted to analyze the quality of our local geoid model. As many as 11 points spanning all over the model's boundaries will be selected as the validation values. Those 11 points should have both ellipsoidal height (h) and orthometric height (H). The distribution of those samples depicted in **Figure 5**, while the data about those validation points is shown in **Table 1**.

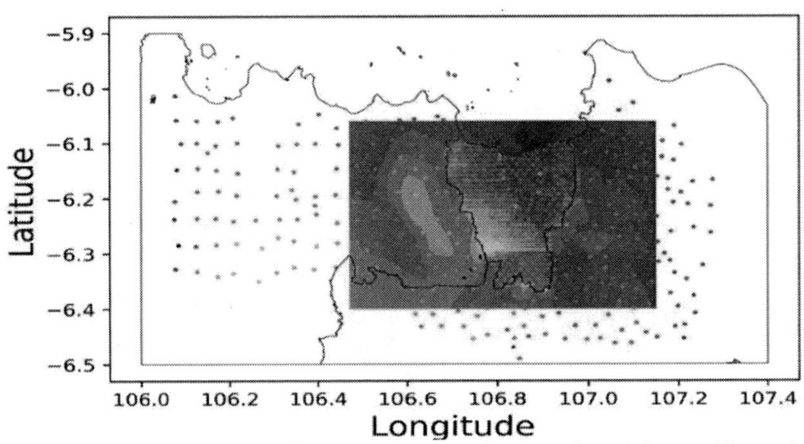

Figure 3. The contour of gridded gravity measurement data produced from trilinear interpolation.

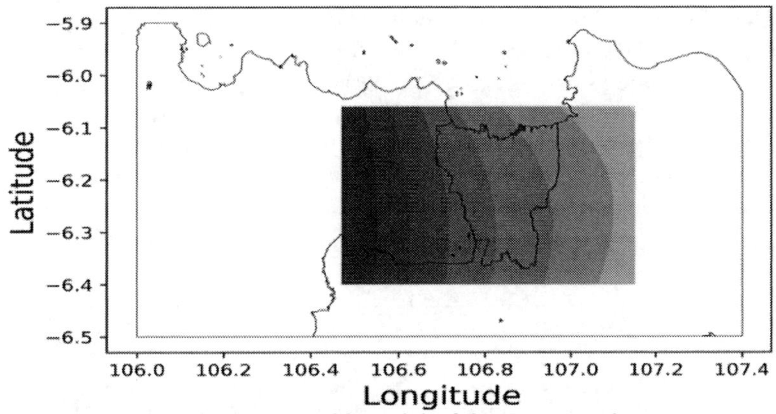

Figure 4. The local geoid model of the Greater Jakarta area.

Table 1. Validation points and their 3D values (position and elevations)

No	Lon	Lat	Height	H_Ortho
1	-6,2172259	106,8749299	34,576	15,499
2	-6,3186845	106,9101893	60,365	41,233
3	-6,2204172	106,7223984	34,627	15,807
4	-6,2190019	106,7293226	34,562	15,912
5	-6,2361289	106,7574098	38,923	20,050
6	-6,3019064	106,8366012	59,962	40,871
7	-6,2367149	106,7515594	40,502	21,762
8	-6,2073957	106,8588015	32,523	13,628
9	-6,2392342	106,7848056	35,204	16,427
10	-6,2882639	106,8707579	51,110	32,018
11	-6,2268139	106,8762478	31,517	12,623

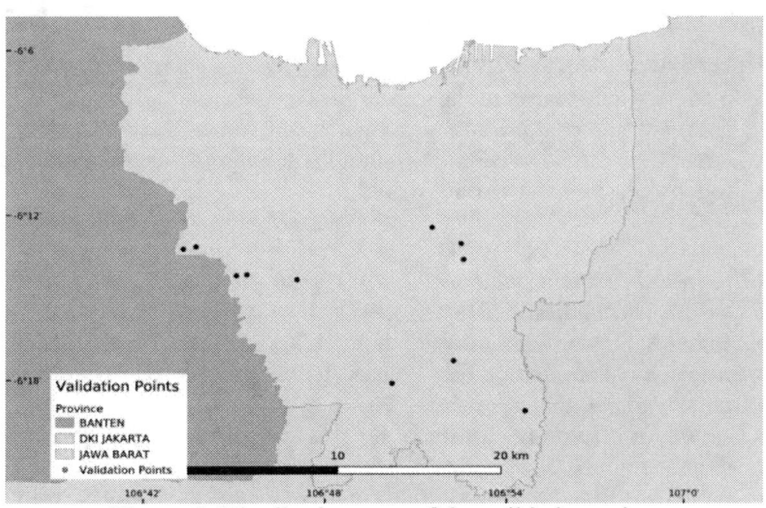

Figure 5. Distribution map of the validation points

Based on those validation points, our local geoid model has a standard deviation of 0.095 m and an RMS of 0.359. This standard deviation value is worse than the previous result by Ramdani et al. [9]. On the other hand, the RMS value of this research's result is somewhat smaller than that result above. The complete verification results are shown in **Table 2**. For this reason, terrestrial gravimetric measurements are carried out in this region by adding a point outside the previously measured area. This study was used to determine the accuracy of the results of terrestrial gravimeter measurements when using an interval of 2.5 km because, in previous studies, the measurement interval was larger.

Table 2. Details of verification results.

No	Lon	Lat	N_{val}	Geoid(2,5km)	d_grv
1	-6,2172259	106,8749299	19,077	18,649	-0,428
2	-6,3186845	106,9101893	19,132	18,668	-0,464
3	-6,2204172	106,7223984	18,819	18,399	-0,420
4	-6,2190019	106,7293226	18,650	18,417	-0,234
5	-6,2361289	106,7574098	18,873	18,489	-0,384
6	-6,3019064	106,8366012	19,091	18,615	-0,477
7	-6,2367149	106,7515594	18,740	18,476	-0,265
8	-6,2073957	106,8588015	18,895	18,605	-0,291
9	-6,2392342	106,7848056	18,777	18,553	-0,224
10	-6,2882639	106,8707579	19,092	18,698	-0,394
11	-6,2268139	106,8762478	18,894	18,663	-0,232
				Std, Dev.	0,095
				RMS	0,365

4. Conclusion

The local geoid model of the Jakarta area based on 2.5 x 2.5 km gridded gravity measurement data has a standard deviation of 0.095 m and RMS of 0.365 m. These results are somewhat ambiguous compared to our previous local geoid model based on 1.5 x 1.5 km gravity data. The ambiguity itself was possibly caused by natural factors such as land subsidence and road condition, which affected the gravimeter readings. Another possible ambiguity source is instrumental factor such as GNSS and gravimeter measurement precision.

Acknowledgements

This research was made possible thanks to the Center for Geodetic and Geodynamics Control Network, BIG, who provided us with the CG-5 digital gravimeter and some GNSS receivers. The authors also acknowledge the Head of Research Division, BIG, and Prof. Fahmi Amhar for their support and advice during the acquisition, processing, and writing works.

References

[1] BIG 2013 *Peraturan Kepala Badan Informasi Geospasial No.15 Tahun 2013 Tentang Sistem Referensi Geospasial Indonesia 2013* vol 1995
[2] Triarahmadhana B, Heliani L S and Widjajanti N 2014 Pemodelan Geoid Lokal D . I . Yogyakarta menggunakan Metode Fast Fourier Transformation dan Least Square Collocation *Conf. Geospatial Inf. Sci. Eng.*
[3] Kotsakis C 2008 Transforming ellipsoidal heights and geoid undulations between different geodetic reference frames *J. Geod.* **82** 249–60
[4] Hofmann-Wellenhof B and Moritz H 2005 *Physical Geodesy* (Vienna: Springer-Verlag)
[5] Kiamehr R 2006 *Precise Gravimetric Geoid Model for Iran Based on GRACE and SRTM Data and the Least-Squares Modification of Stokes' Formula: with Some Geodynamic*

Interpretations (KTH)
[6] Ries J C, Bettadpur S, Poole S and Richter T 2011 Mean background gravity fields for GRACE processing, in GRACE science team meeting (Austin, Texas) pp 8–10
[7] Ramdani D, Priyatna K and Andreas H H 2014 EVALUASI GLOBAL GEOPOTENSIAL MODEL UNTUK PERHITUNGAN GEOID DI JAKARTA (The Evaluation of Global Geopotential Model for Geoid calculation in Jakarta) *J. Ilmian Geomatika* **20** 1–8
[8] Ries J C C, Bettadpur S, Poole S and Richter T 2011 Mean background gravity fields for GRACE processing *GRACE science team meeting, Austin, TX* pp 8–10
[9] Ramdani D, Aryadi Y and Widyantoro B T 2015 Geoid of Jakarta *ACRS 2015 - 36th Asian Conference on Remote Sensing: Fostering Resilient Growth in Asia, Proceedings*
[10] Chapman M E 1979 Techniques for interpretation of geoid anomalies *J. Geophys. Res. Solid Earth* **84** 3793–801
[11] van Gelderen M and Rummel R 2001 The solution of the general geodetic boundary value problem by least squares *J. Geod.* **75** 1–11
[12] Tscherning C C, Knudsen P and Forsberg R 1994 *Description of the GRAVSOFT package. Geophysical Institute, University of Copenhagen*
[13] Hartanto P, Huda S, Putra W, Variandy E D, Triarahmadhana B, Pangastuti D, Pahlevi A M and Hwang C 2018 Estimation of marine gravity anomaly model from satellite altimetry data (Case Study : Kalimantan and Sulawesi Waters-Indonesia) *IOP Conf. Ser. Earth Environ. Sci.* **162** 012038

Peak Ground Acceleration (PGA) and Peak Ground Velocity (PGV) Analyze for Microzonation of Earthquake Hazard Area: Case Study in West Nusa Tenggara

T Razin[1], K Khatimah[1*], Y Annisa[1], A Hamzah[1] and M F I Massinai[1]

[1]Geophysics Dept., Hasanuddin University

[*]text.khusnul@gmail.com

Abstract. The Lombok region is located around a complex tectonic zone with an Indo-Australian oceanic crust transition zone with Australian continental crust in the west and Sundanese arc in the east. This complexity makes some area in West Nusa Tenggara have a high level of earthquake vulnerability and to determine the potential level of seismic damage risk this study was conducted by analyzing Peak Ground Acceleration (PGA) and Peak Ground Velocity (PGV) using earthquake data since 2000 - March 2020 with an intensity more than M4.5. Earthquake data are analyzed using the Yin-Min Yu formula to get the relationship between Peak Ground Acceleration (PGA), Peak Ground Velocity (PGV), and earthquake intensity, so areas with risk level of earthquake damage can be mapped as preliminary earthquake mitigation schemes. The results of the study show that the highest PGA value in West Nusa Tenggara is 74.73 gal at the bedrock and when it on the surface, the PGA value can increase due to amplification of local soil conditions. Likewise PGV value about 32.21 gal where this maximum value is located in East Lombok Regency and North Lombok Regency. According to the classification of PGA and PGV values, the study area has a potential high-risk level of earthquake damage.

1. Introduction

Based on the Central Statistics Agency (BPS), the population of West Nusa Tenggara reached 5.320.092 people per 2020 [1]. This condition is one of the high-risk vulnerability factors which at any time causes huge losses in the event of a natural disaster. The possibility of an earthquake disaster affecting West Nusa Tenggara includes two parts there are, the subduction zone in the south of West Nusa Tenggara and the Back-Arc thrust in the north of West Nusa Tenggara. This is the cause of frequent earthquakes in West Nusa Tenggara. Considering the high earthquake risk, earthquakes Hazard analysis in Mataram area is very necessary [2]. Peak Ground Acceleration (PGA) is the highest ground acceleration value that appears on the surface of the area within a certain period of time due to seismic vibration [3]. Whereas, Peak Ground Velocity (PGV) is the value of the maximum ground velocity on the surface that has occurred in an area with a certain time period due to earthquake vibrations [4].

The level of damage caused by the earthquake depends on the strength and quality of the building, geological, and geotectonic conditions as well as peak ground acceleration (PGA) in the area where the earthquake occurred [5]. Also, the empirical measurement of PGA and PGV is crucial in spatial planning, particularly in infrastructure development. Therefore, the microzonation of earthquake-prone areas can be utilized to minimize the occurrence of damage in these areas [6]. Accordingly, this research expects to provide input for the design of earthquake resistant buildings in West Nusa Tenggara. In addition, this earthquake disaster update is an important step in earthquake disaster mitigation in Indonesia.

1.1. Data Collection
The data used in this study are seismic data obtained from the official website of the United States Geological Survey (USGS) from January 2000 to March 2020. The seismic database is limited to $116{,}067^0$ - $119{,}332^0$ E and 8^0-9^0 S, with magnitude ≥ 4.5 The seismic data used in this study is 321 events.

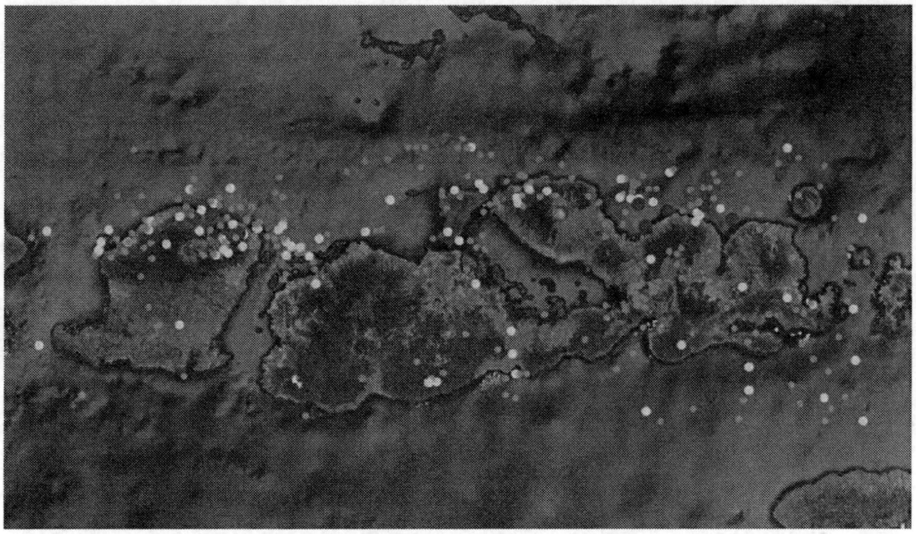

Figure 1. West Nusa Tenggara as study area from Google Earth.

1.2. Calculation of Peak Ground Acceleration
The value of Peak Ground Acceleration (PGA) in this study was calculated using the empirical equation, as follows [7] :

$$\text{PGA} = exp\,(\frac{I-0{,}7}{2}) \quad (1)$$

where,

$I = I_0 exp^{-b\Delta}$
$I_0 = 1{,}5 * (M\text{-}0{,}5)$
I = Intensity at the epicenter distance (observation station)
I_0 = Intensity of the earthquake source
Δ = Epicenter distance
b = 0,00051
M = Earthquake magnitude

1.3. Calculation of Peak Ground Velocity
The empirical equation that correlated Peak Ground Velocity (PGV) and earthquake intensity, as follows [7]:

$$\text{PGV} = exp\,(\frac{I-1{,}89}{2{,}14}) \quad (2)$$

where,

$I = I_0 exp^{-b\Delta}$
$I_0 = 1{,}5 * (M\text{-}0{,}5)$

I = Intensity at the epicenter distance (observation station)
I_0 = Intensity of the earthquake source
Δ = Epicenter distance
b = 0,00051
M = Earthquake magnitude

1.4. Data Analysis

Data analysis was carried out in three stages, are the analysis of the PGA and PGV values, an analysis contour map of PGA and PGV, and determining the relationship between the peak ground acceleration, velocity, and earthquake intensity. The peak ground acceleration and velocity then made into a contour map with an interpolation system using the *Inverse Distance Weighted* (IDW) method [8]. Then to obtain the relationship between the PGA and PGV, it is analyzed with excel by plotting the result data of PGA and PGV.

2. Results and Discussion

The calculations result shown the relationship between PGA and earthquake epicenter intensity in equation (1) and the relationship between PGV and earthquake epicenter intensity in equation (2). Based on the seismicity map of West Nusa Tenggara that we determined, it can be seen that the event occurred in the study area were scattered with various magnitudes. Where events with M4.5 - 4.9 are scattered in almost all regions of West Nusa Tenggara both on land and sea, but for events with M5.7 - 6.9 it is only found in coastal areas to sea parts of West Nusa Tenggara. In addition, the study area is surrounded by active faults wherein the tectonic arrangement of West Nusa Tenggara is affected by the Flores back arc thrust in the north and the subduction zone in the south. Subduction often produces deep events. On the contrary, the back-arc thrust tends to produce shallow seismic events [9].

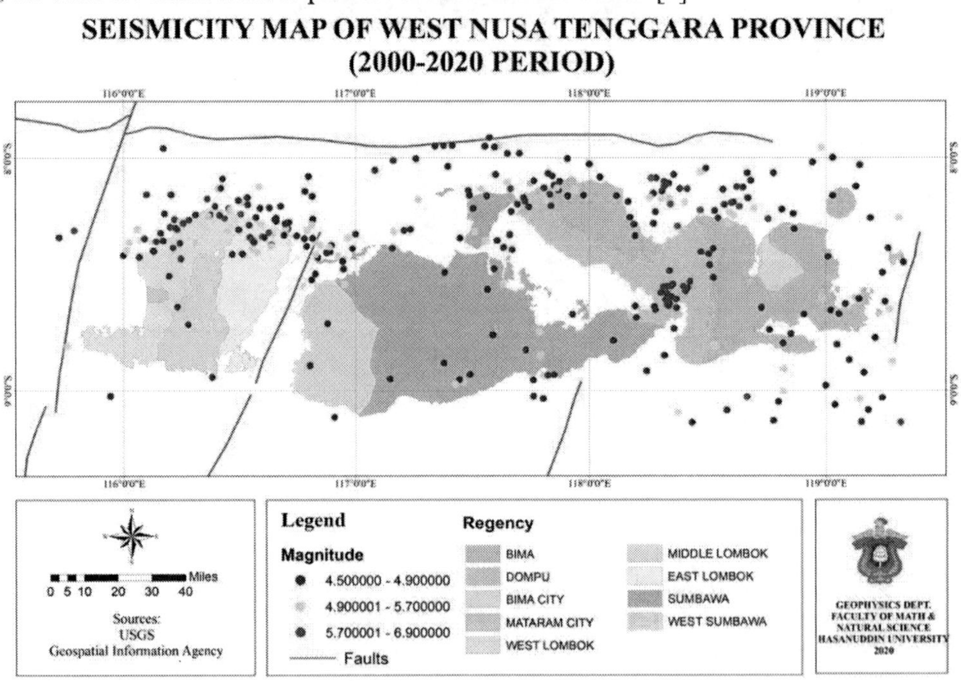

Figure 2. Seismicity Map of West Nusa Tenggara

An empirical approach is necessary for further earthquake studies due to the lack of acceleration and velocity of ground data on the *accelerograph*. So that from the maximum ground intensity and acceleration data for each coordinate, the contour map can be obtained for the microzone of the area.

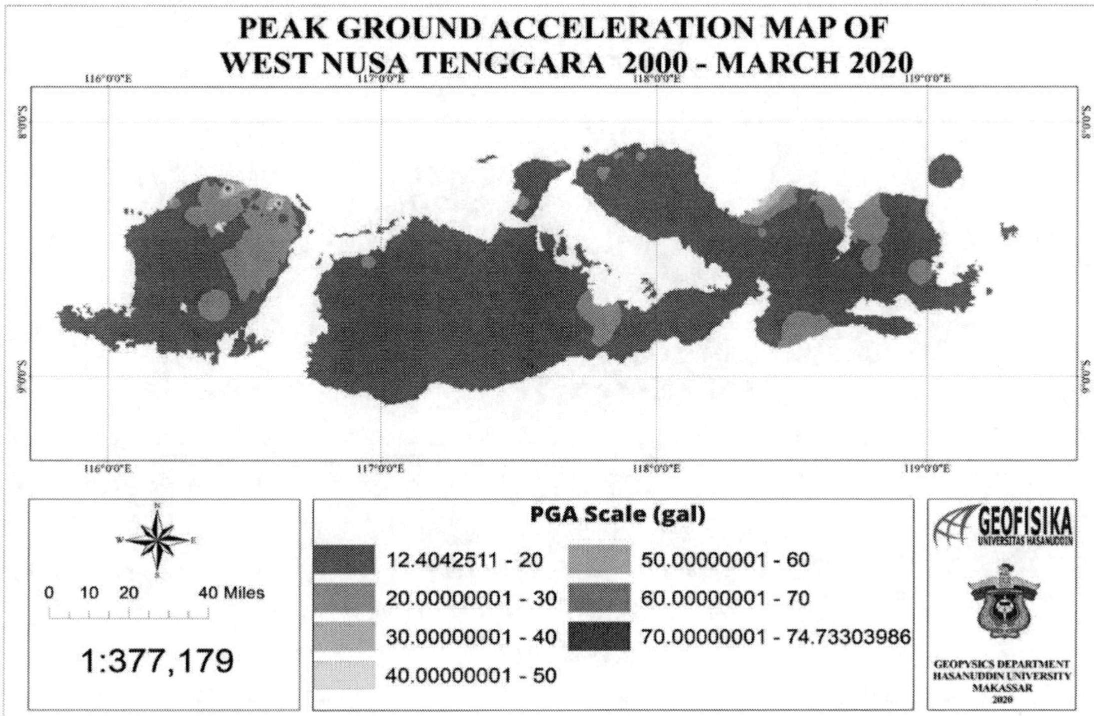

Figure 3. Peak Ground Acceleration Map of West Nusa Tenggara.

In Figure 3, it can be seen that the Peak Ground Acceleration value from West Nusa Tenggara is divided into 7 classes, with the first class with the lowest PGA value scale of 12.404 - 20 gal and the PGA value of this class relatively dominates in West Nusa Tenggara. Then for the areas that were found to have the largest PGA value with 70.001 - 74.733 gal, the coverage area is still wide, therefore microzonation is carried out by dividing into two maps (Figure 5 and Figure 6), is the PGA map of the districts of North Lombok and East Lombok, so that the mapping of distribution of ground acceleration is more specific.

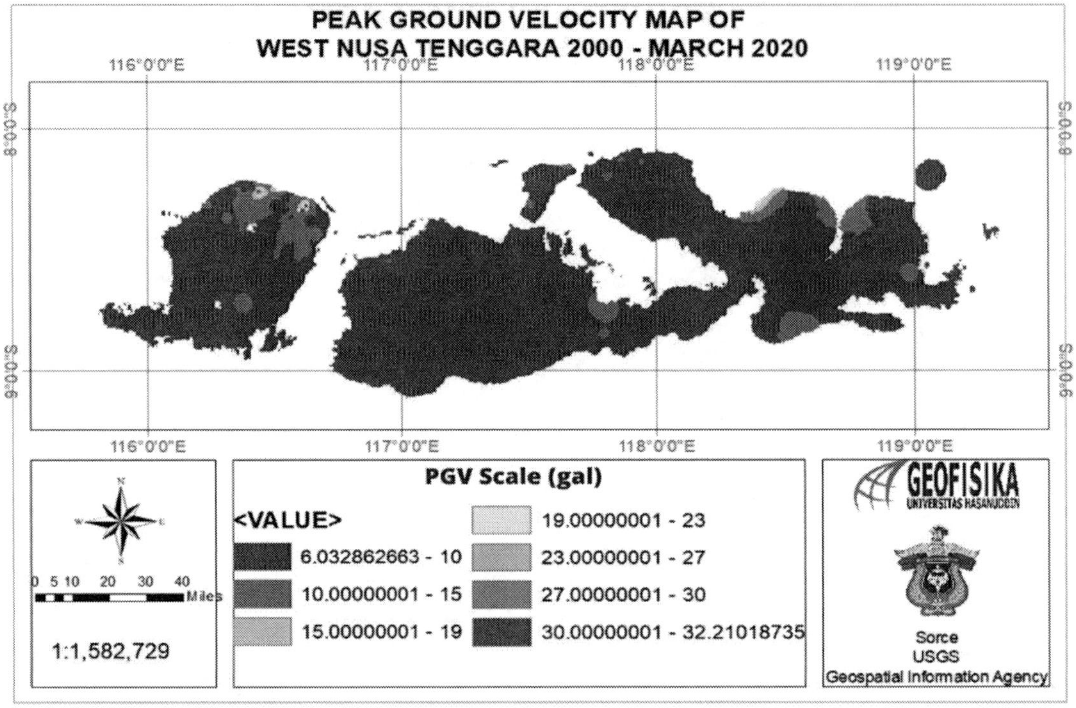

Figure 4. Peak Ground Velocity Map of West Nusa Tenggara.

The results of the modeling of the Peak Ground Velocity (PGV) value obtained from the interpolation using the Inverse Distance Weighted (IDW) method, it is known that the PGV value in West Nusa Tenggara with the smallest classes is 6.0328 - 10 gal and the highest with a value of 30.001 – 32.210 gal, so that microzonation was also carried out in areas with the highest PGV value, are North Lombok and East Lombok districts (Figure 7 and Figure 8).

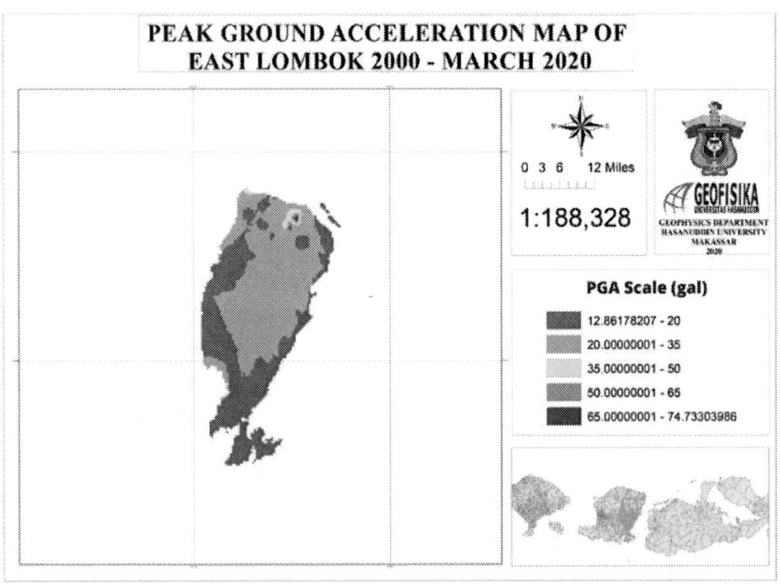

Figure 5. Peak Ground Acceleration Map of East Lombok

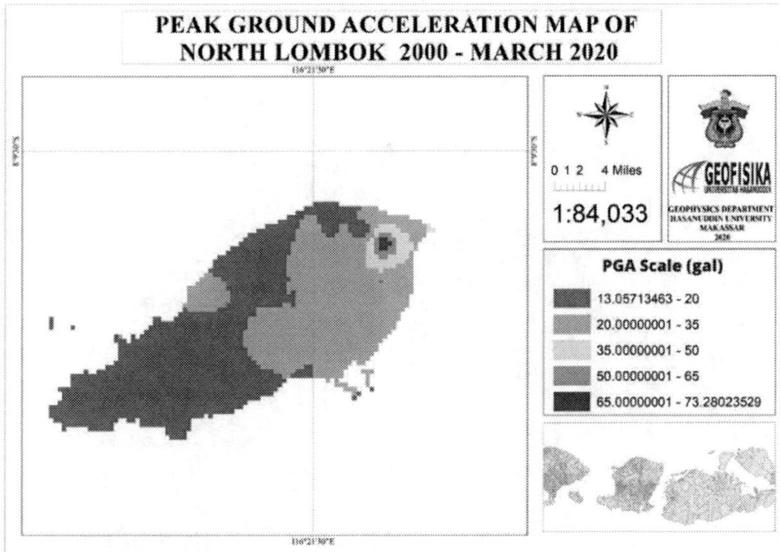

Figure 6. Peak Ground Acceleration Map of North Lombok

The microzonation map shows both East and North Lombok PGA values are quite high, are 65 - 74.733 gal also with PGV values wheres both have 25 – 32.313 gal. When viewed from the tectonics, both areas do have a direct impact when there is an activity from the Flores back arc thrust, for example during the 2018 Lombok earthquakes measuring M_w7.0 and caused a lot of material loss and lives [10]. As we have known before, due to the compression between the Australian continental plate and the Eurasian plate[11], the tectonic model of Lombok is formed by the large thrust of the south subduction and the back-arc thrust, so it causes the south subduction to tend to produce deeper earthquakes and the rear thrust of North Flores tends to produce shallow earthquakes. Then, the earthquake ground motion will be affected by the source, propagation, and site effects [12]. The 2018 Lombok's earthquake seems to influence by the seismic activity of Flores back-arc thrust [13].

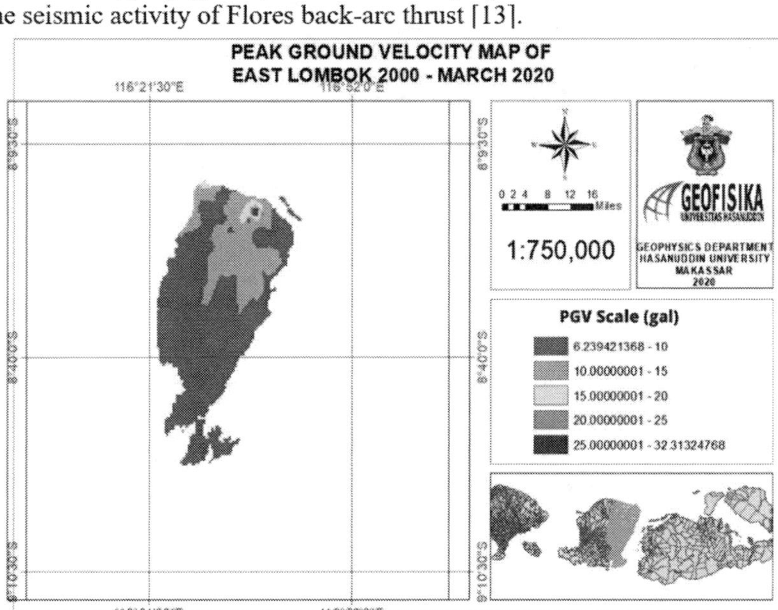

Figure 7. Peak Ground Velocity Map of East Lombok

IOP Conf. Series: Earth and Environmental Science **873** (2021) 012046 doi:10.1088/1755-1315/873/1/012046

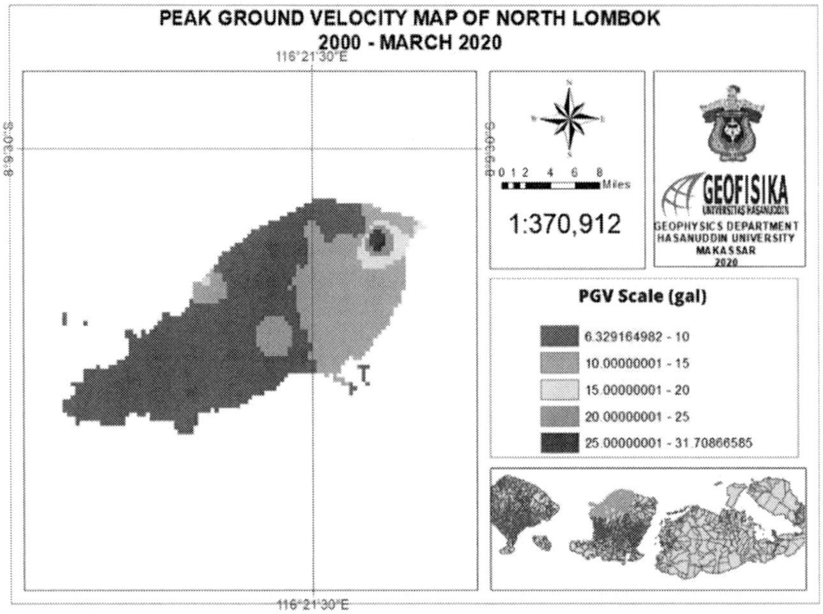

Figure 8. Peak Ground Velocity Map of North Lombok

The curve shows a linear relationship between PGA and PGV. This shows the relationship between the two parameters in determining earthquake-prone areas. The greater the PGA value and the PGV value indicates the higher earthquake risk level [3], it can be seen that PGA and PGV are directly proportional (Figure 9).

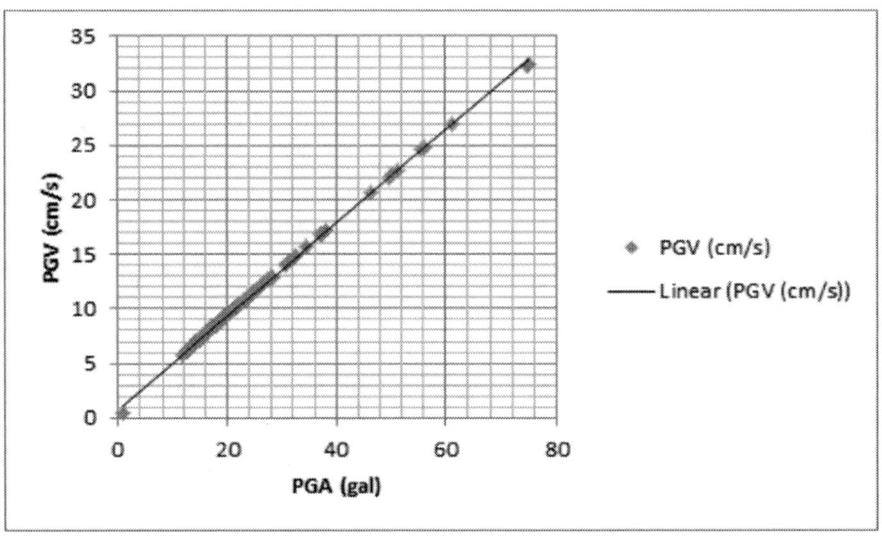

Figure 9. The relationship curve between PGA and PGV

3. Conclusion

1. The Highest value of Peak Ground Acceleration (PGA) and Peak Ground Velocity (PGV) in West Nusa Tenggara in January 2000 to March 2020 with the intensity of ≥ M4.5 are in East Lombok Regency and North Lombok Regency, are PGA is 74.733 gal and PGV is 32.210 gal.

2. The minimum Peak Ground Acceleration value (PGA) in West Nusa Tenggara in 2000 - March 2020 with intensity ≥ M4.5 is 12.404 gal and the value and Peak Ground Velocity (PGV) is 6.032 gal.
3. Regions that have a high PGA value also have a large PGV value. The greater PGA value and PGV value, also have the greater intensity value, then, greater the intensity value makes higher earthquakes risk level.
4. The areas that have the highest earthquake risk level are East Lombok Regency and North Lombok Regency.

Acknowledgment

This research was supported by Geophysics Department, Faculty of Mathematics and Natural Science of Hasanuddin University.

References

[1] Badan Pusat Statistik. https://sensus.bps.go.id/main/index/sp2020, accessed on 17th September 2020

[2] Taruna, R.M, Banyunegoro,V.H, and Daniarsyad G 2018 Peak Ground Acceleration at Surface for Mataram City with a Return Period of 2500 Years using Probabilistic Method *MATEC Web of Conferences* **195** 03019. doi: 10.1051/matecconf/201819503019

[3] Massinai, M.A, Lantu, and Virman 2016 Analisis Percepatan Tanah Maksimum, Kecepatan Tanah Maksimum Dan MMI Di Wilayah Sulawesi Utara *J. SNF.* **5** 33-36. doi: 10.21009/0305020407

[4] Boatwright, J., Thywissen,K. and Seekins, L 2001 Correlation of ground motion and intensity for the 17 January 1994 Northridge, California earthquake. *Bull. Sism. Soc. Am.* **91** 739-752

[5] Prabowo, U.N, and Amalia, A.F 2018 Analisis Percepatan Getaran Tanah Maksimum Untuk Memetakan Resiko Bencana Gempa Bumi Di Kab Pemalang, Jawa Tengah *J. Sci. Tech* **4** 80-83

[6] Jamal, R.J, Lantu, Aswad S, and Sulaiman C 2017 Mikrozonasi Kawasan Rawan Bencana Gempabumi Dengan Studi Peak Ground Acceleration Menggunakan Metode Boore Atkinson Dan Data Mikrotremor Di Daerah Kupang *J. Geocelebes* **1** 5-12

[7] Wu, Y. M., Teng, T. L., Syin, T, C., Hsiao, N. C 2003 Relationnships Between Peak Ground Acceleration, Peak Ground Velocity and Intensity in Taiwan. *Bull. Sism. Soc. Am.* **93** 386-396

[8] Soebagio, Anwar N, Edijatno and Widodo B 2018 Reliability Test of Inverse Distance Weighting (IDW) Method by Linier Limits *J. Of Engineering and Applied Sciences* **13** (20) 8476-8482

[9] Agustawijaya, D.S, Sulisstiyono H, and Elhuda I 2018 Determination of the seismicity and peak ground acceleration for Lombok island: an evaluation on tectonic setting *MATEC Web of Conferences* **195** 03018. doi: 10.1051/matecconf/201819503018

[10] Sunarti, E., et al. 2021. Critical Point on Housing Construction, Resilience and Family Subjective Welfare After Disaster: Notes from the Lombok, Indonesia, Earthquake Sequence of July-August 2018. *Geomatics, Natural Hazards and Risk*, 12:1, 922-938, doi: 10.1080/19475705.2021.1910576

[11] W. Hamilton 1974 Earthquake map of Indonesian region, *U.S. Geo. Surv., Misc. Inv.* Serial Map 1-875-C, scale 1:5.000.000

[12] D. Hadley, D. Helmberger. J. Orcutt 1982 *Bull. of the Seismol. Soc. of America* **72** (3) 959-979

[13] Agustan, Hanifa R.N, Anantasena Y, Sadly M and Ito T 2019 Ground Deformation Identification related to 2018 Lombok Earthquake Series based on Sentinel-1 Data *IOP Conference Series: Earth and Environmental Science* **280** 012004. doi:10.1088/1755-1315/280/1/012004

IOP Conf. Series: Earth and Environmental Science **873** (2021) 012047 doi:10.1088/1755-1315/873/1/012047

Experimental Study of Flow Characteristics in A Rectangular Gravel Bed Channel

Robby Yussac Tallar, Dea Teodora Ferninda, Efferiki , Prabu Mandvi Hafiz Anjar Suhendar and Frankie Pandapotan Purba

Universitas Kristen Maranatha (Maranatha Christian University),
Jln. Surya Sumantri No. 65, Bandung 40641

robbyyussac@yahoo.com or robby.yt@maranatha.edu

Abstract. In an open channel flow, the characteristics of flow resistance are greatly affected by the roughness of the base and the walls of the channel. The existence of an object or other material, including gravel, also influenced the resistance of flow, therefore the purpose of this study is to examine the flow characteristics (flow velocity and flow resistance) in gravel open channel by using experimental study. A laboratory study to explore the effect of channel bed in terms of roughness of types of sediment on the hydraulics flow in 8 m length x 40 cm width a rectangular channel is presented. The study consists of an extensive set of rectangular flume experiments for flows with certain slope and sediment bed. The study was using the Before After Control Impact (BACI) method by set up five different scenarios. The results show that the lowest flow velocity (v=0.3041 m/sec) was occurred in the scenario 3 (50%sand and 50% gravel). Based on the Manning' coefficient (n), it was also found that at the 100% discharge flow condition, the highest value of friction factor (f=0.0780) within 5 scenarios was scenario 3 with the sediment consisted of 50%sand and 50%gravel. Whereas the value of the lowest friction factor(f=0.0652) was scenario 1 with the sediment only gravel within. It concluded that the results gave the lower value of Manning' coefficient (n) compared to the table of Manning's coefficient (f= 0.04) for the channel with gravel base condition.

1. Introduction

In recent years, rapid development and the changing land-use are major factors causing frequent floods and become an issue of growing concern around the world[1–3]. Floods generally can be caused by natural and human factors[4–6]. Natural factors included heavy rainfall intensity and the other hydrologic responses[7]. It also depends on the physiographic characteristics (slope, soil texture, cover land, rocks permeability, and profile curvature) of the affected area. Meanwhile, human factors included blocking of channels or aggravation of drainage channels, improper land-use, deforestation in upper watersheds, etc. An open channel both in natural channel and human-made channel is a waterway, drainage channels or conduit in which a liquid flow with a free surface. An open channel flow describes the fluid motion in open channel such as rivers or streams. In natural channels, the characteristics of flow resistance are greatly affected by the roughness of the base and the walls of the channel[8–10]. The existence of an object or other material, including gravel, also influenced the

IOP Conf. Series: Earth and Environmental Science **873** (2021) 012047 doi:10.1088/1755-1315/873/1/012047

resistance of flow, therefore the purpose of this study is to examine the flow characteristics (*flow velocity and flow resistance*) in gravel open channel by using experimental study.

2. Methods

In this study, we addressed this problem by proposing experimental study that set up at Hydraulic Laboratory, Civil Engineering Department, Maranatha Christian University. A laboratory study to explore the effect of channel bed in terms of roughness of types of sediment on the hydraulics flow in 8 m length x 40 cm width a rectangular channel is presented (Figure 1 and 2). The study consists of an extensive set of rectangular flume experiments for flows with certain slope and sediment bed. The study was using the Before After Control Impact (BACI) method by set up five different scenarios. (Table 1).

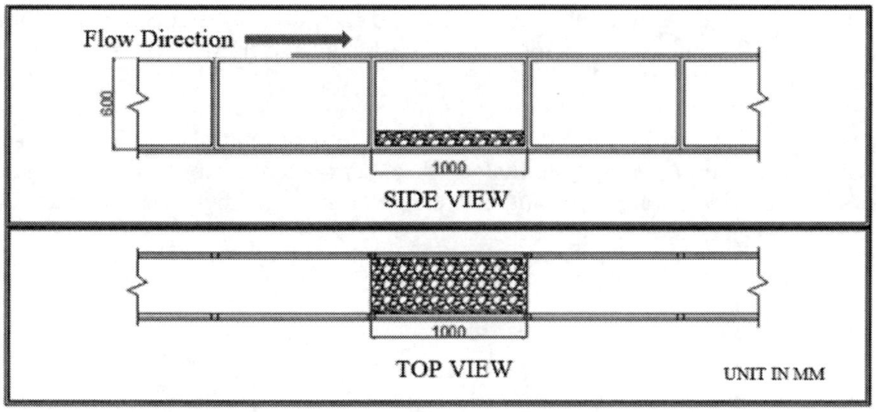

Figure 1. Rectangular Channel: Physical Model (a)

Figure 2. Rectangular Channel: Physical Model (b)

Table 1. Results of Sieve Analysis

Scenario	Type of material
1	100% coarse
2	100% sand
3	50% coarse 50% sand
4	30% coarse 70% sand
5	70% coarse 30% sand

2.1. Preliminary Study

There are two main analysis in preliminary study that should have been completed for analyzing the flow characteristics (Manning coefficient, Chezy Coefficient and friction factor):

2.1.1. Discharge Rating Curve.

The purpose of drawing the discharge rating curve is to determine the maximum discharge from a channel and to obtain a defined discharge in experimental study. A rating curve is a graph of discharge versus stage for a given point on a stream or open channel. The results of discharge rating curve in this study can be seen in Figure 3.

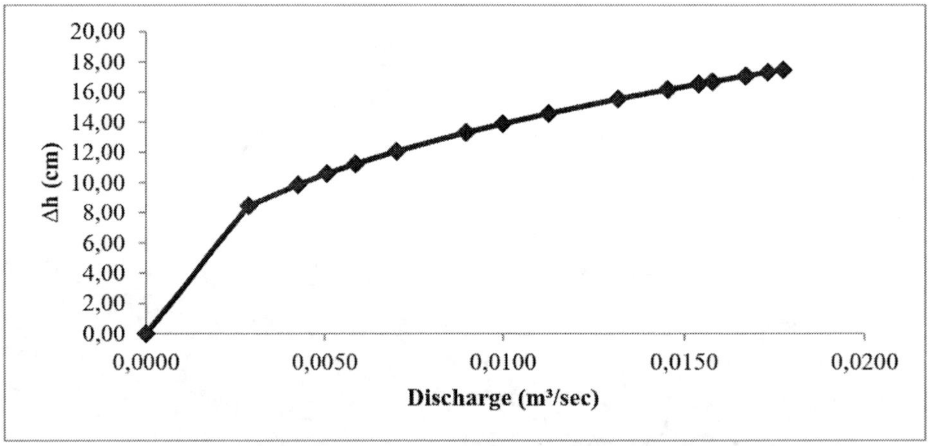

Figure 3. Discharge Rating Curve

2.1.2. Sieve Analysis.

The purpose of conducting the sieve analysis is to determine the grain size of the soil, to classify the soil, and to obtain the uniformity coefficient (Cu) and gradient coefficient (Cc) from the grain size curve distribution. This research uses gravel and sand aggregates which are then placed in the rectangular channel. To classify the aggregate of gravel and sand, it is necessary to do the sieve analysis process first. The sieve analysis is carried out by set up the weighed aggregate into a set of pre-arranged sieves. Then a set of filters is inserted into the vibrating device for 15 minutes. After the sieve analysis test is carried out, the weight data needs to be processed in order to obtain the retained percentage and D_{10}, D_{30}, D_{60} which will be used for sediment classification. Initial aggregate weight is 1000gr. The results of sieve analysis can be seen on Table 2.

Table 2. Results of Sieve Analysis

Sieve No.	Sieve Opening (mm)	Weight Soil Retained (gr)	Percent Retained (%)	Percent Cum Retained (%)	Percent Finer (%)
4	4,750	378	40,170	40,170	59,830
10	2,000	189	20,085	60,255	39,745
20	0,850	108	11,477	71,732	28,268
35	0,500	70	7,439	79,171	20,829
100	0,150	86	9,139	88,310	11,690
200	0,075	54	5,739	94,049	5,951
pan		56	5,951	100,000	0,000
	Total	941	100,000		

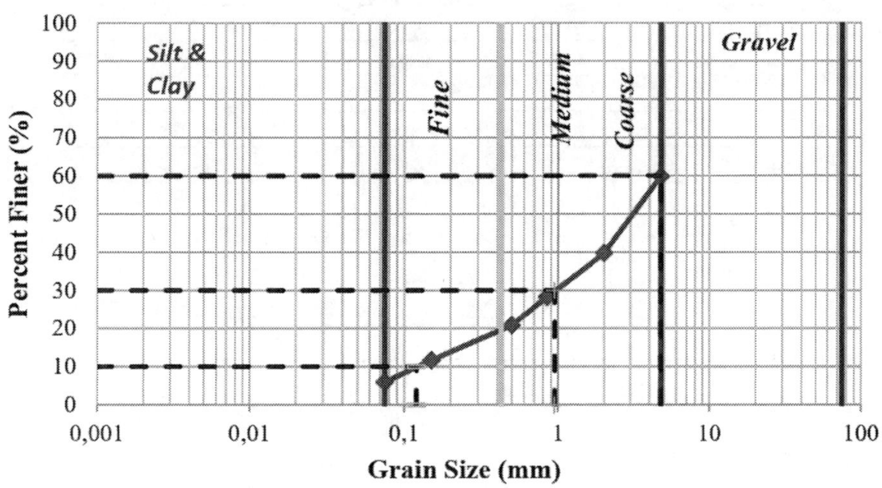

Figure 4. Grain Size Distribution

From Graph in Figure 4, it is D_{10}, D_{30} dan D_{60} to calculate uniform coefficient (C_U) and gradation coefficient (C_C):

$$C_U = \frac{D_{60}}{D_{10}} \qquad C_C = \frac{D_{30}^{\ 2}}{D_{10}\ x\ D_{60}}$$

with: D_{60} = 60% finer (mm), D_{10} = 10% finer (mm), and D_{30} = 30% finer (mm). Therefore, the analysis of C_U and C_C:

$$C_U = D_{60}/D_{10} = 4{,}75/0{,}12 = 39{,}583$$

$$C_C = D_{30}^2/(D_{10}*D_{60}) = 0{,}95^2/(0{,}12*4{,}75) = 1{,}583$$

From the graph in Figure 4 shows that the relationship between grain size and percent finer so that the results is D_{60} = 4,75mm; D_{30} = 0,95mm; D_{10} = 0,12mm. Furthermore, C_U = 39,583 and C_C = 1,583. By using Soil Classification Chart, it is classified that sediment is categorized in *Poorly Graded Sand.*

2.2 Flow Rate Velocity Analysis.

The flow rate velocity is taken by several positions both x and y direction. Position A, B, and C were located in cross section area with the distance between the points is 100mm and B1, B2 and B3 were

IOP Conf. Series: Earth and Environmental Science **873** (2021) 012047 doi:10.1088/1755-1315/873/1/012047

located in horizontal direction of the channel with (Figure 5). Furthermore, considering the flow rate velocity in depth, the study used 0.2h, 0.6h, and 0.8h.

Table 3. Results of Flow Rate Velocity Based on Scenarios

Flow Velocity (m/sec)									
Scenario 1									
Depth	B1			B2			B3		
	Position			Position			Position		
	A	B	C	A	B	C	A	B	C
0,2h	0,326	0,335	0,330	0,337	0,340	0,337	0,333	0,345	0,337
0,6h	0,337	0,345	0,340	0,340	0,345	0,343	0,337	0,343	0,333
0,8h	0,333	0,325	0,333	0,287	0,292	0,297	0,311	0,321	0,325
Scenario 2									
Depth	B1			B2			B3		
	Position			Position			Position		
	A	B	C	A	B	C	A	B	C
0,2h	0,352	0,357	0,347	0,340	0,338	0,330	0,331	0,335	0,330
0,6h	0,337	0,342	0,343	0,321	0,316	0,323	0,304	0,304	0,297
0,8h	0,282	0,256	0,251	0,258	0,227	0,248	0,273	0,248	0,258
Scenario 3									
Depth	B1			B2			B3		
	Position			Position			Position		
	A	B	C	A	B	C	A	B	C
0,2h	0,313	0,318	0,308	0,309	0,319	0,316	0,319	0,325	0,316
0,6h	0,321	0,328	0,321	0,313	0,309	0,313	0,319	0,328	0,313
0,8h	0,277	0,280	0,256	0,275	0,243	0,215	0,285	0,285	0,255
Scenario 4									
Depth	B1			B2			B3		
	Position			Position			Position		
	A	B	C	A	B	C	A	B	C
0,2h	0,326	0,328	0,323	0,330	0,330	0,326	0,335	0,335	0,337
0,6h	0,328	0,335	0,330	0,326	0,331	0,333	0,335	0,342	0,330
0,8h	0,302	0,302	0,287	0,277	0,284	0,251	0,251	0,239	0,229
Scenario 5									
Depth	B1			B2			B3		
	Position			Position			Position		
	A	B	C	A	B	C	A	B	C
0,2h	0,331	0,335	0,331	0,338	0,338	0,330	0,342	0,343	0,337
0,6h	0,338	0,345	0,345	0,335	0,340	0,325	0,330	0,333	0,338
0,8h	0,311	0,292	0,244	0,301	0,277	0,256	0,296	0,260	0,226

This study used the current-meter to measure the flow rate velocity. The propeller type is 2-141579 with diameter 50mm and pitch 0.10. The equation is $V = 0{,}1023 * n + 0{,}016$ for $n \leq 9.33$, or $V = 0{,}1008 * n + 0{,}030$ for $9{,}33 \leq n \leq 19{,}54$, with n = lap/time. In the beginning, the hypothesis of this study is the characteristics of flow resistance are greatly affected by the roughness of sediment due to type and diameter measurement. Therefore, the average of flow rate velocity can be seen on Table 4.

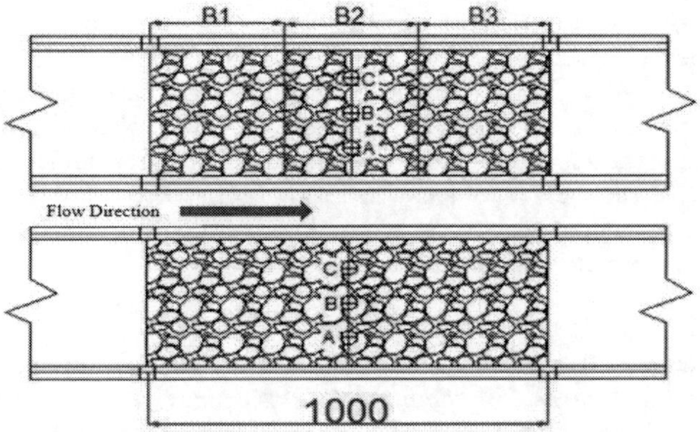

Figure 5. Location of velocity points

Table 4. Results of The Average of Flow Rate Velocity Based on Scenarios

Discharge	Position	Flow Rate Velocity (m/sec)				
		Scenario 1	Scenario 2	Scenario 3	Scenario 4	Scenario 5
100%	B1	0,3354	0,3240	0,3077	0,3212	0,3252
	B2	0,3289	0,3051	0,2956	0,3149	0,3199
	B3	0,3331	0,2989	0,3088	0,3115	0,3171
Average		0,3325	0,3094	0,3041	0,3159	0,3207

2.3 Flow Resistance Analysis.

The further analysis is flow resistance analysis to find out the value of roughness of sediment. Several equations were performed to calculate the flow resistance analysis. The results can be seen Table 5.

Table 5. Results of Flow Resistance Analysis

Scenario	Flow Velocity V (m/sec)	Koef. Manning n	Koef. Chezy C	Friction Factor f
1	0,3325	0,0194	34.6272	0,0652
2	0,3094	0,0208	32.2965	0,0754
3	0,3041	0,0212	31.6871	0,0780
4	0,3159	0,0204	32.9298	0,0723
5	0,3207	0,0201	33.4212	0,0701

3. Discussions

Based on the Manning' coefficient (n), it was also found that at the 100% discharge flow condition, the highest value of friction factor (f=0.0780) within 5 scenarios was scenario 3 with the sediment consisted of 50%sand and 50%gravel. Whereas the value of the lowest friction factor(f=0.0652) was scenario 1 with the sediment only gravel within. It concluded that the results gave the lower value of Manning' coefficient (n) compared to the table of Manning's coefficient (f= 0.04) for the channel with gravel base condition.

4. Conclusions

This study aims to to examine the flow characteristics (flow velocity and flow resistance) in gravel open channel by determining flow resistance in environmentally acceptable channels. Special emphasis was placed on addressing the hydraulic effects of gravel as sediment. For this reason, laboratory flume studies with scenarios were employed. The most notable finding was that, when compared to 5 scenario conditions, the presence of sediment increased the friction factor. This was strongly dependent on the flow rate velocity. In addition, the linkage between flow resistance, channel properties, and other physical condition was investigated. Further recommendation, in this study the parameters for the cross-sectional geometry and flow resistance were found to be simple but nonetheless valuable in evaluating the success of a field project which aims to restore local hydraulics. It is necessary to modify geometric parameters in the future study.

References

[1] Tallar R Y and Suen J-P 2015 Identification of waterbody status in Indonesia by using predictive index assessment tool *International Soil and Water Conservation Research* **3** 224–38
[2] Naef F, Scherrer S and Weiler M 2002 A process based assessment of the potential to reduce flood runoff by land use change *Journal of hydrology* **267** 74–9
[3] Saghafian B, Farazjoo H, Bozorgy B and Yazdandoost F 2008 Flood intensification due to changes in land use *Water resources management* **22** 1051–67
[4] Tallar R Y and Suen J-P 2016 Aquaculture Water Quality Index: a low-cost index to accelerate aquaculture development in Indonesia *Aquaculture international* **24** 295–312
[5] Tanuwidjaja G, Widjaya J M and Tallar R Y 2010 Creative collaboration in urban polder in Jakarta, in the framework of integrated water management
[6] Brody S, Blessing R, Sebastian A and Bedient P 2014 Examining the impact of land use/land cover characteristics on flood losses *Journal of Environmental Planning and Management* **57** 1252–65
[7] Yang L, Tian F and Niyogi D 2015 A need to revisit hydrologic responses to urbanization by incorporating the feedback on spatial rainfall patterns *Urban Climate* **12** 128–40
[8] Anon 2021 Experimental study of boulder concentration effect on flow resistance in gravel bed channels *CATENA* **205** 105458
[9] Nikora V, Goring D, McEwan I and Griffiths G 2001 Spatially averaged open-channel flow over rough bed *Journal of Hydraulic engineering* **127** 123–33
[10] Cameron S M, Nikora V I and Stewart M T 2017 Very-large-scale motions in rough-bed open-channel flow *Journal of Fluid Mechanics* **814** 416–29

Identification of Ransiki fault segment in South Manokwari Regency, West Papua Province, Indonesia based on analysis of a high-resolution of global gravity field: Implications on the Earthquake Source Parameters

Richard Lewerissa[1*], Nur Alzair[2], Laura Lapono[3]

[1] Department of Physics, Faculty of Mathematics and Natural Sciences, Papua University; Manokwari, Papua Barat, Indonesia.

[2] Department of Geological Engineering, Faculty of Mining and Petroleum Engineering, Papua University; Manokwari, Papua Barat, Indonesia.

[3] Department of Physics, Faculty of Sciences and Engineering, Nusa Cendana University; Nusa Tenggara Timur, Kupang 85148, Indonesia.

r.lewerissa@unipa.ac.id

Abstract. The province of West Papua in Indonesia is an area crossed by three major faults, including Sorong, Koor, and Ransiki, leading to the collision of Australia, the Pacific, and Eurasia. In the past, there have been strong and damaging earthquakes on these faults, manly Ransiki fault in the South Manokwari regency. Identification of the Ransiki fault segment was conducted by geological subsurface modeling using the earth gravity field of the Global Gravity Map (GGM) based on satellite measurements implicates for earthquake source parameters. The GGM is seen as a solution for the unavailability of direct measurements in the region. The gravity field analysis begins with data reduction using SRTM2gravity as modern terrain correction to obtain a complete Bouguer anomaly. Furthermore, the gravity gradient approach through vertical and horizontal gradients, analytical signal, and the tilt angle are applied to emphasize a contact or fault structures that are not visible using a 2D fast Fourier transform. Overall, the gravity gradient analysis obtained results that were compatible with the alignment of the Ransiki fault segment which direction of the northwest to south. The gravity inversion produces a geological subsurface model that clearly shows the Ransiki fault segment, associated with a low rock density distribution, thought to the Befoor formation and quaternary sediments, located between high-density rocks correlated to Arfak volcanic rocks as a basement.

Keywords: *Ransiki, Gravity field, GGMplus, ERTM2160, SRTM2gravity, active fault parameters.*

1. Introduction

West Papua Province of Indonesia is situated in the western part of Papua Island, tectonically one of the most active and complex areas in the world due to extreme deformation in the Australian, Eurasian and Pacific Plate convergence zone [1]. The geological setting at the plate boundary involves the formation and movement of microplates, ocean-forming lithosphere fractures, basins, continental-arc collisions, and so on [2]. The plate activity has consequences for the development of various major faults of West Papua, including the Sorong, Koor, and Ransiki faults, which have contributed to many earthquakes. On October 10, 2002, an earthquake with a magnitude of 7.6 occurred at a shallow depth at the southern part of the Ransiki fault, which had a dextral strike-slip focal mechanism along the fault [1]. A large and destructive earthquake also occurred near the Ransiki region on 21 April 2012, with a magnitude of 6.7 at a shallow depth, with moderate and severe damage to buildings and casualties [3].

The Ransiki region is part of the South Manokwari regency administrative area, the Manokwari regency's expansion area in 2012. This place is a district capital situated in lowlands at an altitude of 0-100 m elevation and near the coast, which is vulnerable to natural disasters, such as the earthquake and tsunami, since the place is on major fault lines the Ransiki fault. Several related studies have been conducted in West Papua, especially in the Ransiki district of South Manokwari. Milsom, (1991) was carried out measurements of gravity and terrane tectonics in the field of New Guinea (Irian Jaya) and reported that the gravity technique was an effective instrument for the study of terranes but could not be used for isolated structures. Most structural analyses of formations in western Papua have been done without regard to evidence of gravity, and there is a disparity of some degree [4]. Serhalawan & Sianipar, (2017) have researched the mechanism of earthquake sources to determine faults based on the Ransiki earthquake on the Mw 6.7 scale of 2012, which states that the Yapen Fault caused the earthquake with a sliding fault mechanism trending west-east, not caused by the Ransiki fault [3]. A fault system study at the triple junction of East Indonesia was undertaken to assess the quaternary activity and its consequences for seismic hazards. The Sorong, Koor, and Ransiki faults are used as a model for the West Papua area [1].

In line with regional planning and development, more studies are required to research geological structures such as faults as with details on the potential for geohazards in Ransiki, South Manokwari regency. Fault identification is generally limited to observing features on the surface, but fault traces may have been lost by geological processes or are covered by layers above them [5]. Our research uses the approach of the gravity method to describe this possibility, which is related to the gravity anomaly correlated with the distribution in subsurface rock density. The fault and the folding system are not only controlled by the current stress field, but also by recent stress. Hence it is crucial to define the features of these systems to obtain an understanding of the tectonic activity that is currently occurring. Geological and geomorphological elements are capable of providing valuable knowledge on the properties of geological formations. However, geophysical surveys, such as gravity and/or seismic methods, are approaches to determine subsurface structures, particularly in areas where active formations are no longer visible on the surface [6]. Furthermore, geophysical methods are considered more desirable for geohazard investigations due to its discrete and more economical nature [7], [8]. The geological structure in the region is occupied by quaternary sediments [9], which is an essential topic for the study of disasters or natural hazards.

Our study utilizes a combination of earth gravity field models with high resolution from the Global Gravity Model plus (GGM) [10], Earth Residual Terrain Model (ERTM) 2160 [11], and the SRTM2gravity [12] for modern terrain correction of disturbance or free-air gravity anomalies in the research place. The use of earth gravity models is used as a solution because there is no direct gravity measurement data available on Ransiki at a dense scale. Geological structure identification is performed by vertical and horizontal derivative (gradient) analysis of the gravity field as well as by inversion modeling of gravity anomaly data in the study area. Geological structure identification is performed by vertical and horizontal gradients (derivative) analysis of the gravity field as well as by inversion modeling of gravity anomaly data in the study area. The derivative analysis uses a first or second degree of vertical or horizontal gradients to enhance edge effects, and also to outline possible boundaries from a gravity source [13]. Gravity inversion modeling is applied to obtain subsurface geological models capable of delineating the existence of the Ransiki fault formation as well as the bedrock density in the study region.

This research is a preliminary study using a qualitative and quantitative evaluation of the corrected gravity anomaly of the GGMplus field of gravity data, and also an analysis of the gravity gradient in the form of vertical, horizontal, analytical signal and tilt angles from the complete Bouguer anomaly using 2D fast Fourier transform (FFT). Gravity modeling relies on Singular Value Decomposition (SVD) and Occam techniques. The goal of the results is to improve and define the boundary of the geological system of the fault covered by the layer of sediment above it so that the planning and construction of the site can pay attention to potential natural hazards.

2. Geology, Tectonic, and Gravity

According to the geological map of the Ransiki sheet issued by the Indonesian Geological Agency in 1989 [9], it was found that the geological formations of the South Manokwari regency were composed of Alluvials, Arfak volcanic rocks, Maruni limestone, Befoor formation, Kemum formation, Wai formation, Anggi granite, and alluvial (Figure 1). For the study area, the Ransiki area is dominated by tertiary alluvial (Qa), Arfak volcanic rock (Tema), and alluvial (Tqb).

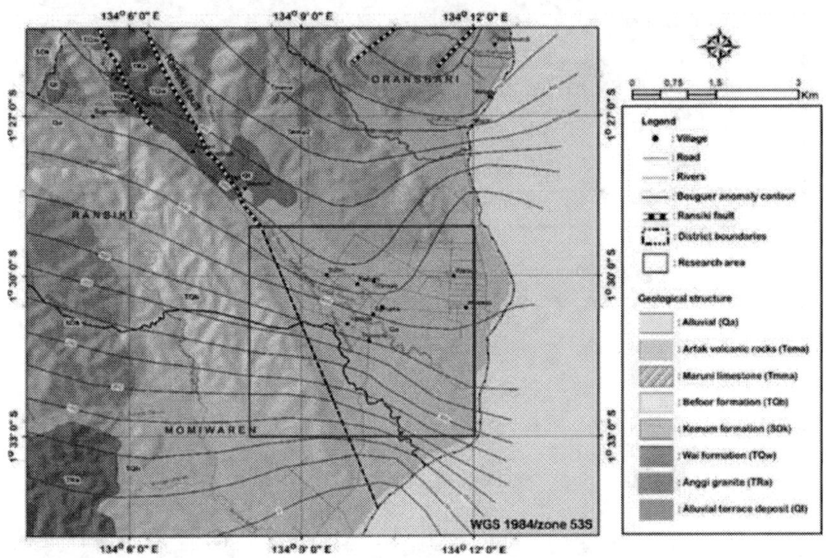

Figure 1. Geological map of the Ransiki area, South Manokwari regency [9]

The research work is included in the bird-headed zone of the Papua islands, a deformed area of active plate tectonics due to impacts between the Eurasian, Pacific, and Australian plates, and also many smaller microplates occupied by major strike-slip faults [14]. The Ransiki fault in the eastern side of the bird's head is believed to be the result of a collision between the archipelago and the Australian continent, spreading into the Bay of Cenderawasih, east of Wandamen, and linked to the Weyland overthrust in the Central Mountains [4], [14], [15]. Today, arc rock is mainly formed by the bedrock on the northern part of Papua Island, such as the eastern west of Papua, the Cenderawasih Bay, and the reefs. The North-Northwest (NNW) direction of Ransiki Fault is believed to be a dextral shear zone connecting to the Sorong fault again with Yapen Fault, which is considered to be inactive [16]. The Ransiki Fault in West Papua has a typical segment length ranging from 20 -50 km with a maximum length of 100 km [1]. Earth's gravity field measurements have been carried out on the bird's head of Papua. Milsom, (1991) conducted gravity and terrane tectonics measurements in the New Guinea region obtained during the geological mapping project of Irian Jaya by Indonesia - Australia (IJGMP), and also involved re-mapping the geology in the south and west of the Cenderawasih Bay, as well as some areas further to the east, as shown in Figure 2 [4]. Figure 2 shows that the study area is on the path of the Ransiki fault having positive Bouguer anomaly values ranging from 0 g.u to 600 g.u (1 g.u = 0.1 mGal). The highest anomaly is associated with Arfak volcanic rocks, while low anomalies are correlated with quaternary alluvial sedimentary rocks.

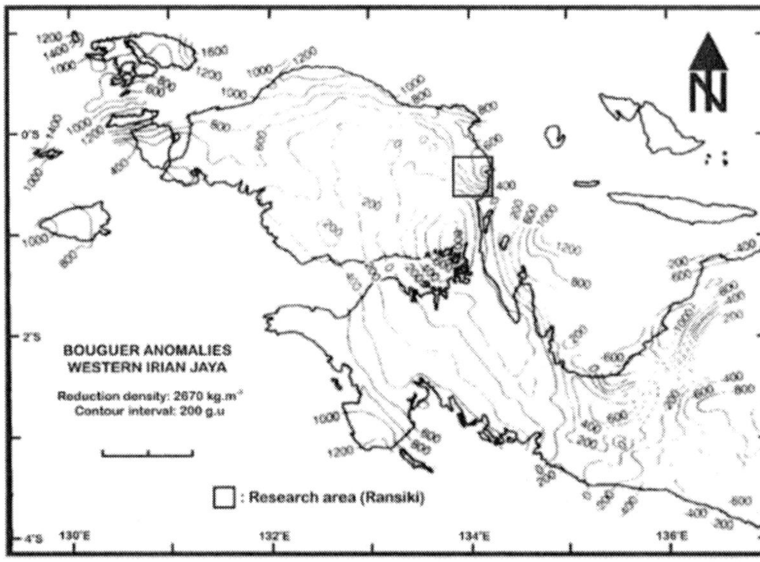

Figure 2. Bouguer anomaly in West Irian Jaya (West Papua) in gravity units (g.u) with an interval of 200 g.u = 20 mGal [4]

3. Methodology

3.1. GGMplus Gravity Data Reduction

This research examines the major strike-slip fault in Ransiki district, South Manokwari, West Papua, Indonesia at coordinates: 134.133° E - 134.199° E and 1.547° S - 1.477° S (Figure 3).

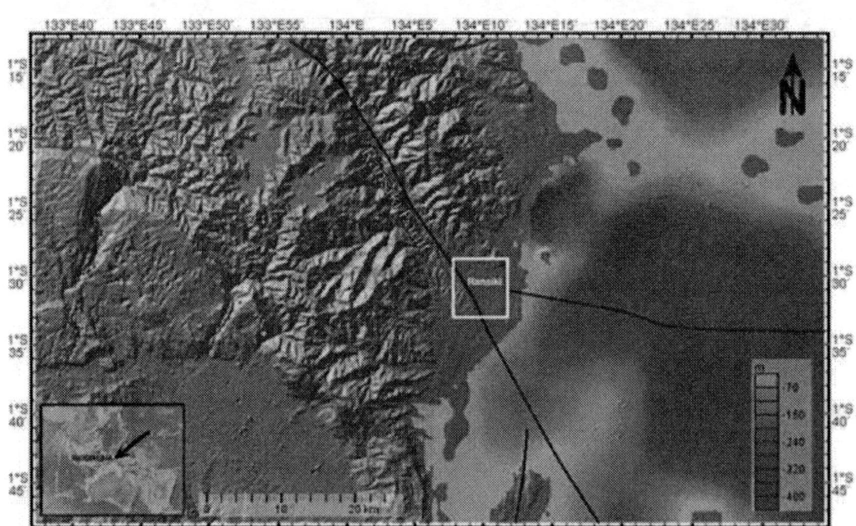

Figure 3. The study area of Ransiki district in South Manokwari regency, West Papua, Indonesia

The reduction of the earth-gravity field starts with the extraction of gravity data for the West Papua area from GGMplus (S05E130), ERTM2160 (S05E130), and SRTM2gravity (S02E134) developed by Curtin University, Perth, Australia. The global high-resolution field of gravity is very useful in several areas, such as exploration, potential field of geophysics, climate, and marine level change analysis. GGMplus is a high-resolution earth-gravity model for local scales on land and islands encompassing an area of ± 60° latitude with a spatial grid of 7.2" (~220 m) [10]. This model results from three gravity measurements, including the GOCE/GRACE gravity satellite, the 2008 EGM model, and topographic gravity. GGMplus is useful in applying geophysics and exploration as a new data source for in-situ reduction of detailed gravity surveys, mineral searches without calculating needed, and reduction of time-consumption [10], [17].

Earth Residual Terrain Model (ERTM) 2160 is a short-scale earth gravity field generated by forward gravity modeling using the SRTM of the global topographic approach. The model has a spatial scale equal to the harmonic spherical coefficient of up to 2160 degrees used to create a small range of GGMplus gravity map from 10 km to 250 m [11]. SRTM2gravity is a modern gravity field correction reflecting earth gravity force from the global topographic mass of ~28 billion computational points. The territory is covered by all land in the -60° to 85° geographic latitude range with a spatial resolution of 90 m. This model specifically incorporates Bouguer shell effect and all residual gravitational field effects on the Bouguer shell. As a result, the model reflects the cumulative signal of gravity provided by the global topographic mass [12]. Gravity anomaly data extraction in the Ransiki region was performed using the Matlab listing software developed by the developer. In addition, we use the GNU Octave program to modify and extract data. The GGMplus model used is a disturbance anomaly model equivalent to free air anomaly data, also ERTM2160 in the form of digital elevation model (DEM) data with a spatial grid of 7.2 "(220 m). The SRTM2gravity model is a full-scale gravity for modern terrain correction with a 3" (~ 90 m) spatial scale.

Data processing begins by verifying the satellite data used, which compars with the gravity field measurement data carried out in West Papua province by the Indonesian geological agency and the IJGMP project [4]. Furthermore, the three types of data are processed to obtain the complete Bouguer anomaly by correcting a free-air anomaly in the form of reduction with full-scale gravity data (terrain correction). SRTM2gravity model contains the total gravity effect of the global topographical mass, so the full-scale gravity (terrain correction) product can be directly reduced to disturbance or free air anomalies to obtain the complete Bouguer anomaly value in the study area. Furthermore, it does not require correction of the spherical Bouguer shell, because it is implicitly included in the product of full-scale gravity [12].

3.2. Regional and residual anomalies separation

This study focuses on the Ransiki fault zone, which preserves sediment associated with residual anomalies so that the complete Bouguer anomaly includes the difference between regional and residual anomalies. Regional anomalies are commonly affected by large and deep structures, while residual anomalies are caused by small and shallow formations [18], [19]. The separating process for regional and residual anomalies is calculated using a frequency filtering technique, using a low-pass filter to pass low-frequency anomalies associated with regional anomalies, while high-frequency anomalies are eliminated. A single filter ring with a boundary susceptible inner and outer ring radius is used in low-pass and high-pass filters.

3.3. Gravity gradients

Vertical and horizontal gradients are intended to describe the boundaries of geological structures and source objects buried in gravity or magnetic field map [20]. Vertical gradient (derivative) is applied to delineate the edges of the gravity anomaly. This technique is commonly used to highlight near-surface geological features, and to raise the large wavenumber element of spectrum, where zero values of vertical gradient (VG) normally corresponds to the geological boundary [21]. The mathematical equation for vertical gradient can be written as:

IOP Conf. Series: Earth and Environmental Science **873** (2021) 012048 doi:10.1088/1755-1315/873/1/012048

$$VG = \frac{\partial g}{\partial z} \tag{1}$$

A horizontal gradient is used to calculate the rate of change in the potential plane in *x* and *y* directions [22]. In addition, a horizontal gradient is used to detect the difference in density or susceptibility to potential field model. This technique is useful for delineating shallow or deep sources relative to vertical gradients, adequate only for shallow structures [23]. The horizontal gradient of gravity data in the directions *x* and *y* is expressed as follows:

$$HG\ (x,y) = \sqrt{\left(\frac{\partial g}{\partial x}\right)^2 + \left(\frac{\partial g}{\partial y}\right)^2} \tag{2}$$

g is a gravity anomaly, which in this study used the complete Bouguer anomaly data in the Ransiki area, South Manokwari regency. The vertical and horizontal gradients of Bouguer anomaly are powerful for the determination of subsurface formations including faults or contacts body represented by the highest value of horizontal gradient and the zero of vertical gradient [6].

To delineate geological structures such as contacts and faults in the research work, an analytical signal of gravity data was used according to Poisson's relation between magnetic and gravity fields. This limitation can be calculated for the maximum amplitude magnitude of the analytical signal [24], [25]. The analytical signal of observation gravity data is generated by 3D sources can be written as [26], [27]:

$$A_s(x,y) = \sqrt{\left(\frac{\partial g}{\partial x}\right)^2 + \left(\frac{\partial g}{\partial y}\right)^2 + \left(\frac{\partial g}{\partial z}\right)^2} \tag{3}$$

where $A_s(x,y)$ is amplitude of analytic signal at position (*x*,*y*), *g* is the gravity observed in the plane (*x*,*y*), and $\left(\frac{\partial g}{\partial x}, \frac{\partial g}{\partial y}\right)$ is horizontal gradient or derivative, and $\frac{\partial g}{\partial z}$ is vertical derivatives in the *z*-component. The analytical signal from the gravity field is obtained using FFT approach in the frequency domain [28].

The tilt angle (*TA*) method was introduced by Miller and Singh, 1994, used to identify structural boundaries from various sources with varying depths [29]. The tilt angle is a ratio between vertical and horizontal derivatives with a range from -90° to 90°. The equation of *TA* can be expressed as [21]:

$$TA = tan^{-1}\left(\frac{VG}{HG\ (x,y)}\right) \tag{4}$$

The value of *TA* is positive above the source, zero or close to zero when it is at the source boundary, and negative on the outside of sources [20], [21]. The process of calculating a vertical, horizontal, and tilt angle gradients is displayed in the frequency domain through a 2-D FFT approach. The gravity gradient is equal to the multiplication of gravity with the wavenumber along the derivative direction in a Fourier domain [30], [31]. The Fourier transform represents the sum of the sine and cosine at different spatial frequencies or wavenumbers (k_x and k_y) which are based on a data range (Dx = max (x) - min (x)) and (Dy = max (y) - min (y)) and data sampling (dx and dy) along the *x* and *y* directions.

3.4. 3D gravity inversion modeling

Subsurface modeling according to the gravity data inversion in the Ransiki area was performed using Grablox and Bloxer tools for model editing and visualization. Grablox is built on a 3D rectangular block, where the main blocks represent the subsurface volume of the gravity survey site, which is separated into small blocks with small volumes. The procedure of inversion was followed the Singular Value Decomposition (SVD) and Occam methods. The SVD is an unconstrained inversion with adaptive damping, while Occam performed to smoothing model using Lagrange multiplier and Roughness parameters [32]. The Lagrange multiplier is a function that minimizes data errors, while the Roughness is a discontinuity parameter. Occam inversion provides a smoot model similar to SVD inversion [32]. The RMS error between the response of the model and the observation data is written as [33]:

$$RMS_d = \sqrt{\frac{1}{M}\sum_{i=1}^{M}\left(\frac{d_i - y_i}{\Delta d}\right)^2} \tag{5}$$

where Δd = maximum gravity data (d_{max}) - minimum gravity data (d_{min}) used for data scale. The initial reconstruction of the model starts with an area of ± 50 km^2. The initial model was established by spatial discretion including 30 small blocks in W-E, 30 small blocks in N-S, and 5 layers with a maximum depth of 2 km. Additional margins are used to eliminate edge effects during the inversion process.

4. Results and Discussion

4.1. Free air and Complete Bouguer anonalies in Ransiki area

The result of GGMplus data extraction for the West Papua zone indicates that the free-air anomaly in the Ransiki area of South Manokwari is positive, ranging from 49.10 mGal to 92.80 mGal. High anomalies are found in the northern and western areas of the study, while moderate to low anomalies are usually observed in the central and southern areas of the study (Figure 4a). The distribution of free air anomalies primarily follows the topographic elevation model, varying from -4 m to 254 m based on the ERTM2160 model (Figure 4b), where high anomalies are correlated with high topography, and vice versa. The SRTM2gravity full-scale gravity model for terrain correction in the Ransiki area produces anomalies between 14.39 mGal and 39.92 mGal. This terrain correction model also follows the elevation pattern, where it is high value in the mountains or hills of the north and west, low in the middle to the south (Figure 4c). The effect of terrain in mountainous areas greatly affects the anomalous value of free air and makes it very varied and irregular [17].

The difference between the value of free-air anomaly and the full-scale gravity (terrain correction) produces a complete Bouguer anomaly in the Ransiki, which is a positive values ranges from 28.82 mGal to 53 mGal. In general, the anomaly pattern is southwest to the northeast with a low anomaly in the southwest, increasing to the east of the Ransiki area (Figure 4d). According to the Ransiki sheet of geological map [9], we compared with Bouguer anomaly results from field measurements by the geological research and development center in 1989, anomaly value and pattern match are obtained, where the resulting regional Bouguer anomaly ranges between 200 μm.s^{-2} (20 mGal) to 600 μm.s^{-2} (60 mGal) with contour intervals of 50 μm.s^{-2}, trending southwest-northeast (Figure 1). Bouguer anomaly of Ransiki area also corresponds to the results of gravity and terrane tectonic measurements in the New Guinea region with a contour interval of 200 g.u (20 mGal), as shown in Figure 2 [4]. Geologically, the high Bouguer anomaly in the northern part is correlated with the Arfak Volcanic Rock (Tema), the moderate anomaly aligned with the Befoor Formation (Tqb), whereas the low anomaly indicates the quaternary sedimentary rock (Qa). The Ransiki fault on the complete Bouguer anomaly map is not clearly described, so further analysis is needed to confirm the structure in the form of anomaly separation and gradient analysis of the anomaly.

4.2. Residual and regional anomalies

This research aims to study and enhance the Ransiki fault zone associated with sedimentary layers that may close the fault path in the study area so that anomalous separation of residual and regional anomalies is required in the complete Bouguer anomaly data. Low pass filter technique is used to achieve regional anomalies, then the differential between the complete Bouguer anomaly and the regional anomaly is a residual anomaly. Regional anomalies is positive with a range between 29.31 mGal and 56.63 mGal, which increased in value from southwest to northeast (Figure 5a). The regional anomaly pattern is close to the complete Bouguer anomaly, then it is presumed that deep and large structures below the surface have a strong effect on the gravity anomaly in the Ransiki region. The location of the fault structure of Ransiki is not visible in the regional anomaly.

IOP Conf. Series: Earth and Environmental Science **873** (2021) 012048 doi:10.1088/1755-1315/873/1/012048

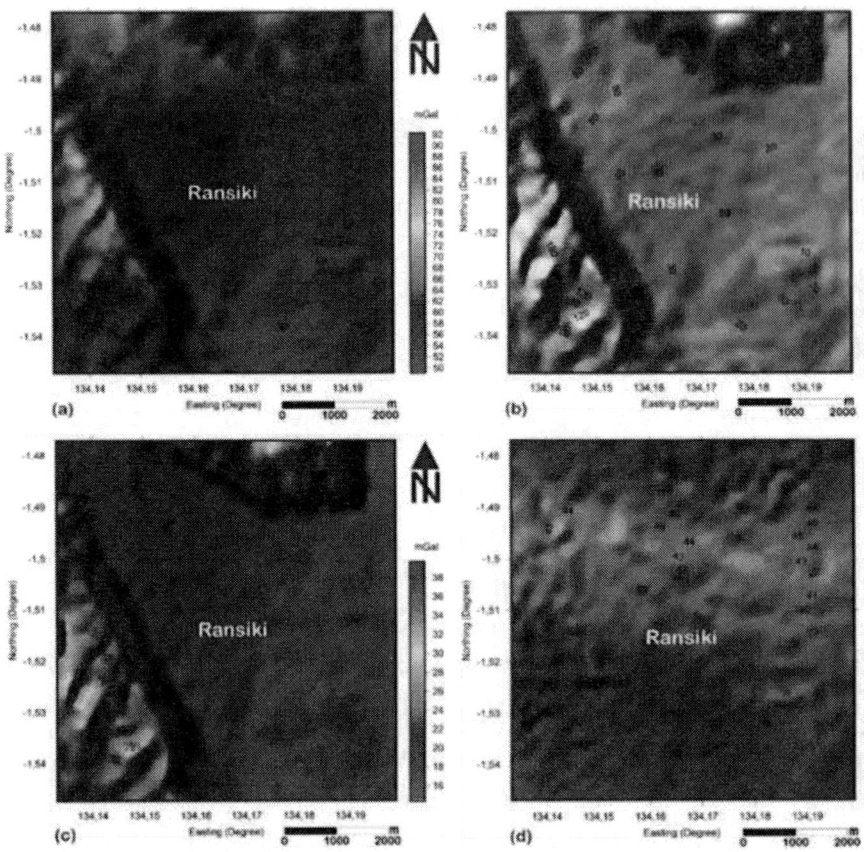

Figure 4. (a) Free air anomaly based on the GGMPlus model; (b) Digital Elevation Model (DEM) based on ERTM 2160 model; (c) terrain correction based on the SRTM2GRAVITY model (d) Complete Bouguer anomaly in the Ransiki area, South Manokwari Regency, Indonesia

Residual anomalies are often more complex relative to complete Bouguer and regional anomalies with values varying from -2.88 mGal to 2.28 mGal, high anomalies in the form of circular spots distributed all across the research location. In this anomaly, the Ransiki fault began to appear and was identified to be associated with a low negative value trending northwest to the south (Figure 9b). In this anomaly, the Ransiki fault began to appear and was identified as being associated with a low negative value in the northwest direction to the south (Figure 5b). Low anomaly values may correlate with weak or crushed zones associated with low density distribution of rock, because of the effect of tectonic activity in West Papua. Watkinson and Hall, (2017) state that the asymmetrical location of the Ransiki River near the Ransiki fault gap supports extensional activity along the fault [1]. To emphasize the structure of the Ransiki fault near the surface, further analysis was provided in the form of an interpretation of the Earth's gradient of gravity from the complete Bouguer anomaly.

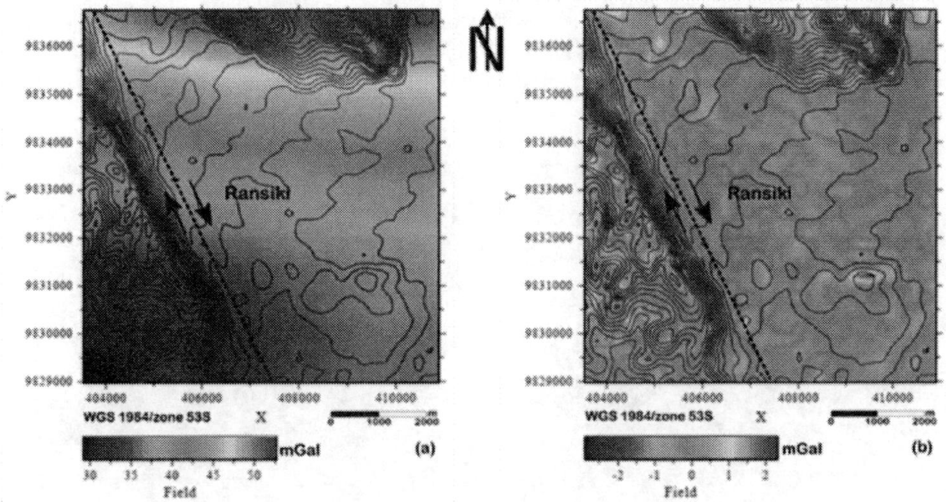

Figure 5. (a) Regional anomaly of the Ransiki area; (b) Residual anomaly of the Ransiki area

4.3. Gravity gradients analysis

Analysis of the gravity gradient of vertical, horizontal gradients, analytical signal, and tilt angles is carried out to sharpen and increase the boundaries of the anomaly source and geology structures in Ransiki. The vertical gradient is negative to positive between -0.012 mGal/m to 0.010 mGal/m (Figure 6a). The fault structure or geological contact is visible on the vertical gradient map, associated with zero values on the anomaly map. The horizontal gradient is positive of 0.00016 mGal/m to 0.01019 mGal/m (Figure 6b).

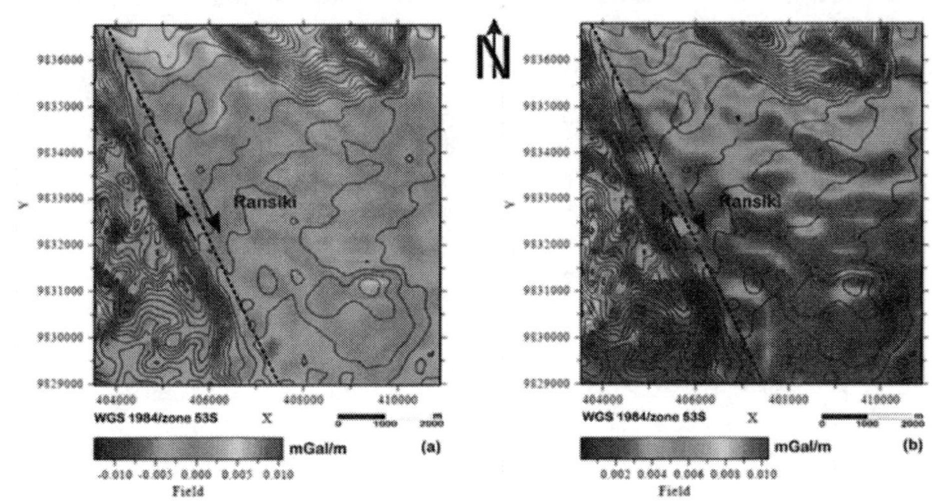

Figure 6. Gradient analysis of Complete Bouguer Anomaly in Ransiki area; (a) Vertical gradient, (b) Horizontal gradient.

On this gradient, the boundaries of the geological structure are correlated with the moderate or intermediate amplitude values. The analytical signal of Bouguer anomaly data in the Ransiki region was positive of 0.00036 mGal/m to 0.1274 mGal/m (Figure 7a). The gradient tilt angle provides more complex information when compared to the vertical and horizontal gradient anomaly maps. The tilt

IOP Conf. Series: Earth and Environmental Science **873** (2021) 012048 doi:10.1088/1755-1315/873/1/012048

angle values ranged from -87.513° to 84.416°, with a predominance of positive angles in the north, and negative in the southern part (Figure 7b). The Ransiki fault structure in the tilt angle model corresponds to zero or close to zero. The zero contour of angle is usually located near to the boundaries of the sources body [29].

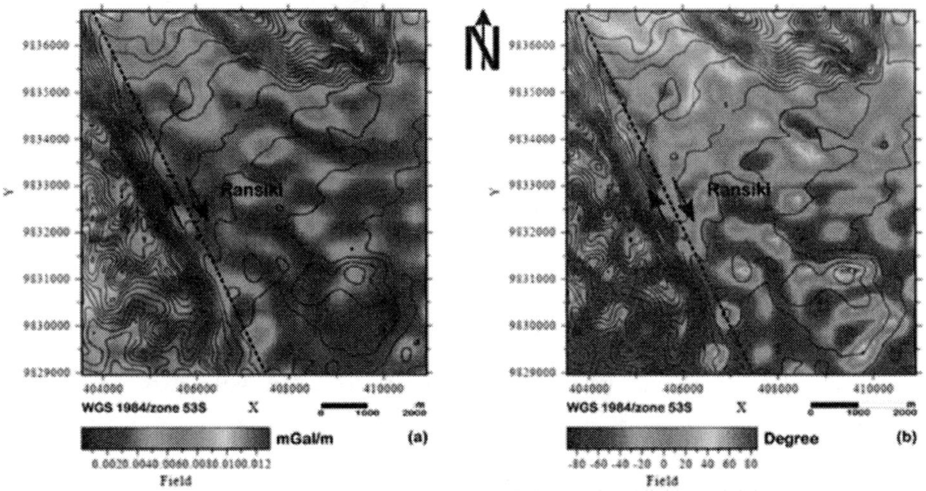

Figure 7. Gradient analysis of Complete Bouguer Anomaly in Ransiki area; (a) The analytical signal, (b) Tilt angle.

4.4. Gravity data inversion and Fault Parameters Interpretation

The purpose of the gravity inversion modeling of global gravity data in the Ransiki Area is to classify subsurface structures associated with the main Ransiki fault line in the field. The inversion process for the complete Bouguer anomaly is performed by optimizing three essential physical parameters, including base, density, and geometry of the block. Base optimization is performed to increase the coefficient of second-order polynomials associated with regional anomalies. Density value and block geometry optimization are carried out to obtain a density distribution model of subsurface rock, which is expected to be capable of describing the Ransiki fault model which is filled by a layer of sediment on the surface. According to the SVD approach, optimization results in an RMS error value of 0.06. Occam density and block optimization generally produce smoother results than SVD with a data RMS error of 0.02 and model RMS error of 0.006. These findings show a very high degree of fitting between measurement and calculation of gravity anomaly data.

The horizontal subsurface model regarding the effect of the inversion of gravity in the Ransiki region is shown in Figure 8a, in the depth of 750 m (Figure 8b), and also the basement (Figure 8c) while the 2.5D model intersecting the Ransiki fault is shown in Figure 8d. The total rock density of the gravity inversion modeling is 2.67 $gr.cm^{-3}$. Figure 8a shows that the study area is dominated by high-density rock (blue color), which is spread evenly over almost the entire region. The high rock density is related to the Arfak volcanic rocks as a basement created during the Upper Eocene to the Middle Miocene [9], [14]. The composition of Arfak volcanic rocks consists of tuff, agglomerate, lava, breccia, andesite, basalt, and gabbro, which are parts of igneous rock with an average density ranging from 2.61 $gr.cm^{-3}$ to 3.03 $gr.cm^{-3}$ [34], [35]. The layer above the basement is thought to be a Befoor formation with a composition of sandstones, mudstones, conglomerates, marl, and volcanic rocks that occurred during the Pliocene [9]. In general, this layer has a rock density that is relatively lower than that of Arfak volcanic rock (light blue color).

The younger layers above the Befoor formation are thought to be alluvial and littoral deposits of quaternary sediments consisting of mud, sand, gravel, peat, and plant matter. Figure 8b shows a 2.5D cross-section profile that intersects the Ransiki fault segment in the South Manokwari district.

According to the model, it can be seen clearly that the alignment of the Ransiki fault segment from the northwest to the south is associated with the distribution of low rock density (in green) between high-density rocks. This zone is thought to be a zone of destruction due to the extension or compression of tectonic plates in the birds head of Papua [1], [14].

Watkinson and Hall in 2017 conducted research on quarternary faults in eastern Indonesia and found that the Ransiki fault zone has a segment length of between 20 – 50 km and a width less than 1000 meters with a class of tectonic activity from maximum to minimum [1]. Moment tensor analysis from the global CMT catalog of large earthquakes in the Ransiki fault zone in 2002 with a magnitude of 7.5 Mw obtained strike and dip angle in the first nodal plane of 60° and 83°, while in the second nodal plane, 329° and 86° [36].

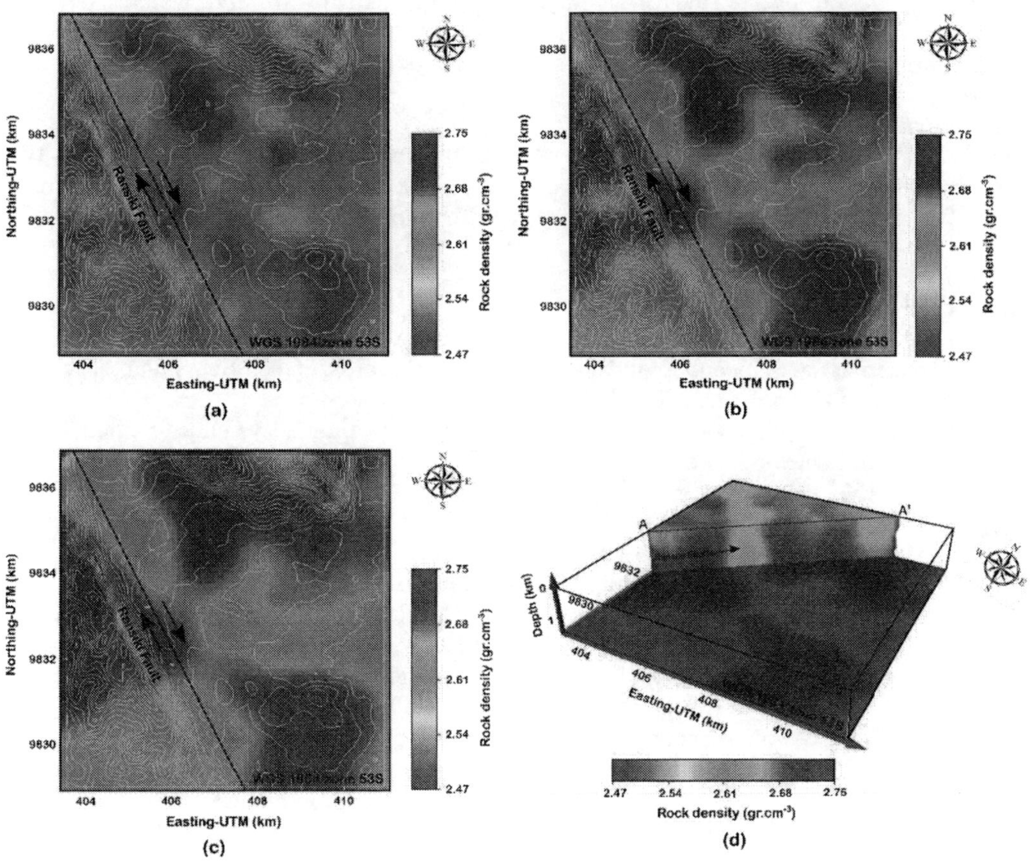

Figure 8. Rock density distrbution according to 3D gravity inversion in Ransiki area (a) horizontal layer at the surface; (b) horizontal layer at the depth of 750 m; (c) horizontal layer at the 1.5 km as the basement; (d) 2.5D cross section perpendicular to the Ransiki fault. Fault wide zone is less than 1000 meters.

5. Conclusion

The use of a high-resolution GGM global gravity field corrected by the SRTM2gravity model to identify the Ransiki fault structure in the South Manokwari Regency, West Papua, Indonesia has shown significant results defining the subsurface layer as a fault line in the study area. Qualitative interpretation through the study of gravity gradients (derivatives) in the form of a vertical, horizontal, analytical signal, and tilt angle gradients of complete Bouguer anomalies indicates an increase in the visibility of the boundaries of anomalous sources of fault systems in the Ransiki region of South Manokwari with northwest to south orientation. The Ransiki fault is correlated with zero values in the vertical gradient and maximum values in the horizontal gradient and analytical signals, whereas the tilt angle provides zero or near-zero degree of fault structures. Quantitative analysis based on gravity inversion models offers a subsurface model of the Ransiki fault segment according to the Arfak volcanic rock with a high rock density as a basement, while a fault line coincides with a low density that is usually correlated with the Befoor formation and Alluvial deposits, which are the extension zone in the birds head of Papua island.

Acknowledgments
We would like to thank Curtin University for providing GGMplus, ERTM2160, SRTM2gravity, and free software extraction for this study. I would also like to thank Mr. Pirtijarvi, who has developed a Fast Fourier Transformation (FFT) Fourpot software to analyze earth gravity Gradient Components, as well as Grablox and Bloxer software for gravity inversion modeling.

Author Contributions
R.L and L.L have formulated concepts and designs for the collection and processing of gravity data. A.N has evaluated the geological of Ransiki area, South Manokwari regency. R.L has also examined the principle of gravity and its gradients, then gravity inversion modeling.

References
[1] I. M. Watkinson and R. Hall, "Fault systems of the eastern Indonesian triple junction: evaluation of Quaternary activity and implications for seismic hazards," Geol. Soc. Lond. Spec. Publ., vol. 441, no. 1, pp. 71–120, 2017, doi: 10.1144/SP441.8.
[2] S. L. Baldwin, P. G. Fitzgerald, and L. E. Webb, "Tectonics of the New Guinea Region," Annu. Rev. Earth Planet. Sci., vol. 40, no. 1, pp. 495–520, May 2012, doi: 10.1146/annurev-earth-040809-152540.
[3] Y. R. Serhalawan and D. S. J. Sianipar, "PEMODELAN MEKANISME SUMBER GEMPA BUMI RANSIKI 2012 BERKEKUATAN MW 6,7," vol. 6, no. 1, p. 10, 2017.
[4] J. Milsom, "Gravity measurements and terrane tectonics in the New Guinea region," J. Southeast Asian Earth Sci., vol. 6, no. 3–4, pp. 319–328, Oct. 1991, doi: 10.1016/0743-9547(91)90077-B.
[5] A. K. Benson and A. R. Floyd, "Application of gravity and magnetic methods to assess geological hazards and natural resource potential in the Mosida Hills, Utah County, Utah," GEOPHYSICS, vol. 65, no. 5, pp. 1514–1526, Sep. 2000, doi: 10.1190/1.1444840.
[6] S. Wada et al., "Continuity of subsurface fault structure revealed by gravity anomaly: the eastern boundary fault zone of the Niigata plain, central Japan," Earth Planets Space, vol. 69, no. 1, p. 15, Jan. 2017, doi: 10.1186/s40623-017-0602-x.
[7] J.-H. Kim, M.-J. Yi, S.-H. Hwang, Y. Song, S.-J. Cho, and J.-H. Synn, "Integrated geophysical surveys for the safety evaluation of a ground subsidence zone in a small city," J. Geophys. Eng., vol. 4, no. 3, pp. 332–347, Sep. 2007, doi: 10.1088/1742-2132/4/3/S12.
[8] Z. Bakhshipour, B. B. K. Huat, S. Ibrahim, A. Asadi, and N. U. Kura, "Application of Geophysical Techniques for 3D Geohazard Mapping to Delineate Cavities and Potential Sinkholes in the Northern Part of Kuala Lumpur, Malaysia," Sci. World J., vol. 2013, pp. 1–11, 2013, doi: 10.1155/2013/629476.
[9] S. Atmawinata, "Peta geologi lembar Ransiki, Irian Jaya : Geological map of the Ransiki sheet, Irian Jaya," 1989.
[10] C. Hirt, S. Claessens, T. Fecher, M. Kuhn, R. Pail, and M. Rexer, "New ultrahigh-resolution picture of Earth's gravity field: NEW PICTURE OF EARTH'S GRAVITY FIELD," Geophys. Res. Lett., vol. 40, no. 16, pp. 4279–4283, Aug. 2013, doi: 10.1002/grl.50838.

[11] C. Hirt, M. Kuhn, S. Claessens, R. Pail, K. Seitz, and T. Gruber, "Study of the Earth's short-scale gravity field using the ERTM2160 gravity model," Comput. Geosci., vol. 73, pp. 71–80, Dec. 2014, doi: 10.1016/j.cageo.2014.09.001.
[12] C. Hirt, M. Yang, M. Kuhn, B. Bucha, A. Kurzmann, and R. Pail, "SRTM2gravity: An Ultrahigh Resolution Global Model of Gravimetric Terrain Corrections," Geophys. Res. Lett., vol. 46, no. 9, pp. 4618–4627, May 2019, doi: 10.1029/2019GL082521.
[13] Y. L. Ekinci and E. Yiğitbaş, "Interpretation of gravity anomalies to delineate some structural features of Biga and Gelibolu peninsulas, and their surroundings (north-west Turkey)," Geodin. Acta, vol. 27, no. 4, pp. 300–319, Oct. 2015, doi: 10.1080/09853111.2015.1046354.
[14] D. P. Gold, P. M. Burgess, and M. K. BouDagher-Fadel, "Carbonate drowning successions of the Bird's Head, Indonesia," Facies, vol. 63, no. 4, p. 25, Oct. 2017, doi: 10.1007/s10347-017-0506-z.
[15] J. Milsom et al., "The Manokwari Trough and the western end of the New Guinea Trench," Tectonics, vol. 11, no. 1, pp. 145–153, 1992, doi: 10.1029/91TC01257.
[16] T. R. Charlton, "The Pliocene-Recent Anticlockwise Rotation of the Bird's Head, the Opening of the Aru Trough – Cendrawasih Bay Sphenochasm, and the Closure of the Banda Double Arc," 2010, Accessed: Aug. 12, 2020. [Online]. Available: http://archives.datapages.com/data/ipa_pdf/081/081001/pdfs/IPA10-G-008.htm
[17] W. Jacoby and P. L. Smilde, Gravity interpretation: fundamentals and application of gravity inversion and geological interpretation ; with CD-ROM. Berlin: Springer, 2009.
[18] J. Nouraliee, S. Porkhial, M. Mohammadzadeh Moghaddam, S. Mirzaei, D. Ebrahimi, and M. Rahmani, "Investigation of density contrasts and geologic structures of hot springs in the Markazi Province of Iran using the gravity method," Russ. Geol. Geophys., vol. 56, pp. 1791–1800, Dec. 2015, doi: 10.1016/j.rgg.2015.11.011.
[19] R. Lewerissa, S. Sismanto, A. Setiawan, S. Pramumijoyo, and L. Lapono, "Integration of gravity and magnetic inversion for geothermal system evaluation in Suli and Tulehu, Ambon, eastern Indonesia," Arab. J. Geosci., vol. 13, no. 15, p. 726, Aug. 2020, doi: 10.1007/s12517-020-05735-7.
[20] A. Eshaghzadeh, A. Dehghanpour, and R. A. Kalantari, "Application of the tilt angle of the balanced total horizontal derivative filter for the interpretation of potential field data," BGTA, vol. 59, no. 2, pp. 161–178, 2018, doi: 10.4430/bgta0233.
[21] I. M. Ibraheem, M. Haggag, and B. Tezkan, "Edge Detectors as Structural Imaging Tools Using Aeromagnetic Data: A Case Study of Sohag Area, Egypt," Geosciences, vol. 9, no. 5, p. 211, May 2019, doi: 10.3390/geosciences9050211.
[22] M. Hicheri, B. Ramdhane, S. Yahyaoui, and T. Gonenc, "New Insights from Gravity Data on the Geodynamic Evolution of Northern African Passive Margin, Case Study of the Tajerouine Area (Northern Tunisian Atlas)," J. Geol. Geophys., vol. 8, no. 1, pp. 1–9, Jun. 2019, doi: 10.4172/2381-8719.1000454.
[23] J. Abderbi, D. Khattach, and J. Kenafi, "Multiscale analysis of the geophysical lineaments of the High Plateaus (Eastern Morocco): structural implications," p. 10, 2017.
[24] S. Hsu, J. Sibuet, and C. Shyu, "High-resolution detection of geologic boundaries from potential-field anomalies: An enhanced analytic signal technique," GEOPHYSICS, vol. 61, no. 2, pp. 373–386, Mar. 1996, doi: 10.1190/1.1443966.
[25] R. O. Hansen, R. S. Pawlowski, and X. Wang, "Joint Use of Analytic Signal And Amplitude of Horizontal Gradient Maxima For Three-dimensional Gravity Data Interpretation," presented at the 1987 SEG Annual Meeting, 1987. Accessed: Aug. 13, 2020. [Online]. Available: https://www.onepetro.org/conference-paper/SEG-1987-0100
[26] I. Marson and E. E. Klingele, "Advantages of using the vertical gradient of gravity for 3-D interpretation," GEOPHYSICS, vol. 58, no. 11, pp. 1588–1595, Nov. 1993, doi: 10.1190/1.1443374.

[27]H. Saibi, J. Nishijima, S. Ehara, and E. Aboud, "Integrated gradient interpretation techniques for 2D and 3D gravity data interpretation," Earth Planets Space, vol. 58, no. 7, pp. 815–821, Jul. 2006, doi: 10.1186/BF03351986.

[28]R. J. Blakely, "Potential theory in gravity and magnetic applications.," Potential Theory Gravity Magn. Appl. Blakely R J Camb. Univ. Press Camb. UK 1995 XIX 441 P ISBN 0-521-41508-X, 1995, Accessed: Aug. 13, 2020. [Online]. Available: http://adsabs.harvard.edu/abs/1995ptgm.book.....B

[29]Z. Chen, L. Mou, and X. Meng, "The horizontal boundary and top depth estimates of buried source using gravity data and their applications," J. Appl. Geophys., vol. 124, pp. 62–72, Jan. 2016, doi: 10.1016/j.jappgeo.2015.11.003.

[30]K. L. Mickus and J. H. Hinojosa, "The complete gravity gradient tensor derived from the vertical component of gravity: a Fourier transform technique," J. Appl. Geophys., vol. 46, no. 3, pp. 159–174, Mar. 2001, doi: 10.1016/S0926-9851(01)00031-3.

[31]H. Grandis and D. Dahrin, "Full Tensor Gradient of Simulated Gravity Data for Prospect Scale Delineation," J. Math. Fundam. Sci., vol. 46, no. 2, Art. no. 2, Jul. 2014, doi: 10.5614/j.math.fund.sci.2014.46.2.1.

[32]G. W. Hohmann and A. P. Raiche, "8. Inversion of Controlled-Source Electromagnetic Data," in Electromagnetic Methods in Applied Geophysics: Volume 1, Theory, 0 vols., Society of Exploration Geophysicists, 1988, pp. 468–504. doi: 10.1190/1.9781560802631.ch8.

[33]D. L. B. Jupp and K. Vozoff, "Stable Iterative Methods for the Inversion of Geophysical Data," Geophys. J. Int., vol. 42, no. 3, pp. 957–976, Sep. 1975, doi: 10.1111/j.1365-246X.1975.tb06461.x.

[34]W. M. Telford, L. P. Geldart, and R. E. Sheriff, Applied Geophysics. Cambridge University Press, 1990.

[35]J. M. Reynolds, An introduction to applied and environmental geophysics. Chichester ; New York: John Wiley, 1997.

[36]Global CMT Project, "Moment Tensor for October 10, 2002 MW 7.5 (GCMT) IRIAN JAYA REGION, INDONESIA," 2012, doi: 10.17611/DP/923555.

IOP Conf. Series: Earth and Environmental Science **873** (2021) 012049 doi:10.1088/1755-1315/873/1/012049

Gravity observation of field camp geophysics in Karangsambung using Data Acquisition 2005-2019

E J Wahyudi[1], A Laesanpura[1] and D Sukmayadi[1]
[1]Institut Teknologi Bandung

Abstract. The study of field camp geophysics in Karangsambung has been done since 1996 until 2019 by geophysical engineering ITB. During the field activities, students was assigned with several data acquisition using various geophysical methods. One of the most common method to conducted alongside with surface geological mapping is gravity. Compilation of gravity data during the activities will be presented in this work. There are two categories of data compilation during 24 years: data compilation 1996-2004, and 2005-2019. The observation conducted using relative gravimeter with data distribution already cover geological surface map in the study area (Luk-Ulo Melange Complex, Karangsambung Formation, Totogan Formation, and Diabas Intrusion). The pattern of gravity observation shows correlated with topographic variation. Range gravity observation from this study is about 62 mGal..

1. Introduction

Field camp activities becomes an important for geoscience students, and it has been conducted worldwide in many universities. Several field camp course reports [1-3] describe many advantages for students because the souls of exploration with curiousity started to be waken-up. More experience for students in the field will help their improvement (technical skills and personal confidence) as exploration expert in the future. The activities in the field will set up the opprtunity for students to show good friendship, team work, and communication skill.

Geophysical engineering of ITB have been conducted field camp course in Karangsambung since 1996. The implementation of geophysical methods for more than two decade in Karangsambung along side with surface geological mapping. In this work, we would like to show our summary of gravity data in the study area. The map compilation will shows coverage of all data and can be use to provide sugestion for next survey in field camp course. This study also part of the work of compilation data that started initially in 2015 and we already published for geomagnetic data only [4]. In this work, the summary of the activites was completed during the covid-19 pandemic period and the merge of gravity data from 24 years of annual field camp course also supported with gravity observation that can be link-up to the refference value of gravity from absolute gravimeter that previously observed in ITB campus.

2. Gravity Observation

Gravity data are collected using relative gravimeter (LaCoste & Romberg (LCR) and Scintrex CG-5). In this work, a compilation data will be categorize into two groups as follow: (a) data 1996-2004 and (b) data 2005-2019. The main data for this work is only group (b) because the data in group (a) insufficient and lack some of the essential information in mapping. We cannot find any reports from the data in 1996 to 1998, so the summary of field camp activities started in 1999 is shown in the table 1. From table 1 we can see the surveys conducted for each year, most of them in June. Only in 2015 the Scintrex CG-5 used in the data acquisition for undergraduate programe.

Table 1. Summary of Gravity Data Acquisition

Date	Month	Year	LCR G-504	LCR G-1158	SC CG5	Loop	#obsv.	tide (μGal)
1st	Jun	1999	v			1	47	(-73 to +106)
11th	Jun	2000	v			1	38	(-58 to +110)
6th	Jun	2001	v			1	47	(-45 to +107)
5th	Jun	2002		v		1	34	(-60 to +150)
5th	Jun	2003	v			1	35	(-34 to +75)
5th	Jun	2004		v		1	34	(-60 to +150)
14th-16th	Jun	2005	v			3	79	(-33 to +91)
18th-20th	Jun	2006	v	v		5	141	(-47 to +105)
19th-21st	Jun	2007	v	v		6	142	(-49 to +92)
10th-12nd	Jul	2008		v		3	97	(-28 to +88)
14th-17th	Jun	2009		v		4	157	(-40 to +57)
21st-25th	May	2010		v		5	148	(-51 to +124)
30th-3rd	May-Jun	2011		v		5	135	(-68 to +111)
20th-24th	Jun	2012	v	v		10	151	(-101 to +141)
4th-8th	Jun	2013	v	v		10	175	(-11 to +125)
13th-17th	May	2014	v	v		9	236	(-78 to +148)
7th-10th	May	2015	v		v	8	149	(-64 to +150)
25th-28th	Apr	2016	v	v		8	257	(-56 to +113)
15th-18th	Jun	2017	v	v		8	281	(-49 to +87)
16th-20th	Apr	2018	v	v		8	176	(-89 to +170)
31th-3rd	Jul-Aug	2019	v	v		8	258	(-107 to +193)

For this work, we have 106 of total loops of relative gravimeter surveys and 2817 of total gravity observations (figure 1). Some statistics of gravity data acquisition from 2005-2019 surveys shows that 2592 observations conducted with LCR G1058 (1478 observations), LCR G504 (1061 observations), and Scintrex CG5 (53 observations). The variation for tide correction for each year can be seen in figure 2. Brief evaluation of data quality from students observation using relative gravimeters can be seen in figure 3 and figure 4. In figure 3 we can see the drift correction for each loop assumed as linear (with zero at the start and (±) maximum stretch/shrink of the spring gravimeter sensor at the end of loop). In figure 4, the coefficient correlation value for each loop shows the relation of gravity data and elevation data. From figure 3 and figure 4, we have 54 loops in the range of maximum drift (-100 to 100 μGal) and 83 loops in the range of negative and strong coefficient correlation (-0.96 to -1.00) between gravity and elevation data.

Map of gravity data distribution in the study area shown as dot-plot in figure 5. Local gravity or relative gravity value with respect to the BASE station show in figure 5A and elevation data from hand-held altimeters shown in figure 5B. Qualitatively, the maps in figure 5 shows negative correlation (with quantitative value of coefficient correlation already mention in figure 4).

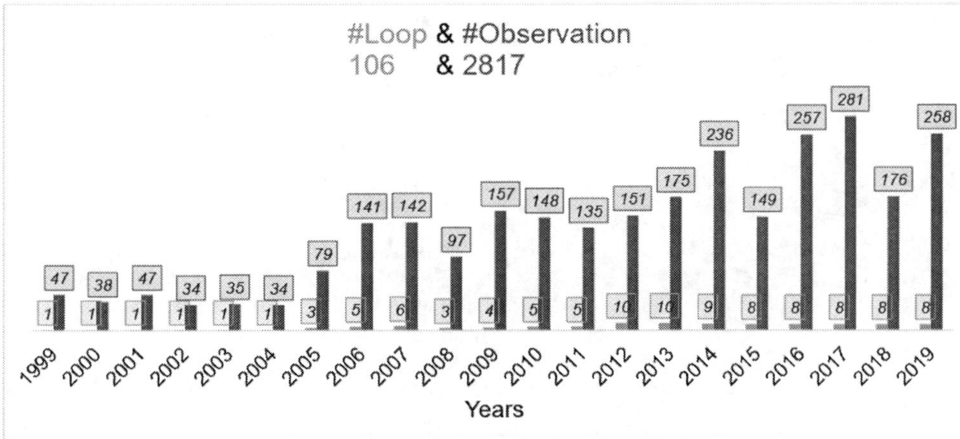

Figure 1 Bar-plot of loop and observation number for each year.

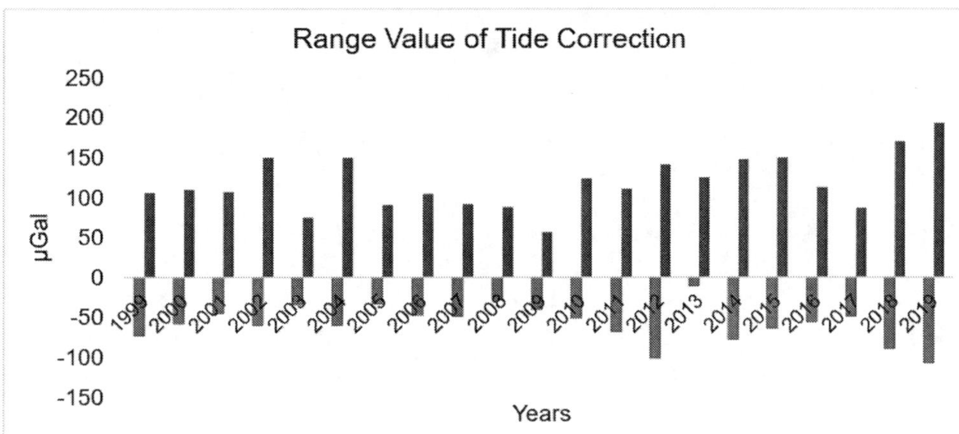

Figure 2 Bar-plot of minimum and maximum value of tide correction for each year.

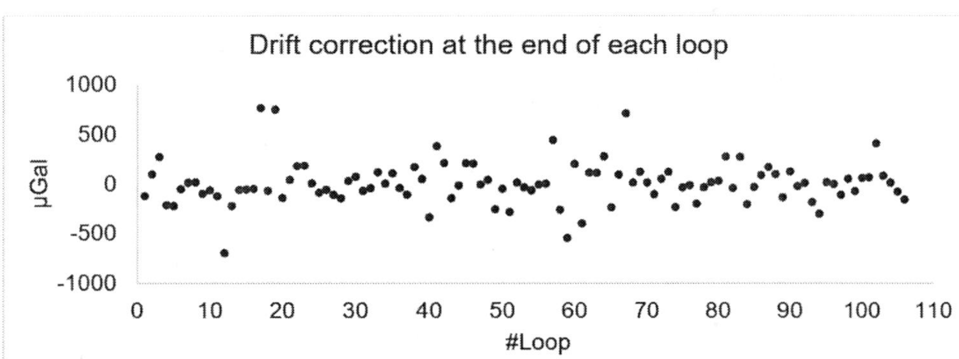

Figure 3 Dot-plot of drift correction (maximum stretch/shrink) for each loop.

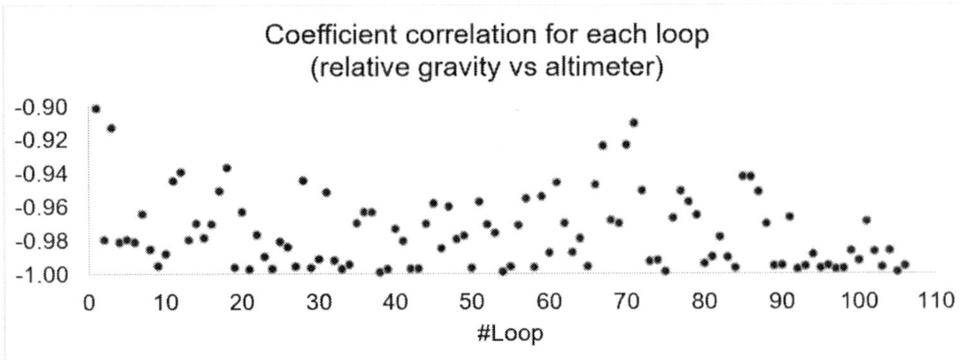

Figure 4 Dot-plot of coefficient correlation value (between gravity and elevation data) for each loop.

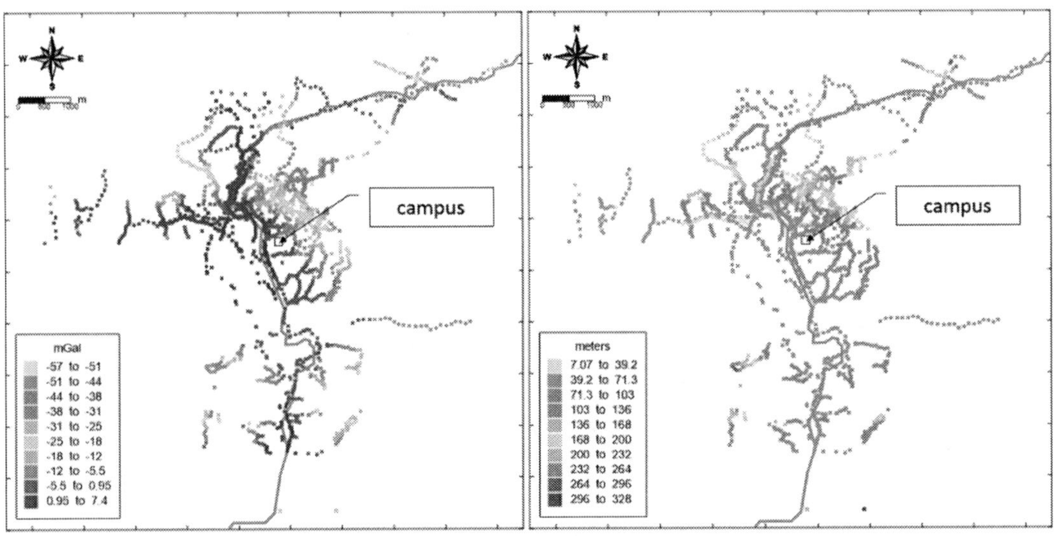

A B

Figure 5 Map of (A) gravity and (B) elevation data distribution in the study area.

3. Gravity Observation Map

We already have the data of absolute gravimeter in Bandung (ITB) that obtained from collaborated study with JICA in 2013 and 2014. The documentation of gravity observation using absolute gravimeter can be seein in figure 6. In September 2019, we conducted the gravity survey from Bandung-Karangsambung to get the value of gravity observation. The observation activity was planed to connect ITB stations (TG and GD) with BASE station in Karangsambung using relative gravimeter Scintrex CG-5. As ilustrated in figure 7, we set up the trasportation from Bandung-Banjar-Gombong-Karangsambung. In figure 7, we also can see the quality of gravimeter leveling using dot plot of digital data of tilt-X and tilt-Y Scintrex CG-5. Most of the data observation in the activity September 2019 was plotted inside the tolerance area (inside the red circle in figure 7). During mobilization in September 2019, we have 11 gravity observation in Bandung, Malangbong, Banjar, Ciamis, Cimanggu, Kebasen, Kebumen, and Karangsambung (table 2).

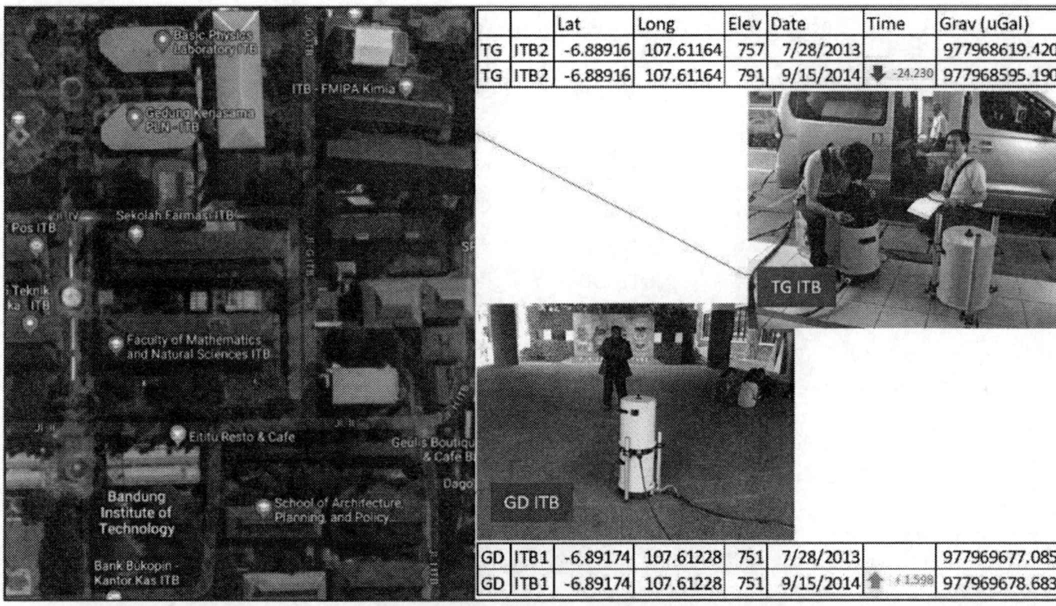

Figure 6 Documentation of activity in 2014 using absolute gravimeter data in ITB.

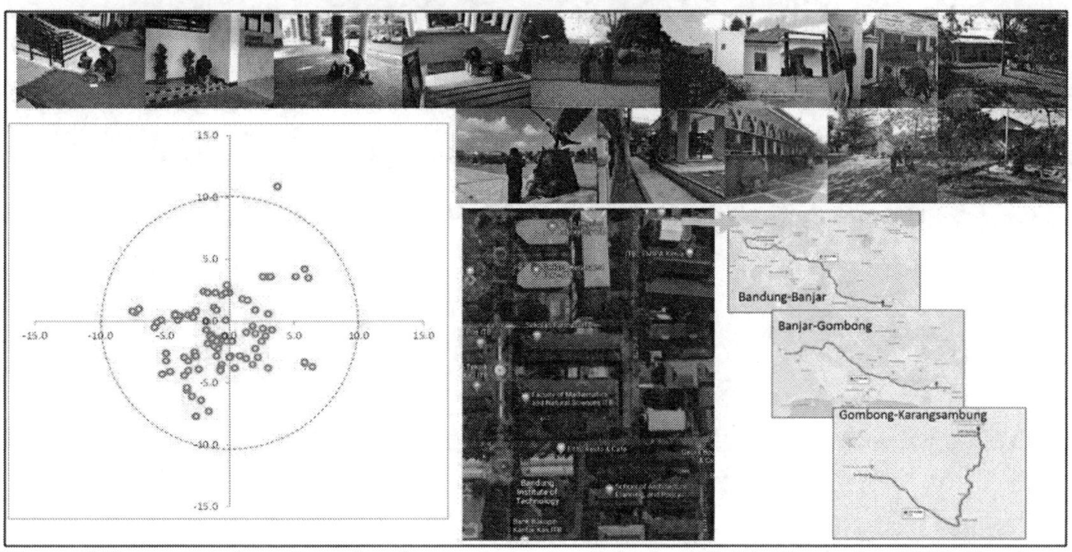

Figure 7 Documentation of gravity observation using Scintrex CG-5 in September 2019 (Bandung-Karangsambung).

In Karangsambung (our study area), we use the value from ID station of LKS (table 2) as our BASE with gravity observation 978201385.048 μGal. The interpolation contour for gravity observation map shown in figure 8. The gravity observation map overlayed with outcorp locations in Karangsambung area (relatively similiar as coverage in previous study [4]). Range gravity observation from the map is about 62 mGal (978144.303 to 978208.783 mGal).

.

Table 2 Gravity Observation from September 2019 activity

ID STS	#Obsv.	StDev. (μGal)	G obs (μGal)	Location
MBJR	12	87.950	978,201,481.760	Banjar
MCMS	24	103.436	978,144,228.506	Ciamis
MCMG	24	104.712	978,144,365.123	Cimanggu
BSCB	42	104.302	977,968,675.235	ITB
TM	18	123.078	977,968,481.796	ITB
BTSK	12	104.826	978,201,062.633	Karsam
LKS	24	102.564	978,201,385.048	Karsam
TBKS	12	104.895	978,201,188.185	Karsam
KBSN	12	89.643	978,179,605.977	Kebasen
KKEB	96	158.300	978,217,162.603	Kebumen
MLBG	24	105.355	978,020,603.567	Malangbong

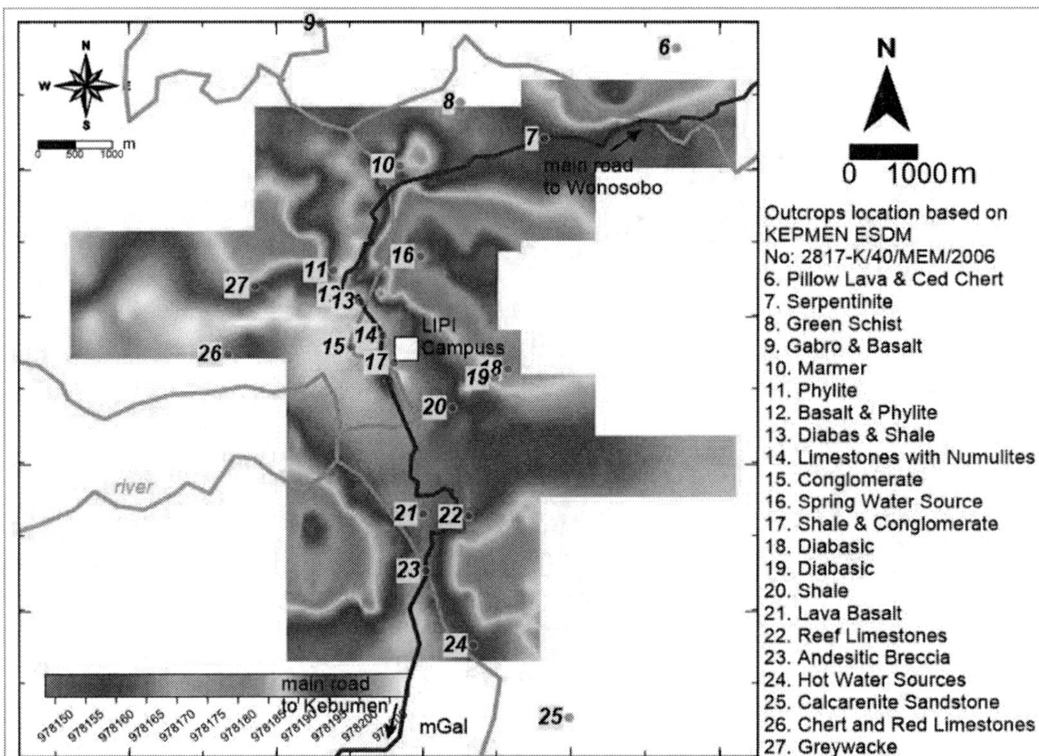

Figure 8 Gravity observation map from data 2005-2019.

4. Conclusion

The pattern of gravity observation shows negative correlation with topographic data variation. Low gravity observation can be seen in the picture correlate to high topographic area. Data distribution from field camp surveys already cover some geological surface map (Diabas Intrusion, Totogan Fm., Karangsambung Fm., and Luk-Ulo Melange Complex). Range gravity observation from this study is about 62 mGal.

IOP Conf. Series: Earth and Environmental Science **873** (2021) 012049 doi:10.1088/1755-1315/873/1/012049

5. References

[1] Stewart R R, Khan S, Hall S, Liner C and Wong J 2010 Geophysics field education: better learning by doing *The Leading Edge* **29** 546
[2] CSM 2016 Geophysical Investigation of the Geothermal System in Pagosa Springs, *CO Colorado School of Mines 2016 Geophysical Field Camp*
[3] McQueen KG, Taylor G, Brown M C and Mayer W 1990 Integration of Teaching and Research in a regional geological mapping project Journal of Geological Education **38** 8
[4] Wahyudi E J and Sukmayadi D 2019 Geomagnetic Map of Field Camp Geophysics in Karangsambung using Data Acquisition 2005-2017 *IOP Conf. Series: Journal of Physics: Conf. Series* **1204** 012050

IOP Conf. Series: Earth and Environmental Science **873** (2021) 012050 doi:10.1088/1755-1315/873/1/012050

Kutei Basin oil bearing sand reservoirs mapping using integrated seismic methods

I M Easwara[1], D S Ambarsari[1], S Sukmono[1], S Winardhi[1], E Septama[2], P D Wardaya[2], V I Rossa[2], B S Murdianto[3], R Raguwanti[2]

[1]Geophysical Engineering Program Study, Bandung Institute of Technology, Jl. Ganesha 10 Bandung, Indonesia 40132
[2]PT Pertamina (Persero) Upstream Reseacrh & Technology Innovation, SopoDel Office Tower 50th Floor, Jl. Mega Kuningan Barat III, Jakarta, Indonesia 12950
[3]Pertamina Hulu Kalimantan Timur, CIBIS NINE Tower 9th Floor, Jl. TB Simatupang, Jakarta Selatan, Indonesia 12560

E-mail: sukmono@office.itb.ac.id

Abstract. The Lower Kutei Basin which contains several giant oil and gas fields is located on the East Kalimantan, Indonesia. This paper discusses the identification and mapping of oil-filled reservoirs and their depositional facies by integrating seismic stratigraphy, attributes, and AI (Acoustic Impedance) inversion methods. The log data cross-plots show that AI can be used to distinguish oil-sands from wet sands and shale, and to derive the total porosity of the sands. However, AI and amplitude values are greatly affected by the oil, porosity and tuning effects, hence they cannot be used to identify the facies containing the oil-bearing sands. Therefore, to map the facies containing the oil-filled sands, the AI map is combined with the variance and sweetness maps. It can be seen clearly from the variance and sweetness maps that the oil-sands suggested by the AI map are contained in a narrow and elongate meander-like geometry which is typical of channel facies. The variance and sweetness maps suggest that there are two channels in the study area. To determine which channel is thicker, spectral decomposition RGB map was made. The result suggests that the right channel is more prospective as it associates with thicker sand deposits. The combination of variance, sweetness and RGB maps strongly indicate that the channels in the study area are in upper-slope environment, and the thicker oil-sands are located in the eastward of the study area.

1. Introduction

The integration of seismic methods can become a powerful tool in hydrocarbon reservoirs identification and mapping. However, the approaches can be effective if the selected seismic attributes to be integrated match with the geological characteristics of the reservoirs. For examples, the study in [1] shows that the combination of amplitude and variance attributes can be used to map channel deposits in Melandong, NW Java basin area. On other studies for the same facies but located in different geological settings, the neural network-based attributes and the integration of amplitude and complex seismic attributes analysis were used [2], [3]. Meanwhile, for the Talang Akar channel deposits in south Sumatra basin, Indonesia, the study in [4] strongly indicates shows that seismic anisotropy is important parameters in seismic analysis. For exploration area with no well data, there are also several possibilities on seismic attributes integration to identify rocks facies and physical properties as discussed in [5]. As a common guidance, study in [6] discussed the general seismic interpretation workflow for interpreting various reservoirs in Indonesia.

The Kutei Basin which contains several giant oil and gas fields is located on the East Kalimantan, Indonesia (Figure 1). The prospective sand reservoirs are Eocene to present in age, and mostly deposited in upper shelf channels to basin floor fan facies. The dominant hydrocarbon trap in the basin is structural traps. The stratigraphic traps are harder to be identified as they often associate with thin layers. The study in [1]) shows that seismic inversion can be used to identify thick gas-filled reservoirs in the Sadewa field, Kutei basin. This paper discusses the identification and mapping of oil-filled reservoirs and their depositional facies by integrating seismic stratigraphy, attributes, and AI (Acoustic Impedance) inversion methods.

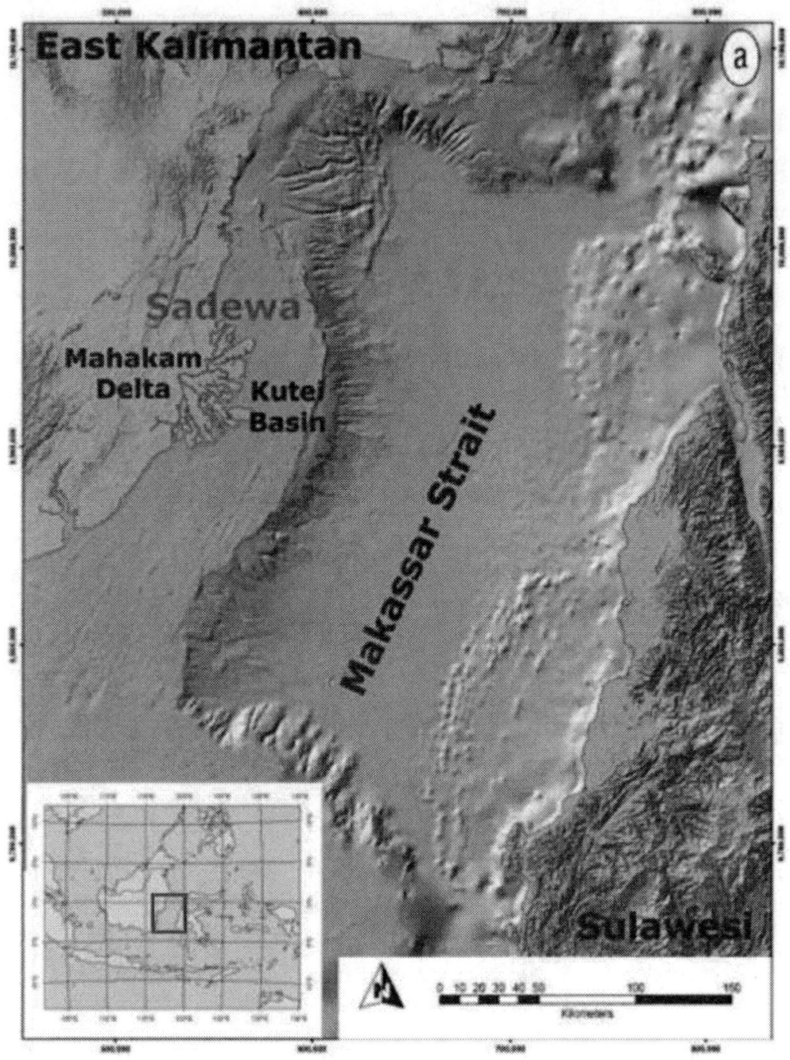

Figure 1. Location of the Kutei basin [1]. The study in [1] used seismic method to identify gas-sands in Sadewa field. This study focuses on the application of seismic methods to identify and map oil-sands in Kutei basin.

2. Data and Methods

In this study, the main data used are the 3D post stack time migration seismic and 3 wells data. Figure 2 displays a sample of seismic section showing the studied interval which lied between light green and white horizons. The Two-Way Time (TWT) map of the studied horizon and the wells' locations are shown in Figure 2' inset map. In the studied interval, the 3 wells discovered oil which associate with typical high seismic amplitude anomalies. From the log data, cross-plots of AI versus gamma-ray, oil saturation and total porosity can be made. The results show that AI can be used to distinguish oil-sands from wet sands and shale, and to derive the total porosity of the sands (Figure 3). It also explains why the porous oil-sands are reflected as high amplitude anomalies in the seismic sections.

The AI inversion was done by using the 3 wells. However, the weaknesses of using AI and amplitude attributes only to map the oil-sands are that they are not sensitive to the facies as their values are greatly affected by the oil saturation, porosity, and tuning effects. Therefore, to map the facies containing the oil-sands, AI is combined with the variance and sweetness attributes. The variance attribute measures of seismic traces similarity; therefore, it is effective to identify channel deposits. The sweetness attribute is the ratio of amplitude envelope maps to the frequency; thus, it will amplify the high-amplitude anomalies associated with oil-sands. Figure 4 shows the AI, variance, and sweetness maps of the studied interval. The variance map clearly shows the presences of narrow and elongate meander-like geometry features which are typical of channel facies. However, both AI and sweetness maps indicate that the presumably oil-filled sands reflected by low-AI and high-sweetness areas are spotty and they are very-likely associate with point-bar facies of the channels.

The variance and sweetness maps suggest that there are two channels in the study area. To determine which sand-filled channel is thicker, spectral decomposition RGB map was made. The result is shown in Figure 5 and suggesting that the right channel is more prospective as it associates with thicker sand deposits. The combination of variance, sweetness and RGB maps also strongly indicate that the channels in the study area are in upper-slope environment. Therefore, it can be expected that thicker oil-sands were deposited in the eastward of the study area.

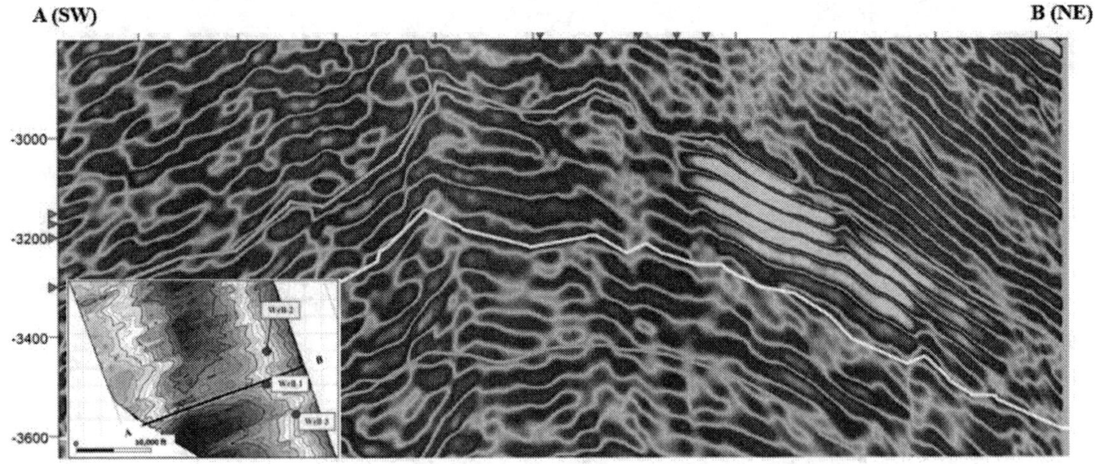

Figure 2. A sample of seismic section showing typical high seismic amplitude anomalies of the oil-sand reservoirs. The studied interval lied light green and white horizons. Inset map shows the location of the section, wells and TWT contours of the light green horizon. Blueish and reddish colours in the TWT map related respectively to the shallower and deeper contours.

IOP Conf. Series: Earth and Environmental Science **873** (2021) 012050 doi:10.1088/1755-1315/873/1/012050

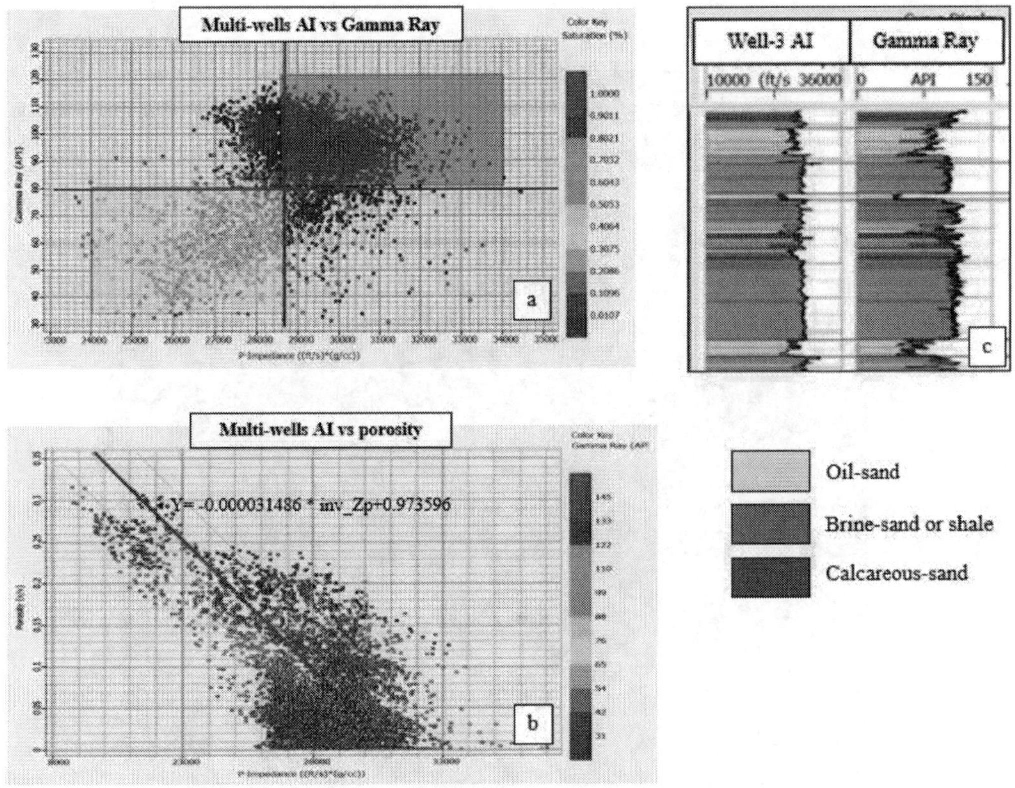

Figure 3. Cross-plots of AI versus (a) Gamma-ray, and (b) Total porosity from the studied interval in the three wells (Depths around 11000 – 13000 ft). Figure c shows a sample of AI and Gamma-ray log in Well-3

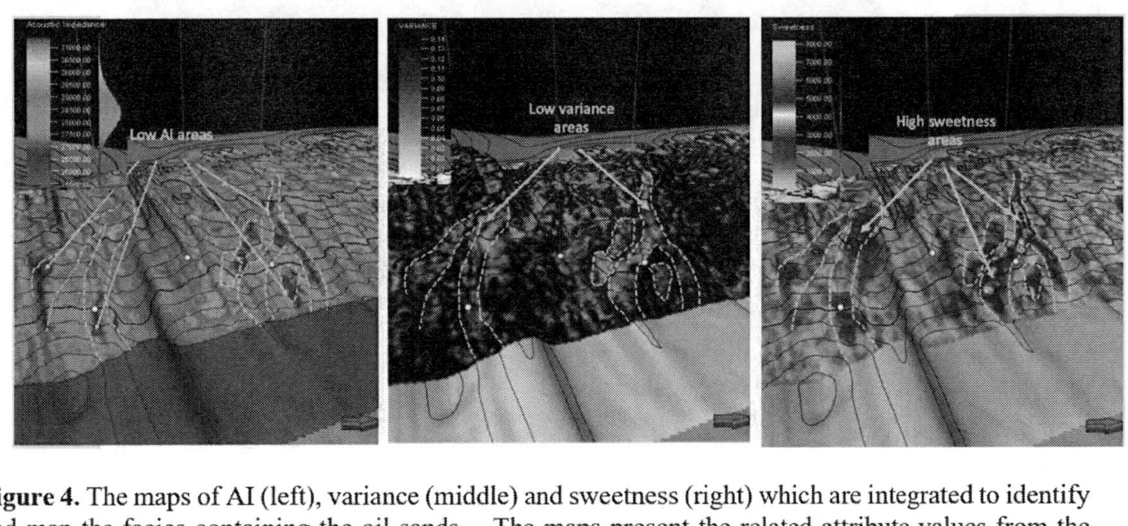

Figure 4. The maps of AI (left), variance (middle) and sweetness (right) which are integrated to identify and map the facies containing the oil-sands. The maps present the related attribute values from the studied interval (light green to white horizons) window. The integrated maps shows that the presumably oil-filled sands reflected by low-AI and high-sweetness areas are spotty and they are very-likely associate with point-bar facies of the channels.

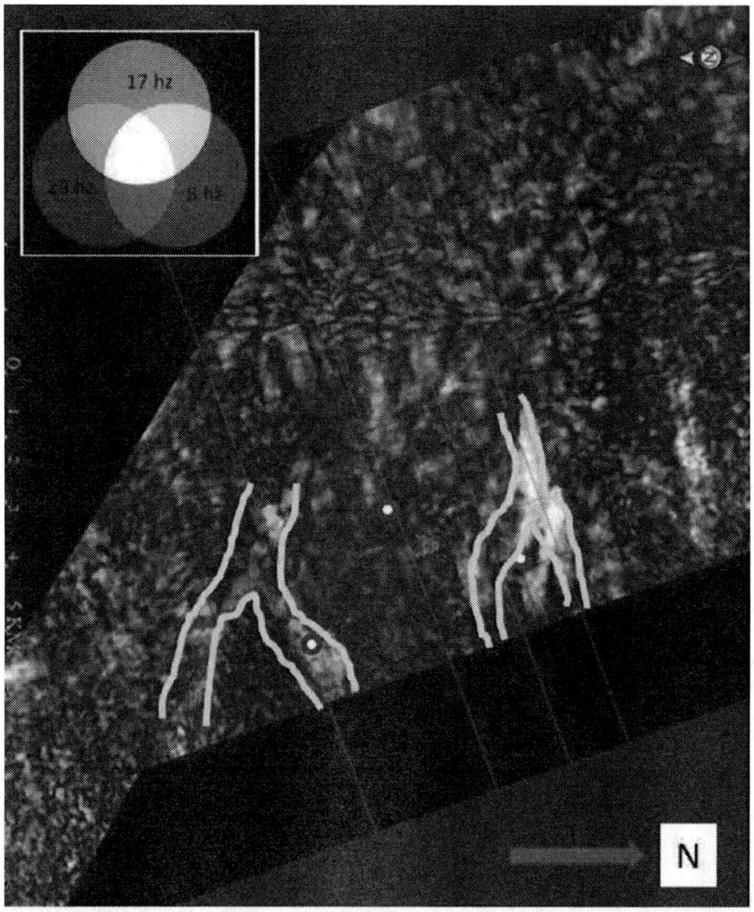

Figure 5. The spectral decomposition RGB map showing that the sand in the right channel is relatively thicker than the sands in the left channel. The maps present the related attribute values from the studied interval (light green to white horizons) window.

3. Conclusions and Recommendations

Previous studies on the application of seismic methods for hydrocarbon identification in the Kutei basin were mostly focused on gas-sands cases. This paper shows that by using the appropriate attributes, the identification of oil-sands reservoirs and their facies are possible. In the studied area, the oil-sands were deposited in the upper slope channels. The AI and sweetness attributes can be used to identify and map the high amplitude anomalies associated with oil-filled sands. The sweetness and spectral decomposition are applicable for mapping the related facies and identifying relative thickness of the sands. The integrated attributes maps suggest that thicker oil-sands were deposited in the eastward of the study area.

References

[1] Thompson P, Hartman J J, Anandito M A A, Kumar D, Magill J, Noguchi K and Gunawan B K 2009 *The Leading Edge* **26** 312
[2] Sukmono S, Samodra A, Sardjito and Waluyo W 2006 *The Leading Edge* **25** 532
[3] Sukmono S 2007 *The Leading Edge* **26** 126

[4] Alamsyah M N, Sukmono S, Marwosuwito S, Sutjiningsih, W and Marpaung L 2008 *The Leading Edge* **27** 260
[5] Sari, V M, Sukmono S, Sanny, T A and Sapiie, B 2018 *First Break* **36** 61
[6] Sukmono and Ambarsari 2019 *Practical seismic interpretation for petroleum exploration* (Bandung: ITB Press)

IOP Conf. Series: Earth and Environmental Science **873** (2021) 012051 doi:10.1088/1755-1315/873/1/012051

Seismic reservoir characterization of gas-sand deposits in the Kutei basin, Indonesia

M Iqbal[1], D S Ambarsari[1], S Sukmono[1], W Triyoso[1], T A Sanny[1], E Septama[2], P D Wardaya[2], R R Pratama[2], B S Murdianto[3], R Raguwanti[2]

[1]Geophysical Engineering Program Study, Bandung Institute of Technology, Jl. Ganesha 10 Bandung, Indonesia 40132
[2]PT Pertamina (Persero) Upstream Reseacrh & Technology Innovation, SopoDel Office Tower 50th Floor, Jl. Mega Kuningan Barat III, Jakarta, Indonesia 12950
[3]Pertamina Hulu Kalimantan Timur, CIBIS NINE Tower 9th Floor, Jl. TB Simatupang, Jakarta Selatan, Indonesia 12560

Abstract. Kutei Basin has the second largest hydrocarbon reserve in Indonesia. In addition to the Miocene inversion related structural traps, slope-fan and channel stratigraphic traps are also important traps in this basin. To guide stratigraphic traps explorations in the basin, the seismic stratigraphy, attributes, and AI inversion methods are integrated to identify and map the reservoir seismic facies, porosity, and pore-fluid. Well data indicates that the studied reservoirs are filled by gas. Seismic data shows that there are two main gas-sand reservoirs corresponding to strong amplitude anomaly. Seismic stratigraphy analysis, guided by seismic attributes, shows that these gas-sand reservoirs were deposited in the channel and local fan facies. The AI inversion is applied to identify and map the porosity and pore-fluid of these two sand reservoirs. Future well locations are identified by integrating the facies, porosity, and pore-fluid maps.

1. Introduction

Kutai Basin, which lied in East Kalimantan – Indonesia (Figure 1), is the country's second largest hydrocarbon basin with oil reserves of 2.47 MMBO and 28.1 TCF of gas [1]. The basin has been dominated by deltaic deposition during the Neogene, but carbonate buildups are also common [2]. During the late Pleistocene, the channel-levee and incised valleys/canyons were developed on slopes, while downslope of the slope the strata are dominated by mass-transport complexes [3]. The study in [3] shows that seismic inversion can be used to identify porous and gas-filled sand reservoirs but only for thick reservoirs close to or above tuning thickness, while for the thin reservoirs, seismic amplitude is more reliable for seismic reservoir characterization.

The most instrumental hydrocarbon traps in the basin are structural traps located mainly in the south-western part of the study area and is related to the Miocene inversion (Figure 1), while the stratigraphic traps in the north-eastern part of the study area (Figure 1) are more difficult to be identified as the potential reservoirs are often associate with thin layers. This paper discusses the integration of seismic stratigraphy, attributes, and AI inversion methods to characterize the gas-sand reservoirs which stratigraphically trapped in the north-eastern part of the study area. The main reservoir parameters to be characterized are depositional facies, porosity, and pore-fluid. The results can be used to guide the exploration of stratigraphic traps in the study area.

The selection of appropriate seismic methods is the main key for successful hydrocarbon reservoirs characterization using integrated seismic data. The study in [4] discussed the application of seismic attributes to map channel deposits in Melandong, NW Java basin area. The study in [5] proposed the application of neural-network based attributes analysis to map the fluvio-deltaic facies for supporting water-flood program. Meanwhile, the integration of amplitude and complex seismic attributes were used to assist exploration well program in the Talang Akar channel deposits in SW Betara field, Sumatra [6]. The study in [7] emphasizes the importance of seismic anisotropy analysis to guide the channel facies characterization in south Sumatra basin, Indonesia. For exploration area with no well data, there are also several possibilities on seismic attributes integration to identify rocks facies and physical properties as discussed in [5]. As a common guidance, study in [6] discussed the general seismic interpretation workflow for interpreting various reservoirs in Indonesia.

2. Data and Methods

In this study, the main data used are the 3D post stack time migration seismic and 5 wells data as shown in Figure 1b. In the studied interval, the 5 wells discovered gas only, while oil is discovered in the deeper sand intervals. The combination of seismic stratigraphy, RMS amplitude and variance attributes is used to deduct the depositional facies of the gas-sand reservoirs. Typical seismic expression of the gas-sand reservoirs in the study area is shown in Figure 2. The shallower reservoir associates with discontinuous reflectors and mounded external shape, while the deeper one has continuous reflectors and wedge shape. Figure 3 displays the cross-plot between well's AI and total porosity. The RMS amplitude maps in Figure 4 shows that both reservoirs are reflected by high amplitude anomalies, but the shallower reservoir has a wider and lobate geometry, in contrast with the narrower and elongate geometry of the deeper reservoir. The combination of typical seismic stratigraphy expression and RMS amplitude maps strongly suggests that the upper reservoir associates with local-fan facies, while the lower-one with the channel-fill facies.

To verify the facies type and their porosities, Acoustic Impedance (AI) inversion was processed using five (5) wells. The related AI maps of both reservoirs overlain to their depositional surface's where AI map shown in Figure 5. The variance maps in Figure 6 shows a more detail lateral geometry and interpreted facies of both reservoirs. It can be concluded that the shallower reservoir is related with slope-fan facies, while the deeper reservoir with channel-fill facies. By using the cross-plot of total porosity, AI and water saturation as shown in Figure 6, then it's possible to identify the porous and gas-filled sands in both reservoirs. From the cross-plot it can be seen also that the ranges of porosity and water-saturation in the reservoirs are between 15-35% and 30-60% respectively.

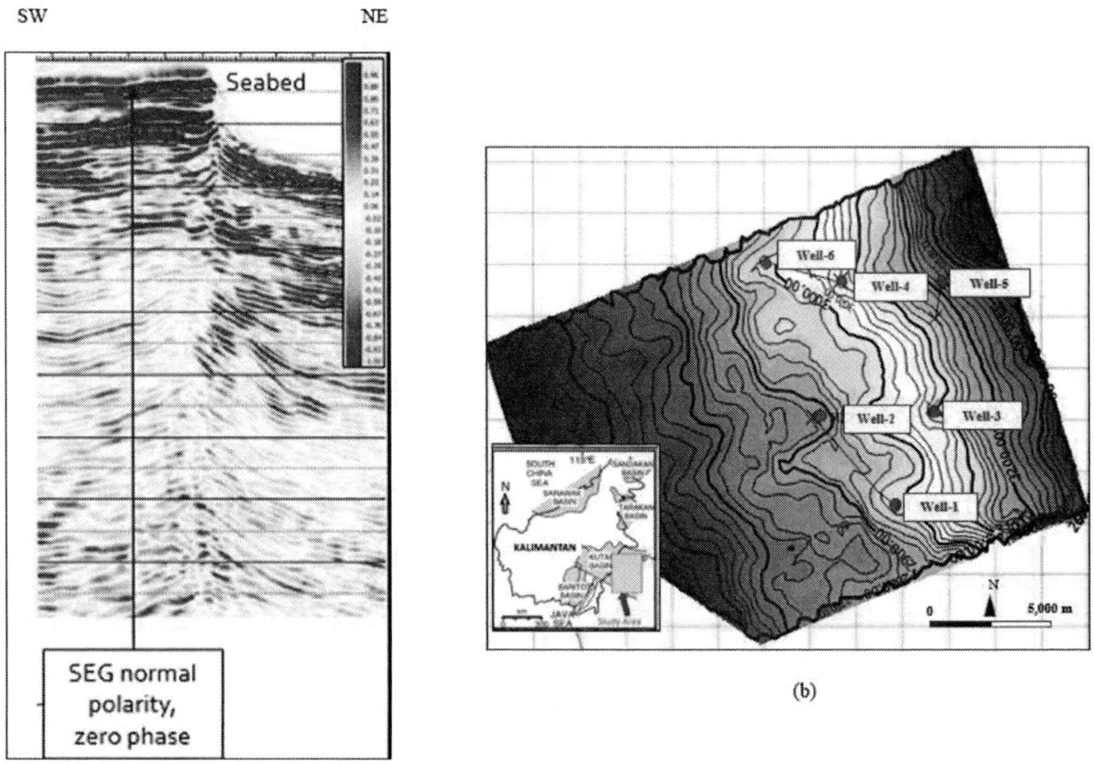

(a)

(b)

Figure 1. (a) A sample of SW-NE seismic section in the study area showing structural high related to the Miocene inversion, (b) TWT map of the studied interval which showing also the 5 wells locations used in the study. Contours in the TWT map is msec; blueish and reddish colors are respectively related to shallower and deeper contours. Inset map in Figure b shows the study area location.

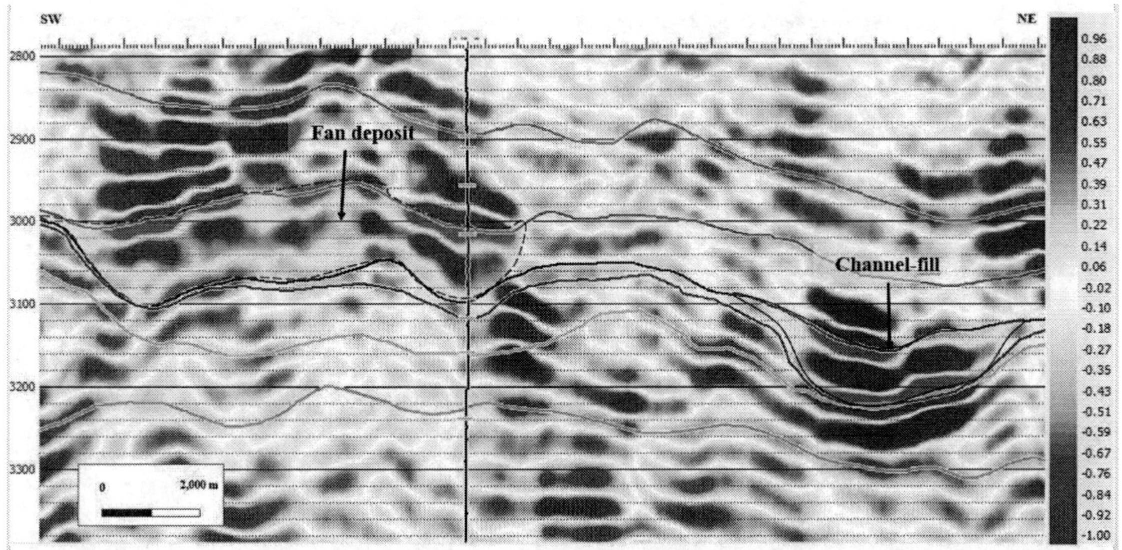

Figure 2. A sample of seismic section showing typical seismic expression of the shallower and deeper gas-sand reservoirs: shallower reservoir associates with local-fan deposit, while the deeper one related with channel-fill deposit. The TWT map of the red horizon is given in Figure 1b.

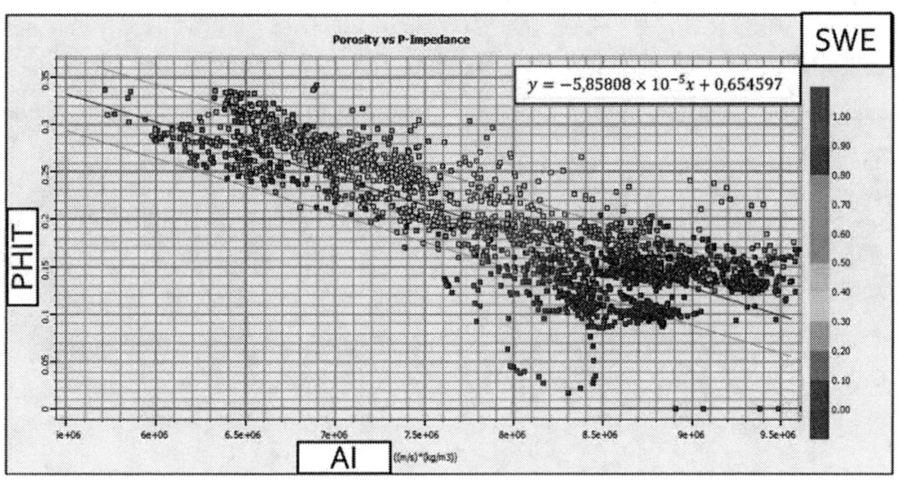

Figure 3. The cross-plot of well's AI versus total porosity with water saturation as the colour legends. The shallower slope-fan and deeper channel reservoirs intervals are integrated in this cross-plot.

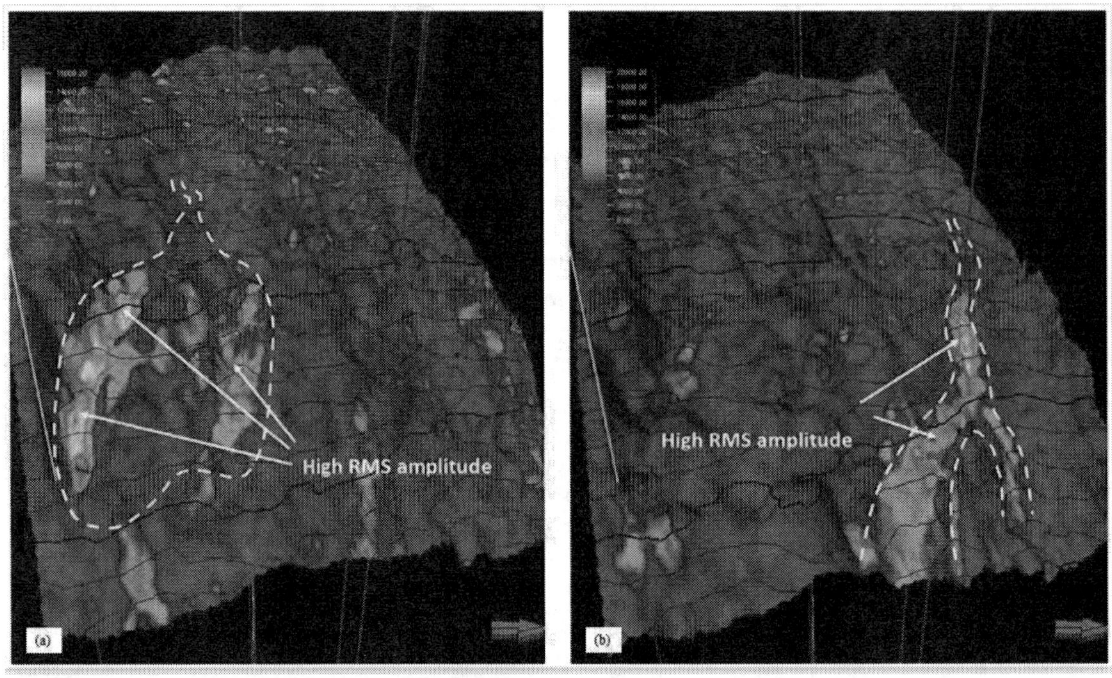

Figure 4. The RMS amplitude maps showing the lateral geometry of the gas-sand reservoirs in the (a) shallower-fan deposit between red and black horizons in Figure 2, and (b) deeper-channel fill deposit between black and blue horizons in Figure 2.

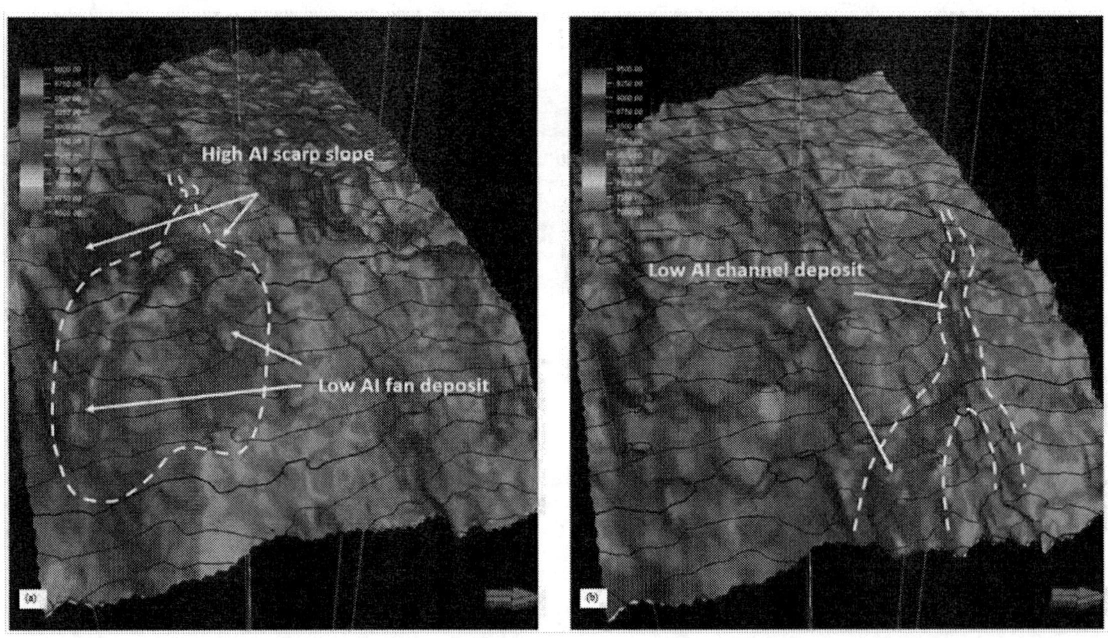

Figure 5. The Acoustic Impedance (AI) maps showing the lateral geometry of the gas-sand reservoirs in the (a) shallower-fan deposit between red and black horizons in Figure 2, and (b) deeper-channel fill deposit between black and blue horizons in Figure 2.

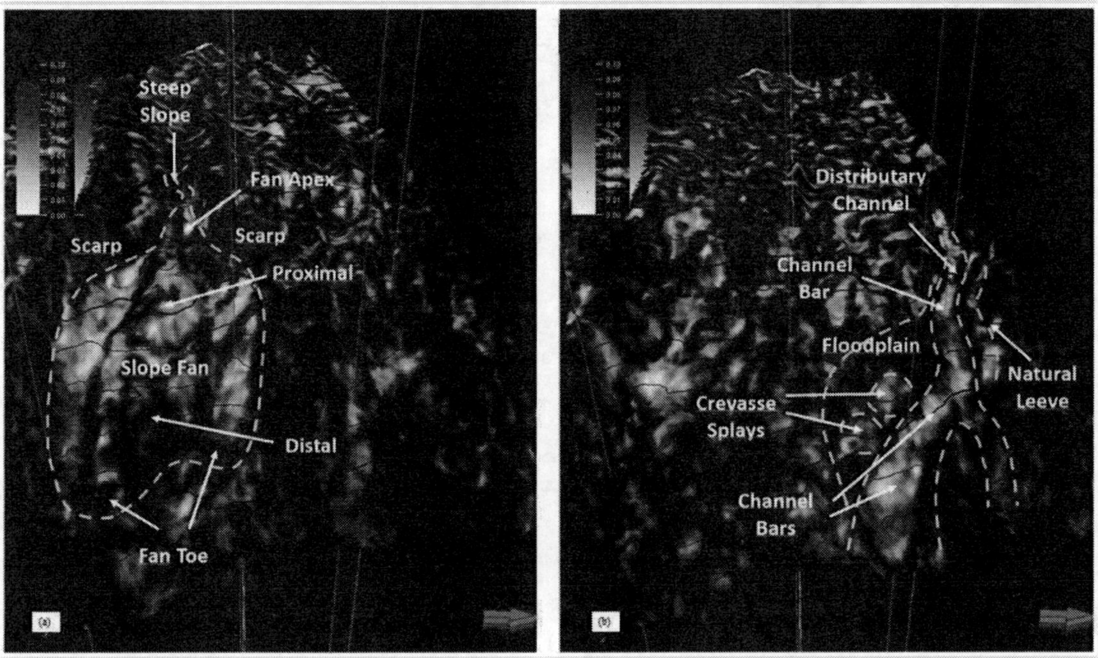

Figure 6. The variance maps showing a more detail lateral geometry and interpreted facies of the (a) shallower-fan deposit between red and black horizons in Figure 2, and (b) deeper-channel fill deposit between black and blue horizons in Figure 2.

3. Conclusions

Previous studies suggest that thick gas-sands in the Kutei basin can be identified by using pre-stack elastic impedance inversion. This paper shows that by using the integration of acoustic impedance, RMS amplitude and variance attributes, the facies containing the thick and thin gas-sands can be identified and mapped. There are two gas-sand reservoirs identified in the studied area. Both reservoirs are reflected by low AI anomalies, but the shallower reservoir was deposited in a higher AI scarp slope environment, in contrast with the deeper reservoir which deposited on much gentler slope and lower AI environment with narrower and elongate geometry. The variance attribute shows a more detail lateral geometry and interpreted facies of both reservoirs. It can be concluded that the shallower reservoir is related with local slope-fan facies, while the deeper reservoir with channel-fill facies. The total porosity and water saturation of both reservoirs can be mapped using AI inversion.

References

[1] Rose R and Hartono P 1978 *Proc. of Indonesian Petroleum Association 7th Annual Convention* p225
[2] Saller A H, Reksalegora S W and Bassant P 2010 *Society of Sedimentology Special Publication* **95** 147
[3] Saller A H, Noah J T, Ruzuar, A P and Schneider R 2004 *American Association of Petreoleum Geologist Bulletin* **88** 21
[4] Thompson P, Hartman J J, Anandito M A A, Kumar D, Magill J, Noguchi K and Gunawan B K 2009 *The Leading Edge* **26** Thompson P, Hartman J J, Anandito M A A, Kumar D, Magill J, Noguchi K and Gunawan B K 2009 *The Leading Edge* **28** 312
[5] Sukmono S, Samodra A, Sardjito and Waluyo W 2006 *The Leading Edge* **25** 532
[6] Sukmono S 2007 *The Leading Edge* **26** 126

[7] Alamsyah M N, Sukmono S, Marwosuwito S, Sutjiningsih, W and Marpaung L 2008 *The Leading Edge* **27** 260
[8] Sari, V M, Sukmono S, Sanny, T A and Sapiie, B 2018 *First Break* **36** 61
[9] Sukmono and Ambarsari 2019 *Practical seismic interpretation for petroleum exploration* (Bandung: ITB Press)

IOP Conf. Series: Earth and Environmental Science **873** (2021) 012052 doi:10.1088/1755-1315/873/1/012052

Fault rupture in Baribis Fault possibly related to the 1847 major earthquake event in the Cirebon area

Mudrik Rahmawan Daryono[1], Danny Hilman Natawidjaja[1], Anggraini Rizkita Puji[1], Sonny Aribowo[1,2]

[1] Research Center for Geotechnology, Indonesian Institute of Sciences, Bandung, Indonesia.
[2] ISTerre, Universite Grenoble Alpes -France

*Corresponding e-mail: mudr001@lipi.go.id or van_mudrik@yahoo.com

Abstract. Baribis Fault is a recently identified active fault known to have thrust movement which located along the northern part of the West Java area. This E-W striking fault runs across high-populated areas, including Cirebon, Indramayu, Sumedang, and Subang area (with a probability of continuing to Jakarta and Banten areas). The last major historical earthquake occurred on November 16th, 1847 around the fault line with a radius of shaking area up to 400 km. The available high-resolution Digital Elevation Model from Geospatial Information Agency, called DEMNAS, has about 7.5-m grid data resolution but still not adequate to be used for identifying fault ruptures of this event. Hence, we conducted an Unmanned Aerial Vehicle (UAV) 3D Photogrammetry survey flown in the lower latitude (~100-m high) in the suspected sites. This study identified clear fault scarp associated with stream-valley offsets indicating strike-slip movement in the Ujung Jaya subdistrict, Sumedang. The trace of fault rupture has a 5±1-meter sinistral offset. This sharp fault deformation feature is possibly related to the 1847 earthquake in this area. This fact is different from regional morphology, which shows that the Baribis Fault is a thrust. Further study is necessary to get more detailed and precise information.

1. Introduction

Baribis Fault is located in the northern part of the West Java area. The Fault is the result of the subduction of the Hindia-Australia Plate with Eurasia Plate with movement about 57 mm/yr. In Sumatra the boundary between the sliver accommodated by the Sumatran Fault. However, in Java the movement of the Sunda Megathrust could be transferred into inland faults in Java such as Cimandiri Fault, Lembang Fault, and this Baribis Fault in Western Java (Figure 1).

On November 16th, 1847 major earthquake occurred around Cirebon city with a radius of shaken up to 400 km [1]. However, there are still many questions about the evidence from the earthquake geology point of view, as the fault rupture line, recent surface deformation, and displacement are also unknown.

We use the Digital Elevation Model of DEMNAS with a 7.5 resolution grid combined the Lower latitude (~100-m high) 3D photogrammetry of the Unmanned Aerial Vehicle (UAV) image to get the 1847 earthquake evidence in this area. The hillshade analysis and coloring technic help improve the topographic scarp view of the suspected fault. We did the Digital Elevation Models' (DEM) surface analysis under the Geographic Information System (GIS) [2].

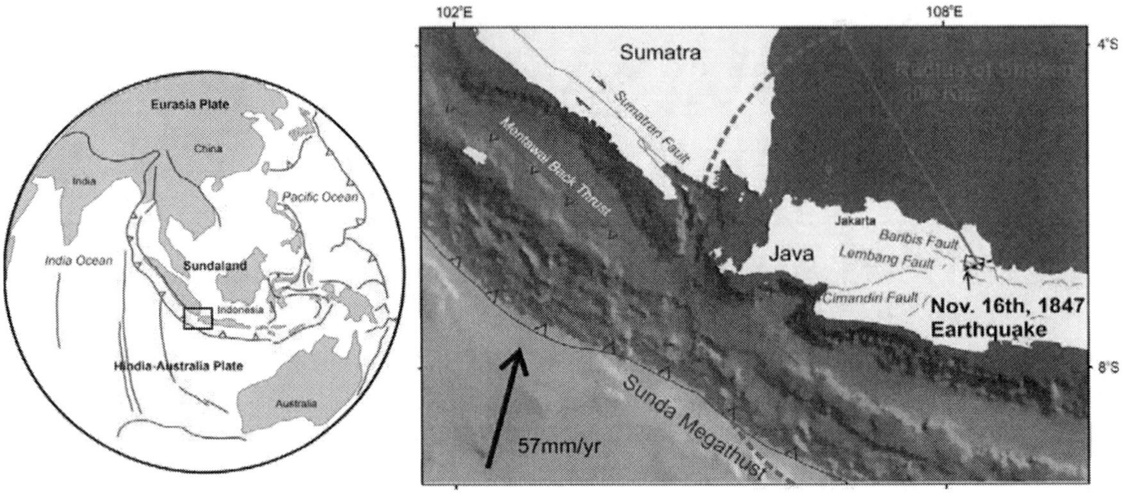

Figure 1. (Left) Baribis Fault location in the globe shows in a black rectangle. (Right) Baribis Fault in between Sunda Megathrust, Sumatran Fault, Cimandiri Fault, and Lembang Fault. Location of November 16th, 1847 earthquake with a radius of shaken about 400 km. The black dashed line is the earthquake source from the Visser Catalog [1]. The black rectangle is the research site.

2. Result and Discussion

Our observation of the DEMNAS image shows a clear NW-SE oriented fault scarp on the hills area. On the Ujung Jaya area has an NW-SE orientation lineament of the fault scarp. This part/segment has a 22 km length until the scarp feature is not existed (Figure 2).

In this area, we identified a left-movement/sinistral river offset. In the eastern area of this part/segment, the fault scarp's trace not clear on the 7.5m DEM images. Hence, we fly a UAV drone to better view the area's topography around Ujung Jaya subdistrict, Sumedang. Our observation produces a 0.3 m resolution of the DEM. With this high resolution DEM, we identified obvious fresh fault scarp deformation. We measure the offset of the river is about 5±1 m left-movement / sinistral (Figure 3). This sense movement is in agreement with earthquake geology study in Lembang Fault [3] and Cimandiri Fault [4]; and geodetic study [5]. Further geodetic research shows a large strain rate related to the strike-slip fault movement in this area (Gunawan & Widyantoro, 2020).

This size of 5±1 m sinistral movement possible correlated to the latest earthquake event in this area since the offset size is capable produce by Mw 7+ class earthquake [6]. The distribution of damage area shows by a line is focus to this area with a shaken radius up to 400 km [1]. A recent historical earthquake study shows that the shake radius is 700 km and possible produce by Mw7.6 [7].

The regional scale of the Baribis Fault is thrust fault [8, 9]. Here we show the local scale of the sinistral river offset. The 5±1m sinistral river offset possibly related to the 1847 earthquake. Further study is necessary to get more detailed and precise information.

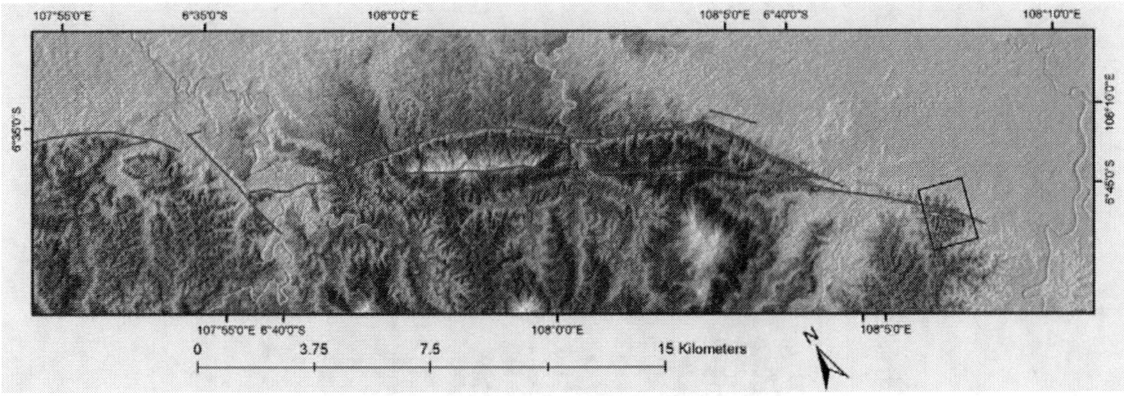

Figure 2. A 22 Km length of part/segment of Baribis Fault. This fault part/segment identified as a fault scrap lineament. The right rectangle is best fresh deformation of this segment.

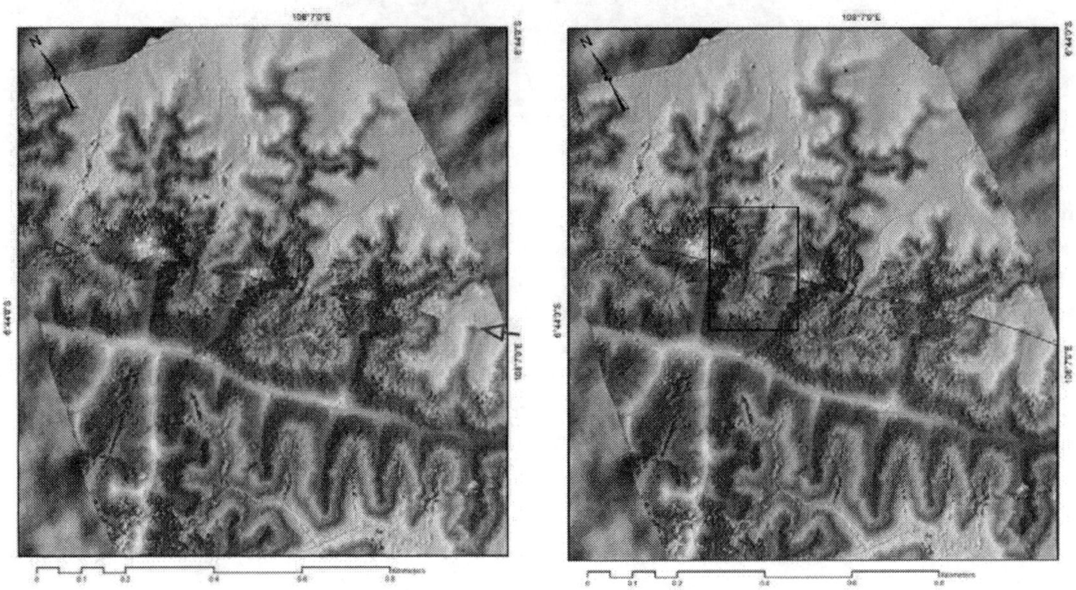

Figure 3. Obvious fresh surface deformation of the last earthquake rupture in this fault part/segment of Baribis Fault. (Left) Drone DEM image original and (right) with fault interpretation and detail site of next figure.

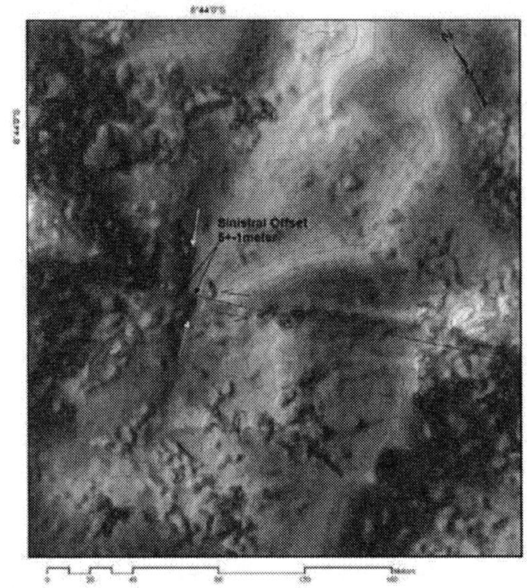

Figure 4. The obvious sharp fault scarp is cut and move to the left about 5±1 m (sinistral/left-lateral).

Conclusion Remark

Baribis Fault is a recently identified active fault known to have thrust movement which located along the northern part of the West Java area. The last major historical earthquake occurred on November 16th, 1847 around the fault line with a radius of shaking area up to 400 km. Unmanned Aerial Vehicle (UAV) 3D Photogrammetry survey flown in the lower latitude (~100-m high) identified clear fault scarp associated with stream-valley offsets indicating strike-slip movement in the Ujung Jaya subdistrict, Sumedang. The trace of fault rupture has a 5±1-meter sinistral offset. This sharp fault deformation feature is agree with the Mw 7+ intensity historical study and possibly related to the 1847 earthquake in this area.

Acknowledgement

This study is funded by the CHL\R1\180173 – GCRF Grant from The Royal Society, UK.

References

[1] S.W. Visser, Inland and Submarine epicentra of Sumatra and Java earthquakes, Koninklijk Magnetisch en Meteorologisch Observatorium te Batavia, Verhandelingen no.9 (1922) 1-14.

[2] M.R. Daryono, Paleoseismology Tropis Indonesia (dengan studi kasus di Sesar Sumatra, Sesar Palukoro-Matano, dan Sesar Lembang) - Tropical Paleoseismology of Indonesia (cases study in Sumatran Fault, Palukoro-Matano Fault, and Lembang Fault), in: Earth Science Doctoral Programme, Institut Teknologi Bandung, Bandung, 2016.

[3] M.R. Daryono, D.H. Natawidjaja, B. Sapiie, P. Cummins, Earthquake Geology of The Lembang Fault, West Java, Indonesia, Tectonophysics, 751 (2019) 180-191.

[4] G.I. Marliyani, J.R. Arrowsmith, K.X. Whipple, Characterization of slow slip rate faults in humid areas: Cimandiri Fault zone, Indonesia, Journal of Geophysical Research: Earth Surface, 121 (2016) 2287-2308.

[5] A. Koulali, S. McClusky, S. Susilo, Y. Leonard, P. Cummins, P. Tregoning, I. Meilano, J. Efendi, A.B. Wijanarko, The kinematics of crustal deformation in Java from GPS observations: Implication for fault slip partitioning, Earth and Planetary Science Letters, 458 (2017) 69-79.

[6] D.L. Wells, K.J. Coppersmith, Updated Empirical Relationships among Magnitude, Rupture Length, Rupture Area and Surface Displacement, (to be submitted to) Bulletin of the Seismological Society of America, (1991).
[7] N. Nguyen, J. Griffin, A. Cipta, P.R. Cummins, Indonesia's Historical Earthquakes Modelled examples for improving the national hazard map, Geoscience Australia, Canberra, (2015).
[8] R. Van Bemmelen, The Geology of Indonesia, Government Printing Office, The Hague, Netherlands, 1949.
[9] T.O. Simandjuntak, A.J. Barber, Constrasting tectonic styles in the Neogene orogenic belts of Indonesia, Geological Society of London, 106 (1996) 185-201.

IOP Conf. Series: Earth and Environmental Science **873** (2021) 012053 doi:10.1088/1755-1315/873/1/012053

Two earthquake events on the Pamsoa Segment of the Matano Fault, Sulawesi

Mudrik Rahmawan Daryono[1], Budhi Kumarawarman[2], Irpan Hidaya Muslim[2] Rr. Triwurjani[3], Reza Permadi[4], Sukmandaru Prihatmoko[4], Setiyo Wibowo[2], Gde Handojo Tutuko[2]

[1] Research Center for Geotechnology, Indonesian Institute of Sciences, Bandung, Indonesia.
[2] PT.Vale Indonesia.
[3] National Research Center for Archeology (ARKENAS)
[4] Indonesian Association of Geologists (IAGI)

*Corresponding e-mail: mudr001@lipi.go.id or van_mudrik@yahoo.com

Abstract. Sulawesi Island crossed by the importance active faults of the Palukoro Fault in the western part and the Matano Fault in the eastern part. Pamsoa Segment, the fourth of six-segments of the Matano Fault (from west to east), located near importance Sorowako mining city, Sulawesi. Pamsoa segment has a 38 km length, oriented N295°E, and has an obvious 475 m sinistral river offset. A reconnaissance survey conduct in this segment along the fault crosses the mining dirt road. Two fault rupture existed on the Holocene river terrace. Accelerated Mass Spectrometry (AMS) carbon dating shows age about 5218 BP and 4446 BP on the soil samples. Nice slicken-side on red soil shows as a fresh fault rupture. This rupture shows a possible two earthquake event on the Pamsoa Segment. The Matano fault has potency similar to the 2018 Palu earthquake. If the fault has similar multi fault rupture earthquake characteristic, the Matano Fault is also classified as a Magnitude 7+ class earthquake. Further research is needed to get precise earthquake geology parameters and ideal charcoal samples.

1. Introduction

Sulawesi Island crossed by importance active faults of the Palukoro Fault in the western part and the Matano Fault in the eastern part [1, 2] with roughly 4 cm/yr sliprate [3, 4]. The Matano Fault is important to be studied since a significant Mw 7.4 Palu tsunami earthquake occurred on the segments of the Palukoro Fault [5]. Matano Fault has six segments, they are (from west to east): Kuleana, Pewusai, Matano, Pamsoa, Ballawai and Geresa (Figure 1). Pamsoa Segment is located near the important Sorowako mining city, Sulawesi. Pamsoa segment has 38 km length, oriented N295°E, and has an obvious 475 m sinistral river offset [6, 7].

We reanalyzed back-matching of Pamsoa Rivers with 8 m resolution of DEMNAS dataset. We also conduct a short reconnaissance survey in the Pamsoa segment (three days survey). With the active fault line from the previous study of Paleoseismology on the Tropical Indonesia of Mudrik Daryono [6], we visited on the small river crossed the fault line. We studied the outcrops and in the same time flew Unmanned Aerial Vehicle (UAV) 3D Photogrammetry survey flown in the lower latitude (~100-m high). We documented the outcrops with 3D Photogrammetry using manual Digital Single Lens Reflector (DSLR) Camera. We also conducted Accelerated Mass Spectrometry (AMS) carbon dating.

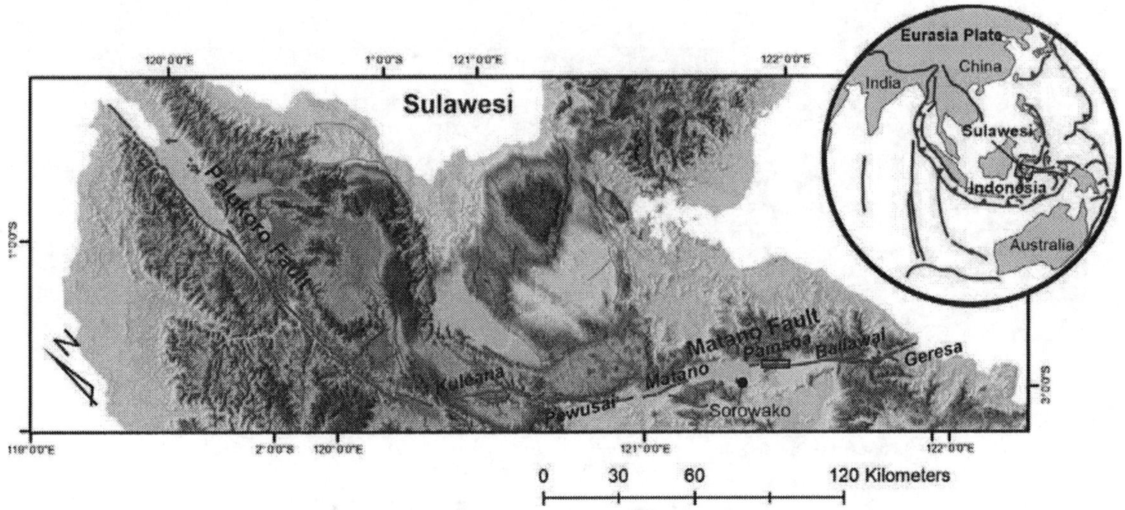

Figure 1. Central Sulawesi faults system. N-E Palukoro Fault trough E-W Matano Fault. Matano Fault consists of Kuleana, Pewusai, Matano, Pamsoa, Ballawai and Geresa Segment. This study is in Pamsoa Segment (black rectangle).

Result and Discussion

Pamsoa Segment consists of ~1 km wide an active fold and faults (Figure 2). Back-matching of Pamsoa Rivers shows about 402m (xx'), 488m (yy'), and 491m (zz'). It is in agreement with previous study with higher 5m resolution dataset of 475±70m sinistral offset [6](Figure 3). On the 2.529° S, 121.500° E coordinate close to the mining dirt road that cross a small river and the fault line, we found obviously two fault ruptures outcropping and cutting a river terrace (Figure 2). Detailed UAV Digital Elevation Model shows the outcrop is in a river branch in which the landscape has already changed due to mining activity (Figure 4).

The outcrops show Rupture A with fault plain N283°E/84°S– slicken side striation trend is horizontal (about N285°E) plunge 5° and Rupture B with N270°E/90°- slicken side striation N270°E plunge 28°. Slicken side striation is in thin red clay-filled fault space indicated as a fresh ruptured (Figure 5).

Detailed 3D photogrammetry shows detail outcrops logging (Figure 6). Those ruptures cut gravel-rich sands layer (yellow) and grey coarse sand layer (grey). We took two bulk samples for each unit due to no charcoal (Figure 6). AMS carbon dating in Radiocarbon Dating Laboratory of Waikato University shows an age about 5218 BP on the yellow unit and 4446 BP on grey unit (Figure 7). This has proved that the river terrace is younger than Holocene age and also prove that this fault is an active fault. But this age result is not correlatable to the co-seismic event. Commonly, bulk sample has older result compare to a pure charcoal sample due to mixing of re-sediment process.

IOP Conf. Series: Earth and Environmental Science **873** (2021) 012053 doi:10.1088/1755-1315/873/1/012053

These ruptures show possible two earthquake events on the Pamsoa Segment. The Matano fault has potency similar to the 2018 Palu earthquake. If the fault has similar multi fault rupture earthquake characteristic, the Matano Fault is also classified as a Magnitude 7+ (bigger than Magnitude 7) class earthquake. This magnitude need to be clarified by measured offset ruptures for each rupture. In the last earthquake event, how much sliprate and how long the recurrent interval are still unknown. Further research is needed to get precise earthquake geology parameters and an ideal charcoal sample.

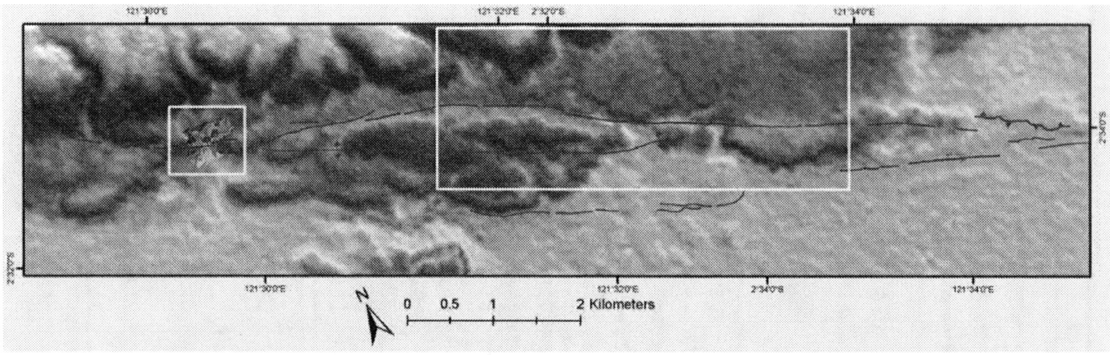

Figure 2. Detail of Pamsoa Segment on DEMNAS hillshade. The white rectangle is detail survey site and back-matching of Pamsoa rivers offset.

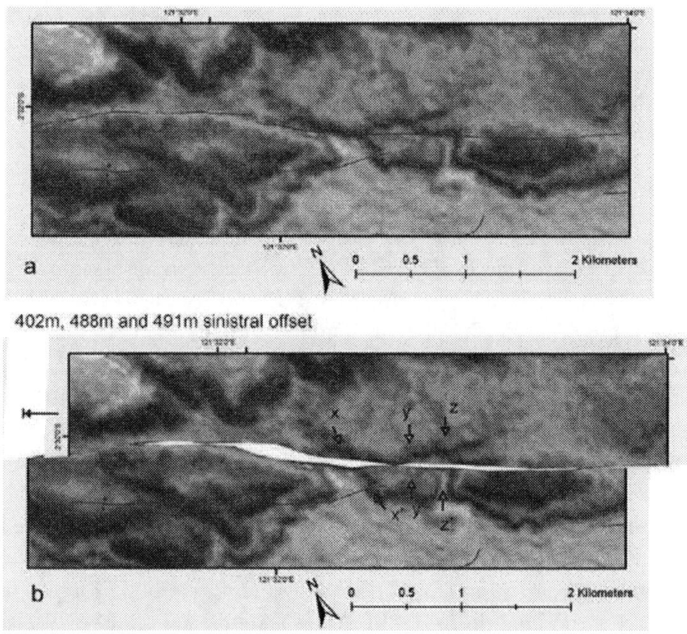

Figure 3. Back-matching of Pamsoa rivers with DEMNAS dataset with 402m to 491 m sinistral offset of xx', yy' and zz'.. Countur lines is 2 m interval.

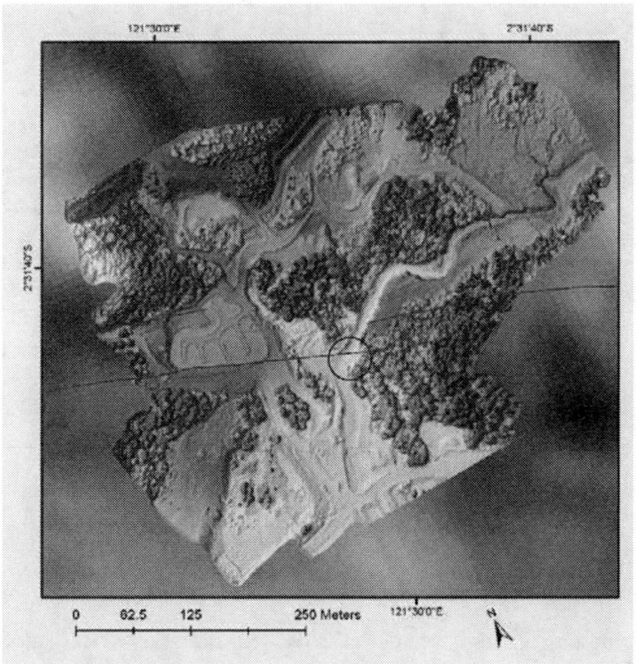

Figure 4. Detail location of Pamsoa Segment line and the fault outcrops (black circle).

(a) (b)

Figure 5. (a) River outcrops and the fault Rupture A and Rupture B. String grid is 1 m space. (b) Detail of Rupture A with shovel head as a scale. Follows the Rupture A with arrows sign.

IOP Conf. Series: Earth and Environmental Science **873** (2021) 012053 doi:10.1088/1755-1315/873/1/012053

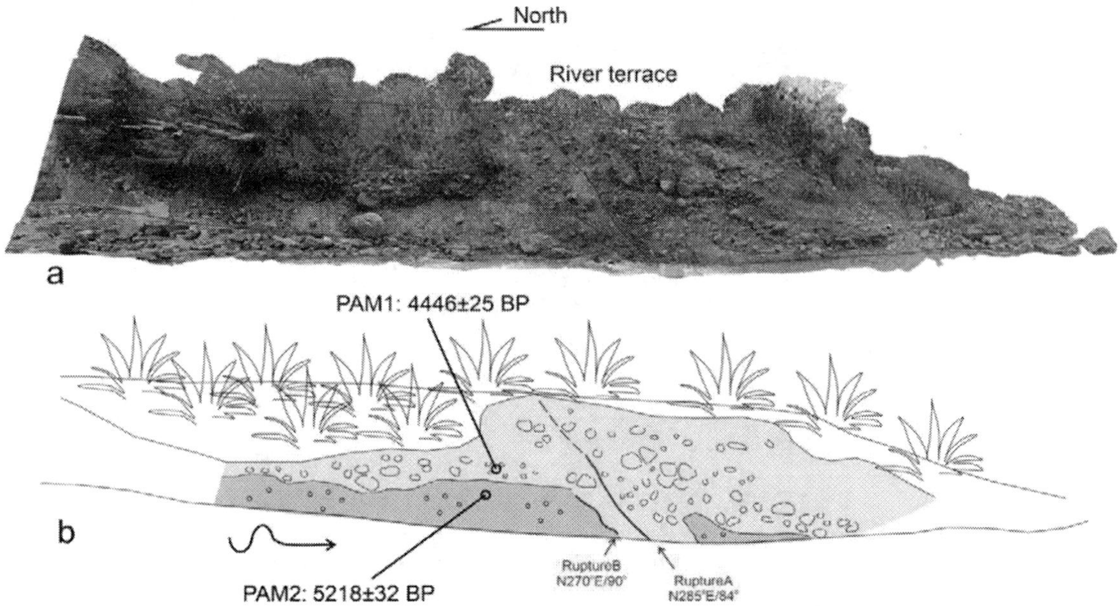

Figure 6. Outcrops logging using manual Digital Single Lens Reflector (DSLR) camera 3D photogrammetry (a) and the interpretation (b). Gravel rich sands (yellow) unit has 4446+-25 BP AMS age and Coarse Sands (grey) unit has 5218+-32 BP AMS age.

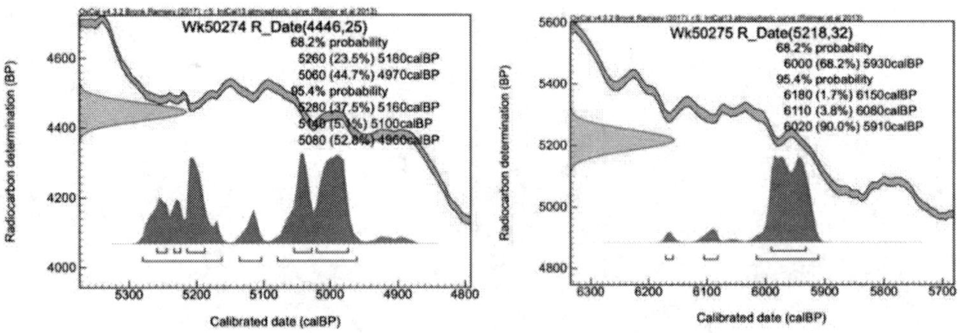

(a) (b)

Figure 7. Detail result of Accelerated Mass Spectrometer (AMS) carbon dating of Yellow Unit (a) and Grey Unit (b).

Conclusion Remark

Pamsoa segment has a 38 km length, oriented N295°E, and has an obvious 475 m sinistral river offset. Two fault rupture existed on the Holocene river terrace. Accelerated Mass Spectrometry (AMS) carbon dating shows age about 5218 BP and 4446 BP on the soil samples. This rupture shows a possible two earthquake event on the Pamsoa Segment. The Matano fault has potency similar to the 2018 Palu earthquake as a Magnitude 7+ class earthquake.

Acknowledgement
The study and publication are funded by the CHL\R1\180173 – GCRF Grant from The Royal Society, UK. Fieldwork is funded and supported by Indonesian Association of Geologists (IAGI), National Research Center for Archeology (ARKENAS) and PT. Vale Indonesia (PTVI). Accelerated Mass Spectrometer (AMS) carbon dating is funded by ARKENAS. DEMNAS dataset is public domain from Indonesian Information Spatial Agency (BIG).

References
[1] J.A. Katili, Large transcurrent fault in Southeast Asia with special reference of Indonesia., Geol. Rdsch., 59 (1970).
[2] I.M. Watkinson, R. Hall, Active faults of eastern Indonesia, Geological Society of London Special Publication, (2016).
[3] A. Socquet, W. Simons, C. Vigny, R. McCaffrey, C. Subarya, D. Sarsito, B. Ambrosius, W. Spakman, Microblock rotations and fault coupling in SE Asia triple junction (Sulawesi, Indonesia) from GPS and earthquake slip vector data, Journal of Geophysical Research, 111 (2006).
[4] O. Bellier, M. Sebrier, T. Beaudouin, M. Villeneuve, R. Braucher, D. Bourles, L. Siame, E. Putranto, I. Pratomo, High slip rate for a low seismicity along the Palu-Koro active fault in central Sulawesi (Indonesia), Terr. Nova, 13 (2001) 463-470.
[5] D. Natawidjaja, M. Daryono, G. Prasetya, Udrekh, P.L.-F. Liu, W. Kongko, W. Triyoso, A. Riskita Puji, I. Meilano, E. Gunawan, P. Supendi, A. Pamumpuni, M. Irsyam, L. Faizal, S. Hidayati, B. Sapiie, M. Kusuma, S. Tawil, The 2018 Mw7.5 Palu supershear ruptures geological fault's multi-segments separated by large bends: Result from integrating field measurements, LiDAR, multibeam bathymetry and seismic-reflection data Geophysical Journal International, (2020 (under review)).
[6] M.R. Daryono, Paleoseismology Tropis Indonesia (dengan studi kasus di Sesar Sumatra, Sesar Palukoro-Matano, dan Sesar Lembang) - Tropical Paleoseismology of Indonesia (cases study in Sumatran Fault, Palukoro-Matano Fault, and Lembang Fault), in: Earth Science Doctoral Programme, Institut Teknologi Bandung, Bandung, 2016.
[7] D.H. Natawidjaja, B. Sapiie, M.R. Daryono, G.I. Marliyani, A. Pamumpuni, M.M. Mukti, Supartoyo, S. Hidayati, A. Solikhin, Earthquake Geology of Indonesia, in: M. Irsyam (Ed.) 2017 Indonesia Earthquake Sources and Hazard (Peta Sumber dan Bahaya Gempa Indonesia Tahun 2017), PUPR, Indonesia, 2017.

Source reconstruction of the 1969 Majene, Sulawesi earthquake and tsunami: A preliminary study

I.R. Pranantyo[1,4,*], **A. Cipta**[2], **H.A. Shiddiqi**[3], **M. Heidarzadeh**[1]

[1] Department of Civil and Environmental Engineering, Brunel University London, Uxbridge UB8 3PH, United Kingdom
[2] Geological Agency of Indonesia, Bandung, Indonesia
[3] Department of Earth Science, University of Bergen, Bergen, Norway
[4] Present: Bandung, Indonesia

E-mail: *ignatiusryanpranantyo@gmail.com

Abstract. We studied the February 23rd, 1969 M7.0 Majene, Sulawesi earthquake and tsunami. It was followed by tsunami reported at five locations. At least 64 people were killed and severe damage on infrastructures were reported in Majene region. Based on damage data, we estimated that the maximum intensity of the earthquake was MMI VIII. Focal mechanisms, derived using first motion polarity analysis, indicated that the earthquake had a thrust mechanism. Furthermore, we built hypothetical earthquake scenarios based on a rectangular fault plane of 40 km × 20 km with a homogeneous slip model of 1.5 m. We run the Open Quake and the JAGURS code to validate the macroseismic and tsunami observation data, respectively. Our best-fitted earthquake model generates maximum intensity of 8+ which is in line with the reported macroseismic data. However, the maximum simulated tsunami height from all scenario earthquakes is 2.25 m which is smaller than the 4 m tsunami height observed at Pelattoang. The possibility of contribution of another mechanism to tsunami generation requires further investigation.

Keywords: Eastern Indonesia, Sulawesi, Earthquake, Tsunami, Numerical Simulations

1. Introduction

Sulawesi Island was composed by few terranes, both detached from Australian and Eurasian plates, making the island tectonically very active, with the most prominent active fault is the Palu Koro Fault Zone [1] (Figure 1). As a result, the island is seismically active and is exposed to tsunami hazard (e.g. Cipta et al., [2], Prasetya et al., [3], Horspool et al., [4]). A deadly earthquake and tsunami occurred on 23rd February 1969 that devastated Majene region, west coast of Sulawesi (Figure 1). The Mw7.0 quake intensely jolted the region of Majene and generated a 4 m tsunami wave at Pelattoang (Figure 1c) [5, 6]. Here we conducted preliminary study on this event to understand the main generation mechanism of the earthquake and tsunami. This study would encourage further investigation to increase earthquake and tsunami resilience for Sulawesi.

2. Historical accounts

Historical accounts in Indonesia were mostly available on a description data-types and incomplete (e.g. [7, 8, 9]). For the 1969 Majene, Sulawesi event, historical accounts were available from

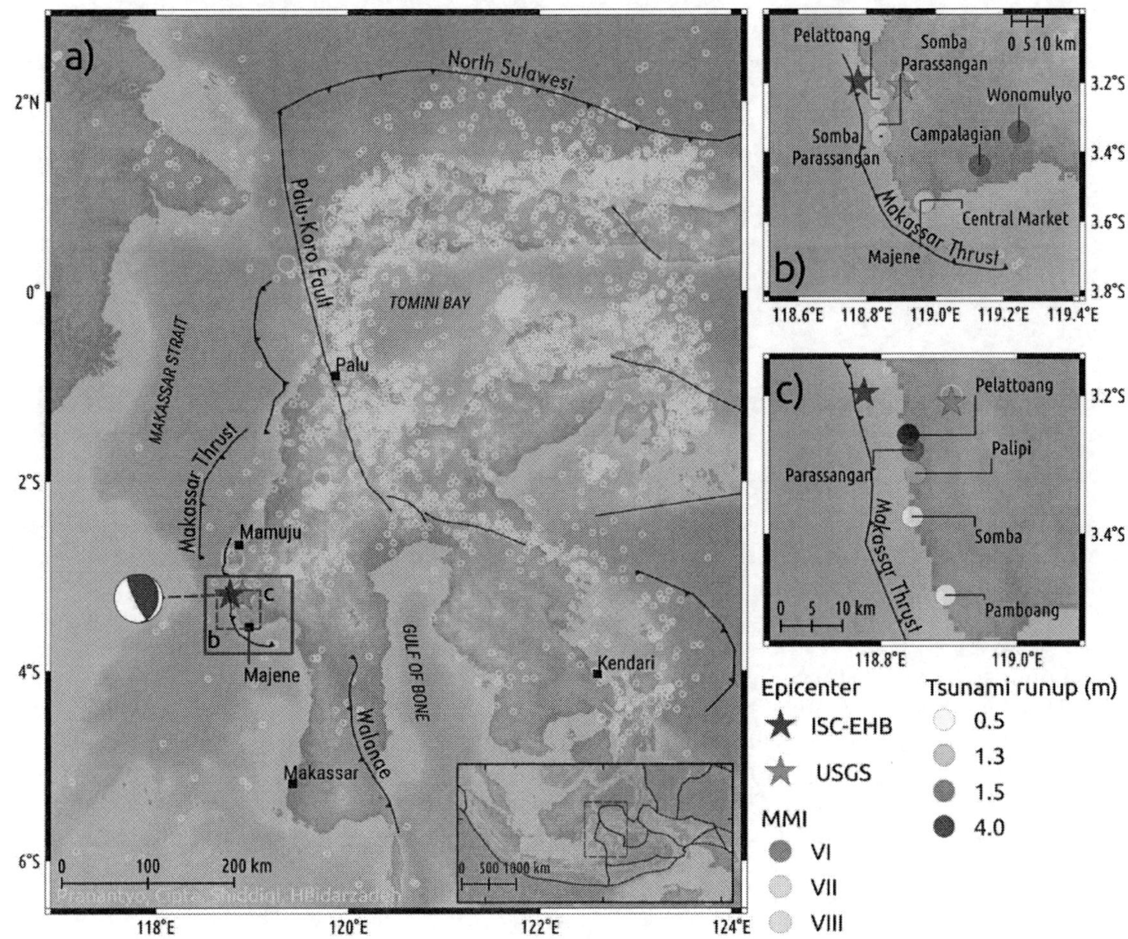

Figure 1. a) Tectonic setting of Sulawesi Island. Red and blue stars indicate the epicentre of the 1969 Majene, Sulawesi earthquake taken from the ISC-EHB and USGS catalogue, respectively. Focal mechanism is based on the first motion polarity analysis (Section 3). The red and blue boxes are inset maps for b) interpreted macroseismic data in Modified Mercalli Intensity (MMI) scales and c) tsunami height observations, respectively.

Soloviev et al., [10], Soetardjo et al., [11], Fikrie and Hardiansya, [12], and International Tsunami Information Center (ITIC) [5, 6]. Further, we converted pieces of information on infrastructure damage, land morphology changes (e.g. rockfalls, landslides, ground cracking), and ground shaking intensity felt by eyewitnesses from several locations into a Modified Mercalli Intensity (MMI) scale (Figure 1b). The data indicate that the earthquake generated massive strong motion with MMI up to VIII at Somba-Parassangan. Tsunami height observations were available at five locations ranging from 0.5 m at Pamboang to 4 m at Pellatoang (Figure 1c). Unfortunately, neither earthquake accounts from locations that were further away from the epicentre nor the estimated tsunami arrival time were available.

3. First motion polarity analysis

Earthquake location and its mechanism are important parameters in this study. According to the ISC-EHB [13] and the USGS earthquake catalogues, their earthquake epicentres are located

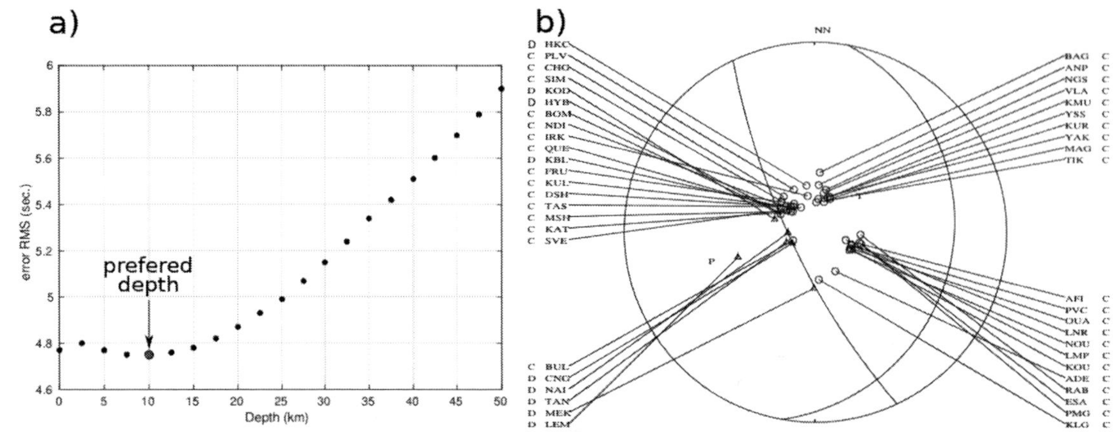

Figure 2. a) The plot of error RMS for various focal depth. The depth with lowest RMS error is marked by the red circle. b) Focal mechanism of the M7.0 1969 Majene, Sulawesi earthquake. Polarities are plotted within the lower hemisphere projection where the compression (C) and dilatation (D) are marked as small triangles and small circles, respectively.

at about 15–20 km distance apart (Figure 1a) with the earthquake depth is between 15 km and 50.4 km. Since this area is situated far from active tectonic margin, it is most likely that the Majene 1969 event was sourced from a shallow crustal fault. Therefore, we re-analysed the earthquake depth using arrival time data from ISC-EHB catalogue. We used direct P- and S-phases, as well as depth- (pP) and mantle-phases (PP, SS, and PcP) from stations at regional and teleseismic distances. Observations with very large residuals are removed. We performed a grid-search by varying the earthquake depth and fixing its epicentre by using the RMSDEP program from the SEISAN package [14, 15]. We used depth range from 0 km to 50 km, with an interval of 2.5 km. The travel times are computed using the iasp91 velocity model [16]. The travel-times error in the root-mean-square (RMS) for each depth is calculated. We found out that the depth of 10 km gives the lowest error RMS (Figure 2a). Furthermore, we utilized the FPFIT software [17] to determine the earthquake focal mechanism based on its first motion polarities reported in the ISC-EHB catalogue. A thrust mechanism was obtained with two possible focal mechanism; Strike1/Dip1/Rake1 (FM1): 10°/18°/124° and Strike2/Dip2/Rake2 (FM2): 155°/75°/80° (Figure 2b).

4. Ground motion and tsunami modelling

4.1. Scenarios and model setup

From the results above, we tested five selected scenarios (Figure 3a, Table 1 to validate the MMI (Figure 3b). We conducted ground motion (GM) modelling using the OpenQuake[1] software [18]. The software calculates peak ground acceleration (PGA) according to the ground motion prediction equation (GMPE) of Boore et al., [19] and Vs30 of Cipta et al., [2], estimated based on geological and topographic parameters. Earthquake intensity (MMI) is obtained by converting the estimated PGA using the Worden et al., [20] formula. We also performed 2D numerical tsunami propagation modelling using the JAGURS[2] code [21] to verify the tsunami height data (Figure 3b, c). We solved the non-linear shallow water wave equations. For this study, we built

[1] https://www.globalquakemodel.org/openquake, last accessed 5 October 2020
[2] https://github.com/jagurs-admin/jagurs, last accessed 10 September 2020

IOP Conf. Series: Earth and Environmental Science **873** (2021) 012054 doi:10.1088/1755-1315/873/1/012054

a digital elevation model (DEM[3]) by combining the national bathymetry (DEMNAS), contours of shallow coastal area (LPI) as well as topography (RBI) then interpolated with a resolution of 0.00225° (~250 m). For this study, we did not conduct inundation modelling.

We assumed a rectangular fault plane model of 40 km × 20 km (based on equations from Wells and Coppersmith, [22]) with the earthquake epicentre was assumed to be the centroid of the fault plane. Firstly, we used the earthquake epicentre from the USGS catalogue with two different solutions from the first motion polarity analysis (the FM1 and FM2), referred to as USGS-FM1 and USGS-FM2. Then we also tested the epicentre from ISC-EHB catalogue, EHB-FM1 and EHB-FM2. Lastly, we applied a trial-and-error approach by shifting the earthquake location as well as changed the mechanism, Majene1969, to find the best-fit model that describes the MMI and tsunami observation data.

Table 1. Selected earthquake and tsunami scenarios to validate the 1969 Majene, Sulawesi event

Scenario	Epicentre	Coordinate*	Strike	Dip	Rake
USGS-FM1	118.9040°E, 3.2010°S	118.8041°E, 3.2751°S, 6.91 km	10°	18°	124°
USGS-FM2	118.9040°E, 3.2010°S	118.8871°E, 3.1091°S, 0.34 km	155°	75°	80°
EHB-FM1	118.7790°E, 3.1960°S	118.6761°E, 3.2701°S, 6.91 km	10°	18°	124°
EHB-FM2	118.7790°E, 3.1960°S	118.7591°E, 3.1041°S, 0.34 km	155°	75°	80°
Majene1969	118.5760°E, 3.1960°S	118.6012°E, 3.4676°S, 2.00 km	13°	15°	67°

Fault length × width = 40 km × 20 km; Homogeneous slip = 1.5 m
*top-right coordinate (longitude, latitude, top-depth) of the fault plane

4.2. Results and analysis

Firstly, by considering the tectonic setting of Sulawesi, the FM1 is the most plausible solution of the earthquake mechanism due to its North-South orientation. The closest structure to the earthquake epicentre is the Makassar Thrust (Figure 1). Therefore, we could eliminate the possibility of the USGS-FM2 and EHB-FM2 scenarios.

Second, the USGS earthquake epicentre was located on land. The GM model of the USGS-FM1 indicates the maximum MMI is a little bit further east compared to the data. Moreover, with an epicentre on land, it caused little earthquake deformation on the sea (Figure 3). Therefore, the maximum tsunami amplitude on the coastline is just a little more than ~0.5 m, which is much lower than the 4 m observation height. On the other hand, the ISC-EHB earthquake epicentre was located on the sea. The GM model of the EHB-FM1 shows a relatively good fit for MMI compared to the data. With half of the earthquake deformation occurred on the sea, it increases the maximum tsunami height to ~1.5 m. Unfortunately, it is still 2.5 m less than the maximum observation.

The USGS-FM1 and EHB-FM1 models produced extremely high MMI in coastal areas near fault and extending further eastward on land. On the contrary, the USGS-FM2 and EHB-FM2 scenarios, the high MMI area was suffered only at the near-fault coastal area, as expected by macroseismic data. However in the southern part of Majene, the predicted intensity was rather lower than expected. In order to obtain better model for MMI data and tsunami

[3] DEM is combination of DEMNAS (https://tides.big.go.id/DEMNAS, last accessed 10 September), LPI and RBI (https://portal.ina-sdi.or.id/downloadaoi, last accessed 10 September 2020)

IOP Conf. Series: Earth and Environmental Science **873** (2021) 012054 doi:10.1088/1755-1315/873/1/012054

height, we modified EHB-FM1 model. Based on the ISC-EHB earthquake epicentre, then we subjectively conducted a trial-and-error approach to find the most plausible earthquake location and mechanism. In this study, we found that the Majene1969 scenario is the best model to reproduce the MMI data (Figure 3). The earthquake location is slightly further west than the EHB-FM1 model with a shallower depth. The Majene1969 scenario increases the maximum simulated tsunami height to ~2.25 m. Likely, the previous scenarios, this model still cannot fit the maximum tsunami height observation.

The observed tsunami height data has a narrow band of maximum tsunami height over 5 km distance along the coastline (Figure 1, 3). The tsunami height then significantly decreases to southward direction. It is likely that the 4 m tsunami height could be due to a local phenomenon such as an underwater landslide induced by the earthquake. However, it would need further investigation to locate where, when, and how large such a phenomenon would be.

5. Summary

We have shown preliminary result of the earthquake and tsunami source reconstruction for the 1969 Majene, Sulawesi event. We collected historical accounts of earthquake intensities and tsunami heights observation from various references. First motion polarity analysis indicates it was a thrust earthquake with the Makassar Thrust as the closest fault to the earthquake epicentre. Majority of the earthquake rupture need to be located on the sea to fit the MMI data and increase the maximum simulated tsunami height. However, all scenarios considered in this study are still unable to reproduce the 4 m tsunami height observed at Pellatoang. We assume that a local phenomenon such as a submarine landslide triggered by the earthquake was involved in the tsunami generation process. Further investigation is needed to determine the parameters of a potential landslide.

Acknowledgment

This study is supported by the Royal Society, UK (grant number CHL\R1\180173)

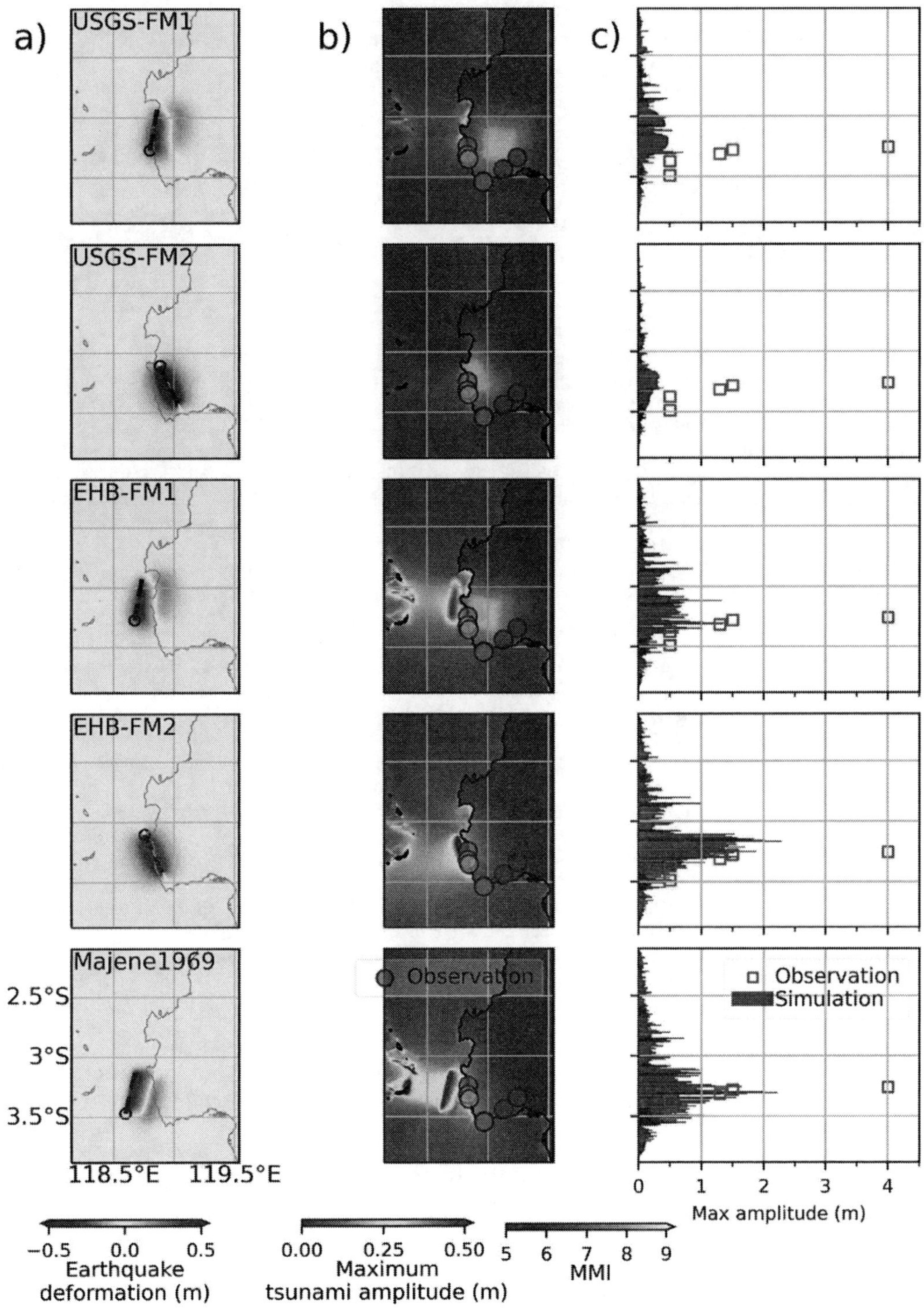

Figure 3. Ground motion (GM) and tsunami modelling of selected scenarios. a) Earthquake deformation model; dashed line and hollow circle indicate fault plane surface projection and coordinate shown in Table 1, respectively. b) GM and maximum tsunami modelling results. c) Maximum simulated tsunami coastal height and tsunami observation.

References

[1] Hall R. Australia–SE Asia collision: plate tectonics and crustal flow. *Geological Society, London, Special Publications* 2011;**355**:75–109. https://doi.org/10.1144/SP355.5.

[2] Cipta A, Robiana R, Griffin J D, Horspool N, Hidayati S, Cummins P R. A probabilistic seismic hazard assessment for Sulawesi, Indonesia. *Geological Society, London, Special Publications* 2016;**441**:133–152. https://doi.org/10.1144/sp441.6.

[3] Prasetya G, De Lange W, Healy T. The Makassar Strait tsunamigenic region, Indonesia. *Natural Hazards* 2001;**24**:295–307. https://doi.org/10.1023/a:1012297413280.

[4] Horspool N, Pranantyo I, Griffin J, Latief H, Natawidjaja D, Kongko W, Cipta A, Bustaman B, Anugrah S, Thio H. A probabilistic tsunami hazard assessment for Indonesia. *Natural Hazards and Earth System Sciences* 2014;**14**:3105–3122. https://doi.org/10.5194/nhess-14-3105-2014.

[5] ITIC. International Tsunami Information Center Newsletter. vol. 2, PO.BOX 3887, Honolulu, Hawaii 96812 USA: ITIC; 1969, p. 8.

[6] ITIC. International Tsunami Information Center Newsletter. vol. 3, PO.BOX 3887, Honolulu, Hawaii 96812 USA: ITIC; 1969, p. 8.

[7] Griffin J, Nguyen N, Cummins P, Cipta A. Historical Earthquakes of the Eastern Sunda Arc: Source Mechanisms and Intensity-Based Testing of Indonesia's National Seismic Hazard Assessment. *Bulletin of the Seismological Society of America* 2018;**109**:43–65. https://doi.org/10.1785/0120180085.

[8] Pranantyo I R, Cummins P R. The 1674 Ambon tsunami: Extreme run-up caused by an earthquake-triggered landslide. *Pure and Applied Geophysics* 2019;**177**:1639–1657. https://doi.org/10.1007/s00024-019-02390-2.

[9] Cummins P R, Pranantyo I R, Pownall J M, Griffin J D, Meilano I, Zhao S. Earthquakes and tsunamis caused by low-angle normal faulting in the Banda Sea, Indonesia. *Nature Geoscience* 2020;**13**:312–318. https://doi.org/10.1038/s41561-020-0545-x.

[10] Soloviev S, Go C, Kim X. *Catalog of tsunamis in the Pacific Ocean*, 1969-1982. Moscow, MGC, 163p 1986.

[11] Soetardjo, Untung M, Arnold E P, Soetadji R, Ismail S, Kertapati E K. *Indonesia*. vol. V. U.S. Geological survey; 1985.

[12] Fikrie M, Hardiansya A. Lembong tallu dan gempa bertubi-tubi di Mamasa 2018. https://lokadata.id/artikel/lembong-tallu-dan-gempa-bertubi-tubi-di-mamasa, last accessed 5 October 2020.

[13] Weston J, Engdahl E R, Harris J, Giacomo D D, Storchak D A. ISC-EHB: reconstruction of a robust earthquake data set. *Geophysical Journal International* 2018;**214**:474–484. https://doi.org/10.1093/gji/ggy155.

[14] Havskov J, Ottemöller L. SeisAn Earthquake Analysis Software. *Seismological Research Letters* 1999;**70**:532–534. https://doi.org/10.1785/gssrl.70.5.532.

[15] Havskov J, Voss P H, Ottemöller L. Seismological Observatory Software: 30 Yr of SEISAN. *Seismological Research Letters* 2020;**91**:1846–1852. https://doi.org/10.1785/0220190313.

[16] Kennet B L N. IASPEI 1991 Seismological Tables. *Terra Nova* 1991;**3**:122. https://doi.org/10.1111/j.1365-3121.1991.tb00863.x.

[17] Reasenberg P A. FPFIT, FPPLOT, and FPPAGE: Fortran computer programs for calculating and displaying earthquake fault-plane solutions. US Geol. Surv. Open-File Rep. 1985:85–739.

[18] Pagani M, Monelli D, Weatherill G, Danciu L, Crowley H, Silva V, Henshaw P, Butler L, Nastasi M, Panzeri L, Simionato M, Vigano D. OpenQuake Engine: An Open Hazard (and Risk) Software for the Global Earthquake Model. *Seismological Research Letters* 2014;**85**:692–702. https://doi.org/10.1785/0220130087.

[19] Boore D M, Stewart J P, Seyhan E, Atkinson G M. NGA-West2 equations for predicting PGA, PGV, and 5% damped PSA for shallow crustal earthquakes. *Earthquake Spectra* 2014;**30**:1057–1085.

[20] Worden C B, Gerstenberger M C, Rhoades D A, Wald D J. Probabilistic Relationships between Ground-Motion Parameters and Modified Mercalli Intensity in California. *Bulletin of the Seismological Society of America* 2012;102:204–221. https://doi.org/10.1785/0120110156.

[21] Baba T, Takahashi N, Kaneda Y, Ando K, Matsuoka D, Kato T. Parallel Implementation of Dispersive Tsunami Wave Modeling with a Nesting Algorithm for the 2011 Tohoku Tsunami. *Pure and Applied Geophysics* 2015;**172**:3455–3472. https://doi.org/10.1007/s00024-015-1049-2.

[22] Wells D L, Coppersmith K J. New empirical relationships among magnitude, rupture length, rupture width, rupture area, and surface displacement. *Bulletin of the seismological Society of America* 1994;**84**:974–1002.

Early Results of Time Domain Receiver Function Data Processing in Mt Merapi and Mt Merbabu

A K Ilahi[1], M F R Auly[1], D A Zaky[1], A Abdullah[1], R P Nugroho[1], S K Suhardja[1], A D Nugraha[2], Z Zulfakriza[2], S Widiyantoro[2,3], J P Métaxian[4], M Ramdhan[5]

[1] University of Pertamina, Jl. Teuku Nyak Arief, Simprug, Kebayoran Lama, Jakarta 12220, Indonesia
[2] Global Geophysics Research Group, Faculty of Mining and Petroleum Engineering Institute of Technology Bandung, Jalan Ganesa No. 10, Bandung 40132, Indonesia
[3] Faculty of Engineering, Maranatha Christian University Bandung, Jalan Surya Sumantri No 65, Bandung 40164, Indonesia
[4]ISTerre, IRD R219, CNRS, Université de Savoie Mont Blanc, Le Bourget-du-Lac, France
[5]Agency for Meteorology, Climatology and Geophysics, Jalan Angkasa I, No. 2, Kemayoran, Jakarta, Indonesia

*Corresponding author: s4ndy104@gmail.com

Abstract. The receiver function method is a method to image the earth subsurface by utilizing Ps conversion waves that are formed due to the velocity boundary. In general, the receiver function method estimates depth of structures such as the mantle-crust boundary by deconvoluting the vertical component from the horizontal component. Typical receiver function data processing is done in the frequency domain where the deconvolution process can be seen as a division between two components. In this study, we tried to reprocess the data using a deconvolution technique in time domain, popularly known as iterative time-domain deconvolution. The principle of iterative time domain deconvolution consists of iterative cross-correlation between the horizontal and vertical component. We use data from the DOMERAPI seismic station network located in the vicinity of Mt Merapi and Mt Merbabu. Mt Merapi is one of the most active volcanoes in the world with frequent eruptions and located at the ring of fire chain volcano in Indonesia. Note that the previous receiver function study in this region showed complex signals at some stations that may be related to sediment at shallow sediment and possible layers of low velocity zone that interfering main signal for a crust-mantle boundary. Our current results show iterative time domain RFs have clearer and smoother signal than the frequency domain that help interpreting the waveform signals. We estimate a range of crust thickness between 26-31 km near Mt Merapi. However, we noticed that iterative time domain calculation requires longer computation time and input signal.

Keywords: Time domain Receiver Function, Merapi, Volcano

1. Introduction

Central Java's location is in the middle part of Java island, Indonesia. This area has an active tectonic activity where the Indian-Australian plate subducts the southern part of the Sundaland. The subduction of Indo-Australian plate beneath Sundaland began after SE Asia evolution in the middle Miocene about 42 MA [1] and existed to date with a convergence rate of 67 mm per year. Central Java is characterized by many volcanic activities and arc magmatism due to the subduction setting's influence. There are two main volcanic arcs in this area: The Southern Mountain Arc (SMA) and Modern Volcanic Arc (MVA). Southern Mountain Arc (SMA) formed in the island's southern coastal region started at the middle Eocene about 45-20 Ma [2]. The subduction in this area ceased in the Cretaceous but then resumed in the late Miocene and formed Modern Volcanic Arc (MVA), which located about 50 km north of SMA [3]. On the edge of Sundaland, a large basin formed in the north of MVA, named as the Kendeng Zone. In this study, the focus area is Mt. Merapi and Mt. Merbabu which are part of MVA.

Mt. Merapi is one of the most active volcanoes in the world and has an eruption frequency between 2 to 6 years [4]. The type of Mt. Merapi is andesitic basaltic volcanoes with a crater containing a lava dome and has pyroclastic flows [5]. [4] studied seismicity discovered a non-seismic zone within the volcano cone and suggested two magma reservoirs at shallow (1-2 km) and deeper depth (5 km). This hypothesis was supported by the petrological composition study [6].

Interestingly, Mt. Merbabu, located 9 km to the North of Mt. Merapi shows lower activities compared to Mt. Merapi. In this study, we tried to reprocess data from DOMERAPI network with receiver function analysis. Receiver function analysis is a method to image the Earth's distinct layer such as Crust-Mantle boundary using 3- components waveform information from far-distant events. [7] showed complex RFs signals in this area by processing the data in a frequency domain. Alternatively, the deconvolution method in RF can also be done in time domain that could enhance the signal. The objective of this study is to reprocess the same data set using time domain deconvolution to better RFs signal. We compare the results between the frequency domain and time domain process to investigate the benefit between each method and finally help the interpretation.

2. Data and Method

We used recorded waveform event from DOMERAPI temporary seismic network carried out from October 2013 to April 2015. The network consisted of 53 three component seismometers with 4 km station spacing spread out near Mt. Merapi and Mt. Merbabu (Figure 1). We selected recordings of teleseismic earthquake with magnitude higher 6 with epicentral distance between 30-90° and gather more than 130 events. In general, after removing some events that has lower to signal ratio, we collected 50 good RFs per station.

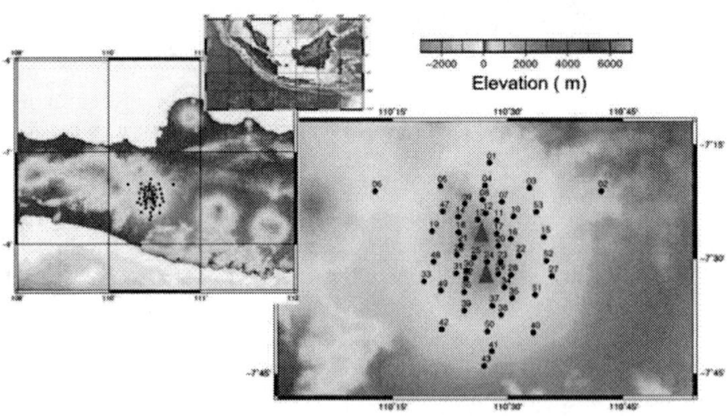

Figure 1. Station Location of Domerapi Network. Red triangle is Mt Merapi and Blue triangle is Mt Merbabu [7]

IOP Conf. Series: Earth and Environmental Science **873** (2021) 012055 doi:10.1088/1755-1315/873/1/012055

Receiver Function

Receiver function technique is a method to generate P-S converted waves from teleseismic data. Ps converted wave formed when the P wave pass through velocity boundaries or discontinuities underneath seismic station. To process the data in RFs, we rotated vertical, North-South and East-West components into vertical, radial and tangential components using back azimuth information between source to events. To improve the Ps signals, further rotation can be done into P, SV and SH component using incidence angle information estimated using ak135 velocity model. In RFs, to isolate Ps waves, the radial component is deconvolved from the vertical component or SV components is deconvolved from the P component.

The deconvolution method can be done either in time or frequency domain. In this study we used the deconvolution technique called iterative time domain deconvolution. The basic of this method is cross-correlation process between radial and vertical component to analyze similarity from both components. Before the deconvolution, first the data will be filtered to reduce the noise. After that, the filtered vertical component is cross-correlated with the radial component to estimate the lag of the first and largest spike in receiver function [8].

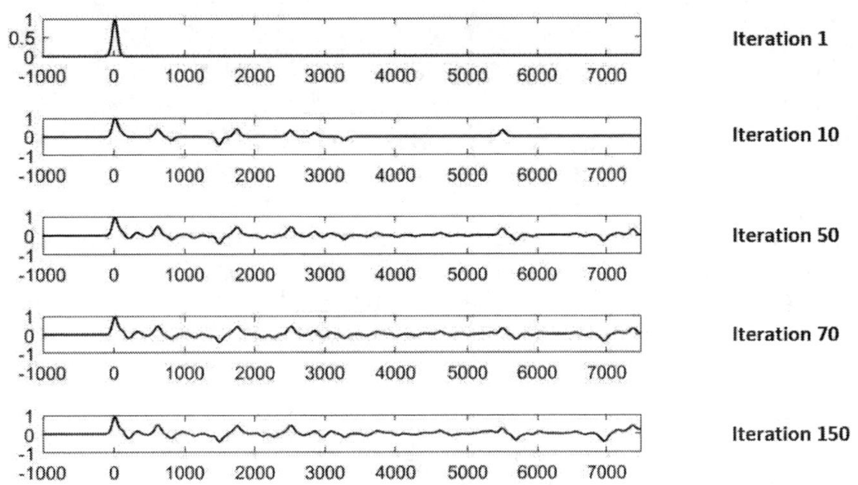

Figure 2. Receiver function prediction result each iteration. Horizontal axis is time at 0.01 s per tick. Zero time is the expected P wave time arrival.

From the cross-correlation result will be obtained the first iteration spike which is a sign of similarity between the signal of the vertical component and the radial component. Then the spike will be filtered with Gaussian width factors that will control the bandwidth of the signal to be estimated (Figure 2). Figure 2 display that typically the first iteration in deconvolution will create the first spike at zero second, and followed by later signals. The result of the filter process will be convoluted with vertical components to produce the prediction of the first iteration radial component. Figure 5 show the difference between radial component estimated from RFs and observation data. Depending of the complexity of the signals, Most RFs show good results after 50 iterations. The whole process will be repeated to estimate spike lag and other amplitudes through iterations until they get results with small errors.

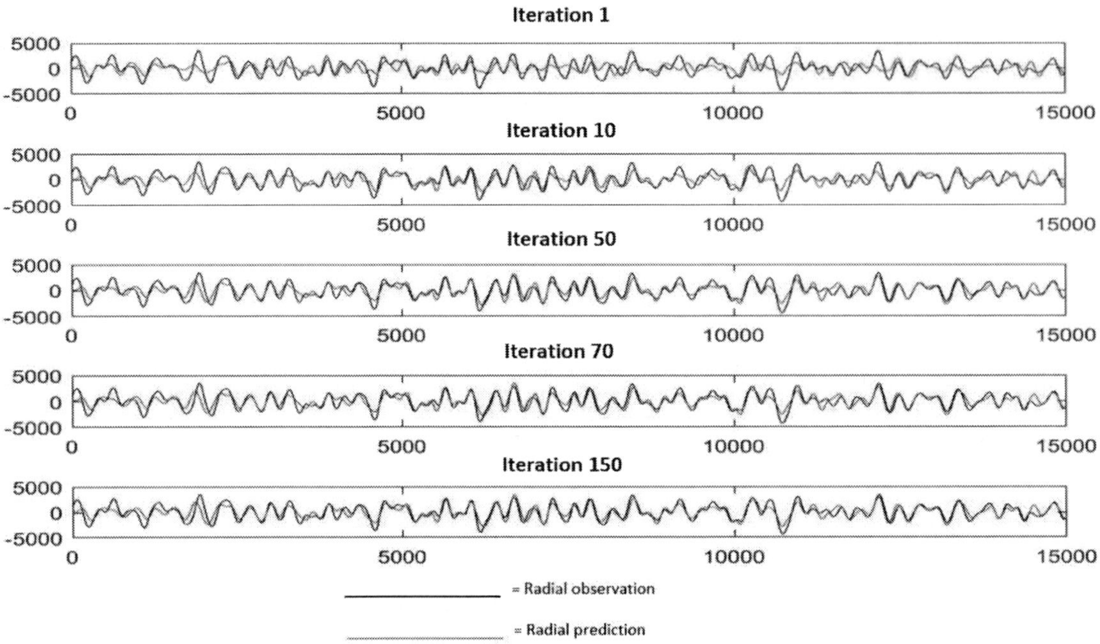

Figure 3. Radial observation compared to radial prediction. Horizontal axis is time at 0.01 s per tick. Zero time represents the arrival of P wave phase.

In this study we also performed H-k stacking method [9]. H-k stacking approach is grid search technique from Zhu and Kanamori to estimate the thickness of crust as well as Vp /Vs ratio below a station. Theoretical delay times of Ps, PpPs and PpSs phases are calculated for each pair of H, k according to a given mean P-wave velocity above the discontinuity of interest. The best estimation of the depth to the discontinuity and the Vp/Vs ratio is expected to coincide with the maximum of the amplitude stack.

3. Result and Discussion

Each receiver function from event was stacked to image subsurface beneath the station. The RF stacked result makes determining the arrival times of p and s waves easier to analyse. H-k stacked was applied to final receiver function stack to observe crustal thickness and Vp/Vs ratio around study area.

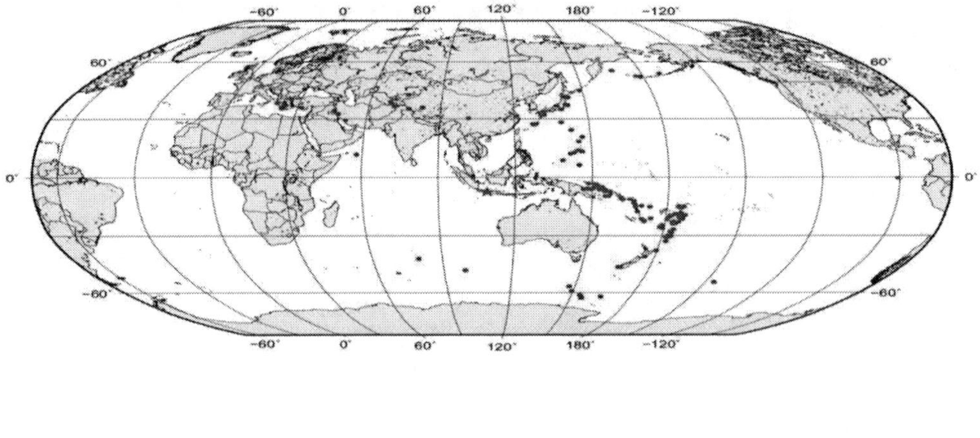

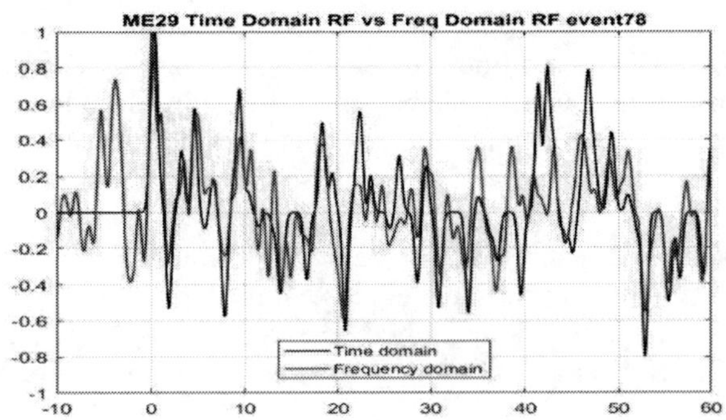

Figure 4. (a) Map of Events distribution in this study. (b) is Plot of RF at station ME29 processed with time domain vs frequency domain.

Figure 4(a) shows earthquake event use in this study. Most of good data come from Japan, Papua New Guinea and New Zealand. An example of one RF signal at station ME29 can be seen in figure 4(b). Black line represents time domain RF. P wave arrive at 0 second is the first signal recorded and have largest amplitude followed by the second recorded signal is Ps converted wave at 4.6177 s. The multiple converted wave which are PpPs recorded at 12.2030 s and PpSs at 16.8207 s. Red line represent frequency domain RF. P wave arrive at 0 second is the first signal recorded and have largest amplitude followed by the second recorded signal is Ps converted wave at 4.75785 s. The multiple converted wave which are PpPs recorded at 12.2403 s and PpSs at 15.8188 s.

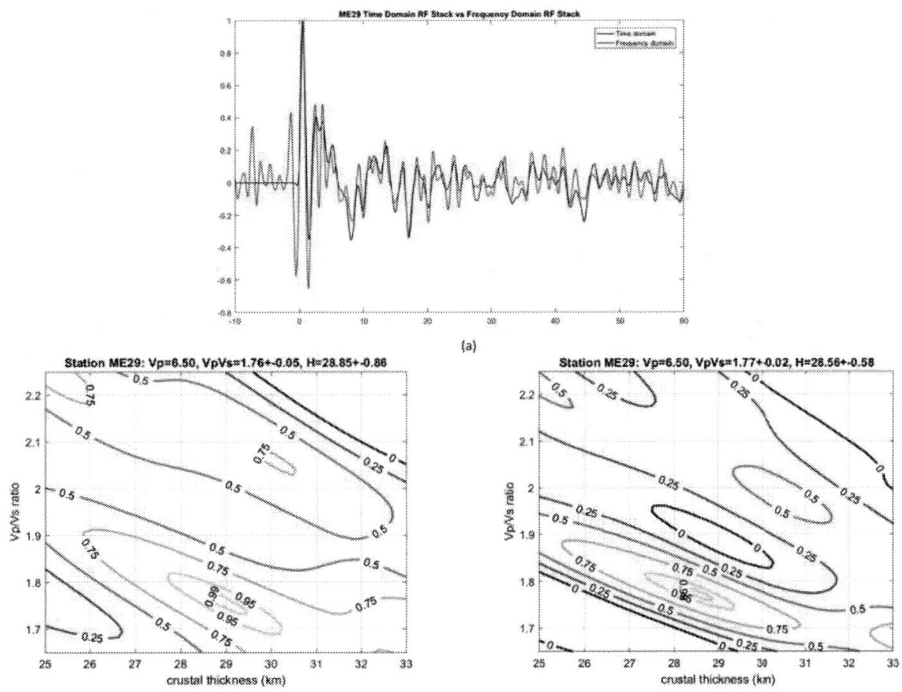

Figure 5. (a) Receiver function stack ME29 time domain vs frequency domain. (b) H-k contour stack RF time domain. (c) H-k contour stack RF frequency domain.

Figure 5(a) shows a comparison between stacked RFs s in the time domain and frequency domain. Black line represents time domain RF and red line for frequency domain. RF comparison from both methods obtain 0.9277 correlation coefficient. In general, stacked RFs in time domain show a simpler signal compared to frequency domain RFs. The frequency domain results also show more wiggles with higher amplitude in later arrival signals.

H-k contour stack from two methods showed by figure 5b (time domain) and 5c (frequency domain). RF time domain method estimate crustal thickness beneath ME29 station is 28.85 km with 1.76 Vp/Vs ratio. In other hands the RF frequency domain method estimate crustal thickness at 28.56 km with 1.77 Vp/Vs ratio. Crustal thickness generated from both methods have fairly similar result.

4. Conclusion

As the conclusion, we proposed that the receiver function time deconvolution method can be used as alternative method for deconvolution data processing since it can generate similar results with frequency domain deconvolution and show simpler signals with less wiggles in the later time. However, still visual inspection is recommended at each RFs to enhance overall data quality.

References

[1] Hall R 2018 *Plate tectonic reconstructions of the Indonesian region* https://doi.org/10.29118/ipa.464.71.84

[2] Smyth H R, Hall R, & Nichols G J 2008 *Cenozoic volcanic arc history of East Java, Indonesia: The stratigraphic record of eruptions on an active continental margin* Special Paper of the Geological Society of America **436**(10) 199–222. https://doi.org/10.1130/2008.2436(10)

[3] Clements B, Hall R, Smyth H R, & Cottam M A 2009. *Thrusting of a volcanic arc: A new structural model for Java* Petroleum Geoscience **15**(2) 159–174 https://doi.org/10.1144/1354-079309-831

[4] Ratdomopurbo A, & Poupinet G 2000 *An overview of the seismicity of Merapi volcano (Java, Indonesia), 1983-1994* Journal of Volcanology and Geothermal Research, **100**(1–4) 193–214 https://doi.org/10.1016/S0377-0273(00)00137-2

[5] Hidayati S, Ishihara K, Iguchi M, Ratdomopurbo A, Mitigation G H, Volcano S, & Yokoyamacho S 2008 *Focal Mechanism of Volcano-tectonic Earthquakes at Merapi Volcano, Indonesia* Indonesia Journal of Physics **19**(3)

[6] Koulakov I, Bohm M, Asch G, Lühr B G, Manzanares A, Brotopuspito K S, Fauzi P, Purbawinata M A, Puspito N T, Ratdomopurbo A, Kopp H, Rabbel W, & Shevkunova E 2007 *P and S velocity structure of the crust and the upper mantle beneath central Java from local tomography inversion* Journal of Geophysical Research: Solid Earth **112**(8) 1–19 https://doi.org/10.1029/2006JB004712

[7] Suhardja S K, Widiyantoro S, Métaxian J P, Rawlinson N, Ramdhan M, Budi-Santoso A 2019 *Crustal thickness beneath Mt. Merapi and Mt. Merbabu, Central Java, Indonesia, inferred from receiver function analysis*

[8] Kikuchi M, & Kanamori H 1982 *Inversion of complex body waves - III* Bulletin - Seismological Society of America **81**(6) 2335–2350

[9] Zhu L, & Kanamori H 2000 *Moho depth variation in southern California from teleseismic receiver functions* J. Geophys Res. **105**(B2) 2969–2980

IOP Conf. Series: Earth and Environmental Science **873** (2021) 012056 doi:10.1088/1755-1315/873/1/012056

Preliminary Results of Receiver Function Forward Velocity Modelling at Merapi Volcano

M F R Auly[1], A K Ilahi[1], I Madrinovella[1], S Widyanti[1], S K Suhardja[1], D Y Fatimah[1], A D Nugraha[2], Z Zulfakriza[2], S Widiyantoro[2,3], J-P Métaxian[4], M Ramdhan[5]

[1] Faculty of Exploration and Production Technology, University of Pertamina, Jl. Teuku Nyak Arief, Simprug, Kebayoran Lama, Jakarta 12220, Indonesia
[2] Global Geophysics Research Group, Faculty of Mining and Petroleum Engineering, Institute of Technology Bandung, Jalan Ganesa No. 10, Bandung 40132, Indonesia
[3] Faculty of Engineering, Maranatha Christian University Bandung, Jalan Surya Sumantri No 65, Bandung 40164, Indonesia
[4] ISTerre, IRD R219, CNRS, Université de Savoie Mont Blanc, Le Bourget-du-Lac, France
[5] Agency for Meteorology, Climatology and Geophysics, Jalan Angkasa I, No. 2, Kemayoran, Jakarta, Indonesia

*corresponding author's e-mail: s4ndy104@gmail.com

Abstract. The tectonic setting of Java island, located at southwestern edge of the Eurasia continent, is dominated by the subduction of Indo-Australia plate. One of the characteristics of active subduction is active seismicity, the generation of arc magmatism and volcanic activity. Mt. Merapi is one example of active volcano related with the subduction process. It is one of the most active volcanoes with location close to high population area. To better understand this area, we employed the Receiver Function technique, a method to image sub surface structure by removing the vertical component from horizontal component. First, we collected high magnitude events and processed RF with water level deconvolution method. Then, we constructed synthetic model with initial velocity input from previous tomography model. Note that we used reflectivity method in generating synthetic model with input parameters matched with parameters from real data processing. Next, we adjusted velocity inputs mainly on tops sediments (1-3 km) to include sediment layers and volcanic rocks, mid-depth low velocity zone that may be related with magma chamber and depth of crust-mantle boundary. Current forward velocity models show a relatively good agreement from 3 stations (ME25, ME32 and ME36). We estimate a thin layer of sediments followed a zone of velocity layer at a depth of 10-15 km and crust-mantle boundary ranging from 26-29 km. In this study, simulated that the signal of sediments layer and low velocity layers interfere main crust mantle boundary that supposed to be highest signal after the P wave in the typical receiver function study.

Keywords: Synthetic Receiver Function, Mount Merapi, Magma Reservoir, Low Velocity Zone

IOP Conf. Series: Earth and Environmental Science **873** (2021) 012056 doi:10.1088/1755-1315/873/1/012056

1. Introduction

Java island is located at the southern part of the Eurasian plate, where the Australian plate is subducting the Eurasian plate. [1] explained that the convergence rate between Australia and Indonesia is 67 mm per year and the slab angle of the slab is getting steeply downward (70°-80°) from a depth of 50 km to the north of Java [2]. Based on the distribution of earthquakes in the global catalogue [3], the plate's depth under the southern part of Java Island is around 100-120 km. The magmatism and volcanic activity that characterize Central Java are primarily affected by the subduction process. Altogether there are two main volcanic arcs, the Southern Mountain Arc (SMA) and the Modern Volcanic Arc (MVA) [4],[5]. Mount Merapi is part of the MVA and the focus of this study.

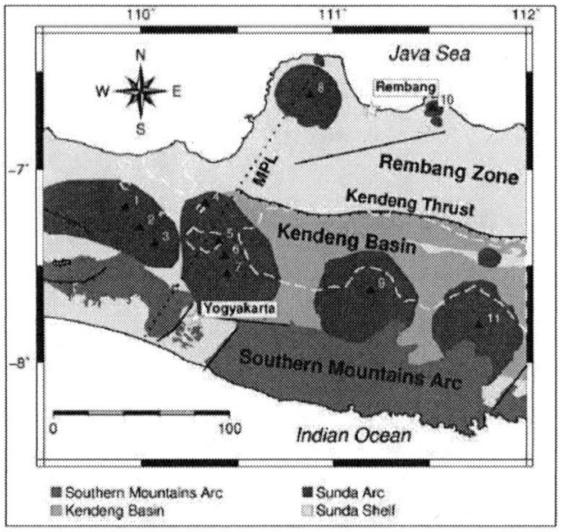

Figure 1. Geological map of the study area [4], with information on volcanoes numbered as, 1: Dieng Volcano Complex; 2: Mt. Sindoro; 3: Mt. Chippy; 4: Mt. Ungaran; 5: Mt. Telomoyo; 6: Mt. Merbabu; 7: Mt. Merapi; 8: Mt. Muria; 9: Mt. Lawu; 10: Mt. Lasem; 11: Mt. Wilis.

Studies on Mt. Merapi have been done using various geophysical and geological methods. Studies using the gravity method with high resolution show evidence of high-density bodies under the crest of Mt. Merapi, Merbabu, and Telomoyo, which can be interpreted as magma reservoirs [6]. [7], using GPS and slope data, concluded that the primary magma reservoir is located between 6-9 km depth below the summit. The result of a tomography study [8] shows there is an anomaly of high Vp/Vs ratio beneath the peak of Mt. Merapi at a depth of ~5 km and ~15 km, which is interpreted as shallow and intermediate magma bodies. Local earthquake tomography was deployed using the data from the same network as this study and found three active zones with a high Vp/Vs ratio beneath the summit [9].

The closest related study is the RFs study that estimates the crust thickness from the DOMERAPI network [10]. The receiver function (RF) is one of the methods used to determine the reflectivity that occurs in the Earth structure. This technique uses the principle that if the P wave encounters a boundary that separates rock layers with high-velocity contrast, an S wave will then be created in the boundary. The converted S wave will then be recorded at the recording station after the P wave because the S wave velocity is always slower than the P wave velocity. If the layer below the boundary has a higher velocity than the value above it, then a positive pulse will be created, and vice versa, if the layer above the boundary has a higher velocity value than the layer below it, the negative pulse will be created. The receiver function is then calculated through deconvolution between the radial component and vertical component. [10] showed complex RFs in this area with lower amplitude signal from crust-mantle boundary and a pattern of negative RFs signals beneath Mt. Merapi. In this study, forward RF modeling has been carried out by generating synthetic RFs from a series of velocity model. Synthetic models are then matched with stacked observed RFs. We focused on the three layers, the first is the sediment of top of layers, a low velocity zone representing a possible magma chamber and crust-mantle boundary depth. This study attempts to provide more detailed information about the structure of the earth's crust beneath Mt. Merapi.

2. Data and method

This study uses recorded data from the DOMERAPI seismic network installed in the area around Mt. Merapi. A total of 53 broad-band seismometers were installed during October 2013 to mid-April 2015, with an average distance between stations is ~ 4km. Earthquake data collected are 150 with magnitude > 5.5 and have epicentral distances of 30 - 100 degrees. The location of the recording station is shown in figure 2.

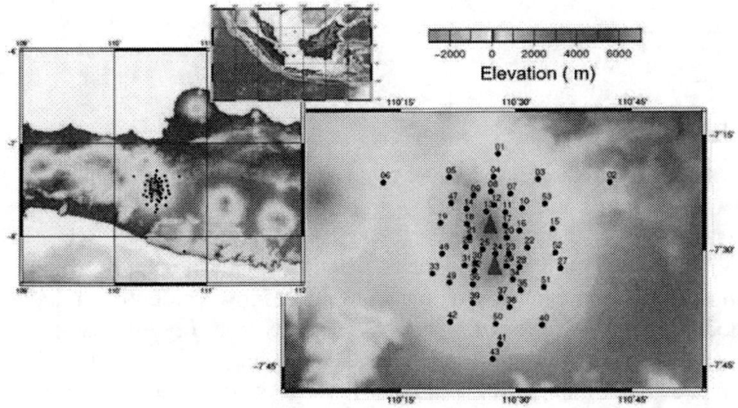

Figure 2. Location of the seismometer installed for this study. The red and blue triangles are Mt. Merapi and Mt. Merbabu [10]

To better understand geological structure beneath the mountains, we performed forward modelling to create synthetic RFs from the determined velocity model. Then, the synthetic seismogram is matched with the observed RFs until they showed good agreement. We used the reflectivity code [11]. Note that the forward modelling carried out in this study used the assumption that the subsurface conditions are isotropy. Figure 3 is an example of synthetic RF result from synthetic velocity model based on ak135 velocity model.

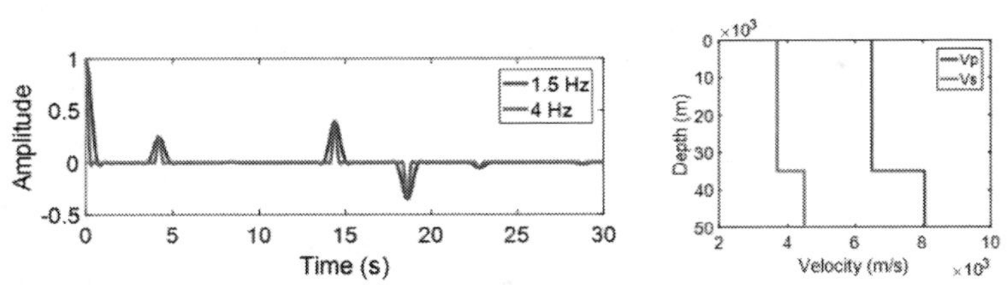

Figure 3. Synthetic RF results (left) from a synthetic crust and upper mantle velocity model (right) based on the ak135 velocity model. In the synthetic RF results, the red and blue curves are the results using a filter frequency of 1.5 Hz (low frequency) and 4 Hz (high frequency).

In the example of synthetic RF, we used 2 different frequency which are low frequency and high frequency, although in reality, the frequency filter used is low frequency because it is more representative of the frequency of the earth. The purpose of using a high-frequency filter in this study is to simplify the RF synthetic modelling that causes a high-frequency response in RF readings such as thin layers and multiples. In a simple model that has a change in Vp and Vs at a depth of 35 km, we can observed 4 signals : The first positive signal at 0 second showed the incoming P wave arrival, the second positive signal at 4.6 second showed the converted Ps phase from the boundary, and the third positive and negative signal are the multiples. We modified the velocity model input by adding two layers: the first is sediment layer and the second is low velocity layer representing possible magma chambers beneath the Mt. Merapi.

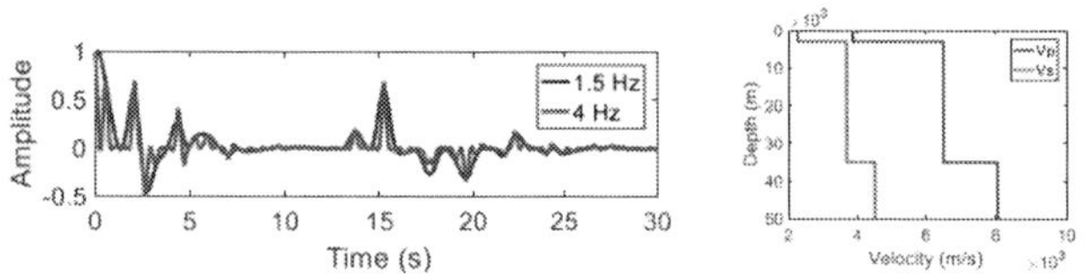

Figure 4. Synthetic RF for velocity models with a 3 km thick sedimentary layer near the surface and Moho at depth of 35 km.

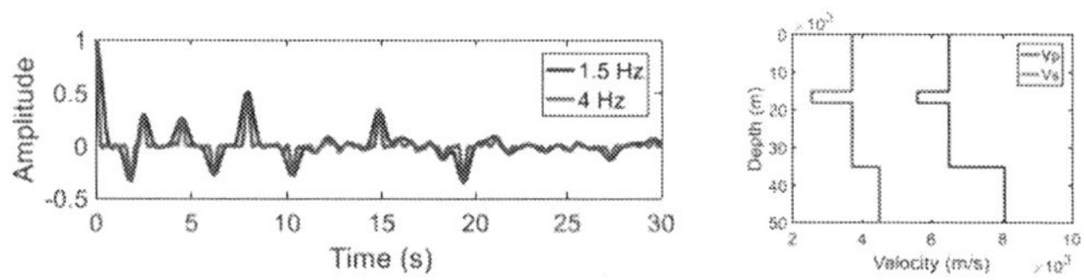

Figure 5. Synthetic RF for velocity model with a 3 km thick of low velocity layer at a depth of 15 km and Moho at depth of 35 km.

. Figure 4 is a synthetic RF for a model with a sediment layer that has a thickness of 3 km and Moho. The sediment layer used has Vp = 3864 m/s and Vs = 2273 m/s, we used velocity number from Christensen [12]. Figure 4 also showed that sediment layer at top will add complexity the RFs signal especially at later arrivals. Additional sediment at top of the model will also lower the amplitude signals from the Moho signal. Figure 5 shows that synthetic RFs with 3 km thick of low velocity at a depth of 15 km. RFs response from this model showed a series of signals that comes from reverberation at low velocity layer and Moho. We noticed from this simple modelling that the negative amplitude before Moho signals can be caused by the presence of a layer of sediment near the surface and the presence of a layer with low velocity at a certain depth though full waveform modelling is necessary.

3. Result and discussion

As a preliminary understanding for constructing a synthetic velocity model, we used a 1-D velocity model from a tomography study by Ramdhan et al. (2019) to estimate the depth of the low velocity zone and Moho at ME25, ME32, and ME36 station.

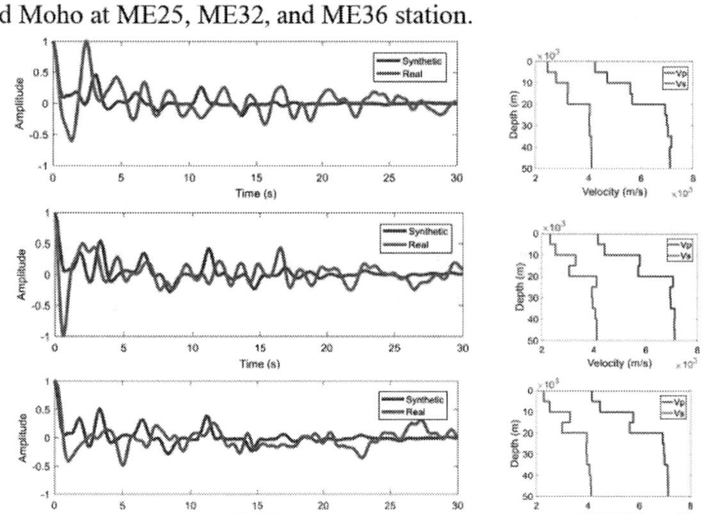

Figure 6. Synthetic RF results (blue line) from the tomography velocity model [8] and the results of RF observations (red line) from Suhardja et al. (2019) for stations ME25, ME32 and ME36 (top to bottom)

Figure 6 is a comparison of the synthetic RF results based on the tomography velocity model from tomography study [8] and the results of the observed RF results from previous study [10]. Visually, the similarity of the resulting signals do not show good agreement. The RF results from tomography are also unable to respond to signals that have negative amplitudes at the time of arrival of 1-2 seconds as in the results of observed RF, which may be the result sediments at shallow depths below Mt. Merapi.

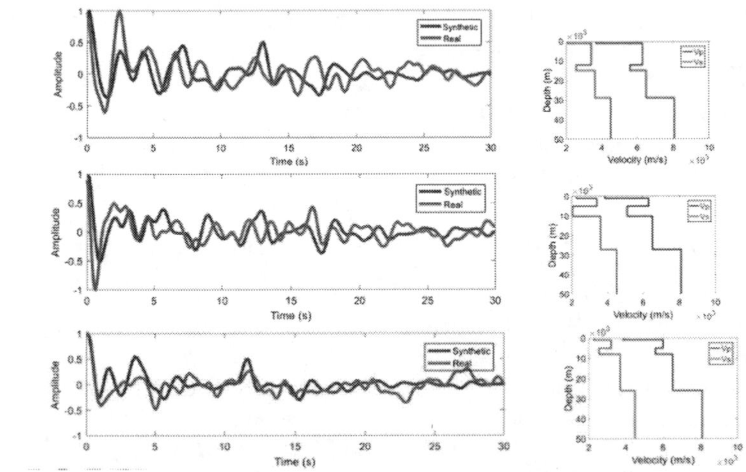

Figure 7. Final synthetic RF results compared with observed RF results (left) and final synthetic velocity models (right) for ME25, ME32, and ME36 station.

Figure 7 show the best velocity model that we can match with stack RFs by finding the best combination of shallow sediment thickness and depth of low velocity layer as magma chamber, dan Moho depth. Station ME25 and ME32 station have shown a good visual matching at the arrival time of 1-10 s, but the RF result at ME36 station seems to lack good visual matching. We observed for arrival time after 10 s, the signals have a poor correlation with real data. This might be related with more complex structures and its multiples. The velocity models that has sediment layer, LVZ, and Moho for each station which can then be made a simple subsurface section (figure 8).

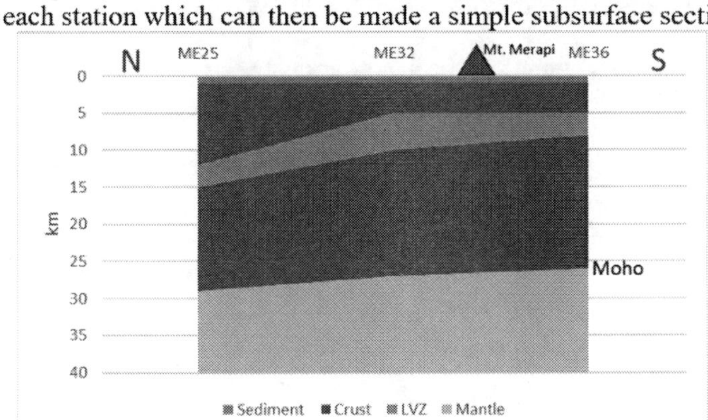

Figure 8. The plot results of the LVZ depth and Moho depth based on the synthetic velocity model of the ME25, ME32 and ME36 station.

In the summary, negative amplitude that appears in the RFs at 0-4 seconds can be caused by two objects, a thin layer of sediment near the surface and the presence of a low velocity zone at a certain depth. Based on the results of synthetic RF at the ME25, ME32, and ME36 stations, the depth of the magma chamber varies. The depth of the magma chamber seems to be getting shallower towards the Mt. Merapi area, as shown by the simple cross-sectional model in figure 8. The crust thickness also getting thin near Mt. Merapi.

4. Conclusion

Based on the results of RF forward modeling, the main signals that appear in the results of RF observations consist of 3 main sources, the sedimentary rock layer with a thin thickness, the low velocity zone which is identified as the magma chamber and the boundary of the earth's crust and the upper mantle, Moho discontinuity. However, identifying the source of the multiples from the observed RF results is quite difficult to do because the multiples response to the RF signal is very complex so that the compatibility of the synthetic RF signal with the observed RF after 10 s does not have a good

IOP Conf. Series: Earth and Environmental Science **873** (2021) 012056 doi:10.1088/1755-1315/873/1/012056

correlation.The results of this study indicate that the depth of the LVZ which is interpreted as a magma reservoir obtained from the synthetic velocity model ranges from 5-15 km with a depth that is getting shallower towards the south, the area around Mt. Merapi. These results support the results of previous studies which explain that the result from tomography study [8] showed there is an anomaly of high Vp/Vs ratio beneath the peak of Mt. Merapi at a depth of ~5 km and ~15 km and study using GPS and slope data which suggests that the main magma reservoir is located between a depth of 6-9 km below the summit [7].

References

[1] Simons W J F, Socquet A, Vigny C, Ambrosius B A C, Haji Abu S, Chaiwat Promthong, Subarya C, Sarsito D A, Matheussen S, Morgan P & Spakman W 2007 *A decade of GPS in Southeast Asia: Resolving Sundaland motion and boundaries* J Geophys Res **112**

[2] Koulakov I, Bohm M, Asch G, Lühr B-G, Manzanares A, Brotopuspito K S, Fauzi P, Purbawinata M A, Puspito N T, Ratdomopurbo A, Kopp H, Rabbel W & Shevkunova E 2007 *P and S velocity structure of the crust and the upper mantle beneath central Java from local tomography inversion* J. Geophys Res. **112** B08310

[3] Weston J, Engdahl E R, Harris J, Di Giacomo D, & Storchak D A 2018 *ISC-EHB: reconstruction of a robust earthquake data set* Geophysical Journal International **214**(1) 474-484

[4] Smyth H R, Hall R & Nichols G J 2008 *Cenozoic volcanic arc history of east Java, Indonesia: the stratigraphic record of eruptions on an active continental margin In: Draut, A. E., Clift, P. D., Scholl, D. W. (Eds.), Formation and Applications of the Sedimentary Record in Arc Collision Zones* Geol. Soc. Am. Spec Paper **436** pp 199–222

[5] Clements B, Hall R, Smyth H R & Cottam M A 2009 *Thrusting of a volcanic arc: a new structural model for Java, Petrol* Geosci **15** 159–174

[6] Tiede C, Camacho A G, Gerstenecker C, Fernández J & Suyanto I 2005 *Modeling the density at Merapi volcano area, Indonesia, via the inverse gravimetric problem, Geochem* Geophys. Geosyst. **6** Q09011

[7] Beauducel F, & Cornet F H 1999 *Collection and three-dimensional modeling of GPS and tilt data at Merapi volcano, Java* Journal of Geophysical Research: Solid Earth **104**(B1) 725-726

[8] Ramdhan M, Widiyantoro S, Nugraha A D, Métaxian J-P, Rawlinson N, Saepuloh A, Kristyawan S, Sembiring A S, Budi-Santoso A, Laurin A & Fahmi A A 2019 *Detailed seismic imaging of Merapi volcano, Indonesia, from local earthquake travel-time tomography* J. Asian Earth Sci. Vol **177** p134-135

[9] Widiyantoro S, Ramdhan M, Metaxian J P, Cummins P R, Martel C, Erdmann S, Nugraha A D, Budi-Santoso A, Laurin A & Fahmi A A 2018 *Seismic imaging and petrology explain highly explosive eruptions of Merapi Volcano, Indonesia* Sci. Rep. **8**(1) 13656.

[10] Suhardja S K, Widiyantoro S, Métaxian J P, Rawlinson N, Ramdhan M, & Santoso B S 2019 *Crustal thickness beneath Mt. Merapi and Mt. Merbabu, Central Java, Indonesia, inferred from receiver function analysis*

[11] Levin V, & Park J 1997 *P-SH conversions in a flat-layered medium with anisotropy of arbitrary orientation. Geophysical Journal International* **131**(2) 253-266

[12] Christensen N I, & Stanley D 2003 *Seismic velocities and densities of rocks* In International Geophysics Vol. **81** pp. 1587-1594

Initial Result of P Wave Tomography Model in Sunda-Banda Arc Transition using FMTOMO

P T Brilianti[1], Haolia[1], M I Sulaiman[1], S S Angkasa[1], S Widyanti[1], I Herawati[1], S K Suhardja[1], A D Nugraha[2], Z Zulfakriza[2], S Widiyantoro[2,3], M Ramdhan[4]

[1] Faculty of Exploration and Production Technology, University of Pertamina, Jl. Teuku Nyak Arief, Simprug, Kebayoran Lama, Jakarta 12220, Indonesia
[2] Global Geophysics Research Group, Faculty of Mining and Petroleum Engineering, Institute of Technology Bandung, Jalan Ganesa No. 10, Bandung 40132, Indonesia
[3] Faculty of Engineering, Maranatha Christian University Bandung, Jalan Surya Sumantri No 65, Bandung 40164, Indonesia
[4] Agency for Meteorology, Climatology and Geophysics, Jalan Angkasa I, No. 2, Kemayoran, Jakarta, Indonesia

*Corresponding author: s4ndy104@gmail.com

Abstract. Our study area is located near island Sumbawa, Sumba, Flores, West Timor, Indonesia and East Timor, popularly known as Sunda-Banda arc transition zone. The tectonic setting is mainly controlled by the movement of the oceanic lithosphere Indo-Australian plate subducting the Eurasian plate and Northward migration of Australian continental lithosphere into western Banda-arc in the region of Flores, Sumba and Timor island. We tried to image velocity structure beneath these regions using regional events and tomography inversion model. We collected 5 years of regional events from the Indonesian Agency of Meteorology, Climatology and Geophysics. In total, we reserved 3186 events recorded on 29 stations. For data processing, we used fast marching method as ray tracing between sources and receiver. We then employed subspace inversion as the tomography procedure to estimate the best velocity model representing the tectonic model in the region. Hypocenter data distribution is concentrated on shallow parts of the region and along the Benioff zone down to a maximum depth of 400 km. One of challenge of this study is that although events are abundance, the stations used are mostly located onshore and does not extend in the south-north direction that leads us to under determined problem in the inversion process. However, checker-board models show most our target area can be retrieved to its initial model with sign of smearing effects shown start from a depth of 50 km. After six iteration and optimized selection of damping and smoothing parameters, we observed low velocity anomaly under Bali, Lombok, Sumba, East Nusa Tenggara at shallow depth that may be related with volcanic activity. Deeper low anomaly can also be seen that may be related with partial melting process. A band of fast velocity is clearly seen that goes deepen to the north depicting subducting slabs own to a depth of 300 km. We also observed a possible of fast velocity in the northern part of our stations at shallow depth that we believe may represent the back arc thrust.

Keywords: Sunda-Banda Arc, Subduction, Back Arc Thrust, FMTOMO.

1. Introduction

The tectonic setting of Indonesia is dominated by the interaction of three major plates. Indo-Australian Plate moves northerly, Pacific Plate which moves westerly, and the Eurasian plate [1]. One example of complex interaction is in eastern part of Indonesia especially in the Sunda-Banda Arc zone and also the subject of this study (Figure 1). The transition zone is affected by the movement of Indo - Australian Plate northerly with a convergence rate of around 7.7 mm/year [2]. The Indo-Australian plate has a variety of subduction ages: in south Java, subduction began from 40 Ma, and in the eastern Indonesia, began from 110 Ma [3]. The differences in subduction age affect the subduction dip. The angle of the slab dipping reaches 55° - 60° in Sunda Arc and 75° - 80° in Banda Arc [4]. One interesting geological feature is the occurrence of Flores back-arc thrust system located in Northeast Bali to Flores [5]. [6] proposed that the mechanism of back-arc thrust is more influenced by the subduction rather than the pressure of volcanic arcs. Based on previous historical earthquakes from BMKG, Flores back - arc thrust generated the 2018 Lombok's earthquake with a 7.0 magnitude that awakens a 50 km tsunami. This area has the potential for tectonic earthquakes with various scales. This study tried to image the geological structure to better understand the geodynamic system of the region such as the relation between subduction and volcanic arc and seismicity pattern.

2. Data and Method

This study is a joint research between Universitas Pertamina and the Agency for Meteorology, Climatology and Geophysics, Indonesia. We selected events data with a ≥ 3 magnitude and 20-400 km depth range within 4 years (January 2015 - December 2018). The total number of earthquake events is 3186. Figure 1 show the seismicity patterns of our data set. Most event shows a gradual depth increase from South to North direction. Interestingly a gap of shallow-mid range depth seismicity is seen between 121° to 123° longitude near the Timor Island. We also observed a series of deep events with depth more than 300 km at the north section of our study. The same area shows a few numbers of shallow events close to geological featured plotted as back-arc thrust.

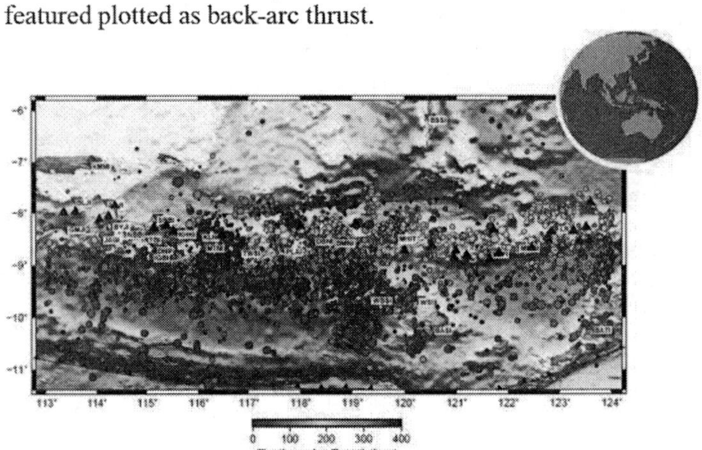

Figure 1. Map of distribution of seismicity and stations in the study area using BMKG data. Magenta inverted triangles represent the seismic stations. Red to blue circles represent the epicenters of earthquakes between January 2015 and December 2018 as a function of earthquake depth. Black triangles are volcanoes.

To illuminate structures in this study, we applied P wave delay time seismic tomography method. We used the fast-marching method for ray tracing reconstruction between sources and receivers [7]. We preferred to select 1-D velocity model from ak135. For the parameterization, we set an interval depth of 56 km in vertical direction, 105.45 km of x-direction (longitude) and 56.61 km of y-direction (latitude). Due to imbalances data set and grid parameterization, we applied damping and smoothing, and the number is set by using the trade-off between data variance and model.

3. The result, Discussion, and Conclusion

3.1. Checkerboard Resolution Test

The checkerboard resolution test is used to investigate an area that has a good resolution. The input model used positive and negative perturbation of 0.2 km/s relative to the 1-D reference velocity model. The size of grid anomaly is adjusted with the expected model size of our geological model and similar with our real model inversion.

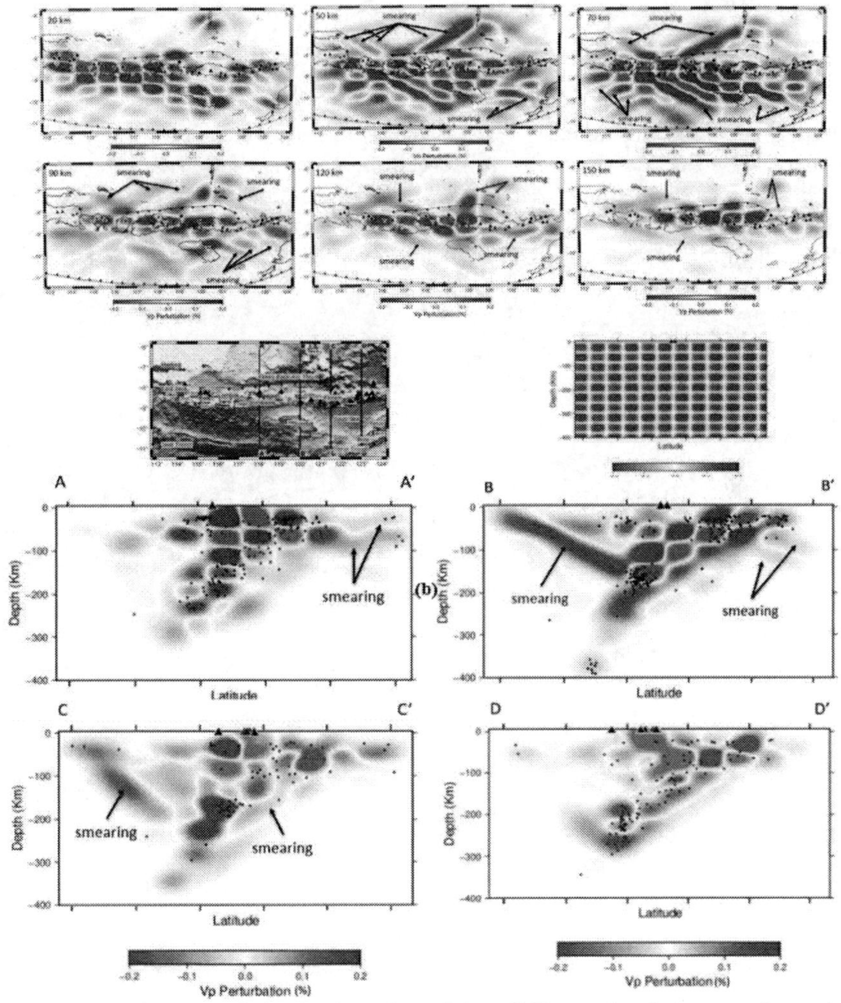

Figure 2. Top is horizontal slices of checkerboard model at different depth. Bottom is vertical slices of checker boards at different cross section from north (left) to south (right)

The checkerboard horizontal section (Figure 2) shows good structures recovery beneath East Java, Bali, Lombok, Sumbawa, Flores, Sumba, Savu Sea, Timor, and Flores back-arc thrust at the depth of 20 to 70 km. We confidence to image our target model within this area though the resolution start to disappear from a depth of 200 km. Vertical cross section at Figure 2 show a possible smearing effect from the North and less resolution in the East, coincidence with the area of seismicity gap near Timor island.

3.2. Vp Tomography Model

Figures 3 and 4 show the final tomography model in horizontal slices and vertical slides. Horizontal slices at shallow section displays a slower or negative velocity anomaly in Bali, Lombok, western Sumbawa, western Flores, and Sumba that coincidence with volcanoes line. Low velocity anomaly also seen beneath western Flores and southern Sumba that extend at deeper depth. We pre-interpreted the low anomaly might be related with a partial melting zone. [8] mentioned that the deeper low anomaly is the result of continent-arc collision related to the northern boundary of the Indo-Australian plate. From the previous study, [9] mentioned that they presented the low anomaly beneath Flores and NE of Sumba at a depth of ~50km. [10] also mentioned that they presented a low shear wave velocity anomaly at a depth of 15-25 km in this area. Vertical section from the west to east shows a similar features of fast velocity model that the dip from south to north direction. They slightly show that the extend of fast velocity anomaly is deeper in the east. The fast velocity most likely represents the subducting Indo-Australian plate to the Eurasian plate. We observed the dipping of slab changes from slightly steep at shallow depth and very steep from a depth of 80 km. A possible break in slab is seen at line D near Timor Island that coincides with a gap of seismicity in this area. A feature of back-arc thrust is slightly seen though not clear in the northern part of the stations. [9] detected of forearc sliver beneath the Savu Sea, central and eastern Flores, and western Timor, however we have not seen that due to less station number in the East.

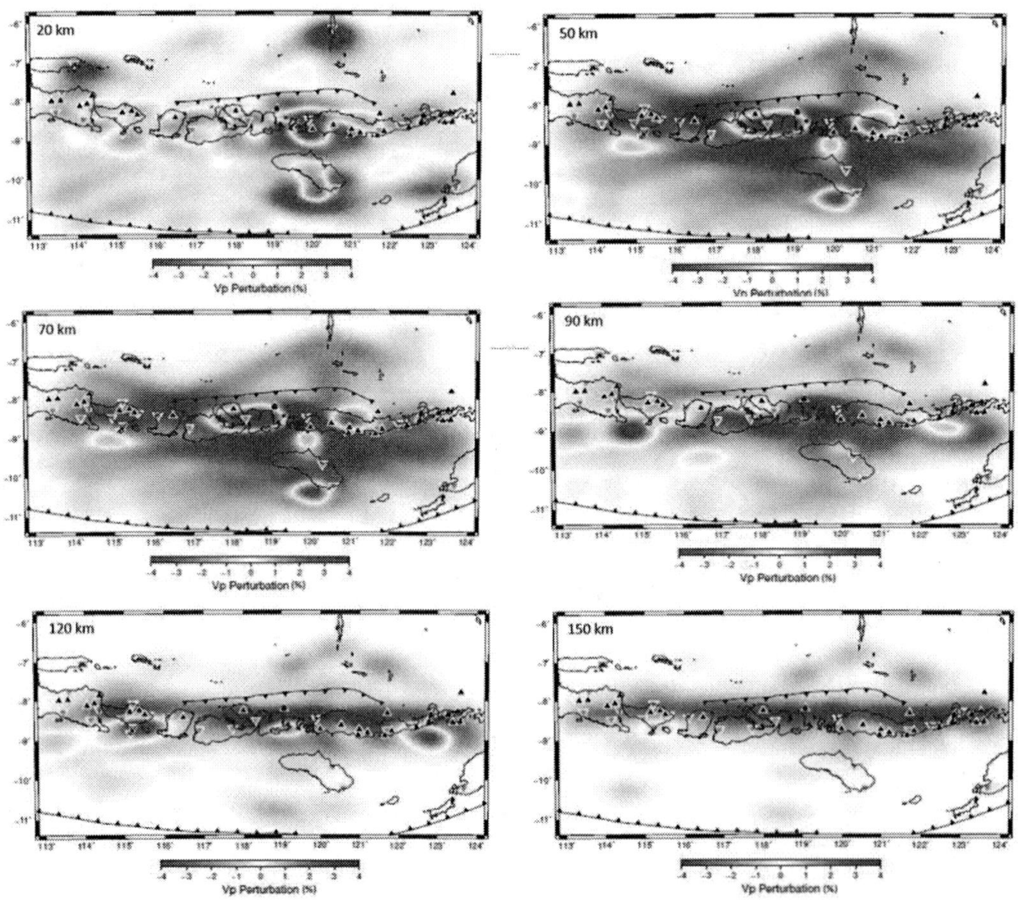

Figure 3. Horizontal slices for Vp models for the depth section at 20-250 km of depth.

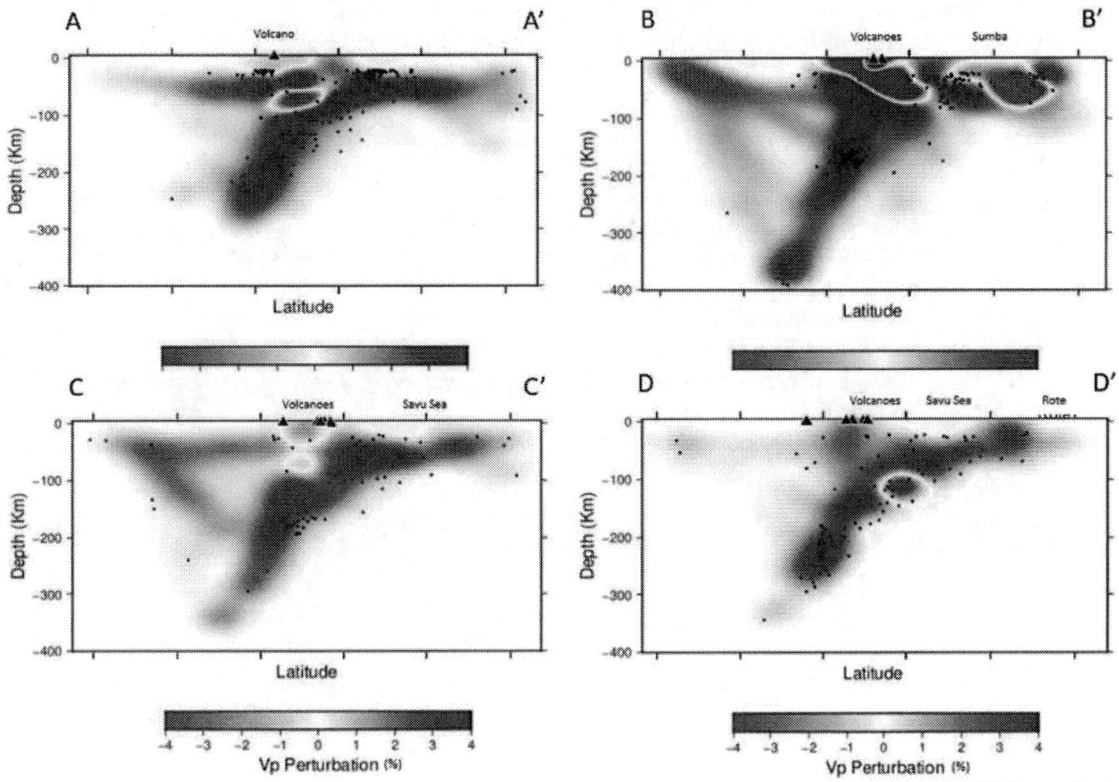

Figure 4. Tomogram Vp models for the vertical section along A-A', B-B', C-C', and D'D' lines. Models are plotted with a perturbation scale of ± 4 km/s. The location of vertical sections is shown in Figure 2

3.3. Conclusion

As the conclusion, this study promises a good tomography model in illuminating subduction slab in the Sunda-Banda arc. A zone of low velocity band is also seen near volcanic line and partial melting area. However, we would like to improve our imaging by adding more regional data set in the near future. We would like to thank the Agency for Meteorology, Climatology and Geophysics, Indonesia for providing data and discussion

References

[1] Hall R 2002 *Cenozoic geological and plate tectonic evolution of SE Asia and the SW Pacific: computer-based reconstructions, model and animations* Journal of Asian Earth Sciences **20**(4) p 353–431

[2] DeMets C, Gordon, R G, & Argus, D F 2010. *Geologically current plate motions*. Geophysical Journal International **181** (1) 1 – 80

[3] Müller R D, Roest W R, Royer J Y, Gahagan L M, & Sclater J G 1997. *Digital isochrons of the world's ocean floor* Journal of Geophysical Research (JGR) **102**, 3211-3214

[4] Royden L H, & Husson L 2009 *Subduction with variations in slab buoyancy: models and application to the Banda and Apennine systems* In Subduction Zone Geodynamics (pp. 35–45) Springer Berlin Heidelberg

[5] Silver E A, Reed D, McCaffrey R, & Joyodiwiryo Y 1983a *Back arc thrusting in the eastern Sunda arc, Indonesia: A consequence of arc-continent collision* Journal of Geophysical Research: Solid Earth **88**(B9) 7429–7448

[6] Silver E A, Breen N A, Prasetyo H, & Hussong D M 1986 *Multibeam study of the Flores backarc thrust belt, Indonesia* Journal of Geophysical Research: Solid Earth **91**(B3), 3489–3500.

[7] Rawlinson N, A M Reading, and B L N Kennett 2006 *Lithospheric structure of Tasmania from a novel form of teleseismic tomography* J. Geophys. Res **111** B02301 DOI:10.1029/2005JB003803

[8] Harris R 1991 *Temporal distribution of strain in the active Banda orogen: A reconciliation of rival hypotheses* Journal of Southeast

[9] Supendi P, Nugraha A D, Widiyantoro S, Abdullah C I, Rawlinson N, Cummins P R, & Miller M S 2020 *The fate of the Forearc lithosphere at arc-continent collision zones: Evidence from local earthquake tomography of the Sunda-Banda Arc Transition, Indonesia* Geophysical Research Letters **47**(6) e2019GL086472

[10] Porritt R W, Miller M S, O'Driscoll L J, Harris C W, Roosmawati N, & da Costa L T 2016 *Continent–arc collision in the Banda Arc imaged by ambient noise tomography* Earth and Planetary Science Letters **449** 246–258

IOP Conf. Series: Earth and Environmental Science **873** (2021) 012058 doi:10.1088/1755-1315/873/1/012058

Comparison of P-Wave Tomography Model in Sunda-Banda Arc by Using Data from BMKG and ISC

P T Brilianti[1], M S Haq[1], Haolia[1], M I Sulaiman[1], R P Nugroho[1], A S B Oka[1], W Pranowo[1], S K Suhardja[1], A D Nugraha[2], Z Zulfakriza[2], S Widiyantoro[2,3], M Ramdhan[4]

[1] University of Pertamina, Jl. Teuku Nyak Arief, Simprug, Kebayoran Lama, Jakarta 12220, Indonesia
[2] Global Geophysics Research Group, Faculty of Mining and Petroleum Engineering Institute of Technology Bandung, Jalan Ganesa No. 10, Bandung 40132, Indonesia
[4]Agency for Meteorology, Climatology and Geophysics, Jalan Angkasa I, No. 2, Kemayoran, Jakarta, Indonesia

*Corresponding author: s4ndy104@gmail.com

Abstract. The tectonic setting of our study area is located between the Island of Java and Timor Leste. The complexity of this region is started with two different plates, The Indo-Australian plate and the Eurasian plate that move with different orientations and convergence rate. This area also shows active seismic activity and has a series of active volcanoes as a product of subduction and collision. To deepen understand this area, we perform delay time tomography using FMTOMO package that includes 3-D finite-difference based ray tracing and sub-space inversion procedure. We used two different sets of data, the first one is 4 years data catalog from the Indonesian Agency of Meteorology, Climatology and Geophysics, and the second one is 47 years of data from the International Seismological Centre. Data from the local Indonesian show agency shows a fewer number of events but more focus clusters. Meanwhile, the data from ISC catalog has more events and evenly distributed data. However, we also noticed that data from ISC has cluster events located at the same depth that can be improved with events relocation for better depth estimation. The Checkerboard models from both data set show a comparable result, though data from ISC show a better recovered model at a deeper depth and shallow part in the eastern area. The checkerboard from the local Agency shows slightly better results in the shallow part. Next, we invert delay time for each data set using we optimized damping and smoothing parameters. Final tomogram models show that data from the local Agency show a more continuous fast velocity band representing a downgoing subducting slab and possible back-arc thrust while results from the ISC data show a more detached fast velocity band that could be contributed from fixed depth problem in the data set. However, we noticed that data from ISC show a higher amplitude low-velocity anomaly especially in the shallow depth

Keywords: Sunda-Banda Arc, Back Arc Thrust, FMTOMO

1. Introduction

The movement of the Indo-Australian plate to the north, subducting the Eurasian plate, has a substantial impact on the tectonic setting of the Sunda-Banda Arc. The Indo-Australian plate is moving northward at a rate of 7.7 mm/year relative to the Eurasian Plate [1]. Based on the modeling of ocean floor age and spreading rate [2] shows that Bali Island to Nusa Tenggara Timur has a variety of subduction ages that the age of the oceanic floor in the East is older than the West. As an additional, back-arc thrust structure formed for the slab's pressure continually made the accumulation of energy in northeast Bali Island to Flores is known as Flores Back-Arc Thrust. This thrust estimate has a depth of up to 60 km and a dip of 30° [3]. Sunda-Banda Arc has various seismic activities. [4] showed that seismicity in Bali to Nusa Tenggara Timur is dominated by the earthquake mechanism of a normal fault with shallow depth. The earthquake that occurred in this area was recorded as a large and destructive earthquake. The objective of this study is to better image geological features in this region using delay time P wave tomography and arrival time recorded by permanent station. We used two different data set, the first ones 4 years data from the Indonesian Agency of Meteorology, Climatology and Geophysics and the second one from the International Seismological Centre (ISC) with 30 years recorded data. The result of the study is expected to better understand the tectonic systems in Bali up to Nusa Tenggara Timur by comparing results from different data set.

2. Data and Method

This study is a collaboration project between Department of Geophysical Engineering, Universitas Pertamina and the Indonesian Agency of Meteorology, Climatology and Geophysics (BMKG). The BMKG data collection period is for 4 years (January 2015-December 2018) with number of 3186 earthquakes. The ISC data collection is for 47 years (November 1969-October 2016) with the number of earthquake events is 8234. Both data set have events with magnitude higher than 3 and a 20-400 km depth range. The station that is used in the study area is 29 for BMKG stations and 37 for ISC stations. Figure 1 shows data distribution in the region for both data set. ISC data set distribution shows a more distributed events especially in the shallow section in the South and deep events in the North. ISC also shows shallow data set di North close to the area where back-arc thrust events located.

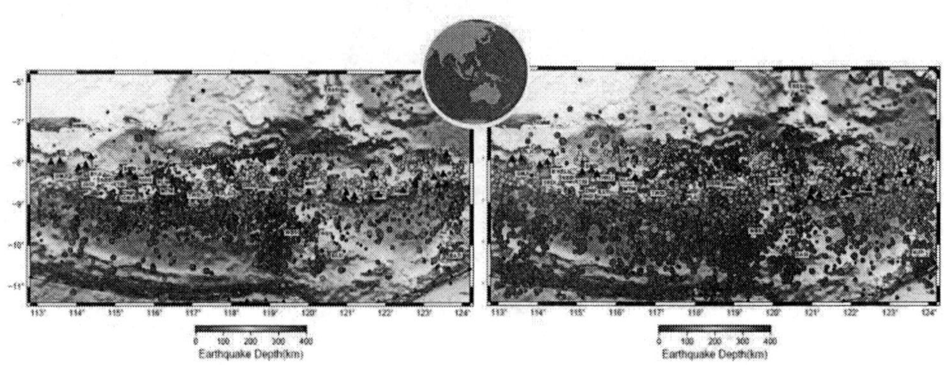

(a) (b)

Figure 1. Map of distribution of seismicity and station in the study area with (a) BMKG data and (b) ISC data. The seismic stations are represented by magenta inverted triangles. The epicenters of earthquakes are represented by red to blue circles as a function of depth. Volcanoes are represented by black triangles.

This study applied delay time seismic tomography to illuminate sub surface structure. We used the fast-marching method to construct ray path between sources and receivers [5]. For parameterization, we set the interval grid size of 56.61 km in latitude direction, 45 km in longitude direction and 40.5 km in vertical direction. The subspace inversion is applied to solve the inversion problem. The damping and smoothing values are chosen based on the trade-off between data and model variance, with the goal of minimizing data variance while maintaining as little model variance as possible.

3. The result, Discussion, and Conclusion

3.1. Checkerboard Resolution Test

We used the checker-board resolution by adding a pattern of positive and negative anomaly next to each other as the velocity model. By using the same combination of source and receiver, we tried to invert synthetic data set and compared with the input velocity model. This technique is performed to check how deep our resolution target, size of geological features, and which area has higher degree of confidence for modeling. We set the input velocity anomaly into ± 0.2 km/s to the 1-D reference ak135 velocity model. The input model also uses the same interval grid as the final model mentioned before. The results of the checkerboard test can be seen in on Figure 2 for horizontal slices and Figure 3 for vertical slices. In general, although the number of events from BMKG catalog are much less than the data from ISC, the pattern and quality of checkboard does not differ a lot. However, better resolutions are seen at deeper depth especially at western part of the model.

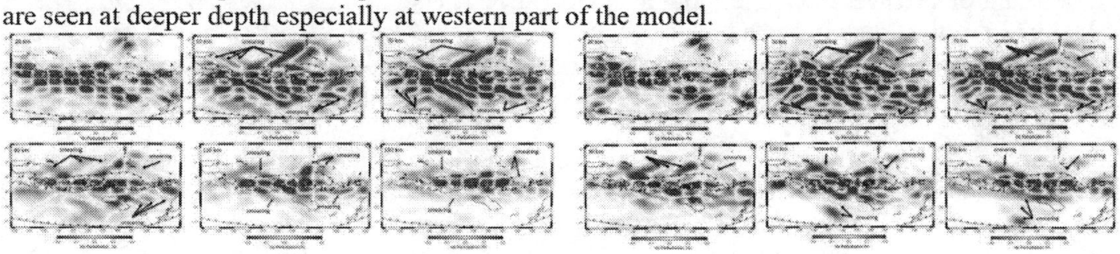

(b) (c)

Figure 2. (a) On the left side is the result of the checkerboard resolution test for the horizontal slice in various depths using BMKG data. (b) On the right side is the result of the checkerboard resolution test for the horizontal slice in various depth using ISC data

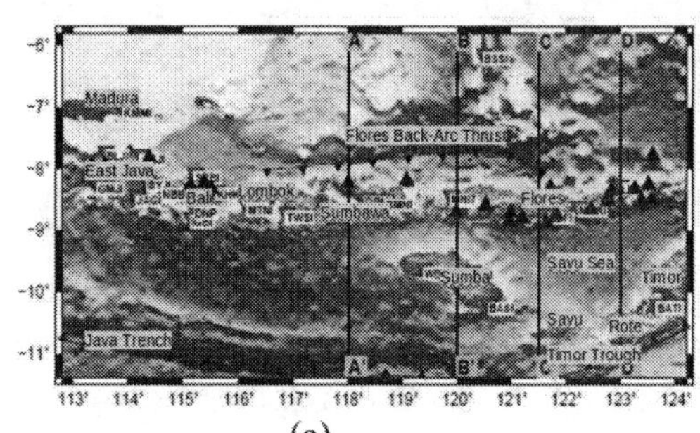

(a)

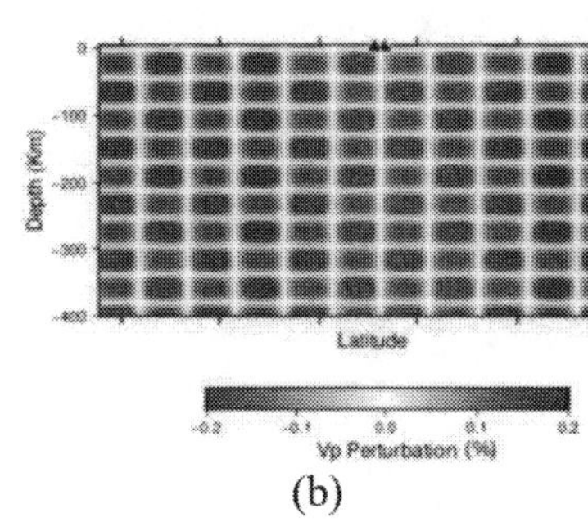

(b)

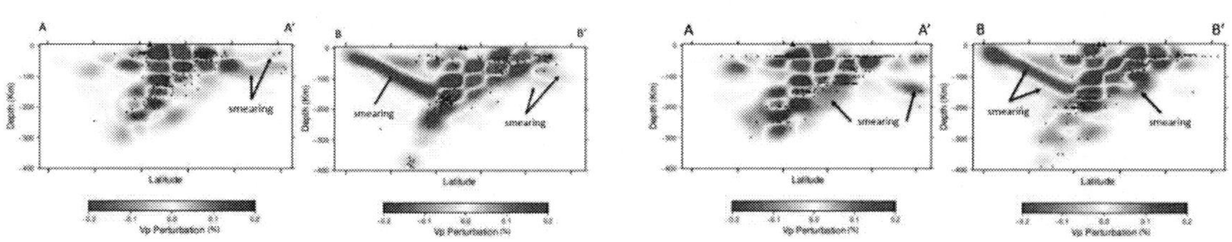

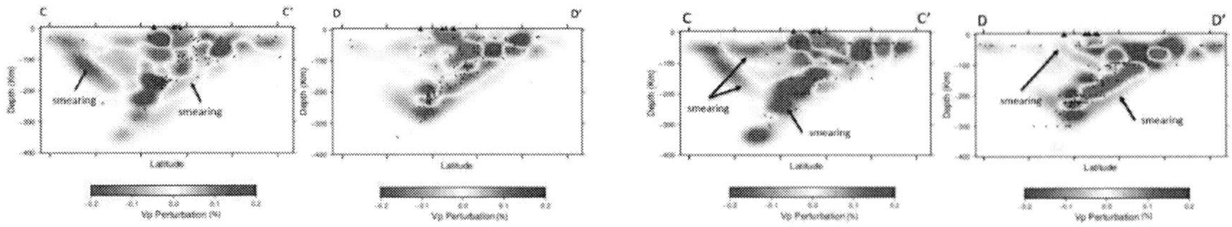

(c) (d)

Figure 3. (a) Map of the cross-section in the study area. Lines A-A', B-B', C-C', and D-D' depict the locations of the vertical section. (b) Vp synthetic input model with perturbation of ± 0.2 km/s for a vertical slice. (c) On the left side is the result of the checkerboard resolution test for the vertical slice in BMKG data. (d) On the right side is the result of the checkerboard resolution test for the vertical slice in ISC data

The results of the checkerboard test for the vertical slice on both data show the same pattern. The Vp model for the vertical slice at the latitude of $-8^o$to $-9.5^o$and the depth of 20-200 km can be recovered well. However, we noticed a stronger amplitude recovery for ISC data set from a depth of 100 km.

3.2. Vp Tomography Model

The travel time residual is plotted as a histogram and is associated with the initial and final models. The value of the initial model residual on BMKG data is 12.9659 ms while on ISC data is 30.03759 ms. The value of the final model residual on BMKG data is worth 8.8201 ms while on ISC data is worth 20.39058 ms. The data from ISC show more numbers due to a higher number of events. Figure 4 show horizontal P wave tomogram at different depth for both data set. The red color is a negative velocity that can interpreted as lower density or higher temperature model and blue color is positive velocity model that can seen as higher density or colder temperature model.

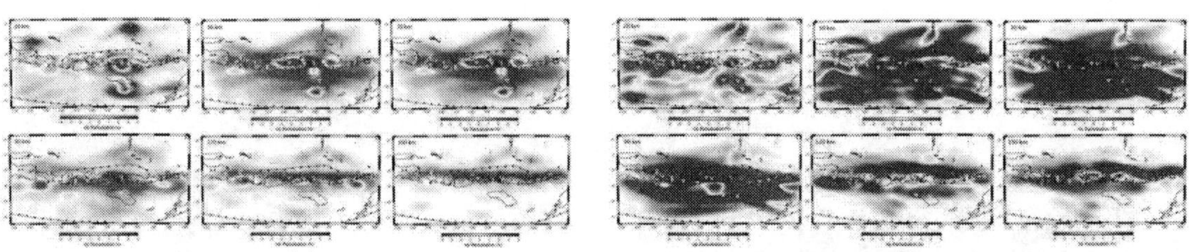

(a) (b)

Figure 4. (a) On the left side is the Vp model of a horizontal slice for BMKG data. (b) While on the left side is the Vp model of a horizontal slice for ISC data. Models are plotted in ± 4 km/s of Vp perturbation. Positive and negative anomalies are marked by blue and red colors

The P wave tomograms show different results for both data set. At a depth of 20 km, the BMKG model shows a presence of low anomaly in a few places such as beneath Bali, Lombok, Sumbawa, central of Flores Island, and Sumba, strong amplitude beneath west of Flores and south of Sumba. The ISC model shows a negative velocity with stronger amplitude with bigger area, such as in Selayar Island, Maluku, Sumba, East Nusa Tenggara, East Java, Bali, Lombok, Sumbawa, and Timor. At deeper depth, results from ISC data set show a domination of positive amplitude compared with model the BMKG. Positive amplitude model is also seen a deeper depth with much thinner area at the BMKG set.

IOP Conf. Series: Earth and Environmental Science **873** (2021) 012058 doi:10.1088/1755-1315/873/1/012058

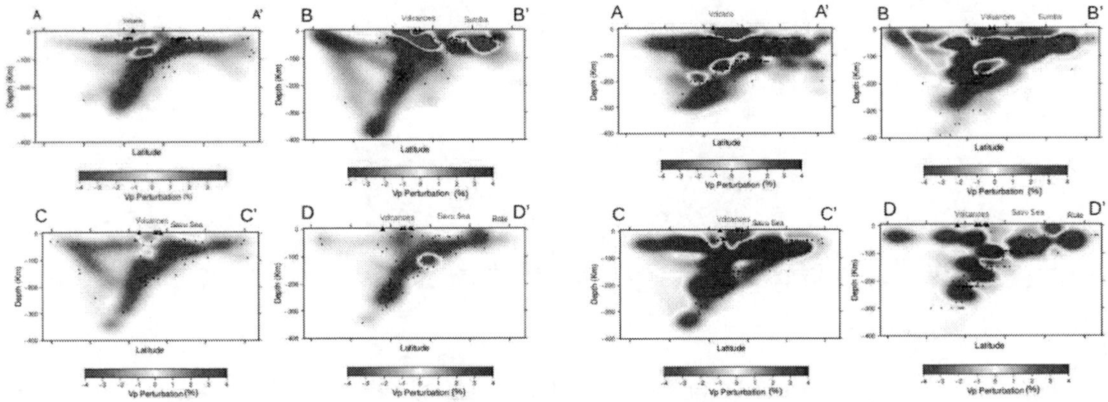

(a) (b)

Figure 5. (a) On the left side is the Vp model of a horizontal slice for BMKG data. While on the left side (b) is the Vp model of the horizontal slice for ISC data. Models are plotted in ± 4 km/s of Vp perturbation. Positive and negative anomalies are marked by blue and red colors

Figure 5 show cross section profiles from south (right) to north (left) for both data set. The subduction slab is clear seen in the BMKG data set as a steep dipping fast velocity model to the north. However, results from the ISC data set, although show a dipping pattern, the fast velocity seem to be oversized and dominated. The fast velocity also seems not as continuous as the model from the BMKG model.

3.3. Conclusion

This study used earthquake event arrival data input and station networks from two different agencies, The BMKG and ISC. Although the ISC data set has more events and arrivals, the hypocenter information may need to be relocated for better modeling and interpretation. A positive amplitude anomaly is clearly seen on the BMKG data set that dipping in northward direction at steep angle. A negative amplitude is seen in some area that coincides with volcanic line. We would like to thank the Agency of the Indonesian Agency of Meteorology, Climatology and Geophysics, and also the International Seismological Centre for providing the data set.

References

[1] DeMets C, Gordon R G, & Argus D F 2010 *Geologically current plate motions* Geophysical Journal International **181** (1) 1 – 80

[2] Müller R D, Sdrolias M, Gaina C, & Roest W R 2008 *Age, spreading rates and spreading asymmetry of the world's ocean crust* Geochemistry Geophysics Geosystems **9**(4)

[3] Silver E A, Breen N A, Prasetyo H, & Hussong D M 1986 *Multibeam study of the Flores backarc thrust belt, Indonesia* Journal of Geophysical Research: Solid Earth **91**(B3) 3489–3500

[4] Dziewonski A M, Chou T-A, & Woodhouse J H 1981 *Determination of earthquake source parameters from waveform data for studies of global and regional seismicity* Journal of Geophysical Research **86** 2825-285

[5] Rawlinson N, A M Reading, and B L N Kennett 2006 *Lithospheric structure of Tasmania from a novel form of teleseismic tomography* J. Geophys. Res. **111** B02301

IOP Conf. Series: Earth and Environmental Science **873** (2021) 012059 doi:10.1088/1755-1315/873/1/012059

An Attempt to Pick Teleseismic P Wave Arrival Using Envelope and Artificial Neural Network Algorithm

R K Lobo[1], Y H L Gaol[1], D Y Fatimah[1], A Abdullah[1], D A Zaky[1], S K Suhardja[1], A D Nugraha[2], Z Zulfakriza[2], S Widiyantoro[2,3], M Ramdhan[4], K H Palgunadi[5], B Mujihardi[6]

[1]Faculty of Exploration and Production Technology, University of Pertamina, Jl. Teuku Nyak Arief, Simprug, Kebayoran Lama, Jakarta 12220, Indonesia
[2]Global Geophysics Research Group, Faculty of Mining and Petroleum Engineering Institute of Technology Bandung, Jalan Ganesa No. 10, Bandung 40132, Indonesia
[3]Faculty of Engineering, Maranatha Christian University Bandung, Jalan Surya Sumantri No 65, Bandung 40164, Indonesia
[4]Agency for Meteorology, Climatology and Geophysics, Jalan Angkasa I, No. 2, Kemayoran, Jakarta, Indonesia
[5]Physical Science and Engineering, King Abdullah University of Science and Technology, Saudi Arabia.
[6]Geothermal Research – Innovation & New Ventures, PT. Pertamina, Jl. Mega Kuningan Barat III, Jakarta Selatan, Indonesia

*Corresponding author: s4ndy104@gmail.com

Abstract. Seismic events detection and phase picking play an essential role in earthquake studies. Typical event detection is done visually or manually on recorded seismogram by choosing a series of higher amplitude signals recorded on at least 4 stations. More sophisticated methods have been used in event detection and picking with additional attributes such as Short Time Average over Long Time Average (STA/LTA). This method is based on average number sampled at multiple predefined windows. However, STA/LTA is dependent on the window size which becomes its drawback. In this study, we explore one derivative attribute, popularly known as envelope or instantaneous amplitude. It has been extensively used in seismic reflection and refraction method. In principle, this method uses the Hilbert Transform to calculate complex seismic trace and take the magnitude of complex seismic trace as envelope amplitude that can be used to analyze P wave arrival time. We employed one of the machine learning methods, Artificial Neural Network (ANN). The ANN method works by analyzing various inputs and training them to recognize patterns in P wave arrival signals. We started our study by applying envelope attribute to synthetic data with noise addition. We found that with noisy data the envelope attribute still gives a clear signal for first-time arrival. Next, we trained 300 seismograms of teleseismic events recorded on IRIS-US networks and tested our trained program on 20 seismograms as a blind test. To compare performance between the two methods, we calculated the difference between the results of automatic picking and manual picking. The final calculation shows an average deviation of 0.355 seconds. Twenty-five percent of testing data (5 samples) has a deviation above 0.5 seconds, and 75% of the remainder (15 samples) already had a deviation under 0.5 seconds. The more significant deviations of the P wave picks are likely due to noisy signals in the data set and complex arrival signals. This study shows that the combination of envelope attribute and machine learning method is promising to distinguish teleseismic P wave arrival and automatically pick them.

1. Introduction.

Arrival times picking of seismic phases plays a significant role in earthquake data since it can provide information about the Earth's interior structure. Seismic waveforms and its arrival time can be utilized with further advance methods such as Receiver Function, Shear Wave Splitting and Seismic Tomography to image Earth sub-surface structures. For the last 10 years, the quantity of seismic stations around the world are quickly increasing, thus did the seismic data volumes. Traditionally, an observer will do arrival time picking by visual observation on the seismogram. The result of arrival time picking depends on the experience of the observer. The manual picking process gives good results, yet this process is time-consuming since the massive volumes of seismic data. Moreover, the manual picking process sometimes has some trouble when it has a low signal-to-noise ratio (SNR).
To perform arrival time picking efficiently, numerous studies have proposed an a automatic arrival time picking algorithms and one of the most broadly utilized methods is the Short-Term Averaging/Long-Term Averaging (STA/LTA) algorithm [1], that simply the ratio of the two averages calculated on sliding/moving windows over the trace. Coppens method, Modified Energy Ratio (MER) method, statistical attributes such as skewness and kurtosis, maximum-likelihood method, and cross-correlation methods are some of the commonly used methods in first arrival time picking algorithm. Most of this method is only effective with high SNR data. For low SNR data, there are possibilities that the first arrival did not detect by those methods [2].

With the target to improve time data processing, we used one of the machine learning technique named neural networks to analyze the first arrival on teleseismic data. STA/LTA (Short Term Averaging over Long Term Averaging) and envelope (instantaneous amplitude) are used as input parameters for neural networks data training. STA/LTA is utilized to clarify the contrast between seismic event signal and noise signal. The purpose of envelope usage is to perceive amplitude fluctuation when an earthquake event is coming so that the patterns of the first arrival on the seismogram could be distinguished. The envelope attribute uses Hilbert transform on a vertical component trace to obtain the complex seismic trace that has two parts, the real and imaginary parts of the trace. This research's primary purpose is to develop an auto-picking algorithm first arrival picking on teleseismic events with an accuracy that is similar with human expert performance. The application of the neural network's method is expected to give good picking results on low SNR data, where the first arrival patterns are not clearly seen.

The data used in this study are vertical component seismograms from teleseismic events from IRIS stations in the United States region. Waveform data were downloaded from the IRIS database using the ISC earthquake catalogue. The teleseismic earthquake event used has a magnitude above 6 (M> 6) with a recording period ranging from 2012-2017. The recording length of each seismogram is two minutes containing the arrival event of the P wave. The seismogram has a sampling rate of 40 Hz, so the data length of each seismogram is 4800 samples. There are 33 teleseismic earthquake events recorded by 61 stations on the IRIS seismological network in the Americas. Most of the earthquake events used are located on the South American Continent with a distance of more than 1000 km from the receiving station (Figure 1). In total, we collected 300 teleseismic events used for the training data process. For the validation process and blind test, 20 earthquake events were used. The blind test process is P-phase picking process on a new set of seismograms.

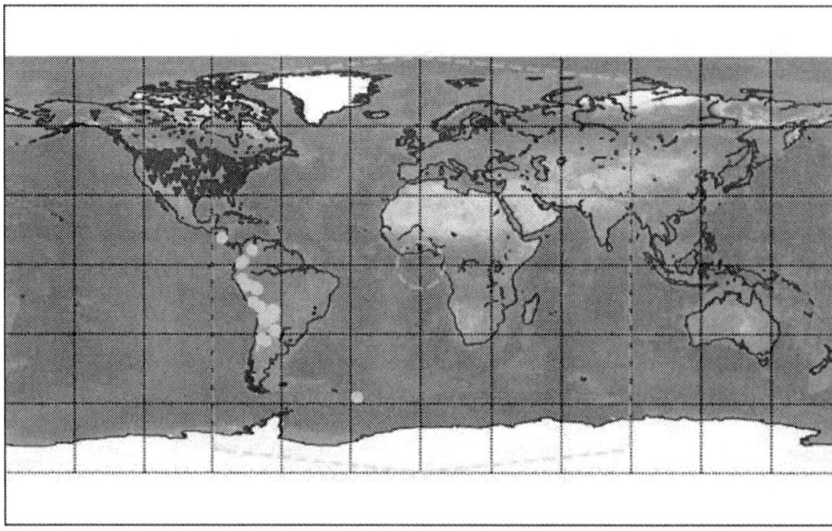

Figure 1. Map of US IRIS seismic network stations (red) and teleseismic event distribution (yellow).

Seismogram pre-processing (filtering and normalization) aims to eliminate the effect of noise and uniform the amplitude value as seen on Figure 2. The filtering process was carried out using a third order low-pass filter with a cutoff frequency of 1 Hz. Less variation of amplitude or a uniform amplitude value will make it easier for the ANN learning process.

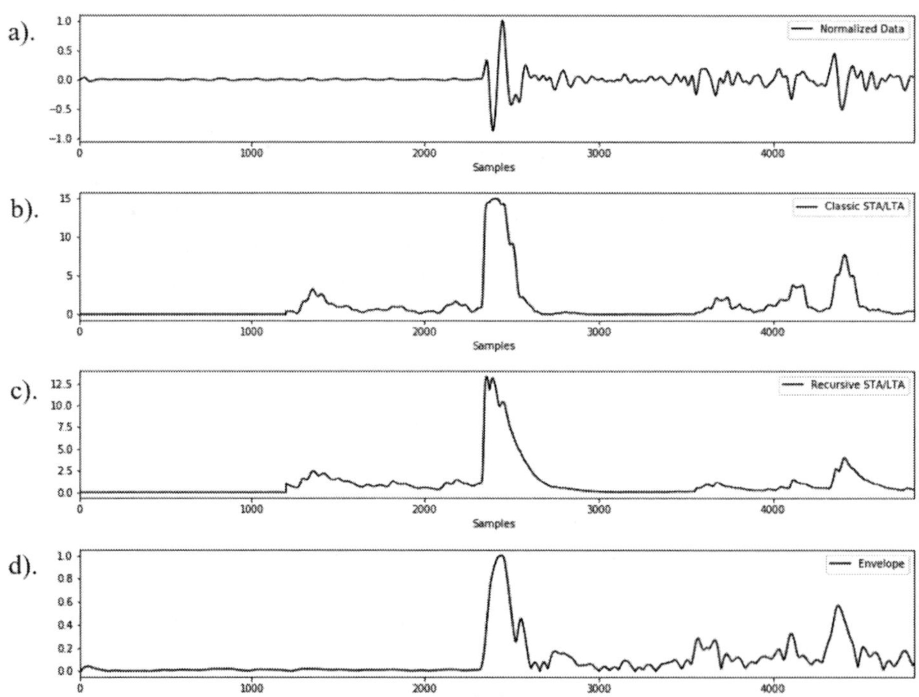

Figure 2. (a) Sample seismogram after pre-processing, (b) classic STA/LTA, (c) recursive STA/LTA, (d) envelope attribute.

IOP Conf. Series: Earth and Environmental Science **873** (2021) 012059 doi:10.1088/1755-1315/873/1/012059

2. Methods

STA/LTA

The short time average/long time average (STA/LTA) has been widely reliable for detecting earthquake events and can be used with other methods to improve picking accuracy. The primary purpose of this method is to clarify the contrast between the earthquake signal and noise. STA/LTA measure the average of the absolute value from a series of seismic signal using two moving time windows – a short time average window (STA) and a long-time average window (LTA). The STA window will be more sensitive to seismic events because it measures the instant amplitude of the seismic signal and keeps track of earthquake events. Meanwhile, the LTA window functions to obtain background noise information, which usually has a longer period [3]. Figure 3 is an example of events and STA/LTA calculation. When clear signal arrived, a spike in STA/LTA attributes is obvious and able to differentiate between real arrival and not. In this research, we used classic STA/LTA [4] and recursive STA/LTA [5]. An example of STA/LTA attributes calculation can be seen on Figure 3.

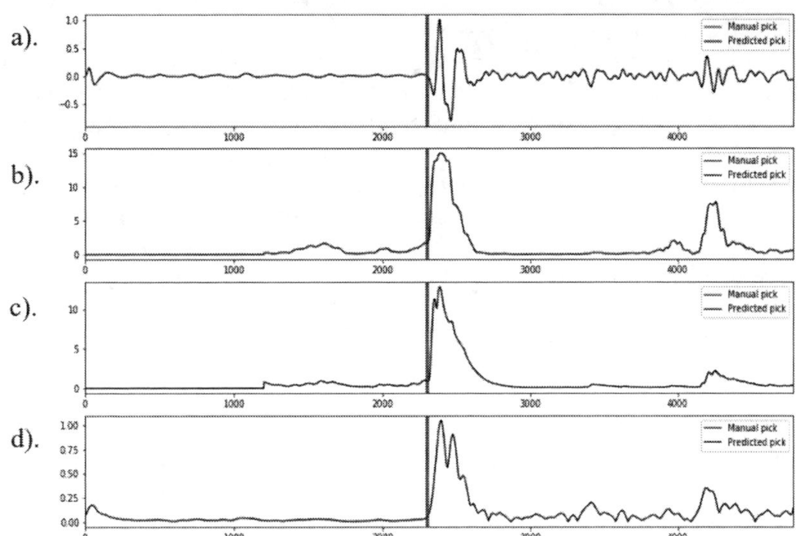

Figure 3. The 10th real data set along with the attributes; (a) Sample seismogram after pre-processing, (b) classic STA/LTA, (c) recursive STA/LTA, (d) envelope attribute.

Envelope Attribute (Instantaneous Amplitude)

Hilbert transform process can used to remove the negative frequency part and double the magnitude of positive frequency. It can be seen as a band pass filter. The outputs of this process are instantaneous amplitude, frequency, and phase attribute information from seismic signal data. A seismic signal is a real part of a complex seismic trace. A complex seismic trace *z(t)* has real *x(t)* and imaginary parts *y(t)* as defined by the equation below:

$$z(t) = x(t) + i\,y(t) \quad (1)$$

The real part *x(t)* is the real seismic trace, and *y(t)* is the imaginary seismic trace obtained from the Hilbert transformation of *x(t)*. The envelope represents the total instantaneous energy of a seismic signal with an amplitude range that varies from zero to the maximum amplitude of a seismic signal. This method is commonly used to detect first breaks in seismic data. A spike or increase in the amplitude value of the envelope can indicate the arrival time of the waves on the seismogram (Figure 3).

Artificial neural Network

Artificial Neural Network (ANN) is one of the machine learning method based on the working principle of human biological neural networks. The machine-learning algorithm focused on studying patterns from a set of training data. Models generated from machine learning have characteristics derived from the training data. After the model studies the patterns in the training dataset, the model can predict some desired output from other data.

The concept of the ANN learning process is learning from the wrong output values and will be updated as iterations increase. At the beginning of the learning step, the output model will have an error value which is the difference between the desired output value and the predicted output value. Changes in output values are based on learning experiences. The data training step aims to get the predicted output value that is close to the actual desired output.

The purpose of using the ANN method is to predict where on seismogram is the P-wave first arrival. The ANN model learns from the dataset input, which consists of the STA/LTA attributes, envelope attribute, and the label. Labels are guidance to differentiate between the first arrival zone and the non-first arrival zone on the seismogram.

The STA/LTA and envelope attribute were generated from the seismogram, these attributes will be used as a guide for the manual picking. From these attributes, the P-wave arrival pattern will be seen from the presence of an amplitude spike. Consequently, manual P-arrival picking is performed on the STA/LTA and envelope attributes. This manual pick intends to guide where the P-wave "arrivals" and "non-arrivals" are located. P-wave "arrivals" will be labeled 1 (true) and the "non-arrivals" zone will be labeled 0 (false). Therefore, this process is also known as labeling. The ANN auto-picker was then trained using the STA/LTA, envelope, and the label as the input parameters. The model generated from the training process is then used to predict the P-wave arrival from new data. Model validation is executed by predicting the P-wave arrival on synthetic data with some given noise. The model must be able to predict P-wave arrivals on synthetic data with results that are close to manual picks by the observer.

3. Results and Discussions.

This section shows the results of predicted P-wave arrival on the real teleseismic data set. ANN auto-picker results will be compared with the manual arrival time picking by the observer. The ANN model was applied to predict arrival time from the blind data set. The model created can recognize the P-wave arrival pattern from the attributes used. The P-wave arrival can be seen at the beginning of the first spike of each attribute. This is following the theory that P-wave is the fastest wave recorded in a seismogram. The ANN model can distinguish between the highest first spike (arrival of the P-wave) and the next spikes, which are the other wave type arrivals.

Figure 4 shows the comparative results between the ANN auto-picker and the manual pick results. Five data samples (25% of the total data) has a deviation of more than 0.5 seconds. Meanwhile, the remaining 15 samples (75% of the total data) already had a deviation of below 0.5 seconds. The average deviation from 20 data samples is 0.355 seconds. This result is quite good because it is quite close to the manual pick results. The greater the deviation is likely due to the more complex attribute patterns from the data and has not yet been identified by the ANN model.

In some situations where the data is a bit noisy (figure 4), the ANN auto-picker will have picking results that are less close to the manual picks. This can occur due to the observer's manual picking errors or the complexity of the P-wave arrival patterns so that the ANN auto-picker cannot appropriately recognize it. More training datasets with more varied attribute patterns such as kurtosis and skewness are needed to make the ANN model better at identifying the correct P-wave arrival.

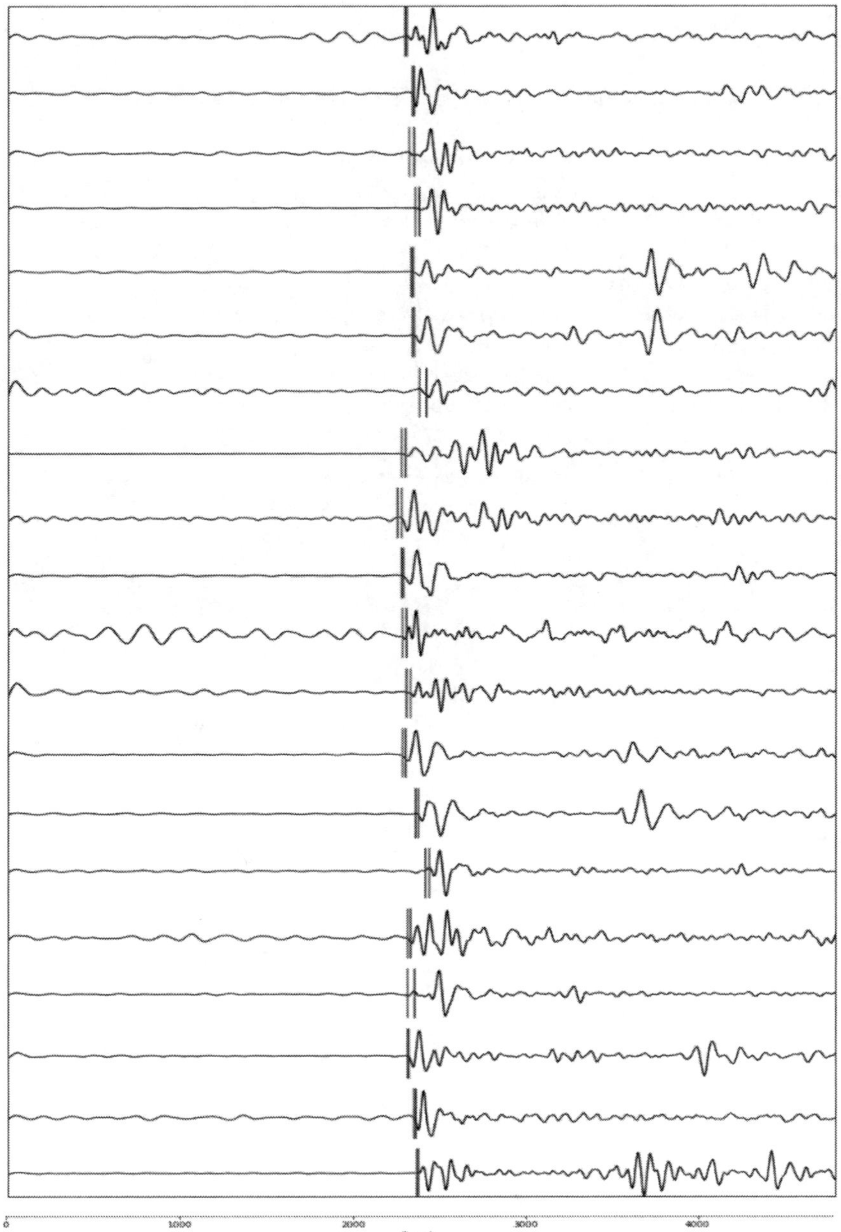

Figure 4. Comparison of manual picking results (red line) and predicted pick results on real seismograms (blue line).

IOP Conf. Series: Earth and Environmental Science **873** (2021) 012059 doi:10.1088/1755-1315/873/1/012059

References

[1] Allen R V 1978 *Automatic Earthquake Recognition and Timing from Single Traces* Bulletin of the Seismological Society of America **68**(5) 1521-1532

[2] Aminzadeh F, Maity D, Tafti T A, Brouwer F 2011 *Artificial neural network based autopicker for micro-earthquake data* In SEG Technical Program Expanded Abstracts pp 1623-1626

[3] Trnkoczy A 2012 *Understanding and parameter setting of STA/LTA trigger algorithm* New Manual of Seismological Observatory Practice (NMSOP) pp. 1-20 Deutsches GeoForschungsZentrum GFZ

[4] Agius M R 2007 *Automatic Earthquake Detection and Localisation from a Three-Component Single-Station* M Sc diss University of Malta

[5] Grigoli F et al. 2014 *Automated seismic event location by waveform coherence analysis* Geophysical Journal International **196**(3) 1742-1753

IOP Conf. Series: Earth and Environmental Science **873** (2021) 012060 doi:10.1088/1755-1315/873/1/012060

Preliminary Results of Automatic P-Wave Regional Earthquake Arrival Time Picking Using Machine Learning with STA/LTA As the Input Parameters

Y H Lumban Gaol[1], R K Lobo[1], S S Angkasa[1], A Abdullah[1], I Madrinovella[1], S Widyanti[1], A Priyono[1,2], S K Suhardja[1], A D Nugraha[2], Z Zulfakriza[2], S Widiyantoro[2,3], M Luqman Hakim[4], K H Palgunadi[5]

[1] Faculty of Exploration and Production Technology, University of Pertamina, Jl. Teuku Nyak Arief, Simprug, Kebayoran Lama, Jakarta 12220, Indonesia
[2] Global Geophysics Research Group, Faculty of Mining and Petroleum Engineering Institute of Technology Bandung, Jalan Ganesa No. 10, Bandung 40132, Indonesia
[3]Faculty of Engineering, Maranatha Christian University Bandung, Jalan Surya Sumantri No 65, Bandung 40164, Indonesia
[4] Agency for Meteorology, Climatology and Geophysics (BMKG), Jalan Angkasa I, No. 2, Kemayoran, Jakarta, Indonesia
[5]Physical Science and Engineering, King Abdullah University of Science and Technology, Saudi Arabia.

*Corresponding author: s4ndy104@gmail.com

Abstract. The Short Term Averaging/Long Term Averaging (STA/LTA) has been widely used to detect earthquake arrival time. The method simply calculates the ratio of moving average of the waveform amplitude at short and long-time windows. However, although STA/LTA signals can distinguish between real events and noise, we still recognize some lack of accuracies in first P wave arrival pickings. In this study, we attempt to implement one machine learning method popularly, Artificial Neural Network (ANN) that employ input, hidden and output layer similar as human brain works. Note that in this study, we also try to add input parameters with another derivative signal attributes such as Recursive STA/LTA and Carl STA/LTA. The processing step started by collecting event waveforms from the Agency of Meteorology, Climatology and Geophysics. We chose regional events with moment magnitude higher than 3 in the Maluku region Indonesia. Next, we apply all STA/LTA attributes to the input waveforms. We also tested our STA/LTA with synthetic data and additional noise. Further step, we manually picked the arrival of P wave events and used this as the output for ANN. In total, we used 100 events for arrival data training in P wave phases. In the validation process, an accuracy of more than 0.98 can be obtained after 200 iterations. Final outputs showed, that in average, the difference between manual picking and automatic picking from ANN is 0.45 s. We are able to increase the accuracy by band pass filter (0.1 – 3 Hz) all signal and improve the mean into 0.15s difference between manual picking and ANN picks.

1. Introduction

Traditionally, the determination of P wave was done manually by an observer/operator through visual inspection. The manual picking method gives us accurate results, but it also requires a lot of time since the massive number of earthquake data. The results will also be different from each operator, depending on their experiences. [1] proposed a method called Short Term Averaging/Long Term Averaging (STA/LTA) to automize the first break picking process that also widely used to replace the manual picking method. However, this method is only effective when the earthquake has a high signal-to-noise ratio (SNR) and will be difficult to identify the first break of earthquake with low SNR. The lack of STA/LTA method in identifying the first break of earthquake with low SNR motivate us to try applying machine learning approach and STA/LTA attributes in predicting the P-wave earthquake arrival time. [2] have tried to improve the STA/LTA method's performance by using a fitness function. [3] used machine learning algorithms to predict the earthquake first-break using 11 input parameters that will be trained to learn its pattern to predict the first break of a new earthquake data. In this paper, we try to apply machine learning algorithm called Artificial Neural Network (ANN) by combining three types STA/LTA (Classic, Recursive and Carl STA/LTA) as the input parameters [3]. These parameters will be trained using Keras, a Python's newest machine learning library. The result expected from this research is a Python-based algorithm that can do automatic first-break picking.

2. Methods

2.1. Artificial Neural Network (ANN)

ANN is one of the machine learning algorithms that work similarly to the human brain system. ANN is a collection of connected units or nodes called neurons that can be separated into three main parts (architecture): *the input layer, hidden layer,* and *output layer (Figure 1).* The general equation of ANN is [4]:

$$y = f\left(\sum_{i=0}^{n}(x_i\, W_i) + B\right) \tag{1}$$

where y is output, x is input, W is weight and B is bias

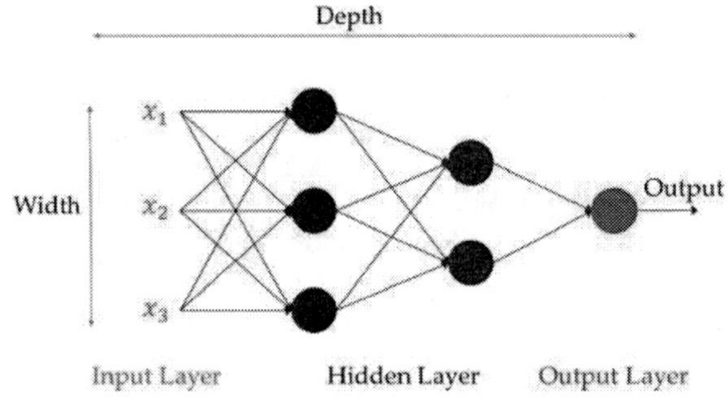

Figure 1. ANN general architecture

2.2. Short Term Averaging/Long Term Averaging (STA/LTA)

Short Term Averaging/Long Term Averaging (STA/LTA) is a method that works similar to the moving average of two particular windows. A possible arrival time is declared when the ratio between a short-term average (STA) of the signal and its long-term average (LTA) exceeds a certain threshold parameter that calculates on a specific sliding/moving windows. Visually, we can quickly identify earthquake arrival time from the STA/LTA curve that will show a higher spike than the overall curve.

In this paper, we calculate three types of STA/LTA (Classic, Carl, and Recursive STA/LTA) methods that are provided by Obspy, a python library for seismic processing. Generally, all of these types of STA/LTA are work similar, and the only difference is their equation that is shown below: *Classic STA/LTA* [5]:

IOP Conf. Series: Earth and Environmental Science **873** (2021) 012060 doi:10.1088/1755-1315/873/1/012060

$$STA_{(i)} = \frac{1}{N_{STA}} \left(\sum_{j=i-N_{STA}}^{i} A_j \right) \tag{2.1}$$

$$LTA_{(i)} = \frac{1}{N_{LTA}} \left(\sum_{j=i-N_{LTA}}^{i} A_j \right) \tag{2.2}$$

$$RSL_{(i)} = \frac{STA_{(i)}}{LTA_{(i)}} \tag{2.3}$$

where *STA* is *short term average*, N_{STA} is STA window length, *A* is earthquake signal amplitude, *LTA* is *long term average*, N_{LTA} is LTA window length, *RSL* is the ratio of STA and LTA.

Recursive STA/LTA [6] [7]:

$$STA_{i+1} = \frac{1}{N_{STA}}(A_i - STA_i) + STA_i \tag{3.1}$$

$$LTA_{i+1} = \frac{1}{N_{LTA}}(A_i - LTA_i) + LTA_i \tag{3.2}$$

$$RSL_{i+1} = \frac{STA_{i+1}}{LTA_{i+1}} \tag{3.3}$$

where *STA* is *short term average*, N_{STA} is STA window length, *A* is earthquake signal amplitude, *LTA* is *long term average*, N_{LTA} is LTA window length, *RSL* is the ratio of STA and LTA
Carl STA/LTA [7]:

$$ETA = star - (ratio \text{ x } ltar) - abs(STA_t - LTA_t) - quiet \tag{4}$$

where: *ETA* is the detector response – a value over 0 means detection, *STA* is the short term moving average of signal, *LTA* is the long term moving average of signal, *star* is the short term moving average of absolute value of signal and *LTA* difference, *ltar* is the long term moving average of star, ratio, and *quiet*-sensitivity parameters.

2.3. Data Processing and General Workflow

In this study, we used 106 local-earthquake waveforms with sampling rate of 20 samples/second and magnitude > 3 that recorded by BMKG station from November 06, 2019 to November 15,2019. We apply a bandpass filter with an average frequency range is 0.5 - 2.5 Hz to increase the signal-to-noise ratio of the waveforms. Next, we manually picked the first break and calculate the ratio of three types STA/LTA that will be used as input parameters. Then, we give a label to the data with 1 and 0; 1 for the first-break and 0 for not first-break. We used 11 samples of each waveform with detail as one sample with label 1 is the first-break, and ten samples with label 0 are non-first-break (five samples before and after the first break). The dataset will be normalized to simplify the calculation process. After that, we train the dataset and do blind testing to check if our model is already good enough or not.

3. Results and Discussions

We trained 106 pairs datasets that consist of kurtosis and skewness of each event. This machine learning model's accuracy has reached 91% after 200 iterations with model loss below 0.25. A good model loss is getting closer to 0 because it means that machine learning's predictive results have no difference to the actual value.

A. Synthetic Data

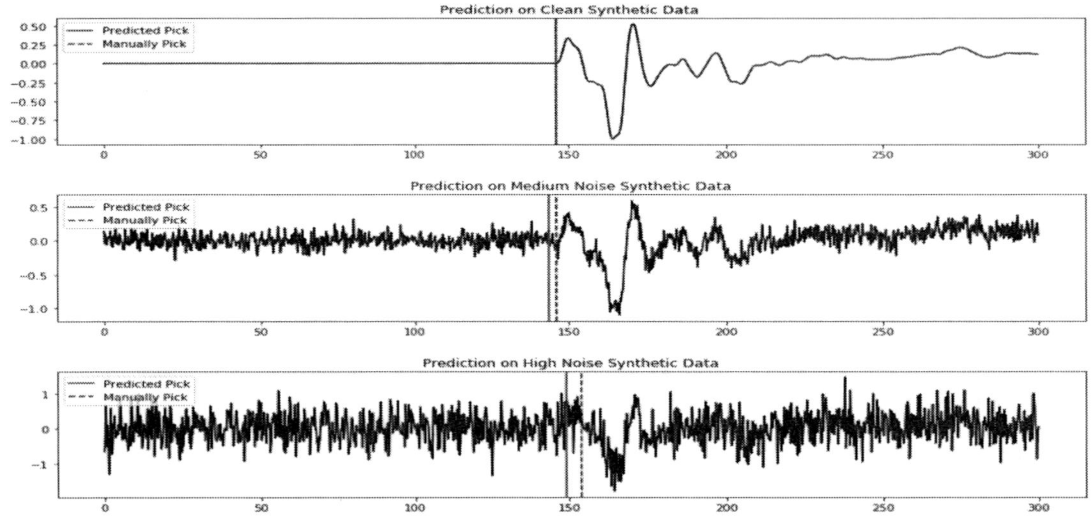

Figure 2. STA/LTA of Clean Synthetic Data

We tested our machine learning model on three types of synthetic data with sampling rate of 4 samples/second **(Figure 2)**. On clean synthetic data, the computer predicted the first break precisely the same with our manual pick. Meanwhile, on medium and high noise synthetic data there are few seconds of differences to our manual pick. (1.25 second and 2.75 second [**Table 1**]) miss to our manual picks on medium and high noise synthetic data. It means that our model is already working well on low noise earthquake waveform and will be less accurate on more noise earthquake waveform.

Table 1. Analysis of first-break picking prediction on synthetic data

	Sample position on manual first-break picking	**Sample position on auto first-break picking**	**Differences**
Clean synthetic data	588th (147 second)	588th (147 second)	0
Medium noise synthetic data	588th (147 second)	593th (148.25 second)	5 samples (1.25 second)
High noise synthetic data	607th (151.75 second)	596th (149 second)	11 samples (2.75 second)

B. Real Data

After that, we tried to apply our machine learning model on two real data or events recorded on GLMI station, one of BMKG stations located on Galela, North Maluku Province. Based on the results on synthetic data, we understand that noise greatly effects the accuracy of our model to determine the first break, so we are going to try our machine learning model on raw and filtered waveform to see how good it works on real earthquake waveform. Event 1 **(Figure 3)** is an earthquake with a magnitude of 5.1 and located in Minahasa Peninsula, Sulawesi. Meanwhile, Event 2 **(Figure 4)** is an earthquake with a magnitude of 6.1 and located in Papua. From the pictures, we can see that our model is working accurately enough to detect the first break on both raw and filtered waveform. The largest error is 0.65 seconds missed to the real first break (manual pick) on Event 1 raw waveform but the error decrease on Event 1 filtered waveform with the 0.15 seconds shift to the real first break **(Table 2)**.

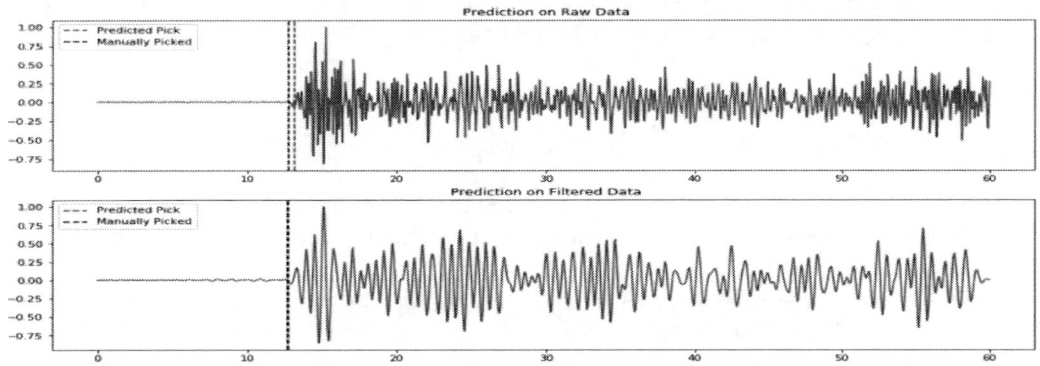

Figure 3. Auto first-break picking on real data 1 (raw and filtered)

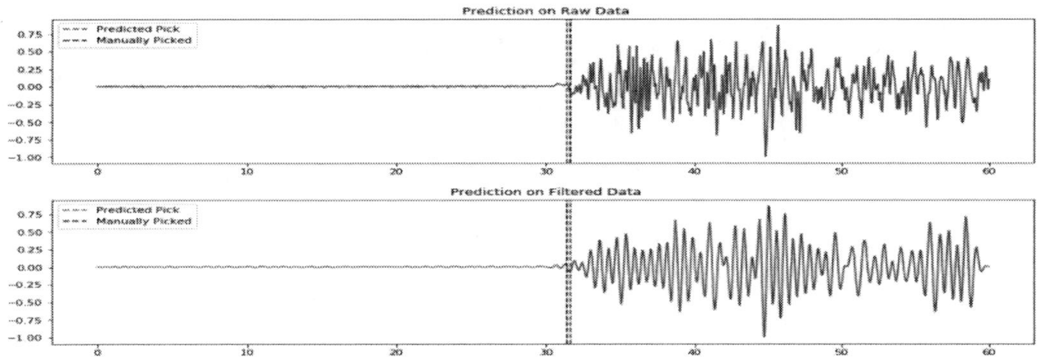

Figure 4. Auto first-break picking on real data 1 (raw and filtered)

Table 2. Analysis of first-break picking prediction on real data

Event	Differences on raw data (s)	Differences on filtered data (s)
Event 1	13 samples (0.65s)	3 samples (0.15s)
Event 2	5 samples (0.25s)	4 samples (0.2s)

4. Conclusion

We have successfully conducted the application of an artificial neural network in predicting the first break of earthquakes. The results show satisfying results on both synthetic and real data. The largest error or difference between our machine learning model and manual picking is 0.65 seconds in raw Event 1 waveform. From the test on synthetic and real data, we also understand that noise greatly affects the accuracy of our model to determine the first break so how effective the waveform is filtered will become one of the most important stages in order to have a satisfying result. We still can improve our machine learning model by adding more data to be trained, and we also need further validation on more data to see how far our machine learning model can work properly.

References

[1] Allen R V 1978 *Automatic earthquake recognition and timing from single traces* Bulletin of the Seismological Society of America **68**(1) 521–1,532

[2] Maurizio V, Satriano C, Lomax A 2012 *Automatic Picker Developments and Optimization: A Strategy for Improving the Performances of Automatic Phase Pickers* Seismological Research Letters **83**(3) 541-554

[3] Mężyk M, & Malinowski M 2019 *Multi-pattern algorithm for first-break picking employing open-source machine learning libraries* Journal of Applied Geophysics 170 103848

[4] Aggarwal C C 2018 *Neural Networks and Deep Learning* In Neural Networks and Deep Learning

[5] Chen Y 2017 *Automatic microseismic event picking via unsupervised machine learning* Geophysical Journal International **212**(1) 88–102

[6] Sharma B K, Kumar A, Murthy V M 2010 *Evaluation of Seismic Events Detection Algorithms* Journal Geological Society of India **75**(3) 533-538

[7] Polkowski M, Plesiewicz B, Wiszniowski J, Wilde-Piórko M, & Group P W 2016 *Local Seismic Events in the Area of Poland Based on Data from the PASSEQ 2006–2008 Experiment* Acta Geophysica **64**(6), 2092–2113

IOP Conf. Series: Earth and Environmental Science **873** (2021) 012061 doi:10.1088/1755-1315/873/1/012061

Preliminary Results of Automatic P-Wave Regional Earthquake Arrival Time Picking using Machine Learning with Kurtosis and Skewness as The Input Parameters

Y H Lumban Gaol[1], R K Lobo[1], S S Angkasa[1], A Abdullah[1], I Madrinovella[1], S Widyanti[1], A Priyono[1,2], S K Suhardja[1], A D Nugraha[2], Z Zulfakriza[2], S Widiyantoro[2,3], M Luqman Hakim[4], K H Palgunadi[5], B Mujihardi[6]

[1]Faculty of Exploration and Production Technology, University of Pertamina, Jl. Teuku Nyak Arief, Simprug, Kebayoran Lama, Jakarta 12220, Indonesia
[2]Global Geophysics Research Group, Faculty of Mining and Petroleum Engineering Institute of Technology Bandung, Jalan Ganesa No. 10, Bandung 40132, Indonesia
[3]Faculty of Engineering, Maranatha Christian University Bandung, Jalan Surya Sumantri No 65, Bandung 40164, Indonesia
[4]Agency for Meteorology, Climatology and Geophysics, Jalan Angkasa I, No. 2, Kemayoran, Jakarta, Indonesia
[5]Physical Science and Engineering, King Abdullah University of Science and Technology, Saudi Arabia.
[6]Geothermal Research – Innovation & New Ventures, PT. Pertamina, Jl. Mega Kuningan Barat III, Jakarta Selatan, Indonesia

*Corresponding author: s4ndy104@gmail.com

Abstract. The traditional method in determining first arrival time of earthquake dataset is time consuming process due to waveform manual inspection. Additional waveform attributes can help determine events detection. One of the widely used attribute is The Short Term Averaging/Long Term Averaging (STA/LTA) which is simply division moving average of waveform amplitude between short time and longer time. Alternatively, waveform attribute can also be built using kurtosis and skewness. The kurtosis attribute is defined as sharpness of data distribution. By definition, the maximum signal should be at or close to the P wave arrival. The skewness is typically used to show normal distribution of the data. The uniqueness of this method is that it has an ability to determine whether the first P wave arrival has positive of negative number. The skewness calculation is similar to kurtosis but it uses the power of 3 instead of 4. With the objective of generating efficient scheme to pick first time arrival, we explore use artificial neural network and a combination of kurtosis and skewness attributes. We use a collection of magnitude events with moment magnitude larger than 3 located close to Moluccas island, Indonesia. We collected all events information from the Indonesian Agency of Meteorology, Climatology and Geophysics. The process is started with manually pick all P wave arrivals using manual inspection. Next, we trained the artificial neural network with attributes numbers as inputs and arrival time we manually picked as the output. In total we used 100 regional events that has clear P wave phases. Then, we validated the results until reaching 0.99 accuracy. In the last step, we tested the once trained procedures on new waveforms contained events. Current result shows an average of 0.4s different between manual pick and trained picked from machine learning. The accuracy can be improved by applying a band pass 0.1-2 Hz filtering with an average of 0.2s.

1. Introduction.

Machine learning is a topic that rapidly developed in the last few years and has already been applied in many different industry branches, also in seismic processing. One machine learning algorithm that can be applied for first-break picking purposes is the Artificial Neural Network (ANN). [1] used this method to predict the earthquake first-break using 11 trained input parameters so the computer can learn its pattern to estimate the first break of a new earthquake data. Another example for first break picking purposes, e.g. support vector machines [2] reinforcement learning [3] or convolutional neural networks [4],[5]. These studies motivated us to try a similar method with a more straightforward yet practical approach.

In this paper, we propose a new innovative and efficient P-wave first break picking method using Artificial Neural Network (ANN) by combining kurtosis and skewness as the input parameters that are effective enough to distinguish a first-break and not first-break. These parameters will be trained using Keras, a Python's newest machine learning library. The result expected from this research is a Python-based program that can do automatic first break picking.

2. Data and Method

2.1. Artificial Neural Network (ANN)

ANN is a machine learning algorithm that works similarly to the human brain system [6]. ANN is a collection of connected units or nodes called neurons that can be separated into three main parts (architecture): *input layer, hidden layer,* and *output layer.* The general equation of ANN is [7]:

$$y = f\left(\sum_{i=0}^{n}(x_i\, W_i) + B\right) \tag{1}$$

where y is output, x is input, W is weight and B is bias

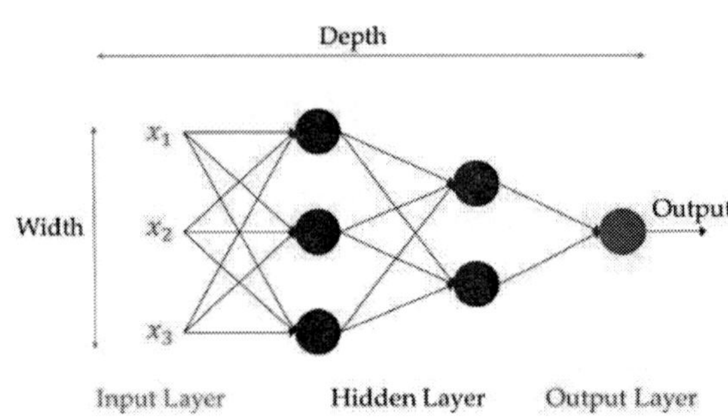

Figure 1. ANN general architecture

2.2. Kurtosis and skewness

Kurtosis

Kurtosis is a degree of peakedness or sharpness from a data distribution. Normal Gaussian distribution will have kurtosis = 0, but earthquake data is a non-Gaussian distribution and has a higher kurtosis value. With this characteristic, kurtosis can be used to identify a non-Gaussian distribution (first-break) that will be characterized by a maximum kurtosis value [8].

$$Kurtosis = \frac{\sum_{i=1}^{M}(x_i - \bar{x})^4}{(M-1)s_x^4} - 3 \tag{2}$$

Skewness

Skewness is a degree of distortion from normal distribution data. In a simple term, skewness measures the lack of symmetry in the data distribution. An earthquake arrival time data is characterized by an amplitude increase, which is higher than its mean amplitude value. When a significant difference in amplitude, the skewness attribute is sufficient to detect a first-break and its type (peak or trough) [8].

$$Skewness = \frac{\sum_{i=1}^{M}(x_i - \bar{x})^3}{(M-1)s_x^3} \tag{3}$$

where M is total samples of amplitude, $x = \{x_1, x_2, \dots, x_M\}$ is amplitude value, $\bar{x}$ is mean of amplitude, and s_x standard deviation [9].

2.3. Mean Absolute Percentage Error (MAPE)

We calculated the error of machine learning auto pick by comparing it to manual pick (considered as true first-break) using MAPE equation [10]:

$$\text{MAPE} = \frac{|P_m - P_o|}{P_m} \text{ x } 100\% \tag{4}$$

where P_m is manual-pick sample's position, P_o is the machine learning auto-pick sample's position.

2.4. Data Processing and General Workflow

In this study, we used 106 local-earthquake waveforms with a sampling rate of 20 samples/second and magnitude > 3 that recorded by BMKG station from November 06, 2019 to November 15, 2019. We apply a bandpass filter with an average frequency range is 0.5 - 2.5 Hz to increase the signal-to-noise ratio of the waveforms. Next, we manually picked the first break and calculated kurtosis and skewness, used as input parameters. Then, we give a label to the data with 1 and 0; 1 for the first-break and 0 for not first-break. We used 11 samples of each waveform with detail as one sample with label 1 is the first-break, and ten samples with label 0 are non-first-break (five samples before and after the first break). The dataset will be normalized to simplify the calculation process. After that, we train the dataset and do testing on synthetic data that we downloaded from the IRIS dataset to check if our model is already good enough or not.

3. Results and Discussions

We trained 106 pairs datasets that consist of kurtosis and skewness of each event. This machine learning model's accuracy has reached 92% after 200 iterations with model loss below 0.22. A good model loss is getting closer to 0, which means that machine learning's predictive results have no difference to the actual value.

A. Synthetic Data

We make first break predictions on three types of synthetic data with sampling rate of 4 samples/second as a benchmark to whether the machine learning model that has been trained is good enough in predicting the first break. The accuracy of predicted results on all three types of synthetic data **(Figure 2)** will be analyzed from MAPE value **(Table 1)**.

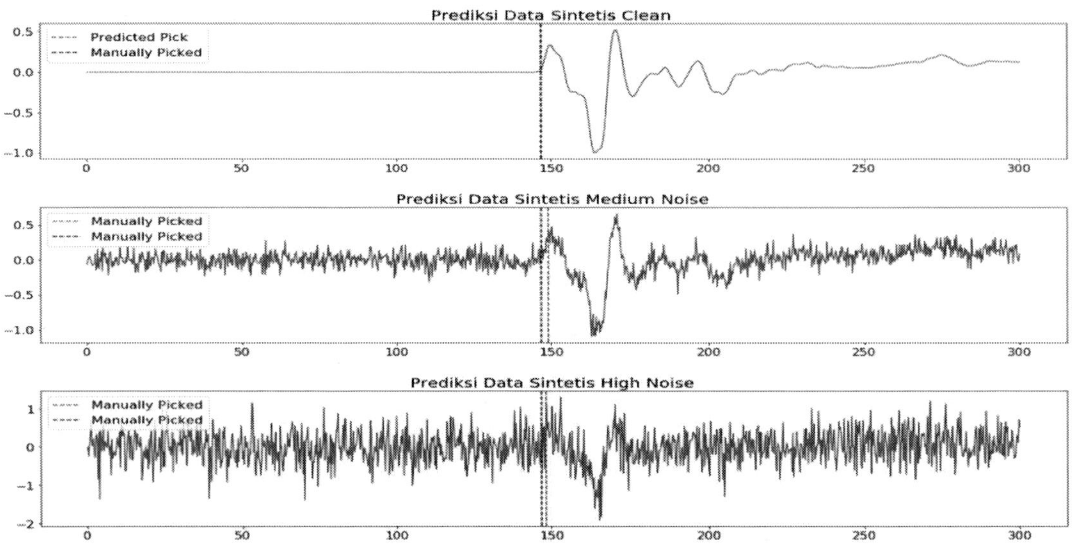

Figure 2. Auto first-break picking on Synthetic Data

Table 1. Analysis of first-break picking prediction on synthetic data

	Sample position on manual first-break picking	Sample position on auto first-break picking	Differences	MAP Error
Clean synthetic data	**588th (147 second)**	**588 (147 second)**	**0**	**0 %**
Medium noise synthetic data	**588th (147 second)**	**597 (149.25 second)**	**9 samples (2.25 second)**	**1.5 %**
High noise synthetic data	**588th (147 second)**	**593 (148.25 second)**	**5 samples (1.25 second)**	**0.8 %**

The machine learning model that has been trained before is good enough in predicting the first break on clean or noisy synthetic data. The error on clean synthetic data is 0% that means the prediction using machine learning is identical to the actual first break. Meanwhile, auto picking results error on medium and high noise data are 0.015% and 0.008%, respectively (2.25s and 1.25s differences from the actual first break).

B. Real Data

Once we are satisfied with the predicted results on the synthetic data, we continued the auto picking prediction on the unfiltered and filtered real data (Figure 3 – Figure 5).

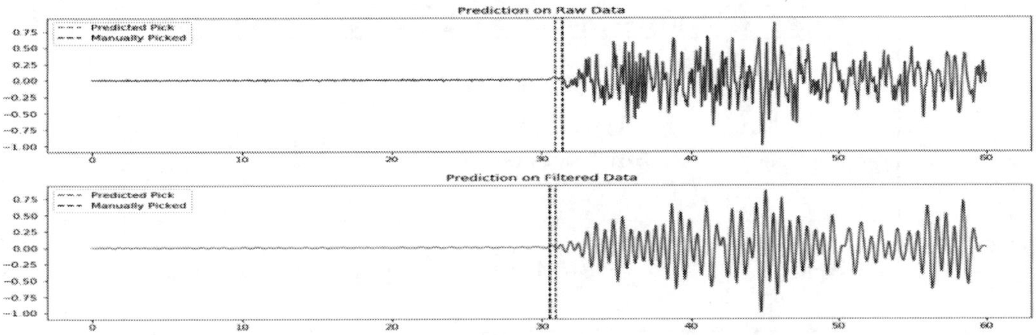

Figure 3. Auto first-break picking on real data 1 (raw and filtered)

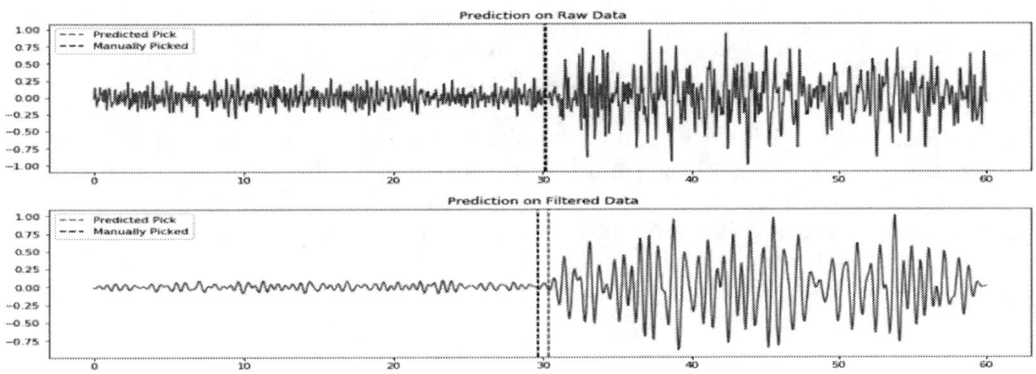

Figure 4. Auto first-break picking on real data 2 (raw and filtered)

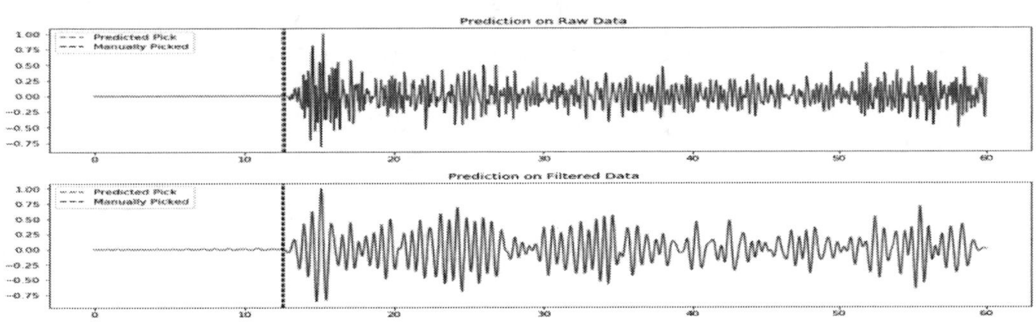

Figure 5. Auto first-break picking on real data 3 (raw and filtered)

Table 2. Analysis of first-break picking prediction on real data

Event	MAPE of Auto-Pick on Raw Data (Differences (s))	MAPE of Auto-Pick on Filtered Data (Differences (s))
Event 1	1.43% (0.45s)	1.14 % (0.35s)
Event 2	0.16% (0.05s)	2.52 % (0.75s)
Event 3	1.18% (0.15s)	0.79 % (0.05s)

Auto-picking was applied to three real earthquake events and show reasonably satisfactory results. On raw (unfiltered) data, the most significant error is demonstrated by Event 1 and Event 3, with the highest deviation of 0.44s and 0.15s, respectively. Moreover, the filtered data results are better, with the highest deviation is only 0.75s on Event 2. Overall, the prediction of filtered data is better than raw data because most of the data we trained is the filtered data. We think that increasing the amount of training dataset can improve accuracy because we have more variations that can be learned by the computer.

4. Conclusions

We have conducted the first break picking of the earthquake using an artificial neural network, a machine learning algorithm. The results showed that our machine learning model is already good enough to predict the first break; however, it still has a few deviation or error in predicting raw data. We can improve our machine learning model by adding more data to be trained so our computer can learn more variation patterns of earthquake's first break.

References

[1] Mężyk M, & Malinowski M 2019 *Multi-pattern algorithm for first-break picking employing open-source machine learning libraries* Journal of Applied Geophysics Vol. **170** 103848

[2] Yalcinoglu L, Millahn K, Fruhwirth R 2011 *Automatic first break detection using support vector machine classification* 73rd Conference and Exhibition incorporating SPE EUROPEC EAGE

[3] Ma Y, Luo Y, 2018 *Automatic first-arrival picking with Reinforcement Learning, CPS/SEG* International Geophysical Conference

[4] Yuan S, Liu J, Wang S, Wang T, Shi P 2018 *Seismic waveform classification and first-break picking using convolution neural network* IEEE Geoscience and Remote Sensing Letters **15** p 272–276

[5] Tsai K C, Hu W, Wu X, Chen J, and Han Z 2018 *First-break automatic picking with deep semisupervised learning neural network* SEG Technical Program Expanded Abstracts : 2181-2185

[6] Brunton S L, Kutz J N 2019 *Neural networks and deep learning* In Data-Driven Science and Engineering pp. 195–226

[7] Aggarwal C C 2018 *Neural networks and deep learning* In Neural Networks and Deep Learning Springer International Publishing AG

[8] Gunawan H et al 2012 *Penentuan waktu tiba gelombang-p secara otomatis dengan metoda skewness dan kurtosis terintegrasi* JURNAL METEOROLOGI DAN GEOFISIKA, VOL. *13* (1) 63–69

[9] Putra J W G 2019 *Pengenalan konsep pembelajaran mesin dan deep learning*. Retrieved from: https://www.researchgate.net/publication/323700644_Pengenalan_Pembelajaran_Mesin_dan_Deep_Learning

[10] Choi Y, Ryu S, Park K, Kim H 2019 *Machine learning-based lithium-ion battery capacity estimation exploiting multi-channel charging profiles* IEEE Access **7** 75143–75152

Pyroclastic Deposits Identification using Near-Surface Seismic Refraction Tomography in Rawa Dano Volcanic Complex, Banten, Indonesia

Maryadi Maryadi[1*], Nanda Aminy Ilahi Bisgar[2], Luthfi Yufajjiru[2], Rifqi Rizqulloh[2], Ghozy El Fatih[2] and Gamma Abdul Jabbar[1]
[1] Geoscience Department, Faculty of Mathematics and Natural Science, Universitas Indonesia, Depok, Indonesia
[2] Undergraduate Program of Geophysics, Faculty of Mathematics and Natural Science, Universitas Indonesia, Depok, Indonesia

*Corresponding author: maryadi@sci.ui.ac.id

Abstract. Rawa Dano is a caldera lake which resulted from Dano Purba Volcano's massive eruption, and it produced a huge amount of pyroclastic deposits that typically formed complex volcaniclastic series. Due to the lack of information regarding the subsurface properties of Rawa Dano area, therefore in this study, a low-energy seismic refraction survey was carried out to identify the distribution of pyroclastic deposits resulted from intensive volcanic eruptions. The data were acquired from two lines in two different sites. Variations of longitudinal velocity in the seismic vertical cross-section suggest that there are more than one type of deposits existed in the area. The results show two main refractors which are related to the deposition of different facies. The seismic velocity shown in the upper part of the seismic tomography model indicates that the pyroclastic deposit has a great thickness. This finding suggests that the eruptions happened massively. By combining the results from both sites, it could be inferred that the preceding one is even bigger in magnitude. The result is in agreement with the earlier surface geological study, which explains a similar conclusion. This research demonstrates the capability of seismic refraction tomography to map the distribution and condition of volcanic deposits around Rawa Dano Volcanic Complex.

Keywords: seismic refraction, velocity, Rawa Dano, volcanic, pyroclastic

1. Introduction

Indonesia is a country located in the Ring of Fire Zone. The fact that it is susceptible to volcanic activities that would create disastrous problems such as pyroclastic fall and flow, which are the products of volcanic eruptions. We can encounter those pyroclastic deposits that were often deposited during different volcanic eruption episodes and typically combined to form complex volcaniclastic series. In some locations, the detailed mechanism of those volcanic events remains unknown due to the lack of evidence and research. The same situation has happened in the area of Rawa Dano. This area is a caldera-like morphology in the middle of Dano Volcanic Complex, which is located in Serang District, Banten Province, Indonesia.

Many international papers mention the utilization of many models to map the distribution of pyroclastic cover deposits, such as slope gradient or elevation [1], and a combination of variety of morphological parameters (for instance [2, 3, 4]). Mainly, these studies have concentrated on analyzing the spatial framework. Research in the Rawa Dano area is still rare, and only focused on surface geological studies. Enhancing our understanding with additional information about the subsurface condition would be beneficial to this study of volcanism. Therefore, this study is aimed to investigate the distribution of the volcanic deposits by identifying their subsurface properties which could differentiate them from soils and basement layer.

Among many geophysical methods in near-surface exploration, seismic refraction tomography is one of the most effective approaches to investigate near-surface geology. This method is generally used to characterize the shallow subsurface layers, which related to the objective of this study is, for example, crustal structure survey [5]; seismic refraction investigations of ignimbrite [6], stratigraphically verified pyroclastic deposits [7], and soil structure [8]. The seismic refraction method in this paper is applied to delineate the distribution of Rawa Dano Volcanic Complex's pyroclastic deposits by using an artificial source of seismic signal, hence making it possible getting the information from depths.

2. Geological Condition

Rawa Dano is a caldera with a lake and swamp with 2,500 hectares coverage [9]. Two significant eruptions caused these features from Dano Purba Volcano during Pleistocene. Upper Banten Tuff deposit (Qvtb) is the vital deposit in this study area, which consists of tuff, pumice tuff, sandy tuff (at the top), crystal tuff, rocky lapilli tuff, glass tuff, and inserts of clay tuff. Dano Muda volcanic rock deposits (Qvd) are in the form of lava flow, composed of andesite or jointed basalt, volcanic breccia, and tuff. Dano Tua volcanic rock deposits (Qpd) consist of lava flow composed of andesite or jointed basalt, volcanic breccia, and tuff. Younger mountain rock deposits (Qhv) consist of volcanic breccia, lava, tuff, lava flows, and other volcanic eruptions (Mount Aseupan, Mount Parakasak) [10].

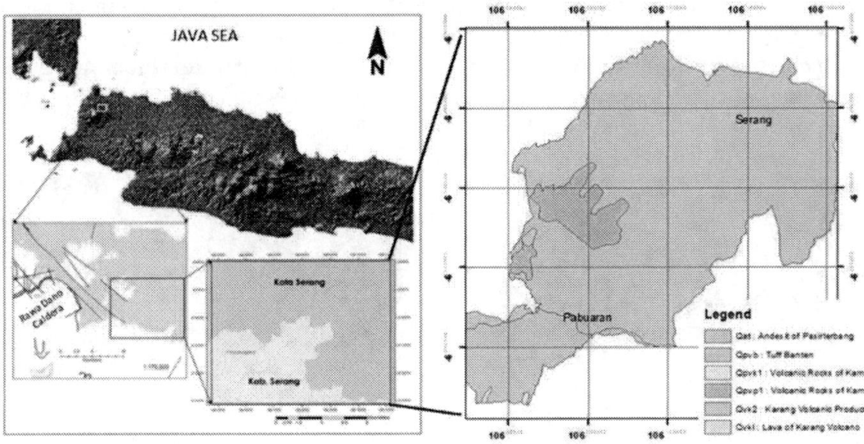

Figure 1. Map of the study area in Rawa Dano (*modified from [11][12]*).

Dano Purba Volcano's first eruption episode is identified as a magmatic eruption. This eruption was likely to have immense explosive power, as indicated by the constituent components, which were more than 5 cm in size (more than lapilli size). This episode's eruption took place with a pyroclastic flow mechanism characterized by lapilli fumarole pipe. After the first phase was completed, there was a hiatus period marked by the presence of a fairly thick paleosoil as indicated by traces fossil of insects [11]. The duration of the eruption is not specific. After the eruption pause is over, another eruption occurs with a magmatic type with two different eruption mechanisms. Based on the geological map of the Serang [12], this eruption happened in the Early Pleistocene. This volcano likely erupted due to the reactivation of subduction in Southern Java during the Plio-Pleistocene period [13, 14]. It is well known that at that time, many volcanoes erupted by producing early quaternary deposits. However, it is not yet significant

IOP Conf. Series: Earth and Environmental Science **873** (2021) 012062 doi:10.1088/1755-1315/873/1/012062

to answer when this eruption occurred. Therefore, it is necessary to analyze age measurements by carbon dating to ensure the hypothesis.

3. Methodology

The study area is located in 2 sites, Site 1 in Pancanegara (6°11'56.1"S 106°05'29.1"E) and Site 2 in Kadu Jaya (6°10'27.5"S 106°08'49.9"E) as shown in Fig. 1. This research uses seismic refraction to identify rock layers based on velocity [15]. Seismic refraction is calculated based on the time it takes for a wave to travel through the rock from the seismic source to the receiver at different distances [16]. We can define the depth of the first (1) and second (2) layer as:

$$Z_1 = \frac{T_i V_1}{2\, cos\alpha} \tag{1}$$

$$Z_2 = \left[T_{i2} - \frac{2Z_i}{V_1 V_3}\sqrt{(V_2)^2 - (V_1)^2}\right]\frac{V_2 V_3}{2\sqrt{(V_3)^2-(V_2)^2}} \tag{2}$$

Like other geophysical methods, seismic refraction survey has three main steps: acquisition, processing, and interpretation. Data acquisition is conducted using seismic refraction instruments to get refraction data. The seismic refraction instrument has three main components: geophones (connected to the main unit), source, and main unit. We use Summit X One as the main unit with a hammer as an artificial low-energy source. Data acquisition took place on Site 1 and Site 2 (Fig. 2). For Site 1, 24 geophones were arranged along 120 m seismic profile in S-N direction. For Site 2, 20 geophones were positioned in S-N direction with the length of 76 m. Each geophone's spacing is 4 m, and shot points are every 8 m, started from 4 m behind the first geophone and ended 4 m in front of the last geophone (Fig. 3). To process the data, we use Easy Refract software to do quality control, pick the first break wave, velocity analysis, and obtain a velocity model of the subsurface layer. Then, interpretation can be done by analyzing the final velocity model to define the subsurface condition.

(a) (b)

Figure 2. Seismic refraction survey location in (a) Site 1 and (b) Site 2.

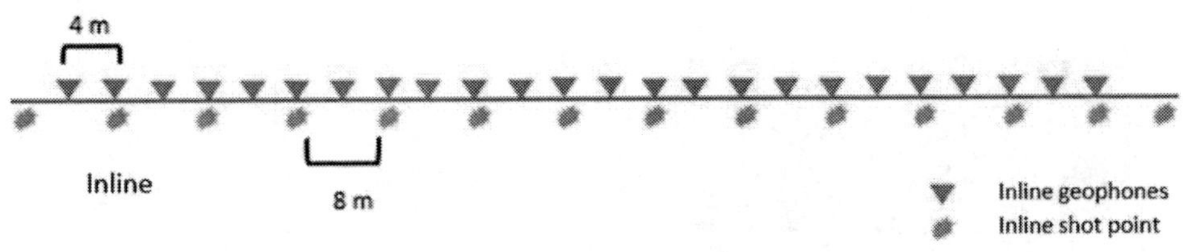

Figure 3. Acquisition geometry

4. Results and Discussions

4.1 Site 1

Seismic refraction survey in Site 1 is aimed to observe the existence of volcanic deposits from different eruption stages, their thickness, as well as the depth of basement layer from their velocity contrast. Based on the data processing results at Site 1, as shown in Fig. 4, there are only two layers of different velocity values were obtained. The first layer, shown in brown, has a primary velocity (Vp) value of 1177.75 m/s and a secondary velocity (Vs) of 565.77 m/s with the shallowest depth of 31 m and the deepest 91 m below the surface. The second layer, shown in green, has a Vp of 1537.14 m/s and a Vs of 738.42 m/s. Both layers have a significant difference in velocity, which apparently describes that it is related to two different types of rock.

Fig. 4 shows that, however, there are no shallow refractors detected. It indicates that neither there are shallow variations of volcanic deposits nor significant velocity differences between layers. The latter could be a more reasonable conclusion by considering the prior geological studies, as stated in chapter 2. From the result of the velocity map and morphological cross-section in Site 1, it can be seen that the first refractor is encountered at a depth greater than 25 m. This layer is getting more in-depth on the right side of the cross-section, showing that this layer is dipping to the north direction. This condition is most probably associated with the deeper pyroclastic deposit, which predates the one on the surface. This layer is having a greater Vp as it got more consolidated after gravitational pressure. This result also suggests that the surface volcanic deposits is having a great thickness. Taking into account the distance from the survey location to the center of the volcanic complex (12 km), this finding also provides further support for the hypothesis that the Rawa Dano eruptions were massive. Unfortunately, this outcome has been unable to give supporting information to explain the eruption stages.

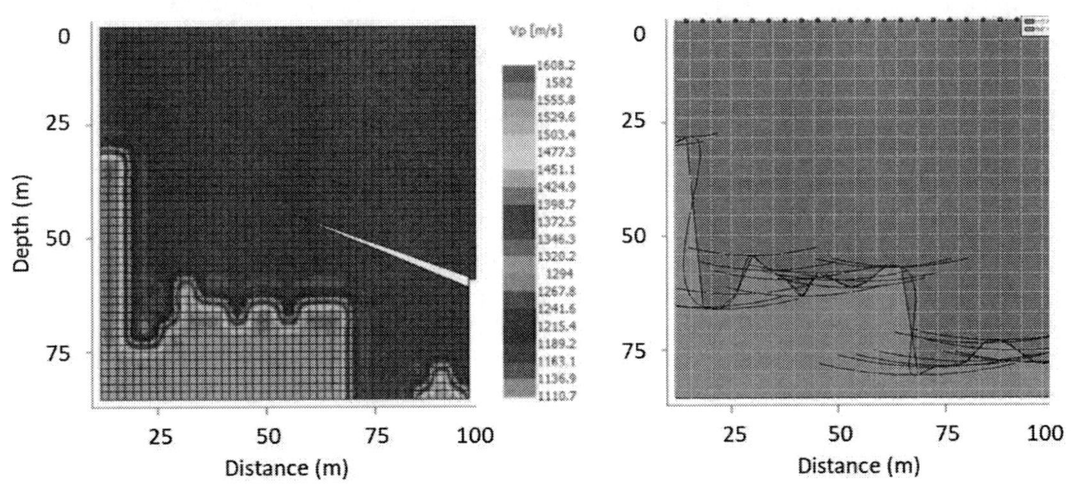

Figure 4. Velocity map (left) and morphology (right) of refractors in Site 1.

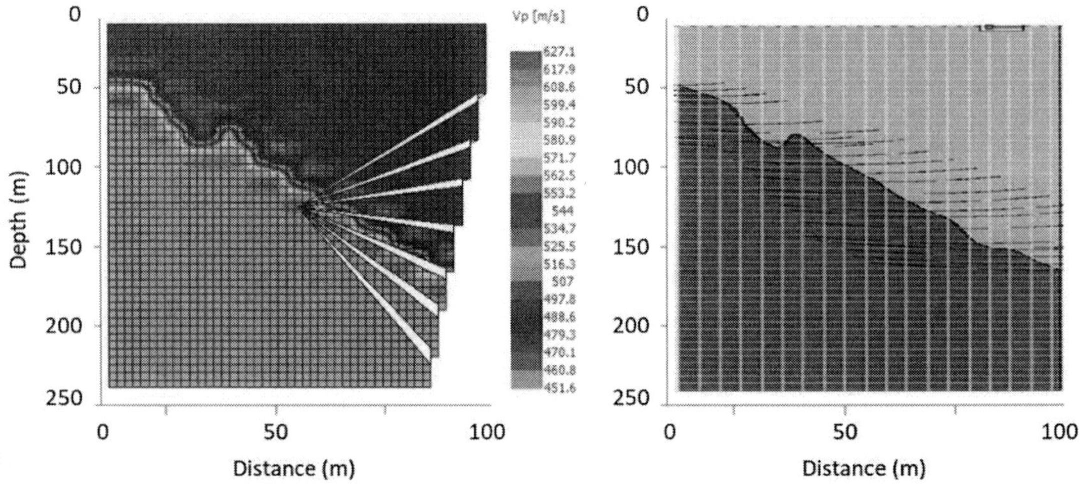

Figure 5. Velocity map (left) and morphology (right) of refractors in Site 2.

4.2 Site 2

The second investigation in Site 2 was carried out to measure the extent of pyroclastic depositions in a farther location which is 7 km to the east from Site 1 or 19 km from the center of Rawa Dano caldera. This investigation is resulting in subsurface velocity map and morphology as seen in Fig. 5. Compared to Site 1, the result of data processing at Site 2 also shows velocity variations with depth. The first layer has a Vp of 473.14 m/s and a Vs of 227.29 m/s from the surface until the deepest of 11 m below the surface. The second layer has a Vp of 603.91 m/s and a Vs of 290.11 m/s. In this cross-section, the second layer is also dipping to the north direction.

Site 2 was the sand mining location which was abandoned for a relatively long time. Measurement was conducted on a dry surface pumiceous volcanic rock which was severely weathered and having bad grain sortation [11]. One interesting result is that the velocity range of the subsurface layers observed in Site 2 differs significantly to the one measured in Site 1. A possible explanation for this is that the layers are made of different facies, although all of them are volcanic products. Different facies here could also suggest that they were formed in different stages. It is interesting to note that the upper layer of Site 2 has a great thickness which indicates another massive eruption happened earlier. The second layer could be related to basement rocks. However, there is not enough information about its geological properties to be correlated to the observed seismic velocity. One unanticipated finding is that the dipping direction of the subsurface layer of Site 1 and Site 2 confirm each other, which could give an additional idea about the paleotopographic condition around Rawa Dano volcanic complex.

5. Conclusions

This study aims to use the seismic refraction method to identify the subsurface geological condition based on seismic velocity. This study has shown that velocity variation with depth and affected by the thickness of layer, compaction, and facies. The results show the deeper position of refractors to the right from the two layers of both sites. The result indicates possible paleotopographic condition around Rawa Dano Volcanic Complex and effect of gravitational pressure caused the greater value of Vp. Facies differences may cause the different value of Vp from both sites. Another interesting result is massive eruption as the product of Dano Volcano's eruption, shown with great thickness in the velocity map. Further studies need to be carried out to explain the eruption stages of Rawa Dano. Additional geophysical method and core data may suit to confirm this study.

IOP Conf. Series: Earth and Environmental Science **873** (2021) 012062 doi:10.1088/1755-1315/873/1/012062

Acknowledgements
This work was financially supported by Universitas Indonesia through PUTI Prosiding Program No. NKB-970/UN2.RST/HKP.05.00/2020. We acknowledge the Energy and Mineral Resources Agency of Banten Province for the cooperation and support for this study.

References

[1] Saulnier G, Beven K and Obled C 1997 *Journal of Hydrology* **202** 158-172
[2] DeRose R C 1996 *Zeitschrift fur Geomorphologie Supplementband* 49-60
[3] Salciarini D, Godt J, Savage W, Conversini P, Baum R and Michael J 2006 *Landslides* **3** 181-194
[4] Segoni S, Martelloni G and Catani F 2013 *Landslide Science and Practice* 127-133
[5] Schlindwein V, Bonnemann C, Reichert C, Grevemeyer I and Flueh E 2003 *Geophysical Journal International* **152** 171-184
[6] Piip V B and Belousov A B 2003 *EAGE 65th Conference & Exhibition*
[7] Gase A, Bradford J and Brand B 2015 *Ground Penetrating Radar and Active Seismic Investigation of Stratigraphically Verified Pyroclastic Deposits* Boise State University
[8] Karslı H, Vanlı S G, Şenkaya M and Güney R 2017 *Eurasian Journal of Soil Science* **6** 134
[9] Adani S, Indra T and Syahputra R 2020 *IOP Conf. Ser.: Earth Environ. Sci.* **538** 012032
[10] Hartanto P, Delinom R and Hendarmawan H 2019 *RISET Geologi dan Pertambangan* **29** 13
[11] Yekini M A 2020 Studi Mekanisme Erupsi Endapan Piroklastik Tuf Banten (Qpvb) di Daerah Pancanegara dan Sekitarnya menggunakan Metode Distribusi Ukuran Butir dan Analisis Komponen *Undergraduate Thesis* Universitas Indonesia
[12] Rusmana E, Suwitodirdjo K and Suharsono 1991 *Peta Geologi Lembar Serang Jawa Skala 1:100.000* (Bandung: Pusat Penelitian dan Pengembangan Geologi)
[13] Katili J 1975 *Tectonophysics* **26** 165-188
[14] Hamilton W B 1979 *Tectonics of the Indonesian Region* (Washington: United States Government Printing Office)
[15] Telford W, Geldart L and Sheriff R 1990 *Applied Geophysics* (Cambridge: Cambridge University Press)
[16] Sabiq H, Rasimeng S and Karyanto 2020 *Jurnal Geofisika Eksplorasi* **4** 58-72

IOP Conf. Series: Earth and Environmental Science **873** (2021) 012063 doi:10.1088/1755-1315/873/1/012063

Rapid Estimation of Earthquake Magnitude using GNSS Data

Susilo[1*], Irwan Meilano[2], Thomas Hardy[3], Muhammad Al Kautsar[1], Dina A. Sarsito[2], Joni Efendi[1]

[1]Geospatial Information Agency, Jl. Raya Jakarta-Bogor Km. 46, Cibinong, Indonesia 16911

[2]Geodesy Research Group, Institut Teknologi Bandung, Jl. Ganesha 10, Bandung, Indonesia 40132

[3]Meteorological, Climatological, and Geophysical Agency, Jl Angkasa 1 No.2, Kemayoran, Jakarta Pusat, Indonesia 10720

E-mail: susilo@big.go.id

Abstract. Earthquake parameters such as the hypocenter location and the magnitude size are important in the development of reliable Earthquake Early Warning (EEW) and Tsunami Early Warning System (TEWS). The Global Navigation Satellitte System (GNSS) data have been used to estimate the earthquake parameters rapidly over the last 15 years. In this study, we present the result and analysis of the scaling properties of Peak Ground Displacement (PGD) as measured by high-rate (sampled at 1 Hz or higher) GPS recordings from Lombok earthquake on August 05th, 2018. The earthquake magnitude from the kinematic solution of CMAT GNSS station is equal to Mw 6.8. This value is achieved between 15 - 20 seconds after the origin time of the earthquake. Our result shows that the displacements from kinematic GNSS data can be used to rapidly determine the earthquake magnitude, typically within the first minute of rupture initiation. Rapid earthquake magnitude determination will be very useful to support EEW and TEWS.

1. Introduction

The determination of earthquake parameters such as location (longitude, latitude, and depth) and magnitude are important in the development of reliable Earthquake Early Warning System (EEWS) and Tsunami Early Warning System (TEWS), and for many projects in earthquake and tsunami mitigation. In the traditional method, these parameters are determined using recording seismic waveform from strong-motion accelerometers or broadband seismometers [1]. Those methods use several seconds of P-wave data to quickly estimate the magnitude, hypocenter, and origin time (OT) of earthquakes [2, 3]. However, the saturation of magnitude is a common problem in the magnitude determination from seismic data for large earthquakes (M $>$ 7) [3–5]. To avoid this saturation problem, several studies [5, 6] have used the Global Positioning Satellite System (GNSS) for the EEWS application over the last 15 years.

Several applications that have used GNSS data for EEWS are ShakeAlert system (developed by the United State Geological Survey (USGS)) [7], Geodetic Alarm System (GlarmS) [8], Bayesian Evidence-based Fault Orientation and Real-Time Earthquake Slip (BEFORES) [9], and Geodetic First Approximation of Size and Timing (G-FAST) [10], and Real-Time on the GNSS Earth Observation Network (GEONET) Analysis System for Rapid Deformation monitoring (REGARD) [11].

On the period of July to August 2018, there were several earthquakes occurred in the North of Lombok Island 1 with a magnitude greater than 6. The first earthquake occurred on July 28th, 2018 (28J) with a magnitude M6.4 continuously with the second earthquake on August 05th, 2018 (05A) with M6.9, and the two last earthquakes occurred on August 19th, 2018, with magnitude M6.3 (19A-1) and M6.9 (19A-2). The magnitude of this event has been modeling using Interferometric Synthetic Aperture Radar (InSAR) data and seismic data [12] and using local seismographic networks [13].

In this research, we use the high-rate data of the CMAT GNSS station to estimate the magnitude of the August 05, 2021 event. We use the Peak Ground Displacement (PGD) formula from GFAST modules to estimate the earthquake magnitude [10]. This event has the biggest coseismic displacement at the CMAT GNSS station than other events [14]. The event has the most significant impact on near volcano [12]. Due to the limitation of GNSS station distribution, we only have one GNSS station which operating continuously in Lombok Island (**Figure 1**). Furthermore, we investigate the estimated magnitude parameter starting from OT to 240 seconds after OT, and to give a comparison of our findings, we used the official magnitude released by Indonesia's Meteorological, Climatological, and Geophysical Agency (Badan Meteorologi, Klimatologi, dan Geofisika / BMKG).

2. Data and method

In this research, we use GNSS data obtained from the Geospatial Information Agency (Badan Informasi Geospasial/BIG) as a part of the Indonesian Continuous Operating Reference System (Ina-CORS). The CMAT GNSS station is located at Mataram city which is $\pm$ 51.89 km from the epicentral location from August, 5th, 2018 event. **Figure 1** shows the location of the CMAT GNSS station. The GNNS data for this research was focusing only on the August, 5th event with Mw 6.9. The GNSS data for another events (28J, 19A-1, 19A-2) were not included in this research because the waveform did not show clearly coseismic displacement.

We use the GPS-Inferred Positioning System and Orbit Analysis Simulation Software (GIPSY-OASIS) version 6.3 from the Jet Propulsion Laboratory (JPL) [15] to generate the kinematic time-series positions of CMAT station at 1 Hz(**Figure 2**). We apply the final precise satellite orbit and clock from JPL 15 and use the IGS14 absolute phase center variations [16]. The FES2004 model from the Onsala Space Observatory (http://holt.oso.chalmers.se/loading/) uses to modeled and corrected the tidal effects (solid earth tide, pole tide, and ocean tide) [17]. The tropospheric wet zenith delays and horizontal gradients were estimated as random-walk parameters [18] and mapped into the tropospheric slant delays down to elevation angle 7 using the Global Pressure and Temperature 2 (GPT2) mapping functions [19]. We apply single-receiver ambiguity resolution to resolve phase ambiguities [20]. We rotate the global coordinates solution (the International Terrestrial Reference Frame (ITRF) 2014) [21] into local coordinates (north, east, up). We produce the displacements waveform in the local reference frame by removing the mean of 10-samples pre-event [1].

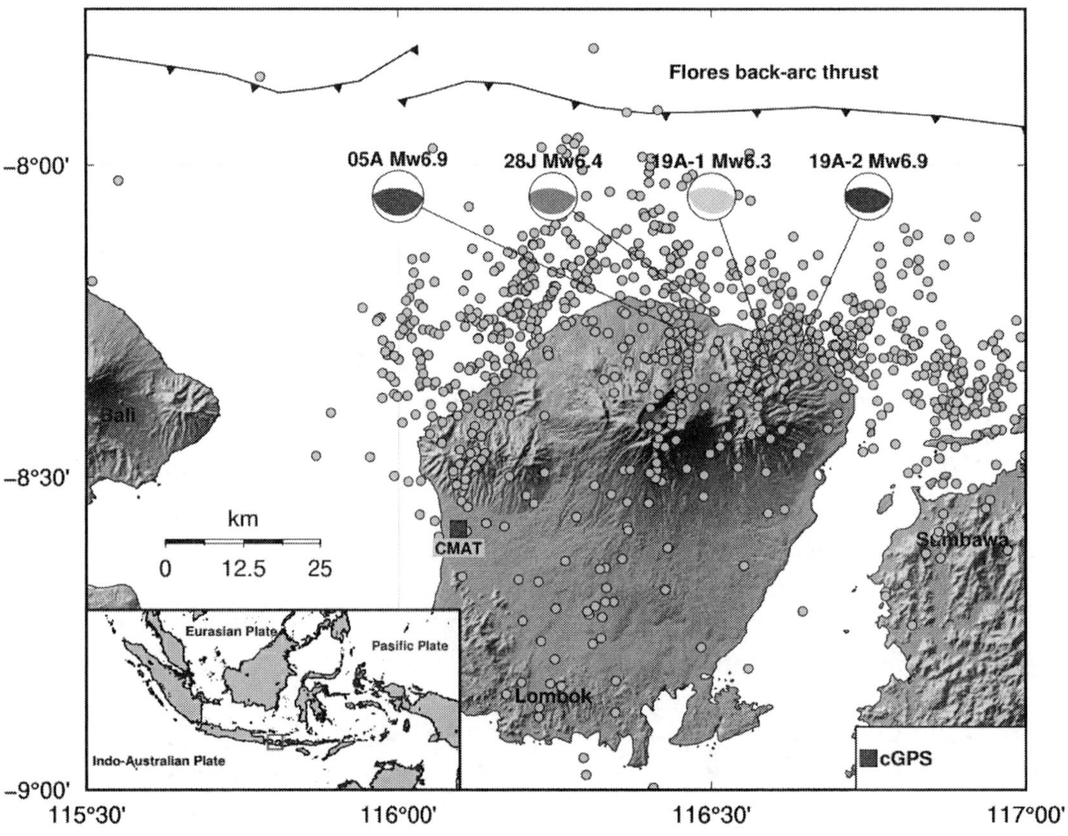

Figure 1. Location of CMAT GNSS station. The focal mechanism and earthquake location were taken from USGS with 28J is event on July 28, 2018; 05A is event on August 5, 2021; 19A-1 is first event on August 19, 2021 and 19A-2 is the second evend on August 19 2018. The light blue circles are aftershock of the event taken from [22]; The fault delineation from Pusat Studi Gempa Nasional (PUSGEN) [23]. The topography data were taken from SRTM15+V2 [24]. Inset shows the position of the study area (red rectangular) area concerning the Indonesian region.

We modeled the GNSS displacements using the G-FAST earthquake early warning packages [10]. We compute the earthquake moment magnitude using PGD scaling. Due to the limitation of station number, we did not apply all the PGD modules of

GFAST. We estimated the earthquake magnitude using epoch-by-epoch of displacement waveform and analyzed the evolution of the earthquake magnitude results.

3. Result and discussion

The displacement waveforms for each component (north, east, up) show in **Figure 2**. The coseismic displacement clearly shows in the North-South (NS) and East-West (EW) components, respectively, while the Up-Down (UD) component did not show any coseismic displacement. The NS and EW components have a maximum coseismic displacement of about 11.25 cm to the North and 6.61 cm to the East directions.

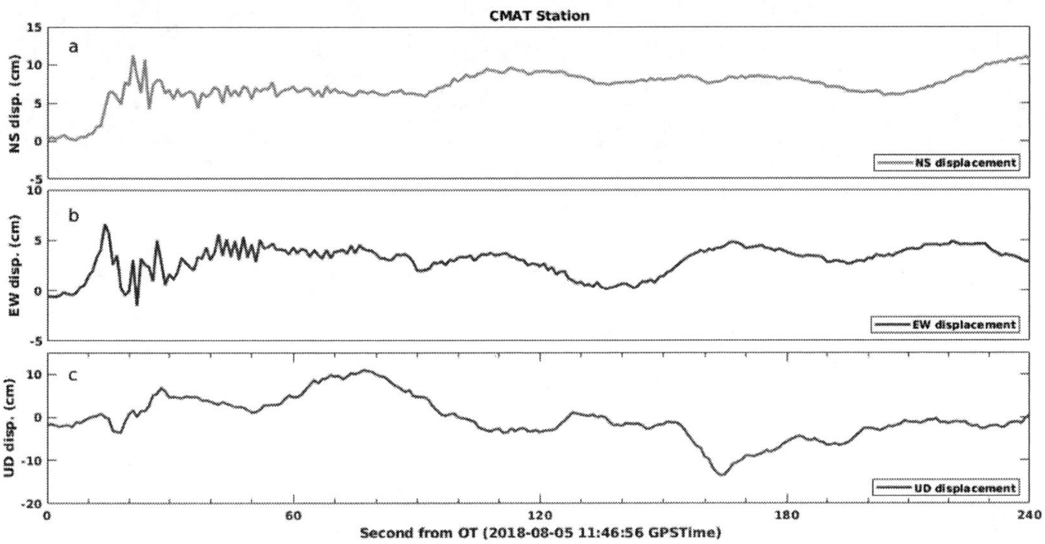

Figure 2. The waveform displacements for each component. (a) is the displacement waveform in the north-south component. (b) is the displacement waveform in the east-west component. (c) is the displacement waveform in the vertical component.

Figure 3 shows the total displacement in two dimensions (2D), three dimensions (3D), the PGD magnitude evolution starting from OT (2018-08-05 11:46:56 GPS time). The 2D displacement was calculated using two components displacement waveform (NS and EW), while the 3D component was calculated using three components of the displacement waveform (**Figure 2**). The 2D and 3D displacements have a similar pattern. This results indicate that the coseismic displacement from vertical component (UD) did not affected to the 3D displacement because there is no significant coseismic displacement in the UD component as saw in the time series 2(c). The total displacement waveform changes quite bigger at 10 seconds after OT. The displacements show the surface wave pattern starting from 10 - 60 seconds after OT.

The PGD magnitude evolution shows in **Figure 3 (c)** and **Figure 3 (d)** for 2D and 3D displacements. In the PGD magnitude calculation, to investigate the evolution of the magnitude, we used all the coseismic displacement to estimate the magnitude. There were two PGD magnitudes in the result which calculates using epicentral distance (Me) and hypocentral distance (Mh), respectively. Our suggestion the vertical coseismic displacement will give a contribution to the PGD magnitude determination. **Figure 3 (c) and (d)** show Mh is relatively higher than Me. The first 10s from OT, the PGD magnitude (2D and 3D) have a value below Mw 6.2 and reach Mw 6.6 - 6.8 at 30s after OT. Overall, from 30 to 240 seconds after OT, the average PGD magnitude has a value between Mw 6.7 - 6.9.

We compare our PGD magnitude reults with the evolution of magnitude from BMKG database. The August, 5th, 2018 event has been detected by the BMKG seismic stations at 90 seconds after OT, and the first magnitude was estimated (FE) at 118 - 119 minutes after OT (green square with red circle in **Figure 3** (c,d)), and first announcement at 150 - 178 seconds after OT (FA).

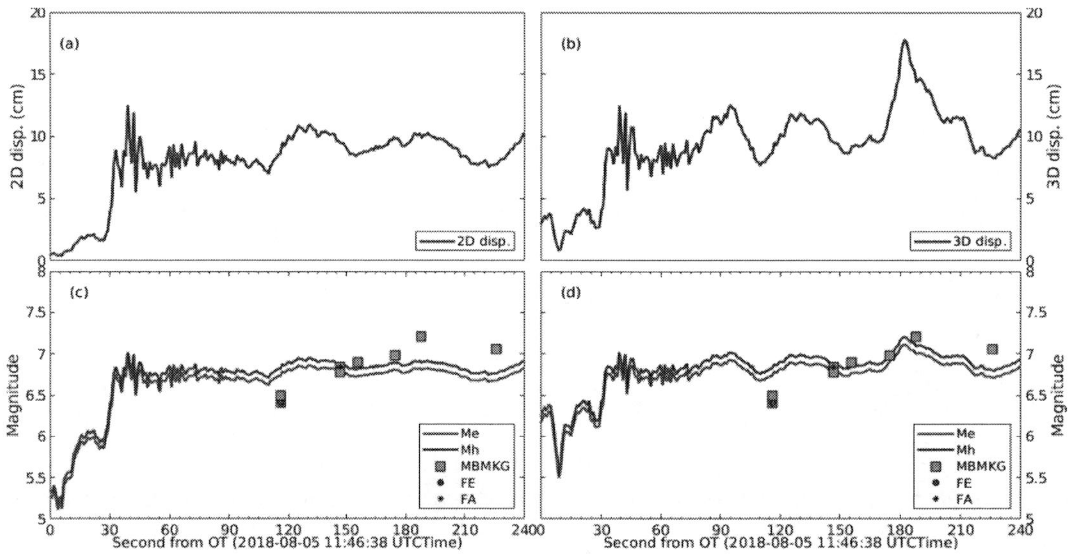

Figure 3. The total displacements and PGD magnitude;Me is the PGD magnitude using epicentral distance; Mh is the PGD magnitude using hypocentral distance; MBKMG is magnitude from BMKG; FE is first magnitude estimation; FA is first magnitude announcement. (a) the 2D total displacement; (b) the 3D displacement; (c) the PGD magnitude from 2D total displacement; (d) The PGD magnitude from 3D total displacement; green squares at (c) (d) are magnitude from BMKG.

4. Conclusions

This research demonstrates the ability to use a high-rate GNSS waveform to determine the earthquake magnitude rapidly. From this research, using GNSS post-processing, the PGD magnitude of the August 05th, 2018 Lombok earthquake with the value Mw 6.6 -

IOP Conf. Series: Earth and Environmental Science **873** (2021) 012063 doi:10.1088/1755-1315/873/1/012063

6.8 was reached at 30 seconds after OT, while the seismic magnitude needs about 188 - 119 s after OT for the first magnitude estimation. The first time seismic magnitude estimation depends on the nearest distance of the seismic stations to the earthquake source location. How fast PGD GNSS has estimated depends on several factors, such as the reliable communication data to transfer GNSS high-rate data to the processing center, the quality of the real-time kinematic positioning, the distance GNSS network stations to the earthquake source location, and the modeling method to estimate PGD magnitude.

Acknowledgments

We thank the Geospatial Information Agency and the Meteorological, Climatological, and Geophysical Agency for providing GNSS data and timestamp magnitude determination. We thank our colleagues from the University of Washington for the discussion about PGD determination. Most figures were generated by Generic Mapping Tools [25].

References

[1] Crowell B W, Melgar D and Geng J 2018 Hypothetical real-time gnss modeling of the 2016 mw 7.8 kaikōura earthquake: Perspectives from ground motion and tsunami inundation prediction *Bull. Seismol. Soc. Am.* **108** 1736–1745 ISSN 0037-1106

[2] Allen R M and Kanamori H 2003 The potential for earthquake early warning in southern california *Science* **300** 786–789 ISSN 0036-8075

[3] Wu Y M and Zhao L 2006 Magnitude estimation using the first three seconds p-wave amplitude in earthquake early warning *Geophys. Res. Lett.* **33**

[4] Brown H M, Allen R M, Hellweg M, Khainovski O, Neuhauser D and Souf A 2011 Development of the elarms methodology for earthquake early warning: Realtime application in california and offline testing in japan *Soil Dynamics and Earthquake Engineering* **31** 188–200 ISSN 0267-7261

[5] Crowell B W, Melgar D, Bock Y, Haase J S and Geng J 2013 Earthquake magnitude scaling using seismogeodetic data *Geophys. Res. Lett.* **40** 6089–6094

[6] Melgar D, Bock Y, Sanchez D and Crowell B W 2013 On robust and reliable automated baseline corrections for strong motion seismology *J. Geophys. Res.: Solid Earth* **118** 1177–1187

[7] Given D D, Cochran E S, Heaton T, Hauksson E, Allen R, Hellweg P, Vidale J and P B 2014 Technical implementation plan for the shakealert production system: An earthquake early warning system for the west coast of the united states *U.S. Geol. Surve. Open-File Rept. 2014-1097* **25** 25 p.

[8] Grapenthin R, Johanson I A and Allen R M 2014 Operational real-time gps-enhanced earthquake early warning *J. Geophys. Res.: Solid Earth* **119** 7944–7965

[9] Minson S E, Murray J R, Langbein J O and Gomberg J S 2014 Real-time inversions for finite fault slip models and rupture geometry based on high-rate gps data *J. Geophys. Res.: Solid Earth* **119** 3201–3231

[10] Crowell B W, Schmidt D A, Bodin P, Vidale J E, Gomberg J, Hartog J R, Kress V C, Melbourne T I, Santillan M, Minson S E and Jamison D G 2016 Demonstration of the cascadia g-fast geodetic earthquake early warning system for the nisqually, washington, earthquake *Seismol. Res. Lett.* **87**

[11] Kawamoto S, Ohta Y, Hiyama Y, Todoriki M, Nishimura T, Furuya T, Sato Y, Yahagi T and

Miyagawa K 2017 Regard: A new gnss-based real-time finite fault modeling system for geonet *J. Geophys. Res.: Solid Earth* **122** 1324–1349

[12] Salman R, Lindsey E O, Lythgoe K H, Bradley K, Muzli M, Yun S, Chin S T, Tay C W J, Costa F, Wei S and Hill E M 2020 Cascading partial rupture of the flores thrust during the 2018 lombok earthquake sequence, indonesia *Seismological Research Letters* **91**

[13] Sasmi A T, Nugraha A D, Muzli M, Widiyantoro S, Zulfakriza Z, Wei S, Sahara D P, Riyanto A, Puspito N T, Priyono A, Greenfield T, Afif H, Supendi P, Daryono D, Ardianto A, Syahbana D K, Husni Y M, Prabowo B S and Sarjan A F N 2020 Hypocenter and magnitude analysis of aftershocks of the 2018 lombok, indonesia, earthquakes using local seismographic networks *Seismological Research Letters*

[14] Susilo S, Wijanarto A, Wibowo S and Abidin H Z 2018 Gps/gnss analysis on lombok earthquakes: Co-seismic deformation *Experimental Findings*

[15] Zumberge J F, Heflin M B, Jefferson D C, Watkins M M and Webb F H 1997 Precise point positioning for the efficient and robust analysis of gps data from large networks *J. Geophys. Res.: Solid Earth* **102** 5005–5017

[16] Schmid R, Steigenberger P, Gendt G, Ge M and Rothacher M 2007 Generation of a consistent absolute phase center correction model for gps receiver and satellite antennas *Journal of Geodesy* **81** 781–798

[17] Lyard F, Lefevre F, Letellier T and Francis O 2006 Modelling the global ocean tides: modern insights from fes2004 **56** 394–415

[18] Bar-Sever Y E, Kroger P M and Borjesson J A 1998 Estimating horizontal gradients of tropospheric path delay with a single gps receiver *J. Geophys. Res.: Solid Earth* **103** 5019–5035

[19] Lagler K, Schindelegger M, Böhm J, Krásná H and Nilsson T 2013 Gpt2: Empirical slant delay model for radio space geodetic techniques *Geophys. Res. Lett.* **40** 1069–1073

[20] Bertiger W, Desai S D, Haines, Harvey N, Moore A W, Owen S and Weiss J P 2010 Single receiver phase ambiguity resolution with gps data *Journal of Geodesy* **84** 327–337

[21] Altamimi Z, Rebischung P, Metivier L and Collilieux X 2016 Itrf2014: A new release of the international terrestrial reference frame modeling nonlinear station motions *J. Geophys. Res.: Solid Earth* **121** 6109–6131

[22] Supendi P, Nugraha A D, Widiyantoro S, Pesicek J D, Thurber C H, Abdullah C I, Daryono D, Wiyono S H, Shiddiqi H A and Rosalia S 2020 Relocated aftershocks and background seismicity in eastern indonesia shed light on the 2018 lombok and palu earthquake sequences *Geophysical Journal International* **221** 1845–1855 ISSN 0956-540X

[23] Irsyam M, Widiyantoro S, Natawidjaya D H, Meilano I, Rudyanto A, Hidayati S, Triyoso W, Hanifa N R, Djarwadi D, Faizal F and et al 2017 *Peta sumber dan bahaya gempa Indonesia tahun 2017* (Pusat Penelitian dan Pengembangan Perumahan dan Permukiman, Kementerian Pekerjaan Umum dan Perumahan Rakyat (in Indonesian))

[24] Tozer B, Sandwell D T, Smith W H F, Olson C, Beale J R and Wessel P 2019 Global bathymetry and topography at 15 arc sec: Srtm15+ *Earth and Space Science* **6** 1847–1864

[25] Wessel P, Smith W H F, Scharroo R, Luis J and Wobbe F 2013 Generic mapping tools: Improved version released *Eos, Transactions American Geophysical Union* **94** 409–410

IOP Conf. Series: Earth and Environmental Science **873** (2021) 012064 doi:10.1088/1755-1315/873/1/012064

Preliminary Model of P-Wave Tomography Beneath Central Java using FMTOMO

M Yasir[1], P T Brilianti[1], S S Angkasa[1], S Widyanti[1], I Herawati[1], W Pranowo[1], S Satiawan[1], S K Suhardja[1], A Priyono[1], S Widiyantoro[2,3], M Ramdhan[4]

[1] Faculty of Exploration and Production Technology, University of Pertamina, Jl. Teuku Nyak Arief, Simprug, Kebayoran Lama, Jakarta 12220, Indonesia

[2] Global Geophysics Research Group, Faculty of Mining and Petroleum Engineering Institute of Technology Bandung, Jalan Ganesa No. 10, Bandung 40132, Indonesia

[3] Faculty of Engineering, Maranatha Christian University Bandung, Jalan Surya Sumantri No 65, Bandung 40164, Indonesia

[4] Agency for Meteorology, Climatology and Geophysics, Jalan Angkasa I, No. 2, Kemayoran, Jakarta, Indonesia

Abstract. The tectonic setting of Java Island is mainly controlled by the collision of Indo-Australian plate subducting the Eurasian plate. The high collision activity of Eurasian and Indo-Australian plates often causes megathrust earthquakes and the rise of arc magmatism that includes volcanic eruption. This study aims to determine the tectonic pattern beneath Central Java based on P-wave tomography inversion. We used the fast-marching method as ray tracing and subspace inversion to image subsurface velocity model to a depth of 150 km. The data used in this study are catalogue events data derived from a temporary seismometer network MERAMEX installed around central Java and DOMERAPI installed surround Mt. Merapi and Mt. Merbabu. We also include events collected from the International Seismological Centre. In total, we processed 563 earthquake events to illustrate velocity structures under central Java. The checker-board model shows that good resolutions can be identified at shallow depth, including offshore south Java contributed from Ocean Bottom Seismometer data. In vertical axis, good resolution models can be expected down to a depth 150 km following rich events from the Benioff zone. Current P wave model show a distinct low velocity zone under Mt Merapi that can be seen down to a depth of 40 km, suggesting a possible separated deep magma reservoir. To the south of Mt Merapi area also shows a low-velocity band that may be related with the southern mountain arc. Additionally, the northern part of Mt. Merapi displays a band of strong low-velocity anomaly to the East and West with the anomaly in the Eastern Part seems to have a deeper extension to a depth of ~50 km. We related this anomaly with Merapi Lawu Anomaly and Kendeng basin. Our results show a similar result with the previous tomography models in this region.

1. Introduction

Indonesia is an archipelago that its tectonic setting is affected by the interaction between 3 large tectonic plates, namely the Indo-Australian Plate, the Eurasian Plate, and the Pacific Plate. The high activity of the movement of these three plates has resulted in many volcanoes forming along the island of Java to Sumatra so Indonesia is called the ring of the fire area. [1] stated that the high collision activity of the Eurasian and Indo-Australian plates often causes megathrust earthquakes which result in the high activity of many volcanoes in the Java Island region, especially Central Java. One of the most active volcanoes today is Merapi. In this study, we conducted a tomographic study in the area of Merapi volcano by using the fast-marching ray tracing method and sub-space inversion. Previous local tomography studies [2] show a region of lower Vp and high Vp/Vs ratios inside the Merapi Lawu Anomaly (MLA) region, located to the north of Mt. Merapi. They suggested that this may be related to the presence of elevated fluids and partial melts. [3] performed local earthquake tomography and found lower velocity anomaly in the back-arc crust north of the active volcano, down to upper mantle area. [4] applied Receiver Function method to estimate crust thickness in this area and proposed that the Moho is shallower beneath Mt. Merapi with higher Vp/Vs ratio. The objective of this study is to image the subsurface structure beneath central Java using alternative fast marching ray tracing method and inversion method.

Data and method

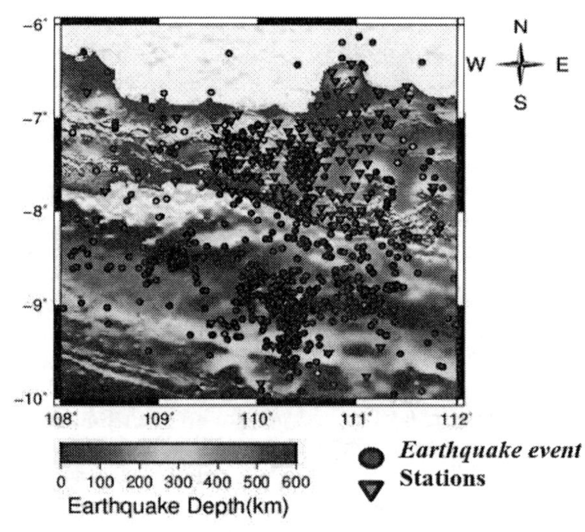

Figure 1. The distribution of earthquakes and stations in the Merapi Volcano area

In this study, we used the same data set as in [5], [2]. The data comes from several networks: 53 stations from DOMERAPI network, 17 from local BMKG network and 134 from MERAMEX network covering a region of 150 x 200 km (Figure 1). The total duration of data is from 2004-2016 although most stations are installed in short duration time. We collected 563 events with a maximum depth of 160 km. This depth limit was chosen based on previous research regarding the depth of subduction seen at a depth of 50-100 km [2].

Events distribution is dominated by less than 100 km depth located offshore at southern part of the island of Java (Figure 1). Most event show a down-going dip to the north and most likely are tectonic event from the subduction slab between the Indo-Australian plate and the Eurasian plate. We used the fast-marching method to reconstruct ray path between source and receivers. The 3-D ray tracing is a grid based eikonal solver that uses a modified finite difference method.

The inversion problem is determined by the subspace inversion method with the ak135 velocity model [6] as the initial and reference model. We used a package named tomo3d [7] to reproduce a 3-D velocity model of crust and upper mantle beneath Central Java, Indonesia. For the parameterization, we set 30 layers in depth direction with 5 km thick, 50 grids in latitude direction with 10 km length and 50 grids in longitude direction with 11 km length. The damping parameter and smoothing parameter is set to be 10 and 15 to resolve underdetermined problem in the inversion.

2. Result

One way to check if the perturbation velocity model is reliable, is by observing the difference between calculated travel time and observation time. A good model can be achieved by observing less number in residuals data between observed arrival time and model arrival time. Figure 2 shows the histogram of travel time between initial model and final model. Before inversion, most residuals data are negative with means about -0.4s and residual data after inversion show a more centered histogram on 0.1s, suggesting the velocity model has been optimized and move closer to real velocity number. Note that in this study, to optimize our model, we performed trafe off analysis in the damping and smooth parameters. The optimal damping and smooth parameters are 15 and 20 consequently.

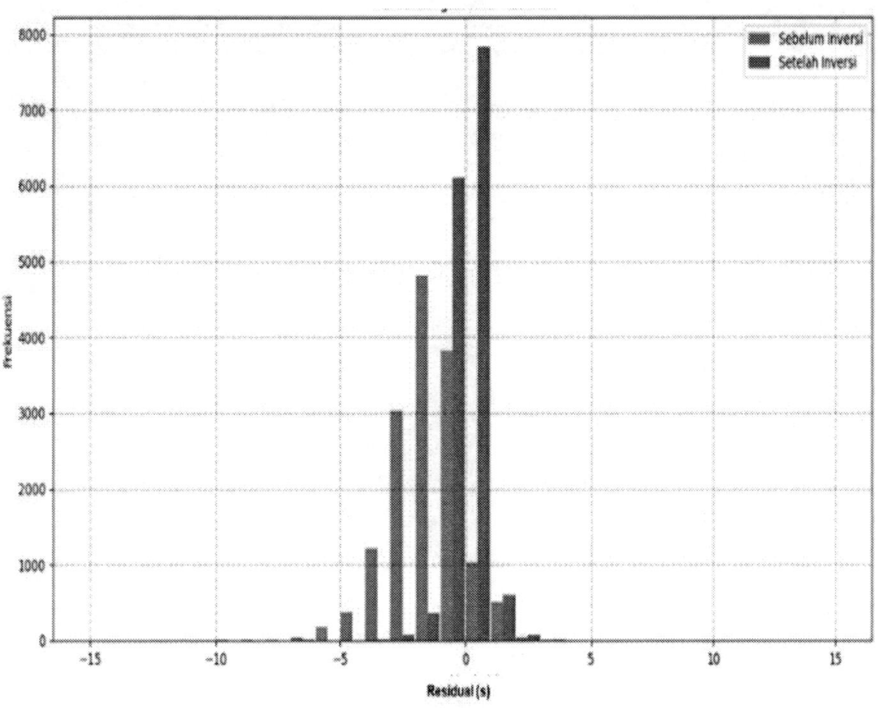

Figure 2. Comparison of residual values

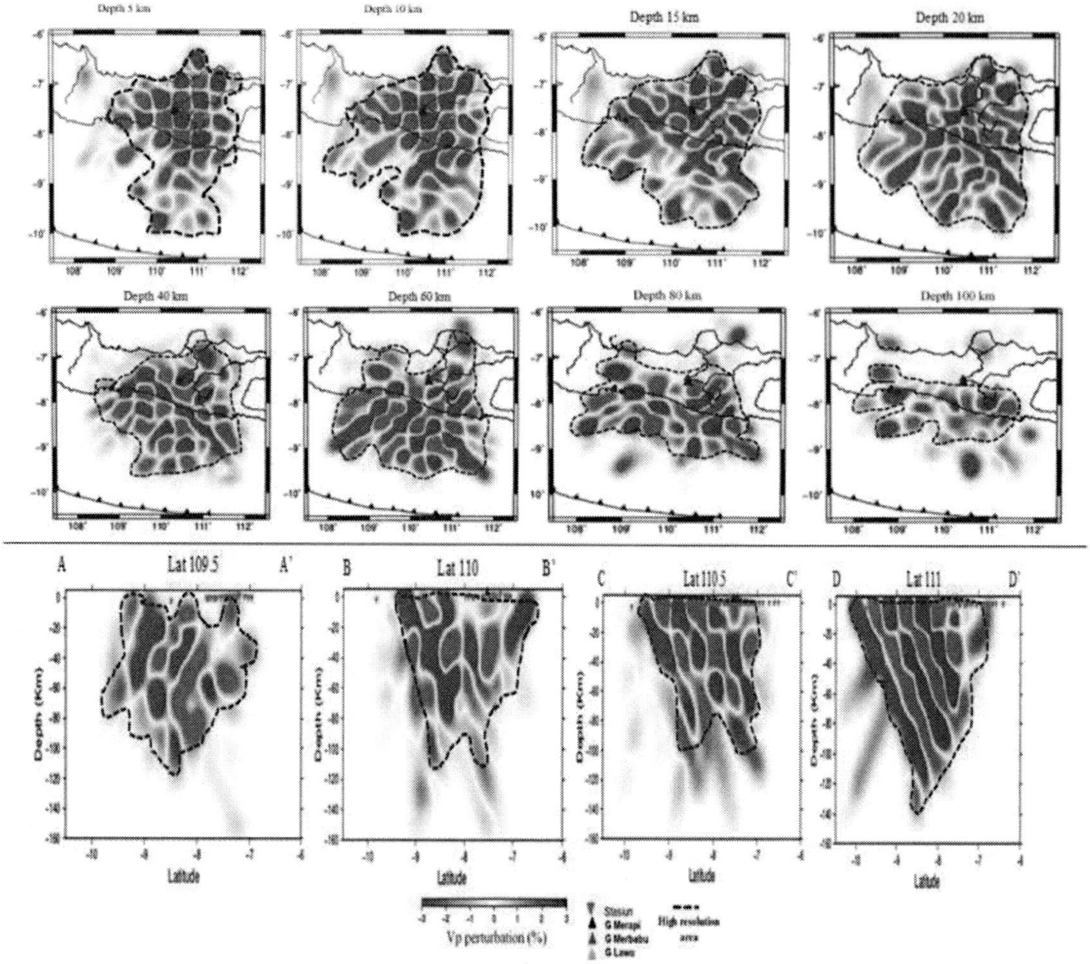

Figure 3. Checker-Board model for horizontal and vertical slices. The vertical slices are taken on latitude 109.5, 110, 110.5 and 111 degree.

To test the robustness of our model, we perform checker-board resolution test. One challenge in tomography inversion is un-balanced condition between unknown parameters and the available arrival data set that can give a non-unique solution in the inversion process. The checker-board simply arrange velocity model into alternating regions of high and low velocity. Then, we tried to invert the velocity using the same source and receiver configuration in the observation data set. The inconsistency between the input checker board model and the recovered structures reveal which regions of the model can be retrieved. Figure 3 shows the results of the checker-board test in this region. Most areas show good recovering down to a depth of 60 km although some smearing start to appear from a depth of 25 km. Resolution quality started to decrease from a depth of 80 km with less area recovered and reach maximum depth of 100 km. Figure 4 and 5 display all recovered P wave tomograms in this region in horizontal and vertical slices. Most shallow depths model show a domination of low velocity zone and deeper than 40 km show high velocity patterns.

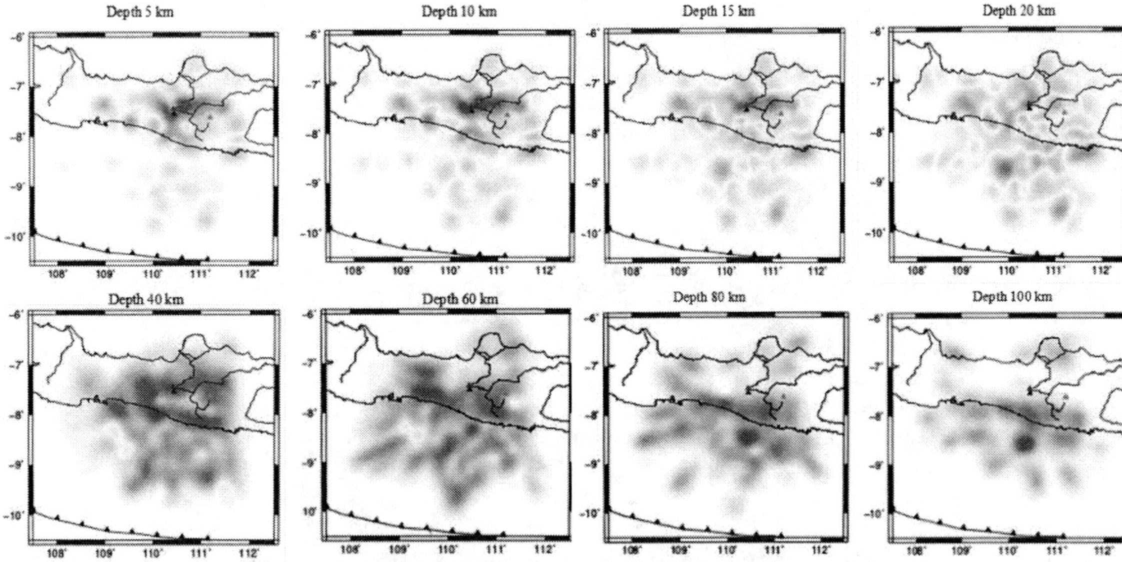

Figure 4. Horizontal tomography model

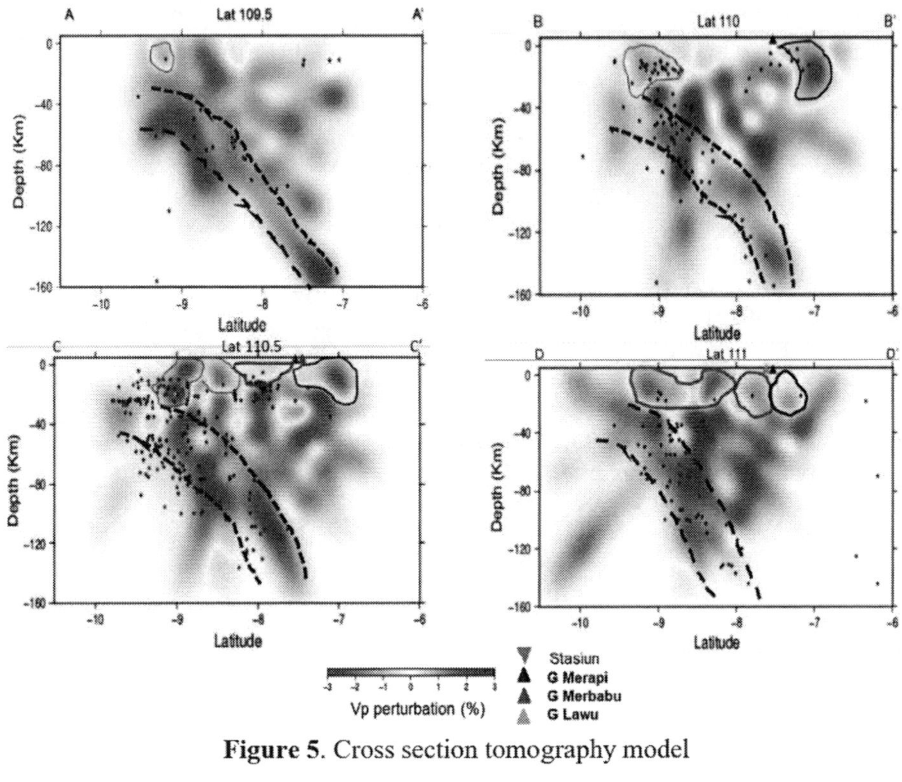

Figure 5. Cross section tomography model

Figure 5 shows a series of cross section from west to east with interesting results. A band of slow velocity can be seen at some areas such as between latitude -9° to -8° from a depth of 5 to 30 km at line B, C and D. Another region of low velocity can also be seen on latitude -7° at shallow depth at line B, C, and D. A band of fast velocity is also seen dipping to the north although not that clear. Plot of seismicity on top of tomogram seems to show a pattern of dipping to the north.

3. Discussion

The high velocity region that dips to north may represent the subducting oceanic plate underneath the Eurasian plate. We see a pattern of steep subduction that increase its angle to the east. Steep angle of subduction may be related to the subduction rollback from north to south which resulted in a change in the subduction path which affected volcanic activity in Central Java.

In line A there is a negative anomaly in the southern part of Java Island from a depth of 5-15 km, which is associated with a marine forearc anomaly which is associated with marine sediment [1]. In line B and C, there is a negative anomaly to the north of Mount Merapi at a shallow depth of 20 km which is interpreted as a partial melting zone [2] while for a shallow depth it is associated with the Kendeng basin zone which is filled with sedimentary material.

In line C, just below and the northern part of Mount Merapi also shows a low anomaly zone to a depth of 30 km associated with a partial melting zone [2] as well as a layer of sediment originating from the Kendeng basin. Besides, there is also a negative anomaly in the southern part of Mount Merapi at a shallow depth of up to 15 km which is associated with the southern mountain arc zone where this zone is associated with volcanic rock and igneous intrusion [5].

4. Conclusion

Tomography model show Vp perturbation anomaly results showing MLA zone / partial melting zone beneath Merapi volcano, Kendeng Basin in northern part of Merapi which contains sedimentary material, igneous intrusion in southern mountain arc, and marine fore arc anomaly in southern Java island. High velocity anomaly and distribution of earthquake events at a depth of more than ± 40 km may represent down-going oceanic plate and increase its angle to the east.

References

[1] Bohm M, Haberland C & Asch G 2013 *Imaging fluid-related subduction processes beneath Central Java (Indonesia) using seismic attenuation tomography* Journal of Tectonophysics **590** (2013) 175–188

[2] Koulakov I et al. 2007 *P and S velocity structure of the crust and the upper mantle beneath central Java from local tomography inversion* JOURNAL OF GEOPHYSICAL RESEARCH, VOL. **112** B08310

[3] Wagner D, Koulakov I, Rabbel W, Luehr B-G, Wittwer A, Kopp H, Bohm M, and Asch G 2007 *Joint inversion of active and passive seismic data in central Java* Geophys. J. Int. **170**(2) 923–932

[4] Suhardja S K et al. 2020 *Crustal thickness beneath Mt. Merapi and Mt. Merbabu, Central Java, Indonesia, inferred from receiver function analysis* Physics of the Earth and Planetary Interiors **302**:106455

[5] Ramdhan M et al. 2019 *Detailed seismic imaging of Merapi volcano, Indonesia, from local earthquake travel-time tomography* Journal of Asian Earth Sciences **177** 134–145

[6] Kennett B L N, Engdahl E R, Buland R, 1995 *Constraints on seismic velocities in the Earth from traveltimes* Geophysical Journal International vol **122** p108-124

[7] Rawlinson N et al 2005 *The fast-marching method: an effective tool for tomographic imaging and tracking multiple phases in complex layered media* Exploration Geophysics **36** 341–350

IOP Conf. Series: Earth and Environmental Science **873** (2021) 012065 doi:10.1088/1755-1315/873/1/012065

Early Results of P Wave Regional Tomography Study at Sunda-Banda Arc using BMKG Seismic Network

M S Haq[1], Haolia[1], M I Sulaiman[1], I Madrinovella[1], S Satiawan[1], D A Zaky[1], S K Suhardja[1], M Ramdhan[4], A Priyono[1], S Widiyantoro[2,3], A M S Nugraha[5]

[1] Faculty of Exploration and Production Technology, University of Pertamina, Jl. Teuku Nyak Arief, Simprug, Kebayoran Lama, Jakarta 12220, Indonesia
[2] Global Geophysics Research Group, Faculty of Mining and Petroleum Engineering Institute of Technology Bandung, Jalan Ganesa No. 10, Bandung 40132, Indonesia
[3] Faculty of Engineering, Maranatha Christian University Bandung, Jalan Surya Sumantri No 65, Bandung 40164, Indonesia
[4] Agency for Meteorology, Climatology and Geophysics, Jalan Angkasa I, No. 2, Kemayoran, Jakarta, Indonesia
[5] Earth Observatory of Singapore, 50 Nanyang Ave., Nanyang Technological University, 639798 Singapore

*Corresponding author: s4ndy104@gmail.com

Abstract. The plate movement, geological structure, magmatism, and seismic activity in the area of Bali to East Nusa Tenggara are mainly related with the subducting of Indo-Australian Plate underneath the Eurasian plate. The complexity is added with the recent collision of Australian continent lithosphere with the western Banda arc, along the islands of Flores, Sumba and Timor island. Our study area is known as the Sunda-Banda arc transition. With the aim of imaging subsurface structure, we perform seismic tomography inversion using regional events. We collected 5 years of earthquake data (January 2015 – December 2019) from the Indonesian Agency of Meteorology, Climatology, and Geophysics (BMKG). The output of our data processing is not limited to only P wave velocity model, but also relocated seismicity pattern in the region. In general, seismicity pattern shows dominant shallow events in the south that progressively shift into deeper events in the north down to a few 500 km, marking a dipping subduction zone in this region. A group of shallow events down to a depth of 50 km is also seen at the norther region that may relate to back-arc thrust activity. P wave tomogram model show a lower velocity perturbation at a depth of 30 km that could be associated with magmatic activity along the volcanic front line. Higher P wave perturbation model are spotted at two different zones, the first one is marking a dipping Indo-Australian plate down to depth of 400 km. We noticed that the angle of dipping is steeper in the Eastern part compared to the Western part. The second a relatively flat at shallow depth at the northern region from the island of Lombok to Nusa Tenggara Timur that may mark the back-arc thrust region

IOP Conf. Series: Earth and Environmental Science **873** (2021) 012065 doi:10.1088/1755-1315/873/1/012065

1. Introduction

The area of Bali to East Nusa Tenggara is located on the ring of fire route which has high earthquake activity due to the subduction zone between the Indo-Australian plate and the Eurasian plate. The Indo-Australian plate is moving at a rate of 6 - 7.3 cm/year to the north [1]. A study by [2] proposed a microcontinental fragments generation when the Indo-Australian plate is moving to the north (Figure 1).

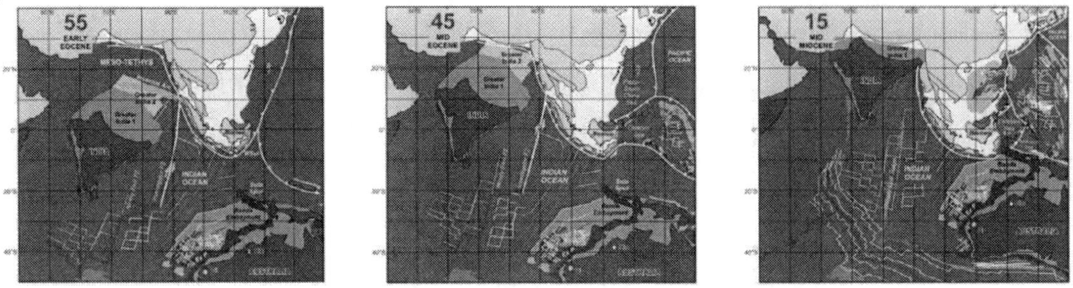

Figure 1. Reconstruction plate at 55, 45 and 15 Ma [2].

[3] estimated that the age of the plates in the southern part of the island of Java to Flores or the eastern part of the Sunda Arc has an age of 100 - 120.4 Ma. The difference in plate age from the subduction zone affects the deformation and seismicity in the Sunda Arc [4]. Interestingly, a series of shallow earthquakes is seen at north of Bali to East Nusa Tenggara [5] and [6] that was interpreted as the back arc thrust events. Tomographic studies have been conducted previously by [7] and [6] to determine the tectonic patterns that developed in the area, which focus on the subduction zone of the eastern Sunda Arc. Tomogram model by [7] displayed a high-velocity zone extending northward interpreted the down-going oceanic plate and also a high velocity between the two low-velocity anomalies, interpreted as forearc sliver from the continent. This study was conducted to image the tectonic structures that developed in this area of Bali to East Nusa Tenggara, by displaying the results of the P wave tomogram in that area. The objective is to better image the structure and understand more about complex tectonic in this region.

2. Data

The data used in this study taken from the BMKG earthquake catalogue for Bali to East Nusa Tenggara with coordinates 112o to 125o E and -5o to -13o S at the period 2015-2018. It has 8875 events with depth 0-700 km and magnitude from ≥3 SR). Majority of events have shallow depth around 1-100 km, particularly to the southern of regions although a dipping down to the north is also seen. A region of shallow events is also spotted in the northern part of the station network (Figure 2). In total, we used 27 stations distributed from east to the west of regions, mainly around Bali to Nusa Tenggara, and a few in the East Java.

3. Method

We performed our processing data in this study by using simulps12 software [8]. This software uses a simultaneous inversion algorithm to relocate earthquake and determine the P-wave velocity structure each layer. For parameterization, we set the center of grid at 9°S and 118.5°E where the x-axis has a 60 km interval in the inner grid and a 110 km interval for the for outer grids. Meanwhile, the y-axis has a 30 km interval in the inner grid and 70 km interval for the outer grids, yet the z-axis has 10 km, 30 km, 60 km, and 100 km intervals. The velocity model used is 1-D global velocity model AK-135 by [9] with depth over 720 km divided into 4 layers.

The simultaneous tomography inversion uses a pseudo-bending method for ray tracing. The purpose of ray tracing is to estimate the ray path and calculate travel times between hypocenter and station based

IOP Conf. Series: Earth and Environmental Science **873** (2021) 012065 doi:10.1088/1755-1315/873/1/012065

on the velocity model. The software will also invert the optimum velocity and sources location based on the kernel matrix and different between travel time and observation time.

The tomography inversion is a non-linear problem that is linearized where the amount of data is not enough to cover all unknown variable parameters so that the matrix becomes under-determined. To avoid the inversion problem, regularization with damping is generally used. The value of damping is acquired from a trade-off curve between data variance and model variance by empirical iteration. Optimum damping value is expected to increase model variance and decreasing data variance. In this study, the optimum damping value is 70 based on the trade-off curve from various damping around 10 until 80. The optimum damping value pointed out variance model value is 0.03739173 and the variance data value is 0.28614.

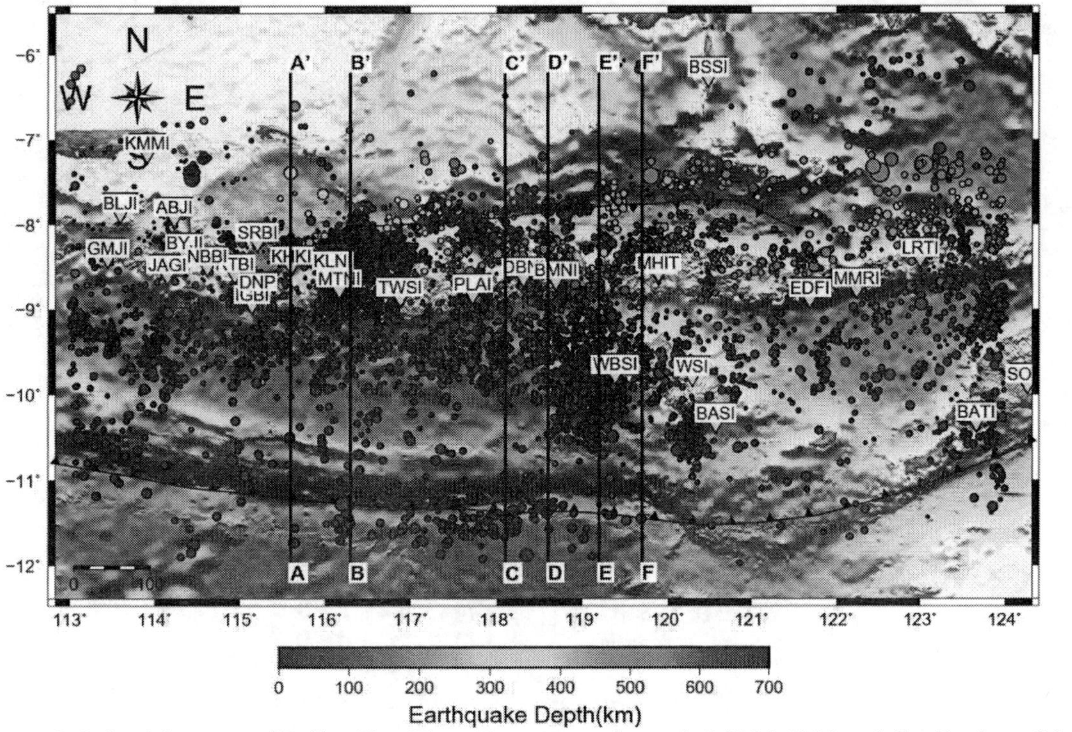

Figure 2. Seismicity map of Bali to East Nusa Tenggara at the period 2015-2018 and distribution of the recording stations used. Black lines indicate the position of the vertical cross sections displayed in Figures 3.

4. Result and Discussion

To validate the results of the tomographic inversion that has been done, a resolution test is required. The resolution test conducted is the checkerboard resolution test, where the resolution test is carried out by alternating positive and negative anomalies such as a chessboard both horizontally and vertically, where each positive and negative anomaly is ±6% of the initial velocity model. On the results of the checkerboard resolution test, the depth limit of the resolution is 240km, the area that has a good resolution to this depth is the northern area of Sumbawa. East Nusa Tenggara, Lombok and the western areas of the study area were only well resolved to a depth of 150km (Figure 3).

Figure 3 also show the results from P wave tomography in horizontal dan vertical slices. The perturbation value is expressed in colour, blue and red. Positive perturbation with a wave velocity value that is higher than the initial velocity model is expressed in blue. Faster velocity can be associate with higher density and colder temperature. Meanwhile, negative perturbation with a lower wave velocity value than the initial velocity model is expressed in red that be seen an area with lower density, weaker structure or higher temperature region. A band of lower velocity is seen a depth of 30 km not too far from the volcanoes line. We also observed a pattern of fast velocities that dip down to the north although this is not as continuous as we expected.

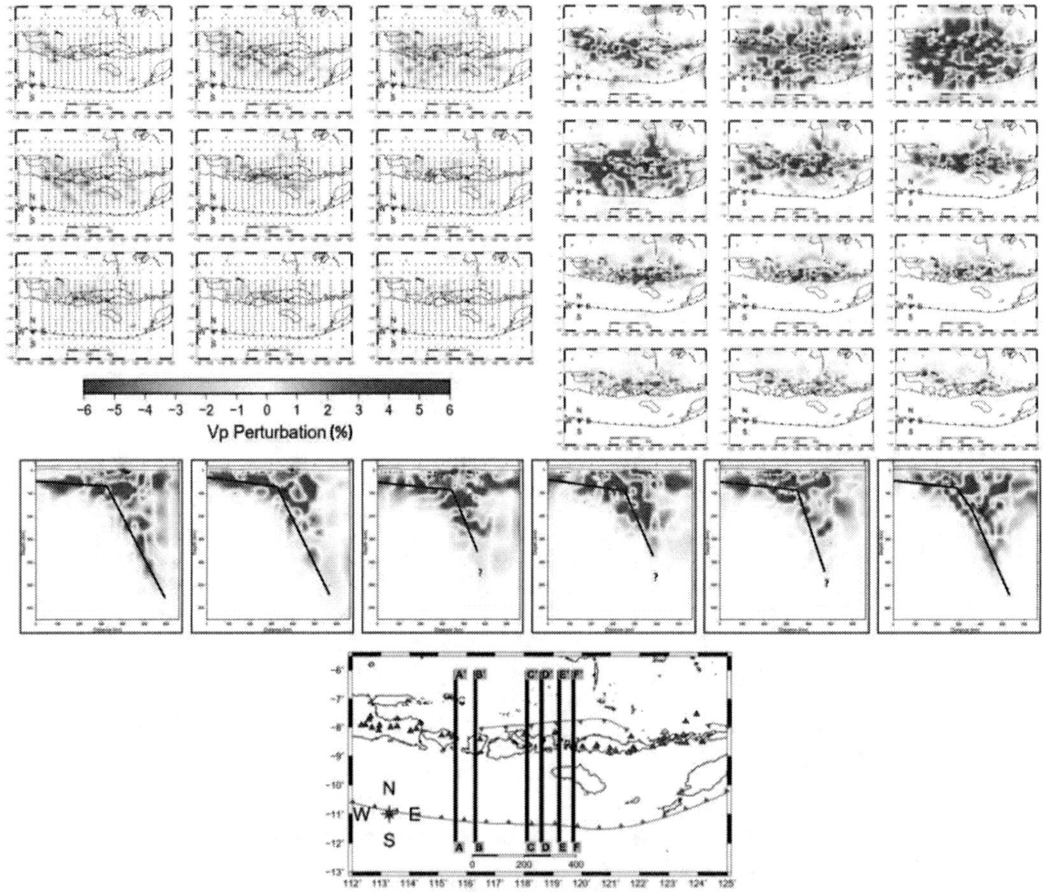

Figure 3. Plot a checkerboard resolution test based on depth (left), Horizontal P wave tomogram plot based on depth (right) and vertically in the study area (bottom). The black line represents the subduction zone modelling in the study area. Bottom figure is the location of cross sections

The fast velocities band may represent the oceanic plate that subducting the continental plate. We see a pattern of low angle dip at shallow depth that turn into steep angle at a depth of 60 km. We have confidence of our resolution down to a depth of 300 km. Below this depth, we have not seen good resolution on our checkerboard model so that it would be difficult to interpret where there is a break in the subducting slab as previous claim by [10]. One interesting feature that is previously targeted in this study is the features of back arc thrust previous imaged as a dipping positive velocity model to the south at shallow depth. We might have seen this on cross section E in our model, however, we see the tomogram model is not convincing.

5. Conclusions

We have conducted tomographic imaging of subduction zones in the area of Bali to East Nusa Tenggara using regional data. A dipping fast velocity to the North is seen to the north that is interpreted as the oceanic plate subducting the Eurasian plate. This model is supported by previous geophysical studies. We see a pattern of lower angle of subduction at shallow depth that increase its angle into around 55o

IOP Conf. Series: Earth and Environmental Science **873** (2021) 012065 doi:10.1088/1755-1315/873/1/012065

at lower depth. A pattern of lower velocity is also seen beneath volcano that might explain the origin of magmatism in the region. Further analysis from the checker-board model and seismicity suggest several improvements to improve the image. We suggest lowering down the standard during event relocation to capture more events since we see too many events removal. Next, we also think of adding more data set and damping tests to improve the resolution at deeper depth

References

[1] DeMets C, Gordon R G, & Argus D F 2010 *Geologically current plate motions*. Geophysical Journal International **181**(1) 1–80.

[2] Hall R 2002 *Cenozoic geological and plate tectonic evolution of SE Asia and the SW Pacific: computer-based reconstructions, model and animations* Journal of Asian Earth Sciences **20**(4) 353–431

[3] Müller R D, Roest W R, Royer J-Y, Gahagan L M, & Sclater J G 1997 *Digital isochrons of the world's ocean floor* Journal of Geophysical Research (JGR), 102,3211

[4] Cloetingh S, & Wortel R 1986 *Stress in the Indo-Australian plate* Tectonophysics **132**(1–3) 49–67

[5] Hamilton W 1979 *Tectonics of the Indonesian Region* Geological Society of Malaysia Bulletin **6**(July) 3–10

[6] Widiyantoro S, & Fauzi 2005 *Note on seismicity of the Bali convergent region in the eastern Sunda Arc, Indonesia* Australian Journal of Earth Sciences **52**(3) 379–383

[7] Supendi P, Nugraha A D, Widiyantoro S, Abdullah C I, Rawlinson N, Cummins P R, Harris C W, Roosmawati N, & Miller M S 2020 *Fate of Forearc Lithosphere at Arc-Continent Collision Zones: Evidence From Local Earthquake Tomography of the Sunda-Banda Arc Transition, Indonesia* Geophysical Research Letters **47**(6).

[8] Evans J R, Eberhart-Phillips D, & Thurber C H 1994 *User's manual for SIMULPS 12 for imaging Vp and Vp/Vs: A derivative of the Thurber tomographic in version SIMUL3 for local earthquakes and explosion* Open-File Report 94-431 101

[9] Kennet B, Engdahl E R 1995 *Constraints on seismic velocities in the Earth from traveltimes* Geophysical Journal International 122 108-124

[10] Widiyantoro S, Pesicek J D, & Thurber C H 2011 *Subducting slab structure below the eastern Sunda arc inferred from non-linear seismic tomographic imaging* Geological Society Special Publication **355**(August 2015) 139–155

Preliminary Results of Regional P and S Wave Tomography at Seram, Indonesia

P A Subakti[1], M I Sulaiman[1], D Y Faimah[1], I Madrinovella[1], I Herawati[1], H M Lubis[1], S K Suhardja[1], A Priyono[1], S Widiyantoro[2,3], M Ramdhan[4], A S Sembiring[4], A M S Nugraha[5]

[1] Faculty of Exploration and Production Technology, University of Pertamina, Jl. Teuku Nyak Arief, Simprug, Kebayoran Lama, Jakarta 12220, Indonesia
[2] Global Geophysics Research Group, Faculty of Mining and Petroleum Engineering Institute of Technology Bandung, Jalan Ganesa No. 10, Bandung 40132, Indonesia
[3] Faculty of Engineering, Maranatha Christian University Bandung, Jalan Surya Sumantri No 65, Bandung 40164, Indonesia
[4] Agency for Meteorology, Climatology and Geophysics, Jalan Angkasa I, No. 2, Kemayoran, Jakarta, Indonesia
[5] Earth Observatory of Singapore, 50 Nanyang Ave., Nanyang Technological University, 639798 Singapore

email : s4ndy104@gmail.com

Abstract. The Seram Trough is located in the northern part of Indonesia and has a complex tectonic setting. The uniqueness of these regions lies in the U-shape subduction system. Several models have been proposed in this region, such as one subduction system that has been rotated 90° or 180^0, two subduction systems, and one subduction that having a slab roll-back that causes extension systems. In this study, we try to invert velocity and seismicity using double-difference tomography with the target of better imaging the sub-surface structure in the region. We use data catalogue collection from the Indonesian Agency of Meteorology, Climatology, and Geophysics. The length of data is 4 years from January 2015 to December 2019 from 16 permanent stations. Earthquake relocations show a focused hypocenter distribution at shallow depth, and we interpreted some of these shallow depth events are related to the magmatic activity. Event distribution also displays a steep angle of seismicity pattern that represents the dipping subduction slab. Inverted Tomography models show a band of faster velocity models that dip from North to South, suggesting a subductions slab. We also observe a possibility of a tear in the slab from the seismicity pattern and tomogram model. The slower velocity perturbation is seen at shallow depth that may associate with magmatic and frequent shallow seismicity. A possibility of partial melting is also seen with low-velocity zone at a depth of 70 km next to the fast dipping velocity.

Keywords: P-wave tomography, S-wave tomography, Double Difference Tomography, Seram

1. Introduction
Seram is one of the areas in the Maluku archipelago. The tectonic setting of this area is affected by 3 main plates, namely the Australian Plate, the Pacific plate and the Eurasian plate. This area is known for having a pattern of curved subduction (Figure1).

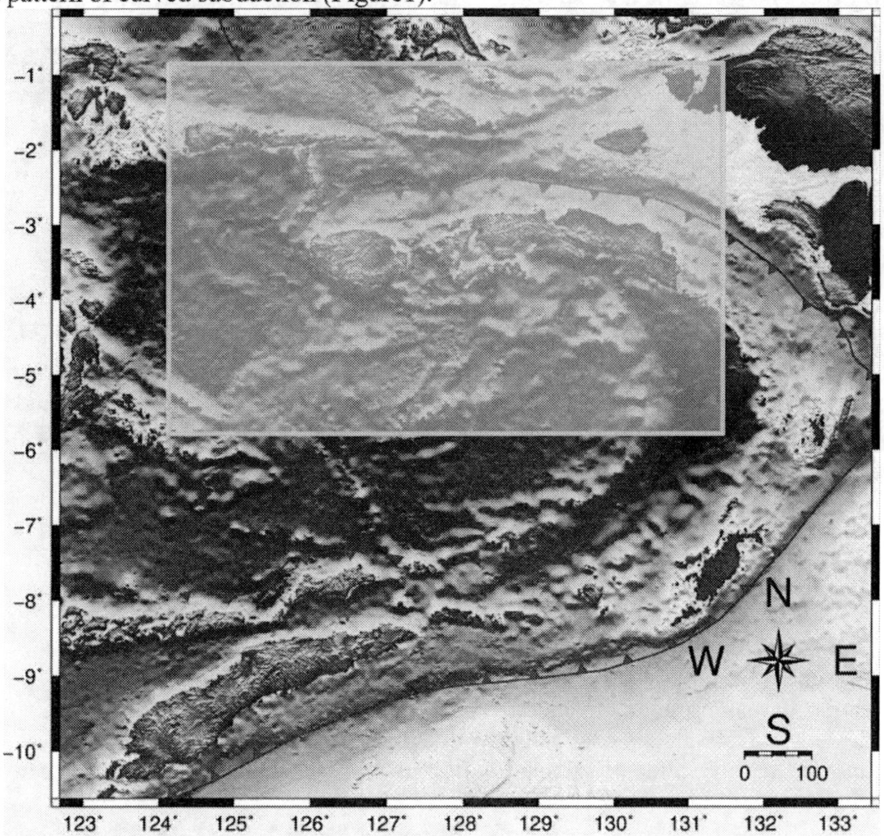

Figure 1. Study area (grey colour) and a pattern of curved subduction (red line)

To the present, there are several models to explain the interesting pattern of curved subduction Spakman & Hall [1] proposed a one slab system that rolled back. Sri Widiyantoro et al. [2] also found that seismic tomography successfully depicts Banda Slab as a single system that twisted and forming spoon-shaped structure towards Banda Sea as well as conforming the slab rollback scenario. In the other hand, Cardwell, R. K. et al [3] with seismicity and fault plane study proposing Banda Slab constructed by two different system. The first one is the Timor – Aru Trough system from south and the second one is the Seram Trough system in the north that joined the Banda Slab because New Guinea Tarera – Aiduna Fault Zone extension. The objective of this study is to image subsurface feature in this region especially at Seram Trough using regional data set and double difference tomography method. We are interested in knowing the seismicity pattern and understand the geodynamic of Seram Trough.

2. Data and Method
This research used data on the arrival times of P and S waves from the Agency of Meteorology, Climatology and Geophysics Agency (BMKG), Indonesia. The data consists of 1238 earthquake events and 16 observation stations from 2015 to 2019. Note that we used velocity model data from previous research of Madlazim [4] as the starting model. The number data before being relocated was

IOP Conf. Series: Earth and Environmental Science **873** (2021) 012066 doi:10.1088/1755-1315/873/1/012066

1238 earthquake events and after being relocated the number of data was reduced to 1225 earthquake events. We do not see a significant difference in hypocenter location between before and after inversion although some improvements in fixed depth events (Figure 2). The data is distributed evenly at shallow depth and some events dipping to south-west direction that can reach a depth of 500 km.

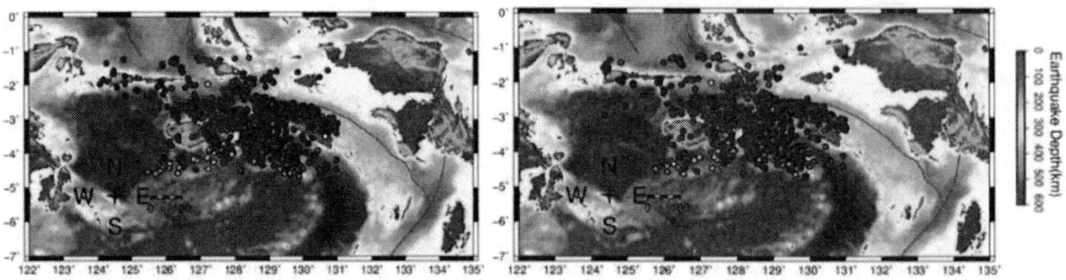

Figure 2. Seismisity map before relocation (left) and after relocation (right)

We used double difference tomography method known widely as TomoDD [5]. The double difference method minimizes the remaining time from the calculated travel time and the observed travel time on two earthquake events that are interconnected with the same number of earthquakes recording stations [5,6,7]. Note the travel time and ray path is estimated using the pseudo bending method. For model parameterization, we set the grid into 40 km cube in all directions. To validate the results of the inversion, a resolution test was performed using the Derivative Weight Sum (DWS) method. DWS is carried out to test the results of the tomogram and the results of the tomographic inversion by weighting the total length of the rays through a node in the inversion grid.

3. Results and Discussion

One indication of having good inversion model is calculating the residual value before and after relocation. The residual value is obtained from the difference in travel time between the travel time of observation and the travel time of calculation. Figure 3 shows that majority of time residual has been removed closer to zeros.

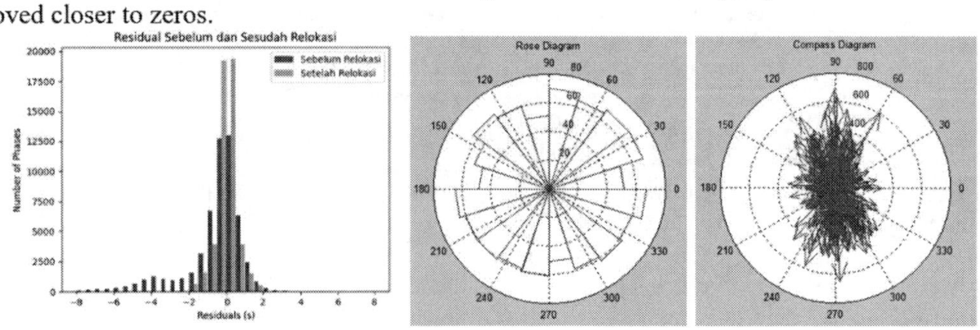

Figure 3. Histogram residual (left), rose diagram (centre), and compass diagram (right)

In Figure 3 shows a diagram of a rose and a compass resulting from a hypocenter relocation. The direction of the hypocenter spacing is indicated by an arrow on the compass diagram and rose diagram to the hypocenter shift in km. Overall, we do not see any relocation weigh in one direction. The wave velocity structure model is obtained by the tomographic inversion process using tomoDD software with 20 iterations. Damping parameter is added in regularization process to solve underdetermined problem with value of 150 estimated from trade off curve between model variance and data variance.

IOP Conf. Series: Earth and Environmental Science **873** (2021) 012066 doi:10.1088/1755-1315/873/1/012066

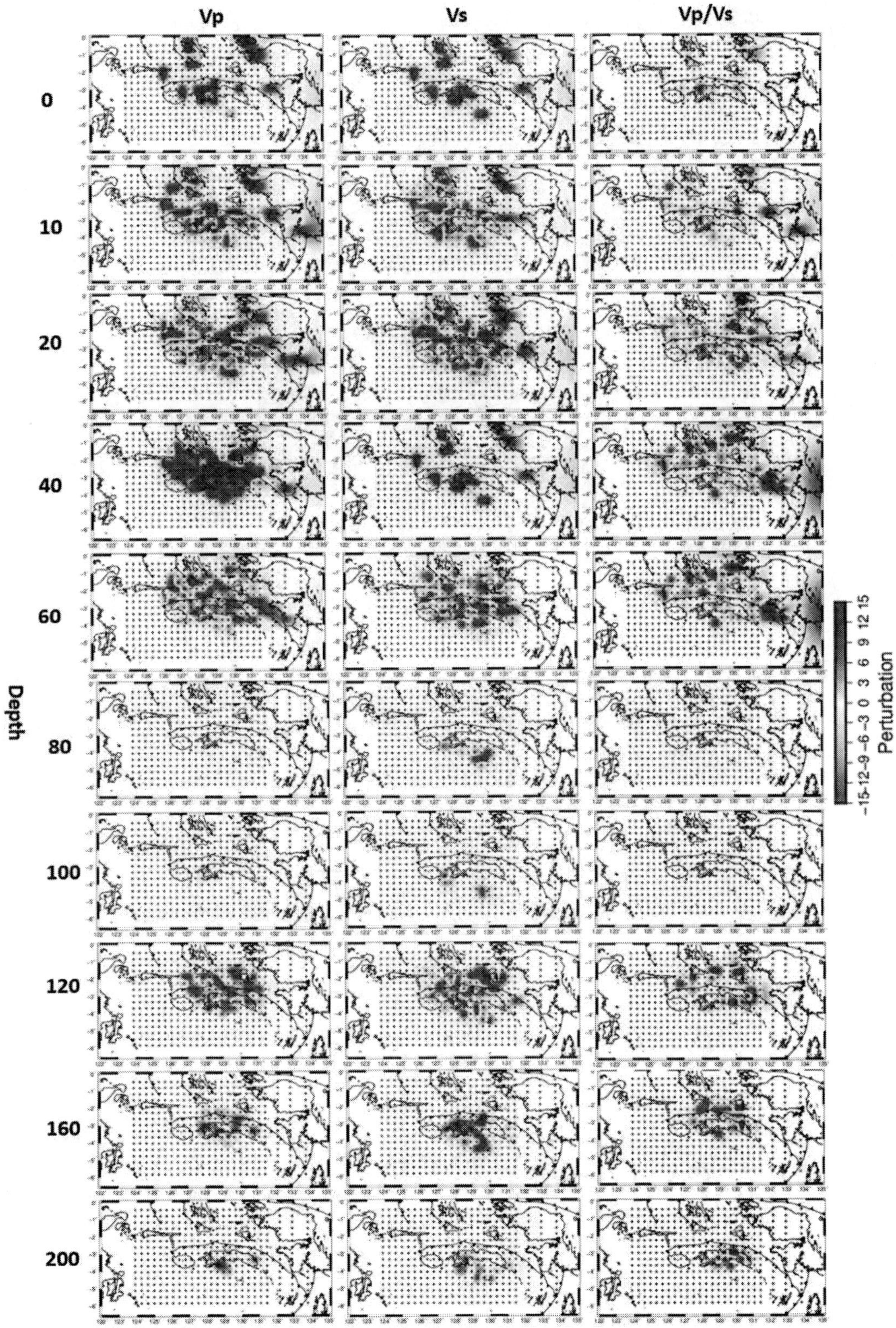

Figure 4. Horizontal tomogram Vp (left), Vs (centre), and Vp/Vs (right)

Figure 4 shows the results of the tomogram model for horizontal P and S wave velocity at each depth. The results of the tomogram show the difference in red and blue, where the blue colour is a positive perturbation with a higher wave velocity value than the initial velocity model and the red colour is a negative perturbation with a lower wave velocity value than the initial velocity model. Positive anomaly shows the density of the structure in the area is denser or colder which is associated with subduction slabs. Negative anomaly shows the density of the structure in the area is more fragile or hotter which could be associated with the local fault zone or partial melt zone The horizontal plotting of perturbation values obtained an anomaly pattern at shallow depths <20 km dominated low velocity, then a high velocity is seen that is stifling from north to south from a depth of 40 km and follows a slab pattern. Some area at the point of volcanic area show velocity model of Vp with low velocity, Vs with high velocity, and Vp/Vs with high velocity indicated there are partial melting in this area.

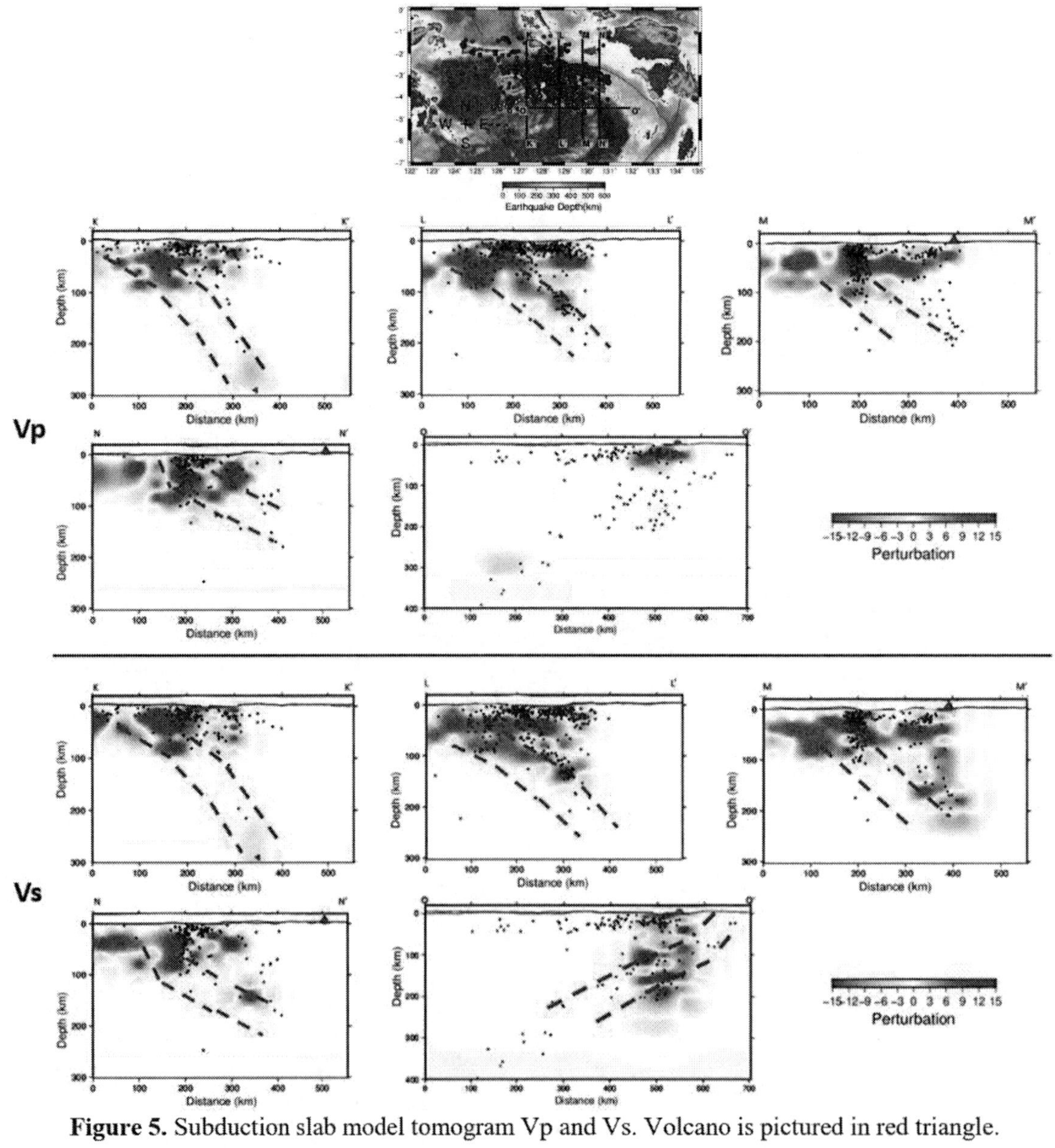

Figure 5. Subduction slab model tomogram Vp and Vs. Volcano is pictured in red triangle.

Figure 5 show different cross section profiles at different longitude from north to south direction. Note that our red dash line interpretation as subduction slab is mainly based on seismicity depth. Cross section L is a good representation of subduction slab that coincides with seismicity. Some sections also display a relationship between low velocity and beneath volcano.

4. Conclusions

Based on the results of relocation earthquake data recorded from the January 2015 - December 2019 period in the study area, the earthquake distribution pattern shows a slab pattern from north to south and east to west. A deep earthquake can be caused by several sources, including the subduction slab plunging to a depth of more than >450 km. Then shallow earthquakes can be caused by local faults close to the surface. The relocation results show good results where the residual travel time histogram data, the residual value of the earthquake after relocation, has increased close to the value of 0 and the value has decreased away from 0.

The tomographic inversion results have succeeded in providing a new 3D velocity model from the initial 1D velocity model. Some sections show convincing down-going fast velocity coincides with seismicity. A low velocity is also found on beneath volcano. Next works would like to include more data set for better interpretation in this area.

Acknowledgments

The data used in this study is from Indonesian Agency of Meteorology, Climatology and Geophysics Agency (BMKG). We would like to thank Indonesian Agency of Meteorology, Climatology and Geophysics Agency (BMKG) which has permited us to collect the data BMKG Jakarta. We also would like to thank University of Pertamina which has support this study.

References

[1] Spakman W and Hall R 2010 Surface deformation and slab-mantle interaction during Banda arc subduction rollback *Nat. Geosci.* **3** 562–6

[2] Widiyantoro S, Pesicek J D and Thurber C H 2011 Complex structure of the lithospheric slab beneath the Banda arc, eastern Indonesia depicted by a seismic tomographic model *Res. Geophys.* **1** 1

[3] Cardwell R K and Isacks B L 1978 Geometry of the Subducted Lithosphere Beneath the Banda Sea in Eastern Indonesia From Seismicity and Fault Plane Solutions melange sediments **83**

[4] Jannah, Supardiyono, Madlazim 2013 Analisa model kecepatan lokal gelombang primer 1-D dan koreksi stasiun di Kepulauan Maluku *Inovasi Fisika Indonesia*. **2** 2

[5] Zhang H and Thurber C H 2003 Double-difference tomography: The method and its application to the Hayward Fault, California *Bulletin of the Seismological Society of America* **93** 1875–89

[6] Waldhauser F and Ellsworth W L 2000 A Double-difference Earthquake location algorithm: Method and application to the Northern Hayward Fault, California *Bull. Seismol. Soc. Am.* **90** 1353–68

[7] Naamin K N, Sahara D P, Nugraha A D and Ramadhan I 2018 Application of Double Difference Tomography Method to Determine The 3D Seismic Wave Velocity Structure of GoLF Geothermal Field *J. Geofis.* **16** 27

IOP Conf. Series: Earth and Environmental Science **873** (2021) 012067 doi:10.1088/1755-1315/873/1/012067

Preliminary Results of Double Difference Tomography at Sunda-Banda Arc

Haolia[1], M. I. Sulaiman[1], P. T. Brilianti[1], R. P. Nugroho[1], I. Madrinovella[1], A. Abdullah[1], D. A. Zaky[1], S. K. Suhardja[1], A. Priyono[1], S. Widiyantoro[2,3], M. Ramdhan[4], A. M. S. Nugraha[5]

[1] Faculty of Exploration and Production Technology, University of Pertamina, Jl. Teuku Nyak Arief, Simprug, Kebayoran Lama, Jakarta 12220, Indonesia
[2] Global Geophysics Research Group, Faculty of Mining and Petroleum Engineering Institute of Technology Bandung, Jalan Ganesa No. 10, Bandung 40132, Indonesia
[3] Faculty of Engineering, Maranatha Christian University Bandung, Jalan Surya Sumantri No 65, Bandung 40164, Indonesia
[4] Agency for Meteorology, Climatology and Geophysics, Jalan Angkasa I, No. 2, Kemayoran, Jakarta, Indonesia
[5] Earth Observatory of Singapore, 50 Nanyang Ave., Nanyang Technological University, 639798 Singapore

*Corresponding author: s4ndy104@gmail.com

Abstract. The Sunda-Arc transition to the Banda Arc is located on the south of the Flores Island, Indonesia, where the Australian lithosphere is moving to the north direction. On-going subduction process dictates the tectonic setting though some studies also suggest a collision and obduction may occur in the past due to of plate buoyancy. This area has active seismicity with frequent large magnitude events. To better understand the tectonic system in this region, we performed double-difference tomography inversion using regional events. We obtained the data catalog from the Indonesian Agency of Meteorology, Climatology, and Geophysics ranging from 116° to 125° east longitude and -6.5° to 12.5° latitude. We collected 4312 events data, detected from 15 stations from January 2015 to December 2019. Final relocated hypocenters showed a reduced fixed-depth problem and a more clustered event, although some deep events disappear. Most events are related to the subducting Benioff zone with some clustered events in the northern area may be related to back-arc thrust. We also observed clustered events near active volcano region and reduced shallow seismicity region to the west of the Timor Island. Resolution test using the checkerboard and Derivative weigh Sum (DWS) shows that fair P wave resolution can be achieved until 300 km, although a smearing start to show at a deeper depth. However, due to lack of arrival S wave data, the resolution test suggest good resolution can only be seen until a depth of 100 km. Tomogram P and S wave models show a clear dipping subducting slab from south to North down to a 250 km. We also spot a fast velocity band near the Timor Island area that similar to the previous tomography study, interpreted as sliver forearm. We spotted a band of lower Vp, lower Vs and higher Vp/Vs at shallow depth close to the volcanic line and we interpreted this as a zone of higher temperature, that may relate to magmatic activity in this region. We also noticed a zone of low velocity and higher Vp/Vs that may relate with dehydration and partial melting. However, we feel this still uncertain due to low Vs resolution.

1. Introduction

Our study area is located at the Eastern part of Indonesia where the Indo-Australian plate is subducting underneath the Eurasian plate. We focused our study at south of the Flores Island where the Australian continental crust moves near the trough and a transition zone from the Sunda Arc to the Banda Arc. The transition area of the Sunda-Banda arc is characterized by changes in tectonic order from the subduction of the Indo-Australian oceanic plate along the Sunda Arc to the collision of the continental arc along the Banda Arc (Shulgin et al., 2009). The change has brought implications for the complexity of the geological and geodynamic arrangements in this transition area. This area shows a high magnitude events such as one that occurred in the south of Sumba Island known as the 1977 Sumba earthquake zone. Earthquakes with a magnitude above 6 with a thrust mechanism also occurred in the north of this region. These events were related to Flores and Wetar back-arc thrust. Both thrusts occurred due to the accommodation of forces from the subduction in the south of the Nusa Tenggara archipelago (Supendi et al, 2020). Supendi et al (2020) used local tomography method and found a positive anomaly model between slow velocity at a depth on 70 km that has 10-20 km thickness and interpreted as forearm sliver. This study tried to use regional events recorded on a permanent station near the area and simultaneous inversion to invert velocity perturbation. We are targeting a better imaging of subsurface structure in the region and seismicity pattern.

2. Data and Method

This study used the BMKG earthquake catalog limited from 116° to 125° east longitude and -6.5° to 12.5° latitude. We used the arrival time phase of P and S waves from 15 stations, with a data period of January 2015 to December 2019. In total, we obtained 3140 event data with most events occur in the western part of Sumba Island, south of Sumba Island, the eastern part of Alor, and the southern part of the eastern island (Figure 1). We observed that most events are shallower than 60 km and presumed these events are related with subduction process in this region. We also noticed a lower scattered number of events in the Savu Sea. We employed the double difference tomography method (Zhang et al, 2003) to determine the velocity structure and to relocate the hypocenter to the study area.

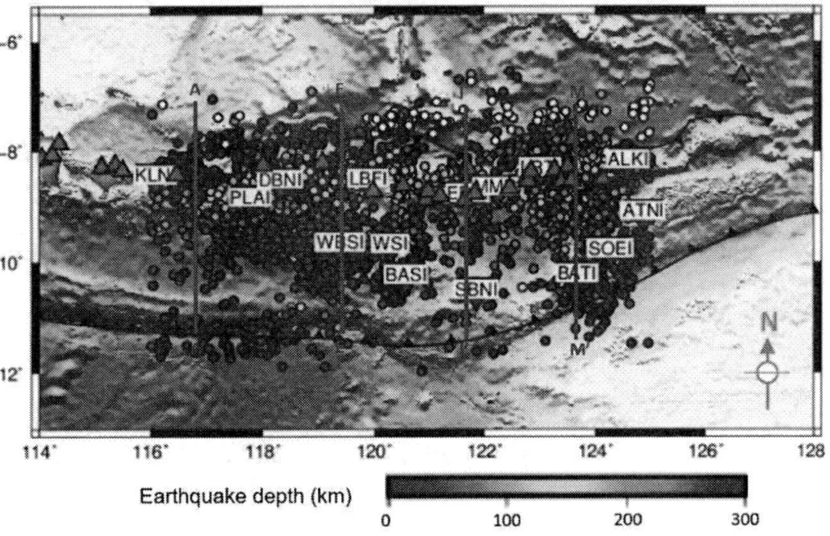

Figure 1. Map of the study area. Red to blue circles represent the initial hypocenter between January 2015 and December 2019. Magenta triangle are volcanoes and blue inverted triangles show the stations. Vertical lines are cross section line used in figure 3 and 4.

IOP Conf. Series: Earth and Environmental Science **873** (2021) 012067 doi:10.1088/1755-1315/873/1/012067

The initial and reference velocity model in this study is the 1-D velocity model AK135 (Kennett et al., 1995). For the parameterization, we considered several factors such as station location and event distribution. One challenging problem in our study is low numbers of receiver stations compared to the numbers of source events. Figure 2 shows the grid parameterization in this study. We set the inner grid to have 40 km space in all direction and larger distance (200 km) on the outer grid to include all data in our study.

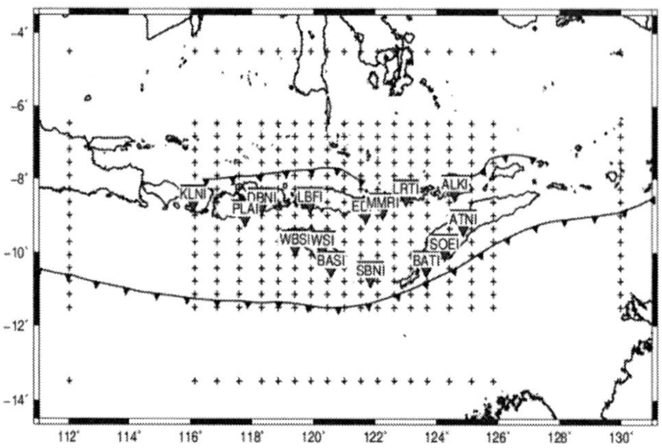

Figure 2. Model Parameterization. the plus symbol indicates the grid points and the blue inverted triangle indicates the recording station.

The optimum damping value can be determined empirically by carrying out a series of inversion process that will generate different variance model value and variance data. The optimum damping value is associated with a reduction in data variance and at the same time does not significantly increase the variance of the model (Eberhart P, 1986). The determination of this attenuation value is obtained by plotting the trade off curve between the data variant and the model variant. We set the appropriate damping factor to be 230 for the Vp dan Vs inversion.

3. Results, Discussions and Conclusions

To investigate the recovered anomalies, we carried out the checkerboard recovery tests. We added 10% velocity anomaly relative to the 1-D reference velocity. Next, we used the same configuration of sources and receiver as our observation data and invert to see how well the checker-board model recovered. The results of checkerboard tests for horizontal slices and cross sections are presented in Figures 3. In general, all Vp anomalies are reproduced well both in horizontal and vertical directions. However, resolution in Vs anomalies is questionable due to less S wave arrivals. Depth resolution for Vp is quite reliable, down to a depth of 150 km.

Figures 4 shows inverted model both for Vp, Vs and Vp/Vs ratio in horizontal slices. Note that the Vp/Vs model are not joint inverted in this study, instead we calculated by dividing the number of Vp and Vs on the same grid. At a depth of 30 km, we noticed a low velocity area for both Vp and Vs at West Nusa Tenggara, Eastern part of East Nusa Tenggara, and the Timor Island. Interestingly, the location of low velocity is close to the Volcanic Line that extend horizontally from Bali between 8° to 9° south latitude, though not any volcano seen on the Timor Island. The low velocity at West Nusa Tenggara and Eeast Nusa Tenggara seems to extend to a depth of 50 km and not any low velocity on the Timor Island at greater depth. We interpreted the low velocity below West and East Nusa Tenggara may be related with partial melting zone that feed current active volcano. We further observed the horizontal slices and found that below a depth of 70 km, most inverted velocities are dominated by higher velocity.

Figure 5 shows inverted velocity model cross section from West to East. We observed a band of dipping down high velocity (blue color) from south to north direction. The location of dipping high velocity is coincidence with the dipping down hypocenter trend from south to north. We interpreted this band of high velocity as the Indo-Australian plate that is subduction Eurasian plate, though the Eastern part of this study at Timor Island is known as the Sunda-Banda Arc transition, represents the subduction from Australian lithosphere that move to the North. The cross section in the East (cross section M in Figure 5) show two interesting features. The first is the band of horizontal low velocity that extends horizontally at a depth of 30-40 km. We are still not sure about this and will investigate further. The second is a zone of low velocity between fast velocity. This feature, although not so identical, resemble structures known as fore arc sliver as previously suggest by Supendi et al, (2020).

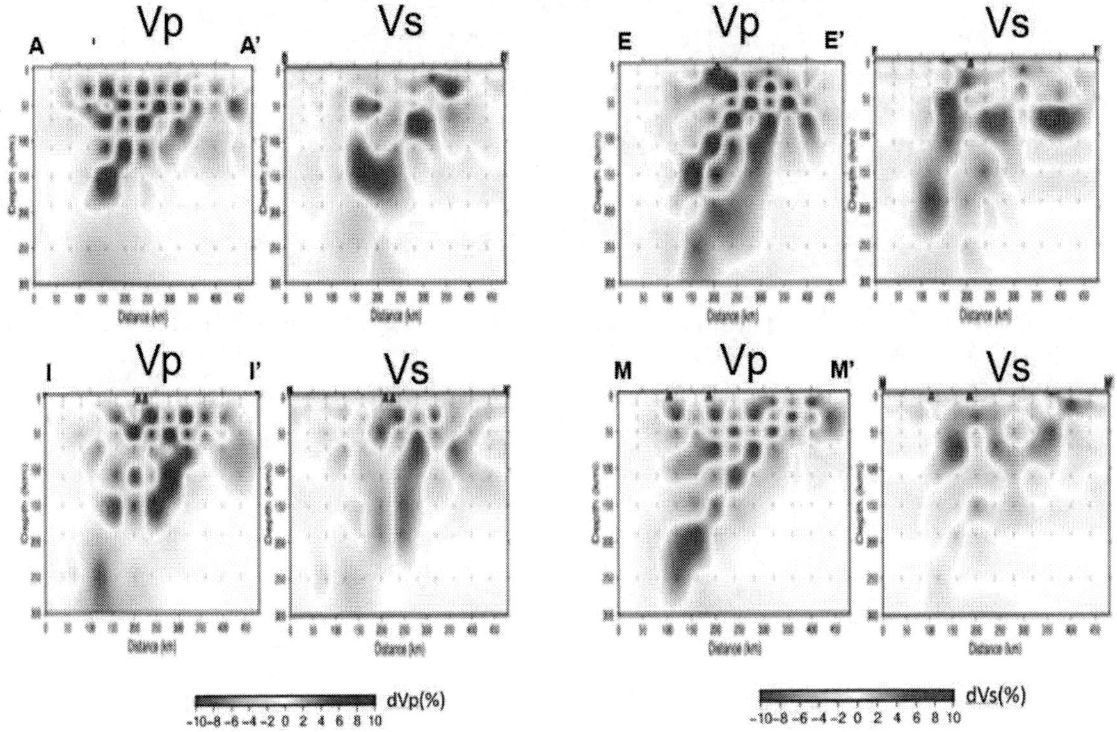

Figure 3. The checkerboard inversion model. Blue color is positive anomaly and red color is negative velocity anomaly. Line for cross section is shown on Figure 1.

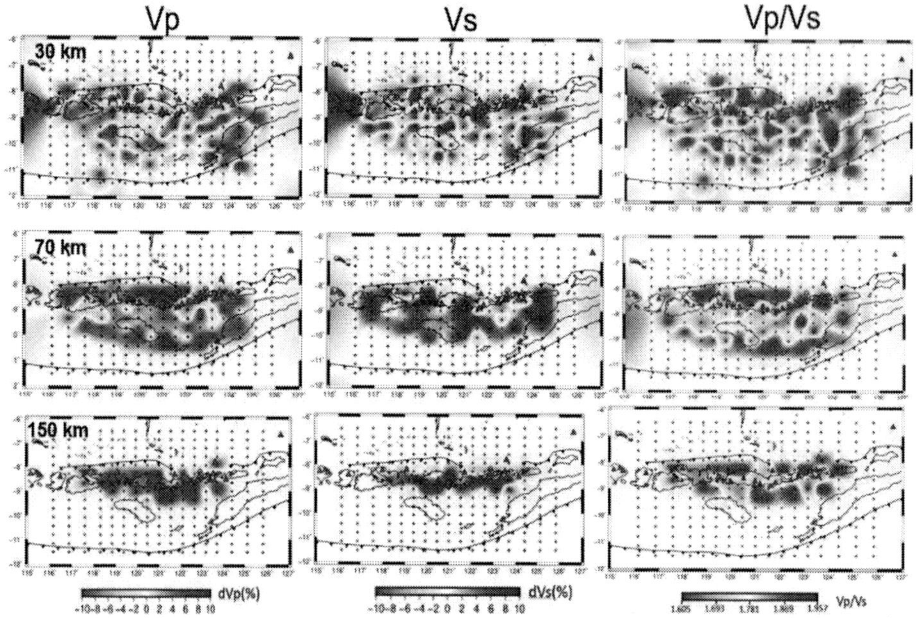

Figure 4. Inverted Vp, Vs and Vp/Vs ratio horizontal slices. Note the color blue is positive anomaly in Vp, Vs and high Vp/Vs Red Color is the opposite.

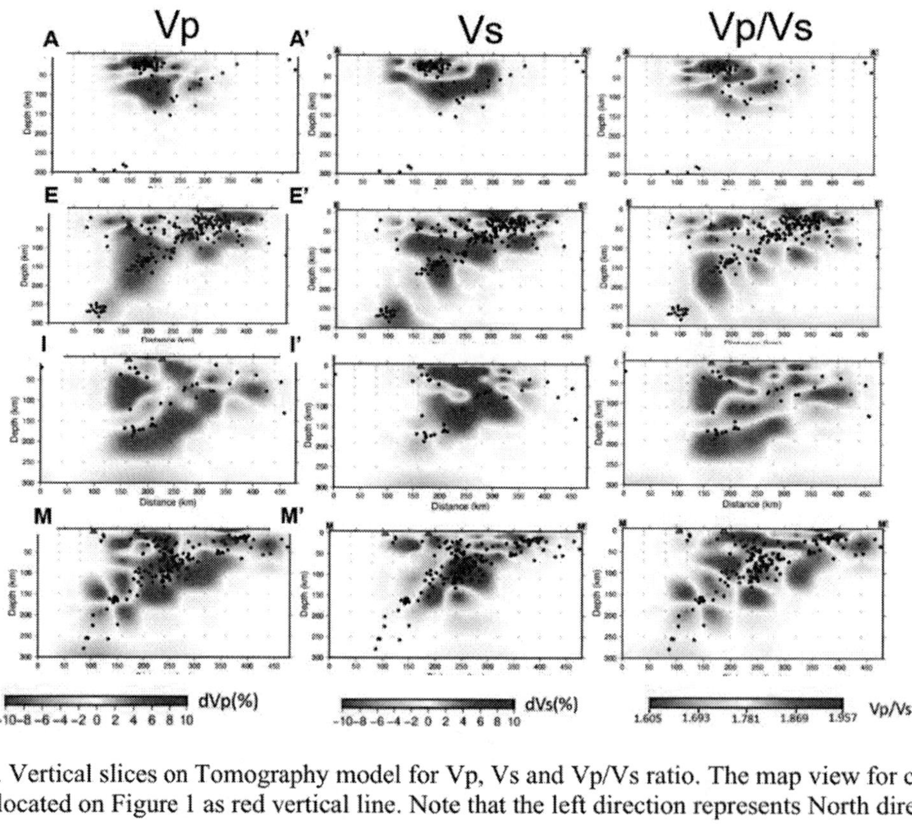

Figure 5. Vertical slices on Tomography model for Vp, Vs and Vp/Vs ratio. The map view for cross-section is located on Figure 1 as red vertical line. Note that the left direction represents North direction.

As for the conclusion, Our preliminary results showed promising subsurface structures beneath Sunda-Banda arc through delay time tomography model, especially in Vp model. However, due to lack of S wave arrival time, the quality of Vs and Vp/Vs ratio need to be improved by adding more stations and possibly longer duration data set. A combination of data that has more stations such as the one from Supendi et al, (2020) would give better image in the eastern part of the model. We would like to thank the the Indonesian Agency of Meteorology, Climatology, and Geophysics for providing data and discussion.

References

[1] Harris, R. (2011). The Nature of the Banda Arc–Continent Collision in the Timor Region. In D. Brown & P. D. Ryan, Arc-Continent Collision (pp. 163–211). Springer Berlin Heidelberg. https://doi.org/10.1007/978-3-540-88558-0_7

[2] Harris, Ron, Vorkink, M. W., Prasetyadi, C., Zobell, E., Roosmawati, N., & Apthorpe, M. (2009). Transition from subduction to arc-continent collision: Geologic and neotectonic evolution of Savu Island, Indonesia. 21.

[3] Kennett, B. L. N., Engdahl, E. R., & Buland, R. (1995). Constraints on seismic velocities in the Earth from traveltimes. Geophysical Journal International, 122(1), 108–124. https://doi.org/10.1111/j.1365-246X.1995.tb03540.x

[4] Kisslinger, C., & Engdahl, E. R. (n.d.). THE INTERPRETATION OF THE WADATI DIAGRAM WITH RELAXED ASSUMPTIONS. 14.

[5] Koulali, A., Susilo, S., McClusky, S., Meilano, I., Cummins, P., Tregoning, P., Lister, G., Efendi, J., & Syafi'i, M. A. (2016). Crustal strain partitioning and the associated earthquake hazard in the eastern Sunda-Banda Arc: Crustal strain in the Sunda-Banda Arc. Geophysical Research Letters, 43(5), 1943–1949. https://doi.org/10.1002/2016GL067941

[6] Liao, P.-C., Lii, C.-C., Lai, Y.-C., Chang, P.-Y., Zhang, H., & Thurber, C. (2014). A Graphics Processing Unit Implementation and Optimization for Parallel Double-Difference Seismic Tomography. Bulletin of the Seismological Society of America, 104(2), 953–961. https://doi.org/10.1785/0120130236

[7] Pusat Studi Gempa Nasional (Indonesia) (Ed.). (2017). Peta sumber dan bahaya gempa Indonesia tahun 2017 (Cetakan pertama). Pusat Penelitian dan Pengembangan Perumahan dan Permukiman, Badan Penelitian dan Pengembangan, Kementerian Pekerjaan Umum

[8] Rutherford, E., Burke, K., & Lytwyn, J. (2001). Tectonic history of Sumba Island, Indonesia, since the Late Cretaceous and its rapid escape into the forearc in the Miocene. Journal of Asian Earth Sciences, 19(4), 453–479. https://doi.org/10.1016/S1367-9120(00)00032-8.

[9] SeismologicalSocietyofAmerica, B. (n.d.). Three-Dimensional Velocity Structure In Northern California Coast Ranges From Inversion Of Local Earthquake Arrival Times By Donna Eberhart-Phillips. 28.

[10] Shulgin, A., Kopp, H., Mueller, C., Lueschen, E., Planert, L., Engels, M., Flueh, E. R., Krabbenhoeft, A., & Djajadihardja, Y. (2009). Sunda-Banda arc transition: Incipient continent-island arc collision (northwest Australia). Geophysical Research Letters, 36(10), L10304. https://doi.org/10.1029/2009GL037533

[11] Spiegelman, M., & McKenzie, D. (1987). Simple 2-D models for melt extraction at mid-ocean ridges and island arcs. Earth and Planetary Science Letters, 83(1–4), 137–152. https://doi.org/10.1016/0012-821X(87)90057-4

[12] Supendi, P., Nugraha, A. D., Widiyantoro, S., Abdullah, C. I., Rawlinson, N., Cummins, P. R., Harris, C. W., Roosmawati, N., & Miller, M. S. (2020). Fate of Forearc Lithosphere at Arc-Continent Collision Zones: Evidence From Local Earthquake Tomography of the Sunda-Banda Arc Transition, Indonesia. Geophysical Research Letters, 47(6). https://doi.org/10.1029/2019GL086472

[13] Toomey, D. R., & Foulger, G. R. (1989). Tomographic inversion of local earthquake data from the Hengill-Grensdalur Central Volcano Complex, Iceland. Journal of Geophysical Research, 94(B12), 17497. https://doi.org/10.1029/JB094iB12p17497.

[14] Von Blanckenburg, F., & Davies, J. H. (1995). Slab breakoff: A model for syncollisional magmatism and tectonics in the Alps. Tectonics, 14(1), 120–131. https://doi.org/10.1029/94TC02051.

[15] Waldhauser, F. and Ellsworth, W.L. 2000. A double-difference Earthquake Location Algorithm: Method and Application to the Northern Hayward Fault, California, Bull. Seismol. Soc. Am. 90, 1353–1368.

[16] Zhang, H. (2003). Double-Difference Tomography: The Method and Its Application to the Hayward Fault, California. Bulletin of the Seismological Society of America, 93(5), 1875–1889. https://doi.org/10.1785/0120020190

[17] Zhang, Haijiang, & Thurber, C. (2006). Development and Applications of Double-difference Seismic Tomography. Pure and Applied Geophysics, 163(2–3), 373–403. https://doi.org/10.1007/s00024-005-0021-y

[18] Zhang, Haijiang, Thurber, C., & Bedrosian, P. (2009). Joint inversion for Vp, Vs, and Vp/Vs at SAFOD, Parkfield, California: JOINT SEISMIC VELOCITY INVERSION AROUND SAFOD. Geochemistry, Geophysics, Geosystems, 10(11), n/a-n/a. https://doi.org/10.1029/2009GC002709

Early Results of Eastern Indonesia P-wave Tomography Study Using Regional Events

M I Sulaiman[1], P A Subakti[1], Haolia[1], D Y Fatimah[1], I Madrinovella[1], A Abdullah[1], D A Zaky[1], A S B Oka[1], M H M Lubis[1], S K Suhardja[1*], A Priyono[1], S Widiyantoro[2,3], M Ramdhan[4], A M S Nugraha[5]

[1] Faculty of Exploration and Production Technology, University of Pertamina, Jl. Teuku Nyak Arief, Simprug, Kebayoran Lama, Jakarta 12220, Indonesia
[2] Global Geophysics Research Group, Faculty of Mining and Petroleum Engineering Institute of Technology Bandung, Jalan Ganesa No. 10, Bandung 40132, Indonesia
[3] Faculty of Engineering, Maranatha Christian University Bandung, Jalan Surya Sumantri No 65, Bandung 40164, Indonesia
[4] Agency for Meteorology, Climatology and Geophysics, Jalan Angkasa I, No. 2, Kemayoran, Jakarta, Indonesia
[5] Earth Observatory of Singapore, 50 Nanyang Ave., Nanyang Technological University, 639798 Singapore

*Corresponding author: s4ndy104@gmail.com

Abstract. The tectonic system of Eastern Indonesia is controlled by several major and minor plates, such as Indo-Australian, Australian plate, and Pacific plates. This area is known for its complexity, and high seismic activity. This study tries to image the complex structures beneath this region by employing regional events data and seismic tomography methods. We used five years of regional events catalog provided by the Indonesian Agency of Meteorology, Climatology, and Geophysics. We have sorted 7336 events recorded between 120° – 136° longitude and 0° – 13°(-) latitude consisting of 46446 P and 15467 S wave arrival data. Relocated hypocenter map shows a better constrain location on seismicity along outer Bandar Arc. A dipping pattern of seismicity is seen that is going deeper to the Banda Sea. The seismicity map also images a steep angle pattern of seismicity that could be related to the subduction slab roll-back model at North of Wetar island. Interestingly, we spotted a seismicity gap in West Seram that could be linked with slab tear zone. The checker-board test suggests a proper resolution is still reliable to a depth of 200 km with a less interpretable model at a depth of 300 km. P-wave tomographic models image the high velocity dipping down going slab. The Banda slab is seen to subduct from south Timor Island to the north, from east Tanimbar and Aru Island to west part, and from north Seram Island to south. We observed the down-going slab meet from all directions at about 300 km beneath the Banda sea. P wave tomogram also shows the Timor Island slab has a steeper dip that agrees with the seismicity pattern. Near the Seram island, we identify a low-velocity anomaly zone infiltrate the Banda slab beneath the shallow part of West Seram, which was previously interpreted as slab tear zone. This study also noticed a higher velocity tomogram model at North of Wetar island that might indicate a

IOP Conf. Series: Earth and Environmental Science **873** (2021) 012068 doi:10.1088/1755-1315/873/1/012068

back-arc thrust. Lastly, a low-velocity band is also exposed at a shallow depth close to the volcano chain along that Banda volcanic arc.

1. Introduction

Indonesia has frequent seismic activity from the West part in Sumatra to East part in Papua. The Indonesian Meteorology, Climatology, and Geophysical Agency reported 44% of earthquakes in Indonesia occur in eastern part of Indonesia and is the subject of this study. Some recent events are: the 26th September 2019 Ambon's earthquake with 6.8 magnitude and 10 km depth causing dozens of people died, thousands of people injured, and damaging public facilities. The 30th September 1899 Seram's earthquake with 7.8 magnitude and 12 km depth took thousands of lives and wreck a lot of buildings. Based on plate reconstruction [1], Indonesia is dominated by three major plate movement which are Indo – Australia, Eurasia, and Pacific Plates. Eastern Indonesia itself mostly be affected by Indo – Australia plate movement toward north and collide with Eurasia plate with convergence rate around 7.7 cm/ year [2] and resulting subduction path along southern part of Indonesia include Banda Arc. The subduction in Banda Arc is unique because it has D – shape geometry or some research said concave spoon/ half-bathtub shape geometry [3]. The curved Banda Arc subduction most likely caused by deformed Australian plate in Banda embayment which give away the unique D – shape and stirred by slab rollback that make the subduction path fitted with Banda embayment. Some study also mentions about slab tear in West Seram Island, to be exact under Buru Island, which indicate because of mantle resistance towards the subduction movement and resulting slab rotation that lead the tearing beneath Buru Island.

As the additional, local structure like back-arc thrust formed in North Wetar Island which happen because the energy accumulation accommodated from subducting slab in Timor Island. Besides that, as the manifestation of subduction, volcanoes are formed along Banda volcanic arc such as Banda Api, Manuk, Serua, Nila, Teon and Damar Volcanoes.

Previous study suggest that Banda Slab is a single system that curved into 180° containing young oceanic crust and mostly formed following the Banda embayment shape and rollback process which lead to crust deformation [4]. [5] also found that seismic tomography successfully depicts Banda Slab as a single system that twisted and forming spoon-shaped structure towards Banda Sea as well as conforming the slab rollback scenario. In the other hand with seismicity and fault plane study [6] proposing Banda Slab constructed by two different system. The first one is the Timor – Aru Trough system from south and the second one is the Seram Trough system in the north that joined the Banda Slab because New Guinea Tarera – Aiduna Fault Zone extension.

This study expected to map the slab geometry along Banda Arc include the tearing zone under Buru Island in West Seram and back-arc thrust structure in North Wetar to validate which structure dominantly generate earthquake with high potential risk of destruction and also the volcanoes activity along Banda Volcanic Arc.

2. Data and Method

The research took place in Eastern Indonesia, to be specific in Banda Arc. This research is a collaboration between Universitas Pertamina with Indonesian Meteorology, Climatology, and Geophysical Agency (BMKG). Earthquake data provided by BMKG from 2015 – 2019 constraints latitude 10.617°S until 0.220°S and longitude 122.695°E until 133.440°E. The data consists of travel time from each event in each seismic station along with its hypocenter parameter such as longitude, latitude, origin time, depth, and magnitude. There are 22 total seismic stations and 7336 total events with 46446 P-wave observation and 15467 S-wave observation.

Based on seismicity map, the possibility of slab subduction shown by the earthquake event pattern from shallow in outer Banda Arc and getting deeper beneath Banda Sea, The east Timor Trough signify the deeper slab area for its earthquake events appear until 600 – 700 km, this area also have probable a steeper dipping slab shown by abrupt change in earthquake depth change from south to north. Besides, seismicity gap likely appears in west of Timor Island, this also confirmed by International Seismological Centre (ISC) and United States Geological Survey (USGS) catalog data event. The Weber Deep area likewise shown rare seismic activity, [7] explain the lack of seismic event due to low angle normal fault, further explanation also mention that the low frequency could indicate energy accumulation which lead to larger event. Moreover, there is a seismic jump pattern found in Seram specifically beneath Buru Island, the seismic event occurs in the shallow area until depth 200 km, then it's unlikely happen between 200 km – 400 km, but then continuously shows seismic event again after depth 400 km. As additional, in the north Wetar Island, revealed a thrusting pattern from north to south in a shallow feature around 100 km.

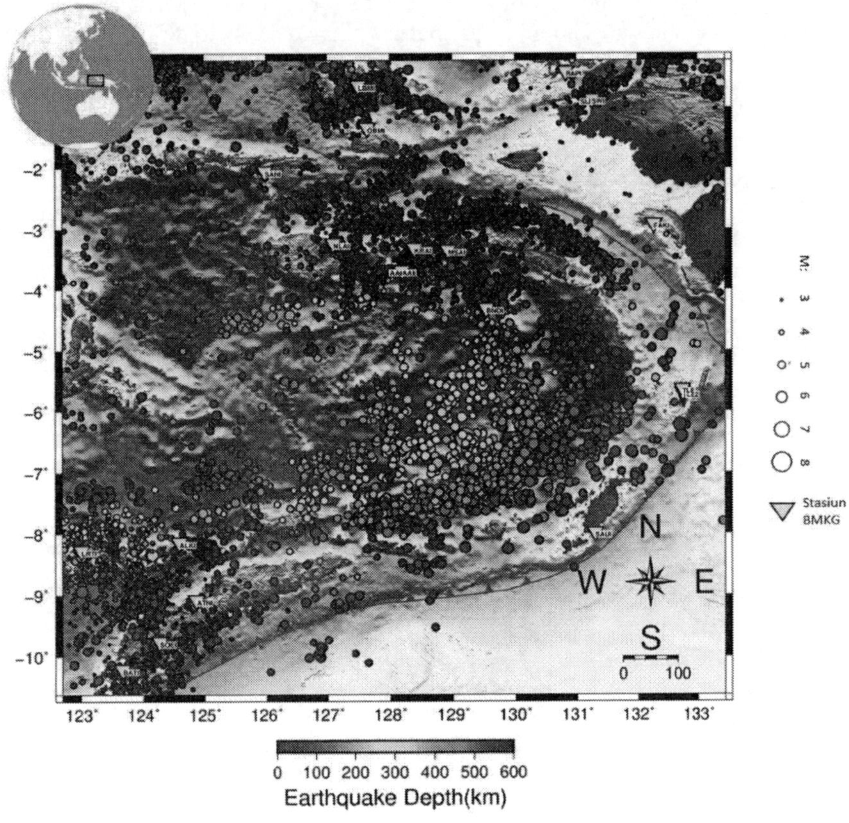

Figure 1. Earthquake and station data distribution

22 total BMKG seismic station utilized in this study distributed from Flores, Timor, Tanimbar, Aru, Papua New Guinea, Seram, Buru and many more surrounding Island. The station distribution is likely uneven due to the installment difficulty in some area, Timor and Seram island is quite dense distributed meanwhile Banda Sea, Aru, Tanimbar still rarely has station around it.

IOP Conf. Series: Earth and Environmental Science **873** (2021) 012068 doi:10.1088/1755-1315/873/1/012068

Travel time seismic tomography applied in this study to image subsurface based on seismic velocity model. This method utilizes earthquake propagation travel time from source to receiver, AK135 velocity [8] as initial model and pseudo-bending ray tracing [9]. Model parameterization be solved by dividing the study area into several grid horizontally (x, y) for longitude and latitude and vertically (z) for depth. Grid interval made in 60 km for x and y axis but 100 – 200 km for outer focus area and varies from 30 – 60 – 120 km for z axis depends on how much earthquake event there to cover our inversion. In the shallower area with denser events applied smaller grid, while deeper area with rare seismic activity applied sparser grid. The determination of grid will affect the ray coverage in inversion process as well as the final resolution.

The process done by SIMULPS12 [10] which simultaneously inverse seismic velocity model and hypocenter location. As a result of using slowness, the inversion process using non-linear inversion with linear approaching to solve the inversion problem. Sometimes, in the inversion process we also face matrix singularity problem due to the lack of ray tracing data in several grids. As the solution for this, we use damping parameter. Optimum damping parameter obtained from trade-off curve between data and model variance [11]. It is associate with significant reduction of data variance and at the same time did not increase the model variance significantly. In this study, we use 70 as optimum damping parameter with 0.34155 s^2 data variance and 0.01118665 km^2/s^2 model variance.

3. Results & Discussions

There is not much difference between hypocenter distribution before and after relocation, the main change is the reduction of earthquake fix depth in shallow area as well as deep event seemingly diminished due to lack of seismic station that record the data. While other part did not change that much and still shows similar patterns. This happen because BMKG data even before relocation process already has an acceptable residual time value for each earthquake in every station, hence we only need to improve the residual error. After relocation, we can enhance the residual time value in 0 s which make the residual error smaller and improved the data quality significantly.

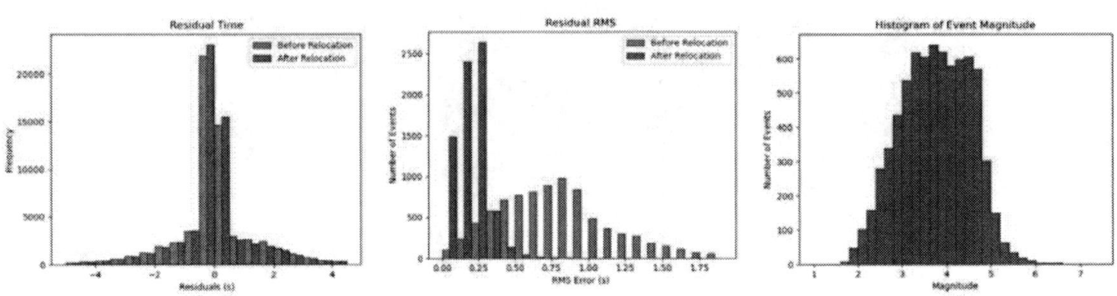

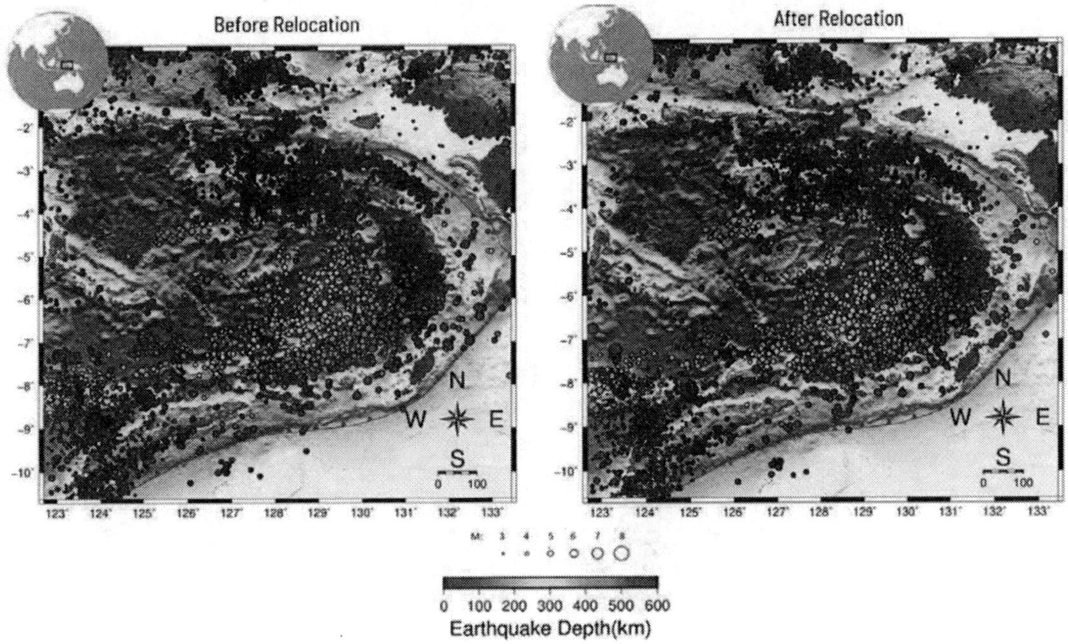

Figure 2. Residual time histogram; Residual RMS error, Earthquake magnitude histogram, and Hypocenter distribution before and after relocation

To determine which tomogram area that has acceptable resolution, therefore we can interpret confidently, we performed checkerboard resolution test (CRT) and Derivative Weight Sum (DWS). We use all the same parameterization but using synthetic velocity model with additional 5% noise. Based on the CRT, for shallower area, Seram and Timor Island shows a high resolution while the deeper area shows a better resolution around Banda Arc to Banda Sea. Besides, the resolution is well-solved and reliable to interpret until depth 200 km. This proved by the DWS which show the density of ray length through the grid, Seram and Timor Island both shows a denser ray coverage in shallower area, while in the depth between 100-300 km shows quite dense ray in northwest of Banda Sea.

IOP Conf. Series: Earth and Environmental Science **873** (2021) 012068 doi:10.1088/1755-1315/873/1/012068

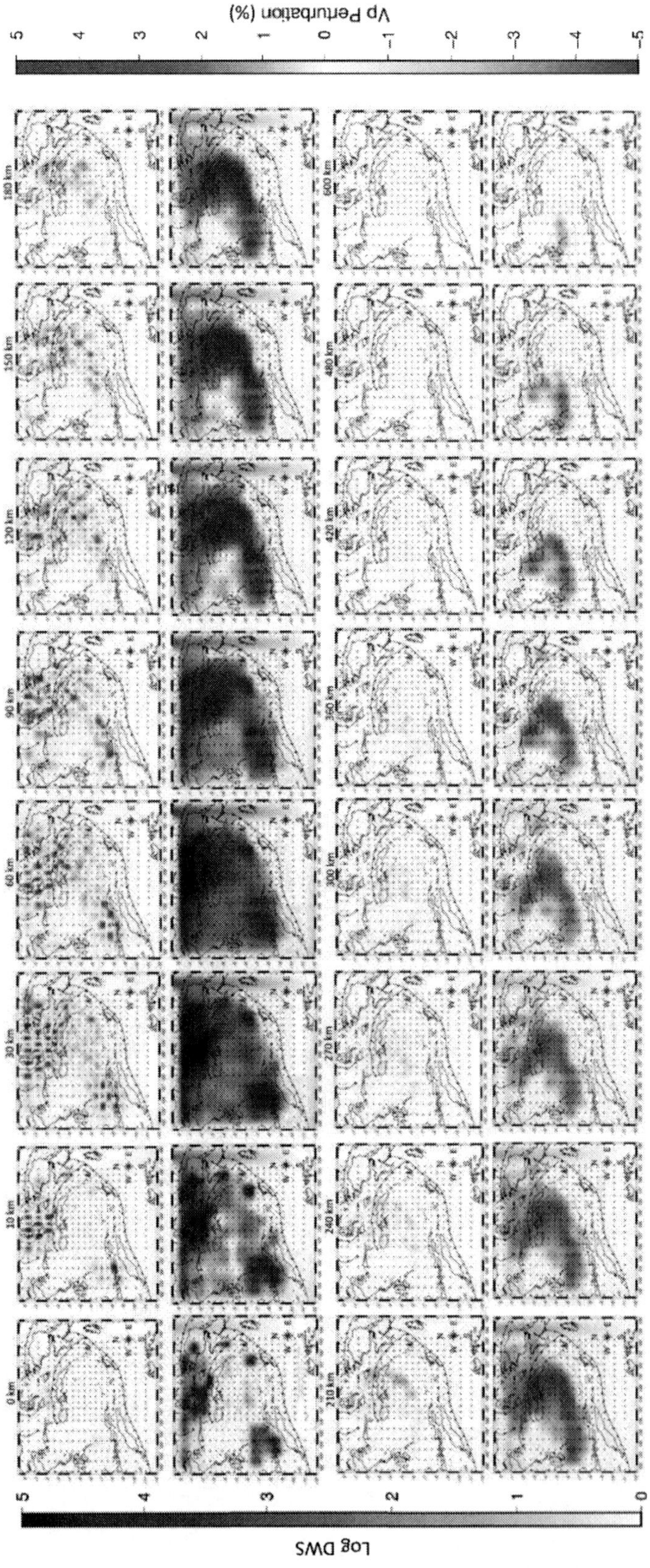

Figure 3. Checkerboard Resolution Test and Derivative Weight Sum result for each depth

IOP Conf. Series: Earth and Environmental Science **873** (2021) 012068 doi:10.1088/1755-1315/873/1/012068

Whereas in depth beneath 300 km is likely less interpretable due to lack of ray coverage. DWS shows the ray only cover some little area in the northwest of Banda Sea. However, it also confirmed that we still have seismic ray down to depth 600 km resulting the possibility of structure reach depth 600 km or more. This resolution is most depending on the seismic station location and earthquake distribution, in this study seismic stations are a lot tighter in the Timor, Seram and West Papua which make this area receive more earthquake ray which make the resolution well resolved and more interpretable than the rest of the area

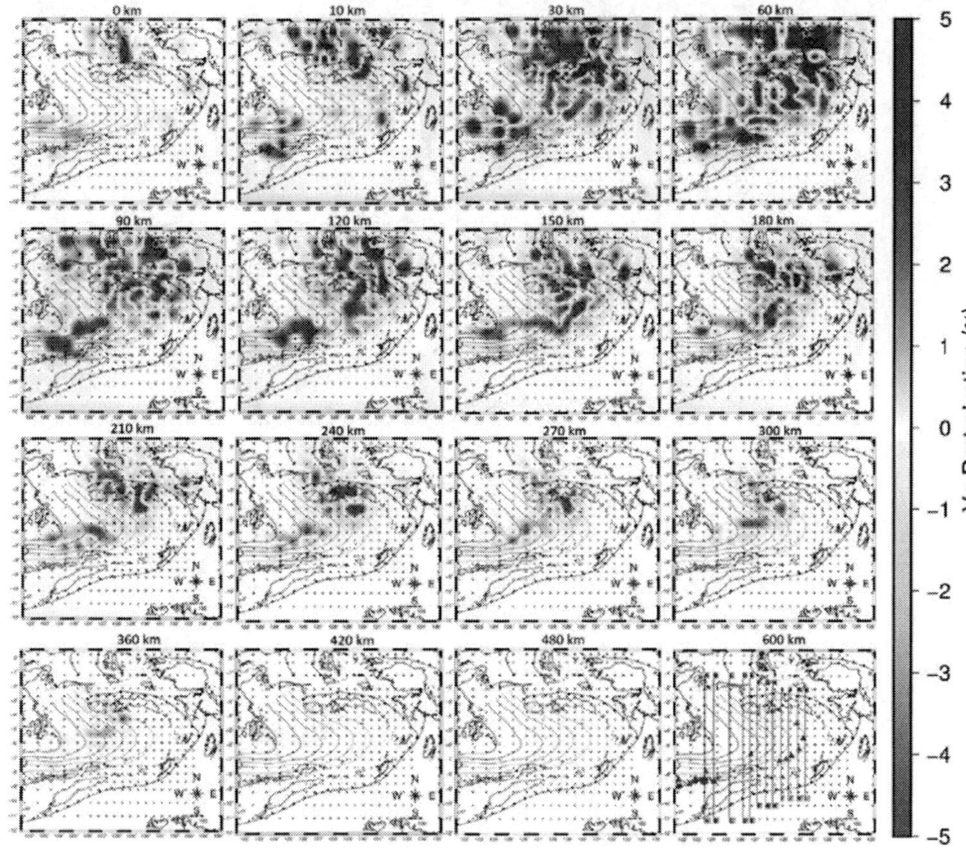

Figure 4. Result of P-wave tomography from depth 0 – 600 km. Blue and red portray positive and negative anomaly respectively

Looking at the horizontal slice tomogram, in Timor, Tanimbar and Aru (Banda Arc Segments) shown positive anomaly depict by blue color move from south to northwest around Banda Sea continuously as depth increases. Moreover, positive anomaly also visible in Seram Segment moving from northeast to southwest as depth increases. Thus, those positive anomalies interpreted as Indo-Australia's subducted slab plate from Banda Arc to Banda Sea forming the spectacular Banda curve slab until depth approximately 600 km. While negative anomaly shown clearly in depth 30 km around Alor (north of Timor Island), and small island along Banda Arc known as "Banda Ring of Fire", this negative anomaly depicts by red color in shallow area till depth approximately 80 – 100 km. Beside that negative anomaly also found in west of West Papua in depth 10 – 30 km.

IOP Conf. Series: Earth and Environmental Science **873** (2021) 012068 doi:10.1088/1755-1315/873/1/012068

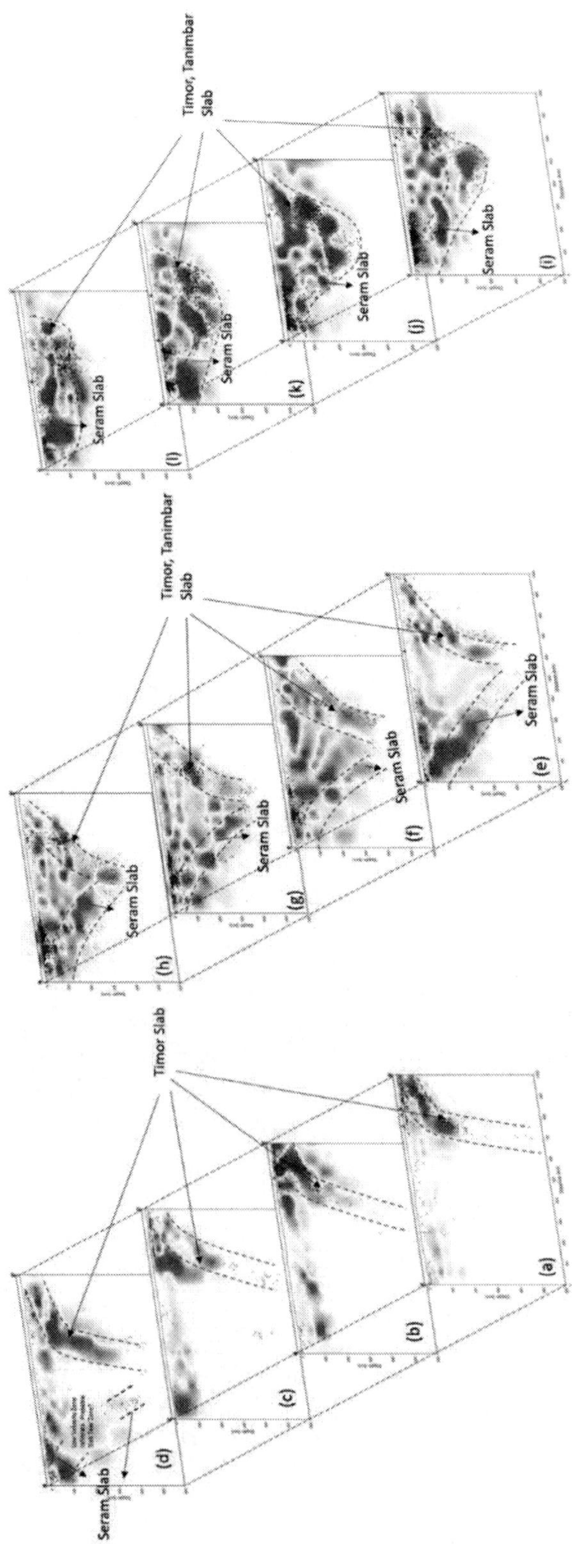

Figure 5. Vertical slice of tomogram from A – L (the cross-section line shown in fig.4 depth 600 km)

In vertical slice tomogram, the positive anomaly distributed fit with the possibility of slab from south of Timor Island and north of Seram Island. In cross-section A, B, C, and D, the positve anomaly shown a specific steep dipping until 600 km in west of Timor Island due to the possibility of rollback extension from Java – Sumatera Slab. Furthermore, in the cross-section D, in the north area near Seram Island, there is a postive anomaly in shallow depth influenced by Seram Slab, but then discontinue until depth 300 – 350 km, the positve anomaly is reappear back in the depth more than 350 km, some of previous research [12],[3] said there is a possibility of break-off or tearing slab beneath west of Seram thus its showing negative anomaly infiltrate the postive anomaly along Seram Slab. In cross-section E, F, G and H, the Slab from both direction Seram in north and Timor in south showing a clear positive anomaly alongside with the hypocenter location subducted and meet in depth approximately 400 – 500 km. These also shows how the slab subducting from east to west proven by the east part is more shallower or sloping than the west part. Morever, the cross-section I, J, K, L shows the same pattern where slab depict in positive anomaly from north and south and increasingly sloping to east where its start subducting.

Local structure like back-arc thrust feature and volcanic activity could be identified by the tomography as well. The back-arc thrust depicted in the north of Wetar Island shown by positive anomaly shown in cross-section D and F. This also supported by hypocenter distribution in that area dominated with shallow earthqauke below 100 km. This back-arc thrust formed as the result of subducting plate in Timor Island which make a pressure accumulation in north of Wetar. Besides, negative anomaly found along Banda Volcanic arc in depth 30 – 80 km known as volcanic front-line zone. Famous volcanoes along that zone such as Banda Api, Manuk, Serua, Nila, Teon and Damar shown negative anomaly beneath them which interpreted as partial melting and magma chamber of those volcanoes.

4. Conclusion

As for the conclusion, Banda geological features are depicted successfully using earthquake travel time tomography. Banda Slab subducted around Banda Arc, from Timor Island in the south and Seram island in the north shown by positive anomaly dipping down until depth 600 km. Wetar back-arc thrust also identified with positive anomaly in depth less than 100 km, and the volcanic activity around the area portrayed as negative anomaly beneath the Banda Volcanic Arc.

References

[1] Hall R 2002 *Cenozoic geological and plate tectonic evolution of SE Asia and the SW Pacific: computer-based reconstructions and animations* Journal of Asian Earth Sciences **20**(4) 353-434

[2] Pairault A A, Hall R, Elders C F 2003 *Structural styles and tectonic evolution of the Seram Trough, Indonesia* Elsevier Marine and Petroleum Geology **20** p1141–1160

[3] Spakman W, Hall R 2010 *Surface deformation and slab–mantle interaction during Banda arc subduction rollback* Nature Geoscience NGEO917

[4] Hall R, Spakman W 2015 *Mantle structure and tectonic history of SE Asia* J.Tectonophysics 658 p14-45

[5] Widiyantoro S, Pesicek J D, Thurber C H 2011 *Complex structure of the lithospheric slab beneath the Banda arc, eastern Indonesia depicted by a seismic tomographic model* Research in Geophysics Vol.1 E.1 PAGEpress Italy

[6] Cardwell R K et al 1978 *Geometry of the subducted lithosphere beneath the Banda Sea in eastern Indonesia from seismicity and fault plane solutions* J. Geophysics **83** 2825–2838

[7] Axen G J 2007 *Research Focus: Significance of large-displacement, low-angle normal faults* J. Geology **35** P 287–288

[8] Kennett B L N, Engdahl E R, Buland R 1995 *Constraints on seismic velocities in the Earth from traveltimes* Geophysical Journal International Vol. 122 p 108-124

[9] Um J, Thurber C H 1987 *A fast algorithm for two-point seismic ray tracing* Bulletin of the Seismological Society of America Vol. 77 No.3 pp. 972-986

[10] Evans J R, Eberhart-Phillips D, Thurber C H 1994 *User's Manual for SIMULPS12 for imaging Vp and Vp/Vs: A Derivative of the Thurber Tomographic Inversion SIMUL3 for Local Earthquakes and Explosions* US Geological Survey USGS OFR 94-431

[11] Eberhart-Phillips D 1986 *Three-dimensional velocity structure in northern California Coast Ranges from inversion of local earthquake arrival times* Bulletin of the Seismological Society of America **76** (4) 1025–1052

[12] Fernández-García C, Guillaume B, Burn J 2019 *3D slab breakoff in laboratory experiments* J. Tectonophyscis Vol. 773

Investigation of Rawa Dano Volcanic Deposits and Its Paleotopography Using Ground-Penetrating Radar

Maryadi Maryadi,[1,*] Fira Mariah Sausan Champai,[2] I Nyoman Triananda,[2] Andi Darmawan,[2] Gamma Abdul Jabbar,[1]

[1]Geoscience Department, Faculty of Mathematics and Natural Science, Universitas Indonesia, Depok, Indonesia

[2]Undergraduate Program of Geophysics, Faculty of Mathematics and Natural Science, Universitas Indonesia, Depok, Indonesia

*Corresponding author : maryadi@sci.ui.ac.id

Abstract. The detailed mechanisms of volcanic eruptions happened around Rawa Dano, Banten, Indonesia, remain undiscovered. One of the key features to this geological event is the presence of a 13.7 km × 6.5 km caldera-like morphology in the middle of Banten tuff deposits. Surface geological investigation in the area indicates that the eruptions are massive and occurred in several periods. Low-frequency ground-penetrating radar (GPR) signals are used as an aid to identify the unexposed part of the deposits in this volcanological study. Common-offset GPR surveys were carried out along three measurement lines traversing over the deposit outcrops. An outcrop which is exposed after sand mining activities at one of the survey locations shows dipping interfaces between the upper pyroclastic flow deposits, pumice-rich deposits, paleosol, and the lower pyroclastic fall deposits. These stratigraphic contacts are detected as well under the surface which are clearly recognizable in radar images. The GPR cross-section also shows some other reflections due to different deposit types. The overall results of the GPR profiles give the idea about the thickness of each type of volcanic deposits and the paleotopography in the surrounding area.

Keywords: ground-penetrating radar, Rawa Dano, pyroclastic, deposit, paleotopography

1. Introduction

Rawa Dano is located in the western part of java, Serang City, Banten Province. Geological map [1] shows that the majority of this area is covered by a pyroclastic deposit namely Banten Tuff formation, the map also shows that there is a caldera structure in Rawa Dano indicating volcanic activities in the past. A recent geological study of the Banten Tuff formation by [2] reveals that in early Pleistosen there has been multiple eruption in Rawa Dano. Eruptions happened in two types of mechanism which are pyroclastic fall and flow mechanism with two gaps of volcanic activities in between those eruptions. [2] In order to find more information regarding these volcanic events especialy information about the geological features in the subsurface, a geophysical approach is needed.

Based on the difference of the physical properties and geometries of subsurface objects, ground-penetrating radar (GPR) is able to differentiate interface between layering of rocks or even objects such as pipes, cables, etc., which enables the application of this method in geology and geotechnical sector

[3]. Several previous research regarding the use of GPR method on volcanic materials can be found. In volcanic setting, research by [4] demonstrates the possibilities of GPR method in characterization of air fall deposits, dykes, structures, and different types of lava flows based on difference in radar signatures and mean EM velocities exhibited by each volcanic material. A similar attempt was done by [5] to characterize pyroclastic fall and flow deposits of Tambora Volcano, the correlation between radar data and volcanic stratigraphy is able to indicates that the reflection might be caused by interface between soil-pyroclastic deposit and of pyroclastic flow and fall deposit. Another research also by [6] aims to identify longitudinal structure of pyroclastic flow deposits at Merapi Volcano in Java Indonesia. Those research shows the reliability of GPR method to be used in identifying volcanic materials in the subsurface.

GPR method with low frequency was used to identify the unexposed part of volcanic deposit and paleosol layering caused by the gap of volcanic activities in this area. Common offset GPR survey was carried out in three lines survey traversing over the deposit outcrop which exposed after sand mining activities in one of the survey location. The outcrop shows a dipping interfaces between the upper pyroclastic flow, pumice rich, paleosol, and the lower pyroclastic fall deposits. The depth and thickness of each layers identified from GPR data interpretation can then be used to estimate paleotopography along with the behavior of past volcanic activities in the area such as duration and magnitude of the eruption.

2. Study Area

The Rawa Dano lies in an ancient caldera of Danau volcanic complex on the north-western tip of West Java, Indonesia at an elevation of around 100 m above sea level (asl) in a caldera of a Plio-Pleistocene volcanic complex on the western tip of Java Island, Indonesia. (Fig. 1) The adjacent high volcanic terrain is mostly between 400–700 m asl. The deepest part in Rawa Dano is a creater which was largely occupied by swamp. The sediment of the swamp consists of fine grained lake deposits (organic silty clay and silty gyttja with shells), with minor intercalations of peat and fluvial sands [7]. Peat deposits prevail in the central swamp area with sequences of up to 4-5m thickness just below the surface.

The Danau volcanic complex is of Plio-Pleistocene age and formed in several of eruptive phases, culminating in the eruption of the voluminous Banten pumice tuffs (also known as `Bantam' tuffs) and formation of the large [8][1]. Banten Tuff (Qpvb) is a pyroclastic deposit that is quite important in the western tip of Java. Banten Tuff (Qvpb) is crucial because it has an extensive distribution which almost covered the entire area of Banten. In the middle of Banten Tuff's (Qvpb) distribution, there is a caldera with a rectangular shape which has an area of about 13.7 km x 6.5 km. [2]

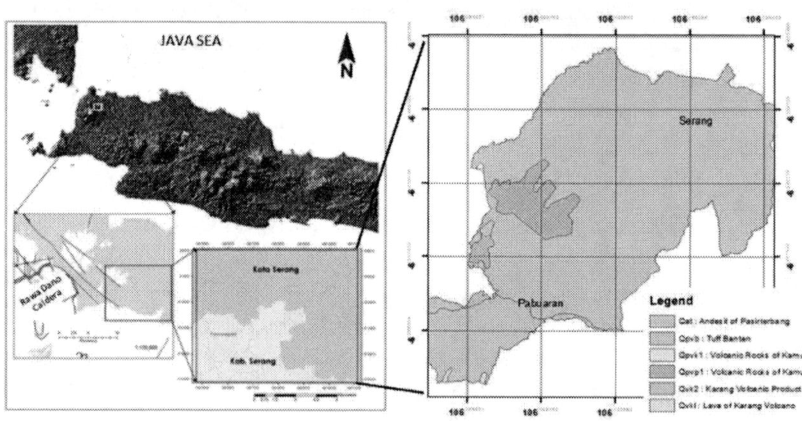

Figure 1. Map of investigation site in Rawa Danau, Banten, Indonesia (modified from [1][2])

This research takes precisely in the area that produces the tephra-stratigraphic correlation. This correlation is identified into four eruptive facies units that exist in one outcrop which called the type locality, each of these facies are then characterized. The four facies have eruption types with different mechanisms. The eruption mechanism of three upper facies sediment is pyroclastic flow. Meanwhile, the mechanism of facies in their subsequent layer is pyroclastic fall [5].

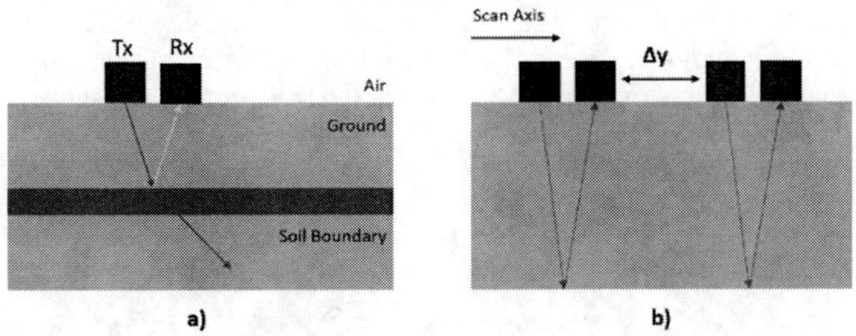

Figure 2. a) GPR Working Scheme, b) Common-Offset Technique (modified from [11])

To identify an unexposed part of the deposits in this volcanological study, we used ground-penetrating radar (GPR) with low-frequency signals. The research sites are located at 6°11'56.1"S 106°05'29.1"E and 6°10'27.5"S 106°08'49.9"E. This survey location was used for sand mining activities, resulting in an exposed outcrop which shows dipping interfaces between the upper pyroclastic flow deposits, pumice-rich deposits, paleosol, and the lower pyroclastic fall deposits.

3. Methodology

Ground Penetrating Radar (GPR) is one of the active geophysical method that used radio wave propagation in range 10 – 2000 MHz for detecting reflected wave from subsurface objects as its responses [9][10]. For the instrument, GPR consists of two antennas which are transmitter and receiver antenna. The wave signals are emitted to the ground through the transmitter antenna, when the signals hit some reflectors, some energy would be reflected to the receiver and the remains are continued to the next layer (Fig. 2).

The magnitude of reflected wave (r) depends on relative dielectric permittivity (RDP) difference from a layer to another layer [10], that can be described with :

$$r = \frac{\sqrt{RDP_2} - \sqrt{RDP_1}}{\sqrt{RDP_2} + \sqrt{RDP_1}} \tag{1}$$

Because radio wave is considered as an electromagnetic wave, its propagation is controlled by some physical properties such as dielectric constant (RDP), conductivity (σ), and permeability (μ), but permeability is less concerned. The quality of GPR data, either the resolution or depth of penetration are determined from antenna frequency and subsurface properties [11]. Soils with high conductivity would produce high levels of attenuation, it caused penetration depth to decreased. Electric conductivity from the soils itself could increase as the amount of water, salt saturation, and clay content increases. Hence, this research should consider whether the sites of investigation can produce an accurate result or not.

This research used 50 & 100 MHz frequency antennas with maximum penetration approximately about 40 m. The frequency was chosen with consideration that the object is estimated to be located at a depth of 50 m. The other reason is the common assumption that this frequency is suitable for soil layer and geological studies. Before running the measurement, it is necessary to set up several parameters on GPR instrument to get maximum result. The parameters used are 9 for medium permittivity, 1200 ns for time range, and 32 for stacking.

There are two sites in this investigation. The first site which located near the main source of eruption consists of 2 lines with profile length about 140 m (named Line 1-1 and Line 1-2). Due to massive attenuation occurred in depth below 25 m, recording duration for one GPR trace was shortened to 400 ns. The first site is an active mining site where the outcrop is exposed after soil extraction. Another site is located in the eastern side of the first site (away from Rawa Dano caldera) and it is lower in elevation. The surface of this area made of more solid layer which was assumed to be the basement. In the location of abandoned sand mining area, 2 lines of GPR survey were carried out with distance of 80 and 50 m for Line 2-1 and Line 2-2 respectively. Location of these GPR profiles are shown in Fig. 3.

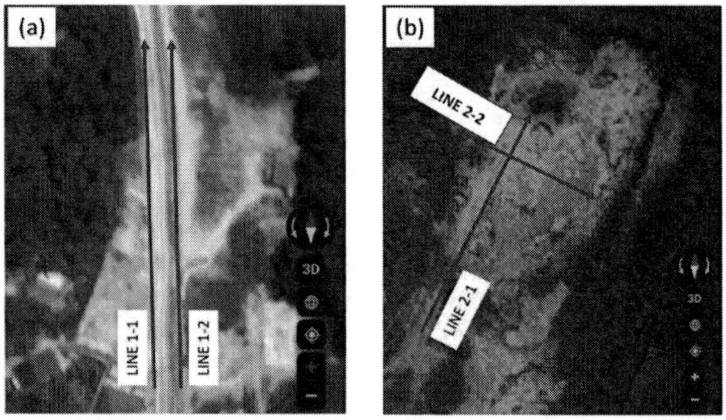

Figure 3. GPR survey profile at (a) site 1 and (b) site 2.

Once all the GPR data has been collected, the data are then filtered using background removal and frequency-based filter. The reflection amplitude was fixed through gain control, so that the results are expected to clearly delineate the soil layer boundaries between tuff or pyroclastic deposits with other soils. All the process (data collection and processing) was carried out using Prism 2.7 GPR software.

4. Result and Discussion

Data processing result from the two measurement lines in Site 1 were shown in Fig. 4. A high-level of unfiltered background noise, indicated by constant straight lines from 0 to maximum distance, makes the reflection data become difficult to interpret. However, the cross-sectional view of Line 1-1 (upper part of Fig. 4) shows some reflectors with various shapes which are visible starting from the depth of 7.5 m. At this depth, there are reflectors in the shape of dunes, which are having a slightly different electrical properties from the overlying soil. One interesting finding is that these morphologies have crisscrossed one another, making it more challenging to analyze. However, by considering such small reflection amplitudes, it can be inferred that this layer consists of the same pyroclastic deposits, but maybe differ in some manners. It could represent the earlier volcanic material deposition than the one exposed on the surface, and it could give the idea of small portion of paleotopography in this area.

The adjacent GPR profile (Line 1-2) shows almost no differences in terms of stratigraphy, except the depth and the position of the reflectors are not the same. It can be seen in the lower part of Fig. 4 that the reflectors are deeper in depth, and farther in distance relative to the first GPR trace. This result indicates that the elevation of this paleotopography is getting low to the East direction. Several high amplitude reflections were detected in the depth about 12 m in Line 1-1 and 15 m in Line 1-2. These reflectors are steeply dipping in the North direction. A possible explanation for this finding might be that the reflection is due to the basement layer which have a clear contrast with the upper volcanic layers. However, due to the limited energy and penetration ability of GPR signal, there could be a potential bias in data interpretation.

In the second part of the GPR survey in Site 2, measurement along the Line 2-1 (Fig. 5) shows a strong reflection with a wavy shape starting from the depth of 6 m below the ground surface. The anomaly is detected along the measurement profile, with some deeper reflector visible at some part where GPR signal is not severely attenuated. There are noticeable signal attenuations below the depth of 20 m which is most probably due to groundwater aquifer. Meanwhile, the GPR cross-section of Line 2-2 shows two main reflectors: the first one is located on its left hand side in a shallow depth; the second reflector is a dipping reflector located on its right hand side after 20 m depth.

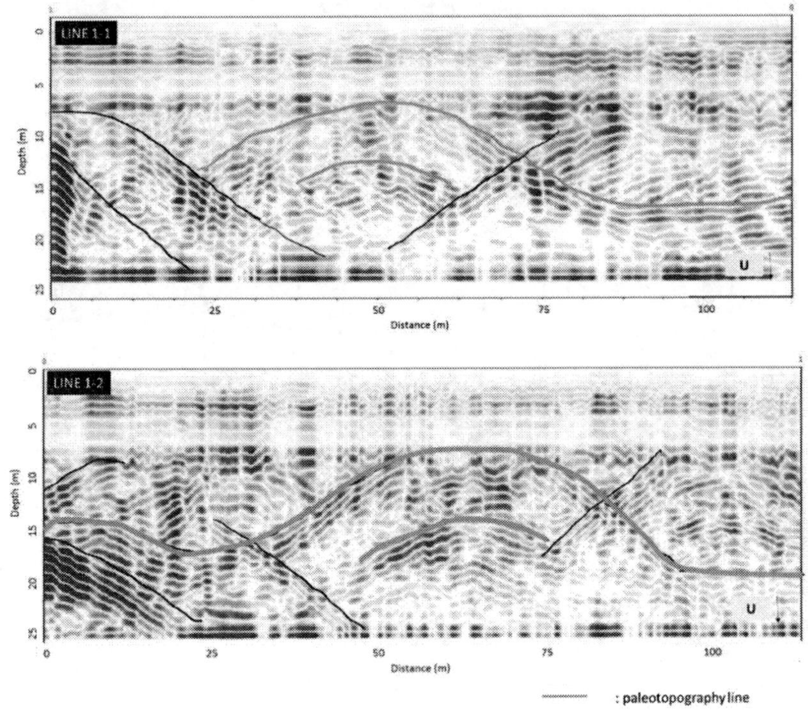

Figure 4. GPR Cross Sectional data in site 1, black lines show the proposed stratigraphic layers.

Comparing the results from Site 1 and Site 2, it can be seen that the strong reflection, which is most likely indicating the basement depth, could be found in shallower depth in Site 2. The thickness of volcanic deposits in this location, is much more reduced compared to Site 1. This finding inferred that the volcanic deposits from volcanic eruptions in Rawa Dano are getting thinner to the east direction. One unanticipated finding is that the second reflector at the end of Line 2-2 is dipping to the west direction, which differs from the rest. This result is probably related to the existence of ancient river, which could also explain the existence of a massive ground water aquifer at the center. However, with a small survey dimension, this interpretation is uncertain, and need to be proven with more reliable data and methodology.

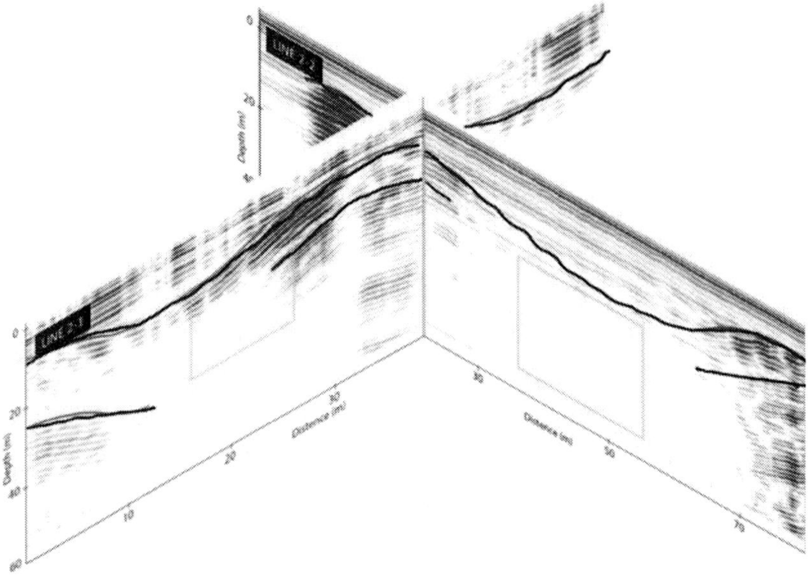

Figure 5. GPR cross-sectional results from Site 2 (Line 2-1 and Line 2-2). Black lines show the layer of proposed stratigraphy and the squares show the electromagnetic signal attenuation.

5. Conclusions

In this investigation, the GPR method was used to gain more information about volcanic deposits and paleotopographic condition around Rawa Dano volcanic complex. This research confirmed that GPR survey was able to differentiate between underground stratigraphic layers by using their contrast in electrical properties. The results from Site 1 shows that the elevation of the general paleotopography around the location is getting low to the East direction. This study has also found that the volcanic deposits such as upper pyroclastic flow deposits, paleosol, and the lower pyroclastic fall deposits determined from the reflection of the GPR signal on several lines. Furthermore, the volcanic eruptions in Rawa Dano are getting thinner to the east direction, which also indicates that the origin of the volcanic deposits is located to the west from both measurement sites. Taken together, these findings strengthen the idea that volcanic eruptions in Rawa Dano was happened in several periods, which was resulting a number of layers of volcanic deposits.

Acknowledgment

We acknowledge the Universitas Indonesia as this study is funded through PUTI Prosiding Program No. NKB-970/UN2.RST/HKP.05.00/2020. We also thank the Energy and Mineral Resources Agency of Banten Province for the cooperation during data acquisition processes.

References

[1] Rusmana E, Suwitodirdjo K and Suharsono 1991 *Geology of the Serang Quadrangle, Java. Scale 1:100,000* Explanatory notesand geological map Geological Research and Development Centre Bandung p 19

[2] Yekini M A 2020 *Studi Mekanisme Erupsi Endapan Piroklastik Tuf Banten (qpvb) di Daerah Pancanegara dan Sekitarnya Menggunakan Metode Distribusi Ukuran Butir dan Analisis Komponen* (Indonesia language) Undergraduate Thesis Universitas Indonesia

[3] C Gomez F L 2009 Block-and-Ash Flow Deposition: A Conceptual Model From a GPR Survey on Pyroclastic-Flow Deposits at Merapi Volcano, Indonesia *Geomorphology* pp 118-127

[4] Bordon P 2019 Automatic Detection of Pipe-Flange Reflections in GPR Data Sections Using

Supervised Learning *Journal of Applied Geophysics* p 170

[5] Lewis J 2007 Characterization of Pyroclastic Fall and Flow Deposits From the 1815 Eruption of Tambora Volcano, Indonesia Using Ground-Penetrating Radar *Journal of Volcanology and Geothermal Research* pp 352-361

[6] C Gomez F L 2008 Longitudinal Structure of Pyroclastic-Flow Deposits, Revealed by GPR Survey, at Merapi Volcano, Java, Indonesia *Journal of Volcanology and Geothermal Research* pp 439-447

[7] Rimbaman 1994 *Penelitian Pelestarian Rawa Danau di Kabupaten Serang, Jawa Barat* (Indonesia language) Center for Geological Research and Development and Regional Development Planning Agency of Dati I Province, West Java 40pp appendices

[8] Van Bemmelen R W 1949 *The Geology of Indonesia* vol. 1, 1A,1B. Government Printing Of®ce, Den Haag, p 7321 maps and appendices

[9] Annan A P 2003 *Ground Penetrating Radar Principles, Procedures, & Applications* Sensors & Software Inc

[10] Jol H M 2009 *Ground Penetrating Radar Theory and Applications* Elsevier Science ISBN: 978-0-444-53348-7

[11] Daniels D J 2004 *Ground Penetrating Radar – 2nd Edition* The Instituion of Electrical Engineers, London ISBN 0 86341 360

IOP Conf. Series: Earth and Environmental Science **873** (2021) 012070 doi:10.1088/1755-1315/873/1/012070

Analysis of M 5.3 Sumbawa, Indonesia earthquake 2020 and its aftershocks based on hypocenter relocation from BMKG seismic stations

M Ramdhan[1,*], Priyobudi[1], A Mursityanto[2], K H Palgunadi[3], Daryono[1]

[1]The Indonesian Agency for Meteorology, Climatology, and Geophysics (BMKG), Jl. Angkasa I No.2, Kemayoran, Jakarta 10720
[2]The Indonesian Agency for Meteorology, Climatology, and Geophysics (BMKG) region III, Jl. Raya Tuban, Kuta, Badung, Bali 80361
[3]Physical Science and Engineering, King Abdullah University of Science and Technology, Thuwal, Saudi Arabia

E-mail: mohamad.ramdhan@bmkg.go.id

Abstract. The 2020 Sumbawa earthquake of moderate magnitude (M 5.3) produced very significant aftershocks. Based on the computation of Utsu's method, those aftershocks would be ended after the 20th day. Those earthquakes along 20 days were relocated using double-difference method. The relocation results show the southwest-northeast orientation and getting deeper into the northwest direction. Those two directions show the strike and the dip from the fault plane of the earthquake which was consistent with the focal mechanism released by the Indonesian Agency for Meteorology, Climatology, and Geophysics (BMKG). Those results showed the majority of earthquakes occurred at a depth of shallower than 20 km. Those earthquake depths were fit with the previous study showing the crustal thickness beneath Sumbawa Island that was about 28 km. We also found that those earthquakes occurred at splay faults propagating to decollement structure. This study is beneficial for earthquake disaster mitigation especially in updating active faults on Sumbawa Island.

1. Introduction

The tectonic system of Sumbawa Island is affected by the Indo-Australian oceanic plate subducting to the Eurasian continental plate. This interaction is caused by the formation of a back-arc thrust fault along the north of Lombok Island to the north of Flores Island (The Flores back-arc thrust) [1]. The series of large earthquakes at the north of Lombok Island in 2018 that occurred in the fault caused many casualties, injured, and infrastructure damages. Besides affected by the fault and the subduction system, earthquake activities around Sumbawa Island were also caused by some active faults, such as the earthquake that occurred on June 13th, 2020. The earthquake was considered as a moderate earthquake (M 5.3) following by a number of aftershocks. The mainshock was felt across Sumbawa, Lombok, and Bali Islands with a maximum intensity felt on Sumbawa Island at IV Modified Mercalli Intensity (MMI) and a minimum intensity felt in

Karangasem Regency, Bali Province at II MMI. The earthquake was hosted by an active fault that had not been previously clearly identified. Thus, this study has a significant contribution in delineating the active fault in the study area. Active fault identification is advantageous in mitigating one area from a disaster caused by a significant earthquake, therefore, devastating effects can be minimized. The active fault delineation had been successfully applied using hypocenter relocation results in the Palu-Koro active fault at the Palu-Donggala fault segment [2].

2. Data and Methods

We used the BMKG region III earthquake catalog for 20 days from June 13th to July 3rd, 2020 within the coordinates of 116^0-120^0E and 8^0-10^0S. We removed unrelated earthquakes with the mainshock. Figure 1 shows the study area and seismic stations used to relocate earthquakes. The time range was obtained from aftershocks decay computation based on Utsu's method [3]. We calculated based on the method using aftershocks for 13 days as shown in figure 2 (a). Based on Utsu's method [3], the aftershocks would be approximately insignificant after the 20th day as depicted in figure 2 (b). We applied double-difference method to relocate the mainshock and aftershocks [4,5]. Each earthquake at least has six phases (P and S phases or P phases only) and maximum distance from earthquake sources to seismic stations was considered 500 km. Those criteria had been applied successfully to detect an interface zone in the Palu-Koro active fault [6]. In this study, we utilized existing models for P-wave speed (Vp) and the ratio of P- to S-wave speed ratio (Vp/Vs) [7,8].

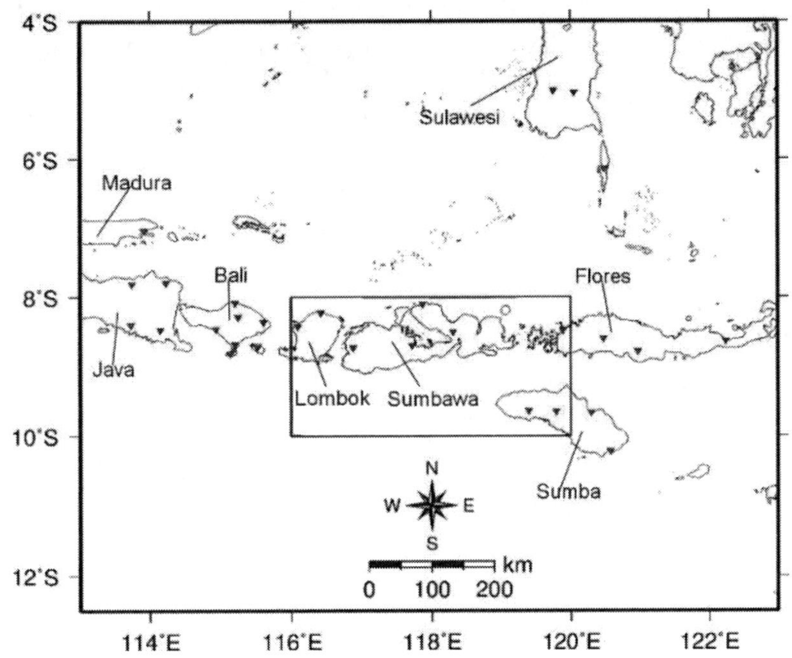

Figure 1. Brown reverse triangles show the seismic station position used in this study. Black rectangle indicates the study area.

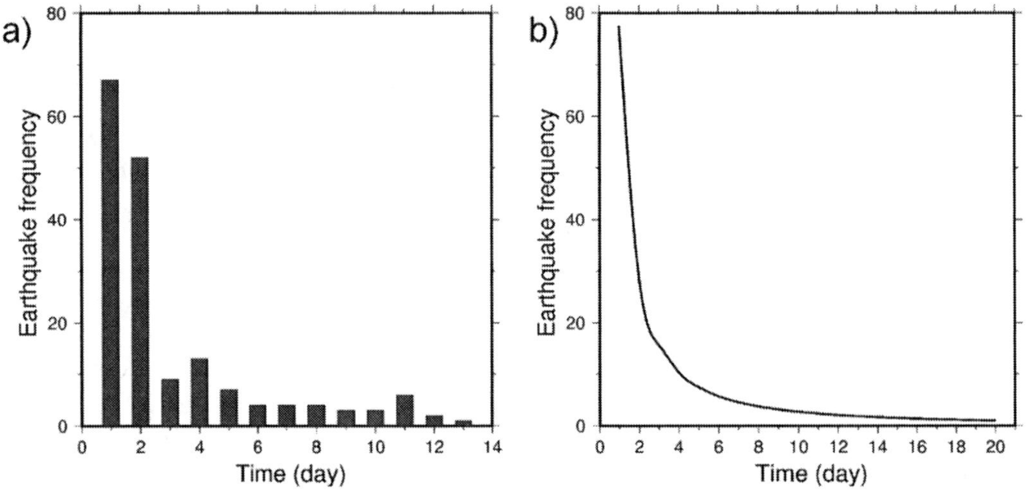

Figure 2. Daily histogram of aftershock distribution for 13 days (a) and aftershock temporal decay based on Utsu's method [3] (b).

3. Result and Discussion

We successfully relocated 135 hypocenters from 221 earthquakes for 20 days. The biggest aftershock magnitude was M 4.5. A number of earthquakes could not be relocated because they were not fit with the inversion criteria. We utilized the bootstrap method to compute the relative error of the hypocenter location (X_0, Y_0, Z_0) [4,9,10]. Those results show a majority of those parameters is less than 1.5 km such as shown in figure 3. The statistic parameter of those errors is presented in table 1.

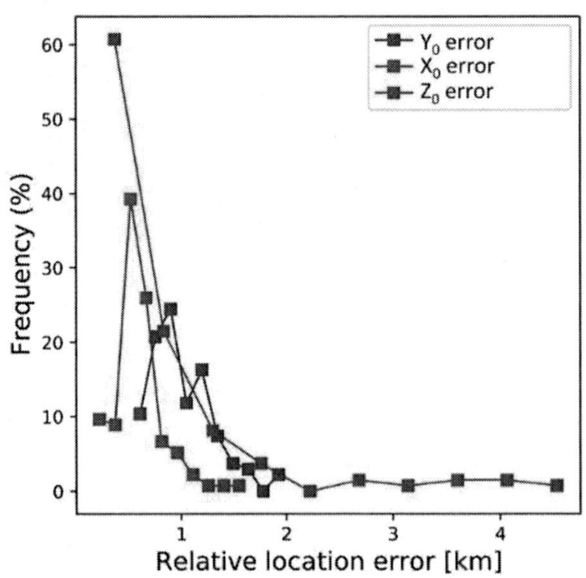

Figure 3. Histograms of relative location errors of mainshock and aftershocks Sumbawa earthquakes 2020 using the bootstrap method. Confidence ellipsoids resulted from a bootstrap analysis for those errors is 95%.

Tabel 1. Statistic parameters of relative error locations computed based on the bootstrap method.

X_0 error (km)				Y_0 error (km)				Z_0 error (km)			
Mean	Min	Max	Stdev	Mean	Min	Max	Stdev	Mean	Min	Max	Stdev
1.09	0.6	2.07	0.29	0.67	0.23	1.69	0.23	0.98	0.38	4.98	0.81

BMKG released a source mechanism with strike orientation N224^{0}E and dip angle 49^0 to northwest direction. The source mechanism was computed using moment tensor inversion [11]. Seismicity map after hypocenter relocation showing the strike orientation is consistent with the source mechanism released by BMKG as shown in figure 4 (b). Before relocation, the strike orientation of aftershock distribution is perpendicular to the strike orientation of the source mechanism as shown in figure 4 (a). This emphasizes the importance of this study in relocating hypocenter, therefore, more reliable aftershocks distribution that is consistent with tectonic stress loading orientation in this area can be useful in understanding an active fault. The results of this study were also consistent with the GPS observation study at the BIMA and LABU stations showing that the local tectonic force on Sumbawa Island was oriented towards the northwest [12]. The tectonic force had the same direction as the mainshock dip direction based on the moment tensor inversion. This showed that the M 5.3 Sumbawa earthquake was caused by the local tectonic force on the island.

The hypocenter relocation results show a significant shift in the hypocenter depth where fixed depths at 10 km could be enhanced to be more reasonable positions as shown in figures 5 (b) and 5 (d). Those fixed depths also caused misoriented dip direction as shown in figure 5 (c). The direction of dip to the northwest is more reliable after hypocenter relocation as shown in figure 5 (d). The fixed depth itself occurred because of the non-convergence during the inversion. The necessity for fast parameters in disseminating earthquake information caused unavoidable fixed depth, thus the relocation process must be applied to obtain a more precise and reasonable hypocenter location. Precise hypocenter parameters are beneficial for further tectonic studies. Meanwhile, the earthquake parameters that are disseminated by the BMKG are accurate enough for emergency response purposes. Stakeholders can use this information to perform rapid rescue and evacuation if there is a potential for a significant earthquake and tsunami disaster because the information can be received in less than 5 minutes. The results of this study show the majority of earthquake depths occurred at a depth of less than 20 km as showed figure 5 (b) and 5 (d). They were consistent with the previous study stated that the crustal thickness beneath Sumbawa Island was about 28 km [13]. The schematic model of fault structures causing the Sumbawa earthquake M 5.3 and its aftershocks are shown in figure 6. The model shows a low-angle thrust fault called a decollement structure at a depth range of 13-15 km. The fault plane became steeper before 13 km depth that indicated splay faults in which the mainshock and majority of aftershocks occurred at those structures. A number of aftershocks also propagated to the decollement structure that is located beneath those faults.

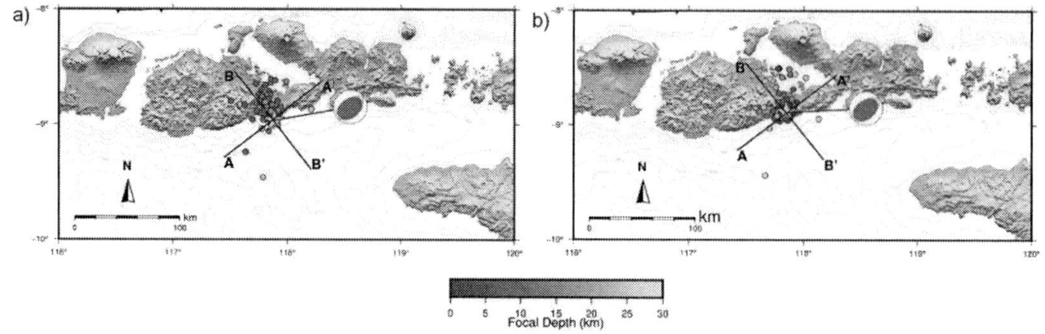

Figure 4. Seismicity map before (a) and after (b) hypocenter relocation. A-A' and B-B' are two vertical section lines following strike and dip directions based on the focal mechanism released by BMKG. The maximum distance of earthquakes to each line is 60 km. The yellow star shows the mainshock of the Sumbawa earthquake M 5.3. The topographic map used SRTM Non-Void Filled [14].

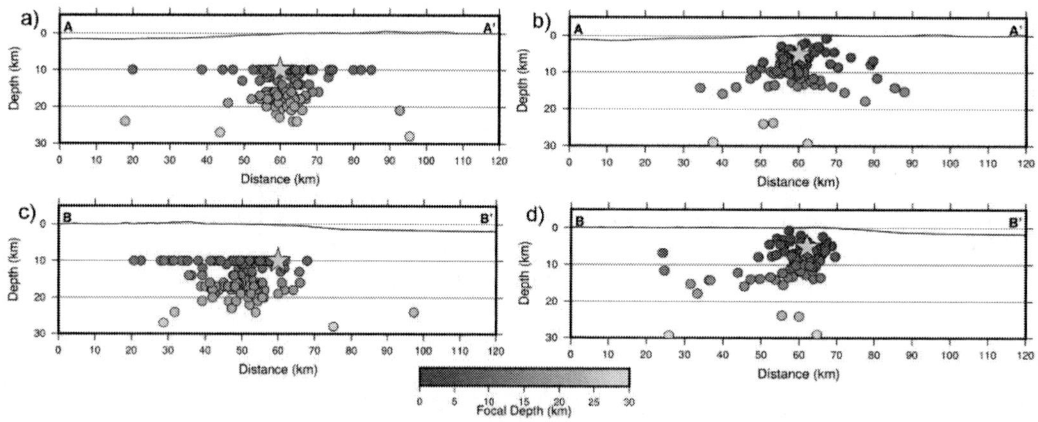

Figure 5. Vertical section before (a) and after (b) hypocenter relocation along strike direction. Vertical section before (c) and after (d) hypocenter relocation along dip direction. The yellow star shows the mainshock.

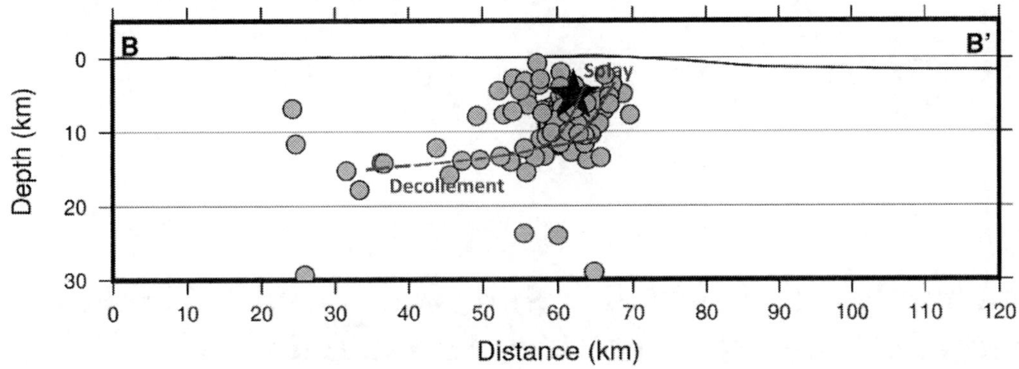

Figure 6. Schematic model of splay faults and decollement structure based on hypocenter relocation along dip direction. The mainshock and aftershocks of the Sumbawa earthquake 2020 occurred at those geology structures. The black star shows mainshock.

4. Conclusions

This study successfully delineates the orientation of strike and dip directions which is consistent with the moment tensor inversion released by BMKG and local tectonic force in the area. Those results show that the mainshock and aftershocks series occurred at splay faults and decollement structure. The Sumbawa earthquake M 5.3 was caused by tectonic force in northwest direction. The study is so worthwhile in advanced tectonic study especially in updating active faults which will be used for disaster mitigation in the area.

Acknowledgement

We are very grateful to The Indonesian Agency for Meteorology, Climatology, and Geophysics (BMKG) region III for providing the earthquake catalog for this research. This study also was supported by *Komite Kajian Gempa bumi dan Tsunami* BMKG 2020.

References

[1] Hamilton W B 1979 (US Govt. Print. Off.)
[2] Priyobudi P and Ramdhan M 2020 *Jurnal Lingkungan dan Bencana Geologi* **11** 1–9
[3] Utsu T 1961 *Geophys Mag* **30** 521–605
[4] Waldhauser F and Ellsworth W L 2000 *Bull Seismol Soc Amer* **90** 1353–68
[5] Waldhauser F 2001 *US Geol. Surv. Openfile report* 01–113
[6] Ramdhan M and Priyobudi 2019 *J. Phys.: Conf. Ser.* **1341** 082009
[7] Koulakov I, Bohm M, Asch G, Lühr B-G, Manzanares A, Brotopuspito K S, Fauzi P, Purbawinata M A, Puspito N T, Ratdomopurbo A, Kopp H, Rabbel W and Shevkunova E 2007 *J Geophys Res: Solid Earth* **112**
[8] Wadati K 1933 *Geophys. Mag* **7** 101–11
[9] Billings S D 1994 *Geophys J Int* **118** 680–92
[10] Efron B 1982 (Society for Industrial and Applied Mathematics)
[11] Minson S E and Dreger D S 2008 *Geophys J Int* **174** 585–92
[12] Nugroho H, Harris R, Lestariya A W and Maruf B 2009 *Tectonophysics* **479** 52–65
[13] Syuhada S, Hananto N D, Abdullah C I, Puspito N T, Anggono T and Yudistira T 2016 *Acta Geophys* **64** 2020–49
[14] Earth Resources Observation And Science (EROS) Center 2017 Shuttle Radar Topography Mission (SRTM) Non-Void Filled

IOP Conf. Series: Earth and Environmental Science **873** (2021) 012071 doi:10.1088/1755-1315/873/1/012071

Potentially active normal faulting zone identified in the eastern margin of Palu City, Central Sulawesi, Indonesia

Anggraini Rizkita Puji[1], Mudrik Rahmawan Daryono[1], and Danny Hilman Natawidjaja[1]

[1] Research Center for Geotechnology, Indonesian Institute of Sciences, Bandung, Indonesia.

*Corresponding e-mail: angg029@lipi.go.id

Abstract. The 2018 Mw 7.5 earthquake in Palu, Central Sulawesi, resulting in ~2,000 fatalities and estimated economic losses of ~22.8 trillion Indonesian Rupiah, according to the report of BAPPENAS and Central Sulawesi Provincial-Government. Therefore, it is necessary to prevent similar disaster in the future by further detailed studies of any other potential sources that are capable of generating such hazards. Palu City is in the vast depression valley bordered by mountains in its eastern and western margins. The 2018 earthquake source is the Palukoro Fault, which runs through the western margin of onshore Palu Valley then continued under the bay. Along the eastern margin of the valley, we also identified a wide zone of many potentially active faults strands orienting N-S and NW-SE, showing predominantly normal faulting. These faults are observed from their normal fault scarps as inspected from Light Detection and Ranging Digital Terrain Model (LiDAR DTM) data with 90-cm resolution and field ground checks. The faults deformed the old terrace sediments (Late Pleistocene, ~125 kya), but it is unclear whether they also cut the Holocene young alluvial like the ruptured fault of 2018 event. Further paleoseismology investigation is then necessary to obtain further information about these potentially-active normal faults, including their slip-rate and the past ruptures.

1. Introduction

Palu City is located in the vast depression valley that is a very dynamic active tectonic region as it is in a triple junction between Australia, Eurasian, and Philippine Sea Plates [1-2]. North of Sulawesi, the Celebes Sea is being subducted beneath Sulawesi [1]. Hence, this region consist of complex arrays of active fault lines throughout the region (Figure 1). The Palukoro & Matano Fault zones, which cut Central Sulawesi, are the most dominant active fault of Sulawesi, and is of particular importance because it is straddled by Palu City (population around 340,000).

The 2018 Mw 7.5 earthquake in Palu City resulting in ~2,000 fatalities and estimated economic losses of about 22.8 trillion Indonesian Rupiahs (~USD 1.5 billion on current exchange rate), according to the report of BAPPENAS and Central Sulawesi Provincial-Government [3]. This event has lead the need for a better hazard risk evaluation of the 2018 earthquake surface ruptures and associated active faults to determine the strategy and policy in mitigating future earthquakes and their associated hazards and risks.

From geological point of view, a fault is evidently active if it cuts or deforms young landscape or sedimentary layers of Holocene to Late Pleistocene period [4]. For example, in New Zealand, the reference (maximum) age for active fault identification is 125,000 years old, which is related to the beginning of the previous warm-interglacial period or correlated with alluvial terraces and the Marine

Isotope Stage 5 (MIS 5) [5]. For an extremely high-risk project, the 'capable or potential' active faults that show any geomorphic-geologic evidences since early Pleistocene (1,8 million years old) should also be taken into considerations.

Active fault strands identification is rather tricky. As an example, in all old-rocks environment, sometimes it is hard to determine whether the identified fault line is active or not since there is no young geological deposits involved. Geological and landscape analysis is also limited by ratio between fault slip rates and rates of erosions and depositions. If a fault slip-rate is very low, lower than rates of deposition and/or erosion, than we may not be able to see surface-landscape expressions of fault movements since they will be erased by erosion or buried under geological deposits. Hence, identifications of active faults heavily depend on resolutions of used topographic/landscape (digital) maps. The higher the resolution will be the smaller morpho-tectonic features can be seen.

We implemented a combination of remote-sensed and field-based method active fault mapping to identify and analyze the earthquake-fault rupture traces of the 2018 earthquake (Mw 7.5) and we have acknowledged that the main source of the 2018 earthquake event is caused by the movement of Palu fault which runs through the western margin of the onshore Palu Valley then continued under the bay water striking relatively N-S direction. However, in this study we focus on fault scarps distributed in the eastern margin of Palu valley. The scarp was originally mapped as a lineament with no sense of fault-slip direction in the seismotectonic maps published by Center for Geological Survey [6]. Our investigation, however, indicates potentially active normal faulting. The fault displaced the Late Pleistocene unit. The distribution of these fault strands is located near the area of relocation housing plan for the 2018 earthquake victims and Tadulako University, highlighting the importance of characterizing these fault strands.

2. Methods

We use a combination of remotely sensed and field-based observations supported with comprehensive desk-study analysis on active fault mapping including evaluating previously published studies and geological maps. The Palukoro Fault is one of the major fault in Indonesia that have been studied rather intensively. However, in general, almost none of them had focus on mapping active fault lines in sufficient details. Most of paper-journal publications only studied the regional (active) tectonics from geological studies and from tectonic geodetic studies [1-2, 6-15].

Our remote mapping is based on LiDAR DTM 1-m grid data resolution developed by Geospatial Information Agency (BIG) after the 2018 Palu Earthquake and DEMNAS 8.5-m grid data resolution. The remotely based mapping was followed by field-based mapping to confirm the identified tectonic features by inputting the selected target sites into handheld Global Positioning System (GPS) and webGIS-based map as a guidance. To help robust and accurate interpretation of the fault line as complimentary to the LiDAR dataset, we also conducted Unmanned Aerial Vehicle (UAV) survey during the fieldwork using small drone (DJI Mavic) to develop Digital Elevation Models (DEM) and Orthophoto. The drone was flown at about 100 m high with overlapping images taken between 8-9 times. The image resolution produced from UAV survey have about 2-4 cm / pixel.

3. Results, Discussions, and Conclusions

In the eastern margin of Palu Valley, we identified numerous potentially active faults predominantly from the observation of normal fault scarps trending N-S and NW-SE in LiDAR, high-resolution orthophoto, as well as on the ground during fieldworks (black lines in Figure 2a, 2b, 3). These faults can be identified cutting and deforming the old Quaternary sediments, labeled as Qp (Quaternary-Pleistocene) in the Geology and Seismotectonic Map [6, 12] (Figure 2b). In some places, they are appeared to have also affected the Quaternary-Holocene landscapes or the young alluvial terraces and sediments. In general, the appearance of their active-tectonic-landscape signatures of the fault lines are rather obscure and spotty, not prominently continuous over long distances like the major active fault in the west side. Hence, it suggests that their rate of activities is significantly lower. Since they need to be further studied in more details, here we assigned them as the potentially active faults, which are colored in black.

The prominent occurrences of the potentially active normal fault orienting N-S is in the Talise Area in the Eastern side of Palu Bay (Figure 2a). These faults are clearly observed by their relatively fresh linear (normal) fault scarps with predominantly west down movements. The trace of the faults gradually disappears northward when passing the young alluvial-fan sediments but reappearing again in the northern hilly areas. The fault seems to run around the west front of the Tadulako University. Hence, the fault is possibly buried under this Holocene sediment. The geological exposure key evidence of the fault was observed on at Site A at the cliffs of the quarry along southern side of the asphalt road (Figure 3). The outcrop clearly shows that the fault cut the old Quaternary sediments (labelled as Qp in the Geological Map and Seismotectonic Map) as shown in Figure 2b. A similar old Quaternary sediment on the western margin has been dated by Bellier [7] using the cosmogenic dating by the measurement of *in situ* produced ^{10}Be concentrations in quartz boulders on top of fan surfaces, which is about 125,000 years old (Figure 4). The maximum fault offsets observed along the main fault is about 2 meters. We called it as Tadulako Fault since it appears to run across the Tadulako University complex (Figure 2a & 2b).

We also plotted the fault map from Seismotectonic Map [6] and its revision by Soehaemi published from Center for Geological Survey (PSG) in 2018 for comparisons (Figure 2b). It is apparent that Soehaemi revised their previous Seismotectonic Map based on the same geological outcrop at Site-A. Hence, it considers the same fault, which is called here as the Tadulako Fault (Figure 2a). However, they did not consider the tectonic geomorphology of the surrounding area (numerous lineaments and fault scarps), and their measurement of the fault strike seems to be a bit off (more westerly).

Based on the available data, we still do not have clear evidence whether these normal faults in the eastern margin also cut or deformed the Holocene sediments (Qh). This is possibly because the slip rates of the faults are significantly lower than the rate of erosions and sedimentations. Consider that, even for the fast-moving Palukoro fault on the western margin, its fault traces on surface were mostly not seen before exposed by the 2018 earthquake ruptures. Hence, further investigations are necessary to fully understand the characteristics and hazard potentials of these normal faults. Nonetheless, considering available data, then the fault slip rate is generally interpreted as between 125,000 (125ka) and 11Ka (early Holocene). However, according to the previous study along the western margin, these secondary faults cut the Holocene alluvial fan, dated ~11,000 years old [7] (see Figure 4); and thus the recurrent interval can be estimated to be about between the 125 ka and 5Ka (~Mid Holocene) or classified as Class IV to Class VI (>5,000 – 125,000 years) [16].

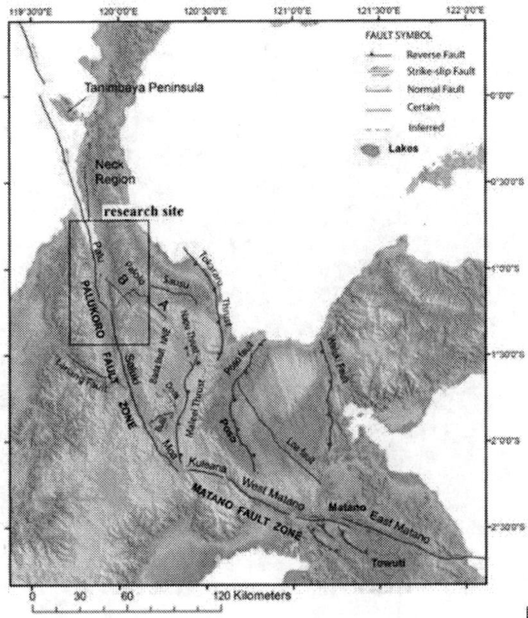

Figure 1. Active Fault Map of Central Sulawesi, modified from Daryono [17] and PuSGeN map [18]. Black rectangle is the research site.

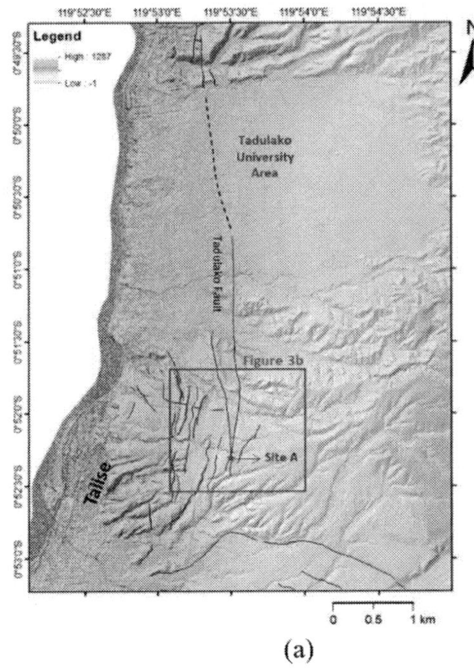

(a)

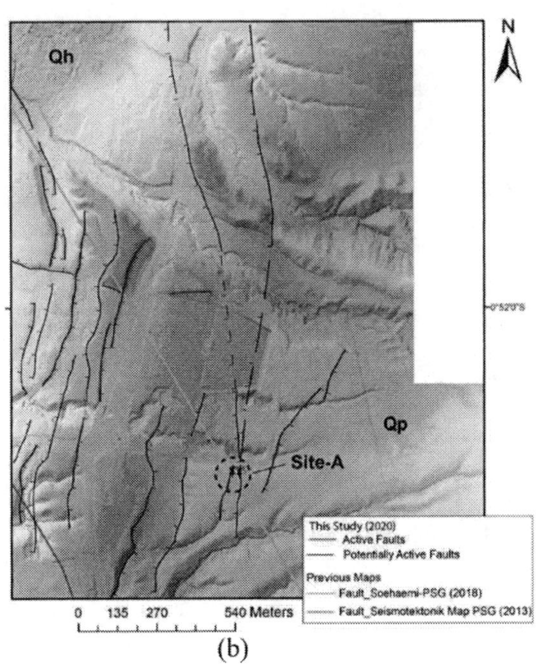

(b)

Figure 2. (a) The potentially active fault trace in the Talise area, eastern side of the Palu City near the bay area. Site A is the key-site observation of the fault exposure on the mining cliff; (b) Comparisons of the potentially active faults trace mapped in this study and the fault trace which published in the Seismotectonic Map and its revision.

Figure 3. Site A in Talise showing the exposure of the potentially active fault (Tadulako Fault) cutting the Late Plistocene sediments (Qp), dated about 125 ka [7]. The fault strike is N350E to N-S. The normal fault displacement on the main fault is up to 2 meters. The traces of the fault lines on the top of ground surfaces have mostly been eroded.

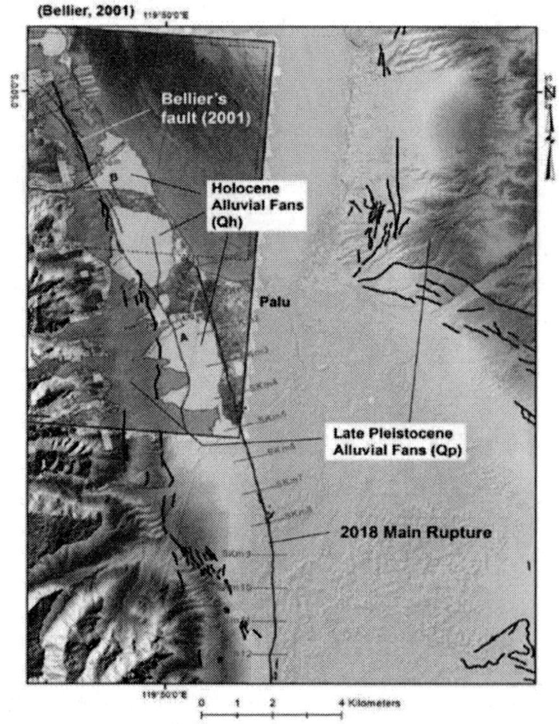

Figure 4. Fault map comparison of this study with Bellier [7]. Note that Bellier's faults are coincided with the secondary normal faulting associated with the negative flower structures.

Acknowledgement

This study is funded by the CHL\R1\180173 – GCRF Grant from The Royal Society, UK.

References

[1] Silver E, McCafrey R, and Smith R 1983 *J. Geophys. Res. Solid Earth* **88** no B11 9407-18
[2] Hall R 2002 *J. of Asian Earth Sci.* **20** no 4 353-431
[3] BAPPENAS and Central-Sulawesi-Local-Government Report 2018 *Rehabilitation and Reconstruction in Palu Area After the 2018 Earthquake (Pemulihan dan Pembangunan Kembali Wilayah Pasca Bencana Provinsi Sulawesi Tengah)* (Indonesia: BAPPENAS, National Agency for Planning and Development)
[4] McClymont B 2001 *Building on the edge - The Use and Development of Land On or Close to Fault Lines* (Wellington: Parliamentary Commissioner for The Environment)
[5] Langridge R and Ries W 2014 *Active Fault Mapping and Fault Avoidance Zones for Central Hawkes Bay: 2013 Update* (New Zealand: GNS Science) 48
[6] Soehaimi A, Marjoyono, and Wahyono W 2013 *Peta Seismotektonik Daerah Palu dan Sekitarnya, Sulawesi Tengah (Seismotectonic Map of Palu and Surounding Area, Central Sulawesi scale 1:150000)* (Bandung: Center for Geological Suvey, Geological Agency, Ministry of Mines and Energy)
[7] Bellier O, Sebrier M, Deauduoin T, Vileneuve M, Braucher R, Bourles D, Siame L, Putranto E, and Pratomo I 2001 *Terra Nova* **13** 463-70
[8] Kadarusman A, Miyashita S, Maruyama S, Parkinson C, and Ishikawa A 2004 *Tectonophysics* **392** 55-83
[9] Katili J 1978 *Tectonophysics* **45**
[10] Socquet A, Simons W, Vigny C, McCafrey R, Subarya C, Sarsito D, Ambrosius B, and Spakman W 2006 *J. of Geophy. Res.* **111**
[11] Stevens C, McCaffrey R, Bock Y, Genrich J, Endang, Subarya C, Puntodewo S, Fauzi, and Vigny C 1999 *Geophy. Res. Let.* **26** 2677-80
[12] Sukamto R 1973 *Reconaissance Geological Map of Palu Quadrangle* (Bandung: Geological Agency, Ministry of Mines and Energy)
[13] Vigny C, Perfettini H, Walpersdorf A, Lemoine A, et al 2002 *J. of Geophy. Res.* **107** 10,1029
[14] Walpersdorf A, Rangin C, and Vigny C 1998a *Earth and Planetary Sci. Let.* **159** 47-55
[15] Walpersdorf A, Vigny C, Subarya, C, and Manurung P 1998b *Geophy. Res. Let.* **25** 2313-16
[16] Kerr J and Nathan S 2003 *Planning for Development of Land on or Close to Active Faults: A guideline to assist resource management planners in New Zealand*, ed Ministry for The Environment (New Zealand: GNS Science) 67
[17] Daryono M 2016 *Paleoseismologi Tropis Indonesia: Studi kasus di Sesar Sumatra, Sesar Palukoro, dan Sesar Lembang (Paleosesimology in Tropical Indonesia. Case Studies in Sumatran Fault, Palukoro Fault, and Lembang Fault)* (Bandung: Institut Teknologi Bandung)
[18] Natawidjaja D, Sapiie B, Daryono M, Marliyani G, and Al, E 2017 *Geologi Gempa Indonesia (Earthquake Geology of Indonesia)* in *Peta Sumber dan Bahaya Gempa Indonesia tahun 2017 (Earthquake Sources and Hazards Map 2017 Version)* ed M Irsyam and S Widiyantoro *et al* (Bandung: Kementrian PUPR (Ministry of Public Works and Housing))

IOP Conf. Series: Earth and Environmental Science **873** (2021) 012072 doi:10.1088/1755-1315/873/1/012072

Seismicity around Cirata Dam, West Java, Indonesia based on BMKG local seismic network

Priyobudi[1], M Ramdhan[1], Daryono[1]

[1]The Indonesian Agency for Meteorology Climatology and Geophysics of Indonesia (BMKG), Jl. Angkasa I No.2 Kemayoran, Jakarta 10720, Indonesia

priyobudi@bmkg.go.id

Abstract. The addition of seismic stations to the seismic network of BMKG in 2019 has successfully located some local earthquakes. In the early 2020 occurred significant earthquakes around Cirata Dam, West Java. During a period of January-March 2020, there have been 5 earthquakes with magnitude ranging from 1.8-3.7. Those earthquakes caused ground shaking up to III MMI intensity scale around the epicenters area. The relocation of the hypocenter using the Teletomo-DD method is applied in this study so that the data can be interpreted to show the fault geometry in this area. The relocated epicenters distribute in the east side of the dam elongated in SSW-NNE direction. Vertical distribution of relocated hypocenters show that the earthquake occurred at 1.1 km down to 14.5 km depth. Hypocenter depths are getting deeper to the north direction, this suggest dip orientation of the fault plane. The reconstructed dip orientation is consistent with nodal plane resulted from moment tensor inversion results, that shown fault planes oriented in N 229^0 –272^0 E direction and dip 49^0–50^0 to the north direction.

1. Introduction

Due to the subduction of the Indo-Australian plate beneath the Eurasian plate along Java Island, it caused stress transfer to the island and produced some active faults. In the West Java area, there are Cimandiri Fault [1], Lembang Fault [2], and Baribis Fault [3]. The latest GPS study indicate an active Kendeng-Baribis fault that extent from Central Java to the south of Jakarta [4]. Beside of those faults, we suspected other active faults in West Java. One of them is Cirata Fault. The fault should also be considered as hazard source in West Java area. Figure 1 showing the area of this study and active faults in West Java. Moreover, the Cirata Dam was built above this hazard zone. Cirata Dam is the main infrastructure for power plants, fisheries, irrigation as well as tourism object.

In the early 2020, between January-March there were five earthquakes occurred around Cirata dam area. Those earthquakes were significant events because they caused earth shocking up to III MMI (*Modified Mercalli Intensity scale*) around the epicenter area. The denser seismic station network of BMKG has provided more proper data to be used for this study. This study is intent to reconstruct fault geometry in this area. Strike and dip orientation of a fault are very important parameters as an input for further seismology studies such as surface rupture estimation, deterministic or probabilistic seismic hazard analysis. This study will also contribute to complete active fault database in Indonesia, especially West Java.

IOP Conf. Series: Earth and Environmental Science **873** (2021) 012072 doi:10.1088/1755-1315/873/1/012072

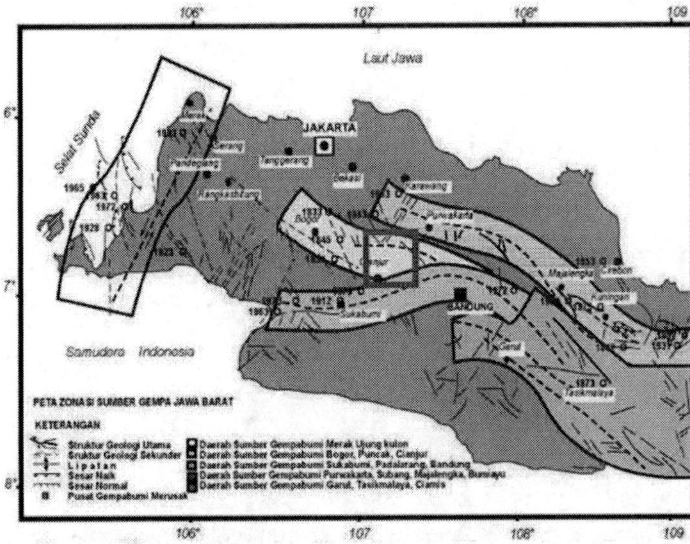

Figure 1 Earthquake hazard zone related to active fault distribution in West Java modified from the previous paper [5]. Research area located in the Bogor-Puncak-Cianjur earthquake zone (red square).

2. Data and Methods

This study used preliminary hypocenter data of the BMKG earthquake catalog. The preliminary hypocenter data were determined using the LOCSAT program [6] and global velocity model IASPEI91 [7], which are integrated in Seiscomp-3 seismic processing program. We then relocated hypocenters using Teletomo-DD [8] using the 3D velocity model for Indonesia region and the 1D velocity model for outside Indonesia [9–11]. Relocation processes were held simultaneously for all seismic events in Indonesia[12].

Moment tensor inversion was applied to selected events using the SCMTV program that included in the Seiscomp-3 seismic processing package. Stable inversion for the complete moment tensor algorithm [13] was used for moment tensor inversion. Green's function was calculated using GEMINI [14] based on the PREM velocity model [15].

3. Results and Discussions

Epicenters of both preliminary and relocated ones distributed around the east side of the dam (Figure 3). Ucertainty of relocated epicenters location varied from 519.4 – 1033 meters for East-West and 280.0 – 437 meters for North-South direction (Table 1). Those results show a majority of those parameters is less than 1.0 km such as shown in Table 1. Relocated hypocenters show a significant shift in vertical distribution (Figure 4). The depth of preliminary hypocenters was ranging from 10 to 17 km. After relocated, hypocenters shifted to shallower crust between 1.1 – 14.5 km depth, with uncertainty ranging from 462.6 – 956.9 meters. Hypocenter parameters are show in Table 1. Relocated hypocenter depth of the two events (M3.5 and M3.7), are also shallower than depth determined by moment tensor inversion (Figure 4). This distinction caused by different velocity model that is used for hypocenters relocation and moment tensor inversion.

Moment tensor inversion were conducted for the two events: M3.5 and M3.7. The results are focal balls that indicate thrust oblique and strike-slip oblique mechanism. Those focal balls show that fault planes are elongated to south west and west direction and dip angle varied from 49^0 to 75^0 (Table 2). The centroid depth of those events are 8 km and 4 km and the resulted magnitude moment are Mw3.3 and Mw3.5. An example of the moment tensor inversion result of the January 14th, 2020 event is shown in Figure 2. The plot of depth vs fitting of observed and synthetic seismogram was best fit at 8 km depth and Mw 3.3.

Tabel 1 Comparison ofhypocenter depth for preliminary, relocated and moment tensor inversion (CMT). EX, EY and EZ are uncertainty of relocated hypocenters for East-West, North-South, and vertical direction.

Event Date	Origin Time (UTC)	Lon	Lat	Error (m)			Depth (km)			Mag
				EX	EY	EZ	PDE	Reloc	CMT	
14-Jan-20	20:16:07	107.33	-6.76	1033.0	713.8	462.6	10	2.2	10	3.5
11-Mar-20	14:51:49	107.34	-6.69	417.7	280.8	727.2	11	5.3	-	2.3
11-Mar-20	15:50:14	107.36	-6.69	488.6	511.2	665.4	17	14.5	-	1.8
16-Mar-20	2:09:17	107.35	-6.73	519.4	437.0	956.9	10	3.3	-	2.4
20-Mar-20	14:41:31	107.35	-6.74	630.5	607.5	656.4	10	1.1	4	3.7

Tabel 2 Moment tensor inversion results of selected earthquakes M3.5 and M3.7

Event Date	OT(UTC)	Lon	Lat	Depth (km)	Mw	Strike, Dip, Rake of NP1	Strike, Dip, Rake of NP2
14-Jan-20	20:16:07	107.33	-6.76	10	3.3	142, 52, 125	272, 49, 52
20-Mar-20	14:41:31	107.35	-6.74	4	3.6	126, 75, 138	229, 50, 20

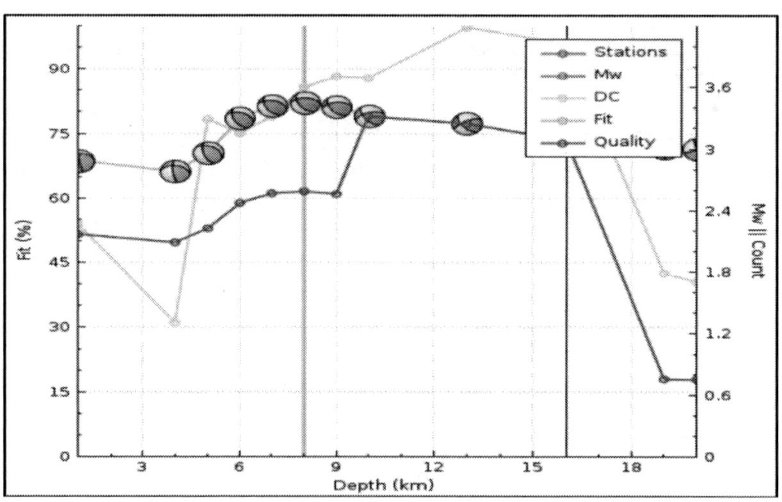

Figure 2 Moment tensor inversion result of the 14 January 2020 event. The plot of depth vs seismogram fitting shows the best result at 8 km depth and Mw3.3.

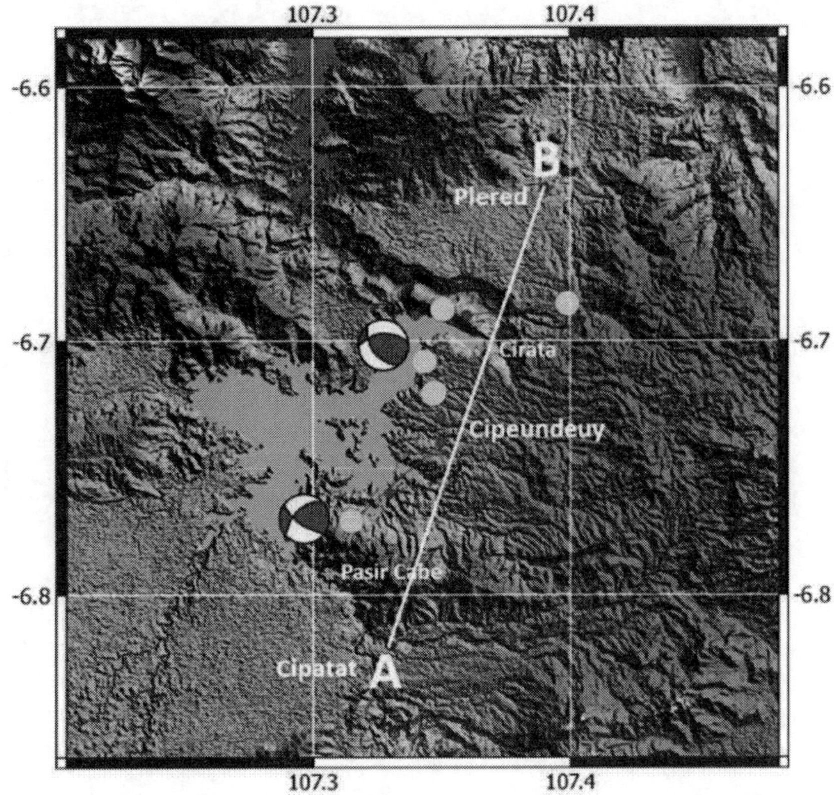

Figure 3 Seismic events distribution around Cirata Dam. Red dots are preliminary results. Yellows dots are relocated epicenters. Beach balls show focal mechanism of M3.7 and M3.5 events. The white line from A to B is section line in Figure 4. Topographic were drawn using DEMNAS 0.27 ArcSecond[16].

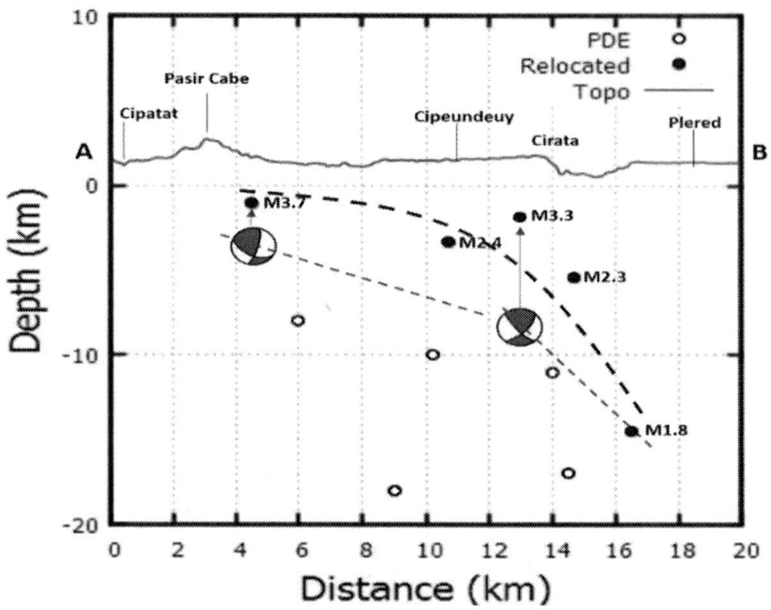

Figure 4 Vertical section of seismicity. Black circles are preliminary hypocenters. Black dots are relocated hypocenters. Beach balls showing focal mechanism of M3.7 and M3.5 events. Black dashed line is reconstructed fault plane interpolated from relocated hypocenters. Red dashed lines are fault plane extrapolated from focal mechanism. Topographic section (blue line) were drawn using DEMNAS 0.27 ArcSecond[16]exaggerated 5 times for its vertical scale.

Epicenters distributed in line from PasirCabe, Cipeundeuy, to Plered along with SSW – NNE direction (Figure 3). The fault plane solution from CMT inversion consists of a couple of planes that elongated to WSW – ENE direction. Therefore, the lineament of epicenters distribution does not represent strike direction of the fault because they areperpendicularto the fault strike. The mechanism of the Cirata Fault was different from the fault mechanism of the Cimandiri Fault as well as the Lembang Fault. The Cimandiri Fault around the Cianjur segment and the LembangFaul showing a left lateral movement[17]while the Cirata Fault shows thrust oblique mechanism.

Previous research showed that there were three microearthquake zones around the Cirata Dam with depth shallower than 10 km. The distribution of those microearthquakes indicated three fault zones, those are the Citarum-Cisomang Fault, the Saguling Fault, and the PasirCabe Fault. Those faults have the potential to generate an earthquake up to M 7.0[18].This study shows that relocated hypocenters below PasirCabe are shallower than hypocenters below Citarum-Cisomang Fault. On the other hand, PasirCabe could be the up-dip part while Citarum-Cisomang could be the down-dip part of the same fault plane. We then interpolated a fault plane along distributed hypocenters (Figure 4). The result is a dashed blackline that is going deeper to the NNE direction. We also extrapolated the fault line from two focal balls (Figure 4) represented by dashed red lines. The extrapolated lineare agree with the fault plane interpolated from hypocenters. Exceptthey are different in-depth, yet the both showing a similar trend and they are represented the plane ofCirata Fault. Based on the cross-section in Figure 4 the length of the interpolated line is around 16 km. This length represents the width of the fault. By using the empirical relation of earthquake and fault size [17]the width of the fault isproportional toMw6.6 earthquake. Magnitude estimation from this study is smaller than M 7 that approximated by the previous study.Even though a Mw 6.6 earthquake with shallow depth can also cause significant damaging.

4. Conclusions

Vertical relocated hypocenter distribution clearly shows a trend related to the structural geology of the region. This trend shows an inclined fault plane to the north direction. The interpretation is also supported by the reconstruction of the fault plane extrapolated from the focal mechanism. Seismic events occurred at the up-dip part, down-dip part, and in between along the fault plane. The seismicity of this region probablycaused by a single fault that is the Cirata Fault rather than three fault zones. Although this study has limitation on hypocenters data along strike direction, this study complements previous studies where the direction of the fault plane has not been resolved.

5. Acknowledgement

We are grateful to The Indonesian Agency for Meteorology, Climatology, and Geophysics (BMKG) supporting data earthquake catalog and waveforms. This research was also funded by *Komite KajianGempabumidan Tsunami BMKG* 2020.

References

[1] Dardji N, Villemin T and Rampnoux J P 1994 *Journal of Southeast Asian Earth Sciences* **9** 3–11
[2] Daryono M R, Natawidjaja D H, Sapiie B and Cummins P 2019 *Tectonophysics* **751** 180–91
[3] Natawidjaja D and Daryono M 2016 *Indonesia Proc. GEOSEA XIV Congress and 45th IAGI Annual Convention*
[4] Koulali A, McClusky S, Susilo S, Leonard Y, Cummins P, Tregoning P, Meilano I, Efendi J and Wijanarto A B 2017 *Earth and Planetary Science Letters* **458** 69–79
[5] Djaja D and Pratomo I 2017 *Jurnal Biologi Indonesia* **5**
[6] Bratt S and Nagy W 1991 *Science Applications International Corporation, San Diego*
[7] Kennet B 1991 IASPEI 1991 *TeNov* **3** 122–122
[8] Pesicek J, Thurber C, Zhang H, DeShon H, Engdahl E and Widiyantoro S 2010 *Journal of Geophysical Research: Solid Earth* **115**
[9] Widiyantoro S and Hilst R van der 1997 *Geophysical Journal International* **130** 167–82
[10] Widiyantoro S and van der Hilst R 1996 *Science* **271** 1566
[11] Kennett B L N, Engdahl E R and Buland R 1995 *Geophys J Int* **122** 108–24
[12] Ramdhan M, Priyobudi, Triyono R, Daryono, Heryandoko N and Imananta R T 2019 *J. Phys.: Conf. Ser.* **1341** 082010
[13] Minson S E and Dreger D S 2008 *Geophys J Int* **174** 585–92
[14] Friederich W and Dalkolmo J 1995 *Geophys J Int* **122** 537–50
[15] Dziewonski A M and Anderson D L 1981 *Physics of the earth and planetary interiors* **25** 297–356
[16] http://tides.big.go.id/DEMNAS/. Accessed 1August 2020
[17] Supendi P, Nugraha A D, Puspito N T, Widiyantoro S and Daryono D 2018 *Geosci. Lett.* **5** 31
[18] Soehaimi A 2008 *Indonesian Journal on Geoscience* **3** 227–240

IOP Conf. Series: Earth and Environmental Science **873** (2021) 012073 doi:10.1088/1755-1315/873/1/012073

Pull-Up Effect Correction and Oil In Place Sensitivity Test by Comparing Velocity Model Method in JAX-Field, Offshore North West Java

Muhammad Yusuf Ibrahim[1], Normansyah[2], Wien Lestari[1], Mariyanto Mariyanto[1]

[1] Geophysical Engineering Department, Institut Teknologi Sepuluh Nopember
[2] Pertamina Hulu Energi Offshore North West Java

Email : myusufibrahim@outlook.com

Abstract. The pull-up effect is the condition of lithology elevated in seismic imaging because of rapid seismic wave propagation through carbonate build-up on it. Pull-up effect conditions can lead to misinterpretation, so it needs to be corrected until the actual geological conditions are obtained. This research was conducted in the JAX-field working area of PT Pertamina Hulu Energi ONWJ. The target reservoirs of this study are in the Main (Upper Cibulakan) Formation under the Carbonate Parigi Formation. The reflectors of the target reservoirs show pull-up effect in time domain seismic data. Thus, building a velocity model for velocity anomaly correction is needed to reduce uncertainty for structure maps and oil in place calculation. The method of correcting the pull-up effect in this study uses three variations of the velocity model: variation structurally controlled model, variation RMS velocity with well control, variation calibrated RMS velocities model. The three variations of the velocity model result can correct the pull-up effect on JAX-Field. Velocity model with variation RMS velocity with well control had the lowest error with 17,31 feet average of depth difference with actual depth from well. Based on three velocity models, the value of original oil in place on the JAX-32 reservoir surface had a range of 59,14-84,59 mmbo, while on the JAX-35A surface has a range of 27,77-31,23 mmbo. These values can be considered in reserve calculation sensitivity.

1. Introduction

The seismic method is an excellent method for subsurface imaging. Seismic method for subsurface imaging has high accuracy and resolution in modelling the geological structure under the earth's surface. In certain cases, 3D seismic images have a pull-up effect that can cause miss interpretation. The pull-up effect is the lifting of lithology in seismic imaging due to the influence of the propagation speed of the build-up on it. The pull-up effect is not a real geological condition and is included in the pitfall/not real [1].

Depth conversion is a technique used to eliminate structural ambiguity inherent in the time domain and verify the structure and present it in a more precise geological senses in the depth domain. Geological and reservoir engineering studies are always in-depth domain, it enables the interpreter to integrate seismic depth with geological, petrophysical, and production data. There are two categories on the depth conversion method. First, the direct time depth conversion method, employs assumed velocities. Second, the velocity modelling method, uses more precise velocities of the subsurface layers, which are obtained by dividing the subsurface layers and defining each layer's velocity, as velocity changes with depth. Both methods can accurately tie wells and effectively predict depth[2].

This study area is in the Northwest Java Basin, more precisely in the Southern Ardjuna Sub-Basin (Figure 1). The South Ardjuna Sub-Basin has long been recognized as a major source area for ONWJ oil and gas fields [3]. In the Parigi Formation, there are carbonate build-ups that impacts a pull-up effect on seismic images. The occurrence of pull-up effect is due to the differences in velocities on shale and carbonate. The objective of this study is to understand how to correct the pull up effects that are formed by the Carbonate Build up on The Main and Massive Formation.

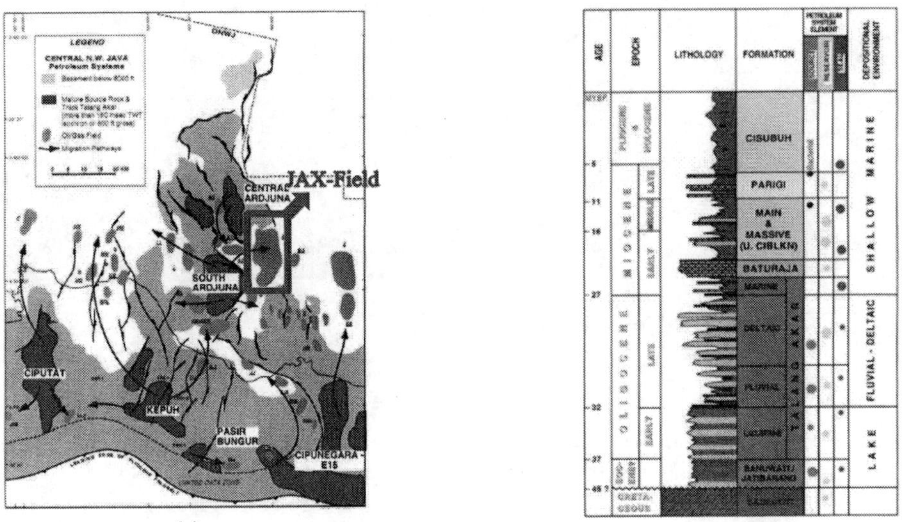

(a) (b)

Figure 1. Location Map of J-Field (a) and Stratigraphic Column (b), Offshore North West Java, Indonesia [3]

2. Data and Methodology

In this study, the data used by researchers are 3D seismic data, seismic surfaces, seismic V_{RMS}, well pick, and well logging data (13 wells). The log data used in this study area are sonic log, gamma ray log, density log, neutron porosity log, and checkshot data.

In this study, researchers made three velocity models that can be seen in the flow chart (Figure 2). Velocity modeling using Depth Team Express software requires multiple data inputs. In the V_{RMS} data a Dix Inversion will be used using the Dix Equation. The Dix Equation calculates interval velocities directly from stacking velocities. Assuming a flat layered velocity and structure model, the interval velocity is related to the stacking velocity by the Dix equation [4],[5]:

$$V_{intn} = \sqrt{\frac{V_n^2 T_n - V_{n-1}^2 T_{n-1}}{T_n - T_{n-1}}} \quad (1)$$

Where:

V_{intn} = interval velocity above the horizon n

T_n = TWT of the normal incidence ray at the horizon

V_n = RMS velocity for the horizon n

surfaces are used for guiding the interpolation of well TD curves, associating with well picks to create pseudo-velocities. Since velocities are often controlled or influenced by geological structure, the application of structural information for model building can often lead to a more accurate velocity model. Surfaces that used in build velocity modeling are Top Parigi, Base Parigi, Top JAX-27, JAX-32, Top Baturaja, and Base Baturaja.

Time to depth curves are converted into instantaneous velocity functions. These functions are then interpolated into a velocity volume. Functions can be interpolated linearly or along surfaces. A well pick is the depth value in a well that corresponds to a specific horizon or geological event. They typically are the most accurate source of depth data available to an interpreter. Well depth picks are associated with surfaces to create pseudo-velocities.

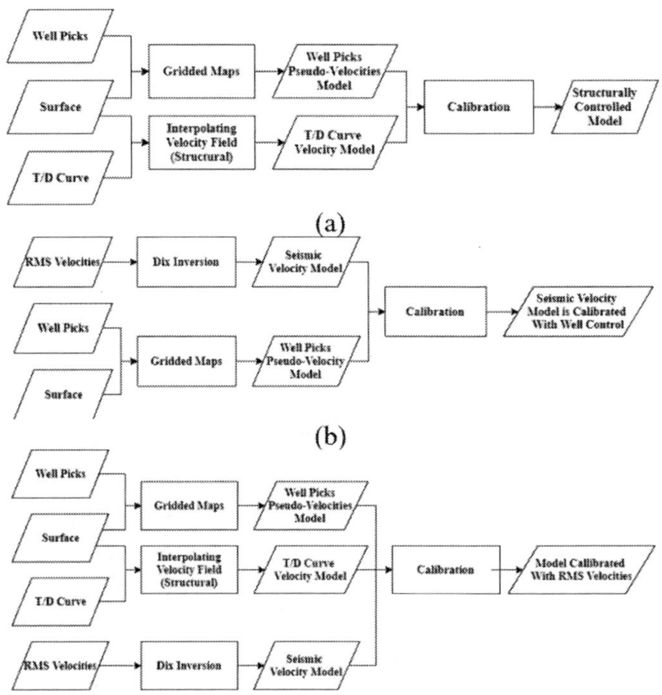

(c)

Figure 2. Velocity Model Development Flow Chart (a) Variation Structurally Controlled Model, (b) Variation RMS Velocity with Well Control, (c) Variation Calibrated RMS Velocities Model

After completing the velocity modeling, researchers do depth conversion on the surface reservoir (JAX-32 and JAX-35A) using TDQ software. Surface with depth domain will be used to calculate Original Oil In Place with the following equation:

$$OOIP = \frac{BRV \ x \ \phi \ x \ (1-Sw)}{Boi} \ x \ 7758 \quad (2)$$

Where,

OOIP	= Original Oil In Place (STB)	Sw	= Water Saturation (%)
7758	= conversion factor from acre/ft to barrel	Boi	= Oil formation volume factor
ϕ	= Porosity (%)		

3. Result and Discussion

The depth conversion results of the three variations in velocity model can be seen in Tables 2 and 3. The true vertical depth (Z) values in the JAX-32 and JAX-35A well picks will be compared with the surface results of depth conversion which will result in a residual value (the differences between the depth of well pick of each well with the depth value of depth domain surface). A positive residual value means that depth conversion results are shallower than well picks depth, while the negative residual value means depth conversion results are deeper than well picks. The calculation of depth residual values shows that the structurally controlled velocity model has the smallest average value for surface JAX-32, while the RMS velocity with well control velocity model has the smallest average value for surface JAX-35A.

Tabel 1. Depth Residual for JAX-32 Using Velocity Model With 3 Variation

Well	Well Picks	Surface Value Variation 1	Surface Value Variation 2	Surface Value Variation 3	Depth Residual Variation 1	Depth Residual Variation 2	Depth Residual Variation 3
J -1	-3321,16	-3354,10	-3344,13	-3357,47	-32,93	-22,97	-36,31
J - 11	-3268,79	-3296,47	-3291,38	-3309,74	-27,68	-22,59	-40,95
J - 15	-3154,90	-3081,29	-3064,91	-3081,41	73,61	89,99	73,49
J - 16	-3318,97	-3277,18	-3288,47	-3298,98	41,79	30,50	19,99
J - 19	-3368,19	-3297,43	-3312,24	-3298,04	70,76	55,95	70,15
J - 3	-3130,49	-3075,80	-3050,80	-3057,47	54,69	79,69	73,02
J - 9	-3302,73	-3254,12	-3238,99	-3253,99	48,61	63,73	48,74
JH - 1	-3186,94	-3116,21	-3098,96	-3108,70	70,73	87,98	78,25
JJ - 1	-3298,19	-3311,93	-3302,84	-3316,48	-13,74	-4,65	-18,29
JK - 1	-3097,25	-3022,44	-3003,42	-3014,26	74,82	93,84	82,99
JL - 1	-3342,52	-3325,33	-3313,15	-3328,18	17,19	29,37	14,34
JM - 1	-3353,75	-3335,97	-3300,50	-3315,40	17,77	53,25	38,34
JN - 2	-3329,79	-3350,44	-3356,54	-3358,79	-20,65	-26,74	-29,00
JN - 4	-3317,87	-3207,31	-3187,94	-3186,55	110,56	129,93	131,32
JNA - 1	-3329,59	-3332,47	-3319,34	-3324,17	-2,88	10,25	5,42
				Average	32,18	43,17	34,10

Tabel 2. Depth Residual for JAX-35A Using Velocity Model With 3 Variation

Well	Well Picks	Surface Value Variation 1	Surface Value Variation 2	Surface Value Variation 3	Depth Residual Variation 1	Depth Residual Variation 2	Depth Residual Variation 3
J -1	-3669,11	-3771,43	-3765,19	-3773,95	-102,32	-96,08	-104,84
J - 11	-3625,68	-3619,41	-3619,87	-3635,85	6,26	5,81	-10,17
J - 16	-3699,45	-3636,24	-3655,06	-3659,69	63,21	44,39	39,76
J - 19	-3784,61	-3773,83	-3787,30	-3771,90	10,78	-2,69	12,71
J - 3	-3478,93	-3506,07	-3482,02	-3480,62	-27,14	-3,09	-1,69
J - 9	-3644,51	-3661,47	-3655,28	-3661,53	-16,96	-10,77	-17,02
JF - 1	-3402,04	-3344,32	-3323,30	-3332,10	57,72	78,73	69,94
JH - 1	-3502,36	-3570,50	-3549,99	-3557,60	-68,14	-47,63	-55,24
JJ - 1	-3681,36	-3703,91	-3704,90	-3709,59	-22,55	-23,53	-28,23
JK - 1	-3430,93	-3388,50	-3373,93	-3376,08	42,43	57,00	54,86
JL - 1	-3704,13	-3755,96	-3746,96	-3758,64	-51,83	-42,83	-54,50
JM - 1	-3761,60	-3822,18	-3774,71	-3795,63	-60,57	-13,11	-34,03
JN - 2	-3672,59	-3795,01	-3787,65	-3795,36	-122,42	-115,06	-122,77
JNA - 1	-3686,56	-3786,84	-3760,11	-3772,16	-100,28	-73,55	-85,60
				Average	27,99	17,31	24,06

On the JAX-32 and JAX-35A surfaces in time domain (Figure 3.a and 4.a), researchers predicted two pseudo anticlines due to carbonate build-up in the Parigi Formation (marked with a red circle). If we look at the depth conversion results using three variations of velocity model, the area of the alleged pull up effect had been fixed because the velocity anomaly in the carbonate build up is included in the making of velocity model.

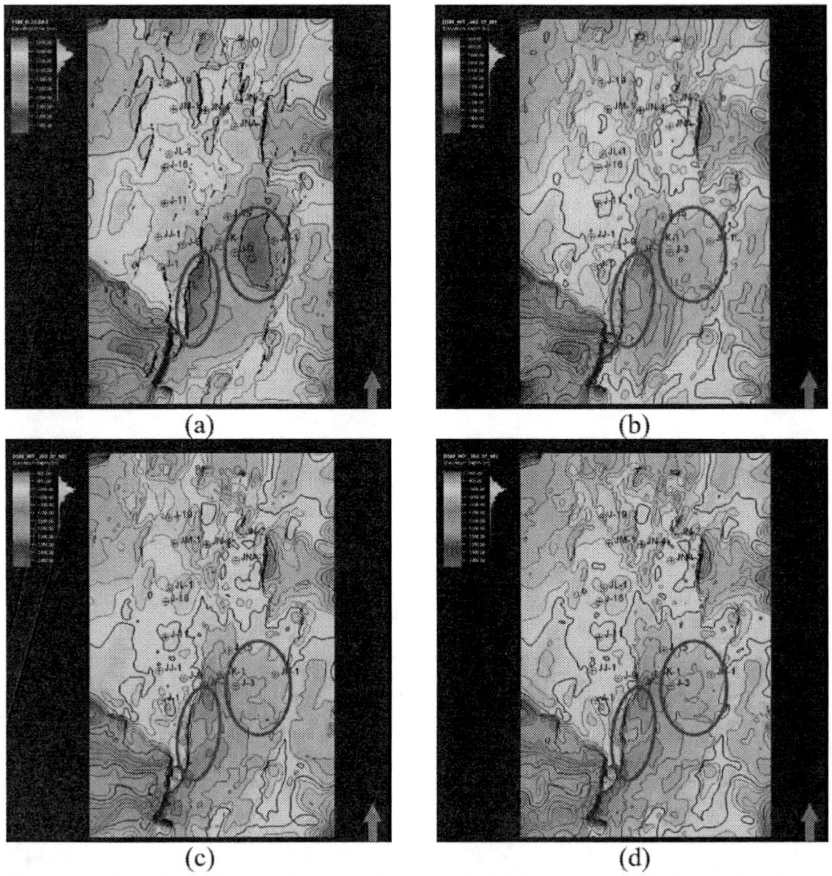

Figure 3. Surface and Result of Depth Conversion at JAX-32 (a) JAX-32 Time Surface, (b) Variation Structurally Controlled Model, (c) Variation RMS Velocity with Well Control, (d) Variation Calibrated RMS Velocities Model

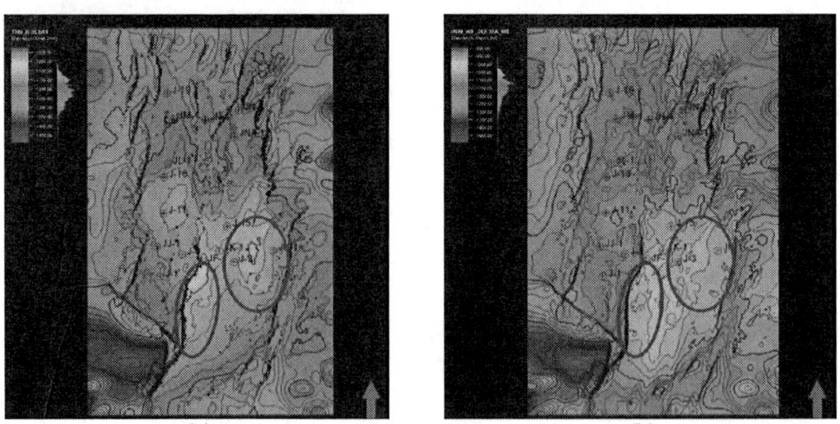

(a) (b)

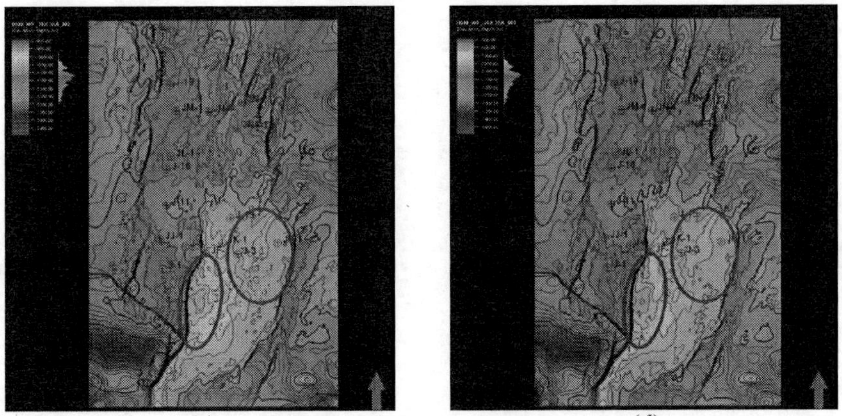

(c) (d)

Figure 4. Surface and Result of Depth Conversion at JAX-35A (a) JAX-35A Surface, (b) Variation Structurally Controlled Model, (c) Variation RMS Velocity with Well Control, (d) Variation Calibrated RMS Velocities Model

In Figure 5 is a comparison graph of the original oil in place (OOIP) value using equation 2 on surfaces JAX-32 and JAX-35A. OOIP calculations use porosity values of 0,18, water saturation of 0,5, and Oil formation volume factor of 1,24. It can be noticed that the largest OOIP value on the JAX-32 surface uses depth conversion results from velocity model variation RMS velocity with well control 84.59 mmbo, while on the JAX-35A surface the largest OOIP value on the result of depth conversion using velocity model variation calibrated RMS velocities model is 31.23 mmbo. On the JAX-32 surface, the differences in OOIP values of the three variations is quite fluctuating compared to the JAX-35A surface, most likely due to the position of JAX-35A is deeper, so that the pull-up effect is smaller than JAX-32, where velocity anomaly decreases with increasing depth.

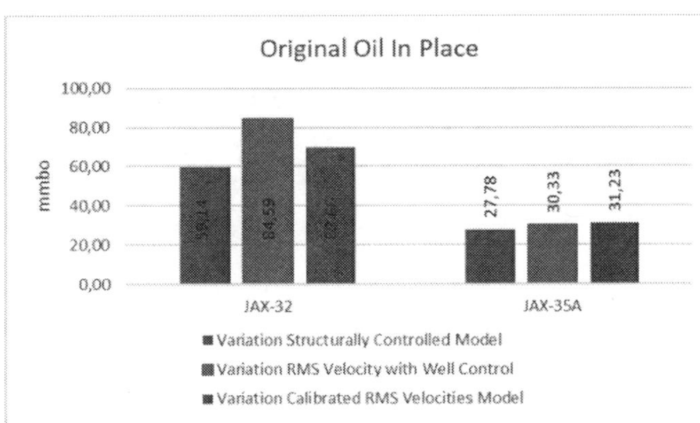

Figure 5. Original Oil In Place Calculation Results

4. Conclusions

This study explained the three variations of velocity model that can be applied to filter the pull-up effect on JAX-Field with range of average error 17,31 – 43,17 feet. In addition, the result defined the sensitivity value of the OIIP around 59,14 – 84.59 mmbo on JAX-32 while on JAX-35A has a range of 27.77-31.23 mmbo. The OOIP range on the JAX-32 had the greater value due to the position of the JAX-32 above the JAX-35A, where the velocity anomaly decreases with increasing depth.

IOP Conf. Series: Earth and Environmental Science **873** (2021) 012073 doi:10.1088/1755-1315/873/1/012073

References

[1] Farline H H, Muksin U and Syukri M 2017 Menghilangkan Efek Pull-up pada Citra Seismik Refleski PSTM Menggunakan Velocity Model pada Lapangan 'MAR', Blok Offshore North West Java **1** 8
[2] A.O S, T D and Jephter D I 2018 Velocity Modelling Using Well Data: Depth Conversion "A Case Study Of K-Field, Onshore Niger Delta Area" *IOSR Journal of Applied Geology and Geophysics* **6**
[3] Noble R A 1997 Petroleum systems of Northwest Java, Indonesia *Proceedings of an International Conference on Petroleum Systems of SE Asia and Australasia* International Conference on Petroleum Systems of SE Asia and Australasia (Indonesian Petroleum Association (IPA))
[4] Dix C H 1955 Seismic Velocities from Surface Measurements *GEOPHYSICS* **20** 68–86
[5] Hubral P and Krey T 1980 Reference *Interval Velocities from Seismic Reflection Time Measurements* General Series (Society of Exploration Geophysicists) pp 199–202

Determination of Earthquake Hypocenter Distribution using the Modified Joint Hypocenter Determination (MJHD) Method throughout the Palu-Koro Fault (Case Study: August-October, 2018)

Dewi Ayu Swastika[1], Harmita Lestari[2]*, Aulia Puji Astuti[1], Sabrianto Aswad[1], Muhammad Fawzy Ismullah Massinai[1]

[1]Department of Geophysics, Hasanuddin University, Makassar, Indonesia
[2]Geophysical Engineering, Faculty of Mining and Petroleum Engineering, Bandung Institute of Technology, Indonesia

*email: harmitalestarymvd@gmail.com

Abstract. The area of Sulawesi, especially along the Palu Koro Fault, is an area that is largely influenced by the confluence and movement of plates as well as regional fault activity pathways with high levels of seismicity. Determining the location of the hypocenter accurately through relocation is required in identifying the detailed tectonic structures in the area. Relocation of the hypocenter using the Modified Joint Hypocenter Determination (MJHD) method using the IASP91 velocity model in the period August to October 2018 with the arrival time data from BMKG catalog. The results of hypocenter relocation using the MJHD method show that from 132 earthquake distribution points to 63 earthquake hypocenter points after the relocation. The change in the location of the hypocenter was much denser along the Palu Koro Fault route than before the relocation as evidenced by the mean value of rms (root mean square) before relocation was 1.31 and after relocation it became smaller (0.61). Changes in parameter values after relocation using the MJHD method caused the distribution of the earthquake hypocenter to be tighter towards the Palu Koro fault than before the relocation, where the distribution had a random and scattered pattern.
Keyword: The Palu Koro Fault; Modified Joint Hypocenter Determination (MJHD); Relocation; Hypocenter; Earthquake

1. Introduction

Sulawesi has two important events that occurred in western Sulawesi at the time kenozoic. The first is rifting and the expansion of the ocean floor in the Strait Makassar in Paleogene, which creates space for the deposition of clastic material originating from Kalimantan. The second is a compressional event started since the Miocene which was influenced by the collision of continents in the west and ofiolite and island arc fragments to the east. Fragments these include the Buton micro-continent, Blacksmith and Baggai Sula. This compression resulting in the West Sulawesi Fold Belt developed in the Early Pliocene. And the effect of the coalition is believed to be the cause of tectonic events in all parts of Sulawesi [1].

In the active regional faults, it is found that the active subduction are still in the Sulawesi area which can trigger activity on large-scale horizontal faults, which are the zone from the source of earthquakes [2]. One of them is along the Palu Koro Fault which is a horizontal fault (strike-slip fault) because it is close to the collision zone of three plates, namely the Eurasian, Indo-Australian and Pacific plates as well as the Philippine micro plate [3].

There are several techniques are developed for relocating the hypocenter. Modified Joint Hypocenter Determination method has been applied in this research. The Modified Joint Hypocenter Determination (MJHD) method inversing the travel time from the group of earthquakes and the amount of station correction used to obtain a hypocenter location is much better than the initial estimate. So that it can provide improvements of errors due to variations in lateral velocity in the previous method.

The Modified Joint Hypocenter Determination (MJHD) method is a development of the previous method, the Joint Hypocenter Determination (JHD) method. The MJHD method simultaneously inverts the travel time of a group of earthquakes and the magnitude of the station correction to get a better hypocenter location. If the station distribution is uneven and the subsurface medium is very heterogeneous, the solution to overcome this problem is to use the MJHD method. The MJHD method is still more profitable because it does not require a major earthquake. This is very effective, especially in cases where no earthquake is well recorded by all stations [8].

1.1 Geological Setting of Central Sulawesi

The Palu region is one of the most seismically active areas in Sulawesi. With the high level of seismic activity in the region. The zone of collision from the junction of these three plates is convergent and colliding in a relative manner and has resulted in the Central Sulawesi and its surroundings becoming one of the areas with a high level of seismicity [4]. The Palu Koro fault, which is active due to pressure from the Flores Sea in the south and at the northern end through the Makassar Strait which intersects the subduction zone of the Sulawesi Sea plate, extends from north of Palu to the south of Malili and ends in Bone Bay along ± 459 km, moving synistrally to left with an average slip of about 32-45 mm/year, and it is a strike slip fault [5].

1.2 Modified Joint Hypocenter Determination (MJHD) Method

The hypocenter is the epicenter of the earthquake which is inside the earth's surface. Generally, the determination of the location of a hypocenter is generated using the results from the seismic phase observations. These results are recorded by three components, namely N-S, E-W, and up-down at each different station [6]. The determination of the hypocenter for earthquake relocation using the MJHD method was first introduced by [7]. The principle is the development of the JHD method, so the algorithm is almost the same. The only thing that distinguishes it is the added limitation to the final flow of the MJHD method. By adding the station correction price, the travel time residual obtained at station i, namely stasiun can be written as follows:

$$r_{ij} = T_{ij}{}^{obs} - (T_{ij}{}^{cal} + s_i) \tag{1}$$

$T_{ij}{}^{o}$ is the travel time of the seismic waves from the epicenter to the station obtained, $T_{ij}{}^{cal}$ is the calculated travel time and s_i is the station correction. With the same assumptions as the Geiger method, equation (1) can be linearised using the Taylor first order expansion as follows:

$$\begin{aligned} r_{ij} &= (t_{ij} - To_{ij}) - T_{ij} \\ &= (\frac{\partial T}{\partial xj}\, dxj + \frac{\partial T}{\partial yj}\, dyj + \frac{\partial T}{\partial zj}\, dzj + dT\, 0j + dsi \end{aligned} \tag{2}$$

Where t_{ij} dan T_{ij} are arrival times and travel times calculated from the occurrence of an earthquake of j at a station of i. It is known that i = 1,…, N indicates the number of observation stations and j = 1,…, M indicates the number of earthquake occurrences. If equation (2) is arranged into a matrix for all stations, the matrix equation is as follows [8]:

$$r_{ij} = A_j\, dX_j + S_j\, d_{si}\,;\; j = 1, ..., M \tag{3}$$

Where r_{ij} is the residual travel time of each earthquake, A_j is the N x 4 matrix of partial derivatives calculated at the source, dX_j is the vector of pertubation or the change in the hypocenter parameter to the predicted (initial) parameter, S_j is the N x N diagonal matrix with the station recording the earthquake earth and d_{si} are station corrections. By adding the station correction price, the travel time residual obtained at station i, namely r_{ij} can be written as follows:

$$r_{ij} = t_{ij obs}(t_{ij cal} + s_i) \tag{4}$$

t_{ij}^{obs} is the travel time of seismic waves from the epicenter to the station which is obtained from the difference between the arrival time of the earthquake waves and the origin time, t_{ij}^{cal} is the calculated travel time and s_i is the station correction. If equation (2) is arranged into a matrix for all stations, the matrix equation is as follows [8]:

$$r_{ij} = A_j dX_j + S_j ds_i; j = 1, ... M \tag{5}$$

Where r_{ij} is the residual travel time of each earthquake, A_j is the N x 4 matrix of partial derivatives calculated at the source, dX_j is the vector of pertubation or the change in the hypocenter parameter to the predicted (initial) parameter, S_j is the N x N diagonal matrix with the station recording the earthquake earth and d_{si} are station corrections. In the iteration of the JHD and MJHD methods, there are 2 ways, the first is calculating the station correction (d_{si}), the second is calculating the hypocenter using station corrections from the first method. Earthquake parameters (x, y, z, t) and station corrections obtained from the previous iteration. So that this is what ultimately becomes the initial parameter for the next iteration [8]. To calculate the value d_{si}:

$$d_{si} = \frac{1}{\sum_{j=1}^{M} w_{ij}^2} \sum_{j=1}^{M} w_{ij}^2\, r_{ij} \tag{6}$$

In the MJHD method, a limitation is added to improve the stability of the JHD results which implies that the station correction used does not depend on the distance and azimuth angle between the center of the study area and the station used. The limitations used in the MJHD method are:

$$\sum_{i=1}^{n} \mathrm{SiDi} = 0$$
$$\sum_{i=1}^{n} \mathrm{Si}\cos\theta i = 0,\; \sum_{i=1}^{n} \mathrm{Si}\sin\theta i = 0,\; \sum_{i=1}^{n} \mathrm{Si} = 0 \tag{7}$$

With S_i is the station correction at the i-th station, D is the distance between the station and the center of the study area, θ_i is the azimuth angle of station i from the regional center and n is the number of observation stations. Determination of the root mean square error (E_{RMS}) with the following equation [9]:

$$E_{RMS} = \sqrt{\frac{1}{N}\sum_{i=1}^{N}(ti_{cal} - ti_{obs})^2} \tag{8}$$

i is the station in one earthquake event and N is the number of stations in one earthquake event. E_{RMS} does not completely improve the data used, but to find out whether the size of the E_{RMS} value approaches the research data model or away from the data model with comparisons from the actual data in the field. The Modified Joint Hypocenter Determination (MJHD) method is more profitable because it does not require a major earthquake. This is very effective, especially in cases where there are no earthquakes that are well recorded by all stations used in the study [7].

IOP Conf. Series: Earth and Environmental Science **873** (2021) 012074 doi:10.1088/1755-1315/873/1/012074

2. Data and Method

2.1 Research Location Overview

This research is located along the Palu Koro Fault, which is at the boundary coordinates of 2° N - 3° S and 119° - 121° E. The location of this research is bordered by East Kalimantan to the west, Sulawesi Sea to the north, Maluku Sea and Seram Sea to the east, and South and Southeast Sulawesi to the south.

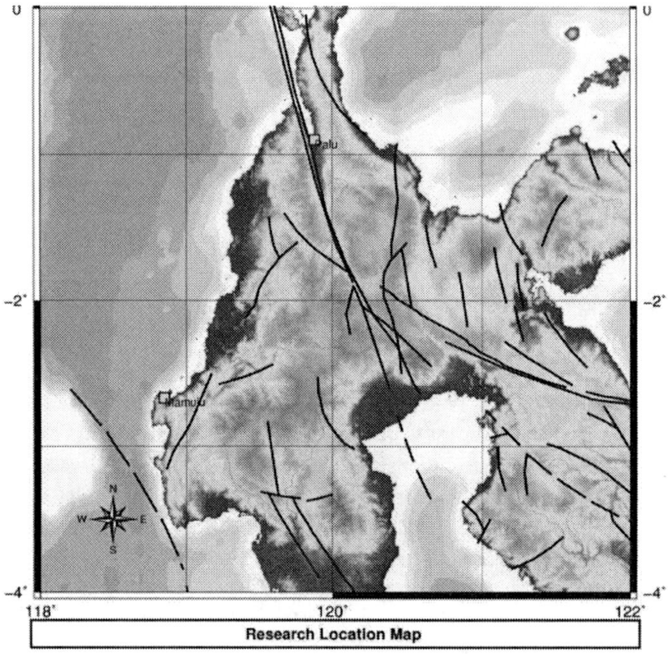

Figure 1 Research Location Map around the Palu Koro Fault, at the coordinates of 2°N-3°LS and 119°E-121°E.

There are 27 total stations used in the calculation and into 21 corrected stations (APSI, BBSI, BKSI, BNSI, BSSI, KMSI, MPSI, MRSI, PMSI, SGKI, SMKI, SMSI, SPSI, SRSI, TMSI, BKB, LUWI, SANI, TOLI2, KAPI, GTOI) after relocation using the MJHD method.

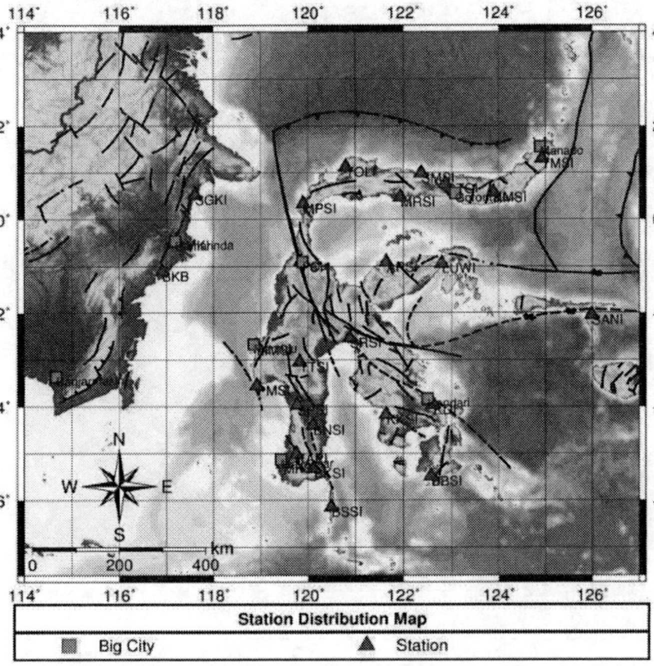

Figure 2 Station Distribution Map, there are a total of 27 stations used and ended up being 21 stations that were corrected

Hypocenter relocation is carried out in several steps, namely taking secondary data from the BMKG earthquake catalog in the form of BMKG arrival time data for the August-October 2018 period to get the final result using the Modified Joint Hypocenter Determination relocation method on linux with the Fortran programming language. The final results of hypocenter relocation in the form of time correction, longitude latitude correction and RMS were processed using the MJHD method. The final result is then plotted and map making using GMT (Generic Mapping Tools) [10] as interpretation and drawing conclusions.

3. Results and Discussion

The MJHD method is expected to minimize the error value so that the hypocenter position is much more accurate. Earthquake relocation data processing uses the MJHD method using earthquake data from the BMKG catalog, namely arrival time data with the IASP91 velocity model. The MJHD method selects earthquakes based on the parameters of the number of MEQ (minimum number of recorded earthquakes), MNST (minimum number of stations recorded), maximum earthquake depth, and maximum residual value determined at the beginning of relocation, so that not all earthquakes can eventually be relocated. The position of the earthquake distribution generated by the MJHD method is considered to be more accurate and fairly dense compared to the previous position which was random and scattered. This is also supported by changes in the rms value or position after the relocation which tends to be better (figure 3).

IOP Conf. Series: Earth and Environmental Science **873** (2021) 012074 doi:10.1088/1755-1315/873/1/012074

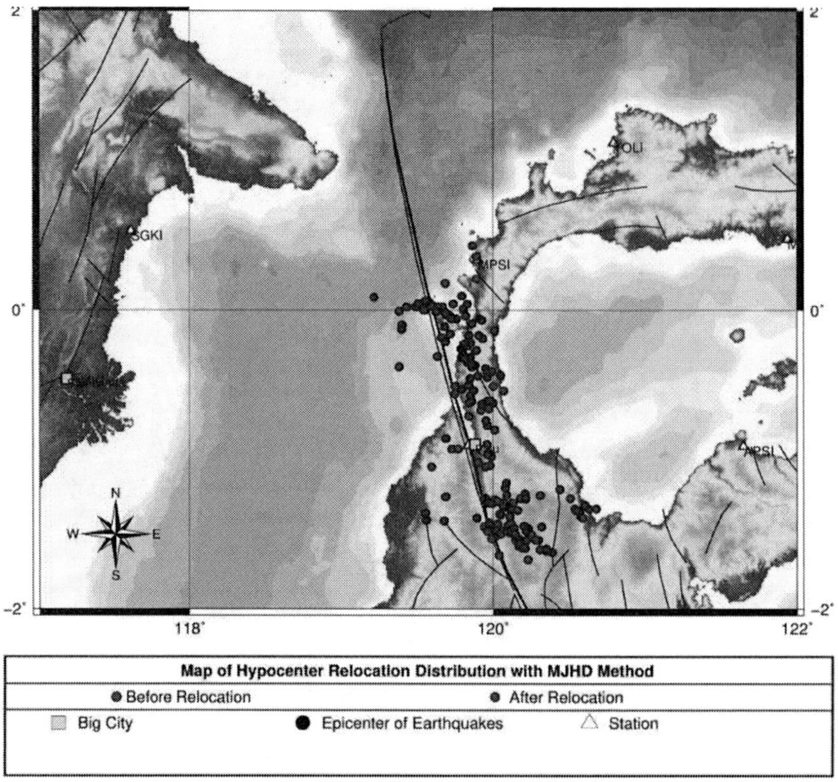

Figure 3 Map of hypocenter relocation before (red) and after (blue) with the MJHD method

In this study, there were a total of 132 earthquake points that were relocated using the MJHD method and turned into 63 earthquake points after relocation with each parameter value used in determining MEQ and MNST being 10. The determination of each parameter value in this study is carried out in stages from the smallest number to obtain a suitable value because these values produce the most ideal and stable results compared to the values before and after the experiment. The relationship between MEQ and MNST with the number of earthquakes relocated is inversely proportional. The greater the MEQ and MNST values, the less earthquakes will be relocated. Because the earthquake was recorded in many stations that will be relocated. Therefore it is necessary to pay attention to the value of MEQ and MNST in order to get ideal results. The ZFIX value or depth is 200 km. The max-ress or residual travel time is set at 2 and the iteration value in 5 consecutive iterations during the MJHD method process is 6,7,8, 9 and 10. With the initial distribution of stations in the study amounting to 27 stations and ending up to 21 stations after relocation. The RMS value obtained before and after relocation is 1.31 before relocation and becomes 0.61 after relocation using the MJHD method.

4. Conclusions

The distribution of the earthquake distribution before the relocation was a total of 132 points and became 63 hypocenter points after relocation using the method. The result of the hyposenter points depending on the input parameters in the form of the number of MEQ, MNST, maximum earthquake depth, and maximum residual value determined at the beginning of the relocation.

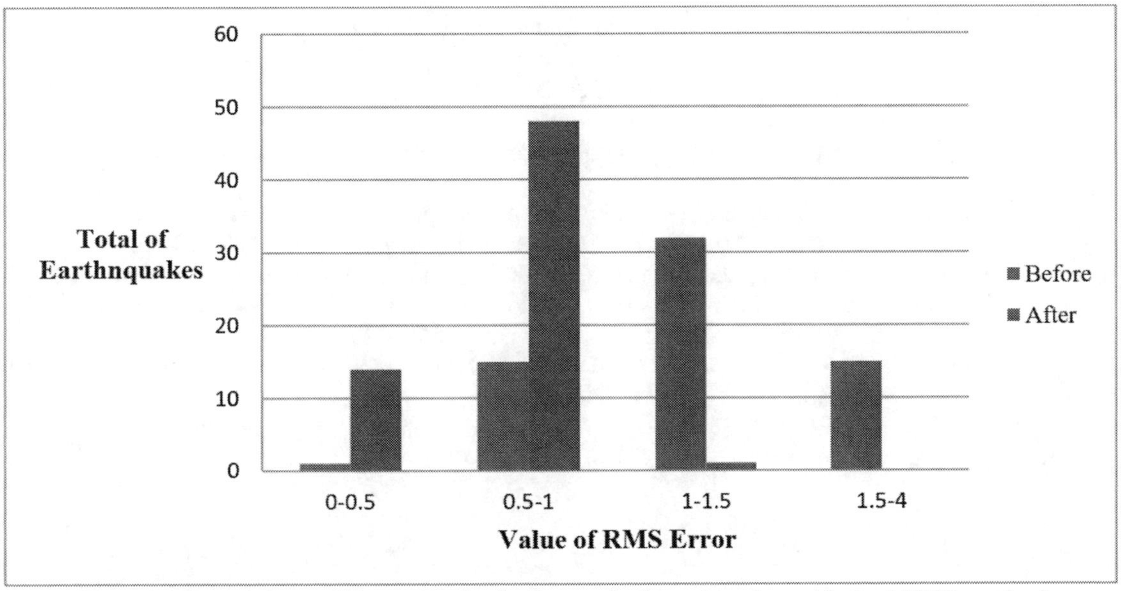

Figure 4 Comparison of the rms value before and after relocation with the MJHD method

MJHD with the Fortran Linux programming language. Changes in parameter values after relocation are origin time, longitude, latitude, depth, magnitude and rms value. The results of the calculation of relocation before and after the earthquake with the greatest magnitude at the time of the incident in the research time period before relocation have sequential parameter values, namely 10:02:43, 119.85, -0.22, 10 km, 7.48 (Mw) and 0.971. The values after relocation were 10: 02.46, 119.84, -0.17, 12.28 km, 7.4 (Mw) and 0.534. The average rms value before relocation was 1.31 and after relocation it was smaller, namely 0.61.

To get a better relocation value, the local velocity model is used in the study area so that the results obtained will be much more accurate. Because in accordance with the existing speed models in the area. We recommend that the earthquake recording station data and earthquake data be used relatively more so that the results of the relocation are much better and able to describe the seismicity of the research area.

Acknowledgment

We would like to thank the Indonesian Meteorology, Climatology and Geophysics Agency (BMKG Region IV, Makassar) for helping and allowing in obtaining the arrival time data used in this research. The maps that used in this research was done using Generic Mapping Tools (GMT) [10].

References

[1] Aswad S 2010 Relokasi Gempa Vulkanik Kompleks Gunung Guntur Menggunakan Algoritma Double Difference *Karya Tulis* Institut Teknologi Bandung.

[2] Calvert S J and Hall R 2003 The Cenozoic Geology Of The Lariang And Karama Regions, Western Sulawesi: New Insight Into The Evolution Of The Makassar Straits Region *Proceeding 29th* Indonesian Petroleum Association.

[3] Koesnama 2014 Horizontal Magnification and Active Support Zone in Sulawesi: The Relationship with Seismicity *JGSM* **15** 75-79.

[4] Watkinson I and Hall R 2011 The Palu-Koro and Matano Faults, Sulawesi, Indonesia: Evolution of an Active Strike-Slip Fault System *Geophysical Research Abstract* **13**, EGU2011-8270 United Kingdom.

[5] Daryono 2011 Tectonic Arrangement and History of the Palu Seismicity, Central Sulawesi.

Geophysical Articles of the Agency for Meteorology, Climatology and Geophysics

[6] Irsyam M et al 2010 *Summary of Study: Development of Seismic Hazard Maps of Indonesia for Revision of Hazard Map* in SNI 03-1726-2002 Bandung.

[7] Havskov J and Ottomoller L 2010 *Routine Data Processing In Earthquake Seismology with sample Data, Exercises and Software* Department of Earth Science University of Bergen. New York

[8] Hurukawa N 1992 Subducting Oceanic Crust of Philippine Sea and Pacific Plates and Weak Zone Normal Compression in Kanto District, Japan *Geophys. J. Int.* **109** 639-652.

[9] Kim K H, Chiu J M, Pujol J and Chen K C 2005 Earthquake relocations, fault zone geometry and constraints on lateral velocity variations using the joint hypocenter determination method in the Taiwan area *Earth Planets Space* **57** 809–823

[10] Grandis H 2009 *Introduction to Geophysical Inversion Modeling* Indonesian Geophysicists Association Jakarta.

[11] Wessel P, Smith W H F, Scharroo R, Luis J, and Wobbe F 2013 Generic Mapping Tools: Improved Version Released *EOS Trans. AGU*, **94** 409–410.

IOP Conf. Series: Earth and Environmental Science **873** (2021) 012075 doi:10.1088/1755-1315/873/1/012075

The Correlation Analysis of Magnetic Susceptibility and Elemental Composition of Very Fine Sand from Anoi Itam Beach in Weh Island, Aceh

T G Pitaloka[1*], S Bijaksana[1], S J Fajar[1], R Nathasa[1] and Z Masrurah[2]

[1] Geophysical Engineering, Faculty of Mining and Petroleum Engineering, Institut Teknologi Bandung, Jalan Ganesa 10, Bandung 40132, Indonesia
[2] Geophysical Engineering, Faculty of Engineering, Universitas Syiah Kuala, Jalan Teuku Nyak Arief 441, Banda Aceh 23111, Indonesia

[*]Corresponding author: taniapita03@gmail.com

Abstract. Ironsand deposits might contain Fe as well as other valuable elements and minerals that could be used in a variety of applications. Often, high Fe content deposits are preferable for exploration. An earlier study shows that the highest Fe content is found in the very fine sand (VFS) size. In this study, seven VFS samples from Anoi Itam were subjected to magnetic measurements as well as X-Ray florescence (XRF), X-Ray diffraction (XRD), and correlation analyses to investigate further characteristics and how magnetic susceptibility correlates with the elemental composition of ironsand. Magnetic susceptibility varies from 2207.77×10^{-8} m^3 kg^{-1} to 4476.68×10^{-8} m^3 kg^{-1}. The main elements contained in the sample are Fe, Ti, Si, and Al. Meanwhile, other elements have small concentrations (<2%). Based on XRD analyses, magnetite and ilmenite are the main minerals with varying concentrations in each sample. The correlation analysis shows that magnetic susceptibility has a weak correlation with Fe probably because Fe forms minerals with very different magnetism, namely magnetite and ilmenite.

1. Introduction

Indonesia has abundant natural resources as a result of volcanic eruptions, one of which is ironsand. Ironsand contains valuable Fe and mineral that can be utilized for various applications. This matter has caused many parties to carry out mining activities. Therefore, the mining process must be selective to obtain potential ironsand. To find out the distribution of ironsand characteristics, research is needed in various regions. Research on ironsand has been carried out, one of which is in Aceh Province [1,2]. Previous studies have shown that the highest Fe content was found in Very Fine Sand (VFS) grain size [3]. In this paper, VFS samples from Anoi Itam Beach become research subjects using the integration of magnetic susceptibility methods, X-Ray Fluorescence (XRF), X-Ray Diffraction (XRD), and rank Spearman correlation analysis to determine the characteristics and correlation between magnetic susceptibility and elemental composition.

2. Geological Setting

Anoi Itam Beach is located in Gampong Anoi Itam, Sukajaya District, Sabang City, Weh Island, Aceh. This beach is located at the coordinates 5°50'15.0" N and 95°22'26.8" E. One of the mountains on Weh Island is Mount Jaboi. Mount Jaboi is a type C volcano and the only mountain that still shows volcanic activity to date on Weh Island. The rocks that make up Weh Island are shown in Figure 1, which consists

IOP Conf. Series: Earth and Environmental Science **873** (2021) 012075 doi:10.1088/1755-1315/873/1/012075

of Alluvium (Qh) and Weh Island Volcano Rocks (QTvw). The alluvium consists of gravel, sand, and mud which is the result of the scraps of various old and young rocks with silt to crust sizes that are deposited on plains such as river estuaries and beaches. Weh Island Volcano Rock (QTvw) is the bedrock that forms Weh Island which consists of andesite-basalt, tuff, and agglomerates. This unit is generally dark gray to blackish. Besides, tuffaceous sandstones are exposed in the Anoi Itam area with a graded bedding structure, coarsely grained to fine, and are thought to be Miocene in age.

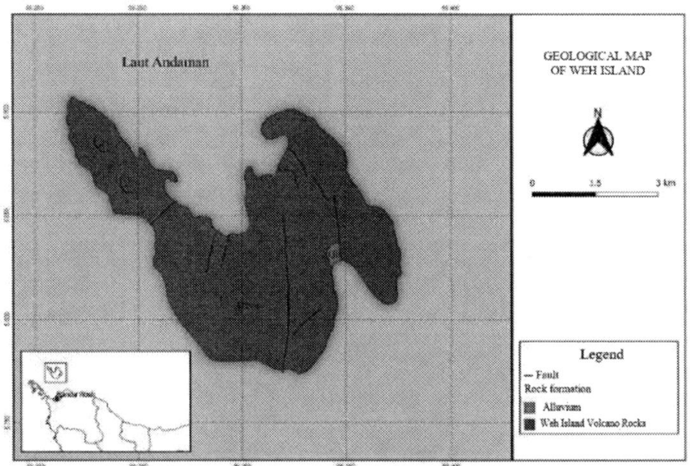

Figure 1. Geological map of the study area (modified from [4]).

3. Experimental Methods

Seven points taken with codes SA-01, SA-02, SA-03, SA-04, SA-05, SA-06, and SA-07 is shown in Figure 2. Ironsand was prepared by sieving at 120 mesh size to get VFS grain. Then, each point was inserted into three holders. Twenty-one samples of VFS ironsand tested for frequency-based magnetic susceptibility using a Bartington MS2 susceptibility meter with two different frequencies, namely 0.46 kHz for low frequencies (χ_{lf}) and 4.6 kHz for high frequencies (χ_{hf}). Seven samples with the most representative magnetic susceptibility values at each point were smoothed using a pestle and mortar and then sieved with a size of 230 mesh for XRF and XRD testing. X-Ray Fluorescence is measured using Supermini 200 X-Ray Fluorescence to determine the elemental and mineral oxide content. X-Ray Diffraction is measured using SmartLab X-Ray Diffraction to determine the mineral content contained in the sample. Furthermore, the Spearman rank correlation analysis was carried out to determine the correlation between element composition and magnetic susceptibility with reference values divided into five parts [5].

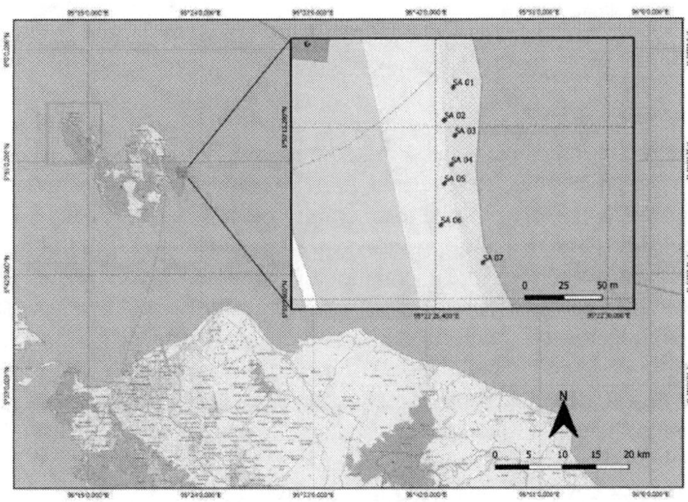

Figure 2. Map of sampling locations on Anoi Itam Beach.

4. Results and Discussion

4.1. Magnetic Susceptibility and Element Composition Characteristics

Based on Table 1, the mean value of χ_{lf} varies between $2207.77 \times 10^{-8}m^3kg^{-1}$ to $4476.68 \times 10^{-8}m^3kg^{-1}$ with $\chi_{fd}\%$ values ranging from 6.65% to 9.57% that indicates VFS ironsand samples have a mixture of superparamagnetic (SP) grain content and not SP [6]. Besides, based on XRF results, the VFS ironsand sample contains the main elements, namely Fe, Ti, Si, and Al as well as the mineral oxide content of them. Other oxide minerals and elements contain small concentrations (<2%). Meanwhile, XRD testing showed that the minerals contained in the samples were quite diverse, with the main minerals being magnetite and ilmenite.

Table 1. Measurement result of magnetic susceptibility of VFS samples.

Codes	Average χ_{lf} ($\times 10^{-8}m^3kg^{-1}$)	Average χ_{hf} ($10^{-8}m^3kg^{-1}$)	Average $\chi_{fd}\%$
SA-01	3648.80	3360.21	7.90
SA-02	3974.35	3708.90	6.65
SA-03	2207.77	2037.50	7.74
SA-04	3949.91	3669.10	7.16
SA-05	2757.55	2561.47	7.06
SA-06	4056.85	3663.92	9.57
SA-07	4476.68	4125.00	7.86

From Table 1, it can be seen that from point SA-01 to SA-07, average χ_{lf} value tends to be an increase. However, there is a decrease in points SA-03 and SA-05. The low χ_{lf} values found at points SA-03 and SA-05 were possible due to the different magnetic minerals in the samples. When associated with XRD results, SA-03-1 contains several diamagnetic minerals such as quartz, dolomite, and rutile. Meanwhile, the SA-05-3 sample contained diamagnetic minerals, namely berlinite and alinite. Previous studies have shown that the presence of berlinite minerals causes a decrease in the susceptibility of Fine Sand (FS) ironsand during an increasing trend of magnetic susceptibility along with decreasing grain size of ironsand [3]. Mineral magnetism level also depends on the amount of Fe contained in it. Sample SA-01-3 has a lower magnetic susceptibility value because the magnetite formula contains less Fe than other samples. This matter indicates that the magnetite contained in the SA-01 is not magnetic enough than the magnetite contained in the other sample points.

4.2. Ironsand Source

The present elements may come from the same source, namely from weathering and the release of minerals by volcanic rock fragments. Besides, the high Fe content in ironsand samples can be caused by the mafic source rock [7]. The mafic source can be determined from the Al_2O_3 / TiO_2 with a low value (<14) [8]. The Al_2O_3 / TiO_2 result calculation shows an average value of 0.72, which indicates the source rock of ironsand is mafic. When viewed from the XRD test results, the types of magnetic minerals contained in each sample have various types of magnetic minerals. Most of the minerals found in ironsand such as berlinite, coesite, sphalerite, rutile, alinite, dolomite, albite, and anorthite are minerals in hydrothermal veins or contact metamorphism. The formation of albite and berlinite minerals in the sample shows that the area where ironsand was formed is a rock containing an andesite-basalt composition [9]. This matter suggests that the minerals contained in VFS ironsand at Anoi Itam Beach probably was formed from the metamorphism of the mafic andesite-basalt hydrothermal veins due to contact with high temperatures in the Jaboi geothermal system.

Figure 3 is a plot between the χ_{lf} and χ_{fd} values from the VFS sample. The magnetic susceptibility of ironsand samples from volcanic rocks will usually increase as the frequency-dependent magnetic susceptibility (χ_{fd}) increases [10]. Meanwhile, for polluted samples, the value of χ_{fd} will decrease as χ_{lf} increases. Besides, the χ_{fd}% result obtained also shows that the magnetic minerals were derived from pedogenesis. This matter is because magnetic minerals derived from pedogenesis have high χ_{fd}% values, whereas lower χ_{fd}% values (<4%) found in contaminated samples or anthropogenic sources [11]. This matter is consistent with the results of other studies that show that the magnetic minerals in ironsand samples from the Cirebon coast are of anthropogenic origin that is indicated by a low value (<2%) [12].

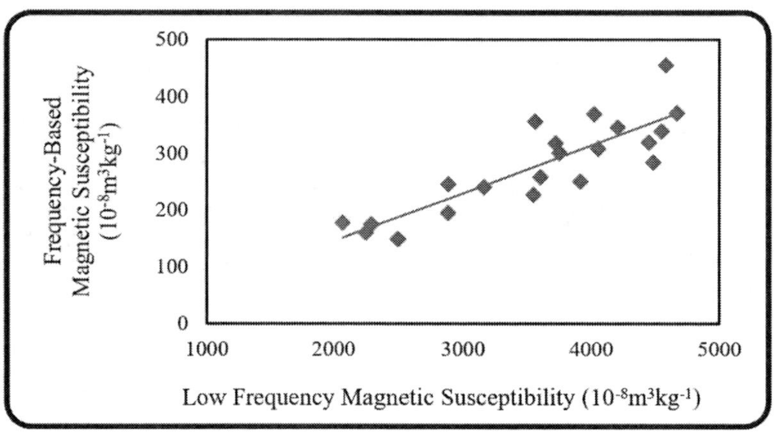

Figure 3. Plot the χ_{lf} value against χ_{fd} from the VFS samples.

4.3. Correlation Analysis between Elemental Composition and Magnetic Susceptibility Value

The correlation coefficient between element composition and magnetic susceptibility values is shown in Table 2. The presence of Ti in the form of rutile minerals causes Ti to have a negative correlation with magnetic susceptibility. Although Ti is also present in ilmenite, the contribution of Fe is more dominant to magnetic susceptibility because the magnetic moment of Fe is higher than Ti [13-14]. Meanwhile, Al is in the minerals anorthite and albite, Ca is in the form of anorthite, dolomite, and alinite, and Mg is in the form of dolomite. Even though it is in the mineral magnetite, the presence of Mg only acts as an impurity element with very little concentration. All of these minerals are diamagnetic minerals. This matter is probably why the element has a weak correlation with magnetic susceptibility.

Besides, most rock-forming minerals that do not contain Fe are diamagnetic. Therefore, in general, the minerals sphalerite, quartz, berlinite, and alinite are diamagnetic. Diamagnetic materials have small magnetic susceptibility values so that Zn, S, Si, P, and Cl have a weak to moderate correlation with magnetic susceptibility.

Table 2. The correlation coefficient between magnetic susceptibility and elemental composition.

	Magnetic Susceptibility
Mg	0.07
Al	-0.16
Si	-0.32
P	0.32
S	0.49
Cl	0.57
K	-0.46
Ca	0.07
Ti	-0.07
Mn	0.61
Fe	0.32
Zn	-0.39
Zr	0.14
V	-0.14

Other elements that appeared during XRF measurements, such as K, Zr, and V, were not identified during XRD measurements. This matter is probably because these elements form minerals with very little concentration. Because it is insensitive to minerals with small concentrations (<4%), XRD is unable to identify these minerals [15]. The correlation analysis results show that their correlation is mostly very weak, so that the minerals formed may be diamagnetic minerals.

Magnetic susceptibility correlates strongly with Mn and weakly with Fe. This matter happens because Fe is in a mineral form that has different magnetic properties. Fe in the form of magnetite will be ferrimagnetic while in the form of ilmenite, it will be paramagnetic. This is probably the reason Fe has a weak correlation with magnetic susceptibility. Previous studies have found that the two samples studied showed negative magnetic susceptibility values even though some paramagnetic materials were at low concentrations [16]. It can be assumed that even though there are ferrimagnetic minerals, the magnetic susceptibility may decrease due to the presence of paramagnetic and diamagnetic minerals which are more dominant.

Mn is an antiferromagnetic metal that can sufficiently influence magnetic susceptibility although it is not as strong as Fe [13,17]. The presence of Mn makes a positive contribution to MnO susceptibility leading to a strong positive correlation between Mn and magnetic susceptibility. Besides, Mn can bind with Fe to form a mineral. Because it has the largest magnetic moment of the transition metal group, the combination of Mn and Fe in one mineral makes a positive contribution to magnetic susceptibility. However, the presence of these minerals is possible in small amounts so that they do not appear in XRD measurements.

5. Conclusion

The VFS ironsand sample at Anoi Itam Beach is classified as magnetic, which is characterized by a high susceptibility value with a mixture of SP and non-SP domains. Based on XRF and XRD testing, VFS ironsand samples have quite diverse concentrations of elements and types of minerals with the largest Fe content and contain the main minerals magnetite and ilmenite. The high Fe content does not always result in high magnetic susceptibility values. This matter is possible due to the presence of paramagnetic and diamagnetic minerals in the sample. Besides, the magnetic minerals were derived from pedogenesis with andesite-basalt source rocks originating from Mount Jaboi. Based on correlation analysis, the weak correlation between element content and magnetic susceptibility is because these elements are in diamagnetic minerals. Meanwhile, a weak correlation between Fe and magnetic susceptibility because Fe forms minerals with very different magnetism, namely magnetite and ilmenite.

References

[1] Saniah, Purnawan S and Karina S 2014 *Depik.* **3**(3) 263–270
[2] Purnawan S, Azizah, Jalil Z and Zaki M 2018 *Jurnal Rekayasa Kimia dan Lingkungan.* **13**(2) 110–119
[3] Masrurah Z 2018 *Characterization of Ironsand Magnetic Properties Depending on The Sand Grain Size in Lampanah Beach and Anoi Itam Beach, Aceh Province* Master's Thesis (Bandung: Institut Teknologi Bandung)
[4] Bennett J D, Bridge D McC, Cameron N R, Djunuddin A, Ghazali S A, Jeffery D H, Kartawa W, Keats W, Rock N M S, Thomson S J and Whandoyo R 1981 *Geologic Map of The Bandaaceh Quadrangle, Sumatra* (Bandung: Geological Research and Development Center)
[5] Momber A W, Buchbach S, Plagemann P and Marquardt T 2017 *Progress in Organic Coatings.* **108** 90–92
[6] Dearing J A 1999 *Environmental Magnetic Susceptibility: Using the Bartington MS2 System* (Kenilworth: Chi Pub)
[7] Yunginger R, Bijaksana S, Dahrin D, Zulaikah S, Hafidz A, Kirana K, Sudarningsih S, Mariyanto M and Fajar S 2018 *Geosciences.* **8**(4) 116
[8] Ali M Y, Hibberd P and Stoikovich B 2018 *Journal of African Earth Sciences.* **140** 60–75
[9] Andreas A dan Putra A 2018 *Jurnal Fisika Unand.* **7**(4) 293–298
[10] Tiwow V A, Rampe M J and Arsyad M 2018 *Sainsmat: Jurnal Ilmiah Ilmu Pengetahuan Alam.* **7**(2) 136
[11] Novala G C, Fitriani D, Susanto K dan Kirana K H 2016 *IOP Conf. Ser.: Earth Environ. Sci.* **29**
[12] Nathasa R 2020 *Magnetic and Geochemical Characterizations of Ironsand Deposits from Cirebon Coastal Area, West Java* Bachelor's Thesis (Bandung: Institut Teknologi Bandung)
[13] Cullity B D and Graham C D 2009 *Introduction to Magnetic Materials* (Hoboken: IEEE/Wiley) p 544
[14] Spaldin N A 2010 *Magnetic Materials Fundamentals and Applications* (Cambridge: Cambridge University Press)
[15] Darmayanti N C E, Manafl A and Briyatmoko B 2000 *Prosiding Seminar Nasional Bahan Magnet I.* 4
[16] Li Y, Zuo R, Bai Y and Yang M 2014 *J. Geochem. Explor.* **146** 17–26
[17] Rosenblum S and Brownfield I K 2000 *Magnetic Susceptibilities of Minerals* 99-529 (Reston: U.S. Geological Survey)

IOP Conf. Series: Earth and Environmental Science **873** (2021) 012076 doi:10.1088/1755-1315/873/1/012076

MAGNETIC AND GEOCHEMICAL CHARACTERIZATIONS OF IRONSAND DEPOSITS FROM CIREBON COASTAL AREA, WEST JAVA

R Nathasa[1*], S Bijaksana[1], S J Fajar[1], T G Pitaloka[1]

[1]Faculty of Mining and Petroleum Engineering, Institut Teknologi Bandung, Jalan Ganesa 10, Bandung 40132, Indonesia

*Corresponding author: reginathasa@gmail.com

Abstract. Cirebon is a densely populated port city which has ironsand deposits at its coastal area. Due to its vicinity to the port and the estuary, these deposits might contain anthropogenic pollutants including, heavy metals. Magnetic measurements, X-Ray diffraction (XRD) as well as X-Ray fluorescence (XRF) analyses were carried out on iron sand samples from three sites along the Cirebon coastal area to identify the anthropogenic pollutants. The samples were separated based on the grain size before the measurement and analyses. Preliminary results show that the ironsand is less magnetic and has smaller frequency-dependent magnetic susceptibility values than ironsand found on Bayuran Beach in Central Java. Combined XRD result and the regional geological map shows that most of the minerals were originated from eroded volcanic rocks. The XRF results show that the samples have a relatively high content of Si. XRF analyses also show relatively high concentrations of Cr and Zn, the Geoaccumulation Index shows that the sediment is moderately to heavily polluted by Cr and Zn indicating the possibility of anthropogenic origin. The Cr and Zn content exceeds the stipulated value in the Sediment Quality Guidelines (SQGS).

1. Introduction

Cirebon is a densely populated port city located in the Province of West Java. Unlike most of the deposits found on the northern coast, this city has iron sand sedimented along the coast. Lots of activities also transpire in the area, such as industrial activity, fishery, trading, and agriculture [1] due to its strategic location. These activities produce pollutants transported and sedimented by the rivers that flow through the city [2]. At a particular concentration, these pollutants might be detrimental to the environment or aquatic life [3].

To sort out this issue, a geophysical study was conducted. The sample collected from three sites along the Cirebon coastal area was subject to geomagnetic as well as geochemical examination. The geomagnetic method was used to identify the susceptibility of the value and the presence of anthropogenic components [4]. Meanwhile, the geochemical method was used to determine the chemical element as well as the oxide compound. This study is expected not only to analyze the pollutant but also to characterize the ironsand.

2. Material and Method

The sample was collected from the Cirebon city along the North Coast of Java on December 17th, 2019. It was collected at three sites along the coastal area as shown in figure 1. Three liters of ironsand were taken at each site as the sample. The sample was washed with water twice, then the contaminant such as shells was. Afterward, it was washed twice with aquabidest and then dried using an oven at 110°C. Thereafter, we took 1.5 kgs samples from each point, for every 250 grams of it was sieved using mesh 60 and mesh 120 to get fine sand (FS) and very fine sand (VFS). From all of the sediment samples, we took nine as the representative samples for geochemical examination. Those chosen samples then were mashed using a mortar and pestle, followed by sieving using the 230 mesh to get the powder sample.

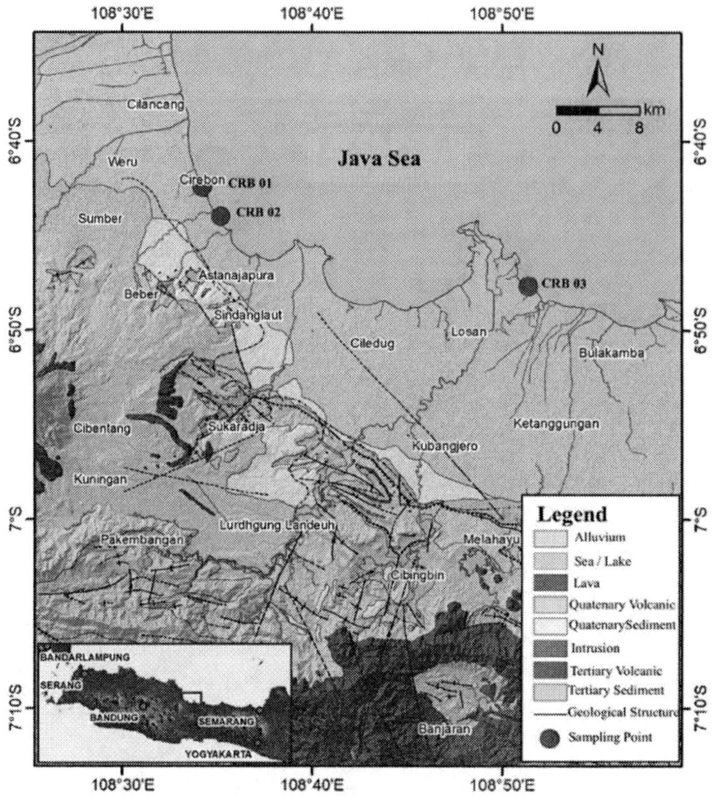

Figure 1. The Cirebon Quadrangle Geological Map (modified from [5]) with the sampling point along the coastal area.

The susceptibility measurement was conducted using Bartington MS2 susceptibility meter both for high and low frequency. Meanwhile, the geochemical examination was conducted using the Supermini 200 X-ray Fluorescence for XRF and SmartLab X-ray Diffraction for XRD. To determine the correlation between each element and the susceptibility value, we used the Spearman rank correlation method since the data is not normally distributed. The correlation was calculated using python.

3. Result and Discussion

3.1. Sediment Source

Based on the XRD data, all of the samples contain andesine and quartz. Other than that, most of the sites contain hornblende, faujasite, muscovite, and pyroxene. Some minerals only appeared in certain samples such as pseudo-malachite, cristobalite, and magnesio-hornblende. Most of these minerals are found in volcanic rocks [6]. The regional geological map supports this observation, since there are Older Cereme Volcanic Products (QTVR) and Younger Volcanic Products of Cereme (QVR) formations at most of the Cirebon area, also there is more formation containing volcanic product such as Ciherang, Gintung, and Halang Formation. High SiO_2 in the sample indicates that the sediment came from an erosion and weathering process [7].

3.2. Ironsand Characteristics

The ironsand sediment from sampling sites is dominated by fine-grained material. It reflects the low current velocity on the deposition area [8]. As shown in Table 1, the susceptibility values for each site and grain size are lower compared to the iron sand from Bayuran Beach, Central Java [9]. The correlation result shows that titanium has a positive very strong relation to susceptibility. Iron and chromium have a strong relation. Meanwhile, silicon and aluminum have a strong negative relation. This negative value means that the elements are inversely related to the susceptibility value. All of the samples have a very low χfd% as presented on Table1 that could indicate the presence of anthropogenic components [10].

Table 1. Susceptibility and χfd% values for each site and grain sizes.

Name	Average χlf (x 10^{-8} m^3/kg)	Average χhf (x 10^{-8} m^3/kg)	Average χfd%
Bulk 1	825.64	816.51	1.08
FS 1	284.60	283.19	0.51
VFS1	2961.95	2947.52	0.49
Bulk 2	354.22	353.07	0.32
FS 2	245.14	243.07	0.84
VFS 2	457.40	455.59	0.32
Bulk 3	2937.79	2936.28	0.05
FS 3	445.13	444.00	0.25
VFS 3	5932.06	5920.55	0.37

The most significant chemicals that composed the sediment are silicon and iron. The iron mineral is most likely to be sourced from hornblende and magnetite minerals. Hornblende minerals present in all grain sizes but magnetite mineral only present in FS. Meanwhile, susceptibility data show that VFS has higher values compared to FS, this indicates that hornblende supplied more magnetic elements, for instance, iron mineral. Other than that, the XRF data shows higher content of silicon rather than iron. There are also several heavy metals present in the samples, for instance, titanium, vanadium, chromium, cobalt, zinc, lead, nickel, manganese, and rhodium.

3.3. Sediment Quality

Based on Bradl [11], some heavy metals including chromium, manganese, and zinc are related to anthropogenic. These heavy metals are present in the XRF data for every sample but not in XRD mineral analyses. It gives the possibility that those elements came from anthropogenic pollutants with the fact that the sampling locations are surrounded by port and fishery industrial. To determine the sediment quality, the heavy metal concentrations are compared to sediment quality guidelines.

Based on Table 2, chromium concentrations exceeded the stipulated value both in the Sediment quality guideline [12] and the Canadian Council of Ministers of the Environment [13]. It's considered to be toxic to the aquatic environment and adverse effects are probable to be observed [14]. Meanwhile, zinc concentrations in all of the samples are above the midrange effects value. Only VFS 3 concentration exceeds the TET value. But based on CCME, zinc concentrations are above the PEL values. In this case, manganese is not stipulated because it's not considered as pollutant since it's a ubiquitous element that composed 0.1% of the earth [15] and also a chemical element that composed ironsand deposits.

Table 2. Chromium, zinc, and manganese concentration in the sample with the Sediment quality guidelines [12-13]

Sample Code		Cr	Zn	Mn
		ppm		
Bulk 1		839	444.4	6100
FS 1		386	426	5960
VFS 1		1630	484	7560
Bulk 2		873	402	5130
FS 2		N/A	426	5180
VFS 2		750	431	5040
Bulk 3		993	417	4970
FS 3		567	326	5820
VFS 3		1310	587	5930
SQG		Cr	Zn	Mn
		ppm		
Midrange Effect	Effect Range Median (ERM)	145	270	-
	Probable Effect Level (PEL)	90	315	-
Extreme Effect	Toxic Effect Threshold (TET)	100	540	-
	Severe Effect Level (SEL)	110	820	-
CCME , 2001	ISQG	53.2	124	-
	Probable Effect Level (PEL)	160	271	-

The Geoaccumulation Index (*Igeo*) values show the comparison between the heavy metal concentration in the samples and the concentration in pre-industrial time [16]. This will give more proof to say if chromium and zinc came from anthropogenic activities. Based on Table 3 *Igeo* level, the samples are moderately to heavily contaminated by zinc and chromium and it has the possibility for those heavy metal to be anthropogenic.

Table 3. Igeo values for chromium and zinc with the classification [16]

Sample Code	Zn Igeo	Classification	Cr Igeo	Classification
Bulk 1	2.08	Moderately to heavily contaminated	2.46	Moderately to heavily contaminated
FS 1	2.02	Moderately to heavily contaminated	1.34	Moderately Contaminated
VFS 1	2.20	Moderately to heavily contaminated	3.41	Heavily contaminated
Bulk 2	1.94	Moderately Contaminated	2.51	Moderately to heavily contaminated
FS 2	2.02	Moderately to heavily contaminated	N/A	-
VFS 2	2.04	Moderately to heavily contaminated	2.29	Moderately to heavily contaminated
Bulk 3	1.99	Moderately Contaminated	2.70	Moderately to heavily contaminated
FS 3	1.63	Moderately Contaminated	1.89	Moderately Contaminated
VFS 3	2.48	Moderately to heavily contaminated	3.10	Heavily contaminated

4. Conclusion

The iron sand deposit on Cirebon coastal is dominated in fine-grained and very fine-grained sand that reflects the low deposition energy in the estuarine area. Geochemically, the deposit contains more silicon (Si) than iron (Fe), causing lower susceptibility values compared to the iron sand deposit in Bayuran Beach, Central Java. The most significant minerals in the sample are andesine and quartz. The mineralogy of the deposit informs that the source of the iron sand is eroded volcanic rocks. The most major heavy metals that are possible to be anthropogenic in the sample are chromium (Cr) and Zinc (Zn). The Cr and Zn concentration in all of the samples exceed the Sediment Quality Guidelines (SQGs) as well as CCME (2001) stipulated values. Hence it is considered to be toxic to the environment and aquatic life. Based on the Geoaccumulation Index, the ironsand deposit on Cirebon coastal is moderately to heavily polluted by chromium and zinc.

References

[1] Ministry of Public Works and Public Housing 2009 Cirebon's Profile Online: http://ciptakarya.pu.go.id/profil/profil/barat/jabar/cirebon.pdf
[2] Nurkhasanah D, Ambariyanto, Suprijanto J, Pusparini N, and Prasetyo B 2019 IOP Conf. Ser.: Earth Environ. Sci. **246** 012010
[3] Canadian Council of Ministers of the Environment 1995 *Protocol for the derivation of Canadian sediment quality guidelines for the protection of aquatic life* (Winnipeg: Canadian Council of Ministers of the Environment)
[4] Sudarningsih S, Bijaksana S, Ramdani R, Hafidz A, Pratama A, Widodo W, Fajar S J, and Agus Santoso N 2017 *Geosciences* **7** 66
[5] Silitonga P, Masria M, and Suwarna N 1996 *Geological Map of The Cirebon Quadrangle, Java* (Bandung: Geological Research and Development Center)
[6] Anthony J W 1990 *Handbook of mineralogy* (Tucson: Mineral Data Pub)
[7] James H 1966 Data of Geochemistry (Washington: United States Government Printing Office)
[8] Nurkhasanah D, Ambariyanto, Suprijanto J, Pusparini N, and Prasetyo B 2019 IOP Conf. Ser.: Earth Environ. Sci. **246** 012024
[9] Miranda R 2019 *Characterization of Magnetic and Geochemistry Properties of Volcanic and Non-Volcanic Iron Sand in Indonesia* (Bandung: Institut Teknologi Bandung)
[10] Crosby C J, Fullen M A, and Booth C A 2014 *Environ. Sci.: Processes & Impacts* **16** 548-557
[11] Bradl H B 2005 *Anthropogenic Sources of Heavy Metals* (Amsterdam: Elsevier)
[12] Burton G A 2002 *Limnology* **3** 65-75
[13] Canadian Council of Ministers of the Environment 2001 *Canadian sediment quality guidelines for the protection of aquatic life: Summary tables* (Winnipeg: Canadian Council of Ministers of the Environment)
[14] Canadian Council of Ministers of the Environment 1999 *Canadian sediment quality guidelines for the protection of aquatic life: Chromium* (Winnipeg: Canadian Council of Ministers of the Environment)
[15] Concise International Chemical Assessment Document (CICAD) 63 2004 (Geneva: WHO)
[16] Muller G 1969 *Geojournal* **2** 108-118

IOP Conf. Series: Earth and Environmental Science **873** (2021) 012077 doi:10.1088/1755-1315/873/1/012077

Early Result for Microtremor Characteristic Observation in Merapi and Merbabu

E. Fernanda[1], A. Bilqis[1], L. Julio[1], K. Nursal[1], Y. H. Christ[1], T. Adiansyah[1], S. K Suhardja[1], M. Ramdhan[2]

[1] Faculty of Exploration and Production Technology, University of Pertamina, Jl. Teuku Nyak Arief, Simprug, Kebayoran Lama, Jakarta 12220, Indonesia
[2] Agency for Meteorology, Climatology and Geophysics, Jalan Angkasa I, No. 2, Kemayoran, Indonesia

Email: yeremia.harvesta@gmail.com

Abstract. The arc magmatism and volcanic activity in Java are dominated by the subducting plate of Indo-Australian into the Eurasian plate. Merapi volcano is located in Central Java and known as one of the most active volcanoes in the world. Several studies have tried to estimate the magma reservoir zone in Mt. Merapi and suggested multiple layers of reservoirs with the shallow one at 1-2 km and a deeper at 6 -9 km or 15 km. The Low-Frequency Passive Seismic is one method to analyze the frequency spectrum below the recording station. Previous related studies show a promising a relation between hydrocarbon reservoir and higher amplitude at vertical component at a frequency between 0.1 – 6 Hz. An observation at the volcano sites have also been reported to display a different spectrum amplitude at the vertical component. This study exploited the same method in LFPS to analyze the frequency spectrum at Mt. Merapi and Mt. Merbabu. We use seismic data from the DOMERAPI temporary seismic network installed in the neighborhood of Merapi and Merbabu volcano. We analyze 53 broad-band seismometers data from October 2013 to mid-April 2015. We also add several stations from MERAMEX network instruments to compare spectrum analysis outside the Merapi and Merbabu volcano. We also removed some tele-seismic and regional events from the data to better analyze the LFPS signal. We have seen a higher amplitude in vertical component near Mt. Merapi and will proceed to analyze all stations.

1. Introduction

The Eurasian continent's southwestern most point is Java Island. The Indo-Australian plate's subduction underneath the Eurasian plate is critical to the region's arc magmatism and volcanic activity. Because of the tectonic setting covered by the Sunda Arc, megathrust earthquakes are common. From 50 km depth to North Java, the subduction rate between Australia and Indonesia is about 67 mm per year (Simons et al, 2007), and the slab dips from near-horizontal to a very steep angle (70°-80°) (Koulakov et al, 2007; Suhardja et al, 2020). Merapi volcano is located in Central Java and known as one of the most active volcanoes in the world, mainly due to the subducting activities in the surrounding area. Merapi volcano has had a total of 22 eruptions since the beginning of the twentieth century. One of the most notable eruptions occurred in 2010, when the lava dome production, gas emissions, erupted lava volume, and seismic energy released exceeded previous eruptions.

IOP Conf. Series: Earth and Environmental Science **873** (2021) 012077 doi:10.1088/1755-1315/873/1/012077

Volcanic earthquake classification has undergone many changes and improvement, the first and most recognized classification proposes that volcanic earthquakes can be divided into four types according to the location of their focus, their relationship to the eruptions, and the nature of the earthquake motion (Minakami, 1960 & 1974). The four types are A-type, B-type, explosion earthquakes, and volcanic pulsation. A continuous seismic signal that is observed during volcanic eruptions that lasts minutes to days in duration is called a volcanic tremor, although it is possible to observe it independently. This tremor which is also called a micro-tremor, mainly consists of surface waves and an irregular sinusoidal wave form. Based on study by Mogi (1963), volcanic earthquake sequences can be categorized into three main types, which are main shock-aftershocks sequence; foreshocks-main shock-aftershocks sequence; and swarm.

Storengy conducted Low Frequency Passive Seismic survey over a geothermal area on 2016, which was proven to detect the geothermal stream saturated area due to its fluid properties. The same response in terms of density and viscosity also was shown in hydrocarbon accumulation. According to Kazantsev et al (2017), since LFPS uses the micro-seismic as the natural diffuse seismic source to analyze the energy spectrum, volcanic tremors can be used as an additional natural source in geothermal context. Furthermore, similar technique used to monitor low frequency tremor associated to the magma chamber activity before eruption (Ferrick et al, 1982; Ripepe and Gordeev, 1999).

The focus of this study is microtremor's characteristics in Mt Merapi and Merbabu. By using local earthquake tomography, it is observed that three active areas with high Vp/Vs are present in this location. The first region is a shallow zone which indicates an intense fluid activity, the second region is thought to be a 20 km depth below MSL pre-eruptive magma reservoir, and the last region is a magma reservoir zone at 30 km depth below MSL which is thought to be the supplier for the main reservoir.

2. Data and Methods

This research is conducted in the vicinity of the volcanoes of Merapi and Merbabu, Central Java. In this area, there is a DOMERAPI station network, consisting of 53 seismometer stations that are installed and scattered around the two volcanoes, with operating times starting from October 2013 to mid-April 2015 (Ramdhan et al, 2017). However, for the data used in this study, only a few recorded events were taken for one week from 10 August 2014 to 17 August 2014. Also, several MERAMEX data networks around the study area were used to compare the results of spectrum analysis outside the Mt. Merapi and Merbabu areas.

In this study, using the passive low-frequency seismic method (LFPS), where the seismometer on the surface will record natural seismic waves originating from below the surface (Tenghamn et al, 2009). The recorded seismic waves will then be analyzed for their low-frequency spectrum to determine the characteristics of the seismic waves.

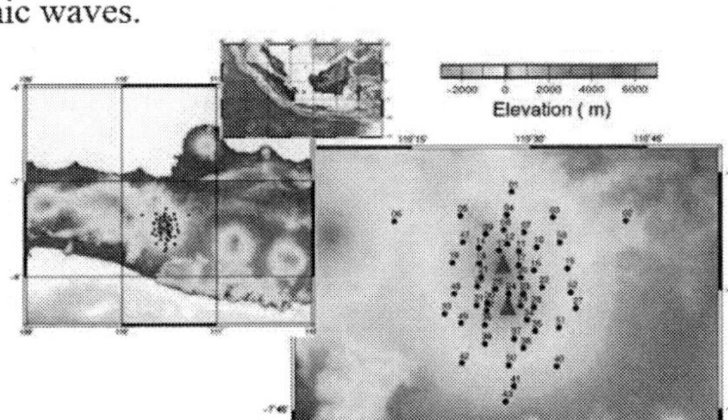

Figure 1. DOMERAPI network around Mt. Merapi and Mt. Merbabu (Suhardja et al, 2020)

We use bandpass filter from 0.1 – 6 Hz to minimize anthropogenic high noise and low-frequency ocean waves. Also, we minimize instrument response by normalizing the recorded signal. STA-LTA, an algorithm that uses short-term and long-term windows for obtaining the ratio of calculated average trace energy to trigger the algorithm activation, is applied to all the 3 components signal of MERAMEX

& DOMERAPI stations in time domain to separate tremor and other wavefield by eliminating teleseismic and regional event. The algorithm requires deep knowledge of the overall trace energy from recorded signal to minimize false triggers and missing weak events.

Fast Fourier Transform is applied to accommodate frequency spectrum of recorded signal on all 3 components for Power Spectral Density technique (PSD). Then, we use the integral value under the PSD on all components to analyze the spectrum attributes. Vaezi & Van der Baan (2015) concluded that PSD technique was compensating the lack of STA-LTA algorithm on detecting weak events and false triggers in frequency spectrum and also helped designing bandpass filter better for analysis. The filter before applying STA-LTA algorithm is not optimal in dynamic noise environment. PSD technique from Vaezi & van der Baan (2014) is based on the concept of stronger spectral over a frequency band of seismic event than total trace energy to detect events in stationary noise conditions.

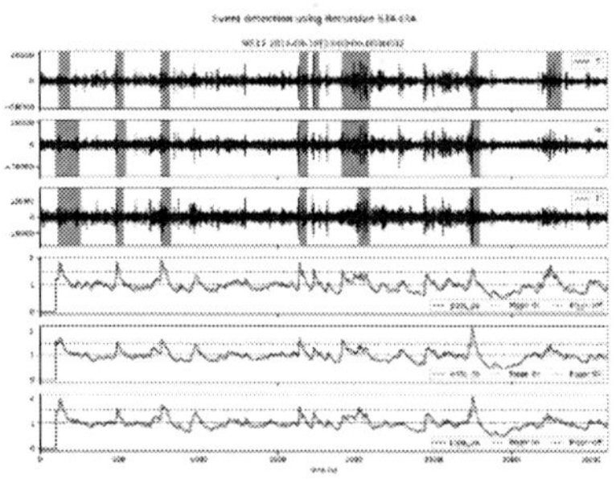

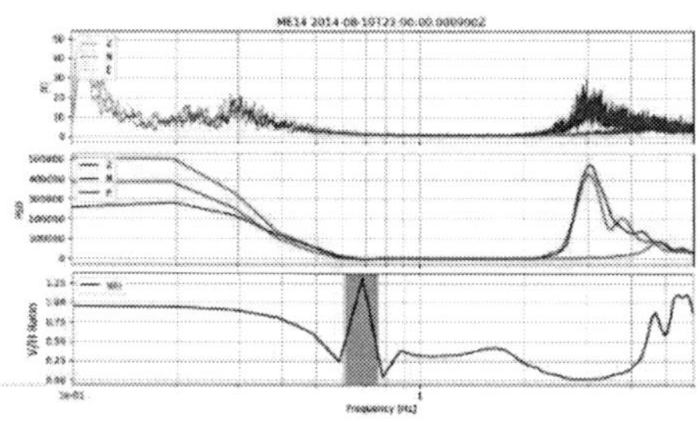

Figure 2. (a) STA-LTA processing from station ME14. The signal was recorded on August 10th,2014 (b) Spectral analysis from station ME14 of August 10th, 2014 recorded signal

Afterwards, we analyze both vertical and horizontal component of PSD attribute to identify the contrast response. Other attributes are frequency and amplitude response. V/H ratio applied on Dangel et al (2003), Saenger et al (2007) & Kazantsev et al (2017) is used in this study. The attributes analyzed on V/H ratio are peak and mean attributes as a comparison to recognize the difference in responses. The V/H ratio is calculated and mapped on limited frequency band. Next, the integral of power spectral density is analyzed both on vertical and horizontal component or we called it component energy.

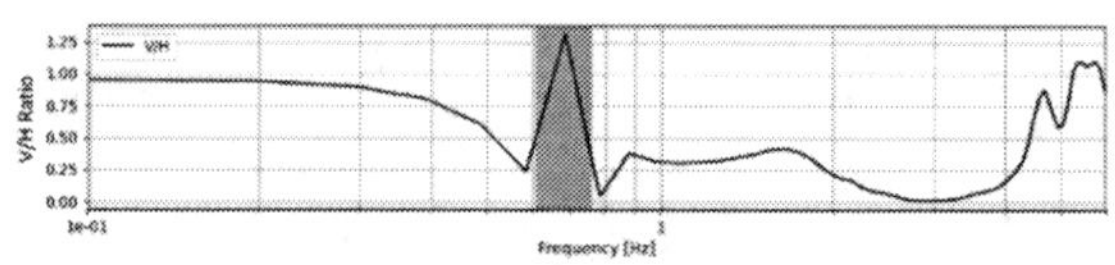

Figure 3. V/H ratio on station ME14 of August 10th,2014 recorded signal

3. Result and Discussion

3.1. Component Signal PSD Analysis

3.1.1. Maximum Amplitude. Based on figure 4, the distributed value of the red polygon has a low mark for range 0 - 0.234 except ME13 and ME23. ME13 and ME 23 has a high value between 0.338 - 1.88. High amplitudes correlate with the top of Merapi and Merbabu, and low amplitude correlates with the body of Merapi. The amplitudes have high attenuation on the body of mountains, in which the body consists of volcanic rocks such as tuff and pyroclastic. Then, the result of component vertical and horizontal has similar interpretation.

IOP Conf. Series: Earth and Environmental Science **873** (2021) 012077 doi:10.1088/1755-1315/873/1/012077

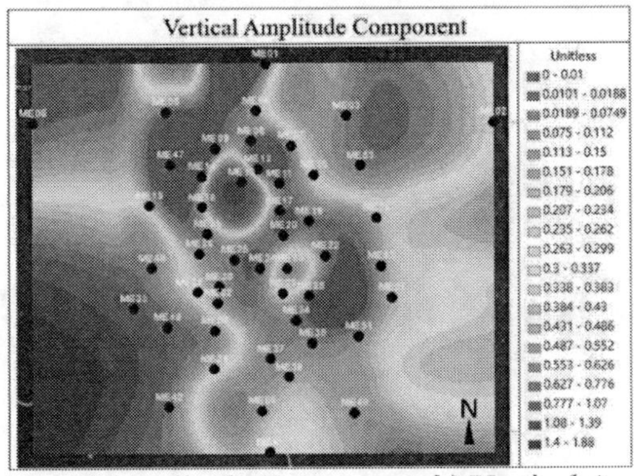

Figure 4. a) Horizontal Amplitude Component, b) Vertical Amplitude Component

3.1.2. Maximum Frequency.

Figure 5 a) is the processing result in form factor of horizontal frequency distribution which located around Mt. Merapi and Mt. Merbabu. In orange box line is the station location around Mt. Merbabu, and the red box line is the stations located around Mt. Merapi. The values are varying from 0 – 8.63 Hz. It distributed evenly around stations. Near the ME16 station, shown as red, the frequency has the highest value than the surrounding areas. ME 16 frequency values is around 7.01 – 8.63 Hz, then the high frequency is distributed to the southern part, which are ME 23 and ME 29. ME 20, ME 17, ME 11, ME 10, ME 15 are surrounding the ME 16 station, and it shows different frequency value, around 0.749 – 1.2 Hz, shown as yellow color. On the Mt Merapi side, ME 22, ME 28, ME 34 and ME 24 surrounds the ME 23 and ME 29 stations. It also shows intermediate frequency values, varying from 0.749 – 1.2 Hz. From the frequency values, it shapes like boundaries surrounding the high values, more precisely around the ME 16, ME 23 and ME 29. Outside the boundary, are distributed thoroughly as low – frequency ranges, which shown as green to blue color.

Figure 5 b) is the result of data processing in the form of the distribution of vertical frequency values around Mount Merapi (green box line) and Merbau (red box line). In the figure, there are frequencies that have values varying from 0 Hz to 2.63 Hz and are evenly distributed throughout DOMERAPI & MERAMEX stations. It can be seen that the distribution of low-frequency values is spreading around Mt. Merbabu and Mt. Merapi area. To the north from Mt. Merbabu, the low frequency ranges from 0-0.241 Hz and surrounds Mt. Merbabu and several stations, ME10, ME11, ME12, ME13, ME16, and ME17. Then, the distribution of intermediate frequencies is spreading to the south and southwest directions of Mt. Merapi, ranging from 0.242-0.377 Hz. It can be shown at the following stations, ME 33, ME 36, ME38, ME39, ME41, ME42, ME49, and ME50. The high-frequency distribution is spreading over the area of Mt. Merapi to the southeast and northeast of the study area, ranging from 0.378-2.63 Hz. Several stations are in this high-frequency range, namely ME 23, ME 29, ME 34, and ME 40.

From the frequency value, it can be explained that the vertical frequency value is classified as low because the frequency is around 0-2.63 Hz, from the frequency value it is included in the thickness of the surface which has very thick sediment, it can be seen that this research area is in a mountainous area has a thick coating. However, it can be seen that in the distribution of vertical frequency values there are differences in frequency changes from low to high ranges, this can be identified as a change in lithology in the area, seen in the frequency change. at stations ME 02, ME 03, ME 15, ME 20, ME 22, ME 24, ME 27, ME 34, ME 35, ME 51, and ME 53. This could be the boundary layer between the sedimentary layer and the volcanic layer surrounding the station.

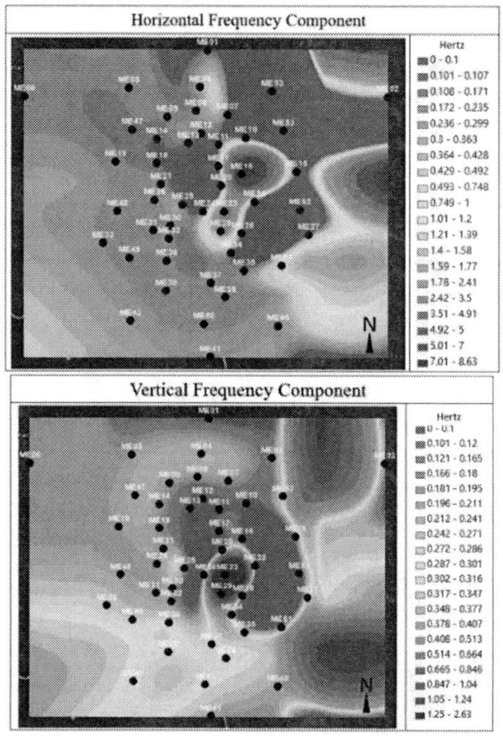

Figure 5. a) Horizontal Frequency Component, b) Vertical Frequency Component

3.2. Signal Energy Analysis

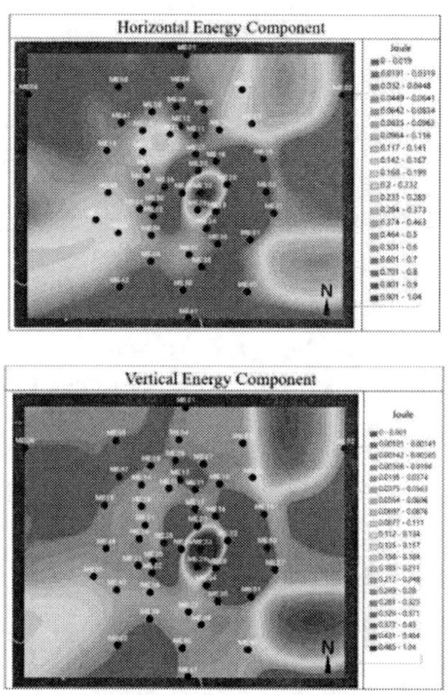

Figure 6. a) Horizontal Energy Component, b) Vertical Energy Component

Figure 6 a) shows the result of data processing in the form of the distribution of horizontal energy values, or the integral of the PSD of the horizontal component around Mt. Merapi and Mt. Merbabu. The map is presented as blue to red color varying low to high values. The integral of power spectral density from horizontal component results varying from 0 - 1.04. The horizontal component is initiated to compare the vertical component result and showed anomalies relatively at the same locations. Unlike the vertical component, the horizontal component presents slightly lower values than the vertical component around station ME23, ME29, ME24, ME20, ME22, and ME28 varying from 0.168 - 1.04 inside the seismic network. Other stations relatively show much lower values outside the previous anomaly mentioned, ranging from 0 - 0.0448, except around ME13 and ME49 stations present slightly higher values than vertical component, varying from 0.168 until 0.373. At the less dense of seismic network, several anomalies are present at the northeast part near ME03 and ME53 stations and the southeast part outside ME51 and ME40 stations while the south western part is closed to ME33 and ME49 stations. The north eastern and south eastern part distributes evenly-values varying from 0.168 - 0.7. Compared to the vertical component, the southwestern to western part distributes higher values at ME33 and ME49 stations than the vertical component varying from 0.168 - 0.373.

Figure 6 b) shows the result of data processing in the form of the distribution of vertical energy values around Mt. Merapi (south) and Mt. Merbabu (north). The distribution of the values of energy varies from 0 – 1.04. Based on the figure, the values in Mt. Merbabu area tend to increase as it goes from west (ME 06) to the east (ME 02) with the highest value distribution in the area (around 0.112 – 1.04) is located in the north eastern part surrounded by ME 02, ME 03, and ME 53 stations. Meanwhile in Mt. Merapi area, the high energy values are distributed in two main parts. The first part is located in the centre, mainly around ME 23 & ME 29 stations and surrounded by ME 22, ME 24, ME 28, and ME 34 stations with distribution varying from 0.372 – 1.04. The second part is located in the south eastern part, close to ME 40 and ME 51 stations with distribution varying from 0.112 – 1.04 and is distributed more evenly than the first part. Aside from the high energy distribution, the area is distributed thoroughly as low – frequency ranges (around 0 – 0.111), which shown as green to blue color.

3.3. V/H Ratio Analysis

In figure 7 a), we could know that the maximum V/H frequency distribution is around 0-6.83 Hz. Distribution in Mt. Merapi and Mt. Merbabu are having low - frequency range values. Low frequency distribution indicates pyroclastic rock lithology in that area, because pyroclastic rock has a high attenuation characteristic.

In figure 7 b), we could know that the maximum V/H ratio value is around 0 – 18.6 Hz. The distribution in Mt. Merapi and Mt. Merbabu has distribution variation from low, intermediate, and high. Based on literature, the sub – surface lithology distribution for the high V/H ratio in Mt. Merapi is caused by solid and dense rocks. For the mean V/H value has similar distribution with V/H ratio.

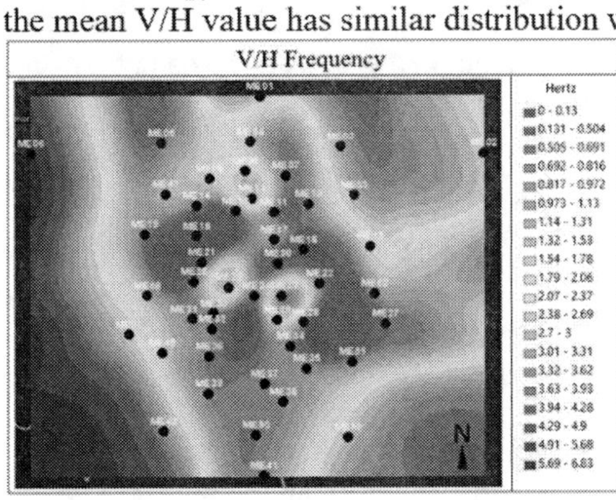

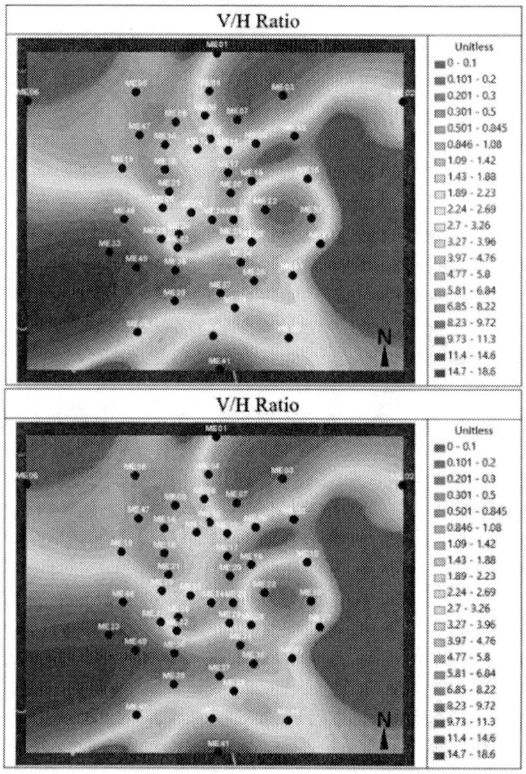

Figure 7. a) Maximum V/H Frequency, b) Maximum V/H Ratio, and c) V/H Mean

4. Conclusion

From the maximum amplitude PSD attribute, we can see similar pattern from the horizontal and vertical component based on qualitative interpretation. There's some kind of boundary consist of low amplitude that surrounds the high amplitude, but the maximum frequency PSD attribute have a different pattern. Qualitatively, the energy signal PSD attribute have similar pattern to the maximum amplitude PSD attribute. The energy signal PSD attribute have similar pattern between horizontal and vertical component on Mt. Merapi area, but on the Mt. Merbabu area, it has different pattern between horizontal and vertical component. From the ratio analysis, there're no pattern similarity between V/H ratio, V/H frequency, and V/H mean ratio. From the result, the best attribute we could purpose for this research is maximum amplitude PSD attribute and maximum energy signal PSD attribute.

Acknowledgement

We would like to thank the Geophysical Engineering Department of University of Pertamina for encouraging us to do this study. We also thank Agency for Meteorology, Climatology and Geophysics for providing the data used in this study.

References

[1] Dangel et al 2003 *Phenomenology of tremor-like signals observed over hydrocarbon reservoirs* Journal of Volcanology and Geothermal Research **128**(1–3) 135–158 DOI: doi.org/10.1016/S0377-0273(03)00251-8

[2] Kazantsev et al 2017 *An innovative methodology based on Low Frequency Passive Seismic data analysis to map geothermal reservoir steam saturated areas* (Jakarta: Proceedings the 5-th Indonesia International Geothermal Convention & Exhibition (IIGCE)) Access: www.storengy.com

[3] Ferrick et al 1982 *Source mechanism of volcanic tremor*. J. Geophys. **87**(B10) 8675-8683 DOI: 10.1029/JB087iB10p08675
[4] Julian BR 1994 *Volcanic tremor: Nonlinear excitation by fluid flow* Journal of Geophysical Research **99** 11 859–11 877 DOI: 10.1029/93JB03129
[5] Koulakov et al 2007 *P and S velocity structure of the crust and the upper mantle beneath central java from local tomography inversion* J. geophys. Res. **112** B08310 DOI: 10.1029/2006JB004712
[6] Martini et al 2013 *A passive low-frequency seismic experiment in the Albertine Graben, Uganda. European Association of Geoscientists & Engineers* Geophysical Prospecting **61**(Suppl. 1) 39-61 DOI: doi.org/10.1111/j.1365-2478.2012.01083.x
[7] Minakami T 1960 *Fundamental Research for Predicting Volcanic Eruptions* Bull. Earthq. Res. Inst. **38** 497-544
[8] Minakami T 1974 *Seismology of volcanoes in Japan. Elsevier: Physical Volcanology* Eds. L Civetta, P Gasparini, G Luongo, and A Rapolla 1–27
[9] Minakami T 1974 *Prediction of volcanic eruptions. Elsevier: Physical Volcanology* Eds. L Civetta, P Gasparini, G Luongo, and A Rapolla 313–333
[10] Mogi K 1963 *Some discussions on aftershocks, foreshocks and earthquake swarms – the fracture of a semi-finite body caused by an inner stress origin and its relation to the earthquake phenomena (Third Paper)* Bull. Earthq. Res. Inst.Tokyo University **41** 615–658
[11] Ramdhan M, Widiyantoro S, Nugraha AD et al 2017 *Relocation of hypocenters from DOMERAPI and BMKG networks: a preliminary result from DOMERAPI project* Earthq. Sci. **30**(2) 67–79 DOI: doi.org/10.1007/s11589-017-0178-3
[12] Ripepe M and Gordev E 1999 *Gas bubble dynamics model for shallow volcanic tremor at Stromboli* J. Geophys. Res. **104**(B5) 10 639-10 654 DOI: 10.1029/98JB02734
[13] Saenger et al 2009 *A passive seismic survey over a gas field: Analysis of low-frequency anomalies* Geophysics **74**(2) O29-O38 DOI: dx.doi.org/10.1190/1.3078402
[14] Saenger et al 2007 *A hydrocarbon microtremor survey over a gas field: Identification of seismic attributes* SEG Technical Program Expanded Abstracts **26** 1277–1281 DOI: doi.org/10.1190/1.2792736
[15] Simons et al 2007 *A decade of GPS in Southeast Asia—Resolving Sundaland motion and boundaries* J. Geophys. Res. **112**(B6) DOI: 10.1029/2005JB003868
[16] Suhardja S K, Widiyantoro S, Métaxian JP, Rawlinson N, Ramdhan M, and Budi Santoso A 2020 *Crustal thickness beneath Mt. Merapi and Mt. Merbabu, Central Java, Indonesia, inferred from receiver function analysis* Physics of the Earth and Planetary Interiors 106455 DOI:10.1016/j.pepi.2020.106455.
[17] Tenghamn S R L and Borresen C N 2009 *Marine Passive Seismic Method For Direct Hydrocarbon Detection. United States, Patent Application Publication* Pub. No. US 2009/0279387 A1 Access: freepatentonline.com/y2009/0279387.html
[18] Vaezi Y and Van der Baan M 2014 *Analysis of instrument self-noise and microseismic event detection using power spectral density estimates* Geophys. J. Int. **197**(2) 1076–1089. DOI: doi.org/10.1093/gji/ggu036
[19] Vaezi Y and Van der Baan M 2015 *Comparison of the STA/LTA and power spectral density methods for microseismic event detection* Geophys. J. Int. **203** 1896–1908 DOI: doi.org/10.1093/gji/ggv419
[20] Widiyantoro et al 2018 *Seimic imaging and petrology explain highly explosive eruptions of Merapi Volcano, Indonesia* Scientific Reports **8**(1) 1–7.
[21] Zobin V M 2012 *Introduction to Volcanic Seismology: Second Edition*. (London: Elsevier) DOI: doi.org/10.1016/C2011-0-06141-0

IOP Conf. Series: Earth and Environmental Science **873** (2021) 012078 doi:10.1088/1755-1315/873/1/012078

Land Subsidence of Java North Coast Observed by SAR Interferometry

T P Sidiq[1], I Gumilar[1], I Meilano[1], H Z Abidin[1], H Andreas[1], A Permana[1]

[1]Geodesy Research Group, Institut Teknologi Bandung (ITB), Indonesia

Abstract. *Land Subsidence became recent issue in environmental management in Indonesia. Large cities in Indonesia, especially in Java Island, are well known to suffer from fast rate land subsidence such as Jakarta, Bandung, and Semarang. However, the phenomena is presumed to also happen in other cities which have large industries and located on the deposit or clay soil layer. The effect of land subsidence may be different in some location, but the coastal zone will have more impact since land subsidence will also induce tidal flood and may cause land area reduction. In this study, we use more than 70 Sentinel-1 data, range from year 2016 to 2020 to map the land subsidence in the Java North Coast. The interferograms are selected based on Small Baseline Subset (SBAS) algorithm. National Digital Elevation Model (DEMNAS) is used in differential InSAR process. Our result shows that many cities along the Java North Coast suffer land subsidence. Jakarta, experience has maximum subsidence more than 5 cm/year. In Central Java, Pekalongan experience up to 7 cm/year of land subsidence, while Semarang and Demak has subsidence rate up to 8 cm/year.*

1. Introduction

Java island is the most populated island in Indonesia, and the location of several main cities of Indonesia such as Jakarta, Surabaya, Semarang, Bandung, and Jogjakarta. Population wise, there are more than 145 million peoples based on 2015 National Survey held by Indonesian Bureau of Statistics. The number depict almost half of the total of Indonesian population, and the fact that both population and development are more concentrated in the North Coast of Java, make this region is very important for Indonesia. However, cities located along the region are prone to the risks caused by environmental changes such as coastal flooding, mangrove loss, and coastal abrasion [1]. Some study also reveal that this region also suffer from land subsidence such as Jakarta [2] and [3], Semarang [4] [5], and some other cities [6].

Geologically, the North Coast of Java is formed by the deposition of sediment or alluvial plain, consist of unconsolidated clay, silt, sand, and gravels of Quaternary age [7].

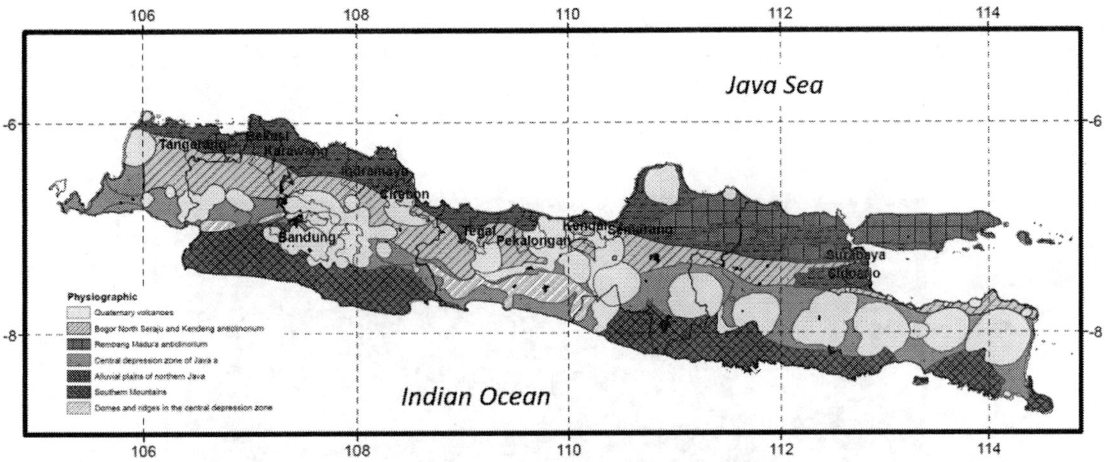

Figure 1 : Physiography of Java Island

USGS explain land subsidence as a gradual settling or sudden sinking of earth's surface which caused by removal or displacement of subsurface material such as water, gas, oil compaction, and loading. The cause of land subsidence can be both natural [8] and anthropogenic, although the study of land subsidence in some cities along the Java North Coast suggest that it mostly driven by anthropogenic activities. Furthermore, land subsidence is considered to be a disaster since it caused both economic loss [9], [4], and natural degradation in large scale [10] [11].

Despite its consequences, many people do not realize the risk because land subsidence affect relatively large area and happened at a slow rate in long term. The risk is realized after the impacts comes to the surface such as infrastructure damage [12], worsen of flood [4], sea water intrusion, and soil quality degradation [13]. The economic loss caused by land subsidence is enormous, such as in Semarang it is estimated to caused loss about 3.5 trillion rupiahs or about 270 million USD [14] [15], and the loss in Bandung is estimated to be around 180 million USD [4].

2. Method

2.1. Data

In this study, we use SAR Sentinel-1 data provided by European Space Agency (ESA) which can be accessed publicly. We choose 70 data from 500 available ascending data in 4 tracks covering Java Island. The average data for each track is about 14 to 16 data, chosen every 3 months of data acquisition to make the computation effective without losing the deformation signal. The acquisition is taken from late 2016 for 2 tracks, while other tracks start from January 2017, to mid-2020.

2.2. InSAR Data Processing

Interferogram generations are done using GMTSAR [16]. We generate 214 interferograms, with average of 51 interferograms for each track. Height correlated tropospheric signal is removed by model the signal by third-order 3D polynomial curve plane [17]. However, due to relatively flat area, the remaining error still affect the interferograms and will be corrected further during time series inversion because the atmospheric error is uncorrelated in time [18]. The Line of Sight (LOS) Displacement and LOS velocity is obtained by interferogram stacking method [19]. Interferogram generation is constrained by perpendicular baseline length below 200 meter, and temporal baseline below 365 days, and then each pixel of every interferogram is evaluated by their coherence value. All pixels have coherence below 0.4 is rejected to make sure its reliability.

Due to temporal decorrelation and rapid seasonal vegetation cover changes in the study area, the coherence value in the area is relatively low with average coherence about 0.1. **Figure 2** below show the coherence value in track 54. The coherence condition in other tracks is similar since the vegetation and surface characteristic is relatively homogeneous in the North Coast of Java. The C-band system also contribute to decorrelation because it is more sensitive to surface changes and vegetation cover compared to L-Band use by ALOS-PALSAR as demonstrated by [6]. However, the access to Sentinel-1 C-band SAR data which made public, prove the dataset to be much more popular to be used more widely.

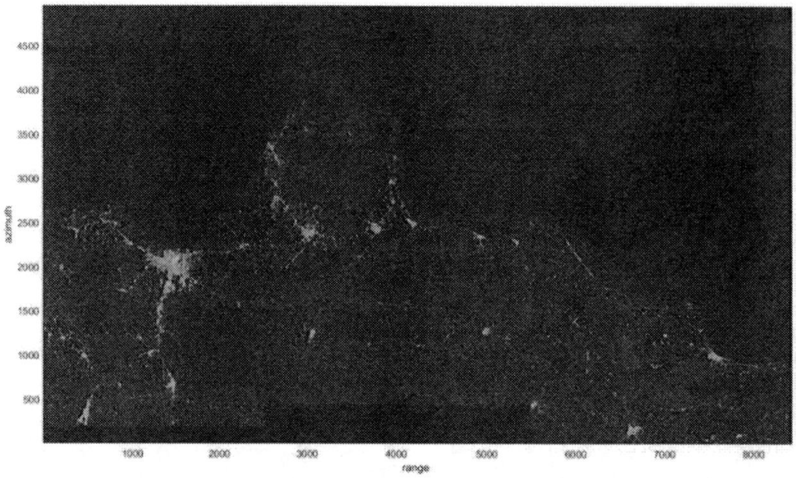

Figure 2 : Average coherence over the area of Track 54

All interferograms are unwrapped using Minimum Cost Flow (MCF) method provided on SNAPHU software [20]. To speed up phase unwrapping process, we also perform nearest neighborhood interpolation on low coherence pixels prior to phase unwrapping process [21]. The unwrapped interferograms are then resampled and geocoded down to a ground resolution of about 60 meters.

3. Result and Discussion

Our result show that there are at least three cities suffer from land subsidence in relatively large area in each city. Those cities are Jakarta, Semarang and Pekalongan (**Figure 3**). The result of each city will be described and discuss on each sub chapter.

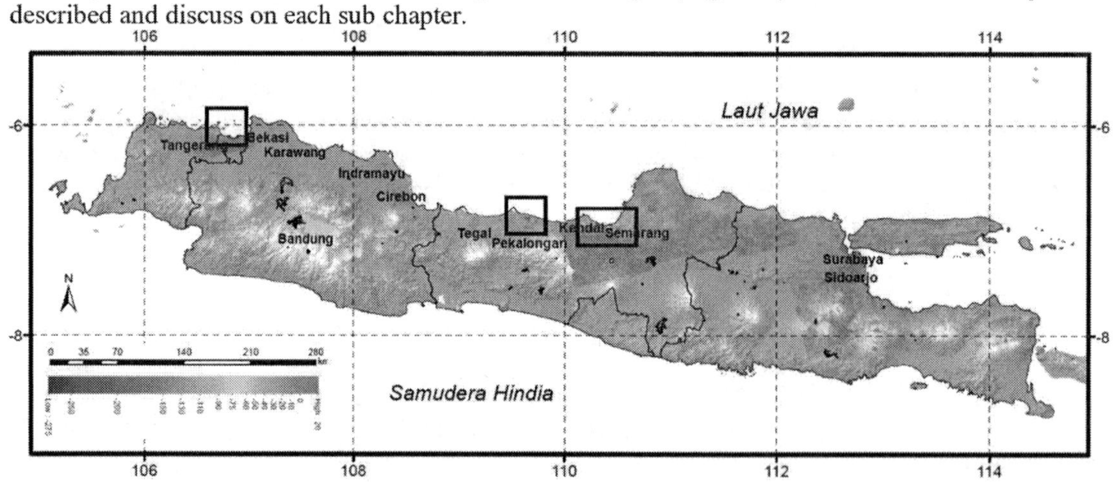

Figure 3 : The Location of the prominent land subsidence in the north coast of java Island as observed from SAR data processing.

3.1. *Jakarta*

Jakarta is the capital city of Indonesia. It is the center for both government and business, as well as the main gate to Indonesia. The rapid development of Jakarta in the last 50 years, make the surrounding area such as Bekasi, Bogor, Tanggerang and Depok was also affected. With population reach more than 25 million peoples in 2016, Jakarta-Bogor-Depok-Tanggerang-Bekasi (Jabodetabek) area is expected to be one of the largest megapolitan city in the world. Beside the population number, Infrastructure development and new industrial area required huge carrying capacity from surrounding area. Access to clean water is one of the most notable problem facing by people in Jakarta and its surrounding area. The limited capability from the government to provide clean water both for household and industry usage, force people to use ground water extensively. Based on observation from 2000 to 2005 in monitoring wells placed in Jakarta, all of them shows the groundwater decreasing level [22].

As can be inferred from figure 1, geologically, Jakarta is dominated by quaternary sediment. This geological condition and excessive ground water exploitation in Jakarta known to caused land subsidence, especially in the northern part of Jakarta. In 1995, [2] report that in the period of 1974 to 1990, the subsidence observed in the northern coast of Jakarta can reach 0.7 meter. The study and monitoring then continued by using geodetic method such as precise leveling, GPS, and InSAR [23] [24] [3]. All study report subsidence rate can reach up to 10 cm/year.

In this study, we report the recent land subsidence going on in Jakarta. We found that the land subsidence continuing with cumulative LOS displacement up to 15 cm, with rate 5 cm/year (**Figure 4**). The spread of subsidence still dominated in the northern coast of Jakarta, and area in West Jakarta. Land subsidence also broaden to the buffer city of Jakarta, such as Tanggerang in the west and Bekasi in the east. We found the rate is much smaller than had been reported in the previous study [6] [24]. However, much recent study provided by JICA [25] shows similar result of subsidence rate using ALOS-PALSAR data.

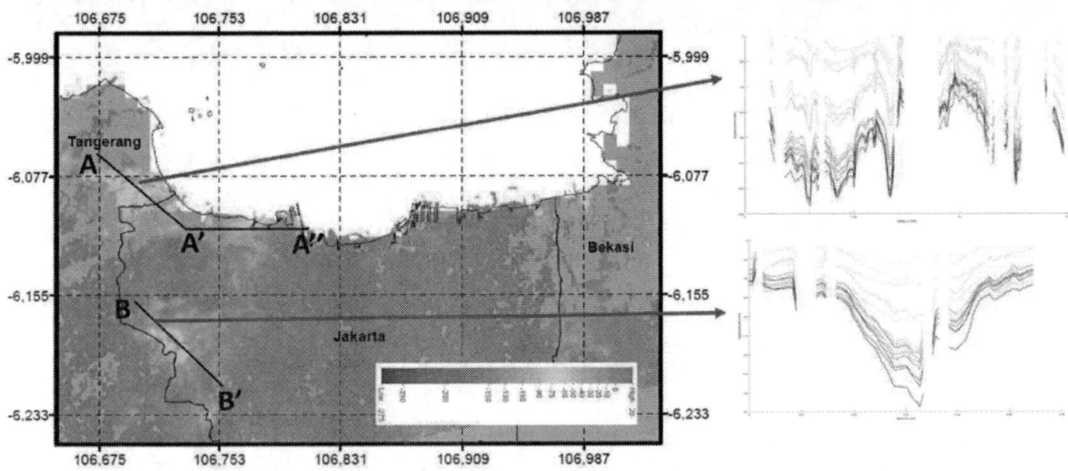

Figure 4 : Land subsidence in Jakarta and its surrounding area (right), and subsidence profile along the line track, lighter color represents earlier data acquisition (2017) and darker color represent recent data acquisition (2020) (left)

3.2. Semarang

As the capital city of Central Java Province, Semarang has become home to 1.6 million peoples. It is one of the largest cities located in the northern coast of Java after Jakarta and Surabaya. Semarang divided into two distinct morphology, coastal lowlands and alluvial deposit in the north-east side, and much more hilly and rough topography in the south-east. Geologically, Semarang formed by three main lithologies, volcanic rock in the southern part, sedimentary rock, and alluvial deposit more in the north coast part (**Figure 1**).

Like Jakarta, Semarang also suffer from land subsidence. This later confirmed by leveling measurement by [26] and reveal the subsidence rate up to 12 cm/year. In 2007, the land subsidence rate in Semarang reported to be as high as 16.5 cm/year [27]. However, different result was also reported by [4] and [5] which report the maximum rate to be around 10-11 cm/year.

The cause of land subsidence in Semarang is triggered by excessive ground water exploitation and damage on groundwater recharge area [26] [4]. Both studies also report the consequences of land

IOP Conf. Series: Earth and Environmental Science **873** (2021) 012078 doi:10.1088/1755-1315/873/1/012078

subsidence, such as damage on infrastructure and sea water inundation. However, for Semarang case, the land subsidence is not solely blame to ground water exploitation, but also from natural compaction which contribute up to 2.2 cm/year [8].

Our result shows that the land subsidence continuing in Semarang, and the cumulative LOS displacement from 2017 to mid-2020 can reach 24 cm, or the average rate of 8 cm/year (**Figure 5**). Interestingly, the maximum displacement not happening in near the coast as in previous study [6], but more to the south. The extension of land subsidence area in Semarang also observed from our result, confirming previous study done by [5]. Outside Semarang city, land subsidence also extends to Kendal City in the west, and Demak in east side of Semarang City.

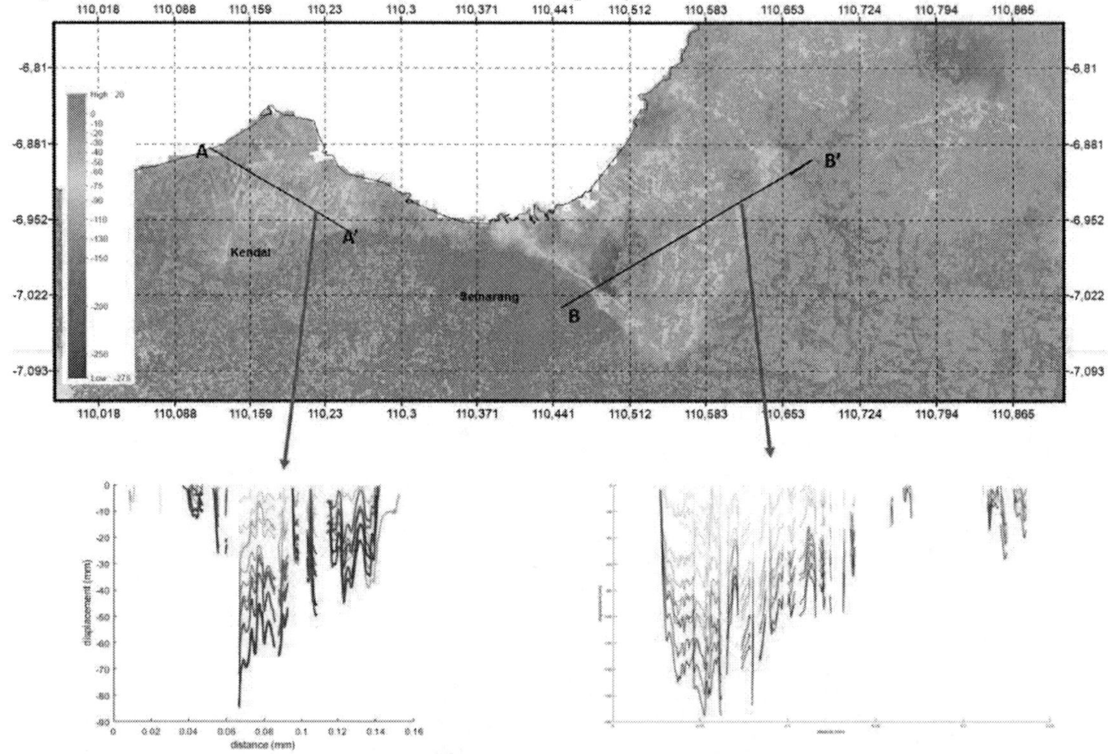

Figure 5 : Land subsidence in Semarang and its surrounding area (top), and subsidence profile along the line track, lighter color represents earlier data acquisition (2017) and darker color represent recent data acquisition (2020) (bottom)

3.3. Pekalongan

Pekalongan located about 80 km in eastern side of Semarang and passed by logistic to and from Jakarta. Administratively, Pekalongan in this context covers both Pekalongan City and Kabupaten (Regency) Pekalongan. It is one of the main industry areas in Central Java, along with history of batik production. Geologically, most of Pekalongan located in alluvial plain in the north, while in the southern part, it divided into quaternary volcano in the southwest, and anticlinorium in the south east part [7].

Environmental issue in the region had drawn more attention from peoples as well as the government since the last decade. The local mass media reported about the worsen of sea water inundation for long time, and later they also report land subsidence phenomena in Pekalongan. The land subsidence Pekalongan was reported to have rate about 3 cm/year [28] to 10 cm/year [6]. The land subsidence in Pekalongan affected tidal flood significantly and projected to cause increasing in flood depth up to 132 cm in 2050 [29].

Our result shows recent subsidence in Pekalongan using Sentinel-1 SAR data. The land subsidence continues with cumulative LOS displacement reach 22 cm from 2017 to mid-2020, or with rate up to 7 cm/year (**Figure 6**). The land subsidence occurs both in Pekalongan city and Pekalongan Regency, although geologically, it occurs in the alluvial plain. The area of affected by land subsidence in Pekalongan like what have been reported by [6].

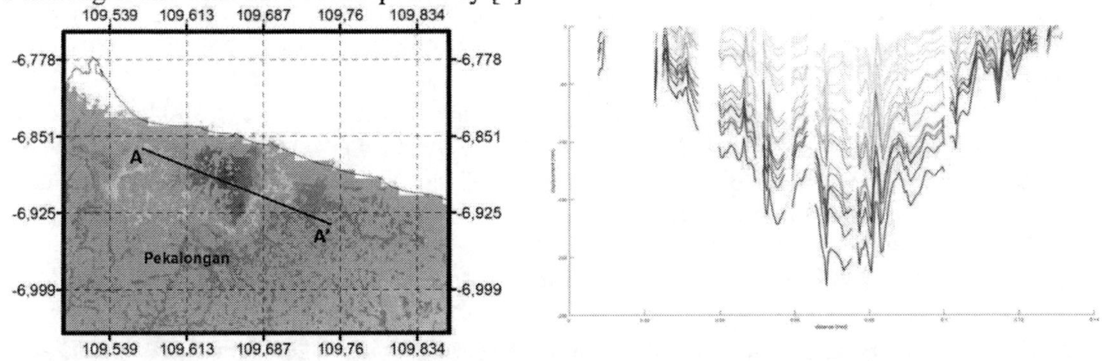

Figure 6 : Land subsidence in Pekalongan Regency (top), and subsidence profile along the line track, lighter color represents earlier data acquisition (2017) and darker color represent recent data acquisition (2020) (bottom)

4. Conclussion

We processed Sentinel-1 SAR data from 2016 to min 2020 in four different tracks to cover all north java coast. Using time-series processing manner, we manage to obtain subsidence rate and land deformation changes both spatially and in time. Our result shows that there are three prominent cities along northern coast of Java suffer from large land subsidence, they are Jakarta, Semarang, and Pekalongan. In Semarang, the area affected by land subsidence widen both to Kendal Regency in the east, and to Demak Regency in the south. Semarang also have fastest subsidence rate up to 8 cm/year, followed by Pekalongan with rate up to 7 cm/year, then Jakarta with rate up to 5 cm/year.

Further study and continuous monitoring of land subsidence in the north coast of Java still need to be continued in the future with better algorithm and more data. Other geodetic method such as GNSS method and leveling is suggested to perform to verify the InSAR method.

Government intervention and strong regulation is needed to slow or stop land subsidence, which from various study conduct in the land subsidence area, the major cause of subsidence is excessive ground water exploitation.

5. References

[1] Syvitski J P 2008 Deltas at risk *Sustainability Science.* **3** 23-32

[2] Hoogeveen R and Leeuwen B V 1995 The large scale development of land subsidence in northwest Jakarta and north Tangerang, Indonesia. *Proc., Symp., Int., 5th on Land Subsidence (The Hague)* pp 433-438

[3] Abidin H Z, Andreas H, Djaja R, Darmawan D and Gamal M 2007 Land subsidence characteristics of Jakarta between 1997 and 2005, as estimated using GPS surveys *GPS Solution* **12** 23-32

[4] Gumilar I, Abidin H Z, Sidiq T.P, Andreas H, Maiyudi R, Gamal M and FukudaY 2013 Mapping And Evaluating The Impact Of Land Subsidence In Semarang (Indonesia) *Indonesian Journal of Geospatial* **2** 26-41

[5] Andreas H, Abidin H Z, Gumilar I, Sidiq T P, Sarsito D A and Pradipta D 2019 On the acceleration of land subsidence rate in Semarang City as detected from GPS surveys *E3S Web of Conferences* **94** 04002

[6] Chaussard E, Amelung F, Abidin H and Hong S H 2013 Sinking cities in Indonesia: ALOS PALSAR detects rapid subsidence due to groundwater and gas extraction *Remote Sensing of Environment* **128** 150-161

[7] Bemmelen R V 1949 *The Geology of Indonesia v. 1* (The Hague: Martinus Nijhoff)

[8] Sarah D, Hutasoit L, Delinom R and Sadisun I 2020 Natural Compaction of Semarang-Demak Alluvial Plain and Its Relationship to the Present Land Subsidence *Indonesian Journal on Geoscience* **7**

[9] Lixin Y, Jie W, Chuanqing D, Guo J W,Yanxiang J and Liu B 2010 Land Subsidence Disaster Survey and Its Economic Loss Assessment in Tianjin, China *Natural Hazards Review* **11** 35-41

[10] Ma K, Zhang Y, Ruan M, Guo J and Chai T 2019 Land Subsidence in a Coal Mining Area Reduced Soil Fertility and Led to Soil Degradation in Arid and Semi-Arid Regions *International Journal of Environmental Research and Public Health* **16** 3929

[11] Howladar M F 2016 Environmental impacts of subsidence around the Barapukuria Coal Mining area in Bangladesh *Energy, Ecology and Environment* **1** 370-385

[12] Abidin H Z, Gumilar I,Andreas H, Murdohardono D and Fukuda Y 2012 On causes and impacts of land subsidence in Bandung Basin, Indonesia *Environmental Earth Sciences* **68** 1545-1553

[13] Keesstra S, Mol G,Leeuw J D,Okx J,Molenaar C, Cleen M d and Visser S 2018 Soil-Related Sustainable Development Goals: Four Concepts to Make Land Degradation Neutrality and Restoration Work *MDPI* **1**

[14] Rosana A 2016 Estimasi Kerugian Ekonomi Pada Bangunan Akibat Penurunan Muka Tanah Menggunakan Rasor di Semarang (Bandung: Institut Teknologi Bandung)

[15] Sarah D, Satriyo N A, Mulyono A and Soebowo E 2014 Kajian Awal Estimasi Kerugian Fisik Akibat Amblesan Tanah di Kota Semarang *Prosiding Pemaparan Hasil Penelitian Pusat Penelitian Geoteknologi LIPI Tahun 2014* (Bandung)

[16] Sandwell D, Mellors R, Tong X, Wei M and Wessel P 2011 Open radar interferometry software for mapping surface Deformation *Eos, Transactions American Geophysical Union* **92** 234-234

[17] Jo M -J, Won J -S, Kim S -W and Jung H -S 2010 A time-series SAR observation of surface deformation at the southern end of the San Andreas Fault Zone *Geosciences Journal* **14** 277-287

[18] Berardino P, Fornaro G, Lanari R and Sansosti E 2002 A new algorithm for surface deformation monitoring based on small baseline differential SAR interferograms *IEEE Transactions on Geoscience and Remote Sensing* **40** 2375-2383

[19] Schmidt D A and Bürgmann R 2003 Time-dependent land uplift and subsidence in the Santa Clara valley, California, from a large interferometric synthetic aperture radar data set *Journal of Geophysical Research: Solid Earth* **108**

[20] Chen C W and Zebker H A 2001 Two-dimensional phase unwrapping with use of statistical models for cost functions in nonlinear optimization *Journal of the Optical Society of America A* **18** 338

[21] Agram P S and Zebker H A 2009 Sparse Two-Dimensional Phase Unwrapping Using Regular-Grid Methods *IEEE Geoscience and Remote Sensing Letters* **6** 327-331

[22] Delinom R M 2008 Groundwater management issues in the Greater Jakarta area, Indonesia. *Proc., Int., Work. on Integrated Watershed Management for Sustainable Water Use in a Humid Tropical Region*

[23] Djaja R, Rais J, Abidin H Z and Kuntjoro W 2004 Land Subsidence of Jakarta Metropolitan Area *Proc., 3rd IG Regional Conference*

[24] Abidin H Z, Andreas H, Gamal M, Djaja R, Subarya C, Hirose K, Maruyama Y, Murdohardono D and Rajiyowiryono H 2005 Monitoring Land Subsidence of Jakarta (Indonesia) Using Leveling, GPS Survey and InSAR Techniques *Int. Association of Geodesy Symposia* Springer-Verlag pp 561-566

IOP Conf. Series: Earth and Environmental Science **873** (2021) 012078 doi:10.1088/1755-1315/873/1/012078

[25] JICA 2019 Project for Promoting Countermeasures against Land Subsidence in Jakarta - Establishment of Implementation Committee and Action Plan, and Capacity Building-

[26] Sutanta H 2002 Spatial Modeling of the Impact of Land Subsidence and Sea Level Rise in A Coastal Urban Setting, case study: Semarang, Central Java, Indonesia

[27] Marfai M A and King L 2007 Monitoring land subsidence in Semarang, Indonesia *Environmental Geology* **53** 651-659

[28] Nashrrullah S, Aprijanto, Pasaribu J M, Hazarika M K and Samarakoon L 2014 STUDY ON FLOOD INUNDATION IN PEKALONGAN, CENTRAL JAVA *International Journal of Remote Sensing and Earth Sciences (IJReSES)* **10**

[29] Pratama M B 2019 Tidal Flood in Pekalongan: Utilizing and Operating Open Resources for Modelling *IOP Conference Series: Materials Science and Engineering* **676** 012029

Laboratory-scaled Azimuthal Resistivity Survey for Fracture Detection

R Salam[1,*] and W W Parnadi[2]

[1]Geography Department, Khairun University, Ternate, Indonesia
[2]Department of Geophysical Engineering, Institut Teknologi Bandung, Jl. Ganesha 10 Bandung 40132, Indonesia

[*]Corresponding author: danipalopo72@gmail.com

Abstract. We conducted azimuthal resistivity survey (ARS) at laboratory scale to study apparent resistivity patterns due to fracture existing in subsurface through physical modeling using test objects buried in a sandbox as well as in a test location outside laboratory building. This survey was divided into 2 experiments, i.e. experiment A and experiment B. In experiment A the survey is implemented on 2.50 m x 1.5 m x 0.81 m sandbox, made of 10mm thick glass plates. Sandstone was used as medium representing quasi homogenous medium. Clay roof tiles as well as steel plates as test objects were buried in the sandbox with three different deep angles: 90°, 45° and 0°. In experiment B this survey was conducted outside laboratory building on the grass field and implemented on 2.50 m x 2.5 m x 1.0 m soil body. Vertical single glass plate as well as vertical double glass plates at 30 cm distance were buried in the soil body. Azimuthal resistivity measurements at 15° angular step using Wenner and dipole-dipole configuration were carried out in both experiments located at 1 point just above anomalous object for experiment A and at 3 points at 15 cm distance from anomalous object for experiment B. As a compliment to ARS we acquired profiling data from two lines parallel and perpendicular to horizontal axis of anomalous object. Our results show that the apparent resistivity pattern can show the direction of anomalous object for both configurations and experiments with a little deviation.

1. Introduction

Fractures are defined as mechanical breaks in rocks originating from strain arising from stress concentrations around the crack, heterogeneity and physical discontinuity [1]. They are formed as a result of tectonic, lithostatic as well as thermal stresses and high fluid pressures. Fractures occur at a wide variety of scales, from microscopic to continental [1]. They form gaps filled with liquid fluid, gas, solid material or their combination. Fracture systems are of great interests for engineer, geotechnical engineer and hydrogeologists. A lot of significant geothermal, water supply as well as petroleum reservoirs form in fractured rocks. They acts as pathways for fluid flow. The existence of fractures is also influence the stability of engineered structures, which are built on surface as well as in subsurface. Studying fracture system are of great importance for many researchers and practitioners. One of geophysical methods suitable for studying fractures is the so-called azimuthal resistivity survey (ARS), a technique belongs to DC resistivity method. The application of ARS technique to fracture problem are widely studied, ranging from practical application in identifying fracture [2, 3, 4, 5] over theoretical development to separate its effect [6] until determining its orientation and characteristics [7, 8, 9, 10].

The existence of fracture system, homogeneous rock layering, oblique layer interface, heterogeneous rock layering as well as their combination in subsurface results in anisotropic effect in the DC resistivity data, that is hard to differentiate. Using only measurement from single azimuth often in inadequate to identify, locate and characterize fracture systems in subsurface [7]. Applying ARS technique based on measurement from multi azimuth is therefore required. ARS method deploys resistivity equipment similar to the equipment used in conventional measurement. This is carried out by measuring the apparent resistivity at various azimuths at a fixed point using a specific electrode configuration. The depth of investigation of this technique can reach 40m – 50 m depending on the electrode spacing used.

Fractures occurred in base rocks cause road and other infrastructures and substructures sitting on weak rocks to collapse. Bumpy road and tilted wall of houses exist. This phenomenon is visibly observed in karstic region in West Java. Since such phenomena are found commonly in Indonesia, investigation of the fracture system in subsurface using ARS technique is extremely needed to be carried out.

This research is aimed at identifying and studying apparent resistivity patterns due to fracture existing in subsurface through physical modeling in laboratory using test objects buried in a sandbox as well as buried in a test location outside laboratory building.

2. Theory

2.1. 1D and 2D electrode configurations

The DC resistivity method studies the variation of rock resistivity beneath the ground by injecting direct current I into the ground through a pair of current electrodes C_1 and C_2 and recording the potential difference ΔV through a pair of potential electrodes P_1 and P_2, resulting an apparent resistivity ρ_a. The apparent resistivity ρ_a is calculated through [3]:

$$\rho_a = K\frac{\Delta V}{I} \quad (1)$$

where K is geometrical factor, depending on electrode position.

Among many electrode configurations available for DC resistivity survey, Wenner and dipole-dipole are two of most used electrode configuration for 1-dimensional (1D) and 2-dimensional (2D) earth model (Figure 1). For Wenner and dipole-dipole, electrode configuration $K = 2\pi a$ and $K = \pi an(n+1)(n+2)$, respectively. a is electrode spacing and n is integer. These K values are valid for 1D and 2D measurements. In the Wenner array the electrodes are uniformly spaced in a line and all electrodes are moved about a fixed center, increasing the spacing a in steps. In the Dipole-dipole configuration, the potential electrodes are closely spaced and remote from the current electrodes, which are also close together and moved about a fixed center between potential electrodes pair and current electrodes pair, increasing the spacing between them.

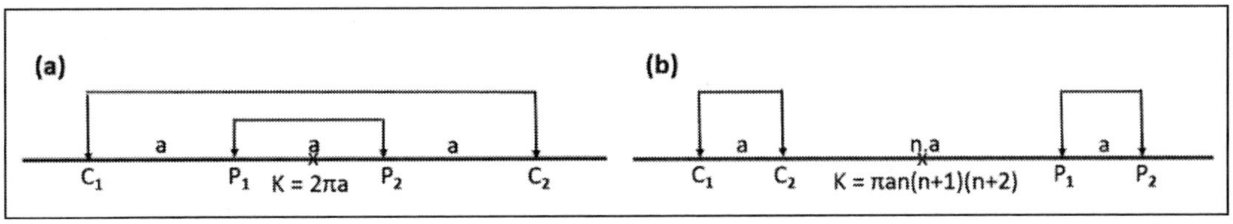

Figure 1. Wenner (a) and dipole-dipole electrode configurations (b) used for 1D and 2D measurements

2.2. Azimuthal Resistivity Survey

Azimuthal resistvity survey (ARS) belongs to a technique in 1D resistivity measurement. The azimuthal Wenner, dipole-dipole as well as square configurations are illustrated in figure 2 where azimuthal data is collected at different azimuths.

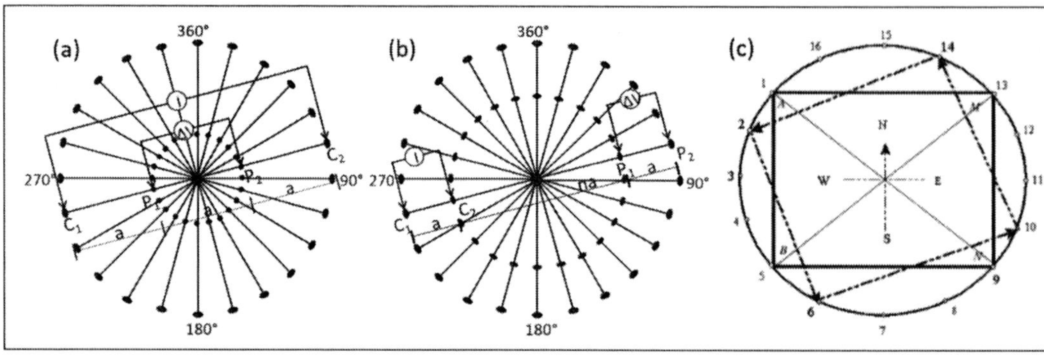

Figure 2. Common electrode configurations used for azimuthal resistivity survey: Wenner (a), dipole-dipole (b) and square arrays (c)

2.3. Azimuthal apparent resistivity plot

The result of azimuthal resistivity measurement is plotted in polar coordinate form as shown in figure 3. An isotropic medium shows a circular geometry whilst an anisotropic medium is elliptical [7], [11]. Best fitted circle and ellipse geometry of the data are shown. To evaluate whether the medium is isotropic or not, the quantity R^2 is used. R^2 is defined as the percentage of the variance σ from the circular model which has been removed by the elliptical model [7]:

$$R^2 = \frac{\left(\sigma^2(circle) - \sigma^2(ellipse)\right)}{\sigma^2(circle)} \tag{2}$$

R^2 values lie between 0 and 1. R^2 values close to 1 indicate very strong anisotropic properties of the rock, while it close to 0 indicate that the medium is isotropic.

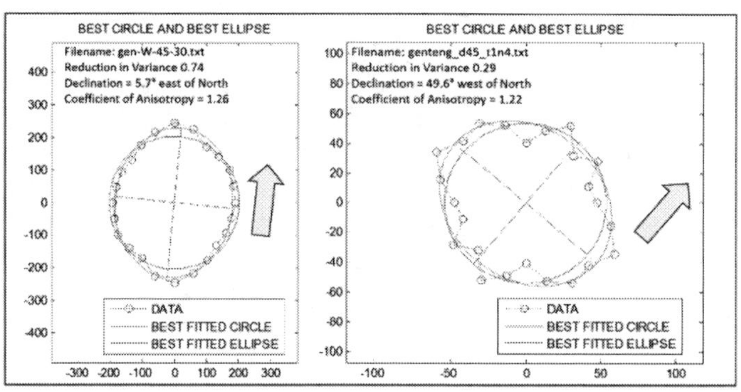

Figure 3. Polar plot of apparent resistivity value versus azimuth. Circles and ellipses of data that fit best are plots on the figure. The orientation of the arrow denotes the strike direction of the fracture-like object.

3. Method

3.1. Experimental setup

We planned and implemented physical modeling through two experiments in this research, i. e. experiment A which was carried out in the laboratory and experiment B which was conducted outside laboratory building.

Experiment A

This survey was conducted in laboratory and implemented on 2.50 m x 1.5 m x 0.81 m sandbox, made of 10mm thick glass plates. Sandstone from a river was used as medium representing quasi homogenous medium. Clay roof tiles as well as steel plates as test objects were buried in the sandbox (figure 4). Clay roof tiles with moderate resistivity values were used that represent a moderately conductive fracture zone in a resistive rocks whereas steel plate was used simulating a highly conductive value in a resistive rocks. The anomalous object was placed with three different dip angles: 90°, 45° and 0°.

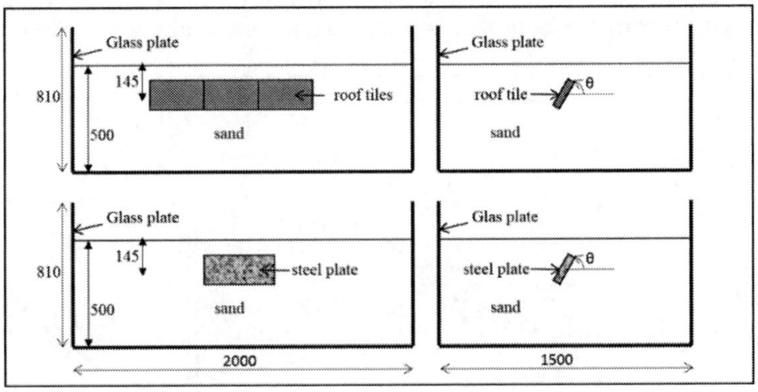

Figure 4. Sketch of physical model of sandbox in broad side (links) and long side views (right). Clay roof tiles (top) and steel plate (bottom) are used as test objects. Θ are varied: 0°, 45° and 90°. The external dimensions of sandbox are 2.00 m x 1.50 m x 0.81 m. Dimensions are in mm.

Experiment B

This survey was conducted outside laboratory building on the grass field and implemented on 2.50 m x 2.5 m x 1.0 m soil body. To make the soil act as a homogeneous medium, the soil dug first dug and crushed in such a way that the soil conditions are close to homogeneous. The soil was then returned to the excavation hole. Glass plates with resistivity values far below that of soil, acting as test objects were buried in the soil body (figure 5). Vertical single glass plate as well as vertical double glass plates at 30 cm distance were buried in the soil body, simulating vertical single fracture as well as vertical double fractures.

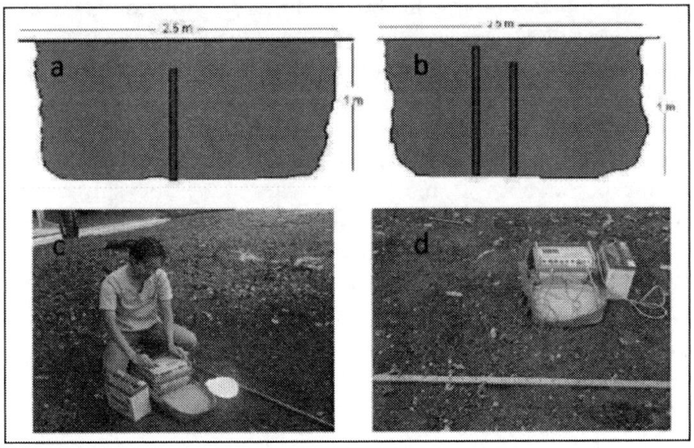

Figure 5. Sketch of physical model of a vertical single glass plate (a) as well as vertical double plates buried in soil environment (b) and their resistivity measurements (c) with Mc OYO resistivity meter (d). The dimensions of physical model are 2.50 m x 2.50 m x 1.00 m. Glass plates are used as test objects representing fracture at small scale.

3.2. Azimuthal resistivity measurement

Azimuthal resistivity measurements were carried out on soil surface above the anomalous objects (figure 6). For experiment A and experiment B with single glass plate, 1 measurement point (point 1) just above a test object is chosen as the location for azimuthal resistivity measurement. For model with vertical double glass plate, azimuthal resistivity measurements was carried out at 3 points locating at 15 cm distance from glass plate as shown in figure 6. For all measurement points in experiment A and B, collinear Wenner as well as collinear dipole-dipole electrode configurations were used and data were acquired at 15° angular step. As a compliment to Azimuthal resistivity measurement, 2D profiling data from two lines parallel and perpendicular to horizontal axis of anomalous object were acquired.

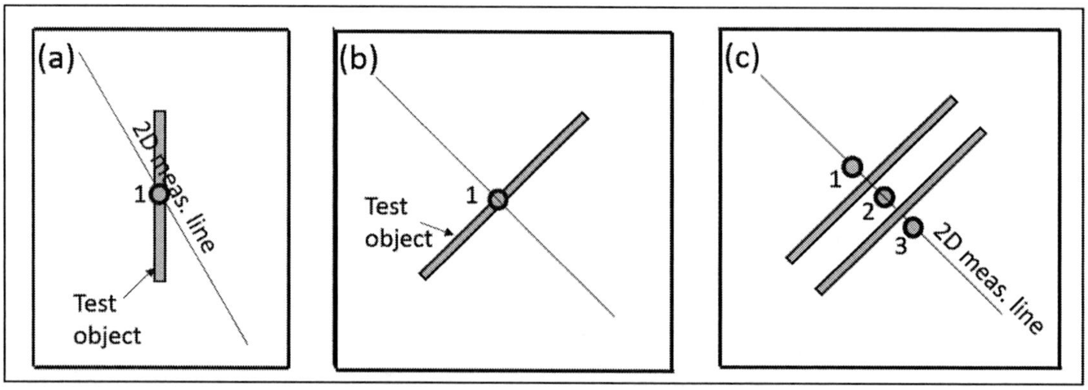

Figure 6. Sketch of measurement point location for experiment A (a), for experiment B with a vertical single glass plate (b) and for experiment B with vertical double glass plates (c).

4. Results and discussion

Figure 7 shows polar apparent resistivity plot at point 1 from model in experiment A using clay roof tiles and steel plate as anomalous objects at 3 different angles: 0°, 45° and 90°. Average calculated orientation of the anomalous objects is shown with grey arrow and the real anomalous object direction is shown with white arrow. All results show consistent anomalous object direction with a slightly average deviation of 12.6° from the real direction. Orientation trend from all data using Wenner and dipole-dipole configuration looks similar. This orientation deviation from the real one is probably due to the sand condition which is not a homogeneous medium and other error occurred during measurement.

Results from azimuthal resistivity measurement from model in experiment B show better results as seen in figure 8. All results showing the direction of anomalous objects approximating 45° that is the direction of both vertical anomalous objects as illustrated in figure 6. Those results are valid for point 1, 2 and 3. Slightly deviation shown in the dipole-dipole data from measurement at point 3. Similar to the results from model in experiment A is most likely caused by the soil condition which is not a homogeneous medium. This condition are obvious from the 2D resistivity profile as seen in figure 9. It seems that azimuthal resistivity measurement using dipole-dipole configuration is more sensitive than that of Wenner configuration in determining the anomalous object orientation. Nevertheless, it is obviously seen that the correct position of anomalous objects in experiment B is visible in the data.

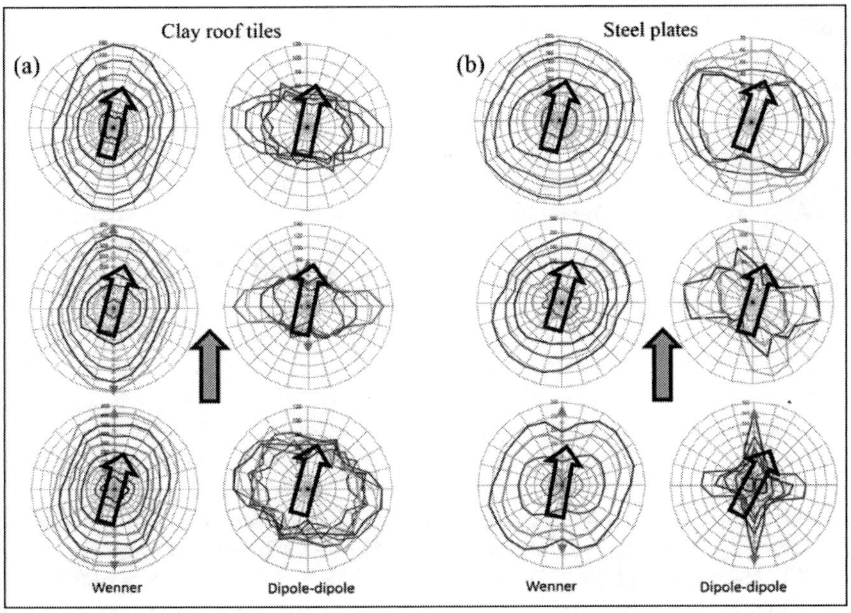

Figure 7. Polar apparent resistivity plot at point 1 from model in experiment A: (a) clay roof tiles and steel plate. Results using Wenner (left) as well as dipole-dipole configurations (right) are shown. Results at different dip angle are shown from top](0°), middle (45°) to bottom (90°). Grey arrow shows the direction of anomalous body in the sandbox and white arrow pinned at the plot is average direction determined from data.

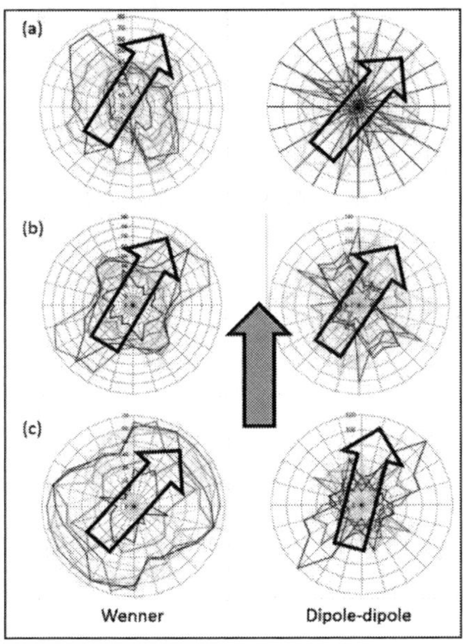

Figure 8. Polar apparent resistivity plot at point 1 to 3 from model in experiment B with vertical double glass plates at point 1 (a), point 2 (b), and point 3 (c).

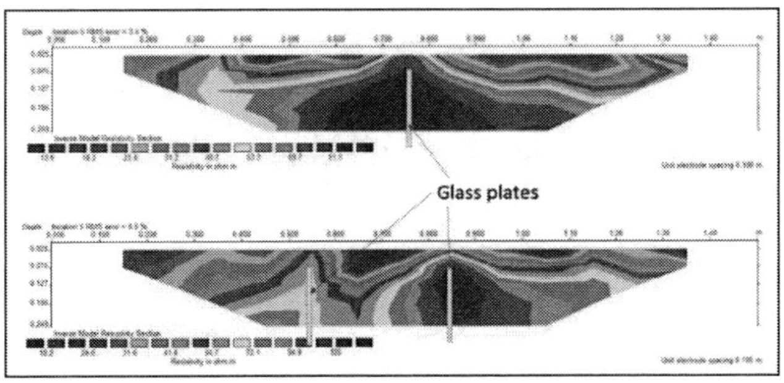

Figure 9. Two-dimensional resistivity profile of model in experiment B with vertical single glass plate (top) and vertical double glass plates (bottom). The position of anomalous objects is shown

5. Conclusions

We conducted azimuthal resistivity survey (ARS) on physical model using test objects buried in a sandbox in our laboratory for experiment A as well as in a test location outside laboratory building for experiment B. Azimuthal resistivity measurements at 15° angular step using Wenner and dipole-dipole configuration were carried out in both experiments located at 1 point just above anomalous object for experiment A and at 3 points at 15 cm distance from anomalous object for experiment B. As a compliment to ARS we acquired profiling data from two lines parallel and perpendicular to horizontal axis of anomalous object. Our results show that the apparent resistivity pattern can show the direction of anomalous object for both configurations and experiments with a little deviation.

Acknowledgments

The authors thank Mr. Ilham Zulkifli for his helpful assistance during data acquisition and processing.

References

[1] Committee on Fracture Characterization and Fluid Flow 1996 *Rock fractures and fluid flow*: National Academy of Sciences 568pp.

[2] Al-Menshed F H and Thabit J M 2016 The use of an azimuthal resistivity survey to detect the flow direction of hydrocarbons in the unsaturated zone at Karbala Governorate, Iraq *Environ Earth. Sci.* **75** 1328 doi: 10.1007/s12665-016-6120-5.

[3] Lane Jr J W Haeni F P and Watson W M 1995 Use of square-array direct-current resistivity method to detect fractures in crystalline bedrock in New Hampshire *Groundwater* **33** 3 pp 476-485

[4] Oladunjoye M A Adefehinti A and Ganiyu K A O 2019 Geophysical appraisal of groundwater potential in the crystalline rock of Kishi area, Southwestern Nigeria *J. African Earth Sciences* **151** pp 107–120

[5] Tepper D H and Phelan D J 2001 *Use of Azimuthal Square-Array Direct-Current resistivity to locate a geologic structure at Fort Detrick, Maryland*: Open-File Report 01-285, US Dept. Interior & US Geological Survey

[6] Watson K A and Barker R D 1999 Differentiating anisotropy and lateral effects using azimuthal resistivity offset Wenner soundings *Geophys.* **64** 3 pp 739-745

[7] Ogunbo J N Mamukuyomi E A Adepoju W S Adebowale H Akinro O and Ukaegbu C R 2018 Panoramic azimuthal Schlumberger Vertical Electrical Sounding for fracture orientation and anisotropy quantification *Heliyon* **4** e00998 doi: 10.1016/j.heliyon.2018. e00998 pp 3-13

[8] Ritzi Jr R W and Andolsek R H 1992 Relation between anisotropic transmissivity and azimuthal resistivity surveys in shallow fractured, carbonate flow systems *Groundwater* **30** 5 pp 774-780

[9] Boadu F K Gyamfi J and Owusu E 2005 Determining subsurface fracture characteristics from azimuthal resistivity surveys: A case study at Nsawam, Ghana *Geophys.* **70** 5 pp B35-B42

[10] Taylor R W and Fleming A H 1988 Characterizing jointed systems by azimuthal resistivity surveys *Groundwater* **26** 4 pp 464-474

[11] Hart D and Rudman A J 1997 Least-squares fit of an ellipse to anisotropic polar data: application to azimuthal resistivity surveys in Karst regions *Computers & Geosciences* **23** 2 pp 189-194

Strong ground motion simulation of the 27 May 2006 Yogyakarta earthquake using Empirical Green's Function

Yeremia Hanniel[1], Ade Anggraini[1], Agus Riyanto[2], Drajat Ngadmanto[2], Wiwit Suryanto[1*],

[1]Seismology Research Group, Department of Physics, Universitas Gadjah Mada, Sekip Utara PO Box Bls 21, Yogyakarta Indonesia
[2]Meteorologi, Climatology and Geophysics Agency (BMKG) Yogyakarta Office, Jl.Wates Km 8, Gamping, Yogyakarta

[*] Corresponding author's e-mail: ws@ugm.ac.id

Abstract. On May 27, 2006, 05:54 am local time, a moderate crustal earthquake of magnitude Mw 6,3 struck the Yogyakarta province, especially in the Bantul regency in the south part of the province. The earthquake damaged or destroyed more than 400,000 houses and buildings and caused more than 5,700 people killed. Several earthquake stations recorded the ground vibration caused by the mainshock very well, except at the stations closest to the earthquake source, namely YOGI in Gamping, West of Yogyakarta, which experienced saturation due to significant vibration. Therefore, information about the maximum ground acceleration near the source is yet not known. We model the ground vibrations near the earthquake source using a stochastic Green's Function approach to obtain information about the ground motions' maximum amplitude. The earthquake source parameters we referred to is the moment tensor solution from the Harvard Moment Tensor. The calculations show that the amplitude is consistent with observations recorded at the BJI Banjarnegara (0.04g) and YOGI Yogyakarta (0.3g).

Keywords: empirical, stochastic Green's function, strong motion, Yogyakarta earthquake

1. Introduction

Yogyakarta Special Region is prone to earthquakes as the collision of the Indo-Australian plate with the Eurasian plate on its south results in high tectonic activity. The most recent major earthquake in Yogyakarta Special Region took place on 27 May 2006 at 05.54 WIB with a magnitude of Mw 6.4. This medium magnitude earthquake claims more than 5,700 lives and destroyed thousands of buildings, including residential house and public facilities [1]. Indonesian Meteorological and Geophysical Agency (BMG, now BMKG) had 27 seismic stations recording the Yogyakarta earthquake on 27 May 2006. However, data from several seismic recording stations were inadequate for analysis due to some instrument problems. Only YOGI (Yogyakarta) and BJI (Banjarnegara) seismic stations located less than 100 km away from the epicentre (Figure 1) could provide relevant data for analysis [2]. However, the mainshock's velocity seismogram recorded by the YOGI station showed the sign of waveform clipping. Typically, a clipped waveform must be excluded from further earthquake parameters estimation [3]. Applying correction procedures to this particular station is necessary as the YOGI station was the closest seismic recording station to the epicentre and damaged areas. Such near-field station potentially holds valuable information of earthquake source and local structure.

Stochastic Green's Function simulation performed in this study is part of the semi-empirical method in strong ground motion prediction. The semi-empirical process is widely known for predicting an earthquake's source, path, and site characteristics. Also, strong ground motion predicted by semi-empirical methods is in the range of critical engineering frequency [4].

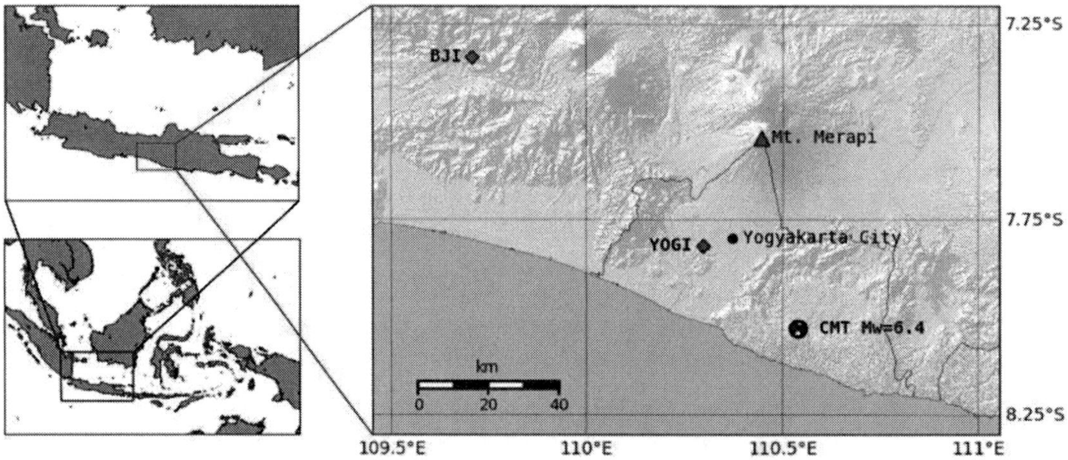

Figure 1. Regional map showing epicentre (blue beach ball) and two nearest seismic recording station from the epicentre (red diamond).

2. Stochastic Green's Function

Ground motion spectrum contains information on how an earthquake physically occurs and how a wave propagates. Stochastic Green's function method simplifies numerous factors of ground motion spectrum to simple equations. The total spectrum of ground motion at a site ($Y(M_0, R, f)$) could be separated into:

$$Y(M_0, R, f) = E(M_0, f)P(R, f)G(f)I(f), \tag{1}$$

where E is earthquake source, P is wave propagation path, G is site effect, and I is instrument or type of motion, M_0 is the seismic moment, R is closest distance to rupture surface, and f is frequency [5].

3. Ground motion spectrum simulation

Firstly, to obtain a simulated ground motion spectrum, we generated a random Gaussian white noise for a given duration based on earthquake event duration [6]. After that, the noise was windowed using an exponential window. We then transformed the windowed noise into the frequency domain using Fast Fourier Transform (FFT) to be normalized by the square root of the mean square amplitude spectrum [6]. Finally, we multiplied the normalized noise spectrum by the ground motion spectrum (Y) and transformed it back into the time domain using FFT (Figure 2).

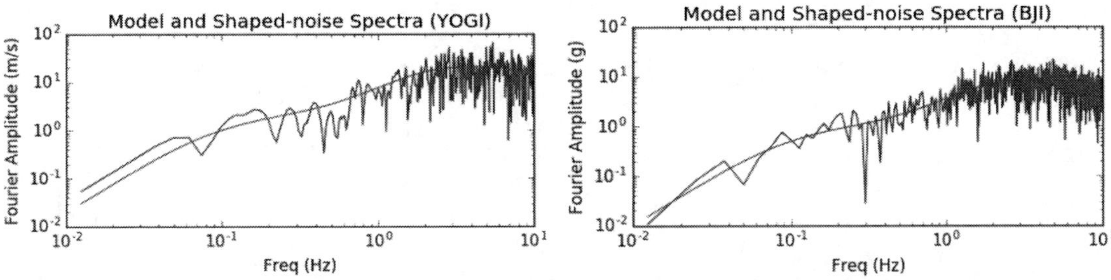

Figure 2. Ground motion spectrum (*Y*; red line) and final spectrum (blue line) from the multiplication of normalized noise spectrum and ground motion spectrum for YOGI and BJI.

4. Results and discussion

We compared the final simulated acceleration waveforms with observed acceleration waveforms from N-S components in the YOGI and BJI case to measure how effective the method is in predicting the ground motion spectrum. Since YOGI records had clipped waveforms, we first compared simulated and observed acceleration waveforms in the BJI case. The bottom plot in Figure 3 shows similarities in waveform shapes and peak ground acceleration (PGA) values between simulated and observed acceleration waveforms at BJI. The simulated PGA value at BJI is around 0.469 g, which is only slightly larger than the observed PGA value at BJI.

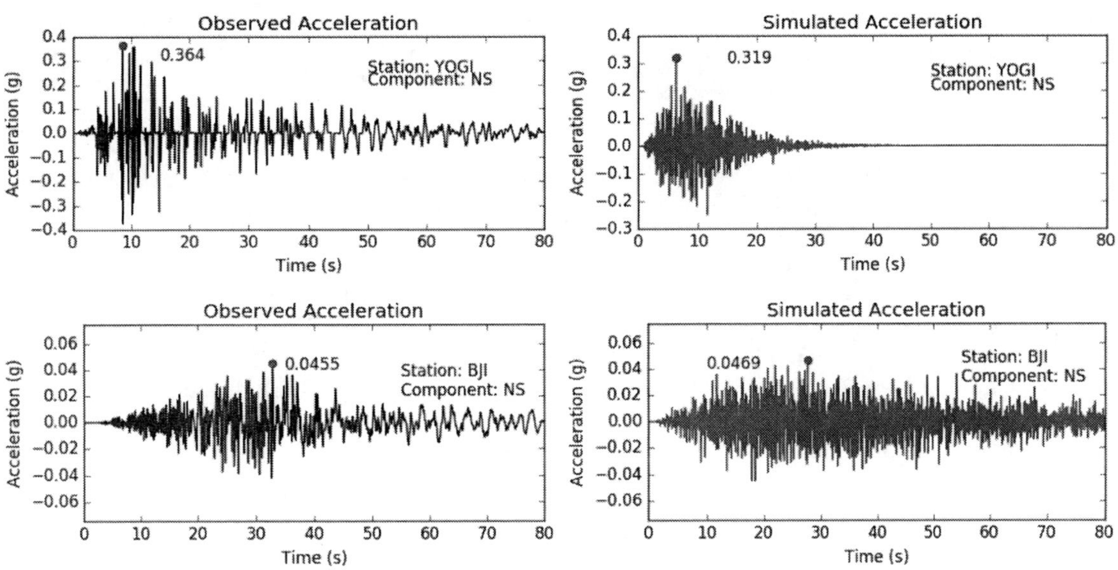

Figure 3. Comparison of observed and simulated acceleration waveforms at YOGI (Yogyakarta) and BJI (Banjarnegara).

The top plot in Figure 3 shows both simulated and observed acceleration waveforms have similar shapes and PGA values. However, the difference of simulated and observed PGA values in the YOGI case is more significant than in the BJI case. Simulated PGA value at YOGI is approximately 0.319 g, while observed PGA values at YOGI is about 0.364 g.

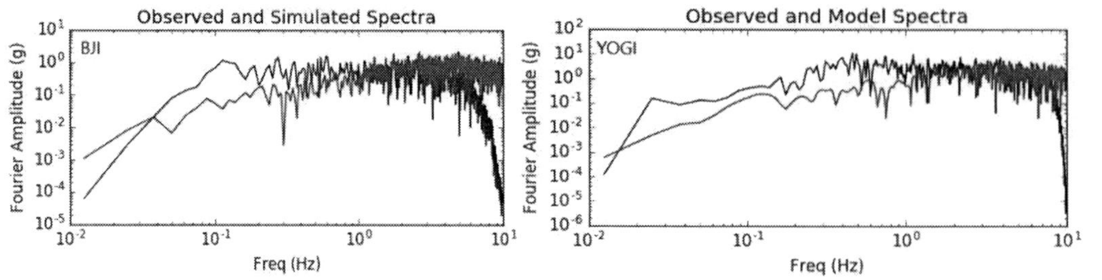

Figure 4. Comparison of simulated and observed acceleration spectrum in the frequency domain at YOGI and BJI.

Figure 4 shows that stochastic Green's function effectively simulates ground motion in a frequency range starting from 0.1 Hz up to 10 Hz. It is the frequency band of interest in engineering seismology ($f > 0.1$ Hz) [5].

5. Conclusion

We have seen that the stochastic Green's function method effectively predicts strong ground motion at a site or a region with limited earthquake data. It is also apparent that the simulated acceleration waveforms fall in a frequency range that becomes the main interest in engineering seismology.

Acknowledgement

BPPTN supported this research through Physics Department Research Grant 2018. Geophysics Laboratory of UGM provided computation facilities for this research.

References

[1] Diambama A D, Anggraini A, Nukman M, Lühr B G, Suryanto W 2019 Velocity structure of the earthquake zone of the M6.3 Yogyakarta earthquake 2006 from a seismic tomography study, *Geophys J Int* **216** (1) pp 439-452
[2] Elnashai A S, Sung J K, Gun J Y and Sidarta D 2007 The Yogyakarta Earthquake of May 27th, 2006 *MAE Center Report* Vol 7 (Urbana-Champaign: the University of Illinois at Urbana-Champaign) pp 15–22
[3] Yang W and Ben-Zion Y 2010 *Seismol. Res. Lett.* **81** 53–61
[4] Tsurugi M 2013 Recipe for Predicting Ground Motion from Crustal Earthquake *Pure and Applied Geophysics* **168** 1 pp 85-104
[5] Boore D M 2003 Simulation of Ground Motion Using the Stochastic Method *Pure Appl. Geophys.* **160** pp 635-676
[6] Boore D M 1983 Stochastic Simulation of High-frequency Ground Motions Based on Seismological Models of the Radiated Spectra *Bull. Seismol. Soc. Am.* **73** pp 1865-1894

Integrating Ground Penetrating Radar, Induced Polarization and Aerial Photograph to Analyze Land Subsidence in Borehole Mining Operation Area: A Case Study from South Bangka Island

Timothy Antonio[1], Wahyudi W Parnadi[2] and Mohammad Nur Heriawan[3]

[1] Master Program of Mining Engineering, Faculty of Mining and Petroleum Engineering, Institut Teknologi Bandung

Email : Timothya.antonio@students.itb.ac.id (Timothy Antonio)
[2] Research Group of Earth Resources Exploration, Faculty of Mining and Petroleum Engineering, Institut Teknologi Bandung

Email : wahyudi@gf.itb.ac.id (Wahyudi Parnadi W Parnadi)
[3] Research Group of Applied Geophysics and Exploration, Faculty of Mining and Petroleum Engineering, Institut Teknologi Bandung

Email : heriawan@mining.itb.ac.id (Mohammad Nur Heriawan)

Abstract. The remaining alluvial tin reserves around large ex-mining area can be an opportunity to optimize the potential reserves using a small-scale mining system. One of the latest breakthroughs in the mining system for the remaining alluvial tin reserves on land is the borehole mining (BHM) method. However, mining operations using BHM have several environmental impacts especially related to its former holes. This study aims to identify the changes in geometry of the former holes both on the surface and below the surface through direct field observations and measurements using geophysical methods. The study site is in the large ex-mining area of Air Nudur Block in South Bangka Regency, Bangka Belitung Islands Province. The geophysical methods used to identify the changes in the geometry of the BHM holes are GPR (ground penetrating radar) and IP (induced polarization). The GPR method has high resolution subsurface visual capabilities at shallow depths and this method can distinguish between stable and unstable layers. The IP method is used to determine the overburden layer and the condition of the alluvial layer where some of it has become a hole. These two methods will mutually reinforce in interpreting the change in the geometry of the BHM holes. Changes in hole diameter and depth were measured directly in the field within 2 (two) months from the completion of mining operations using the BHM method followed by the periodic topographic measurements using UAV (unmanned aerial vehicle). The results of observations on changes in the elevation of the ground surface around the former BHM hole indicated that the land level subsided by about 1 m. Meanwhile, the results of GPR data processing showed changes in the hole geometry with the hummocky radargram pattern into chaotic. Interpretation of IP data processing shows resistivity values of 250 - 1700 ohm.m and chargeability of 15 - 25 ms in the disturbed alluvial layer around the former BHM holes.

.Keywords: *BHM, GPR, IP*, topographical change, former holes

1. Introduction

Air Nudur large mine is no longer operated but there is a plan to remining alluvial tin reserves large ex mining area and optimation . Re-mining is carried out with the aim of optimizing the remaining reserves that have not been mined optimally[1]. One of the latest breakthroughs in the mining system for the remaining alluvial tin reserves on land is the borehole mining (BHM) method. Mining operations using BHM have several environmental impacts especially related to its former holes of BHM[2]. This study is to identify the land subsidence around in BHM area and the changes former of geometry holes both on the surface and below the surface through direct field observation and measurement using geophysical methods is needed to avoid this environmental impact on both surface and subsurface. Meanwhile, subsurface visualization can be measured using the Ground-Penetrating Radar (GPR) method and the Induced Polarization (IP) method[3]. The actual method and the geophysical method are expected to be able to identify and monitor the occurrence of land subsidence in the mine environment[4]. Geophysical methods can be used to identify subsurface and can visualize subsurface conditions. The purpose of the GPR method is to detect objects that are under the ground with a certain level of depth using radio waves that have a frequency 100 MHz[6]. The GPR method has high resolution subsurface visual capabilities at shallow depths and this method can distinguish between stable and unstable layers[4]. The IP method is used to determine the overburden and alluvial layer[4].

IOP Conf. Series: Earth and Environmental Science **873** (2021) 012081 doi:10.1088/1755-1315/873/1/012081

2. Research Method

The GPR and IP acquisition survey was conducted after the presence of aerial photographs to illustrate the target area of the hole that will be targeted for monitoring. Aerial photographs are also used as non-renewable location update data from satellites. In addition, it is used to model the topographic surface that will be validated by the GPS RTK to see the alluvial upper limit. IP measurement was conducted on GPR lines so that the data can be correlated and can connect the two models for interpretation.

a. Ground Penetrating Radar (GPR)

Some cases in geological environmental problems can be solved by GPR with a 100MHz antenna and good resolution can be delivered to provide a shallow layer image[4]. GPR acquisition using MALA X3M system with 100 MHz frequency was carried out for shallow or medium resolution targets [3] Figure 1. Measurements with a trigger as a provider of instructions to the radar system to collect recorded data with an interval of 0.1 m[6]. GPR acquisition was carried out on 5 BHM holes, namely BHM 11 to BHM 15. Trajectories were made of 4 trajectories each hole directed in all directions and clockwise. Data processing is done using Reflexw 6.0.7 software through a process of static correction, substract-mean, gain, bandpass and topographic correction to interpret the radargram data. Interpretation is doing by finding in research area and look the proceesing section which will be linked with theories so we can get the accurate and correct information

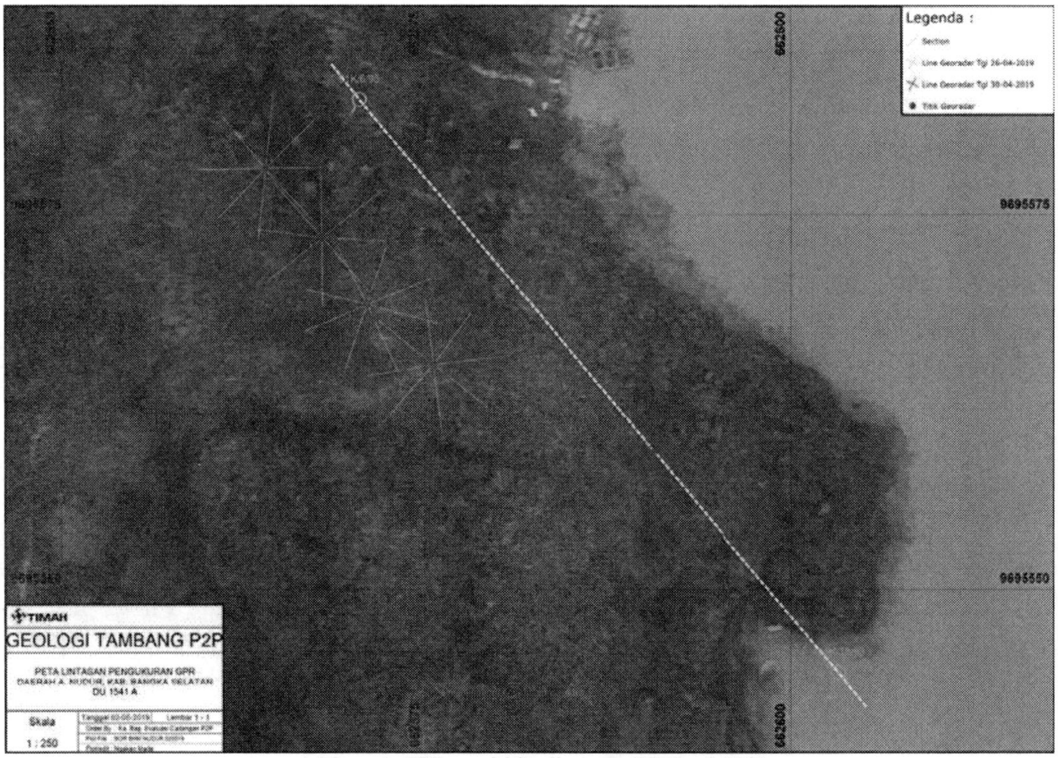

Figure 1. GPR acquisition line map (PT.Timah Tbk)

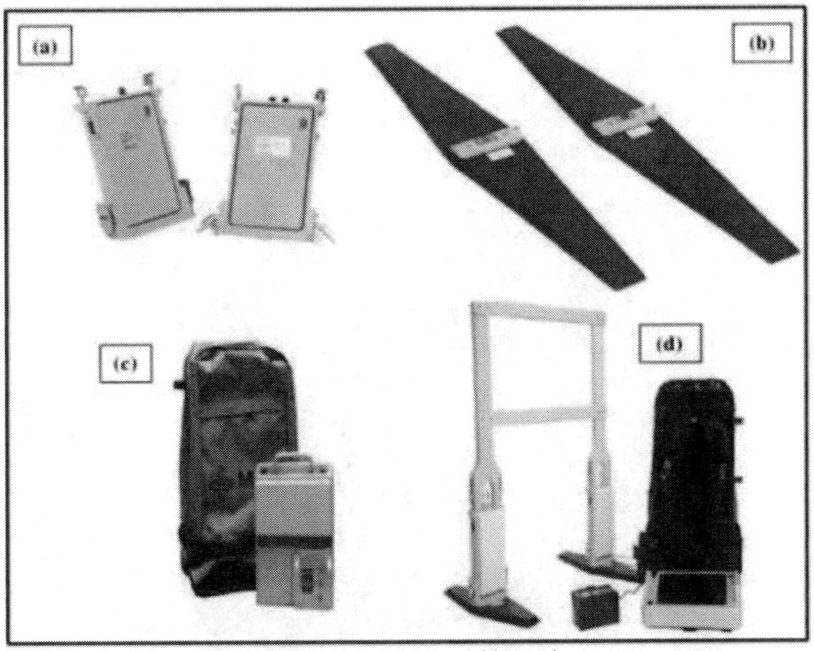

Figure 2. GPR acquisition tools

b. Induced Polarization (IP)

IP acquisition data at 1.5 m electrode spacings was collected. The measurement of IP has a trajectory from northwest to southeast. IP measurement uses a set of 4point light hp 10w of Lipmann tool and switchboxes were mounted on a cable and connected directly to the electrodes Figure 2. This acquisition uses a multichannel dipole-dipole configuration and fast data acquisition measurement. Data processing is done using RES2DINV to determine the resistivity and chargeability distribution to get a 2-dimensional cross section of the inversion model[1].

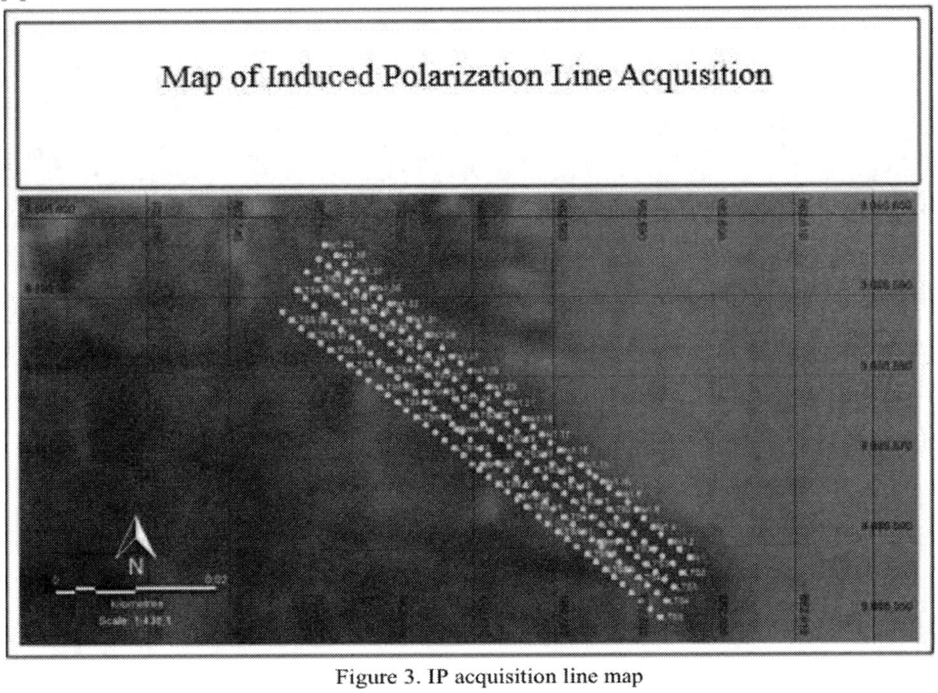

Figure 3. IP acquisition line map

c. ***Topography***

Topography is used for monitoring changes in elevation data. Retrieval of data using drones, GPS. The monitoring was carried out for 4 weeks starting from mid-April 2019 to mid-May 2019. Topographic monitoring was used to see the surface deformation in the BHM hole area and around the BHM hole, which then made a topographic cross section for each week. Topographic section A-A 'represents holes BHM 15 and BHM 14, topographic sections B-B' represent holes BHM 13 and BHM 12, topographic section C-C 'represents holes BHM 11 in Figure 4.

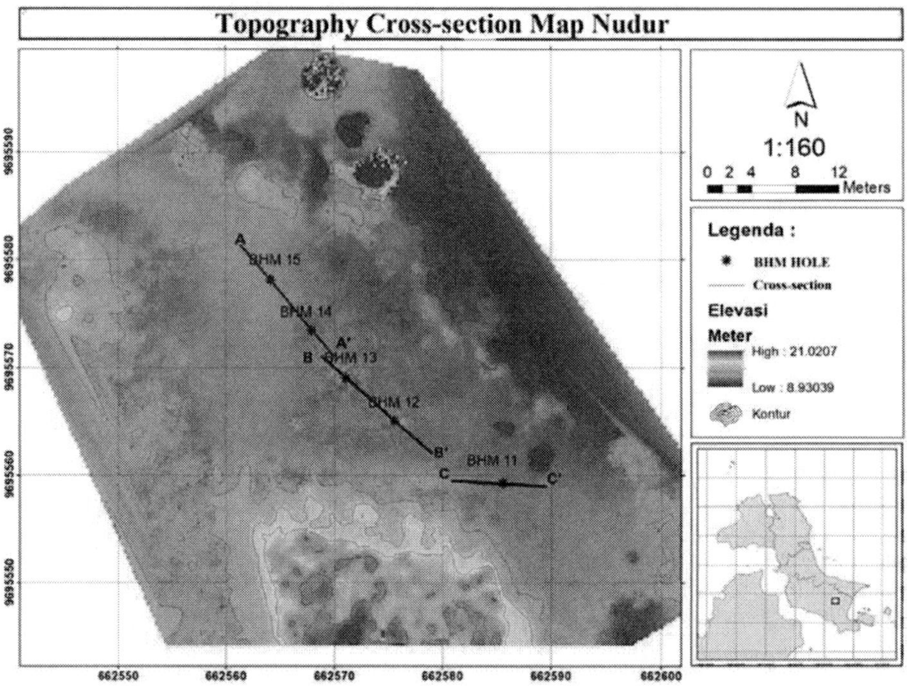

Figure 4. Topography cross-section map

3. Result and Discussion

The results of the three methods of GPR, IP and topography show that there is land subsidence based on table measuring the diameter and depth of BHM holes. The table 1 shows the changes of borehole geometry and land subsidence in BHM holes. The change in diameter from the first week to the 4th week shows that there is material on the surface that has collapsed into the hole, result that subsidence shows around borehole.

Table 1. Table measuring the diameter and depth of BHM holes

NO	BHM Hole ID	Diameter (m)			Depth (m)		
		Week			Week		
		I	II	III	I	II	III
1	BHM 11	0.16	0.2	0.3	7.5	7.5	7.5
2	BHM 12	2.9	3.28	3.42	1.7	1.7	1.5
3	BHM 13	3.2	3.24	3.52	1.29	0.98	0.89
4	BHM 14	0.85	1.38	1.52	1.54	1.36	1.09
5	BHM 15	2.4	2.54	3.1	3.3	0.78	1

a. ***Cross section topography***

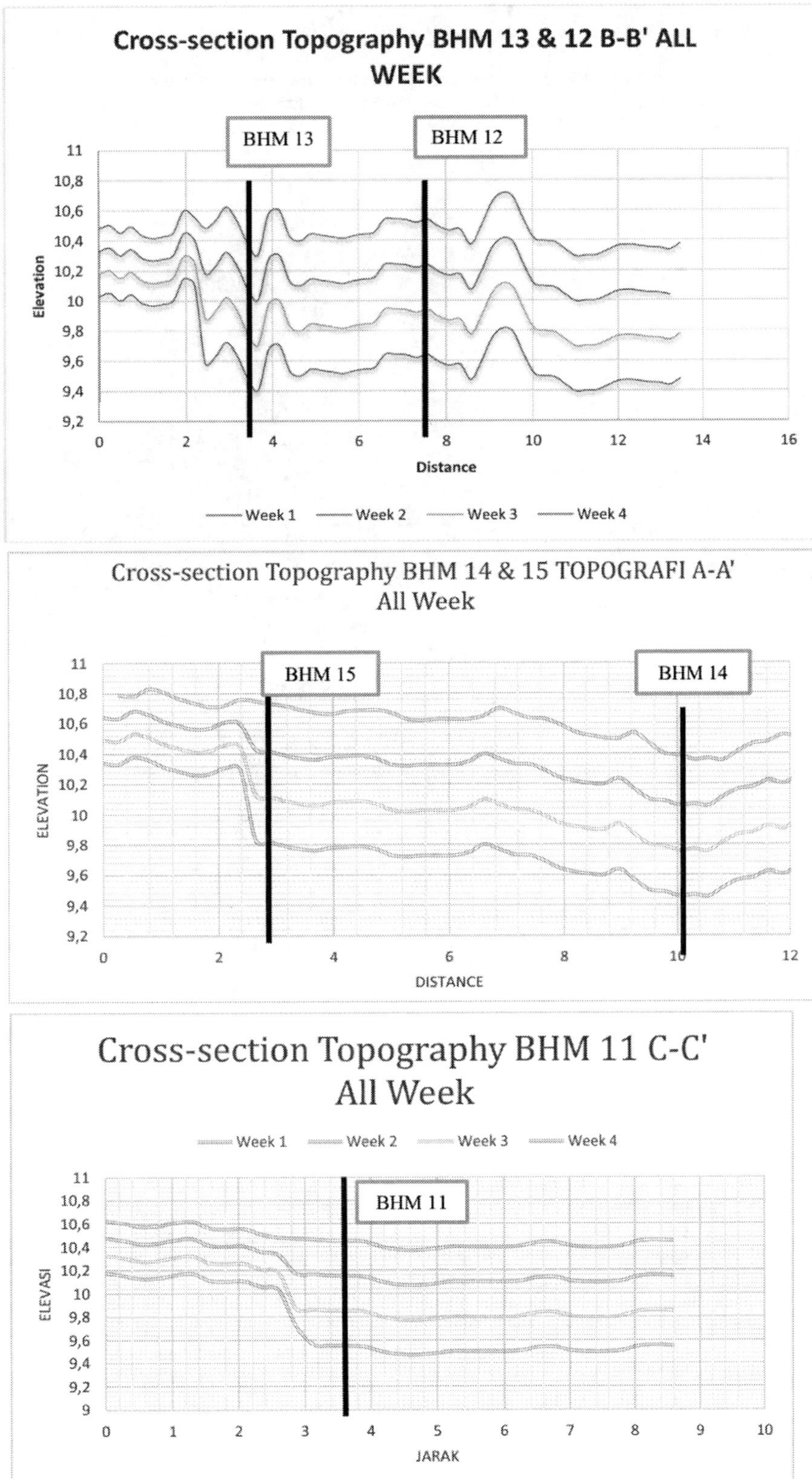

Figure 5. Cross Section Topography BHM All Week

Topography cross-section shows that all topography cross-section shows that BHM Holes experienced land subsidence of 1 m for 4 weeks. In the GPR interpretation, the Radargram used is BHM12 and the line used to interpret is BHM 12_1 – BHM 12_1' to represent another BHM hole. The subsidence in subsurface conditions are explained through GPR interpretation. GPR interpretation shows a change in the radargram pattern in Figure 6 A to Figure 6 B, a change in the radargam pattern from hummocky to chaotic shows a decrease in the subsurface layer. subsidence below the surface occurs to a depth of 1.7 m. subsurface decline occurred during 3 weeks is 0.3 m

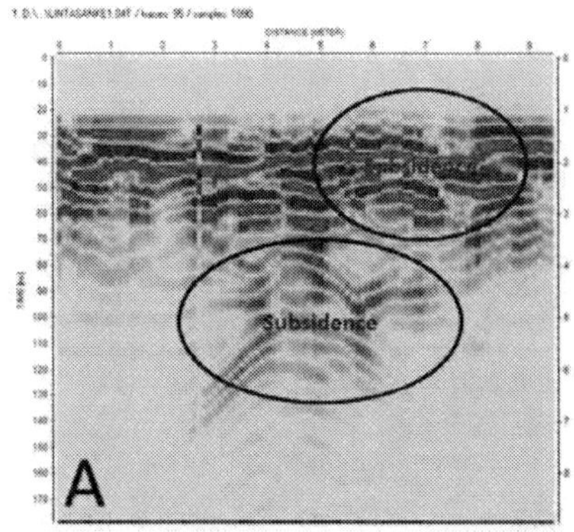

Figure 6A. Result of GPR Monitoring BHM Cross section 12_1-12_1 'April 2019 and (B) BHM Cross section 12_1-12_1' May 2019

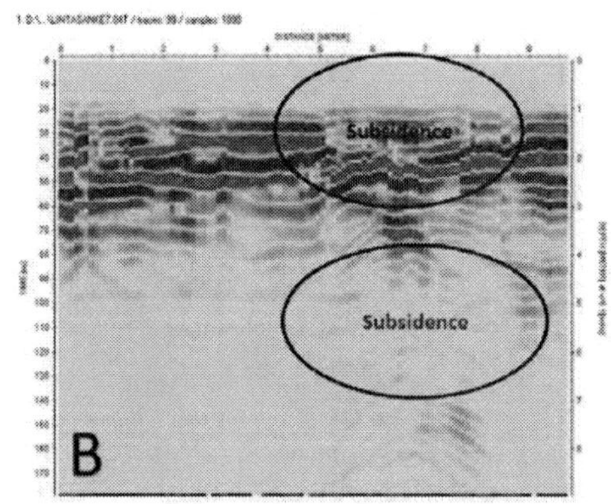

Figure 6B. Result of GPR Monitoring BHM Cross section 12_1-12_1 'May 2019 and (B) BHM Cross section 12_1-12_1' May 2019

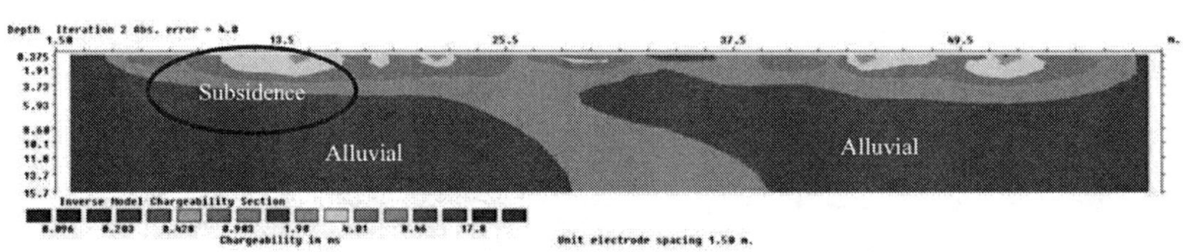

Figure 7. Result of IP Inversion and Topography effect

GPR interpretation shows that the layer subsidence indication occurs in the depth range of 1.7 – 2.0 m at BHM 12 and at a depth of 4- 5 m there is a change in layer shape. Then the GPR interpretation is linked to IP to show subsidence area. IP indicates the presence of holes and alluvial layers with a value range of 2-4 ms which is indicated in yellow and orange color

4. Conclusion

Based on interpretation, The topography shows a decrease in elevation values and land subsidence around the BHM operating area of 1.2 m and the BHM hole of 1 m. subsurface subsidence is indicated by GPR interpretation where the hummocky radargram pattern changes to chaotic and the layer decreases by 0.3 m below the surface. IP shows the alluvial layer which is still dominated by tin minerals under the surface with ranges chargeabillity value 2-4 ms

Acknowledgment

We sincerely thank PT Timah Tbk. For their contribution and support in the field investigation and mapping in Bangka Island and also for this paper.

References

[1] Lubis, I.A. (2006). Filosofi Penambangan Timah dan Konsep Penambangan Endapan Timah Dalam di Laut. PT Timah (Persero) Tbk.

[2] Lubis, I. A., Susilo, R. B., & Romi, S. (2014). A Technological Innovation To Conserve Tin Alluvial Mining : Subsurface Hydraulic Mining by BTM-SR-4 Equipment. Journal of S&T Policy & R&D Management, 12(2), 147–156.Lowrie, W., (2007). Fundamental of Geophysics (2nd ed). New York: Cambridge University Press, Earth's age, thermal and electrical properties, 265-267

[3] Reynolds, J.M., (1996). An Introduction to Applied and Environmental Geophysics, John Willey and Sons, P. 681-749.

[4] Telford, W.M., Geldart, L.P., Sheriff R.E. 1990. Applied Geophysics (2nd ed.), New York: Cambridge University Press, 578-583.

[5] Lunia, O.V. (2015), Deformation style in the damage zone of the Monday fault: GPR evidence (*Tunka basin, southern East Siberia). Russian Geology and Geophysics*, Russian

[6] Malla, Geoscience, Ramac (1997). GPR software manual. Swedia.

Study of Fluid Flow Movement by Using Self Potential Data

Neng E. Jubaedah[1,*] and Wahyudi W. Parnadi[2]

[1]Department of Mineral Mining Engineering, Politeknik Geologi & Pertambangan AGP, Bandung, Indonesia
[2]Department of Geophysical Engineering, Institut Teknologi Bandung, Jl. Ganesha 10 Bandung 40132, Indonesia

[*]Corresponding author: endahne@gmail.com

Abstract. The Self-Potential (SP) method is passive method in geophysics which works based on the natural presence of an electric field on the surface due to anomalies below the surface. SP value on surface can be generated by fluid flow through rock pores or fractures. We study fluid flow movement in subsurface using the velocity value of fluid flow derived from SP values measured on surface. For that purpose, we carried out mathematical modelling, connecting the Helmholtz-Smoluchovsky's electrokinetic potential gradient equation with Darcy's law. The velocity of fluid flow depends on the intrinsic permeability of rocks, electrokinetic potential gradient and electrohydrolic conductivity constant. We tested derived velocity of fluid flow on the SP data from a pilot project test site. Study results show that fluid flow in vertical direction can be identified from SP data at locations where there are significant changes of positive and negative SP values. Fluid flows from a high SP value to a low SP value and this flow is opposite the positive SP gradient. The SP value at study site lie in the range -80mV to -160mV, whereas the value of the water flow velocity lie in the range 0.08 cm/s - 0.21 cm/s.

1. Introduction

There is a lot of environmental pollution in the form of fluids that flows near the ground surface. Such fluid flows cannot be seen directly from surface. The Self-Potential (SP) method is passive method in geophysics which is suitable to be applied in identifying those fluid flows. This technique works based on the natural presence of an electric field on the surface due to anomalies below the surface. SP value on surface can be generated by fluid flows through rock pores or fractures [1]. The SP method is popularly applied by researchers because it is a cheap and efficient method for describing subsurface conditions. This method has many applications, ranging from its use to detect underground caves [1] , over hydrological studies [2] to studies in volcanic environments to investigate multiscale fluid flow [3]. Success in our laboratory examination to study fluid flow velocity determination motivated us to apply the results to a pilot project scale. In this study we investigated the movement of lateral fluid flow using the SP values measured at the surface in the study area.

2. Data and methods

2.1. the Helmholtz-Smoluchovsky's electrokinetic potential gradient equation

Helmholtz developed the theoretical basis of the streaming potential, stating that streaming potentials are present when a conduction current balances the convection current caused by the flow of positive ions as seen in figure 1 [1]. Due to the difference in pressure, fluids will flow in the rock pores and attract positive charges, resulting in a convection current.

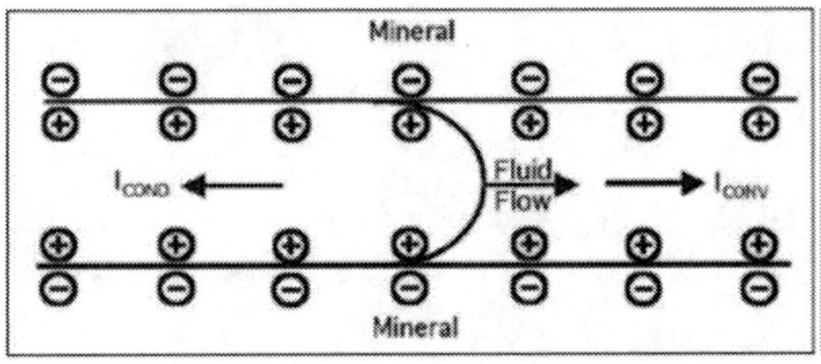

Figure 1. Streaming current in a capillary [1]

The convection current I_{CONV} carried by the flow along the pore is represented by the equation [4]

$$I_{CONV} = \frac{\varepsilon\zeta}{\mu}\Delta P \ . \tag{1}$$

where ε is the dielectric constant of the fluid, ζ is the zeta potential, μ is the viscosity of the fluid and ΔP is the pressure difference.

The conduction current I_{COND} for each unit area caused by the convection current is formulated by Ohm's law [5]:

$$I_{COND} = -\sigma\,\Delta V \tag{2}$$

Where σ is the conductivity of the fluid and ΔV is the difference of streaming potential.

At steady state the convection current I_{CONV} equals the conduction current I_{COND} and this relation leads to the Helmholtz-Smoluchowski equation [1]

$$\Delta V = \frac{\varepsilon\zeta}{\mu\sigma}\Delta P \ . \tag{3}$$

Substituting $P = \rho g h$ from Bernoulli's equation, where ρ is the fluid density; g is the earth acceleration of gravity and h is the height of fluid level, into equation (3) results in

$$\Delta V = \frac{\varepsilon\zeta}{\mu\sigma}.\rho g \Delta h \tag{4}$$

In gradient form, equation (4) becomes

$$\nabla V = \frac{\varepsilon\zeta\rho g}{\mu\sigma}\nabla h. \tag{5}$$

Equation (5) is also called the Helmholtz-Smoluchovsky's electrokinetic potential gradient equation.

2.2. Darci's law

Darcy's law acts as the basic law for fluid flows in porous media. It says that the discharge rate is proportional to the gradient in hydraulic head and the hydraulic conductivity. It is represented mathematically by the following equation [6]

$$q = \frac{Q}{A} = -K\frac{\Delta h}{\Delta L} = -K\frac{dh}{dl} \qquad (6)$$

where q is the discharge rate, Q is the volume of flow in unit time, A is the cross-sectional area normal to the flow and K is the coefficient of permeability. Δh and ΔL are the difference in total head and the length of flow in a medium, respectively.

In gradient form, equation (6) is written as

$$q = -K\nabla h \qquad (7)$$

where

$$K = \frac{k\rho g}{\mu} \qquad (8)$$

q is also known as the discharge velocity v, which is the velocity of fluid flows in porous media. The minus sign in equation (6) means that q is in the direction of decreasing total head. k is intrinsic permeability of fluid.

2.3. Velocity of fluid flows

Velocity of fluid flowing through porous media can be determined from the streaming potential by connecting the Helmholtz-Smoluchovsky's electrokinetic potential gradient equation with Darcy's law.

First, we define Electrohydraulic conductivity constant C as the ratio between ∇V and ∇h [7]

$$C = \frac{\nabla V}{\nabla h} = \frac{\varepsilon\zeta\rho g}{\mu\sigma} \quad . \qquad (9)$$

Substituting equation (9) into equation (5) results in

$$\nabla V = C\,\nabla h \qquad (10)$$

Using C value from equation (9), equation (7) can be written as

$$v = q = -K\nabla h = -K\frac{\nabla V}{C} = -\frac{K}{C}\nabla V \quad . \qquad (11)$$

Thus, the velocity of fluid flow v depends on the intrinsic permeability of rocks K, on the electrokinetic potential gradient or streaming potential gradient ∇V and electrohydraulic conductivity constant C. Equation (11) states that the velocity flow direction is opposite to the SP gradient direction. Based on this equation, fluid flows in the opposite direction to the SP gradient and flows from locations with high SP value to locations with low SP value.

In three dimensional Cartesian coordinate system, with origin O and axis lines X, Y and Z and assuming isotropic media, the velocity component of fluid flow in each direction v_x, v_y as well as v_z is

$$v_x = -\frac{K}{C}\frac{\partial V}{\partial x} \approx -\frac{K}{C}\frac{\Delta V}{\Delta X} \qquad (12)$$

$$v_y = -\frac{K}{C}\frac{\partial V}{\partial y} \approx -\frac{K}{C}\frac{\Delta V}{\Delta Y} \qquad (13)$$

$$v_z = -\frac{K}{C}\frac{\partial V}{\partial z} \approx -\frac{K}{C}\frac{\Delta V}{\Delta Z} \qquad (14)$$

with X, Y and Z are horizontal axis in x direction, horizontal axis in y direction perpendicular to x axis and vertical axis, respectively.

If the fluids flow only in horizontal direction, its total velocity magnitude can be calculated as

$$|v| = \frac{K}{C}|\nabla V| = \frac{K}{C}\sqrt{\left(\frac{\Delta V}{\Delta X}\right)^2 + \left(\frac{\Delta V}{\Delta Y}\right)^2} \quad . \qquad (15)$$

Based on measurement in laboratory using pure water flowing in sand as medium, K and C are correlable each other and have this relation [7]

$$K(\frac{cm}{s}) = 0.0262\ln\left(C(\frac{mV}{cm})\right) + 0.1105 \tag{16}$$

2.4. Data

Pilot project test site of 2400 meters x 1600 meters was used for this study. The topography and morphology of the research area is quite flat with an elevation between 43 - 66 m above sea level. In general, the research area consists of rubber plantations, tall shrubs, swamps, and small forests. In general, the morphology of the study area is a slightly wavy plain, with a maximum slope of 7% which is classified as gently sloping. Overall, the elevation factor does not really affect the measurement results. The lithology in the study area is composed of claystone, siltstone, and tuff sandstone with coal inserted. Water flowing horizontally in subsurface and surface was observed in some parts of the study area. Upflow water was found also near some SP stations.

The equipment and materials used in SP measurements consist of the Fluke Multimeter, switch box, cable, porous pot, CuSO4 solution, meter, CS60 Garmin GPS receiver, copper electrodes, and other field measurement equipment. Figure 2 shows equipments used for data acquisition.

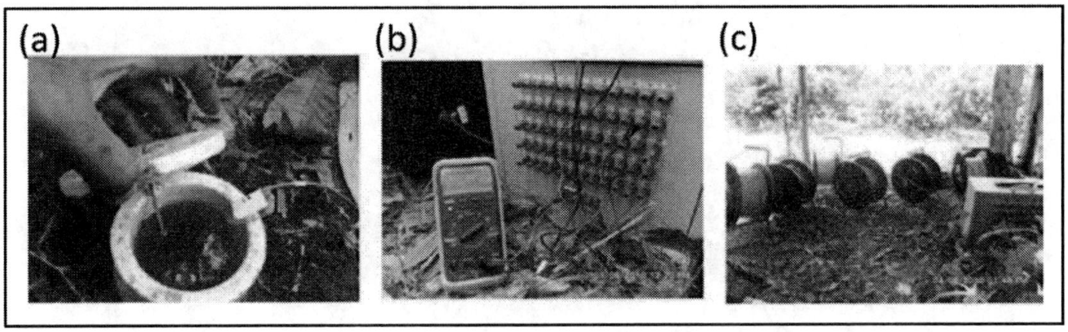

Figure 2. Equipments used in data acquisition: porous pot with CuSO4 in it (a), Fluke Multimeter as resistivitymeter and its switchbox (b) and rolls of cables (c)

SP measurement was carried out in a gridded system at 100 meters station distance in X direction (longitudinal direction) and 50 meters station distance in Y direction (transverse direction). Totally 1155 measurement stations with 4 geological wells were available for this study. Figure 3 show map of measured SP in the study area, SP profile of line A-A' and its corresponding resistivity profile in the same line. SP values lie in the range of -160 mV – 110 mV. Resistivity data was acquired using pole-pole configuration at 20 meters electrode spacing and with wells are used for validation.

3. Results and discussion

The velocity of fluid flowing in the study area is calculated by using equations (12), (13) and (15). Assuming $\frac{K}{C}$ value satisfies equation (16) this relation depends only on the medium properties in the study area and K is calculated as

$$K_{\text{fluid}} = \frac{\rho_{\text{fluid}} \cdot \mu_{\text{pure water}}}{\mu_{\text{fluid}} \cdot \rho_{\text{pure water}}} K_{\text{pure water}} \quad . \tag{17}$$

Using properties value for pure water and that of fluid in the study area, K in equation (17) is easily determined.

The velocity distribution of fluid flowing in X-direction, Y-direction as well as total velocity, all in horizontal direction are shown in figures (4) – (6). Water flowing in subsurface was observed in Wells

IOP Conf. Series: Earth and Environmental Science **873** (2021) 012082 doi:10.1088/1755-1315/873/1/012082

1 – 4 and showed dominantly in lateral direction as seen also in all figures. The magnitude of water velocity in subsurface lies in the range 0.08 cm/s – 0.21 cm/s. Figure (7) shows the distribution of lateral velocity direction overlying SP map. Water velocity direction is shown by arrow sign and its size denotes its magnitude.

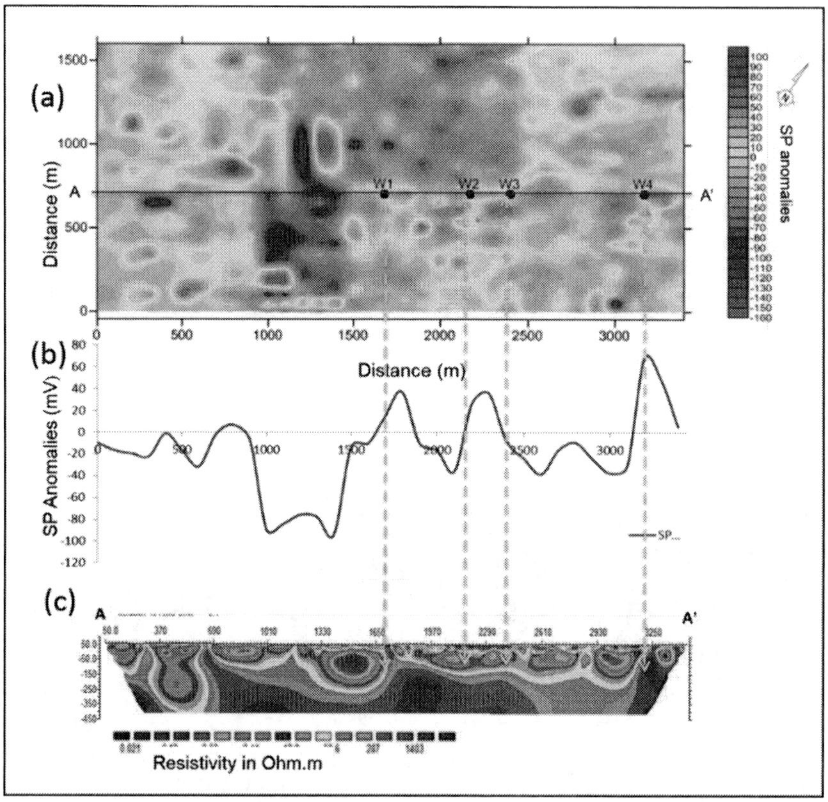

Figure 3. Map of measured SP in the study area (a), SP profile (b) and its corresponding resistity profile of line A-A' (c). W1 – W2 are the location of test wells.

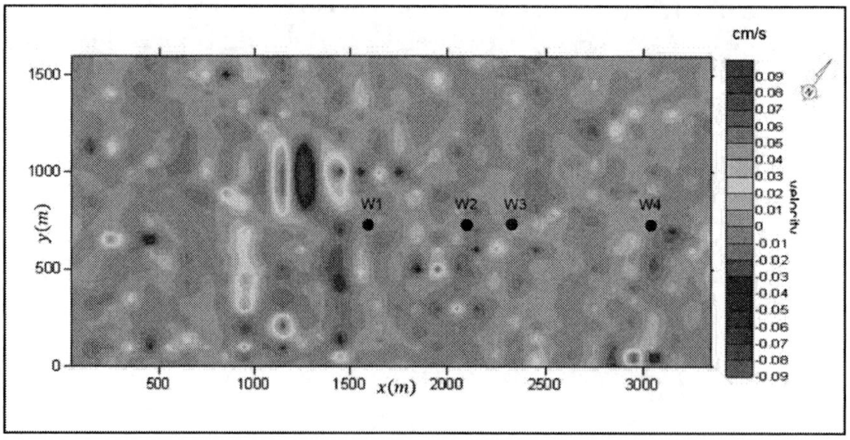

Figure 4. Map of fluid velocity in X direction. W1 – W2 are the location of test wells.

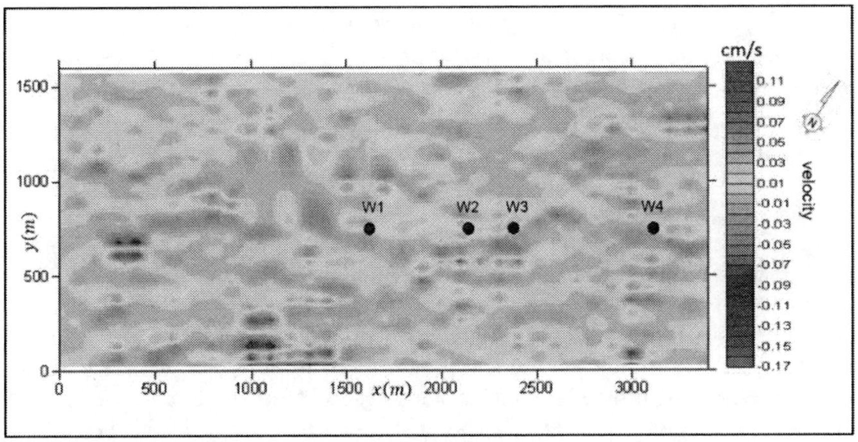

Figure 5. Map of fluid velocity in Y direction. W1 – W2 are the location of test wells.

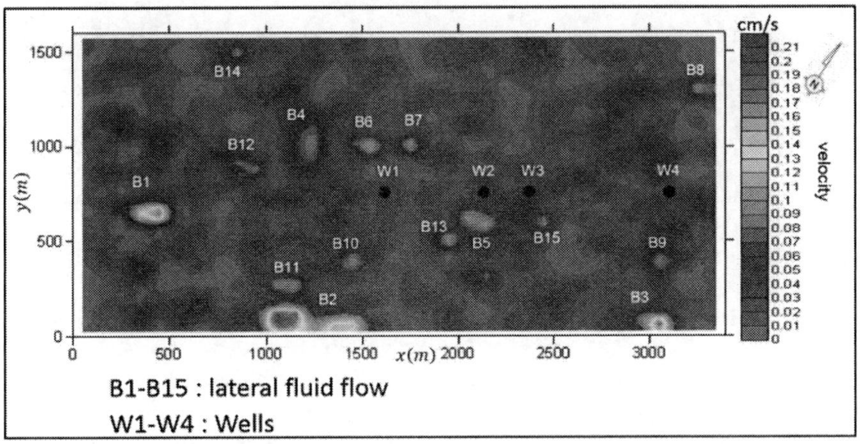

Figure 6. Map of total velocity in lateral direction. W1 – W2 are the location of test wells.

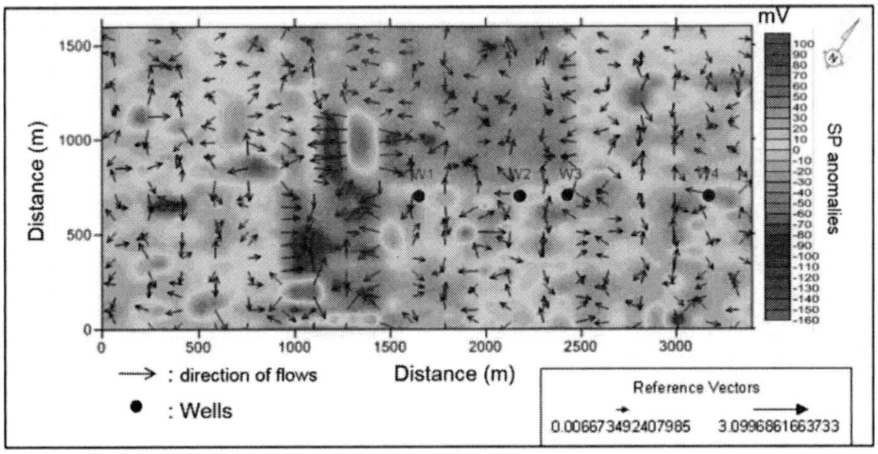

Figure 7. Distribution of lateral total velocity overlying SP map.

As seen also from figure (3), the A-A' line passes through 4 test wells. Those wells coincide with the SP profile which shows a significant change in positive and negative SP signs. A significant change in value from negative to positive or vice versa in the SP profile indicates a vertical fluid flow, because the fluid contains an electric charge greater than the surrounding medium.

4. Conclusion

Our study concludes following results:

1. The location of water seepage near the surface can be well identified using the SP method. The SP value due to water flowing at the study site ranges from: -80mV - -160mV.
2. DC resistivity method is good to be used as a support method to identify the location of water flows, at which the resistivity value for water at the study site is <0.5 Ωm.
3. The formula derived to determine the velocity of the fluid flow works well in modeling the water flow velocity in the study area with the velocity value lie in the range 0.08 cm/s - 0.21 cm/s.
4. A significant change in value from negative to positive or vice versa in the SP profile indicates a vertical fluid flow which is observed in the study area.

References

[1] Vichabian Y and Morgan F D 2002 Self potential l in cave detection: The Leading Edge September 2002 pp 866-871. *Phys.* **35** L74

[2] Lénat J-F 2007 Retrieving self-potential anomalies in a complex volcanic environment: an SP/elevation gradient approach *Near Surface Geophysics* pp 161-170

[3] Grobbe N and Barde-Cabusson S 2019 Self-Potential Studies in Volcanic Environments: A Cheap and Efficient Method for Multiscale Fluid-Flow Investigations *International Journal of Geophysics* 2985824 19pp doi: 10.1155/2019/2985824

[4] Griffiths D J 1995 *Introduction to Electrodynamics 2nd edition*: Prentice Hall of India Private Limited.

[5] Reynolds J M 1997 *An Introduction to Applied and Environmental Geophysics*: John Willey & Sons, New York.

[6] Gudmundsson A 2011 *Rock Fractures in Geological Processes*: Cambridge University Press 578pp

[7] Hamzah M 2009 *Studi Pemodelan Self-Potential (SP) untuk Menentukan Kecepatan Aliran Fluida pada media Berpori*. Dissertation Institut Teknologi Bandung (unpublished)

IOP Conf. Series: Earth and Environmental Science **873** (2021) 012083 doi:10.1088/1755-1315/873/1/012083

Utilization of Gravity Satellite Data to Imaging Subsurface Conditions, Case Study: Lake Toba, North Sumatera

Dadi Abdurrahman[1], Setianingsih[1], Susanti Alawiyah[1], Wawan Gunawan A. Kadir[1], Djoko Santoso[1], Darharta Dahrin[1], Indra Gunawan[1]

[1]Applied Geophysics and Exploration Research Group, Institut Teknologi Bandung , Bandung, Indonesia

E-mail : abdurrahman@gf.itb.ac.id

Abstract. The purpose of this study is to utilize satellite gravity data for free to interpret subsurface conditions. The data can be obtained from the topex site with a latitude range is +/- 80,738. The author tries to interpret the gravity data for Lake Toba, North Sumatra. The area is attractive because it has a sizeable geological object in the form of a large lake and a fault called the Sumatran fault. This gravity data is processed like ordinary gravity processing so that Complete Bouguer Anomaly (CBA) is obtained, then regional-residual separation is carried out using a moving average. From CBA, it can be seen that there are two closures just below Lake Toba. There are two possible chambers under Lake Toba, namely the central and southern parts. In addition, it is also clear that the lineament of the Sumatran fault is also visible, as well as three other lineaments which may be faults that are not identified from the surface. So, we can take advantage of this free gravity satellite data for interpretation of sizeable geological objects that can be applied to other areas. It is helpful to know the general picture of regional geology before conducting a more detailed survey.

1. Introduction

Gravity anomaly represents an image of the subsurface, can be obtained from land, marine, airborne, and satellite. Each of these data has its advantages and limitations. Land data usually have distances between stations ranging from tens of meters to kilometers. Generally, this survey was conducted for preliminary measurements to the detailed stage, both for geotechnical and environmental purposes ([1], [2]), oil and gas industry [3], geothermal [4], and mining ([5], [6]). Marine gravity data is carried out to determine the distribution of gravity values in the ocean area, usually using a boat either directly above it or placing a gravimeter on the ocean floor. Airborne gravity data [7] is generally carried out for difficult areas to reach by land. Generally, this data collection is very fast and wider than land measurements. Gravity satellite data is rarely used because of technological factors that make the data expensive. This data usually has an extensive range, but the spacing between the data is not tight.

Considering the advantage of satellite data that can cover an extensive area, it would be excellent if this data was used to study subsurface conditions that are very regional and deep, such as tectonic studies volcanoes and faults which are very deep and long. In this research, the author tries to use gravity satellite data which can be downloaded from https://topex.ucsd.edu/cgi-bin/get_data.cgi for Lake Toba area, North Sumatra. The data on the site can be downloaded for free. The author tries to use this data in the.

2. Geologic Settings

Sumatra is located in the southeastern part of the Eurasian Plate and northeast of the Indian Ocean Plate. The Indian Ocean Plate has experienced sloping subduction under the island of Sumatra to the Sunda Strait at N20 to the east in a range of 6-7cm ([8], [9]). The tectonic and uplifting activities that occurred from the late Miocene to the Pilo-Pleistocene led to Bukit Barisan and volcanoes along the western part of Sumatra Island. The active strike-slip fault on the left side of the Sumatra fault system indicates that tectonic activity has continued from Mesozoic to now. This tectonic activity resulted in a reversal in the Cenozoic basin along western Sumatra, Central Sumatra Basin, and South Sumatra Basin, as well as two forearc basins namely Sibolga Basin and Bengkulu Basin [10]

3. Gravity Data

The gravity data is obtained by downloading from the site https://topex.ucsd.edu/cgi-bin/get_data.cgi. It contained gravity and topography data. To download both data, we must register. On this site, we can download gravity data for all areas. Downloaded gravity data can have denseness of up to 1-minute. The data output from this site is a free air anomaly (FAA) map.

4. Result and Discussion

By utilizing free air anomaly (FAA) and topography data from this site, we use it for the Lake Toba area and its surroundings, with a range of 30 28 '40.1' '- 10 34' 08.0 " and 97 0 54 '52.7' '- 99 0 57 '35.6' ' with 1-arcsecond resolution. Both data were processed to obtain a Complete Bouguer Anomaly (CBA) map. The CBA map of this area is shown in Figure 1.

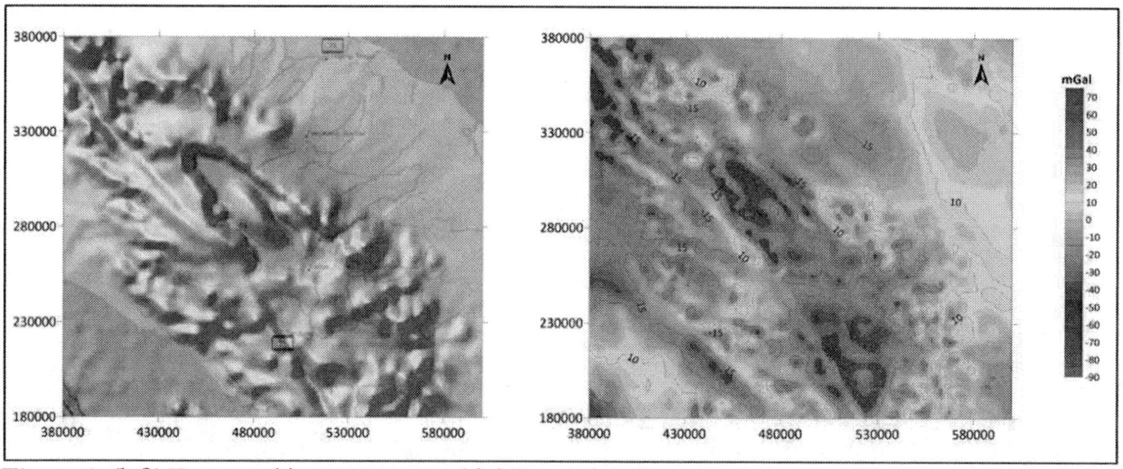

Figure 1. (left) Topographic contour map, (right) complete Bouguer Anomaly (CBA) map of Lake Toba and its surroundings.

From the CBA map, we try to separate regional and residual signals using the moving average technique. By using this technique, we can separate regional-residual gravity anomalies. The regional gravity anomaly is shown in Figure 2. The regional gravity anomaly map shows geological objects that extend wide and deep. From Figure 2, we can see that Lake Toba has a negative gravity response (blue), extending from northwest to southeast. The interesting thing about this regional anomaly map is that the blue anomaly is a collection of two closures: central and south closure. This suggests that the magma chamber of Lake Toba probably consists of two magma chambers.

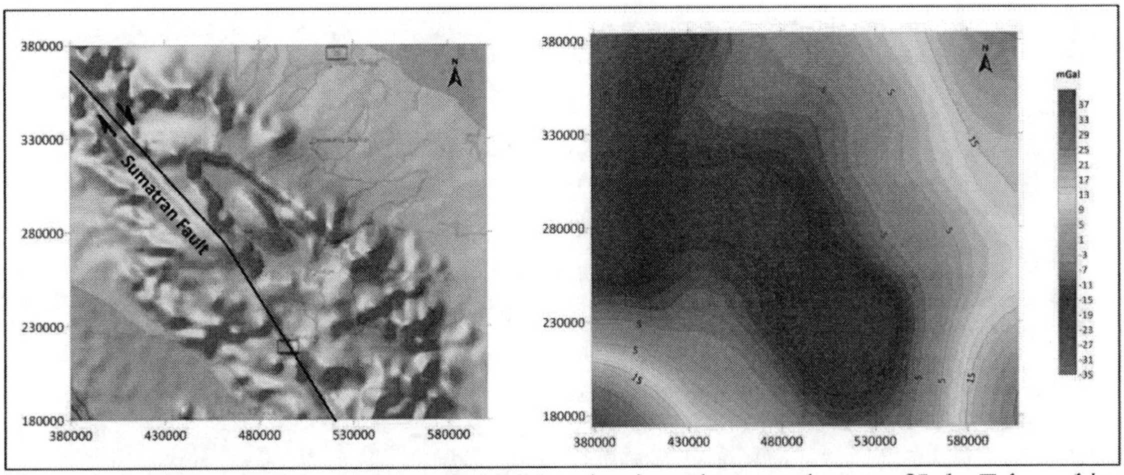

Figure 2. (left) Topographic contour map, (right) regional gravity anomaly map of Lake Toba and its surroundings. There is two closures with low anomalous under Lake Toba.

Reducing the Complete Bouguer Anomaly and regional gravity anomaly maps results in a residual gravity anomaly map. It shows objects narrower and shallower than the regional gravity anomaly. The residual gravity anomaly is shown in Figure 3. In this picture, it can be seen that this anomaly very clearly shows some straightness, as clearly seen in the straightness of the Sumatran fault. So that the residual gravity anomaly can delineate developing faults. We can see another straightness, which is like the mark A, the mark B, and the mark C on Figure 3 (right). Most likely, the straightness is faults that have not been identified.

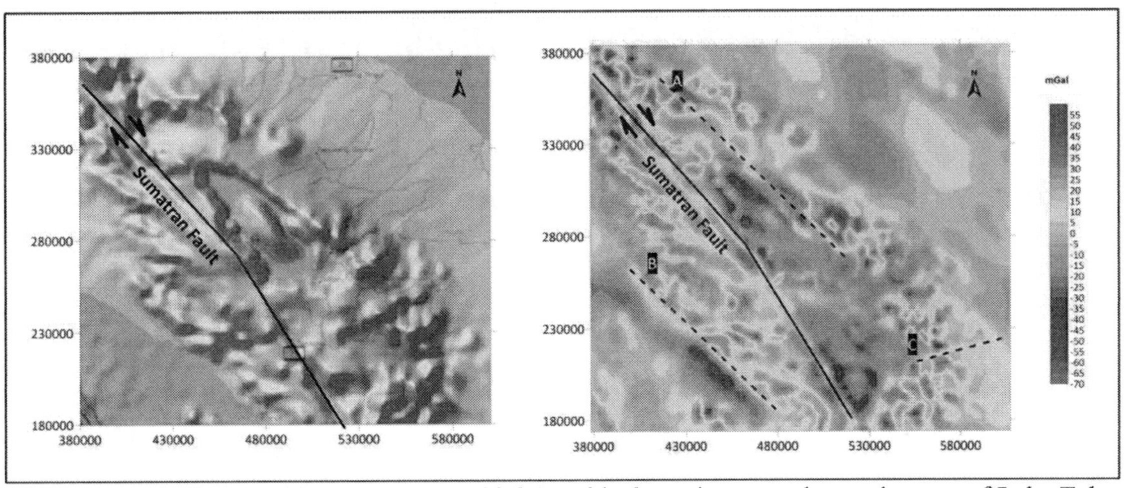

Figure 3. (left) Topographic contour map, (right) residual gravity anomaly gravity map of Lake Toba and its surroundings. Not only the straightness of Sumatran fault, but also there are at least three other straightness that has not been identified.

5. Conclusion

The gravity satellite map can be downloaded for free on the site https://topex.ucsd.edu/cgi-bin/get_data.cgi with an accuracy of up to 1-minute. The output from this site is a free air anomaly. This map can be used for the interpretation of subsurface conditions such as the Lake Toba area. From the results of this data processing, we can see the distribution of Lake Toba's chamber, which is divided into two closures. This is clearly seen on the regional gravity anomaly map. Meanwhile, other developed faults are visible from the residual anomaly map. The straightness of the Sumatran fault is clear. In addition to the Sumatran fault, there are three other straight lines, namely A's, B's, and C's straightness. Most likely, these three straightness are faults that have not been identified.

References

[1] Arzi A A 1975 Microgravity for Engineering Applications: *Geophys Prosp.* **23** 408-425.
[2] Butler D K 1984 Microgravimetric and Gravity Gradient Techniques for Detection of Subsurface Cavities *Geophysics* **49** 1084-1096
[3] Liu L Zhang M Wang P Zhang S and Qiao J 2018 Construction of Geophysical Geological Structural Models for Complex Basins: Application of interpretation of gravity, magnetic, electric and seismic data to exploration in the Chuxiong Basin Chinese *Journal of Geophysics* **61(12)** 4921–4933
[4] Zaher M A Saibi H Mansour K Khalil A and Soliman M 2017 Geothermal Exploration Using Airborne Gravity and Magnetic Data at Siwa Oasis Desert Egypt *Renewable and Sustainable Energy Reviews* Western **82** 3824-3832
[5] Chen G Liu T Sun J Cheng Q Sahoo B Zhang Z and Zhang H 2015 Gravity Method for Investigating the Geological Structures Associated with W-Sn Polymetallic Deposits in the Nanling Range China *Journal of Applied Geophysics* **120** 14-25.
[6] Dransfield M 2007 Airborne Gravity Gradiometry in the Search for Mineral Deposits *Proceedings of Exploration 07: Fifth Decennial International Conference on Mineral Exploration* Milkereit B341-354
[7] Moore D Chowdhury P R Rudge T 2013 FALCON®Airborne Gravity Gradiometry Provides a Smarter Exploration Tool for Unconventional and Conventional Hydrocarbons: Case Study from the Fitzroy Trough, Onshore Canning Basin *10th Biennial International Conference and Exposition*
[8] Sørensen M B Atakan K and Pulido N 2007 Simulated Strong Ground Motions for the GreatM9.3 Sumatra–AndamanEarthquake of 26 December 2004 *Bulletin of the Seismological Society of America* **07** S139–S151
[9] Hamilton W 1979 Tectonic of Indonesia Region *U.S. Geol. Survey* Prof. Papper 1078.
[10] Bachtiar A 2013 Tectono Volcanic Toba Lak: Practical Analogue Sedimentology Model. North Sumatra, Indonesia.

Analysis of Tectonic Plate Velocity Variations in the Sunda Strait Based on GPS Time-series Data

Dhira Y[1], I Meilano[1,2], and Dudy D W[1]

[1]Geodesy and Geomatics Engineering, Faculty of Earth Science and Technology, Bandung Institute of Technology, Ganesha 10, Bandung, Indonesia

[2]Center of Earthquake Science and Technology, Bandung Institute of Technology, Ganesha 10, Bandung, Indonesia

*Corresponding author: dyovenia@gmail.com

Abstract. Indonesia is an earthquake-prone country located in the junction of four tectonic plates, namely the Indo-Australian, Eurasian, Philippine, and Pacific. The convergent boundary between tectonic plates is also called a subduction zone that can produce great earthquakes in the future. One of the subduction zones in Indonesia is the Sunda Strait subduction zone which predicted can release a M7.8 earthquake. Previous research stated that there is a change in tectonic plate velocity after an earthquake ruptured. It is likely that this could happen in the Sunda Strait area which has experienced several large earthquakes. In this study, we conducted research to find out the information on the tectonic plate velocity changes in the Sunda Strait. We used Global Positioning System (GPS) time-series data provided by Indonesia Geospatial Information Agency (BIG). The time series data is used to calculate the earthquake displacement, the changes in GPS velocity of before and after earthquake, and the changes in velocity of each time interval. Our results show that the horizontal displacement due to the earthquake at all GPS stations ranged from 3.34 mm to 7.36 mm in the north-south direction and -27.45 mm to 0.18 mm in the east-west direction. Furthermore, the result of the changes in GPS velocity before and after an earthquake ranged from 2.25 mm/year to 12.60 mm/year and 1.80 mm/year to 13.35 mm/year. The pattern of change in velocity is likely due to post-seismic deformation from the 2012 Indian Ocean earthquake, the 2016 Sumatra earthquake, and also other tectonic factors.
Keywords: earthquake displacement, GPS, subduction zone, velocity changes

1. Introduction

Indonesia is a country located at the confluence of the world's large plates and several small plates or microblocks [1]. The islands in Indonesia are surrounded by four main plates, namely the Eurasian plate, the Indo-Australian plate, the Philippine plate, and the Pacific plate, resulting in rough topography, many earthquakes, and volcanoes (figure 1) [4].

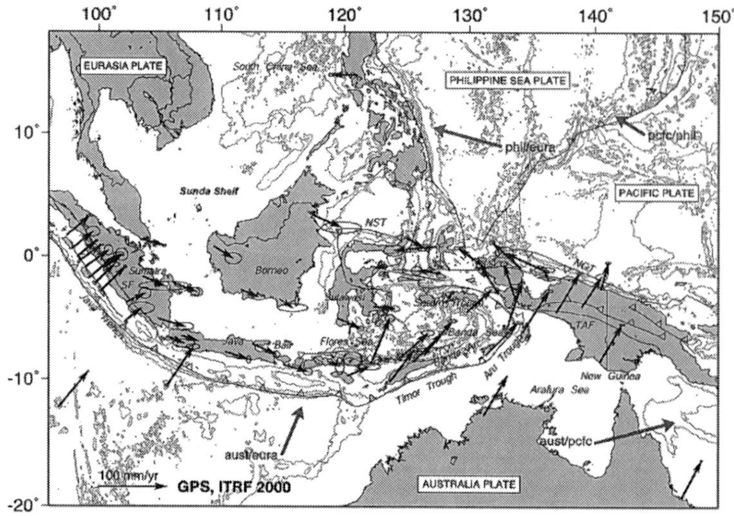

Figure 1. Map of Indonesia's tectonic order from Bock et al. [2]

The Sunda Strait is a transition zone due to changes in direction and velocity of subduction of the Indo-Australian plate against the Eurasian plate. The Indo-Australian tectonic plate pierces the Eurasian plate so that the island of Sumatra moves to the northwest, causing the Sunda Strait to be situated on a fault line and prone to earthquakes [6]. A plate subduction that is accelerated due to a large earthquake and is propagated along the fault by stress [3,5]. On April 11th, 2012 there was an earthquake with a magnitude of 8.6 in the west of the Sumatra Island. This obviously results in changes, such as changes in plate velocity displacement in the area. Therefore, this study aims to determine the effect of that earthquake on GPS observation stations around the Sunda Strait and to determine the pattern of plate velocity changes that occur at GPS observation stations in the Sunda Strait.

2. Data and Methods

8 GPS observation points around the Sunda strait, using the Indonesia Geospatial Information Agency, Continuously Operating Reference System station (CORS BIG) data. Namely CTCN, CLGI, CSBK, CGON, CPSR, CUJK, CRKS, and CMLP (figure 2).

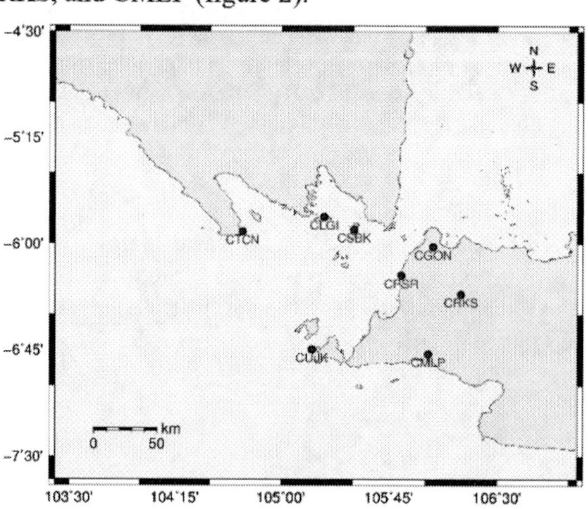

Figure 2. Distribution of observation points

IOP Conf. Series: Earth and Environmental Science **873** (2021) 012084 doi:10.1088/1755-1315/873/1/012084

Each observation point was corrected for Euler pole calculation (EPC). In each time series data from the observation points used, there is an effect of displacement of the Sunda block and seasonal effects [7,8]. Therefore, the process of calculating the influence of the Sunda block is carried out using the Altamimi parameter. The data that has been corrected by the EPC is then further processed with eliminate outliers, and seasonal effects. After that, the time series is used to calculate the velocity vector and change in velocity.

3. Results and Discussion

As a result, time series after cleared from outliers and seasonal effects referring to ITRF 2008 (figure 3).

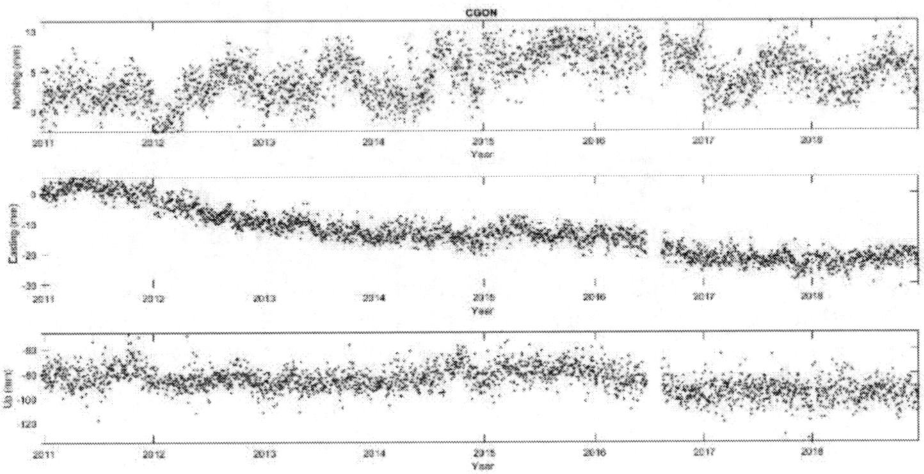

Figure 3. The example time series of CGON observation station point

Then, time series data used to calculate velocity before and after earthquake (figure 4 and figure 5).

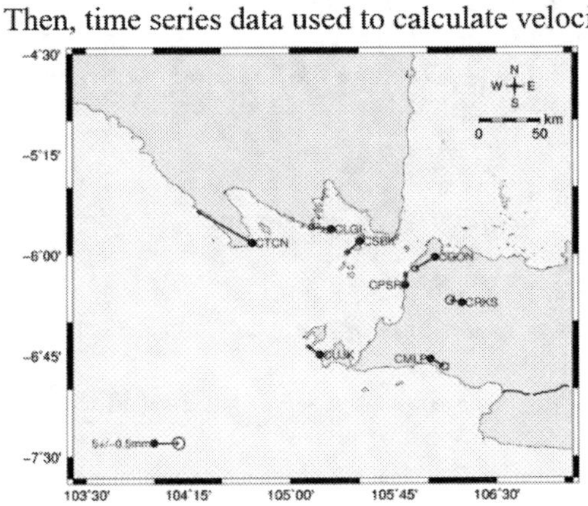

Figure 4. Velocity vector before earthquake

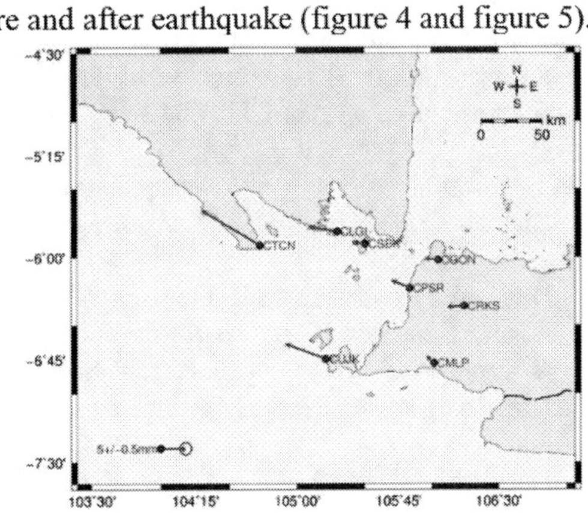

Figure 5. Velocity vector after earthquake

Based on the calculation of the displacement vector and the displacement velocity vector, it is known that the observation station points are not static, but are dynamic or moving. The calculation of changes in velocity for all points is carried out to determine whether the point of the observation station has experienced a change in velocity in its movement (figure 6).

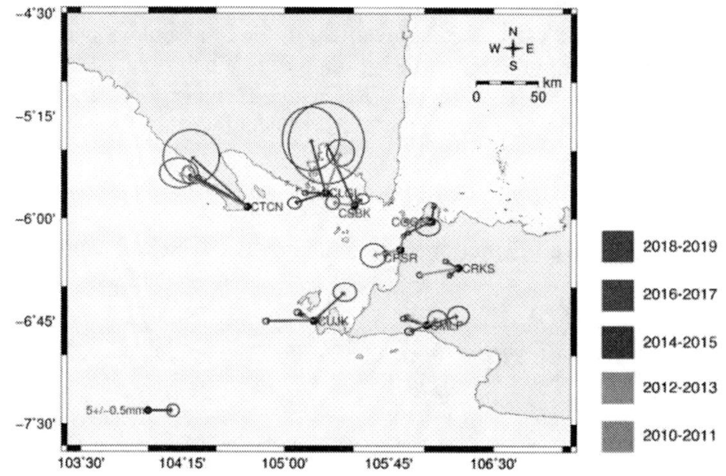

Figure 6. Change of velocity vector all site in five sessions

The trend calculation is carried out based on a predetermined session. If we take one sample observation point (i.e. CGON) (figure 7), the change in velocity caused by the 2012 and 2016 post-seismic earthquake can be analyzed in Figure 6. To summarize the pattern of changes in velocity caused by post-seismic deformation from the 2012 earthquake, M8.6 has an impact the lasts for 2 years and the 2016 M7.8 earthquake had an impact the lasts for 1 year. This shows the effect of the release of energy which causes the plates around the Sunda Strait to return to its equilibrium state.

4. Conclusions

Based on results of calculation, the horizontal displacement value due to the earthquake for all stations range from 3.34 mm - 7.36 mm north and -27.45 mm to 0.18 mm east. The calculation of the horizontal displacement velocity before and after the earthquake ranged from 2.25 mm / year to 12.60 mm / year and 1.80 mm / year to 13.35 mm / year. This shows that there was a change in the velocity of the shift after the earthquake on April 11, 2012 and earthquake on March 2, 2016.

5. Reference

[1] Bird P 2003 *An updated digital model of plate boundaries* (Geochemistry, Geophysics, Geosystems) 4 3

[2] Bock Y, Prawirodirdjo L, Genrich J F, Stevens C W, McCaffrey R, Subarya C, Puntodewo S S O, and Calais, E 2003 *Crustal motion in Indonesia from global positioning system measurements* (Journal of Geophysical Research: Solid Earth) 108 B8

[3] Bott M H P and Dean D S 1973 *Stress diffusion from plate boundaries Nature* 243(5406) 339-341.

[4] Hamilton W B 1979 *Tectonics of the Indonesian region* Vol 1078 (US Government Printing Office).

[5] Heki K and Mitsui Y 2013 *Accelerated pacific plate subduction following interplate thrust earthquakes at the Japan trench* (Earth and Planetary Science Letters 363) 44-49.

[6] Lasitha S, Radhakrishna M, and Sanu T D 2006 *Seismically active deformation in the Sumatra–Java trench-arc region: geodynamic implications* (Current Science) 690-696.

[7] Nikolaidis R *Observation of Geodetic and Seismic Deformation with the Global Positioning System*: *Ph. D Thesis* University of California San Diego 2002 265 p

[8] Susilo, I Meilano, Abidin H Z, and Sapiie B 2015 *A New Definition of Sunda Block* Injoint Convention Balikpapan 2015 HAGI-IAGI-IAFMIIATMI

IOP Conf. Series: Earth and Environmental Science **873** (2021) 012085 doi:10.1088/1755-1315/873/1/012085

Vulnerability Analysis of Earthquake Hazards in Tasikmalaya City Using Horizontal to Vertical Spectral Ratio (HVSR) Method

Najla Insyirah Lianza Bakri[1*], Tedi Yudistira[2] and Yayan Sopyan[3]

[1]Undergraduate Program of Geophysical Engineering, Institut Teknologi Bandung
[2]Global Geophysics Group, Institut Teknologi Bandung
[3]Geological Survey Center, Bandung

*corresponding author's email: najlainsyirah23@gmail.com

Abstract. Tasikmalaya City is one of the regions in West Java Province that is often hit by earthquakes due to its location near the Indo-Australian Plate subduction zone towards the Eurasian Plate. The surface deposits in this city are alluvium and weakly consolidated step deposits which can cause wave amplification during an earthquake. As a mitigation effort, seismic zoning needs to be carried out to map the areas that will experience heavy damage when an earthquake occurs. This study uses the Horizontal to Vertical Spectral Ratio (HVSR) method which is applied to the microtremor recording data to obtain spatial variations in the predominant frequency and amplification values that can explain the characteristics of the geological layer beneath the surface. Based on the obtained results, the predominant frequency ranging from 0.7 to 9.5 Hz with the lowest frequency distribution in the eastern and northwestern parts, which indicates a thicker sediment layer. Amplification ranging from 1.2 to 12.6 with the distribution of higher values in the eastern, southeastern, and northwestern parts. The inversion of the HVSR curves was carried out to determine the value of shear wave velocity (V_s) in order to obtain a more detailed subsurface geological structure that can be used to determine the level of vulnerability of earthquake hazards. The Neighborhood Algorithm is used to find an optimum model. Based on the results of the inversion process, the V_s ranging from 150 - 3054 m/s with lower V_s values in the eastern, southeastern, and northwestern parts at depth of about 25 meters. The average value of shear wave velocity at a depth of 30 meters (V_s^{30}) can also be used to determine the type of soil for geotechnical study. From the obtained V_s^{30}data, the types of soil in the research area are classified into moderate soil, hard soil, and rocks.

1. Introduction

Tasikmalaya City is one of the regions in West Java Province that is often hit by earthquakes due to its location near the Indo-Australian Plate subduction zone towards the Eurasian Plate. The surface deposits in this city are alluvium and weakly consolidated step deposits which can cause wave amplification during an earthquake. Historically, this city was hit by quite destructive earthquakes, including the Pangandaran Earthquake on 2006, the Java Earthquake on 2007, and the West Java Earthquake on 2009. The impact of the earthquake was very massive, from infrastructure damage to fatalities. As an effort to mitigate the earthquake disaster, zoning needs to be carried out to map the areas that will experience heavy damage when an earthquake occurs.

2. Geological Setting

Geologically, Tasikmalaya City is included in the Central Depression Zone or the Bandung Zone [1], a volcanic area that has a relatively depressive shape compared to the zone that flanks it. This zone is mostly filled by alluvial and young volcanic deposits (Quaternary) from the surrounding volcanic products. The surface sediments are alluvium which is deposited in large river flood areas and weakly consolidated step deposits with a thickness of up to 150 m [2]. Lithology in the research area is shown in Figure 1.

3. Data and Methods

3.1. Data

We use microtremor recording data in the Tasikmalaya City collected by Indonesian Geological Survey Center between 15 to 28 May 2019 with recording time around 12.00 – 02.00 (GMT+7) with a measurement duration of ±30 minutes at each point. Measurements in relatively quiet areas are carried out during the day and in busy areas, such as markets, are carried out during midnight. There are 309 measurement locations with the distribution as shown in Figure 2.

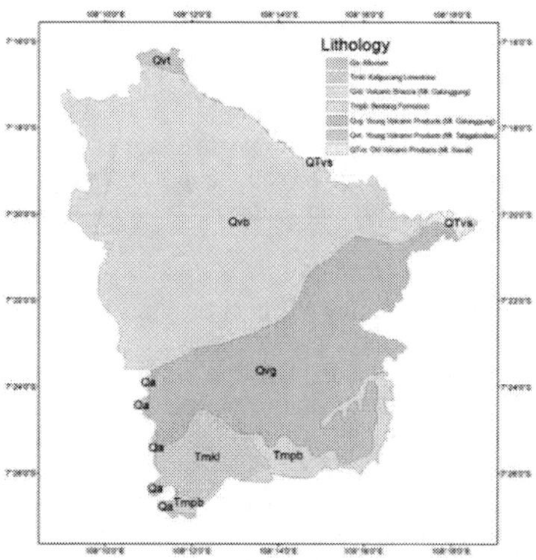

Figure 1. Geological map of Tasikmalaya City [2].

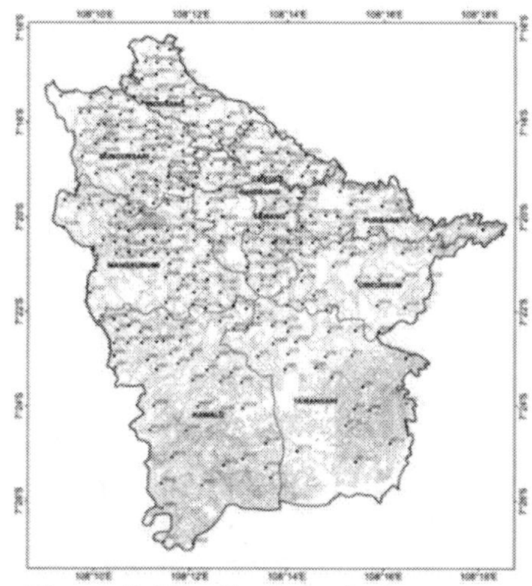

Figure 2. Distribution of microtremor measurement points in Tasikmalaya City.

3.2. Method

HVSR curves were created using Geopsy developed by Wathelet et al. [3], this is a collaboration of various institutions in the SESAME European Project. In order to calculate HVSR curve, we use these parameters: bandpass filter 0.5 – 10 Hz, temporal window length 60 s, STA / LTA 0.1 – 2.5, and smoothing constant 40. These parameters are determined based on literature studies and trial and error to obtain stable and representative HVSR curves. The amplitude spectra of the NS and EW components were averaged by the quadratic mean method. Predominant frequency (f_0) and amplification (A_0) values obtained for each curve then interpolated using Inverse Distance Weighted method (IDW) to produce the predominant frequency and amplification maps.

Inversion was carried out using Dinver package [3] to obtain the values of shear wave velocity (V_s) and layer thickness (h) underneath the stations. This software use Neighborhood Algorithm (NA), a direct search method for non-linear inversion that utilizes Voronoi cells to identify a multidimensional model space and obtain a new model iteratively in the best cells. Searching interval of the model parameters are determined by considering the geological conditions in the study area based on the previous study or using trial and error until a fairly small misfit is obtained in most of the data. In our

IOP Conf. Series: Earth and Environmental Science **873** (2021) 012085 doi:10.1088/1755-1315/873/1/012085

case, the lack of stratigraphic data in the study area makes it hard to determine precise parameters searching range, so that the parameter range is made as wide as possible to represent the variability of sediment and rock types below the surface [4]. In spite of that, the selected V_p and V_s range values are in the limit recommended in [5]. The interval model parameters are set as follows: 5 layers with 100 m depth on the top 3 layers and 2000 m depth on 4th and 5th layer, V_p 200 – 5000 m/s, V_s 150 – 3500 m/s, Poisson's Ratio 0.2 – 0.5, and constant density of 2000 kg/m^3. The V_s model obtained at each station then interpolated to create horizontal section at a certain depth.

4. Results and Discussion

4.1. Predominant Frequency Maps

The obtained predominant frequency ranging from 0.71 to 9.42 Hz with the distribution of low frequency values in the eastern, southeastern, and northwestern part (Figure 3). Comparing with the elevation contour data, areas with low frequencies forming a basin-like geometry which allows thicker surface deposits in this area. These deposits are thought to be weathered Quaternary deposits.

4.2. Amplification Maps

The obtained amplification ranging from 1.15 to 12.59 with the distribution of medium to high amplification values in the eastern, southeastern, southern, and northwestern part (Figure 4). The increasing values to the eastern and southeastern are in accordance with [6]. This increasing amplification is thought to be caused by hard bedrock of volcanic products underlying the weathered Quaternary deposits.

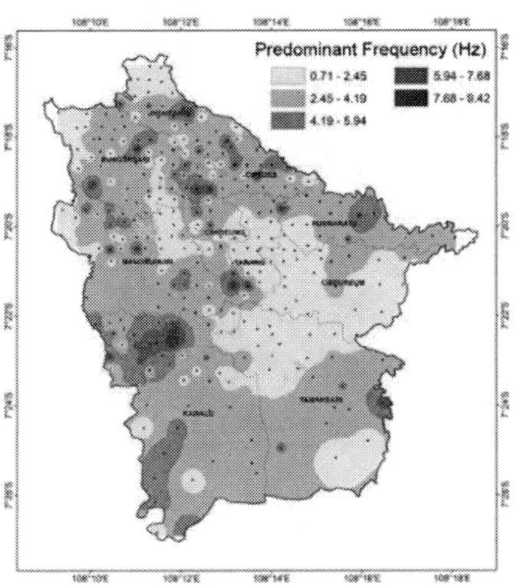

Figure 3. Predominant frequency map in Tasikmalaya City.

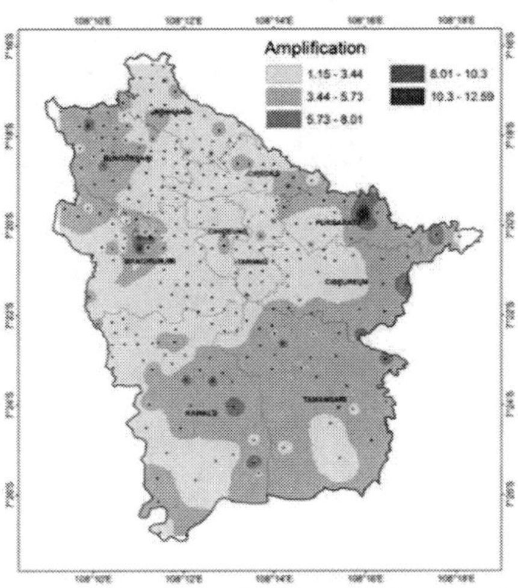

Figure 4. Amplification map in Tasikmalaya City.

4.3. Inversion Results

Figure 5 shows the example of NA result of M235 station. The observed HVSR curve (blue) is approximated by set of calculated curves where the red curve is the calculated data with the best model characterized by lowest misfit (Figure 5.a). The blue vertical line in the observed curve is the standard deviation. Figure 5.b shows the variation of V_s to depth, red stair curve is the best 1-D model related to the red HVSR curve in Figure 5.a. Based on the inversion results the obtained V_s models ranging from 150 - 3054 m/s.

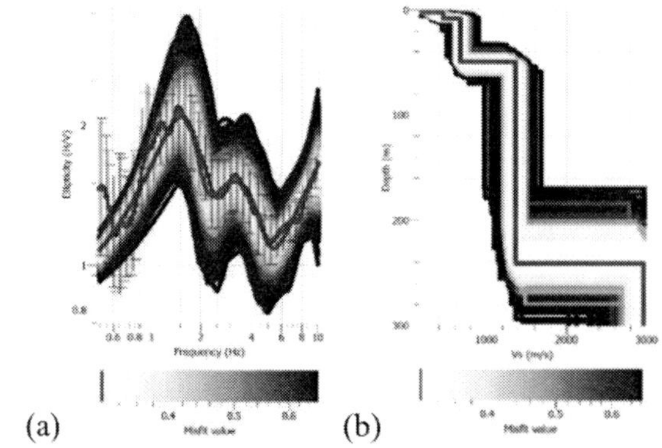

Figure 5. Inversion results on M235 station. (a) HVSR inversion curve, (b) Variation of V_S to depth.

4.4. Horizontal Section of V_S

In order to investigate the lateral variation of shear velocity we create the horizontal slices of several depth as shown in Figure 6. We first classified the shear velocity into two big part, less than and greater than 750 m/s. The greater one dominated begin at 40 m and deeper. The horizontal section at a depth of 25 m (Figure 6.b) most clearly visualizes the boundary areas below and above the engineering bedrock (V_S value of 750 m/s). In the eastern, southeastern, and southwestern the $V_S$value ranging from 0 – 750 m/s which indicates that at depth of 25 meters those area are still a sedimentary layer, while in the southern, western, and northern the $V_S$value ranging from 750 – 1500 m/s which is an engineering bedrock.

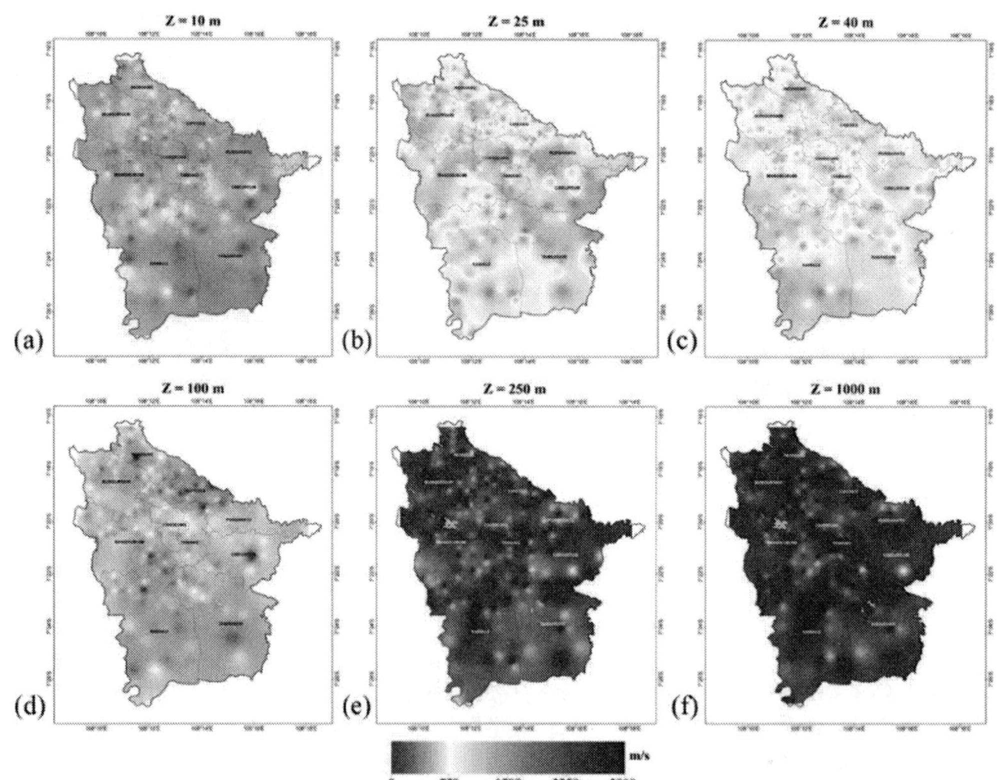

Figure 6. Horizontal section of V_S in depth of (a) 10 m, (b) 25 m, (c) 40 m, (d) 100 m, (e) 250 m, and (f) 1000 m.

4.5. Vertical Section of V_S

Vertical section of V_S were also created and shown in Figure 7. It shows the variation of V_S up to 50 m depth, with red color representing the velocity in the sediment layer (0 – 750 m/s) and the blue color is the velocity in the engineering bedrock (750 – 1750 m/s). In general, the cross section shows the undulating sediment layer structure and dominated by velocities of 250 – 750 m/s. The area with the thickest sediment is on the eastern of the AA' section (Figure 7.b) and on the western of the XX' section (Figure 7.d).

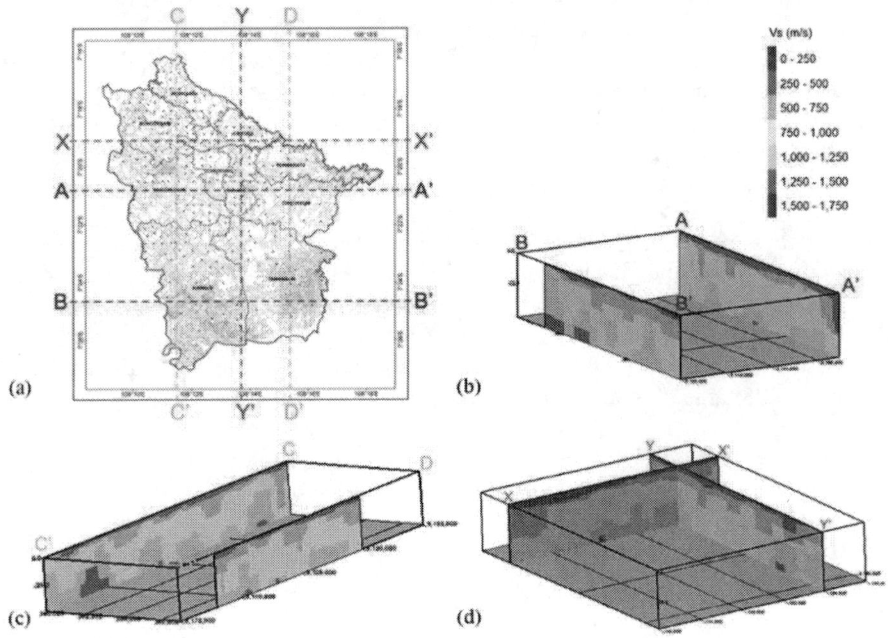

Figure 7. (a) Line to obtain vertical section of V_S, (b) E-W direction, (c) N-S direction, (d) E-W and N-S direction.

4.6. Average Shear Wave Velocity at Depth of 30 m (V_S^{30})

V_S^{30} at each station are calculated from 1-D *Vs* model and then mapped according to the type of soil [7] as shown in Figure 8. V_S^{30} in Tasikmalaya City ranging from 264 - 772 m/s with soil types dominated by hard soil and at several points there are moderate soil and rocks.

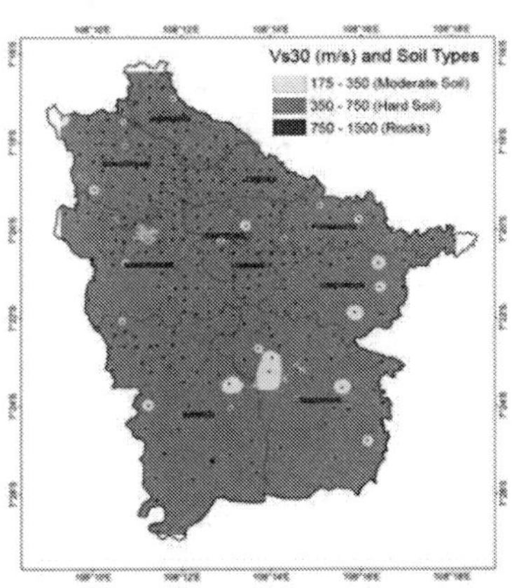

Figure 8. V_S^{30} and soil types in Tasikmalaya City.

4.7. Earthquake Hazard Vulnerability

Large seismic vulnerability correlates with low predominant frequency values and high amplification values [8]. Based on predominant frequency and amplification map, the areas that are more prone to earthquake hazards are the western and northwestern Bungursari Subdistrict; northwestern Mangkubumi Subdistrict; eastern Kawalu Subdistrict; northeastern, eastern, southeastern, southern, and southwestern Cibeureum Subdistrict; and southeastern, northwestern and northern Tamansari Subdistrict. These areas has low predominant frequency values and medium to high amplification values. This is also supported by the horizontal section of V_S at 25 m depth (Figure 6.b) which shows that in that area at 25 m depth the layers are still sediment with V_S below 750 m/s. It can be concluded that the thickness of the sediment in these areas are relatively thicker than in other areas and are more susceptible to damage if an earthquake occurs. There were buildings damage at SMKN 3 Tasikmalaya and the Asy-Syuhada Mosque which is located in the northwestern Tamansari Subdistrict due to the 2015 earthquake [9].

5. Conclusion

Based the obtained results we summarize some finding. The areas that are more prone to earthquake hazards are the western and northwestern Bungursari Subdistrict; northwestern Mangkubumi Subdistrict; eastern Kawalu Subdistrict; northeastern, eastern, southeastern, southern, and southwestern Cibeureum Subdistrict; and southeastern, northwestern and northern Tamansari Subdistrict. These areas has low predominant frequency values, medium to high amplification values, and low V_S values at 25 m depth which are indicates thicker sediment layer.

References

[1] Bemmelen R V 1949 General geology of Indonesia and adjacent archipelagoes, Government Printing Office, The Hague Martinus Nijhoff

[2] Budhitrisna T, Supandjono J B, Panggabean H and Marino 1986 Peta geologi lembar Tasikmalaya, Jawa Barat, skala 1: 100.000 *Pusat Penelitian dan Pengembangan Geologi Bandung*

[3] Wathelet M 2008 An improved neighborhood algorithm: parameter conditions and dynamic scaling *Geoph. Res. Lett.* **35** 9

[4] Cipta A, Cummins P, Dettmer J, Saygin E, Irsyam M, Rudyanto A and Murjaya J 2018 Seismic velocity structure of the Jakarta Basin, Indonesia, using trans-dimensional Bayesian inversion of horizontal-to-vertical spectral ratios *Geophys. J. Int.* **215(1)** 431-449

[5] Sambridge M 1999 Geophysical inversion with a neighbourhood algorithm—II. Appraising the ensemble *Geophys. J. Int.* **138(3)** 727-746

[6] Praja N K, Supartoyo and Sulaiman C 2018 Observasi Dampak *Laporan Cepat Gempa Tasikmalaya Provinsi Jawa Barat 15 Desember 2017 (M6.5)* (Bandung: Puskim PU)

[7] Badan Standardisasi Nasional 2012 *Tata cara perencanaan ketahanan gempa untuk struktur bangunan gedung dan nongedung* (Jakarta: BSN)

[8] Nakamura Y 2008 On the H/V spectrum *Proc. 14th World Conf. on Earthquake Engineering (Beijing)* (China)

[9] Ridwan M, Mulia A Y dan Megantara Y 2018 Observasi Kerusakan, *Laporan Cepat Gempa Tasikmalaya Provinsi Jawa Barat 15 Desember 2017 (M6.5)* (Bandung: Puskim PU)

Crustal thickness estimation in Indonesia using receiver function method

M F Fauzi[1], A Anggraini[1], A Riyanto[2], D Ngadmanto[2] and W Suryanto[1*]

[1]Seismology Research Group, Department of Physics, Universitas Gadjah Mada Yogyakarta, Indonesia
[2]Meteorology, Climatology and Geophysics Agency (BMKG) Yogyakarta Office, Jl. Wates Km.8, Gamping, Yogyakarta, Indonesia

*E-mail : ws@ugm.ac.id

Abstract. The existence of seismic wave velocity difference in the Earth crust and mantle creates the possibility to use earthquake data for estimating the crustal thickness utilizing the Ps conversion phase in the boundary. The radial component signal was deconvolved from the vertical component in the frequency domain to estimate receiver function for Indonesia region. We implemented the water level deconvolution techniques with a Gaussian filter of 2.5 Hz to eliminate the high frequency noise in the receiver function. The H-k stacking technique was performed to all receiver functions from each event to predict the crustal thickness and the Vp/Vs ratio below the stations. We analyzed ten azimuthally distributed teleseismic earthquakes recorded by 108 stations of BMKG. The result shows that the crustal thickness in Indonesia varies from 20 to 39.9 km. The western part of Sumatera, northern part of Sulawesi Island, and North Maluku region show generally thinner crust with value about 20 to 25 km. The North Sumatera, Central Java, and East Java show a considerably thicker crust of up to 36 km. Furthermore, our result also reveals a difference of crustal thickness about 5 km with the previous studies.

1. Introduction

Earth can be separated into three sections based upon chemical composition: the crust, the mantle, and the core. The crustal thickness varies in each area. The existence of seismic wave velocity difference in the Earth crust and mantle introduces the possibility to use the earthquake data for estimating the crustal thickness. Receiver function method uses Ps conversion phase in the crust-mantle boundary and processes the earthquake data to identify the structure of Earth crust and upper mantle. The crustal thickness studies in some areas in Indonesia have been conducted using this method [1-4]. Limited studies about the crustal thickness of Indonesia encourage us to conduct research covering all areas in Indonesia using greater station number. In this study, we are not only estimating the crustal thickness variation in Indonesia, but also comparing the result with the previous studies.

Information about crustal thickness can support the study of tectonic evolution and geological process in certain area. The Moho discontinuity which separates the crust and mantle is an important parameter for many studies, such as gravity modelling, upper-mantle tomographic studies, location of seismic events, and forward seismic wave propagation [5]. Therefore, this study can be used as supporting data to any researches in the future and also provide additional references for the crustal thickness of Indonesia.

IOP Conf. Series: Earth and Environmental Science **873** (2021) 012086 doi:10.1088/1755-1315/873/1/012086

2. Data and Methods

Seismic wave that arrives at seismic station contains information about the seismic source, travel track, and the structure below the station. By eliminating source structure and travel track effect, we obtain the isolated local structure information that is called receiver function. Receiver function method is based upon the fact that the P wave will be refracted not only as P wave but also as S wave when it hits a subsurface boundary (for example The Moho discontinuity). The arrival time and amplitude of P-S converted wave and its reverberation represent the boundary's position.

In this study, we calculate the receiver function through analyzing teleseismic events recorded on the three-component instrument from 108 stations of BMKG. Ten azimuthally distributed events with magnitude more than 6.5 and distance more than 30° are utilized. Location of stations and teleseismic events utilized in this study are shown in Figure 1.

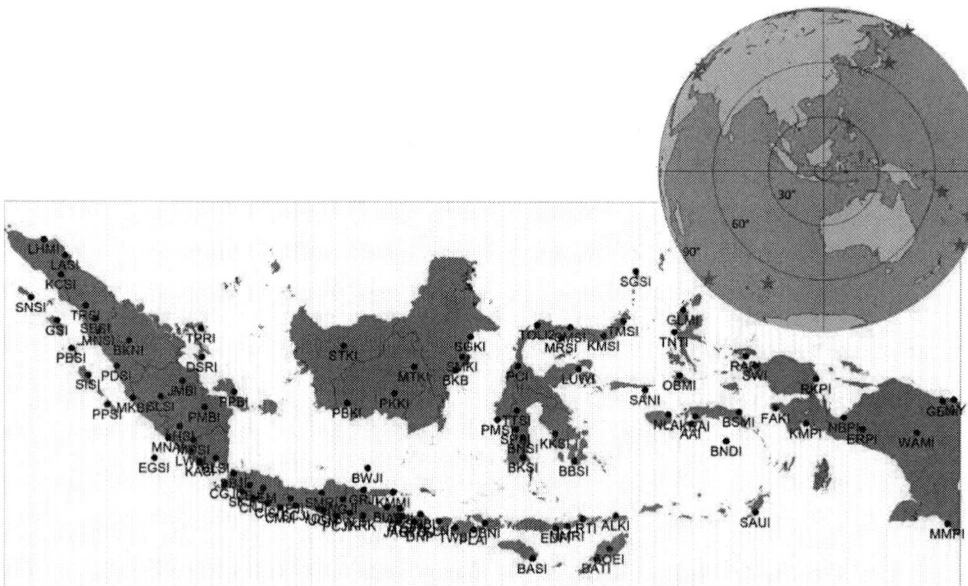

Figure 1. Seismic stations and teleseismic earthquakes utilized in this study. Seismic stations are denoted by black dots, teleseismic events used in this study are shown with red stars and the study area is shown in red colour.

To calculate receiver function, first we windowed the teleseismic waveform from 10 seconds before to 30 seconds after P wave arrival. Then the horizontal components were rotated before frequency-domain water level deconvolution [6] was implemented to generate receiver function. The radial component is deconvolved from the vertical component in frequency domain with the water level of 0.1, 0.01, 0.05, 0.001, and 0.005, adjusted to the smallest misfit obtained. To remove high frequency noises in the receiver function, we carried out a Gaussian filter of 2.5 Hz.

We then implemented H-k stacking method [7] to all receiver functions (RFs) obtained from each event. This method is applied to estimate the crustal thickness and the Vp/Vs ratio below each station. The H-k stacking method involves the summation of all RFs amplitude at times in accordance with Ps, PpPs and PpSs+PsPs. The arrival time of Ps converted phase and its multiple we used are obtained by grid search across a range of crustal thickness (H) and the Vp/Vs ratio (k). Predicted H and k value would yield the largest amplitudes summation. In this study, the grid search range was set from 20 to 40 km for crustal thickness and 1.6 to 2.0 for Vp/Vs ratio with interval 0.01. For this H-k stacking method, we assumed that the mean velocity of P-wave is 6.3 km/s, and the value of ray parameter is 0.06 s/km.

IOP Conf. Series: Earth and Environmental Science **873** (2021) 012086 doi:10.1088/1755-1315/873/1/012086

3. Result and Discussion

Figure 2 shows observed radial component recorded at PMBI station compared to the calculated radial component which is generated from the convolution of the vertical component with receiver function. Observed radial component and calculated signal are quite similar, which indicate the receiver function generated from water level deconvolution has a good quality. The stacked receiver function and all receiver functions ordered by the back azimuth of 4 stations are shown in Figure 3.

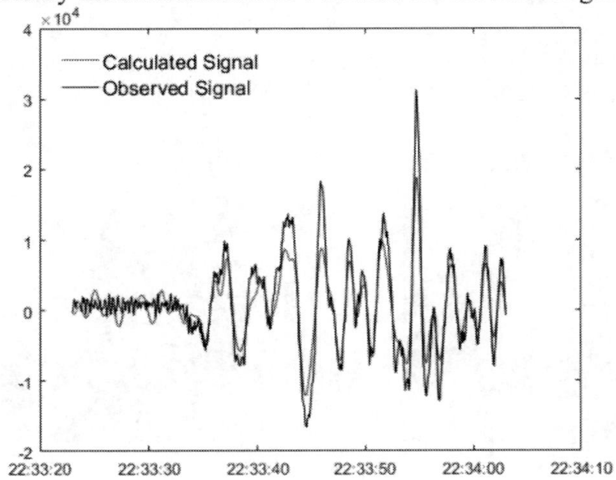

Figure 2. Observed radial component recorded at PMBI station (black line) compared to the calculated signal (red line).

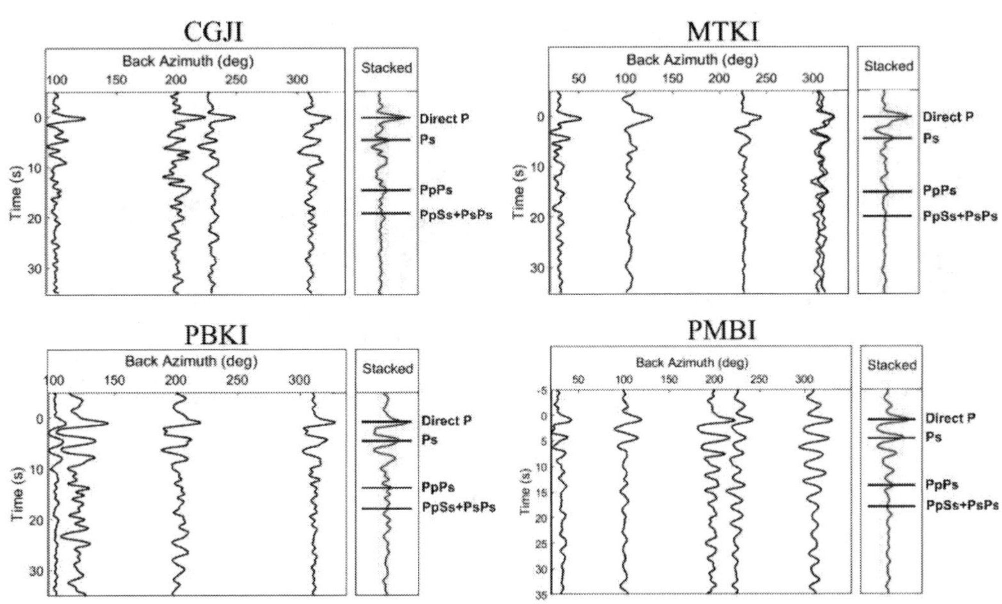

Figure 3. Stacked receiver function (red line) and all receiver functions (black line) ordered by the back azimuth for 4 different stations.

To estimate the thickness of the earth crust (H) and Vp/Vs ratio (k) below each station, we applied H-k stacking method to all receiver functions. Figure 4 shows the H-k stacking results for 4 stations.

Black cross represents the estimated value of H and k. All stacked receiver functions and H-k stacking results for all station can be accessed at bit.ly/hasilRF.

We can see yellow ringing area around black cross in Figure 4, which represents uncertainties in H-k stacking. The uncertainties of H-k stacking could be caused by the mean velocity of P-wave and the value of ray parameter we assumme are not the same as the real condition. If the real value of ray parameter and the mean velocity of P-wave beneath the station are quite close to the value we assume, the ringing structure in H-k stacking could be clear as shown at Station PMBI. Otherwise, some stations exhibit severe ringing which may be caused by the real value of ray parameter are not quite close to 0.06 s/km, and the real velocity of P-wave beneath the station are not quite close to 6.3 km/s.

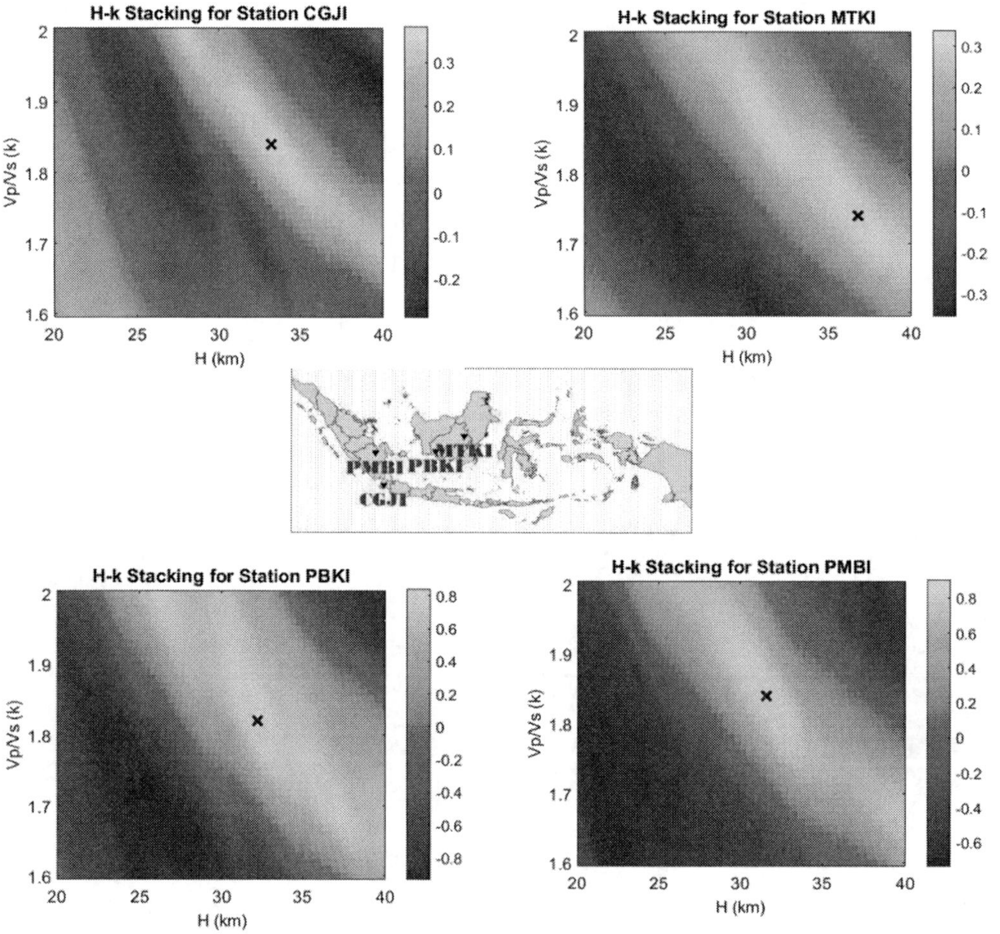

Figure 4. H-k stacked result for 4 different stations. Black cross represents estimated crustal thickness (H) and Vp/Vs ratio (k). Small map shows the location of 4 stations.

We calculate the arrival time of Ps, PpPs and PpSs+PsPs using the estimated value of H and k based on Zhu and Kanamori [7]. Based upon this calculation, we pick the arrival of Ps, PpPs and PpSs+PsPs in the stacked receiver function as shown in Figure 3.

The estimated crustal thickness of each station is then gridded and interpolated to provide an estimate of crustal thickness variation map as shown in Figure 5. The result shows that the crustal thickness in Indonesia varies from 20 to 39.9 km. Some areas in Indonesia have a lower crustal thickness approximately 20 to 25 km which includes the western part of Sumatra, the northern part of Sulawesi

Island, and also the North Maluku region. Meanwhile, the crustal thickness beneath North Sumatra, Central Java, and East Java is considered as the highest with the value up to 36 km thick.

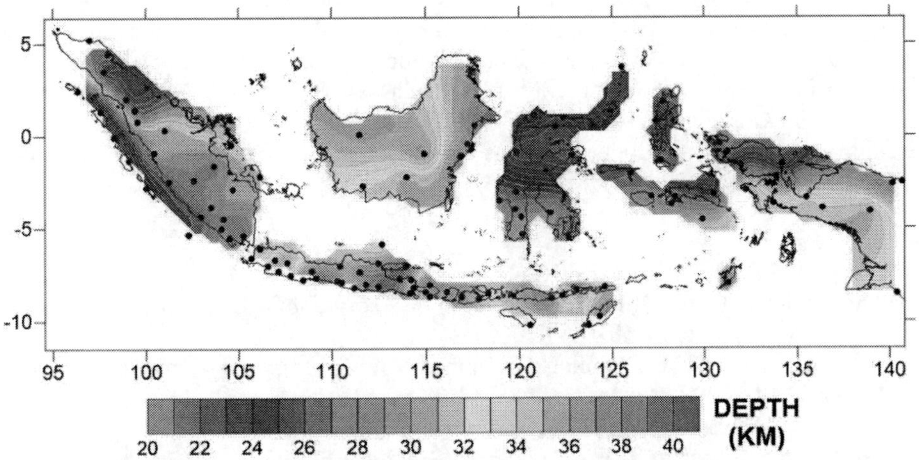

Figure 5. Crustal thickness map of Indonesia. The black dot represents seismic stations.

Compared to Goyal's study [1], our result shows difference ranging from 1 to 4 km for the crustal thickness beneath GSI, LHMI, MNAI, BKNI, and PMBI station. The differences are also found with Syuhada and Anggono's study [2] at 5 stations located in West Java which are CGJI, CISI, CMJI, CNJI, and SKJI, with a range from 1 to 5 km. In comparison to Syuhada's study in Nusa Tenggara [3], our result shows a difference by about 1 to 4 km. Wölbern and Rümpker [4] inferred the crustal thickness ranging from 27 to 39 km in Central Java and East Java, whereas we estimated the crustal thickness of 34 to 38 km in the same area. Table 1 summarizes our result compared to the previous studies.

Table 1. Crustal thickness in this study compared to previous studies.

Station	Crustal thickness in this study (km)	Crustal thickness based on previous studies (km)
GSI	20.39 ± 1.01	19 [1]
LHMI	38.25 ±0.63	35 [1]
MNAI	27.25 ±2.29	31 [1]
BKNI	30.65 ± 4.21	27 – 30 [1]
PMBI	31.60 ±1.94	33 [1]
CGJI	37.2 ± 1.3	33.20 ± 1.52 [2]
CISI	31.9 ± 0.7	32.78 ± 2.70 [2]
CMJI	34.5 ± 2.0	39.96 ± 2.30 [2]
CNJI	35.3 ± 1.4	30.03 ± 2.25 [2]
SKJI	33.6 ± 1.1	31.00 ± 1.58 [3]
SOEI	34.67 ± 0.45	37.98 ± 4.11 [3]
BATI	27.28 ± 0.39	28.79 ± 3.04 [3]
BASI	33.73 ± 4.68	34.06 ± 1.51 [3]
PLAI	28.16 ± 1.95	30.81 ± 4.32 [3]
DBNI	28.35 ± 4.07	32.48 ± 4.13 [3]
MMRI	34.67 ± 1.96	37.61 ± 1.90 [3]
LRTI	28.95 ± 0.88	27.33 ± 2.16 [3]

4. Conclusion

We calculated receiver function for 108 stations of BMKG by analyzing 10 teleseismic events. Then we performed the H-k stacking method to all receiver functions to predict the thickness of the earth crust and Vp/Vs ratio beneath each station. Crustal thickness map of Indonesia is then made based on the H-k stacking results. The result shows that the crustal thickness in Indonesia varies from 20 to 39.9 km. The western part of Sumatera, northern part of Sulawesi Island, and North Maluku region show generally lower thickness with value about 20 to 25 km. The North Sumatera, Central Java, and East Java areas have a significantly thicker crust of up to 36 km. Compared to previous studies, our result shows a difference of about 5 km.

References

[1] Goyal A 2016 *BS-MS Thesis* IISER, Kolkata
[2] Syuhada and Anggono T 2016 *Proc. Int. Sym. on Frontier of Applied Physics* (Bandung) vol 1711 (New York: AIP Publishing) p 070001
[3] Syuhada, Hananto N D, Abdullah C I, Puspito N T, Anggono T and Yudistira T 2016 *Acta Geophys.* **64** 2020
[4] Wölbern I and Rümpker G 2016 *J. Asian Earth Sci.* **115** 69
[5] Abrehdary M and Sjoberg L E 2019 *J. Appl. Geodesy* **13** 291
[6] Langston C A 1979 *J. Geophys. Res.* **84** 4749
[7] Zhu L and Kanamori H 2000 *J. Geophys. Res.* **105** 2969
[8] Mooney W D, Laske G and Master T G 1998 *J. Geophys. Res.* **103** 727
[9] Ligorria J P and Ammon C J 1999 *Bull. Seismol. Soc. Am.* **89** 1395

Implementation of Landslide Susceptibility Model Using Machine Learning for Semi Detailed Map Scale in Mountainous Region of Java Island

Imam A. Sadisun,[1,2,*] Rendy D. Kartiko,[1] and Indra A. Dinata,[1]
[1] Applied Geology Research Group, Faculty of Earth Sciences and Technology, Institut Teknologi Bandung, Indonesia
[2] Research Centre for Disaster Mitigation, Institut Teknologi Bandung, Indonesia

*Corresponding author: sadisun@gmail.com

Abstract. Landslide susceptibility modeling using neural network (ANN) are applied to semi detailed volcanic-sedimentary water catchment. Annually landslide occurred in catchment area frequently in unconsolidated and weathered material combined with uncertainty in rainfall pattern that complicated landslide occurrence. Data used for analysis including landslide inventory, geology, digital elevation related data, distance to stream, and several other available data. Results show that machine learning method yield fair result data based on evaluation on Area under Curve (AUC). Thus, it can be suggested that machine learning methods for landslide susceptibility model could still be develop to produce robust prediction model with different characterization of parameter data and machine learning parameters.

Keywords: landslide, neural network, support vector machine, AUC, SCAI.

1. Overview

Lately in second decade of second millennium, machine learning and artificial intelligence have become hot issues and commodities. A lot of fields try to grab these concepts and adopt it, either successfully or only partially successful. Research in landslide as part of normal science in predicting occurrence of landslide event is also part of this trend. Latest comparative review [1] with paper citation develop exponentially until around 2018. At least 11 machine model architectures are reviewed by Merghadi, with ensemble (combining methods) have higher overall performance. AUC (area under curve) as one indicator of model performance showed variation of 0.816 to 0.894 based on Merghadi's validation. Meanwhile, other researchers [2, 3, and 4] also has quite high (even excellent) AUC value, range from 0.87 to 0.982. This pique our interest to see how well this new trend when applied to Indonesian case.

2. Study Area

Small subset of water catchment in West Java had been picked for study area as seen in Figure 1. This area is part of Cidadap dan Cisokan rivers in West Java. Older volcanic peaks are distributed south of study area. Intense geomorphological process occurred in this area, whether rainfall and also as an active tectonic region.

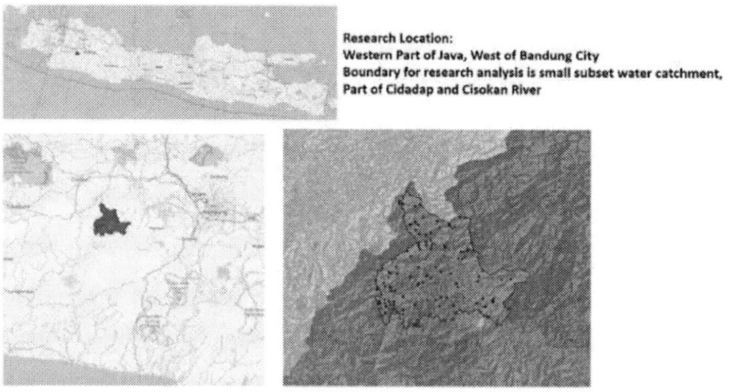

Figure 1. Study area is located in western part of Java Island, along mountainous volcanic region with tertiary sediment exposed in between.

3. Analysis

Input data consisted of scale value data (slope angle, NDVI), ordinal data (distance to structure, distance to stream), and also nominal data such as lithology. Input data distribution can be seen in Figure 2. Landslide in the study area consisted of 212 location which are distributed within 94 km2.

Artificial Neural Network classification are applied using SPSS multi layers perceptron. Two hidden layers are used for classification. All parameters (6 parameters) are included without any statistic test filter to find overall effect on method. Two main scenarios are conducted in this analysis, first with pure data set and second combined with interval scaled data. As an example, lithology is put as nominal scale value in the first scenario, while in second scenario lithology is weighted with frequency ratio value from landslide distribution within lithology zone.

Binary logistic regression is performed for controlling factor to see how accurate artificial neural network compared to other method, in this case with multivariate method. Logistic regression also uses pre-weighted value for nominal parameters, such as lithology.

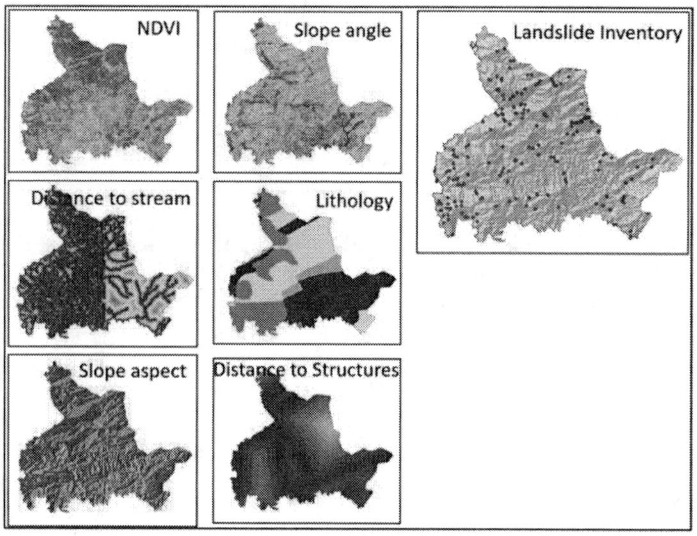

Figure 2. Input data for landslide susceptibility analysis.

4. Result

Area under curve value (AUC) for three scenarios as seen in Figure 3, Figure 4, Figure 5 show that Logistic Regression has the highest AUC with 0.694, followed with pre-weighted ANN (second scenario) with 0.682, and unweighted ANN (first scenario) with 0.612.

Visually, gradation within second ANN scenario is quite jarring as seen in the Figure 4 with distance to structure and lithology has over control for whole region, which made very distinct zones. Gradation in first ANN scenario show graded zone with still distinct zone area easily recognized due to distance from stream parameter. Logistic regression gradation shows more graded zone than both ANN scenario with slope parameters as dominant control of susceptibility zoning.

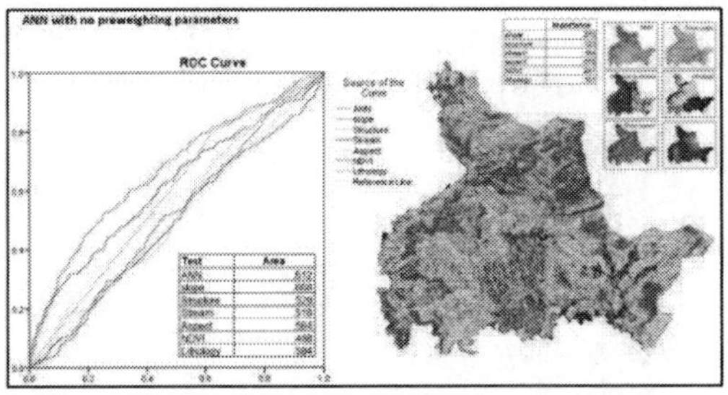

Figure 3. Landslide susceptibility with ANN first scenario (without pre-weighting parameters).

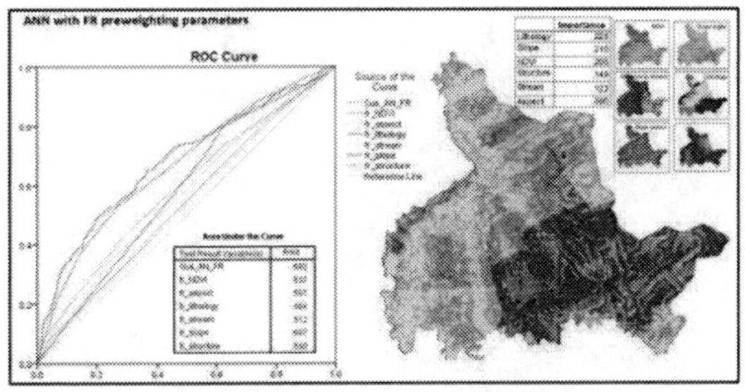

Figure 4. Landslide susceptibility with ANN second scenario (with pre-weighting parameters).

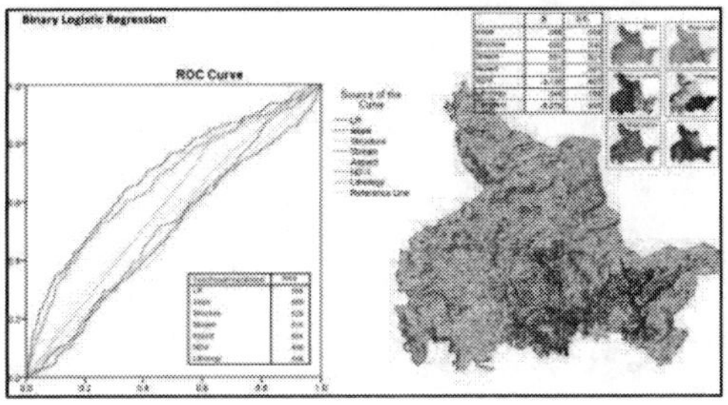

Figure 5. Landslide susceptibility with Binary Logistic Regression.

5. Discussion

Result of landslide susceptibility within this small set watershed show only moderate AUC (or ROC) value, 0.6 to 0.7. Highest AUC is not even ANN method. This does not directly show that ANN are less powerful in predicting landslide than multivariate method.

Some approach should be analysed later on ANN method, especially in preparing data for analysis and internal adjustment within ANN method itself. It seems that data preparation using pre-weighted data will resulted in higher AUC value but must be inspected visually not just statistically. ANN method itself perhaps need several adjustments such as learning time value, number of hidden layers, and also added more parameters. Thus, this preliminary result perhaps can be expanded later on, and will yield in better AUC and visual distribution of landslide susceptibility zones.

6. Conclusion

- The dominant control in landslide susceptibility in three scenarios is slope which has 0.668 AUC value in unweighted ANN (first scenario), 0.667 AUC value in pre-weighted ANN (second scenario), and 0.668 AUC value in Logistic Regression.
- Area under curve value (AUC) for three scenarios show that Logistic Regression has the highest AUC with 0.694, followed with pre-weighted ANN (second scenario) with 0.682, and unweighted ANN (first scenario) with 0.612.

References

[1] Merghadi A, Yunus A P, Dou J, Whiteley J, Thai P. B, Bui D T, Avtar R, and Abderrahmane B 2020 Machine learning methods for landslide susceptibility studies: A comparative overview of algorithm performance *Earth-Science Reviews* **207** 103225

[2] Arabameri A, Pradhan B, Rezaei K, and Lee C W 2019 Assessment of landslide susceptibility using statistical-and artificial intelligence-based FR–RF integrated model and multiresolution DEMs *Remote Sensing* **2019** 11 999

[3] Hong H, Liu J, and Zhu A X 2019 Landslide susceptibility evaluating using artificial intelligence method in the Youfang District (China) *Environmental Earth Sciences* **15** 488

[4] Phong T V, Phan T T, Prakash I, Singh S K, Shirzadi A, Chapi K, Ly H B, Ho L S, Quoc N K, and Pham B T 2019 Landslide susceptibility modeling using different artificial intelligence methods: A case study at Muong Lay district, Vietnam *Geocarto International* 1–24

Landslide Susceptibility Zonation of Rongga District and Surrounding Areas Using Weight of Evidence (WoE) Method

Imam A. Sadisun,[1,3,*] Dika B. Prasaja[2] Rendy D. Kartiko,[1] and Indra A. Dinata[1]
[1] Applied Geology Research Group, Faculty of Earth Sciences and Technology, Institut Teknologi Bandung, Indonesia
[2] Graduate student, Institut Teknologi Bandung, Indonesia
[3] Research Centre for Disaster Mitigation, Institut Teknologi Bandung, Indonesia

*Corresponding author: sadisun@gmail.com

Abstract. Rongga District is located on West Bandung Regency, West Java, which is prone to landslide disaster. Morphological conditions in the form of steep hills become the one of landslide controlling factors. There are many landslide occurrences happen in this area, such as Nyomplong, Cibitung Village on March 23, 2020. The incident was triggered by heavy rain and strong winds. This area was chosen to assess the landslide susceptibility using the Weight of Evidence (WoE) Method. WoE is probabilistic bivariate method which connecting parameters causes landslide against distribution of landslide in research area. Landslide data which generated from direct observation in the field and satellite imagery morphology are 572 landslide events. The data is divided into two groups, the analysis data set (70%) and the validation data set (30%). The parameters used in the analysis are land use, slope, slope direction, curvature, elevation, rainfall, lithology, NDWI, NDVI, distance from road, distance from the river, distance to lineament, flow direction, lineament density, stream density and river density. The parameters validated by determining the value of the area under curve (AUC). AUC value> 0.6 will be used in the landslide susceptibility zonation next analysis. Validation of landslide susceptibility zonation was carried out using 172 landslide events. The result of the WoE validation shows the AUC success rate of 0.70 and AUC prediction rate of 0.76. The value of AUC shows that the modelling is good and acceptable.

Keywords: *landslide, Rongga, area under curve, weight of evidence, bivariate.*

1. Introduction

Landslide susceptibility identification is the most effective and economical way to provide basic data in spatial planning, land use, adaptation, and disaster mitigation [1]. Although the time and location of landslide events are difficult to predict, landslide susceptibility potential evaluation of an area should be carried out. Landslide susceptibility analysis already developed using various remote sensing-based methods and Geographical Information Systems (GIS) [2,3,4,5,6,7].

One of the methods commonly used is the Weight of Evidence (WoE) method. The Weight of Evidence (WoE) method is one of the popular methods used in the Bayesian Theory approach to determine the conditions of the spatial probability between parameters or in the form of layers. This landslide susceptibility analysis requires many landslide events which associated with factors or parameters [8].

2. Location

The research location is located at Rongga District, West Bandung Regency (Figure 1). The research location is bordered by Cianjur Regency in the west and south, Bandung City in the east and Purwakarta Regency in the north. The research location is approximately 61 km from Bandung City or approximately 1.5 hours by motorcycle via the Cililin route.

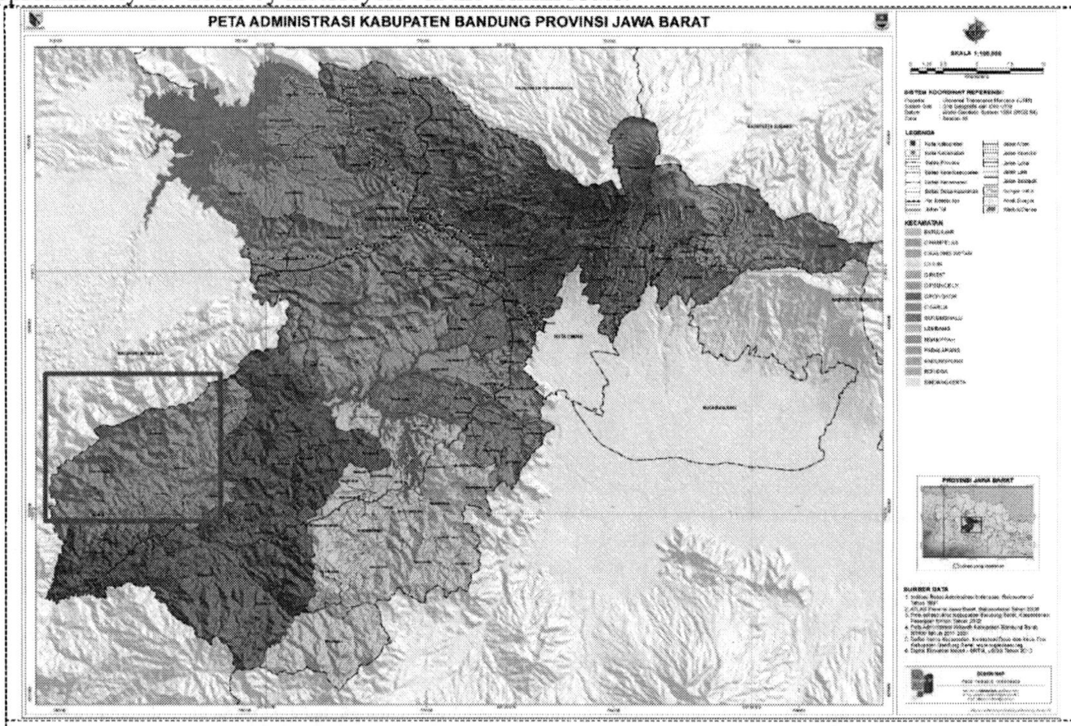

Figure 1. The research location is located at Rongga District, West Bandung Regency.

3. Methods

The WoE method is basically a bivariate method based on the Bayesian probability which is analysed using certain software/GIS [9]. The WoE method is a quantitative technique based on data, using many parameters, and a combination of data to produce a map of weighted data, both continuously and categorically validated using the initial (prior) and after (posterior) probabilities [10]. The WoE weighting system is divided into positive and negative. The positive weight of a particular geofactor class describes the weight of landslide probability in a geofactor class. The negative weight describes the weight of improper landslide in a geofactor class.

4. Data

The data used in this study are primary and secondary data. Primary data were obtained directly from the field and observations of landslides using google earth images (Figure 2), while secondary data were obtained from previous research or related references via internet. The data obtained were used to determine the value of each parameter required in landslide susceptibility zoning map using WoE method. The landslide location divided into two data, namely: train data (ls_train) with a percentage of 70% and test data (ls_test) with a percentage of 30%.

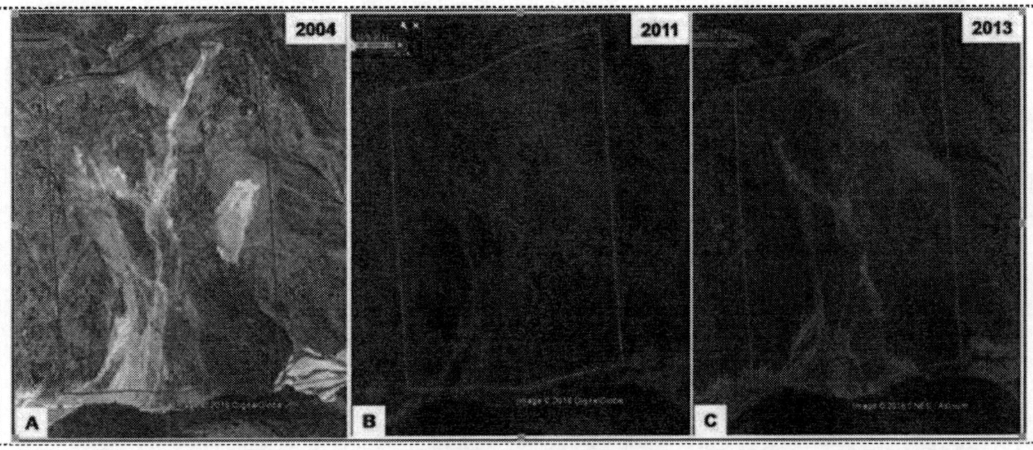

Figure 2. An example of landslide collecting using Google Earth time lapse.

Each parameter will be tested using the train data set (ls_train) to determine the AUC value (area under curve). Parameters with AUC> 0.6 are declared to have passed the validation which will be included in landslide susceptibility zoning map analysis using WoE Method (Table 1).

Table 1. Classification of AUC value [11].

AUC Value	Class
> 0.9	Excellent
0.8 – 0.9	Very good
0.7 – 0.8	Good
0.6 – 0.7	Average
< 0.6	Poor

5. Result and Discussion

Landslide observation data in Rongga area and its surroundings which taken through direct observation in the field (Figure 3) are 66 locations and based on observations through google earth satellite imagery (Figure 4) are 506 locations. A total of 572 landslide events were made into 2 sets of data, they are 70% train data (ls_train; 400 landslide events) for each parameter AUC test and 30% validation data (ls_test; 172 landslide events) for prediction rate validation of landslide susceptibility map (Figure 5).

Figure 3. Example of landslide location which taken direct observation.

Figure 4. Example of landslide location taken through google earth imagery.

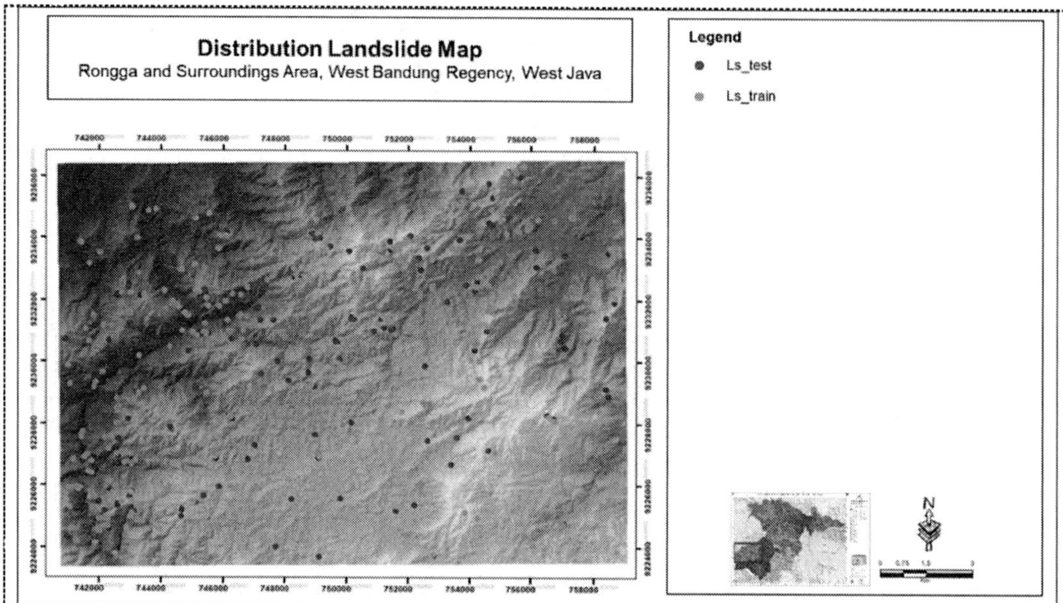

Figure 5. Landslide distribution map for train data (ls_train) and validation data (ls_test) in Rongga area and its surroundings.

The landslide parameters were analysed using commonly used parameters. After reclassification, the parameters are multiplied by the test data (ls_train) using raster calculator on the ArcGIS toolbox to get the table attribute and AUC value. The results of weighting using the WoE method of each parameter that are thought to influence the occurrence of landslide at research location are presented in Table 2.

Table 2. The AUC value of each parameter using the WoE Method.

No	Parameter	AUC Value
1	Land Use	0.60
2	Slope	0.68
3	Slope Aspect	0.58
4	Curvature	0.61
5	Elevation	0.63
6	Rainfall Intensity	0.62
7	Lithology	0.65
8	NDWI	0.54
9	NDVI	0.56
10	Distance from Road	0.58
11	Flow Direction	0.57
12	Lineament Density	0.52
13	River Density	0.60
14	Distance from Lineament	0.56
15	Distance from River	0.60

The parameters that pass the AUC value (AUC> 0.6) are: land use, slope, curvature, elevation, rainfall, lithology, river density and distance from the river then totalled using raster calculator on GIS. After obtaining the zoning of landslide susceptibility using the WoE method from the sum of the parameters, it will be validated and reclassified based on SNI 8291-2016.

Table 3. Distribution of landslide susceptibility zoning according to SNI 8291-2016.

Susceptibility	% LS	% LS Acc	Class WoE	% Area	% Area Acc	WoE Value From	WoE Value To	LS
High	60	60	0-93	47	47	123	153	103
Moderate	20	80	94-130	13	79	-926	-897	34
Low	15	95	131-184	25	87	-2,478	-2,444	26
Very Low	5	100	185-255	15	100	-10,709	-10,125	9

Furthermore, the landslide susceptibility zone is classified into four classes, namely: very low, low, medium, and high landslide susceptibility. The existing colours are adjusted to the colours that have been regulated in SNI 8291 of 2016, namely: very low is blue (RGB: 0 255 255), low is green (RGB: 0 255 0), medium is yellow (RGB: 255 255 0), and high is magenta (RGB: 255 0 255) (Figure 6). The zoning map is strongly influenced by slope and lithological distribution.

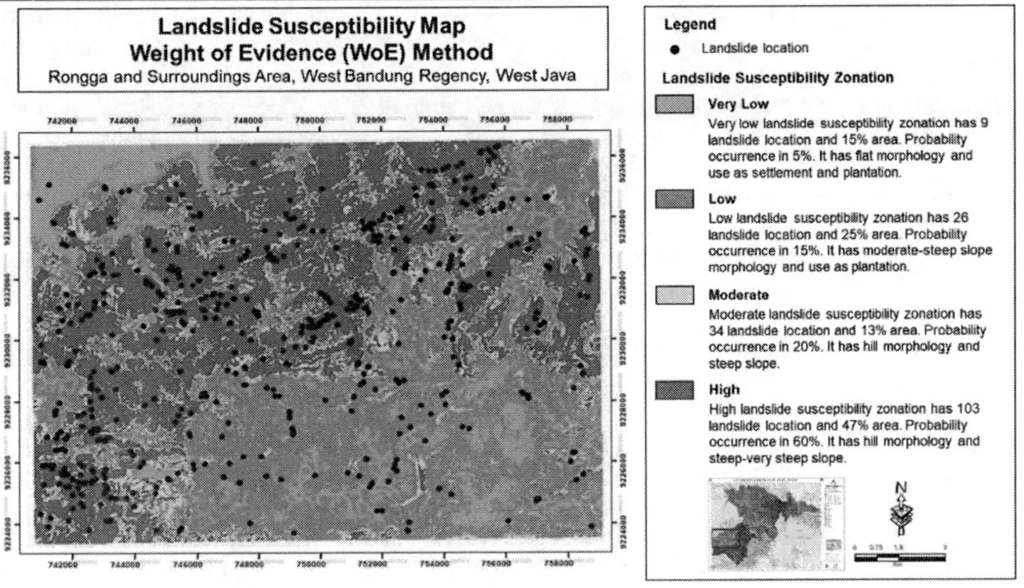

Figure 6. Landslide susceptibility map using WoE method.

Validation of zoning map is to determine the AUC value (area under curve) so that the map can be seen according to existing standards or not. The AUC value calculated is the AUC success rate and AUC prediction rate. The success rate is obtained by combining the sum of WoE data with landslide train data (ls_train). While the prediction rate is obtained by combining the total WoE data with landslide validation data (ls_test). The AUC value of landslide susceptibility zone success rate by WoE Method was 0.69 (Figure 7). The AUC value of landslide susceptibility zone prediction rate by WoE method was 0.66 (Figure 8). This value indicates that landslide susceptibility model (parameter selection and sufficiency of landslide event data) is quite good and acceptable because the AUC value obtained exceeds the recommended limit of 0.6.

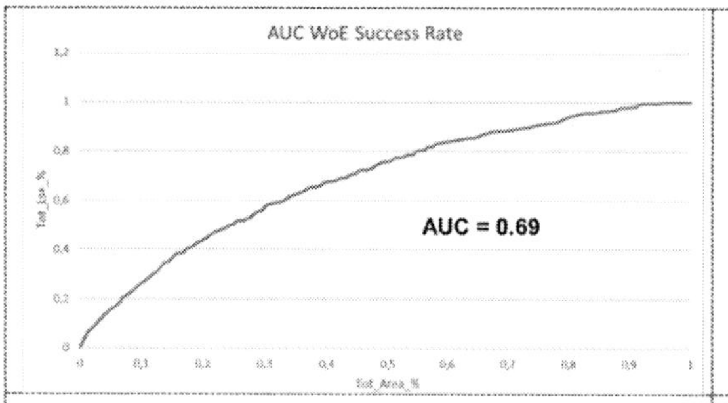

Figure 7. The success rate AUC value of landslide susceptibility map with WoE Method.

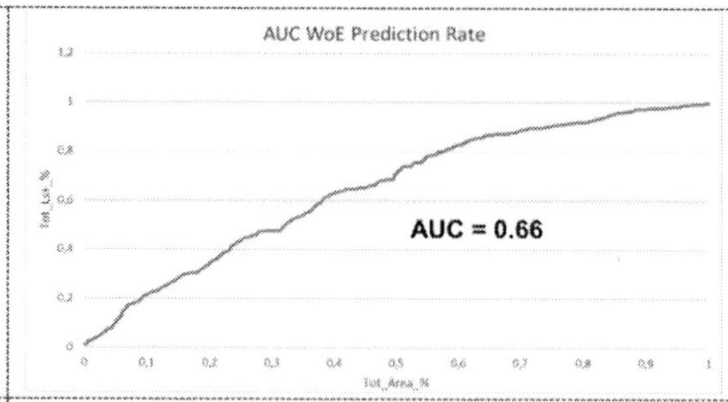

Figure 8. The prediction rate AUC value of landslide susceptibility map with WoE Method.

6. Conclusion

There are 8 parameters that pass the AUC value which are used for the analysis of landslide susceptibility zoning map using Weight of Evidence (WoE) method. After summing, the landslide susceptibility zoning map reclassified into 4 zones, they are: very low susceptibility zone with 15% covering area, low susceptibility zone with 25% covering area, medium susceptibility zone with 13% covering area, and high susceptibility zone with 47% covering area. The obtained AUC success rate is 0.69 and the AUC prediction rate is 0.66, both is reflecting that the resulting map is average and acceptable.

References

[1] Zhou S, Wang W, Chen G, Liu B and Fang L 2016 A Combined Weight of Evidence and Logistic Regression Method for Susceptibility Mapping of Earthquake-induced Landslides: A Case Study of the April 20, 2013 Lushan Earthquake, China (China: Acta Geologica Sinica)
[2] van Westen C J 1993 Application of Geographical Information System to Landslide Hazard Zonation. Enschede, Netherlands (ITC Publication, No.15)
[3] Guzzetti F, Reichenbach P, Cardinali M, Galli M and Ardizzone F 2005 Geomorphology 72 272-299
[4] Lee S and Pradhan B 2006 Journal of Earth System Science 115 661–672
[5] Dahal R K, Hasegawa S, Nonomura A, Yamanaka M, Masuda T and Nishino K 2008 Environ Geology 54 311–324
[6] Yilmaz I 2010 Environment Earth Sciece 61 821-836
[7] Yalcin A, Reis S, Aydinoglu A C and Yomralioglu T 2011 Catena 85 274–287
[8] van Westen C J 2007 Mapping Landslides: Recent Development in the Use of Digital Spatial Information ITC Publication Postbus 6
[9] Mezughi T, Akhir J M, Rafek A G and Abdullah I 2011 EJGE 16 1259–1273
[10] van Westen C J, Rengers N and Soeters R 2003 Natural Hazards 30 399-419
[11] Pourghasemi H R, Moradi H R and Aghda F S M 2013 Natural Hazard 69 749–779

Crustal deformation of Eastern Indonesia regions derived from 2010-2018 GNSS Data

Suchi Rahmadani[1,2], Irwan Meilano[1,2], Dina A. Sarsito[1,2], Susilo[3]

[1] Geodesy and Geomatics Engineering, Faculty of Earth Sciences and Technology, Institut Teknologi Bandung, Jalan Ganesha No. 10, Bandung, Indonesia
[2] Center of Earthquake Science and Technology, Institut Teknologi Bandung, Jalan Ganesha No. 10, Bandung, Indonesia
[3] Indonesian Geospatial Information Agency (BIG), Jalan Raya Bogor Km. 46, Nanggewer Mekar, Cibinong, Indonesia

E-mail: suchi.rahmadani18@gmail.com

Abstract. Eastern Indonesia lies in a complex tectonic region due to the interaction of four major tectonic plates: the Australian Plate, Pacific Plate, Philippine Sea Plate, and Sunda Block. Therefore, this region hosted some destructive seismic activities as well as tectonic deformation, such as the Mw 7.5 Palu Earthquake, the sequences of the 2018 Lombok Earthquake, and the Mw 6.5 Ambon Earthquake in 2019. Our work proposes a recent study on crustal deformation in Eastern Indonesia inferred from Global Positioning System (GPS) velocity field. We used GPS data from the observations of 49 permanent and 61 campaign stations from 2010 to 2018. Here, our velocity field result represents long-term tectonic deformation regions in Eastern Indonesia continuously, from Bali in the west to Papua in the east, demonstrated both in the ITRF 2008 and the Sunda reference frames. The spatial pattern of velocity field map collected from this research will give an initial insight into the present-day tectonic condition in Eastern Indonesia and then can be used to improve our ability to assess this area's earthquake potential.

1. Introduction

Eastern Indonesia is the region with the most complex tectonic configuration in Indonesia. This region is formed by the convergence of four active plates, namely the Australian Plate, Pacific Plate, Philippine Sea Plate, and the Sunda Block [1]. The convergence of the Australian Plate and Sunda Block has partitioned between a megathrust and a continuous zone of back-arc thrusting extending over 2000 km started from east Java to northern Timor [2]. Meanwhile, from the geological study, the Banda Arc itself was formed by a single slab subduction of the Australian plate [3].

This complex tectonic setting has contributed significantly to most of earthquakes happened in Eastern Indonesia. Some of them are the destructive earthquakes, such as the Mw 7.9 1992 Flores earthquake [4], the Mw 7.5 Palu Earthquake which followed by a tsunami [5], the sequences of the 2018 Lombok Earthquake [6, 7], and the Mw 6.5 Ambon Earthquake in 2019 [8], which followed by thousands of aftershocks. Figure 1 shows the historical earthquakes (Mw >5) from 1900 to 2015 in Eastern Indonesia based on the catalog of the National Center for Earthquake Studies (PuSGeN) [9].

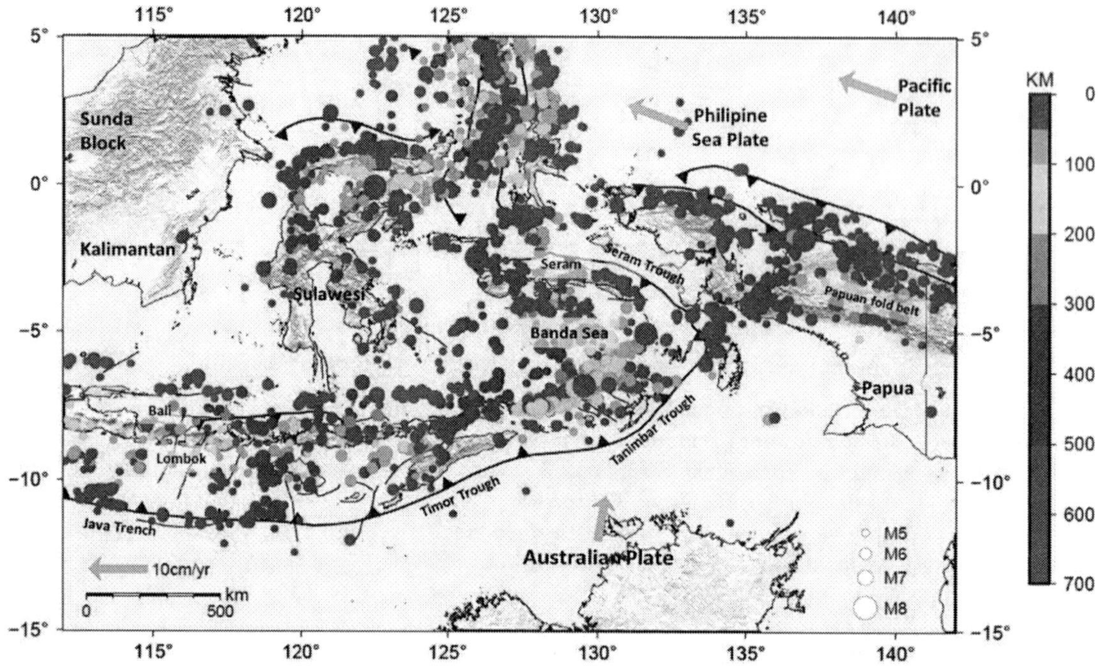

Figure 1. Tectonic setting of Eastern Indonesia. The black lines are active faults, and the colored circles are seismicity data [10]. The color bar shows the depth of the earthquakes. The grey arrow represents the plates velocity with respect to Sunda Block [11].

The previous geodetic studies in Eastern Indonesia focused on the part of the Banda Arc and Sulawesi with lack of GPS observations distribution [10, 12, 13]. This study will be the first one in the last decade of conducting the continuous velocity field in the vast area of Eastern Indonesia. This study area spread from Bali in the west to Papua in the east, with a denser and longer time of Global Positioning System (GPS) observation data. GPS could provide precise three-dimensional relative position up to millimeters fraction over a baseline range of hundreds of meters to thousands of kilometers [14]. Therefore, this technology is very suitable for geodynamics study.

2. Data and Method

We used GPS data from 110 stations provided by the Indonesian Geospatial Information Agency (BIG). The data consist of 49 continuous and 61 campaign stations (Figure 2). The data covered a nine-year time span from 2010 to 2018, although some covered shorter times. But in this study, we only directly processed the GPS data of 31 continuous stations using the GAMIT/GLOBK 10.7 software [15] to get daily time-series, while the rest of the GPS velocities that we used (18 permanent and 61 campaign stations) were from Susilo (2018) [16]. We also incorporated 16 IGS (International GNSS Service) stations (ALIC, AUCK, BAKO, CEDU, COCO, CUSV, DARW, GUUG, HYDE, IISC, KARR, PIMO, TOW2, TWTF, XMIS, and YAR2) into our processing to connect our regional network to the global GPS network, to get better daily solutions. To estimate the linear horizontal velocity, we removed outliers and seasonal variations and corrected for offsets in the time-series [17].

Since our velocity results were discrete, to effectively portray the velocity field throughout Eastern Indonesia, we used the spline interpolation with a tension algorithm suggested by Sandwell and Wessel (1998) to determine the continuous velocity field. After several testing, we implemented the interpolation within a square-based grid of the east and north velocity components with a grid size of 0.3° and applied a 0.3 tension factor, as recommended for topographic interpolation [18].

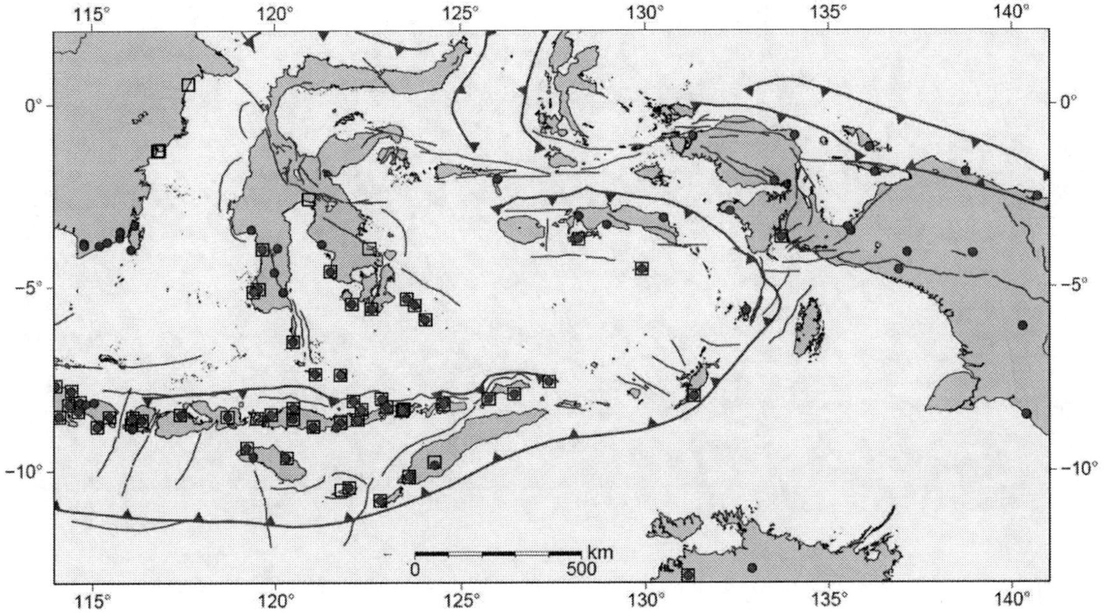

Figure 2. Map of Eastern Indonesia GPS stations used in our analysis. Red and blue circles represent permanent and campaign stations used in this study, respectively. The hollow squares are the GPS sites of the previous study [2]. The brown lines are faults.

3. Result and Discussion

The first result of our study is the interpolated velocity field based on our GPS observations with respect to the 2008 International Terrestrial Reference Frame (ITRF 2008) [19], which is shown in Figure 3. The velocity rates surely indicates that the velocities increase significantly from west to east with a range of ~0.8-72 mm/year. A significant value difference was demonstrated between west and east of Banda Arc (around Tanimbar trough), implying that this area's deformation is characterized mainly by crustal shortening due to the collision of Australian Plate. The velocity field also demonstrates a clear counterclockwise rotation from west to east and up to the north,. The vectors have changed the direction from SE in the southern of Papua to NW in the northern of Papua, with magnitude ~50-70 mm/year.

For more analysis, we tied our GPS velocity to the Sunda reference frame [20]. The result (Figure 4) shows that the west part of Eastern Indonesia (Kalimantan, West Sulawesi, Bali, and Lombok) has minimal velocities compared to the velocity rates with relative to the ITRF 2008 reference frame. It suggests that the west part of the study area is part of Sunda Block and others block different from the east part of Eastern Indonesia [2, 20, 21, 22]. The northern part of Papua reflects higher velocity rates (~95 mm/year) value than the result in Figure 3] and tends to move towards the NW; this is due to the Pacific Plate's convergence to the Papua. While velocities in the southern part of Papua tend to move northward as the result of the Australian Plate. The pattern of the velocity field in Papua has separated around the Jayawijaya mountains, as there is a major fault mechanism in there called Papuan fold belt [15].

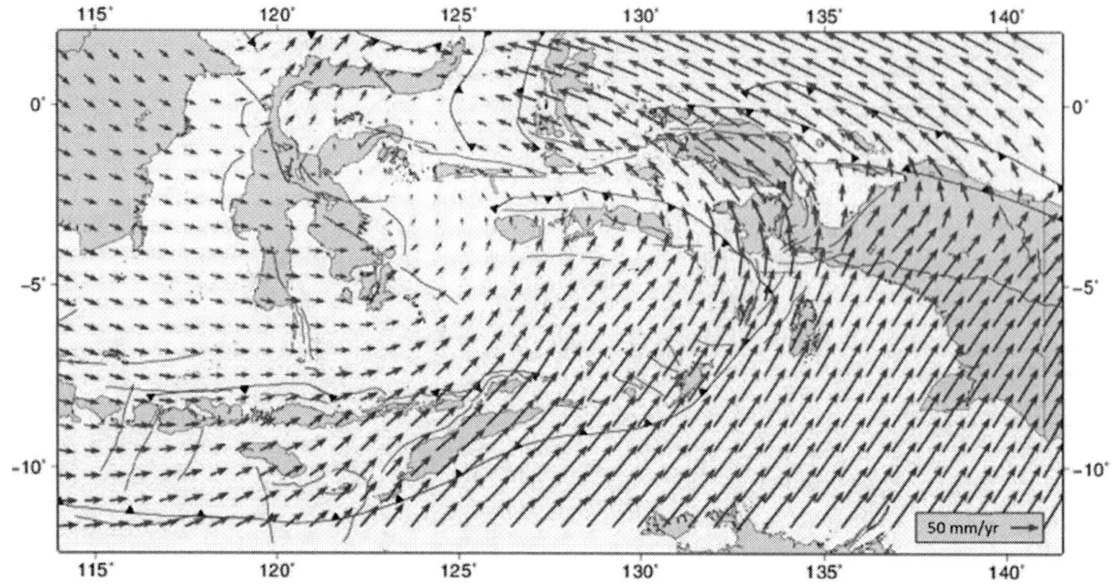

Figure 3. GPS velocity field of Eastern Indonesia with respect to the ITRF 2008 reference frame.

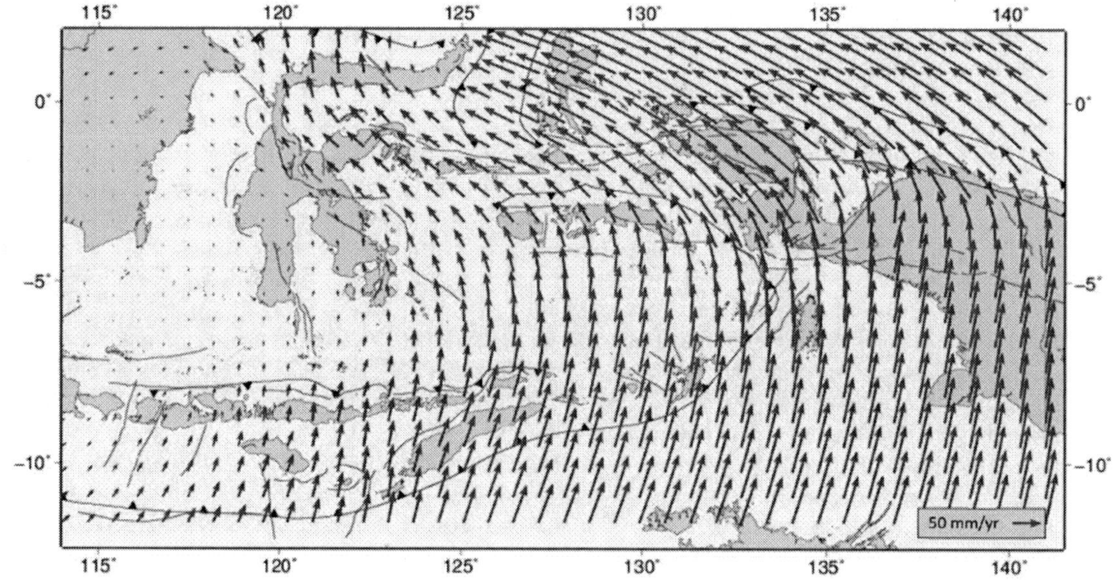

Figure 4. GPS velocity field of Eastern Indonesia with respect to Sunda reference frame.

4. Conclusions

Our denser velocity field result represents long-term tectonic deformation regions in Eastern Indonesia continuously, from Bali in the west to Papua in the east, demonstrated in the ITRF 2008 and Sunda Block reference frame. The results clearly show that the deformation pattern in Eastern Indonesia has partitioned into several blocks that have to be detailed with a comprehensive strain and block modeling in the future study. The pattern of velocity field map produced from this research will provide insight into the recent tectonic condition of Eastern Indonesia and then can be used to improve our knowledge to assess the earthquake potential of this area.

Acknowledments
We are grateful to BIG for providing the GPS data. Figures were generated using GMT software [23]. This research was funded by the 2020 PMDSU research scheme of the Ministry of Research and Technology.

References
[1] Hamilton W B 1979 Tectonics of the Indonesian region. *Tech. Rep.* **1078** U.S. Govt. Print. Off., Wash.
[2] Koulali A, Susilo S, McClusky S, Meilano I, Cummins P, Tregoning P, Lister G, Efendi J and Syafi'i M A 2016 Crustal strain partitioning and the associated earthquake hazard in the eastern Sunda-Banda Arc *Geophys. Res. Lett.* **43** (5) pp 1943-1949
[3] Spakman W and Hall R 2010 Surface deformation and slab–mantle interaction during Banda arc subduction rollback *Nat. Geosci.*
[4] Pranantyo I R, Cummins P R 2019 Multi-Data-Type Source Estimation for the 1992 Flores Earthquake and Tsunami *Pure Appl. Geophys.* **176** 2969–2983
[5] Socquet A, Hollingsworth J, Pathier E, and Bouchon M 2019 Evidence of supershear during the 2018 magnitude 7.5 Palu earthquake from space geodesy *Nat. Geosci.* **12** 192–199
[6] Salman R, Lindsey E O, Lythgoe K H, et al 2020 Cascading Partial Rupture of the Flores Thrust during the 2018 Lombok Earthquake Sequence, Indonesia *Seismol. Res. Lett.* **91** pp 2141–2151
[7] Supendi P, Nugraha A D, Widiyantoro S, Pesicek J D, Thurber C H, Abdullah C I, Daryono D, Wiyono S H, Shiddiqi H A, Rosalia S 2020 Relocated aftershocks and background seismicity in eastern Indonesia shed light on the 2018 Lombok and Palu earthquake sequences *Geophys. J. Int.* **221** pp 1845–1855
[8] Sahara D P, Nugraha A D, Muhari A, et al 2021Source mechanism and triggered large aftershocks of the Mw 6.5 Ambon, Indonesia earthquake *Tectonophysics* **799** 228709
[9] National Center for Earthquake Studies 2017 *Peta Sumber dan Bahaya Gempa Indonesia Tahun 2017*. Pusat Litbang Perumahan dan Permukiman, Badan Penelitian dan Pengembangan, Kementerian Pekerjaan Umum dan Perumahan Rakyat, Bandung
[10] Nugroho H, Harris R, Lestariya A W, Maruf B 2009 Plate boundary reorganization in the active Banda Arc-continent collision: Insights from new GPS measurements *Tectonophysics* **479** (1-2) pp 52-65
[11] DeMets C, Gordon R G, and Argus D F 2010 Geologically current plate motions *Geophys. J. Int. V* **181** no 1 pp 1-80
[12] Herawati Y A, Meilano I, Sarsito D A, Efendi J 2017 Strain analyis in Banda Sea using grid strain based on GPS observation *AIP Conf. Proc.* **1857** 4987075
[13] Pradipta G C, Meilano I, Gunawan E, Efendi J 2017 Deformation analysis in the east nusa tenggara and banda islands based on GPS observation from 2010-2015 *AIP Conf. Proc.* **1857** 4987076
[14] Segall P and Davis J L 1997 GPS applications for geodynamics and earthquake studies *Annu. Rev. Earth Planet. Sci* **1997** pp 301–36
[15] Herring T A, King RW and McClusky S C 2018 *Introduction to GAMIT/GLOBK, Release 10.7*. Cambridge: Massachusetts Institute of Technology
[16] Susilo S 2018 *Realisasi Model Deformasi Geodetik Untuk Datum Semidinamik di Indonesia* Academic Dissertation, Institut Teknologi Bandung
[17] Nikolaidis R 2002 *Observation of Geodetic and Seismic Deformation with the Global Positioning System* Academic Dissertation, Univeristy of California
[18] Wessel P and Bercovici D 1998 Interpolation with splines in tension: a Green's function approach

Math. Geol. **30** 77–93
[19] Altamimi Z, Métivier L and Collilieux X 2012 ITRF2008 plate motion model *J. Geophys. Res.-Sol. Ea.* **117**
[20] Yong C Z, Denys P H, and Pearson C 2017 Present-day kinematics of the Sundaland plate *J. Appl. Geodesy* **11** 169–177
[21] Bock Y, Prawirodirdjo L, Genrich J F, Stevens C W, McCaffrey R, Subarya C, Puntodewo S S O and Calais E 2003 Crustal motion in Indonesia from Global Positioning System measurements *J. Geophys. Res.* **108** 2367
[22] Simons W J F, Socquet A, Vigny C, Ambrosius B A C, Haji Abu S, Promthong C, Subarya, C, Sarsito D A, Matheussen S, Morgan P and Spakman W 2007 A decade of GPS in Southeast Asia: resolving sundaland motion and boundaries *J. Geophys. Res.* **112** B06420
[23] Wessel P and Smith W H 1998 New, improved version of generic mapping tools released. *Eos Trans. Am. Geophys. Union* **79** pp 579-579

Multi frequency Ground-Penetrating Radar method for mapping underground pipes and man-made objects

W W Parnadi[1,*], W Warsa[1], A Laesanpura[1] and D Santoso[1]

[1]Department of Geophysical Engineering, Institut Teknologi Bandung, Jl. Ganesha 10 Bandung 40132, Indonesia

[*]Corresponding author: wahyudi@gf.itb.ac.id

Abstract. Mapping of subsurface infrastructure such as pipelines, pipe ducts as well as man-made objects located below the earth surface is a difficult task in geotechnical engineering due to the need to direct contact to the object. Ground-Penetrating Radar (GPR) method as one of geophysical methods is an appropriate technique to solve such problems due to its capability to effectively provide the existence, the location and the depth of buried shallow objects. Despite many advantages, the application of GPR method is limited by achieved depth and resolution range that is frequency dependent. To overcome such problems, the so-called multi-frequency compositing method with Optimal Spectral Whitening (OSW) technique is studied and applied on real data acquired in the Muarakarang Combined Cycle Power Plant area located in the Northern part of Jakarta, the Capitol of Indonesia. Datasets from GPR measurement with 100 MHz and 250 MHz was used. Each radargram from the same path was processed individually and then joined to become multiple composite radargram. The resulted composite radargram from a GPR line we studied has shown clearer image for interpretation rather than individual interpretation. The applicability of this multi-frequency compositing recommend further study to solve some facing problems in geotechnical engineering.

1. Introduction

Mapping of near surface infrastructures such as pipelines, pipe ducts and other man-made objects located below the ground is difficult task for geotechnical work since its technique needs direct contact to the object to recognize its existence. Ground-penetrating radar (GPR) as a geophysical method is commonly used to overcome such problems due to its non-destructive nature and its implementation in the field can coexist with other activities that do not interfere with each other. This method is able to identify and find subsurface objects effectively. Despite many advantages, this method is limited by achieved depth and resolution range that is frequency dependent [1, 2].

Resolving power of GPR system to find anomalous objects underground is determined by the frequency content of GPR waves emitted by transmitter. Its dominant frequencies deliver information about optimum depth of investigation whilst its bandwidth determine the spatial resolution. Anomalous objects having a variety of size and depths require various GPR antenna frequencies to be deployed in order to image their general condition in the measurement area. Higher antenna frequencies can reach shallower depth with higher spatial resolution and vice versa. To obtain a more detailed subsurface image from surface to desired depth of investigation, a technique to combine radargrams obtained from

IOP Conf. Series: Earth and Environmental Science **873** (2021) 012090 doi:10.1088/1755-1315/873/1/012090

measurement with various antenna frequencies on the same path has been developed. This so-called multi frequency radargram compositing technique [3, 4, 5] was applied on real data, which are acquired in the Muarakarang Combined Cycle Power Plant area in Jakarta, Indonesia to mapping pipeline network and other made-made objects in subsurface. This study demonstrates the applicability of multi-frequency compositing technique to solve a problem in geotechnical engineering.

2. Data and methods

2.1. Multi-frequency compositing

Development of Multiple-frequency compositing technique is aimed at increasing radargram resolution at deeper depth at working frequency reference by conducting a fusion of radargrams acquired with two or more different working frequency antennas from the same path into one radargram. Some different approaches are now exist in the multiple frequency compositing technique. Priska et al. (2019) studied three different approaches, i.e. simple summation; dominant frequency amplitude equalization (DFAE) which comparing average radargram amplitude spectrum of each radargram; and optimal spectral whitening (OSW), an approach including the amplitude ratio but based on the Berlage wavelet analysis as the weighting factor in the radargram compositing technique [5]. From that study, it is concluded that the OSW technique produces the best resolution results among other multiple-frequency techniques used in the study [6].

Based on the above mentioned results, we apply OSW technique. In this method, the weighting value is stated by the equation below as [5, 6]:

$$Aw = \begin{bmatrix} A_{f_1}^{m_1} & A_{f_1}^{m_2} & \cdots & A_{f_1}^{m_M} \\ A_{f_2}^{m_1} & A_{f_2}^{m_1} & \cdots & A_{f_2}^{m_1} \\ \vdots & \vdots & & \vdots \quad \vdots \\ A_{f_N}^{m_1} & A_{f_N}^{m_1} & \cdots & A_{f_N}^{m_1} \end{bmatrix} \begin{bmatrix} w_{m_1} \\ w_{m_2} \\ \vdots \\ w_{m_N} \end{bmatrix} = \begin{bmatrix} s_{f_1} \\ s_{f_2} \\ \vdots \\ s_{f_N} \end{bmatrix} \quad (1)$$

where

Aw : radargram spectral amplitude value

$A_{f_N}^{m}$: the spectral amplitude on m-th data and N-th frequencies

w_m : weighting of m-th data

s_f : spectral amplitude value on N-th frequencies, which is usually a constant

Weighting factor based on the OSW method was applied at some time interval, with the time interval was calculated from the dominant period length of Berlage wavelet from the lowest frequency. After the weighting factor was calculated, then it was scaled based on the RMS amplitude value on that interval before being distributed to all data on the radargram, using linear interpolation by multiplying it with radargram amplitude that being previously normalized. This so-called multi frequency radargram compositing technique was applied on real data, which were acquired in the Muara Karang Combined Cycle Power Plant area in Jakarta, Indonesia to detect pipeline network and man-made objects in subsurface.

2.2. Site conditions

Study area is B-zone in Muarakarang Combined Cycle Power Plant area located in the northern part of Jakarta, the capitol of Indonesia (figure 1). Typical soil profile of the area is shown in figure 2. This soil profile is extracted from boring and SPT data of BH-3 located in the study area. Lithologically the soil profile consists of mixed soft, brown colored gravelly - silty clay with middle plasticity from surface to 2 m depth. Beneath this layer, brown colored, water saturated, degradable fine sand exists. This layer has a very low bearing capacity with only 3-4 N-SPT value from 2m to 12 m depth as seen in figure 2.

2.3. GPR data acquisition and processing

Ramac GPR system with 100 MHz and 250 MHz antenna was deployed. Two datasets from GPR measurement at 10 lines are available for this study, but only resulted data from line 19 is used in this paper. Figure 3 (a) and (c) show GPR measurement in lines adjacent to line 19 by using 100 MHz antenna. That measurement line crosses the location of buried pipe. The appearance of pipes on the surface can be seen in figure 3(b) and (d) and will be used for validation.

Individual GPR data was processed using Reflexw software from Sandmeier [7] with conventional processing steps including dewow, bandpass filtering, DC component subtraction, f-k filtering and background removal. Multiple composite radargram compositing step was then performed to join individual radargram.

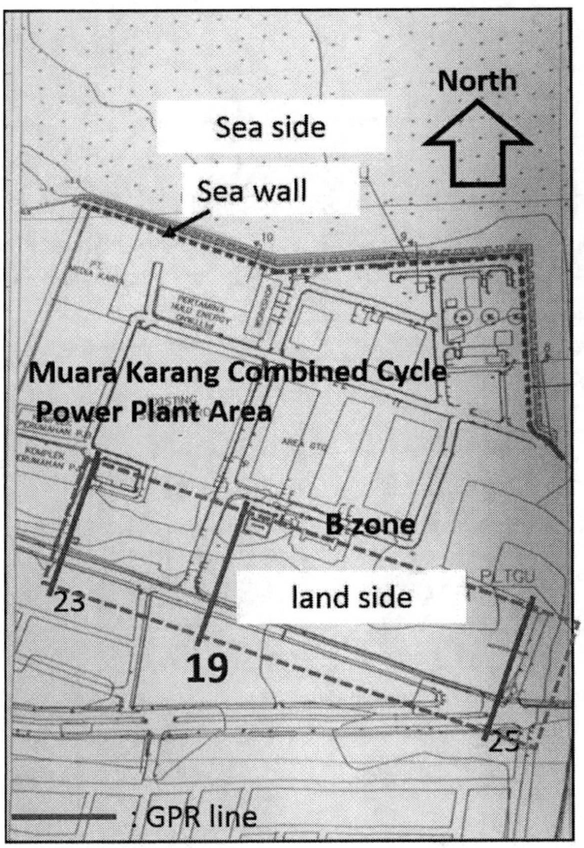

Figure 1. Study area located at Muara Karang Combined Cycle Power Plant in the northern part of Jakarta, the capitol of Indonesia. Some of GPR lines are shown.

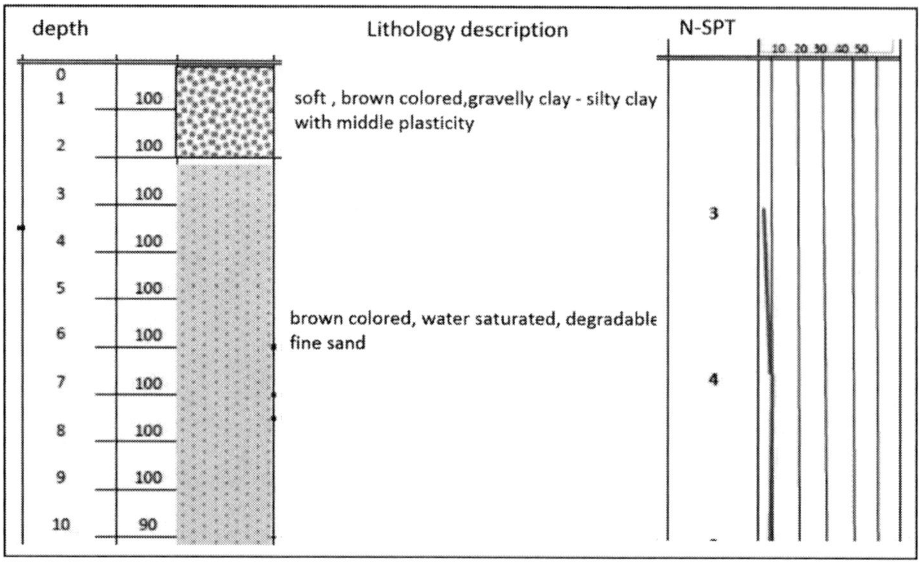

Figure 2. Typical soil profile extracted from boring and SPT data of BH-3 located in the study area. Note very low SPT value of fine sand from the depth of 2 m to 10 m exhibits very low bearing capacity.

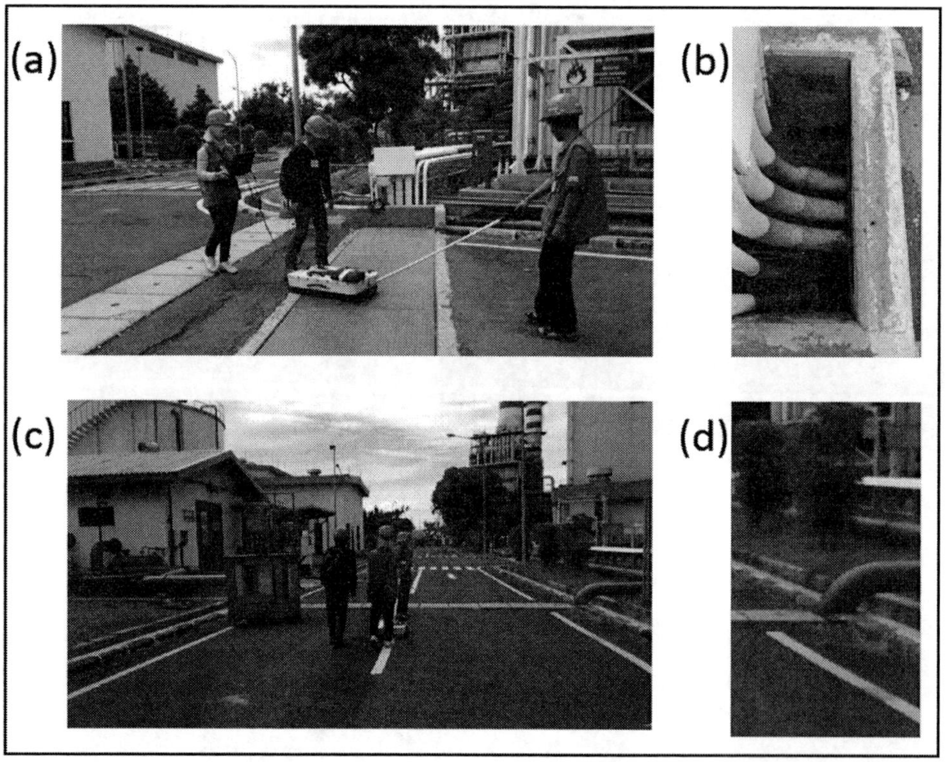

Figure 3. GPR acquisition in the study area: GPR campaign on the road crossing location of buried pipes (a; c). Six pipes of 10cm in diameter at 2 levels that go into the duct (c) and one 50cm diameter pipe that go into the ground (d)

IOP Conf. Series: Earth and Environmental Science **873** (2021) 012090 doi:10.1088/1755-1315/873/1/012090

3. Results and discussion

Soil profile as seen in figure 2 shows soil condition that is relative homogeneous soil from surface until depth of 2m. That soil profile is extracted from Boring at point BH-3 that was carried out until the depth of 30m. SPT within this 2 m thick soft clay was not carried out. SPT value or N-SPT of about 4 from 2 m to 12 m depth for fine sand are relatively very low. That value indicates a very low and homogeneous soil bearing capacity. Such soil conditions make it easier to interpret the data.

We collected GPR data from measurement using 100 MHz and 250 MHz antenna, but only resulted data from line 19 is discussed. Figure 4.(a) and (b) show processed radargrams of line 19 which were acquired with 100 MHz antenna and 250 MHz antenna, respectively. From 100 MHz radargram, only object located at depth of about 1 m is recognized. Produced blurred radargram is most probably caused by a relative water saturated ground.

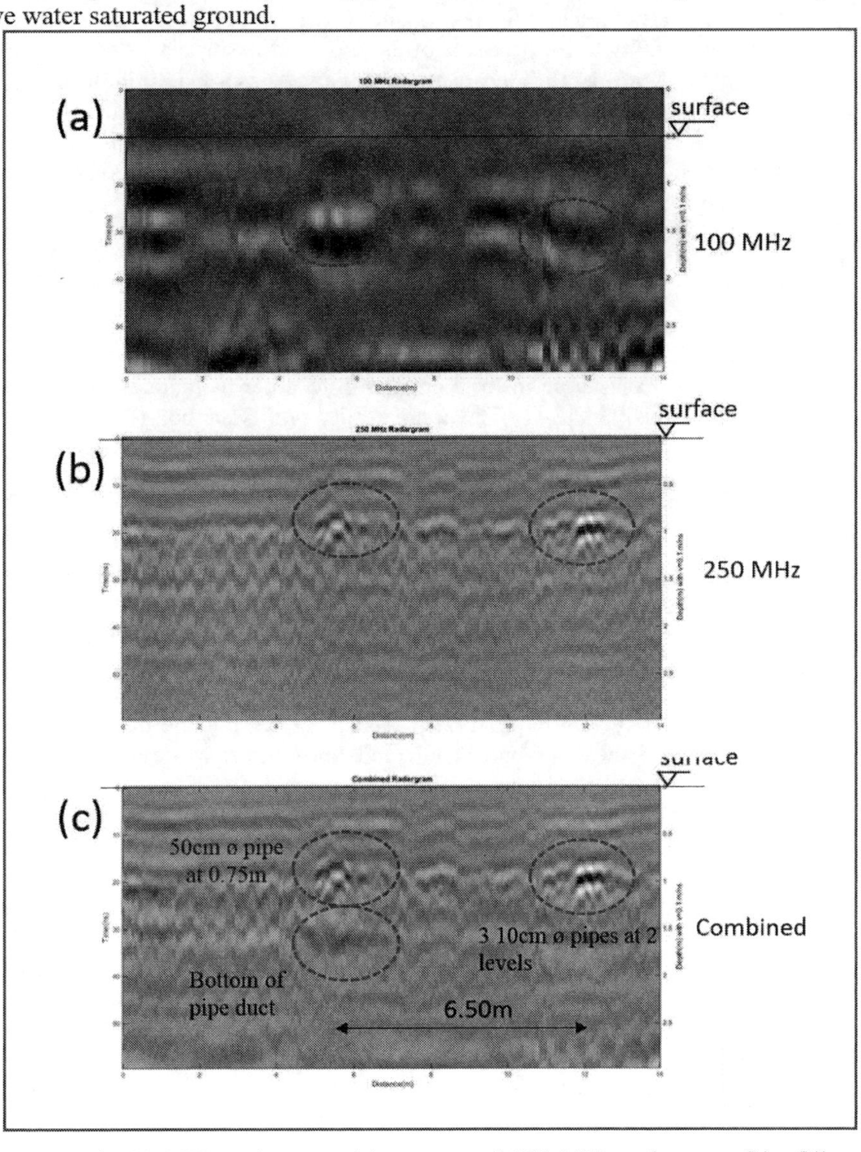

Figure 4. Processed 100 MHz radargram (a), processed 250 MHz radargram (b) of line 19 and their composite radargram using multi-frequency compositing technique. Combined radargram shows clearer images for interpretation rather than individual radargram.

Similar result is also shown in 250 MHz radargram, exhibiting clearer image of subsurface. This result shows that GPR system with 250 MHz antenna is adaptable for measurement on water saturated ground until 2.5 m depth. Combined radargram produced using multiple frequency compositing with OSW technique is shown in figure 4(c).It shows a clearer image of subsurface compared to individual radargram.

Three anomalous objects are recognized in the combined radargram, i.e. object at 6 m distance from origin at depth of about 0.75 m and object at 6.5 m distance from the first pipe at 0.9 m and last object at 1.5 m depth. We interpret the first anomalous object as pipe with 50 cm diameter located at about 75 cm depth, as revealed in the figure 3(d). The second anomalous object is most likely to be a group of 6 pipes with 10 cm in diameter arranged in 2 level at about 90 cm depth. Those type and size of the pipes are validated by pipeline continuity the appearance of the pipe on the surface, as seen in figure 3(b) and (d). The last object located at 6m depth beneath the first object is probably bottom part of the pipe conduit. Bottom part of conduit under second object is hard to recognize in the combined radargram.

The success of this research prompts us to further develop this method in detail for common cases in geotechnics such as detecting underground sewers, the boundaries of the hard layer and the overlying soft ground, and complex pipelines

4. Conclusion

We applied multi-frequency compositing method with the aim at increasing the radargram resolution at deeper depth at a working frequency reference. This technique was done by combining radargram acquired with two or more different working frequency antennas from the same path. Optimal spectral whitening (OSW) technique, that is superior in comparison to other techniques in multiple-frequency compositing we studied previously, is used. The applicability of this method is tested on data acquired in Muarakarang Combined Cycle Power Plant area by using 100 MHz and 250 MHz GPR antenna, the resulted composite radargram joining 100 MHz radargram and 250 MHz radargram show clearer image for interpretation rather than individual interpretation.

Three anomalous objects were identified, i.e. 50 cm diameter steel pipe at 6 m distance from the origin at 0.75 m depth, a group of 6 10cm diameter pipe at 0.9 m and bottom part of duct at 6.5 m distance from the origin at 1.5 m depth. The applicability of this multi-frequency compositing recommend further study to solve some facing problems in geotechnical engineering.

Acknowledgments

The authors thank Dr. Silvia Jannatur Fajar for her assistance during the data acquisition and Ms. Apulina Priska for her help in the data processing and hard effort in developing the method for geotechnical works during her magister program. This work was partly financed by P3MI2020 Program from Institut Teknologi Bandung.

References

[1] Davis J L and Annan A P 1989 Ground-penetrating radar for high-resolution mapping of soil and rock stratigraphy *Geophys. Prosp.* **37** pp 531-551

[2] Parnadi W W Amin S S Heriawan M N Parnadi R G 2019 Study on the use of Ground-Penetrating Radar technique to recognize tin deposits in Bangka Island *IOP Conf. Series: Earth and Environmental Science* **318** 012027. doi: 10.1088/1755-1315/318/1/012027

[3] Endres A L booth A D and Murray T 2004 Multi-frequency compositing of spatially coincident GPR datasets *Proc. the 10th International Conf. on Ground-Penetrating Radar* pp 271-274

[4] Dougherty M.E Michaels P Pelton J R and Liberty L M 1994 Enchancement of ground penetrating radar data through signal processing *Proc. the Symposium on the Application of Geophysics to Engineering and Environmental Problems*, EEGS

[5] Booth A D Endres A L and Murray T 2009 Spectral bandwidth enhancement of GPR profiling data using multiple-frequency compositing *J. Applied Geophysics* **67** pp 88–97

[6] Priska A Parnadi W W Parnadi R G 2019 On the Application of Multi-frequency Ground-Penetrating Radar Method for Mapping Underground Pipe *IOP Conf. Series: Earth and Environmental Science* **318** 012028. doi:10.1088/1755-1315/318/1/012028
[7] Sandmeier K J 2006 *ReflexW Version 4.0* Sandmeier Scientific Software, Karlsruhe

IOP Conf. Series: Earth and Environmental Science **873** (2021) 012091 doi:10.1088/1755-1315/873/1/012091

Sanding Wells on The Island of Lombok due to Earthquake Vibrations

Bagus Endar B. Nurhandoko[1], M. Rizka Asmara Hadi[2], Kaswandhi Triyoso[2], Rio K. Martha[2], Sri Widowati[3], Bakti Sukrisna[4], Syamsuddin[4], Nur Isnaini Romli[5]

[1] Faculty of Mathematics and Natural Sciences, Institut Teknologi Bandung
[2] LPIK, Institut Teknologi Bandung
[3] Telkom University
[4] Mataram University
[5] Formerly FMIPA ITB, Presently Rumah Amal Salman ITB

Correspondence: bagusnur@bdg.centrin.net.id

Abstract. The Lombok earthquake in 2018 was unique, the shocks occurred sequentially. Several major earthquakes were followed by thousands of aftershocks. The earthquake caused a devastating disaster which destroyed many homes, buildings including wells as the main fresh water supply in the Lombok Island. The focal mechanism of main earthquake shows a thrust fault mechanism. Lombok Island is originally a volcano Island which is still growing actively. Therefore surface of Island is dominated by volcanic materials, such as: volcanic rock, volcanic ash, pumice. This paper describes the phenomenon of sanding wells in Lombok, including the physical mechanisms among rock's grains when vibrating earthquake waves. These earthquake waves can eliminate static friction between grains and reduce cohesion between grains of rock. Some subsurface images shows a strong correlation between damage grade and the existence of loose sand and hard rock.

1. Introduction

In 2018, the Lombok earthquake occurred over a long period time. The Lombok earthquake occurred serially for more than one month. The seismic waves caused the disaster that destroyed many homes and buildings. North Lombok was almost completely collapsed. All of the Groundwater wells in northern Lombok are almost silting due to the sanding phenomenon. The sanding phenomenon is the phenomenon where wells produce sand due to loss of rock strength. The sanding phenomenon associated with earthquakes maybe related to the phenomenon of liquefaction. The sanding problems have been published by some authors i.e.: Rissness et al. [1], Bratli and Risnes [2],. Fattahpour et al. [3], Morita et al. [4], Bradford and Cook [5], Nurhandoko et al. [6], Meltcafe et al. [7]. This paper present the massive sanding phenomenon occurred on Lombok Island related earthquake parameters. There is a strong correlation with the damage condition in the area, including the severity of the sanding phenomena on the wells during the earthquake.

2. Lombok Earthquake 2018 and Geology of Lombok

The focal mechanism of the Lombok earthquake showed thrust fault mechanism [8] as shown in **Figure 1**. Analysis of land deformation show the northern part of Lombok Island had been uplifted for

IOP Conf. Series: Earth and Environmental Science **873** (2021) 012091 doi:10.1088/1755-1315/873/1/012091

range 25 cm [9] to 29 cm after the earthquake series [10]. Geological map published by the Ministry of Energy and Mineral Resources of the Republic of Indonesia, the lithology around North Lombok consists of mostly gravel, sand, clay tuff, peat, and a fragment of coral. The surface area in Northern of Lombok Island consists of volcanic products i.e.: volcanic ash, pumice, and loose sand.

3. Grains of rocks parameters of North Lombok

Parameter grains of the samples show low specific mass (about 1.0 to 1.3 g / cc) and contains pumice which has a very low mass density (0.5 g / cc). The grain size measurement showed that the dominant size is 0.4 mm, see in Figure 2.

When the 2018 Lombok earthquake occurred, almost all of the groundwater wells in Northern Lombok suffered shallowing and structural damage due to earthquake vibrations. The earthquake vibration can result the shear stress exceeds the static friction's stress which impact the vibration causes the grains of the matrix to separate from each other. The release of bonds among grains causes a loss of shear modulus. The siltation phenomenon due to the sanding mechanism observed in almost whole groundwater wells in Northern Lombok.

4. Acceleration and focal mechanism of the 2018 Lombok Earthquake

The hypocenter of the 2018 Lombok earthquake was spread over almost all of northern Lombok and formed a half-ring pattern. Based on data from USGS peak ground acceleration (PGA), the main earthquake occurred in Lombok in the range of 20%–50% g. PGA played a dominant role in the occurrence of the liquefaction phenomenon in several areas in North Lombok when the earthquake struck mainland Lombok. Figure 1 shows that focal mechanism of earthquakes in Lombok was the thrust fault mechanism. The existence of the thrust fault mechanism shows that vertical motion during the earthquake was very dominant. Therefore, the waves generated during the earthquake were most likely P waves and Rayleigh waves (P and S waves).

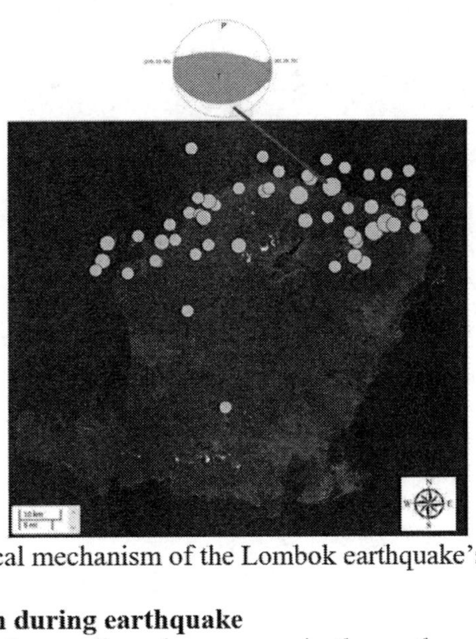

Figure 1. Focal mechanism of the Lombok earthquake's main shock

5. Grain dynamics mechanism during earthquake

The well shallowing caused by the sanding phenomenon in the northern part of Lombok Island was significantly influenced by decreased effective stress contact among grains. When effective stress on the contact area among grains decreases, the grains tend to separate from each other, creating fluid-like movement. Furthermore, the sand flows to reach the water table. These phenomena are predominantly influenced by the dynamic mechanism of pressure induced by earthquake vibrations.

The mechanism of vertical vibration caused by the thrust fault mechanism causes dynamic pressure in grains and pore rocks through vertical movement. Under ambient conditions or conditions of no earthquake vibrations, the static vertical force of each grain can be determined as follows:

$$F_{effective} = (\rho_{grain} - \rho_{water})V_{grain}g + F_{overburden} - F_{pore} \tag{1}$$

$F_{effective}$ indicates the vertical force of each grain, $F_{overburden}$ indicates overburden force, V_{grain} indicates the volume of grain, ρ_{grain} indicates the density of grain, ρ_{water} indicates the density of water, and g indicates gravity acceleration. The earthquake waves are considered to be made up of vertical dominated waves on a set of rock pores with a pore pressure P_0, where $P_{earthquake}$ denotes the maximum pressure generated by waves. The static vertical effective pressure on grains contact is as follows:

$$P_{effective} = -(P_o + P_{earthquake}e^{i(kr-\omega t)}) + P_{overburden} + \frac{(\rho_{grain}-\rho_{water})V_{grain}\ g}{A_{effective}} \tag{2}$$

Therefore, the vertical vector of stress ($P_{vertical}$) at the interface among grains is estimated as follows:

$$P_{effective} = -P_{pore} - P_{earthquake} + \int_{h=0}^{h=depth} \rho(h)_{overburden}g dh + (\rho_{grain} - \rho_{water})\ 2.67\ g.r_{grain} \tag{3}$$

Equation (3) indicates that the liquefaction parameters are the pore pressure (P_{pore}), maximum pressure generated by waves ($P_{earthquake}$), grain density, overburden stress ($P_{overburden}$), grain size (r), and acceleration of earthquake waves. The sanding mechanism will occur if the P-vertical of equation (3) is negative.

$$P_{effective} = -\int_{h=water\ table}^{h=total\ depth} \rho_{water} g dh - \frac{\rho v}{\omega} a_{max} + \int_{h=0}^{h=total\ depth} \rho(h)_{overburden} g dh + (\rho_{grain} - \rho_{water})\ 5.33\ g.r_{grain} \tag{4}$$

When the Peffective + tensile stress of rock was less than zero, the grains were separated from each other or the well under sanding conditions. The well became shallower due to these grains flowed to fill the wells until reaching the water table. The center zone at the bottom of the well suffered the most severe sanding phenomenon due to the absence of rock overburden pressure resulting in the smallest value of effective pressure among the grains. Therefore, when the earthquake with thrust fault mechanism hit northern of Lombok, the greatest upward vertical pressure occurred at the bottom of the well. The distribution of rock grains size is **Figure 2**. The rock were thrown upwards and caused a sanding mechanism, as illustrated in **Figure 3**.

When the earthquake waves shock the grains in the range of peak ground acceleration (PGA) 20% g to 50% g, the sanding phenomenon happened. Sand became like fluid flowing from the bottom of well and reached the water table.

IOP Conf. Series: Earth and Environmental Science **873** (2021) 012091 doi:10.1088/1755-1315/873/1/012091

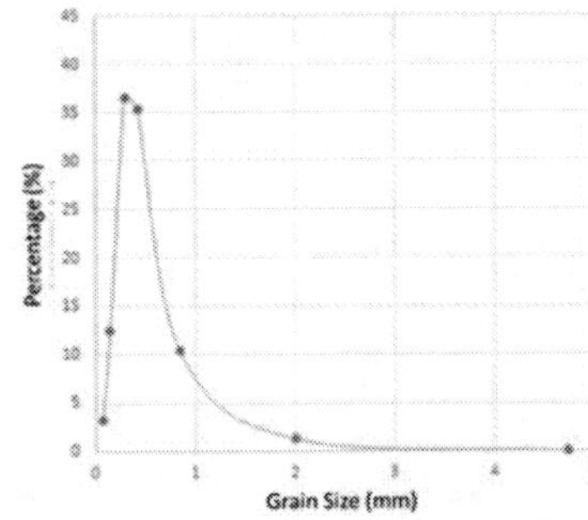

Figure 2. Distribution of the sample size (in mm diameter) as determined by sieve analysis.

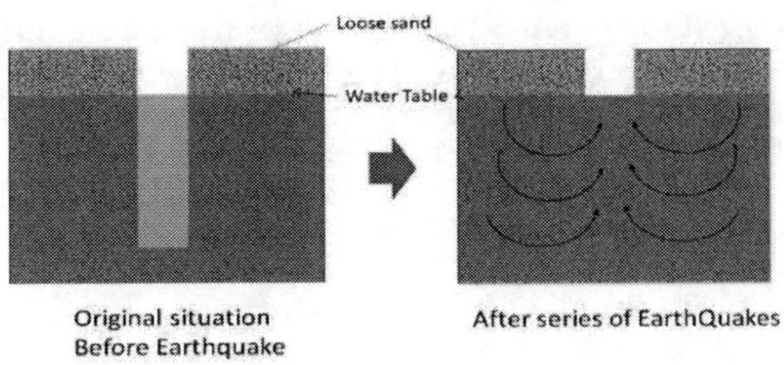

Figure 3. Illustration of the sanding mechanism phenomenon causing siltation on wells in the North of Lombok island

6. Conclusions

The island of Lombok was rocked by a series of earthquakes in 2018 which destroyed large numbers of homes. the earthquake also vibration caused silting of wells due to the sanding phenomenon. The focal mechanism of the Lombok earthquake shows the vertical movement of the thrust fault mechanism. The surface of North Lombok Island dominated by volcanic products and has a very shallow water surface.

Earthquakes can reduce the effective stress between the grains and grains are separated from each other.

We have presented a physical mechanism which correlates wave parameters and saturated water pore pressure, radius, and grain buoyancy pressure, leading to a drop in the effective pressure between grains. The north of Lombok has suffered a sanding phenomenon because the effective pressure between the grains is negative at their original depth conditions. Wells in North Lombok lost their water column, and there was catastrophic water scarcity.

References

[1] Risnes R, Bratli R K and Horsrud P 1982 Sand stresses around a wellbore *Society of Petroleum Engineers Journal 22*(06) pp 883–898

[2] Bratli R K and Risnes R 1981 Stability and failure of sand arches *Society of Petroleum Engineers Journal 21*(02) pp 236–248

[3] Fattahpour V, Moosavi M and Mehranpour M 2012 An experimental investigation on the effect of rock strength and perforation size on sand production *Journal of Petroleum Science and Engineering 86–87* pp 172–189 doi: 10.1016/j.petrol.2012.03.023

[4] Morita N, Whitfill D L, Massie I and Knudsen T W 1989 Realistic sand-production prediction: numerical approach *SPE Production Engineering 4*(01) pp 15–24

[5] Bradford I D R and Cook J M 1994 A semi-analytic elastoplastic model for wellbore stability with applications to sanding *Paper presented at the Rock Mechanics in Petroleum Engineering* Delft Netherlands

[6] Nurhandoko B E B, Susilowati and Listiyobudi M 2018 Thick Walled Core Testing for Sanding Analysis of Chalky Carbonate Reservoir in Production Borehole *The 43rd Annual Scientific Meeting Himpunan Ahli Geofisika Indonesia*

[7] Metcalfe G, Tennakoon S G K, Kondic L, Schaeffer D G and Behringer R P 2002 Granular friction, Coulomb failure, and the fluid-solid transition for horizontally shaken granular materials *Physical Review E 65*(3) pp 031302

[8] USGS Catalog 2020 https://earthquake.usgs.gov accessed 2020 https://earthquake.usgs.gov

[9] NASA Earth Science Satellite-Derived Map of Ground Deformation from Earthquake beneath Lombok, Indonesia [Online] Available: https://disasters.nasa.gov/lombok-indonesia-earthquake-2018/satellite-derived-map-ground-deformation-earthquake-beneath-lombok

[10] Ramdani F, Setiani P and Setiawati D A 2019 Analysis of sequence earthquake of Lombok Island, Indonesia *Progress in Disaster Science 4* pp 100046 doi: 10.1016/j.pdisas.2019.100046

IOP Conf. Series: Earth and Environmental Science **873** (2021) 012092 doi:10.1088/1755-1315/873/1/012092

Preliminary Study: Identifying the Heat Source of Hot spring Discovery at Non-Volcanic Region with Gravity Method

Abi Mawalid[1]* and Andi Mardhotilla[2],
[1]Geophysics Engineering Department, Universitas Pertamina, Jakarta, Indonesia
[2]Geophysics Engineering Department, Universitas Pertamina, Jakarta, Indonesia
*Author to whom any correspondence should be addressed.
mawaleeda@gmail.com

Abstract. Gravity method is one of the geophysical methods that's using variation of gravity value caused by density variation of subsurface. Gravity method has been utilized for many purposes such as understanding the geological information about one area or understanding the subsurface structure. This study is using gravity method to identify the heat source from the geothermal manifestation (hot spring) which is located in the non-volcanic region in North Java. Gravity data was obtained from WGM 2012 satellite data that can be downloaded at BGI website and processed using Oasis Montaj software. Data processed using horizontal derivative method to enhance the boundaries resolution of the possibility geological structures location in the study area that is predicted to be the way of the heat source of hot spring manifestation. The result of this study shows that the heat source of the hot spring in Northern East Java is probably coming from Lasem Volcano's heat source. The result can be proven by the similarity of the Bouguer Anomalies value between the Lasem Volcano region and the manifestations distribution area, which varies in 174-185 mGal. Besides that, Lasem Volcano is the closest volcano from the manifestation with a distance of approximately around 27 km.

1. Introduction

This study is discussing geothermal manifestations that exist in non-volcanic regions. In a non-volcanic region, geothermal manifestation is barely found because it is located too far from the heat source. But geothermal manifestation sometimes exists in the non-volcanic area. An example of this case can be found at Jatirogo, Tuban. Jatirogo is located on Wonocolo Formation that is dominated with limestone [1]. However, the interesting case here is hot spring manifestation can be discovered in this area.

The primary goal of this study is to understand how the heat source can reach the manifestation area. To help us more understand the subsurface characteristic of the manifestation area, gravity method was applied in this study. Gravity method that used in this study is the satellite gravity method. Satellite use to record the wavelength component of the earth as a correction for measurement on the surface. However, this study can be developed and specified for potential studies by increasing the accuracy and spectrum of their waves. The example of study about satellite survey is done by GRACE (German Aerospace Center Gravity Recovery and Climate Experiment) to map temporal variation of gravity to the dynamic mass of water [2].

1.1. Regional Geology

Based on similarities of morphological and tectonic setting, east java area and madura are divided into seven zones [3] such as:

1. Quaternary volcanoes zone
   This zone occupies the middle part along Solo zone, except for Muria mountain which occupies alluvial plain of North Java.
2. North Java alluvium plain
   This zone is divided into two parts, begin from eastern part from Surabaya to the northwest and the western part from Semarang to the east and ends in the Java sea.
3. Rembang zone
   This zone is anticlinorium that extends in the west-east direction. Starting from the east of Semarang to the north of Rembang.
4. Randublatung zone
   This zone is synclinorium that extending from west of Semarang to the east of Wonokromo
5. Kendeng zone
   This zone is anticlinorium that extending from Semarang and then it narrows to the east until north of East Java. This zone is generally formed by volcanic deposits, sandstone, claystone, and marl.
6. Solo zone
   This zone divided into 3 sub-zone (Blitar sub-zone, Middle Solo sub-zone, and north Ngawi sub-zone)
7. Southern mountain zone (East region)
   This zone extends along the southern coast of East Java and Wonosari to the most eastern of Java Island. The topography of this zone is mostly formed by limestone and volcanic material.

Based on the physiographic map division below, the research area is included in the Rembang zone (Figure 1). This zone consists of folded ridges (anticlinorium) that extend with west-east direction start from north of Purwodadi via Blora, Jatirogo, Tuban, and ends in Madura island.

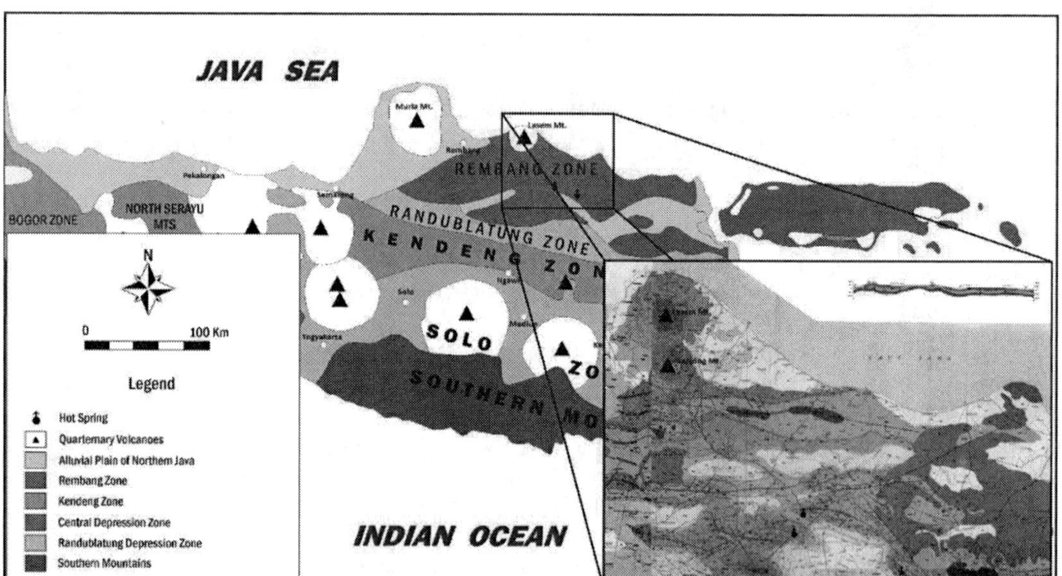

Figure 1. Physiographic Map Division of Central Java and East Java [4]

IOP Conf. Series: Earth and Environmental Science **873** (2021) 012092 doi:10.1088/1755-1315/873/1/012092

The northern part of the East Java Basin (Kendeng Zone, Rembang Zone, Java Sea Shelf Zone, and Randublatung Depression Zone) has two periods of time that caused the relative direction of the magmatic path or its tectonic pattern change [3], i.e.:

1. In Paleogene Era (Eocene-Oligocene), which was oriented northeast-southwest. This pattern causes the northern part of East Java Basin to have a tectonic strain regime, indicated by the pre-tertiary age bedrock lithology which shows a northeast-southwest trending accretion pattern in the form of the orientation of faults in bedrock, horst, and graben.
2. In Neogene Era (Miocene-Pliocene) it changed to the relative east-west, which was a compression tectonic regime, resulting in folds of geological structures, factoring faults, and causing the North East Java Basin to rise.

The condition of the folding structure in the northern part of the East Java Basin is generally oriented west-east, while the fault structure is generally northeast-southwest trending, and there are several upward faults trending east-west.

2. Methodology

This research analysis is based on two approaches: geological and geophysical analysis. The data used for geophysical analysis was obtained from Bureau Gravimetrique International (BGI) website: https://bgi.obs-mip.fr/, and focused on WGM 2012 data type that has 1' resolution and already form in Complete Anomaly Bouguer (CBA) data [5]. The boundary of the research area is around longitude 110.5 until 112.5 and latitude -8 until -6.5, and there are 2806 points of CBA data in that region (Figure 2).

Figure 2. Research Area

According to the geological assumption, the processing that is used in this geophysical analysis is focusing on structure/edge identification. A numerical derivative method used to process the data is Horizontal Gradient Magnitude (HGM). HGM was used to enhance the anomaly contrast resolution. HGM is formed by a bidirectional horizontal derivative (North-South and East-West direction) that is calculated to popping the positive value, which will be categorized as the probability of the geological structure location [2]. Then the HGM equation is:

$$HGM = \sqrt{\left(\frac{\partial g}{\partial x}\right)^2 + \left(\frac{\partial g}{\partial y}\right)^2} \tag{1}$$

Furthermore, that equation is resembling a quadratic addition from the x-direction and y- direction derivative of the anomaly. Inspired by the study about the utility of free software in gravity and magnetic method [6], this research processing will be done using open-source software: Oasis Montaj Viewer and USGS GX version 64.

3. Discussion

3.1. Geological Analysis

The main interest of this case is the location of the manifestation which is not distributed in a non-volcanic zone and formed on the top of limestone. To understand this case, a geological analysis will be used in this study. The geological analysis will be described regionally because information about the real condition of the manifestation is limited. Furthermore, based on the regional geology study of the Northern Java Region, founded that the heat source of the manifestation is probably obtained from the existence of the geological structures around them. 2 assumptions can be considered based on [7] such as:

1. Sedimentary Basins

   Sedimentary basin geothermal system is formed from the heat that appears due to layering sequences of permeable and impermeable strata that alternately. Go along with this research location is around carbonates rock/limestone (permeable), there is a possibility of this type of heat source, but considering that the sediment thickness is quite low due to the subduction process, there is a justification to choose another assumption

2. Off-flow from Volcanic Geothermal Systems

   This system is formed from the off-flow of volcanic geothermal areas (high-temperature), and heating groundwater by direct contact with the hot ground or mixing with deep reservoir water. That assumption (contact and mixing) could happen because of the existence of geological fractures. Resemble with this research problem that there is a discovery of some of the geothermal manifestation in the around the non-volcanic zone, around Northern Java. However; there are high-temperature geothermal areas around Eastern Java (a huge of geothermal prospects distributed around Eastern Java and include the northern region [8], and quite recognizable to correlate with this geothermal system type. Furthermore, in the area also found strike-slip fault (sinistral), and distributions of fractures that can be considered as the pathway of the heat.

Furthermore; to confirm the previous opinion [9] that either Lasem Volcano or Senjong Volcano, both it have lava flow to the southern area and the same magma evolution (partial melting of the mantel due to the subduction process). Then, for the brief conclusion, we can say that the heat source of the manifestation is started from Lasem Volcano, then going to the south to Senjong Volcano, and leaking to the west through the fractures/faults (Off- flow from volcanic geothermal systems).

3.2. Geophysical Analysis

As seen in figure 3, manifestation location is distributed in around slightly low CBA value (174-185 mGal), similar to the anomaly variation around Lasem Volcano and Senjong Volcano. The similarity shows a possibility of the same subsurface condition, such as rocks content or even in the hydrology aspects. Then, according to the previous opinion that the heat source of manifestation was coming from Lasem Volcano and migrated by the faults/fractures, it can be proven by looking at the HGM value pattern around the manifestation and the suspected volcanoes. Figure 4 shows that there is a structured pattern that can be directing the pathway of the heat from Lasem Volcano to the manifestation. This pathway was described with marked by dashes lines that estimated from the high positive value of HGM (indicates the location of the anomaly contrast/suspected structure identification)

IOP Conf. Series: Earth and Environmental Science **873** (2021) 012092 doi:10.1088/1755-1315/873/1/012092

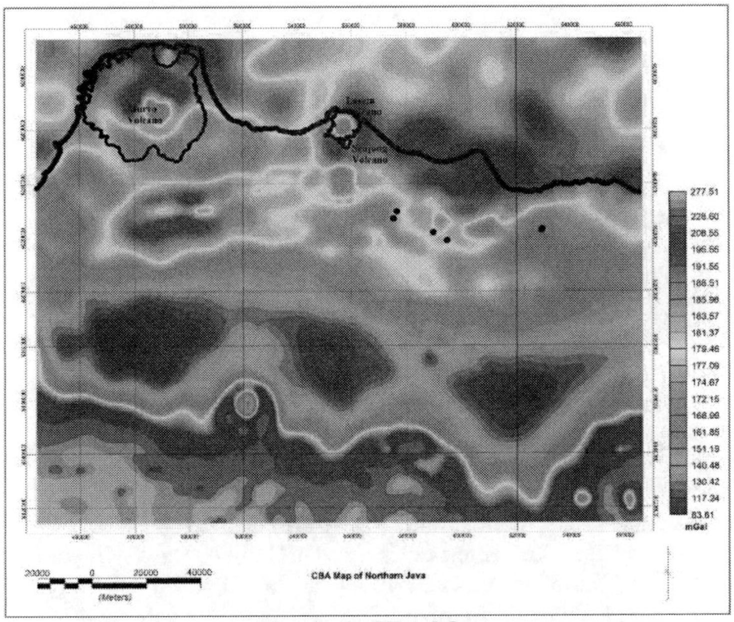

Figure 3. CBA Map

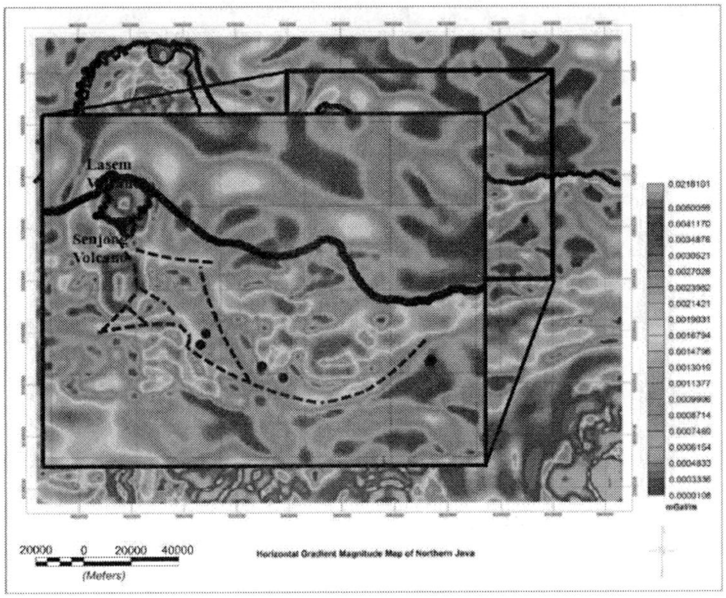

Figure 4. HGM Map

4. Conclusion

The horizontal derivative gravity method has been applied to predict the heat source distribution pathway to the hot spring manifestation in Jatirogo. The response of bouger anomalies shows the similarity of value bouger anomaly values between Lasem Volcano and manifestation distribution which varies between 174-185 mGal. This confirms the assumption of heat source distribution to the hot spring manifestation in Jatirogo, Tuban controlled by Lasem Volcano.

Moreover, another function of the derivative method is given to enhance the boundaries resolution of possible geological structures located in the study area. The geological structure for this study is very important to predict the heat source pathway to the manifestation. Geological structure direction (strike and dip) also useful to help us understand how the heat source transferred from Lasem Volcano to the nearest manifestation in Jatirogo that approximately has 27 km distance.

References

[1] Situmorang R L, Smit R, and Vessem E J V 1992 Geologic Map of Jatirogo Quadrangle, Java. Bandung.
[2] Hinze W J, Frese R R B, & Saad A H 2013 Gravity and Magnetic Exploration Principles, Practices, and Applications. Cambridge.
[3] Bemmelen R W Van 1949 The Geology of Indonesia **1A**, The Hague, Netherland.
[4] Mardhotilla A 2020 Identifikasi Batas dan Estimasi Kedalaman Cekungan Bintuni, Papua Barat Menggunakan Data Gaya Berat WGM 2012 (World Gravity Map 1.0). Jakarta.
[5] Balmino G, Vales N, Bonvalot S, & Briais A 2012 Spherical Harmonic Modelling to Ultra high Degree of Bouguer and Isostatic Anomalies. p 499-520.
[6] Grandis H and Dahrin D 2017 The Utility of Free Software for Gravity and Magnetic Advanced Data Processing. Bandung.
[7] Saemundsson K 2014 Geothermal Systems in Global Perspective Presented at Short Course IX on Exploration for Geothermal Resources. p 3-5.
[8] KSDM (Kementrerian Energi dan Sumber Daya Mineral) 2017 Potensi Panas Bumi Indonesia **1** Direktorat Panas Bumi Kementerian Energi dan Sumber Daya Mineral. Jakarta.
[9] Moktikanana M L A, Harijoko A, and Wibowo H E 2019 Evolusi Magma Gunung Lasem dan Gunung Senjong, Kabupaten Rembang Provinsi Jawa Tengah. p 340-361.
[10] Prihutama, Faiz A, Danistya A, and Wedada S 2018 Karakteristik Geologi dan Skenario Reservoir Hidrokarbon Sebagai Rencana Pengembangan Zona Prospek Lapangan "Tesseract" Cekungan Jawa Timur Utara Pada Zona Rembang Jawa Timur. p 436-454.
[11] Pringgoprawiro H and Sukido 1992 Geologic Map of The Bojonegoro Quadrangle. East Java, Bandung.

Application of P-Wave Moment Magnitude (M_{wp}) and Rupture Time Duration (T_{dur}) to Analyze A Potential Tsunami Earthquake in Sumatra

Andrean V H Simanjuntak [1,4], Noviana Sihotang [1], Afryanti V Simangunsong [1], Buha M M Simamora [1], Djati C Kuncoro [1], Yusran Asnawi [3], Muksin Umar [2,4], I. Irwandi [2,4,*]

[1] Badan Meteorologi Klimatologi dan Geofisika (BMKG), Indonesia
[2] Physics Department of Universitas Syiah Kuala (USK), Banda Aceh, Indonesia
[3] Education of Information Technology Department, Universitas Islam Negeri Ar-Raniry, Aceh, Indonesia
[4] Tsunami Disaster and Mitigation Research Center of Universitas Syiah Kuala (USK), Banda Aceh, Indonesia

Abstract. Tsunami warning is one of many important reports to save lives and reduce the damage for local peoples. A moment magnitude of P-wave (Mwp) and the rupture time duration (Tdur) can be used as the quickly parameters to diseminate the tsunami warning. In this paper, we analyze the seismic waveform from global network to get Mwp and Tdur of South-West Coast of Sumatera earthquake. Mwp was calculated using automatic and manual phase picking of P phase. The results of this study show a well-analyzed relationship between P wave from automatic and manual picking, Mwp and time duration, respectively. The result also give an encouraging studies for the early warning system that will be set up in the future in the region.

Keywords: Mwp, Rupture, P wave, Tdur

1. Introduction

Tsunami is one of nature disaster that cause damage in land by generating massive erosion and loss of life in coastal areas. Based on that, a quick and robust tsunami early warning need to be announced earlier before run-up touch the coastline [1, 2, 3]. By fast calculating of seismic waveform, we can spread a quickly report for earthquake parameters which can generate the tsunami [3].

Specific parameters are important to give a decision about tsunami warning for an earthquake that quickly occurred in the subduction zone [4, 5]. The warning give more time to do a mitigation and evacuation. That time is known as the golden time to reduce and minimalize the damage and victims. These parameters is influenced by window length of seismic waveform that started from origin time of earthquake depend on station distance, arrival time of seismic phase and waveform quality They are calculated to get specific parameters like P-wave Magnitude (Mwp) and rupture duration time.

* correspondent author : irwandi@unsyiah.ac.id

Tsuboi et al [9] analyzed the first arriving of P-waves phase to estimate earthquake size accurately and promptly and derived the P-wave moment magnitude. Mwp. Whitmore et al. [10] make a linear comparison between Mwp with Mw to get a correction about Mwp value. Tsuboi et al. [9]make the Mwp by calculating the P-waves phase from far-field stations, but Mwp also can be applied for local and regional events [11,12 13]

Mwp is useful as the initial estimator of seismic moment from large earthquake that can be inferred from the earliest-arriving seismic phase in the far field. Mwp has been implemented and currently constitutes the procedure in use at both the Pacific Tsunami Warning Center (PTWC) and the West Coast/Alaska Tsunami Warning Center (WC/ATWC) for the purpose of estimating earthquake sources for tsunami-genic potential. In other hand, Lomax and Michelini [14] suggested if tsunami early warning can be reported in 3 – 10 mins after origin time with calculating the earthquake rupture duration time (Tdur). Rupture duration has a correlation with rupture length which the most influential parameter and can describe how big the magnitude of an earthquake that caused the tsunami.

In addition, the duration of the rupture, Mwp can be used for tsunami early parameters because of its methods used faster and more stable [9, 10, 13, 14] where the earthquake used was the tele earthquake (300-900). Both parameters use the P-wave phase so that it does not require time long and contains comprehensive information about earthquake event size and source character [16, 17, 18, 19, 20]. In this paper, we present Mwp and Tdur values determined for one big earthquake with Mw 7.7 that occurred in the subduction zone of South-West Coast of Sumatera. This area is very seismically active and mostly many earthquake is generated by thusting system in subduction system. From the result, we get a good linear regression between our Mwp and Tdur [9, 10].

2. Data and Method

We used seismic waveform from tele stations that downloaded from IRIS in Fig (1). The stations distance (dist > 45°) is used to avoid the effect of the lateral heterogenity.

Figure 1. Seismic Stations Location (Blue inverse Triangle) and earthquake epicenter (Red Stars). This is the waveform recording from all stations in distance (45 ° - 80°) in Fig (2).

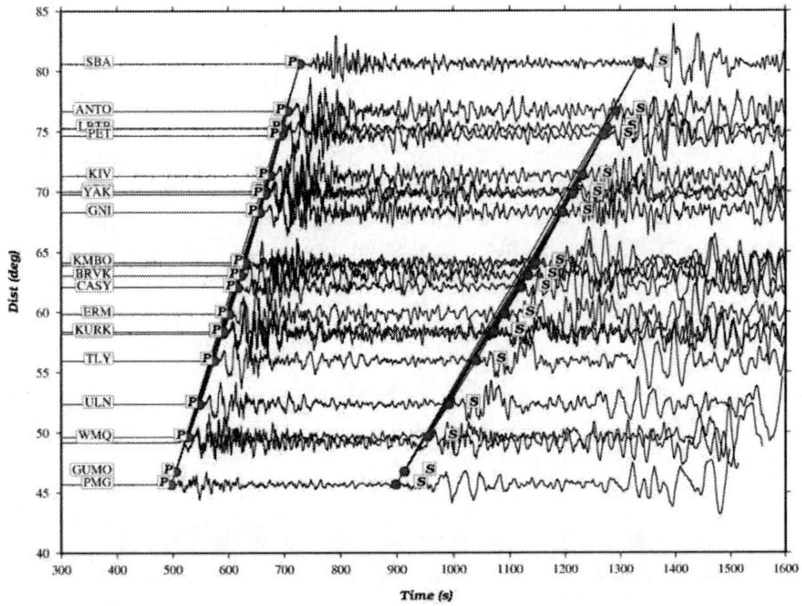

Figure 2. Seismic waveforms were started in 500 – 700 s after the main earthquake occurred and the body-wave phase P (red circle) and S (blue circle).

Mwp is a magnitude that obtained based on recording analysis of P wave. To produce Mwp, we must obtained the moment magnitude. It takes several steps to produce moment magnitude. Seismogram vertical component in velocity must be convert to displacement by integration. Then, the time domain is transformed to frequency domain with operation Fast Fourier Transform (FFT) so it can be known a corner frequency to determine the upper and lower frequency limit. In the processing, Bandpass butterworth filter is used. Seismic Moment (M_o) is calculated by using Tsuboi et al. [9] with Eq. 1.

$$M_0 = \left| max\left(\left(\int U_z(x_r\,, t)dt \right) \right| 4\pi\rho\alpha^3 r \right. \quad (1)$$

Then, the displacement signal as a source time function and the integration of the seismic moment derived from maximum amplitude of the P wave is formulated by equation (1). $U_z(x_r\,, t)$ is value of the shift in the vertical component of the P wave for a far earthquake (farfield) at the station x_r

$$r = \cos - 1|\sin\theta_E \sin\theta_S + \cos\theta_E \cos\theta_S \cos(\varphi_S - \varphi_E)| * 6.371 \quad (2)$$

θE and φE as epicenter latitude and longitude coordinates, θS and φS as station latitude and longitude coordinates. The actual P velocity wave (apparent P velocity, APV3) is expressed as α. Kanjo et al. [21] changed the method relationship of distance and velocity from the P wave based on the earth model table of IASP91 in equation (2).

$$\alpha = 0{,}16\,\Delta + 7{,}6\ km/s \quad (3)$$

Calculation of the moment magnitude using the Kanamori [22] method is mathematically formulated in Eq. 4 as follows :

$$M_w = \frac{1}{1.5}(\log M_o - 9.1) \quad (4)$$

IOP Conf. Series: Earth and Environmental Science **873** (2021) 012093 doi:10.1088/1755-1315/873/1/012093

Mwp can be obtained in equation (5) by Tsuboi (1995):

$$M_{wp} = M_w + 0.2 \tag{5}$$

Adding a value of 0.2 in calculation of the moment magnitude of P wave (Mwp) to correct the radiation pattern of the propagation P wave to the value of Mwp. Tdur is the overall duration signal or set of dominant P waves from a high frequency seismogram. Tdur is one of the additional parameters in issuing tsunami early warnings. Lomax and Michelini [14] have found that the rupture length parameter of an earthquake is the most dominant parameter linear to rupture duration.

3. Result and Discussion

We investigated the rupture duration of each stations with several ways. First, we investigated the noise artifact and the amplitude spectra with spectrogram visualization in Fig (3).

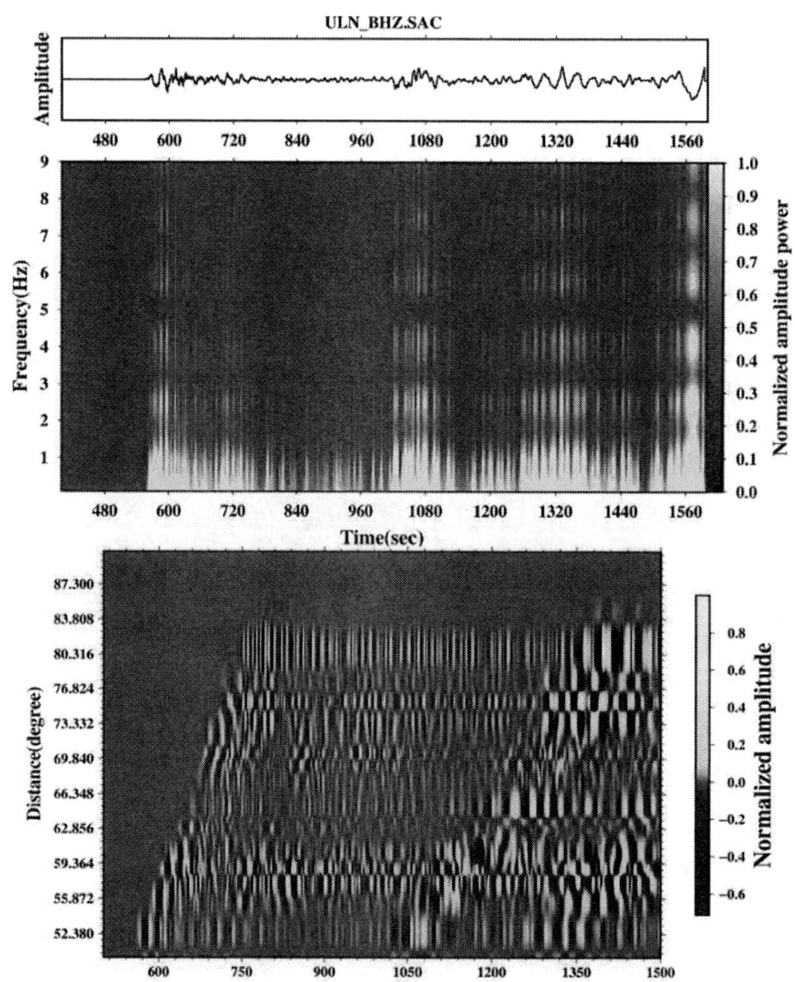

Figure 3. Spectrogram on Z component in ULN station and all stations show a good SNR and amplitude spectra.

From looking and investigating the waveform, we gave a bandpass filter (0.1 – 9 Hz) with assuming the P wave very clear. Then, we enveloped to see the amplitude spectra and from that we get the time duration of earthquake by using seismic analysis code [23]. The example processing is showed in Fig. 4 and 5, we compare the close and far stations distance.

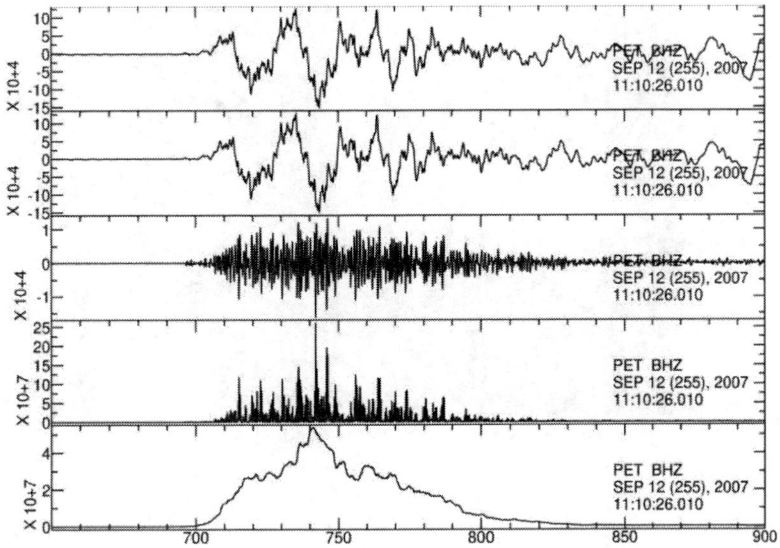

Figure 4. A multiple plot of PET station which show real recording, waveform with bandpass 0.1 – 10 Hz, the envelope and time duration.

The comparison results in Fig (4) on LBTB station have same pattern of PET station in (Fig (5). The different result is addressed on enveloping graph of amplitude spectra and the window length of earthquake time duration which has bigger value than PET station.

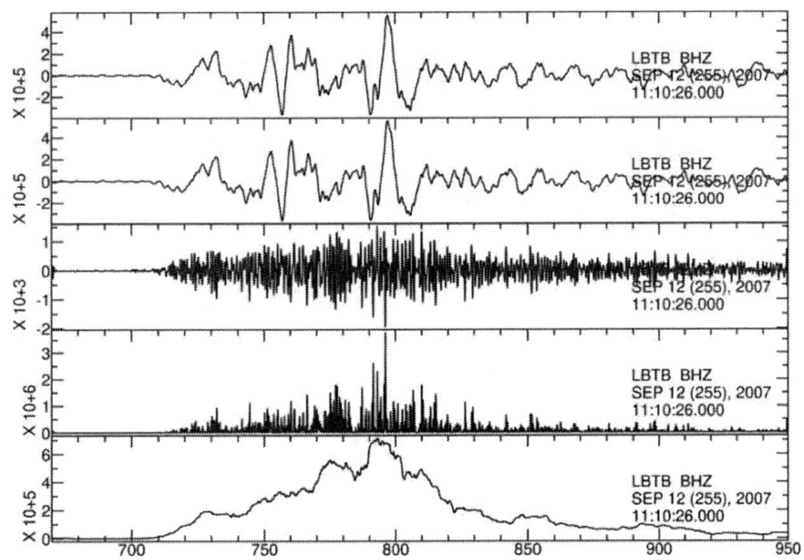

Figure 5. An another multiple plot of LBTB station which show real recording, waveform with bandpass 0.1 – 10 Hz, the envelope and time duration.

After calculating the Mwp and Tdur, we compared all result to see the correlation factor. The correlation is analyzed by linear regression to get pearson values which shown in Fig (6) and Tabel 1.

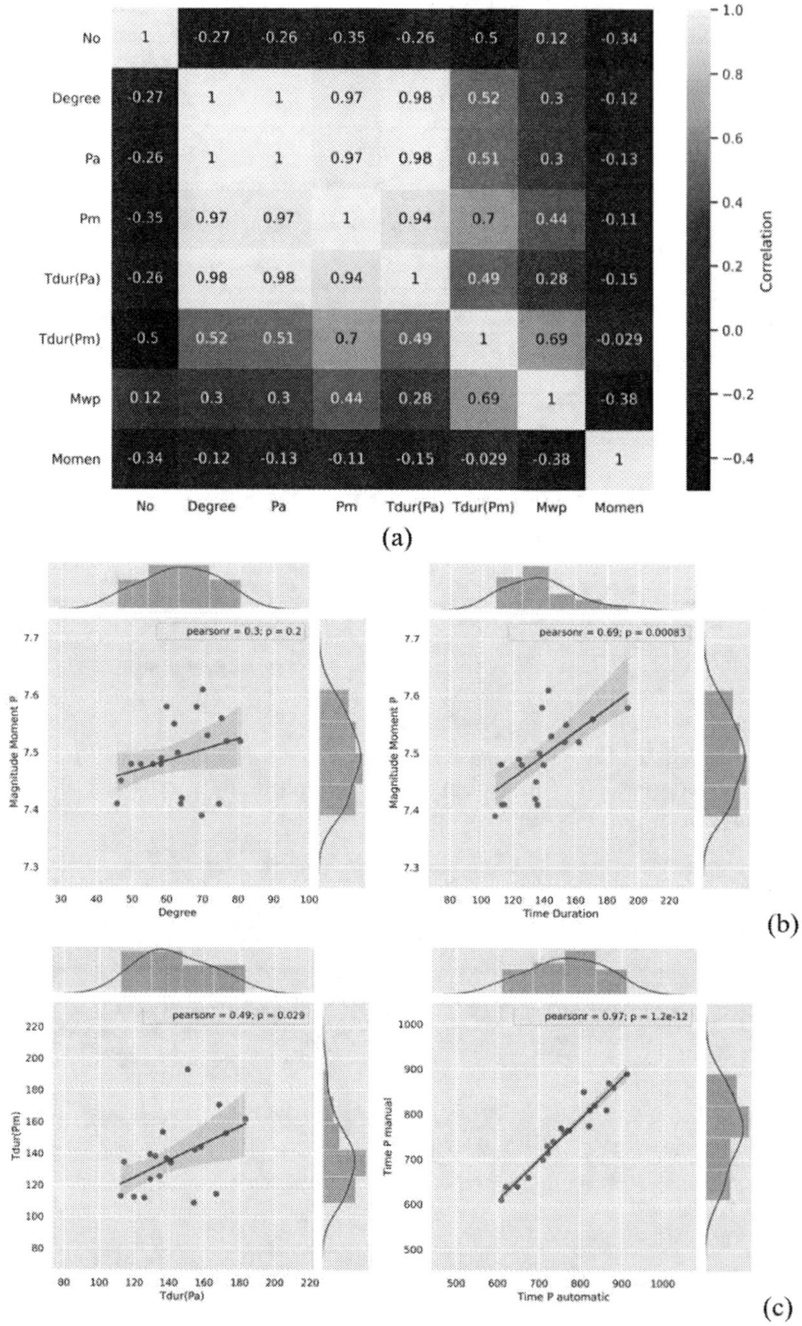

Figure 6. The comparison of all result and regression linear show a relationship between automatic and manual picking of P wave. In Fig 6 (a), the heat matrix correlation show a good and bad values, in Fig 6 (b) and (c), both graph and histogram show the comparison result of linear regression between Mwp and Distance, Mwp and Tdur, Tdur in automatic picking and manual picking, and Pp arrival time with automatic and manual, all relationship perform a well-analyzed result.

Table 1. Result comparison of p-wave time, mwp and mo for all stations.

Stations	Distance (°)	P arrival time	P_p Auto	P_p Manual	T.dur Auto	T.dur Manual	Sensitivity	M_w	moment (m_o)
LBTB	75.2722	699.3	867.81	870	168.51	170.70	7.60E+09	7.46	1.9498E+20
ABKT	58.2227	590.58	719.72	730	129.14	139.42	1.72E+09	7.48	2.0893E+20
CASY	61.9895	616.42	753.01	770	136.59	153.58	1.02E+09	7.51	2.3173E+20
FURI	63.9425	629.41	770.18	765	140.77	135.59	9.56E+08	7.71	4.6238E+20
GNI	68.2565	657.12	807.77	850	150.65	192.88	1.06E+09	7.48	2.0893E+20
GUMO	46.7459	505.51	619.55	640	114.04	134.49	8.67E+08	7.65	3.7583E+20
PET	74.6567	695.74	862.61	810	166.87	114.26	1.06E+09	7.51	2.3173E+20
SBA	80.4926	728.3	911.5	890	183.2	161.70	8.38E+09	7.52	2.3988E+20
KMBO	64.1654	630.87	772.13	765	141.26	134.13	1.03E+09	7.59	3.0549E+20
ANTO	76.6644	707.23	879.55	860	172.32	152.77	9.72E+08	7.46	1.9498E+20
KIV	71.2980	675.85	833.98	820	158.13	144.15	1.31E+09	7.57	2.8510E+20
PMG	45.6840	497.16	609.27	610	112.11	112.84	1.03E+09	7.51	2.3173E+20
YAK	69.7372	666.33	820.56	775	154.23	108.67	1.07E+09	7.54	2.5704E+20
TLY	55.9473	574.47	709.23	700	134.76	125.53	1.60E+09	7.58	2.9512E+20
BRVK	63.0087	623.23	761.98	760	138.75	136.77	1.03E+09	7.47	2.0183E+20
ULN	52.3371	548.14	673.96	660	125.82	111.86	1.06E+09	7.35	1.3335E+20
WMQ	49.6137	527.66	647.41	640	119.75	112.34	9.19E+08	7.55	2.6607E+20
KURK	58.3377	591.38	720.74	715	129.36	123.62	1.63E+09	7.59	3.0549E+20
ARU	69.9802	667.82	822.66	810	154.84	142.18	1.72E+09	7.61	3.2734E+20
ERM	59.8222	601.67	733.88	740	132.21	138.33	1.69E+09	7.68	4.1686E+20

From Fig (6) and Table 1, we perform a well-analyzed relationship between earthquake magnitude moment of P-Wave with time duration. The results of linear regression imply if the Mwp and Tdur can be used as the quickly parameters to report a tsunami warning and both can calculated from automatic and manual pick for P-wave arrival time. As we mentioned before, the tsunami warning issue is the one of important case to reduce and need to combine with seismic hazard assessment to give a vulnerabitlity for potential earthquake that generate a tsunami. The probability and deterministic analysis of seismic hazard can be used to explain the vulnerability effect of tsunami event [24], explain the seismic system in active faults [25, 26, 27],and also the ground motion prediction equation to disseminate a reliable report of shakemap of tsunami earthquake [28, 29, 30] in the future.

4. Conclusion

We perform a well-analyzed results to show an important relationship between Magnitude moment of P-Wave and time duration for quickly report of potential tsunami earthquake. The result of linear regression between Mwp and time duration is very reliable with physical waveform properties and both can be used as the specific parameters to disseminate a quickly report of potential tsunami. Futhermore, seismic hazard anaklysis and ground motion prediction equation can be used to explain the vulnerability level of the area which close to the source of potential tsunami earthquake in the future.

Acknowledgments

The computer facility for this study is funded by the NAS and USAID under the USAID Prime Award Number AID-OAA-A-11-00012 in the room of STEM Research Center. This work is financially supported by "Penelitian Terapan" the Directorate of Research and Community Service, Ministry of Research and Innovation Nomor 215/SP2H/LT/DPRM/2020.

References

[1] Pasari S, Simanjuntak A V, Mehta A and Sharma Y 2021 *Natural Hazards* 1-13.
[2] Kodijat, A. M., Rossel, B. A., Julius, A. M., Octonovrilna, L., & Cahyaningrum, A. P. (2018). In *MATEC Web of Conferences* (Vol. 229, p. 01016). EDP Sciences.
[3] Pribadi S, Puspito N T and Ibrahim G 2013 *Journal of Mathematical & Fundamental Sciences 45*(2).
[4] Pasari S, Simanjuntak A V, Mehta A and Sharma Y 2021 *Pure and Applied Geophysics* 1-18.
[5] Pasari S, Simanjuntak A V and Sharma Y 2021 *Geoscience Letters* **8**(1) 1-13.
[6] Julius, A. M., & Widana, I. D. K. K. (2021, April). In *IOP Conference Series: Earth and Environmental Science (Vol. 708, No. 1, p. 012106)*. IOP Publishing.
[7] Julius, A. M., Pribadi, S., & Muzli, M. (2018, March). In *IOP Conference Series: Earth and Environmental Science (Vol. 132, No. 1, p. 012012)*. IOP Publishing.
[8] Julius, A. M., & Sunardi, B. (2015, April). In *AIP Conference Proceedings (Vol. 1658, No. 1, p. 040001)*. AIP Publishing LLC.
[9] Tsuboi S, Abe K, Takano K and Yamanaka Y 1995 *Bull. Seismol. Soc. Amer.* **83** pp 606-613
[10] Whitmore P M, Tsuboi S, Hirshorn B and Sokolowski T J 2002 Magnitude-dependent correction for Mwp *Science of Tsunami Hazards*.
[11] Hirshorn B, Weinstein S and Tsuboi S 2013 *Pure Appl. Geophys.* **170** pp 975-991
[12] Tezel T and Yanik K 2013 *Seismol. Res. Lett.* **84** pp 991-996
[13] Nurana I, Simanjuntak A V H, Umar M, Kuncoro D C, Syamsidik S and Asnawi Y 2021 *Elkawnie: Journal of Islamic Science and Technology* **7**(1) 131-145.
[14] Lomax A and Michelini A 2011 *Geophysical Journal International* **185**(1) 283-291.
[15] Boatwright J and Choy G L 1986 *Journal of Geophysical Research: Solid Earth* **91**(B2) 2095-2112.
[16] Simanjuntak, A. V., Muksin, U., & Setiawan, Y. (2019, June). In *IOP Conference Series: Earth and Environmental Science* (Vol. 273, No. 1, p. 012021). IOP Publishing.
[17] Simanjuntak, A. V., Muksin, U., & Sipayung, R. M. (2018, December). In *Journal of Physics: Conference Series (Vol. 1116, No. 3, p. 032033)*. IOP Publishing.
[18] Qadariah Q, Simanjuntak A V and Umar M 2018 *Journal of Aceh Physics Society* **7**(3) 127-132.
[19] Simanjuntak A V, Muksin U and Rahmayani F 2018 *MS&E* **352**(1) 012046.
[20] Simanjuntak, A. V., Sihotang, N., Kuncoro, D. C., & Irwandi, I. (2021, March). In *Journal of Physics: Conference Series (Vol. 1811, No. 1, p. 012123)*. IOP Publishing.
[21] Kanjo K, Furudate T and Tsuboi S 2006 *Earth, planets and space* **58**(2) 121-126.
[22] Kanamori H 1977 *Journal of geophysical research* **82**(20) 2981-2987.
[23] Goldstein P, Dodge D, Firpo M and Minner L 2003 *International Handbook of Earthquake and Engineering Seismology* International Geophysics **81** Part B pp 1613-1614
[24] Irwandi, I. (2017, February). In *IOP Conference Series: Earth and Environmental Science* (Vol. 56, No. 1, p. 012007). IOP Publishing.
[25] Muksin U, Bauer K and Haberland C 2013 *Journal of volcanology and geothermal research* **260** 27-42.
[26] Muksin U, Bauer K, Muzli M, Ryberg T, Nurdin I, Masturiyono M and Weber M 2019 *Journal of Asian Earth Sciences* **171** 20-27.
[27] Idris Y, Cummins P, Rusydy I, Muksin U, Syamsidik, Habibie M Y and Meilianda E 2019 Indonesia *Journal of Earthquake Engineering* 1-18.
[28] Rusydy I, Idris Y, Muksin U, Cummins P and Akram M N 2020 *Geoenvironmental Disasters* **7**(1) 1-16.
[29] Irwandi, I., & Jamaluddin, K. (2018, November). In *Journal of Physics: Conference Series* (Vol. *1120, No. 1, p. 012092)*. IOP Publishing.

IOP Conf. Series: Earth and Environmental Science **873** (2021) 012094 doi:10.1088/1755-1315/873/1/012094

Comparison of Airborne Electromagnetic and Ground-based Resistivity Observations: Case Study Banda Aceh Basin

I, Irwandi [1, a)], Syafrizal [2)], Rifqan [1)], H. Naufal [3)], B. Siemon [4)]
[1] Universitas Syiah Kuala, Aceh, Indonesia
[2] Universitas Malikussaleh, Aceh, Indonesia
[3] Institut Teknologi Bandung
[4] Federal Institute for Geosciences and Natural Resources (BGR), Germany
[a)] Corresponding author: Irwandi@unsyiah.ac.id

Abstract. Airborne measurements are very useful to cover very large areas. Nine month after the Aceh Tsunami and Earthquake in 2004, BGR (Federal Institute for Geoscience and Natural Resources) conducted a fresh water supply exploration survey within a project called Helicopter Project Aceh (HELP ACEH). The helicopter-borne electromagnetic (HEM) device operates at five frequencies. The HEM can estimate the 1D resistivity models down to a depth of 150 m for the high electrical resistivity areas and 50 m for low electrical resistivity areas. In this paper, the airborne data of 2005 are compared with resistivity data acquired in Banda Aceh basin in 2018. The HEM output consists of 1D resistivity models derived by inversion of the processed data. These 1D resistivity models are compared with the 2D resistivity models derived from ground-based resistivity measurements. However, the 2D models on the ground are transformed into 1D resistivity models so it can be used for comparison. The transformation is conducted by averaging the resistivity values in the each layer, so every layer only has one resistivity value. Both methods are influenced by many factors. For example, resistivity on the ground is affected by local conditions. The airborne measurements are also influenced by objects that are at the surface of the ground. In some cases, the airborne resistivity models have some differences in absolute resistivity values, but they often have the same structural pattern compared with the ground-based resistivity models.

1. Introduction

Geophysical parameter measurements can be carried out on surfaces, move on surface, above surfaces, and in the subsurface by ground-based, vehicle-based, airborne and borehole measurements, respectively. A common ground-based method to derive the resistivity of the subsurface is direct-current geolectrics with all its variants, which require current injections into the earth and voltage measurements at the surface. Measurements that are not in direct contact with the surface are very useful because there are less logistical problems. Vehicle-based measurement such as VLF [1] can cover very long route from in tens to hundreds km in short time. Airborne measurements such as helicopter-borne electromagnetics (HEM) [2] can cover very large areas in short time. Such airborne electromagnetic measurements have been carried out in the Banda Aceh basin between August 23 and September 12, 2005, within the HELP ACEH (Helicopter-Borne Geophysical Investigations in the Aceh Province, Indonesia) project, which was able to cover an area of approximately 1000 km^2 [3][4].

HELP ACEH covered three quite extensive areas, Banda Aceh basin which included the City of Banda Aceh and the Aceh Besar District, Calang–Meulaboh[3] and Sigli [5]. The purpose of the survey was to find areas with potable water after the tsunami 2004. Therefore, the studies focussed on hydrogeology for subsurface water mapping. However, the large dataset of the covered areas, which also include magnetic and radiometric data, might be very interesting, when used for other applications [6], such as determining soil liquefaction conditions, carrying capacity of the soil, bed rock mapping [7], determining Vs30 for shakemap [8], and structural sediment for Neo-Deterministic Seismic Hazard Analysis (NDSHA) [9]. It might be also used to determine active faults around Banda Aceh Basin [10][11]. In this paper, we discuss a comparison of airborne and ground-based resistivity models.

2. Methodology

HEM resistivities are derived from measured secondary electromagnetic fields caused by the induction of primary fields, which interact with ground material of earth's subsurface as shown in Figure 1a. Each primary electromagnetic field is generated at a discrete frequency by a transmitter coil (T), and a receiver coil picks up the secondary electromagnetic field [12]. A bucking coil cancels out the direct primary field at the receiver and a calibration coil helps to get accurate data. All coils are installed in a tube ("bird") with a length of about 10 m and a diameter of 0.5 m. The shell of the bird is made of Kevlar, a material that has extremely high mechanical stability and poor electric conductivity. This system (Resolve) was manufactured by Fugro Airborne Surveys (FAS), Canada [12] The digital electromagnetic system in horizontal-coplanar coil configuration is operated at five discrete frequencies as shown in Table 1.

Table 1. The frequencies bird instrument and related band name [13] .

Frequency	387 Hz	1820 Hz	8225 Hz	41.550 Hz	133.200 Hz
Wavelength	775km	164.8km	36.47km	7.22km	2.25km
Band name	ULF	ULF	VLF	LF	LF

The bird is connected with the helicopter by a cable of approximately 45 m in length and towed about 41 m beneath the helicopter, depending on the flight speed. This length of cable is chosen to largely reduce the influence of the helicopter on the highly sensitive magnetic and electromagnetic sensors. The average flight altitude of the HEM sensor was 60 m above ground and the normal survey speed was 110–165 km/h.

Three altimeters (radar, laser, barometric altimeter) are used. The radar altimeter is installed beneath the nose of the helicopter to get the distance of the helicopter to the surface. Besides, the laser altimeter is used for getting the altitude of the bird above the ground. The barometric altimeter is used for correction of GPS elevation data and to get air-pressure values. GPS systems are in both the helicopter and the bird to get precise navigation and position data.

The 20 km x 50 km survey area between 95°13'E to 95°44'E and 5°13'N to 5°37'N covers Banda Aceh as well as the valley of Krueng Aceh river,. A flight-line map of the northwestern portion of the survey area is shown in Figure 1a. There are 163 profile lines (red lines) in SW–NE direction and 24 tie–lines flown (green lines) in SE–NW in the entire survey area. The total length of the profile lines is about 3960 line-km [14].

The airborne electromagnetic data are stored per site along the survey lines as shown in Figure 1b. The inversion calculation is done per point providing inverse 1D models by using Levenberg-Marquardt inversion [15][16][17]. The data file with the inversion models [14] contains more than 900,000 1D resistivity models, so it has been necessary to create an AWK [18] script to retrieve the closest 4 sites to the sites on the ground which check for all profile and tie-lines. All four 1D profiles are plotted using gnuplot [19] together with a ground-based resistivity model as seen in Figures 3,4 and 5.

ERT (electric resistivity tomography) or direct-current geoelectrics is a standard method for resistivity measurements directly at the surface of the earth by injecting an electric current and installing other electrodes to measure the subsurface voltage [20]. The data acquisition in the field is carried out using SuperSting R8/IP resistivitymeter (Figure 2a) with Wenner-Schlumberger array. The data are obtained from SuperSting R8/IP are saved to the computer and converted for inversion with res2dinv [21]. An example is shown in Figure 2b. The output is a 2D resistivity model that is converted into an averaged 1D model using an AWK script as shown in Figure 2c. In this paper, 9 ground-based resistivity datasets are discussed as shown in Table 2 which will be compared with the inversion results of helicopter-borne electromagnetic resistivity results.

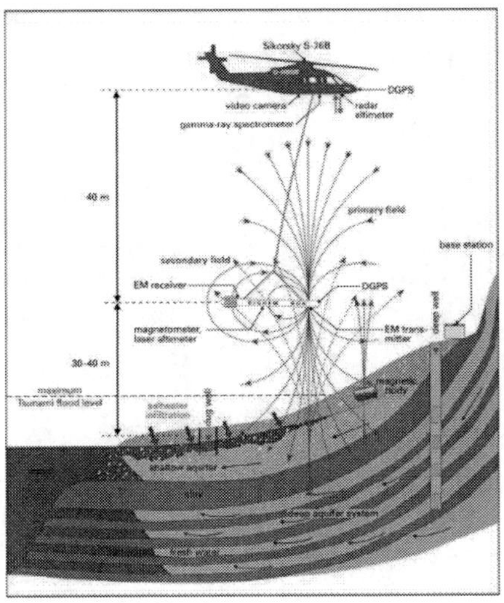

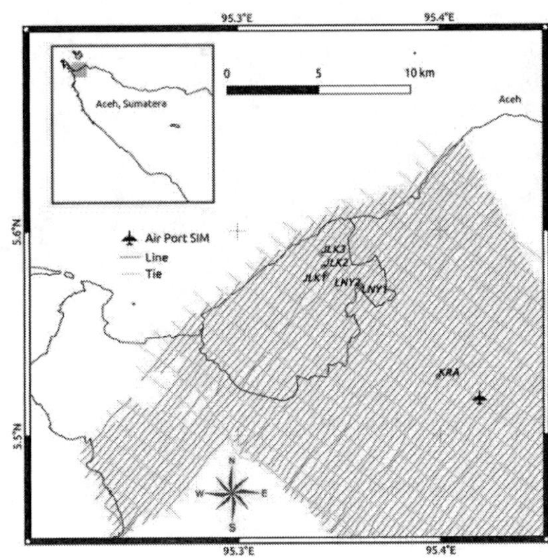

(a) (b)

Figure 1. (a) BGR's airborne geophysical survey system [14] (b) HELP ACEH airborne flight lines for Banda Aceh basin overlaid with the sites of the resistivity measurements on ground.

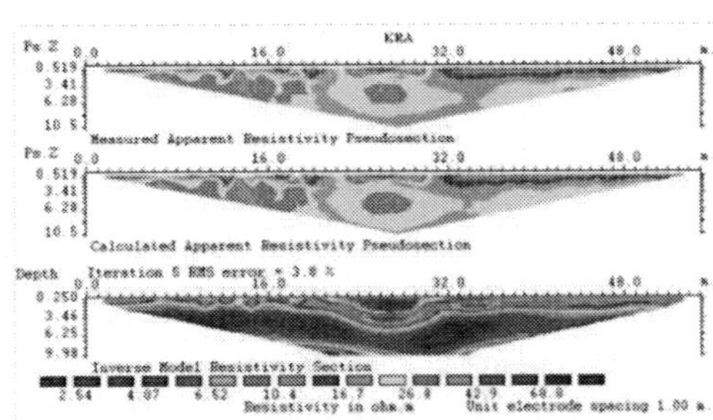

(a) (b)

Figure 2. (a) Resistivity measurement at the KRA site. (b) Processing the resistivity data using res2dinv software generates 2D profiles. (c) 1D resistivity model derived from 2D model.

IOP Conf. Series: Earth and Environmental Science **873** (2021) 012094 doi:10.1088/1755-1315/873/1/012094

Table 2. Location of ground-based resistivity measurements

No	Site Label	Longitude	Lattitude	Note
1	LNY1	95.36052	5.57290	Lamyong Bridge
2	LNY2	95.36004	5.57261	Lamyong Bridge
3	KRA	95.39887	5.52852	Cave Bunker
4	JLK1	95.34335	5.57856	Jeulingke
5	JLK2	95.34235	5.58226	Jeulingke
6	JLK3	95.34104	5.58896	Tibang

3. Result and Discussion

This paper shows the comparison of ground-based resistivity models using direct-current resistivity methods and resistivity models derived from airborne electromagnetic measurements. Figures 3 and 4 show overlaid sections of ground-based and airborne resistivity models (taken from closest four locations).

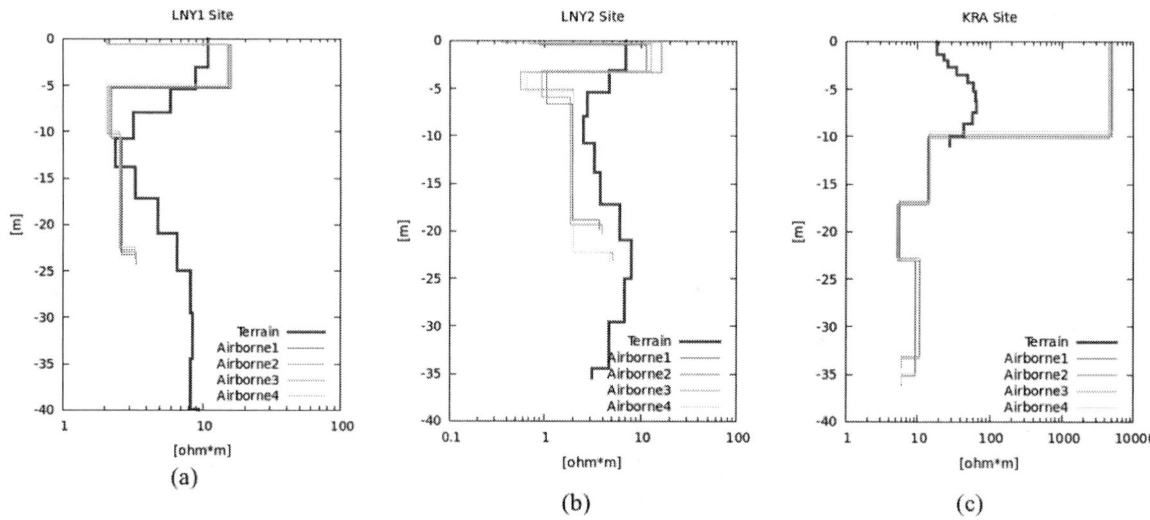

Figure 3. Plotting the ground-based (Terrain) and airborne resistivity models (Airborne1-4) for Lamyong Bridge at sites (a) LNY1, (b) LNY2, and Japan's Bunker Cave at site (c) KRA.

In Figures 3a and 3b, a comparison is shown for the Lamyong Bridge site and in Figure 3c for the Cave Bunker site. The point measurement is near enough, where an airborne profile is fluctuating. The layers have some differences, but the fluctuation of the airborne resistivity is around the value of ground-based resistivity. The penetration of ground-based values can exceed 40 m, but this is not plotted to focus more on the penetration of the airborne data, which is limited here to less than 40 m. Figure 4 shows a comparison for locations that are relatively close at Lingke area. There, the penetration of airborne data is limited to about 10 m depth due to very low (< 1 Ωm) resistivities caused by salt water. Because of the skin depth effect, the airborne electromagnetic method did not successfully probe the area below the surface, so it was clear that the airborne resistivity value was relatively smaller than the value of terrain resistivity. Especially for the JLK2 site, there seems to be a bit of a curve in the pattern formed.

Generally, the comparison shows that the models are relatively close at all sites. Figure 3c shows high resistivity values, because the measurement area of the ground-based measurement is just above the bunker cave. Thus, the local conditions greatly influence the results of the ground-based resistivity measurement, while the airborne resistivity is derived from a larger area.

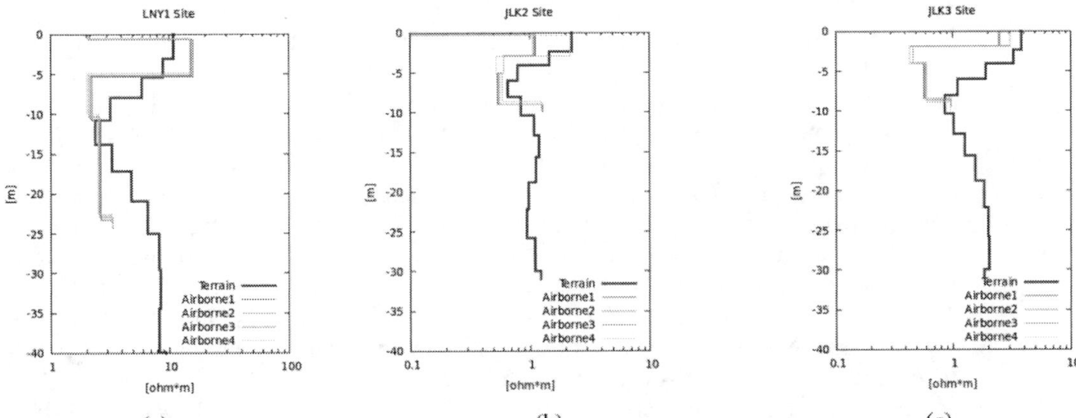

(a) (b) (c)

Figure 4. Plotting the ground-based and airborne resistivity models for Jeulingke Village at sites (a) JLK1, (b) JLK2, and Tibang Village at site (c) JLK3

4. Result and Discussion

The resistivity distribution in the subsurface, particularly at shallow depth, determines the penetration depth of airborne electromagnetic fields. If the resistivity value at shallow depth is low, then the skin depth effect will inhibit the electromagnetic fields from deep penetration and, thus, the results of inversion will produced shallow models. However, if the shallow resistivity values are high, electromagnetic fields can get deeper into the earth and the inversion will resolve deeper parts. The opposite effect is observed for the results of the ground-based resistivity method as low shallow resistivities still enable deep penetration, but high resistivities are problematic. This finding confirms that electromagnetic methods are sensitive to conductivity and direct-current geoelectric methods to resistivity. Local resistivity variations affect ground-based measurements stronger than airborne measurement due to the generally larger footprint of airborne measurements. On the other hand, electromagnetic measurements are always affected by anthropogenic installations containing highly conductive material (metals) such as buildings, fences, power lines, or roads tracks. This cultural noise can be excluded (and interpolated) during HEM data processing, but this was not done in 2005. Geoelectric methods are less sensitive to such anthropogenic effects. In this study, the number of measurement on the ground was limited and needs to be expanded to more sites for a better comparison. Furthermore, one has to pay attention to rainfall factors or ground moisture and differences due to near-surface saltwater effects caused by the tsunami of 2004.

Acknowledgments

Authors wishing to acknowledge the researchers were grateful to BGR (Institute for Geosciences and Natural Resources) for conducting the electromagnetic airborne HELP ACEH and PVMBG (Centre of Volcanology and Geological Hazard Mitigation) for providing the data. This work financially support by Directorate of Research and Community Service (DRPM), Ministry of Research, Technology and Higher Education with project "Penelitian Terapan", contract No.91/UN11.22.1/PT.01.03/DPRM/2020 and PNBP Unsyiah with project "Unggulan Unsyiah", contract 239/UN11.2.1/PT.01.03/PNBP/2020.

References

[1] Irwandi, I., Marwan, M., & Fashbir, 2017b. Applications of the VLF-EM method for rapid Sumatran fault identification in Leuser national park, Aceh. In AIP Conference Proceedings (Vol. 1861, No. 1, p. 030050). AIP Publishing.

[2] Siemon, B., Christiansen, A.V. & Auken, E., 2009. A review of helicopter-borne electromagnetic methods for groundwater exploration. Near Surface Geophys. 7, 629–646.

[3] Siemon, B., Steuer, A., Meyer, U. & Rehli, H.-J., 2007. HELP ACEH – A post-tsunami helicopter-borne groundwater project along the coasts of Aceh, northern Sumatra. Near Surface Geophysics, 5 (4), 231-240, doi: 10.3997/1873-0604.2007005.

[4] Siemon, B. & Steuer, A., 2011. Airborne Geophysical Investigation of Groundwater Resources in Northern Sumatra after the Tsunami of 2004. In: Mörner, N.-A. (Ed.), The Tsunami Threat - Research and Technology. INTECH, ISBN: 978-953-307-552-5, 576-594.

[5] Steuer, A., Siemon, B. & Eberle, D., 2008. Airborne and ground-based electromagnetic investigations of the fresh-water potential in the tsunami-hit area Sigli, northern Sumatra. Journal of Environmental & Engineering Geophysics, 13 (1), 39-48, doi: 10.2113/JEEG13.1.39.

[6] Auken, E., Boesen, T. & Christiansen, A.V., 2017. A review of airborne electromagnetic methods with focus on geotechnical and hydrological applications from 2007 to 2017. Advances in geophysics, 58, pp.47-93.

[7] Oldenborger, G. A., Logan, C. E., Hinton, M. J., Pugin, A. J.-M., Sapia, V., Sharpe, D. R. & Russell, H. A. J. (2016). Bedrock mapping of buried valley networks using seismic reflection and airborne electromagnetic data. Journal of Applied Geophysics, 128, 191–201

[8] Irwandi, I,. Muzli, M. & Jamaluddin, K., 2018, November. GMPE based Shakemap Generation of Peak Ground Motion and Intensity Maps for Pidie Jaya Earthquake. In Journal of Physics: Conference Series (Vol. 1120, No. 1, p. 012092). IOP Publishing.

[9] Irwandi, I., 2017, February. Advantages of Realistic Model Based on computational method: NDSHA versus Standard PSHA. In IOP Conference Series: Earth and Environmental Science (Vol. 56, No. 1, p. 012007). IOP Publishing.

[10] Barber, A.J., Crow, M.J. & Mmsom, J.S. (eds) 2005. Sumatra: Geology, Resources and Tectonic Evolution. Geological Society, London, Memoirs, 31.

[11] Tabei, T., Kimata, F., Ito, T., Gunawan, E., Tsutsumi, H., Ohta, Y., Yamashina, T., Soeda, Y., Ismail, N., Nurdin, I & Sugiyanto, D. (2015). Geodetic and Geomorphic Evaluations of Earthquake Generation Potential of the Northern Sumatran Fault, Indonesia.

[12] Fraser D.C. 1978. Resistivity mapping with an airborne multicoil electromagnetic system. Geophysics 43, 144–172.

[13] ITU-R, 2015. Nomenclature of the frequency and wavelengh bands used in telecommunications. Recommendation ITU-R V.431-8. 2015.

[14] Siemon B., Röttger B. & Rehli, H.-J. 2006. Helicopter-borne geophysical investigation in the province Nanggroe Aceh Darussalam, Northern Sumatra, Indonesia. BGR Report Vol. A-1, Archives No. 0126196 A-1, Banda Aceh, Hannover.

[15] Levenberg, K., 1944. A method for the solution of certain non-linear problems in least squares. The Quarterly of Applied Mathematics, 2, 164–168.

[16] Marquardt, D.W., 1963. An algorithm for least-squares estimation of nonlinear parameters. Journal of the society for Industrial and Applied Mathematics, 11(2), pp.431-441.

[17] Sengpiel, K.-P. & Siemon, B., 2000. Advanced inversion methods for airborne electromagnetic exploration. Geophysics, 65, 1983–1992.

[18] Robbins, A., 2015. Effective AWK programming: universal text processing and pattern matching. " O'Reilly Media, Inc.".

[19] Phillips, Lee. 2017. Gnuplot 5. Packt Publishing (2017). ISBN : 978-0-692-92716-8

[20] Telford. 1990. Applied Geophysics Second Edition, Cambridge University Melbourne. Australia.

[21] Loke, M.H., 2006. RES2DINV ver. 3.55, Rapid 2D resistivity & IP inversion using the least-squares method. Software Manual, 139.

IOP Conf. Series: Earth and Environmental Science **873** (2021) 012095 doi:10.1088/1755-1315/873/1/012095

Support Vector Machine for Land Cover Classification using Lidar Data

M I Hariyono[1,2], Rokhmatuloh[1] , M P Tambunan[1] , R S Dewi[2]

[1]Department of Geography, Faculty of Mathematics and Natural Science, University of Indonesia, Depok City, 16424, Indonesia
[2]Geospatial Information Agency (BIG), Jl. Raya Jakarta-Bogor KM.46 Cibinong, 16911, Indonesia

Email : moch.irwan@ui.ac.id, mochamad.irwan@big.go.id

Abstract. The Lidar technology is widely used in various studies for mapping needs. In this study was to extract land cover using Lidar data by incorporating a support vector machine (SVM) approach. The study was located in the city of Lombok, Nusa Tenggara Barat. Image extraction was performed on single wavelength Lidar data to produce intensity and elevation (Digital Surface Model) features. Feature extraction of Lidar data was implemented by using a pixel-based approach. The extracted features used as an attribute for training data to generate the SVM prediction model. The prediction model to predict the types of land cover in the study area such as buildings, trees, roads, bare soil, and low vegetations. For accuracy assessment purposes, we used topographic map available in shapefile format as the reference map and estimated the accuracies of the resulted classifications. In this study, land cover classification used combination bands which improved the overall accuracy by approximately 20%. The use of the intensity data in this band combination was the reason for the increasing accuracy.

1. Introduction

Currently, many studies use Lidar data as the main data, this data can be used for various applications, such as topographic mapping, DEM modeling [1], building extraction [2,3] vegetation [4] and others. In simple terms, in many researches, Lidar data was used to classify ground and non-ground. Furthermore, Lidar data can be useful for land cover classification [5,6,7,8]. Lidar has high accuracy and resolution for both vertical and horizontal directions. From the Lidar data, features are obtained in the form of height and intensity. The intensity data provides good information in land cover classification processes [9,10] The height features commonly used for classification are DSM, nDSM, DTM and Slope. In this research, we used intensity data from Lidar point cloud. Based on previous study conducted by Hui et al. [11], land cover classification method by combining intensity data with height data provided good results. Hui stated that intensity data has advantages compared to multispectral remote sensing data by avoiding shadows that usually exist in multispectral data [11].

In literature, researches by using Lidar intensity data are increasingly being developed, and carried out in various combinations to improve the accuracy of land cover classification results by combining intensity data with other features [12]. Another study by Yan et al. [13] conducted an evaluation before and after radiometric correction. In their study, land cover homogeneity was also incorporated so that it

IOP Conf. Series: Earth and Environmental Science **873** (2021) 012095 doi:10.1088/1755-1315/873/1/012095

helped in increasing feature extraction or similarity of land cover surfaces which was further increased the accuracy of the classification results after radiometric correction.

Various methods have been used for land cover classification processes, one of which is machine learning support vector machine (SVM). According to Noi and Kappas [14]· this method shows a better accuracy compared to random forest (FR) method, and K-Nearest Neighbor (kNN). SVM has capabilities that have been tested and have a major role in classification processes [15]. Another study using SVM for land cover classification with a combination of Lidar data and ortho image shows excellent accuracy results from the combination of these data [16]. The purpose of this study was to extract land cover using Lidar data by incorporating a support vector machine (SVM) approach.

2. Methodology

2.1 Study Area

The study area was located in a part of the village of Tanjung Karang, Mataram, Nusa Tenggara Barat (NTB), Indonesia. This area was selected because the area has various land cover features on the ground including buildings, vegetations and roads. Several datasets were used in this study including: a) Lidar data (Figure 1a); b) orthophoto (Figure 1b) and c) topographic map (RBI) provided by Geospatial Information Agency (BIG, 2017) at scale 1: 5000. The point cloud density of the Lidar data was approximately 10 points per square meters (ppm). The aerial photo was acquired at the same time with the Lidar data which was recorded in 2016. Figures 1a-b represent the Lidar and the ortho-rectified aerial photos (orthophoto) of the study area.

(a)

(b)

Figure 1. Study area that has various land cover feature presented by using: a). Lidar, and b). Orthophoto

2.2 Data Preparation

The data sets used in this research were prepared by converting Lidar point cloud into raster images with pixel size equals to 15 cm. Lidar data were filtered to obtain ground and non-ground information, the filtering process was carried out using Global Mapper software. New images (bands) were created namely: DTM (digital terrain model), DSM (digital surface model), nDSM (normalized digital surface model), and intensity. In this step, DSM, DTM and intensity data were extracted by using tools provided in Global Mapper software. DSM was extracted from non-ground data while DTM was extracted from ground data. Following these, we extracted nDSM by calculating the differences between DSM and DTM data. Figures 2a-c illustrate the produced Lidar raster images (intensity, DSM, and nDSM).

Furthermore, to evaluate the influence of intensity data on the classification results, seven band combinations were created consisting of one, two, and three band combinations, namely: 1) DSM; 2)

intensity; 3) nDSM; 4) DSM and intensity; 5) DSM and nDSM; 6) Intensity and and nDSM; 7) DSM, Intensity and nDSM. We used these seven band combinations as the input image when performing classification.

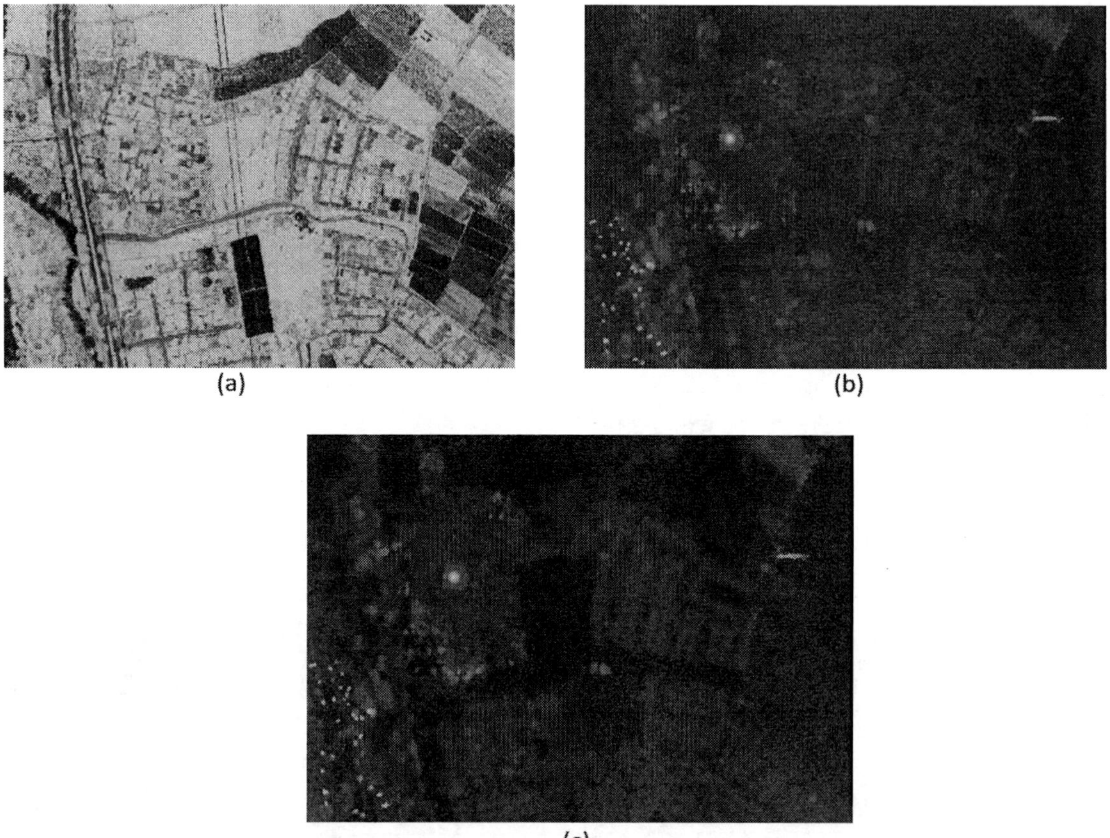

Figure 2. The visualisation of a) Intensity, b) DSM, c) nDSM

2.3 Data Classification

Support vector machine is a supervised algorithm that requires training data before applying the model. For the study area, we identified five land cover classes, namely buildings, tress, roads, bare soil, and low vegetations. SVM was then performed on seven band combinations, and statistical assessment was conducted for the training data. We evaluated the classification accuracies obtained when performing SVM and added more training data if required.

In this research, we used Radial Basis Function (RBF) when performing SVM. From the previous study, the use of these functions affect the efficiency of the classification when performing SVM and increase the accuracy [17], besides using two influential parameters namely cost (C) and gamma (γ) [18].

For accuracy assessment purposes, we used topographic map available in shapefile format as the reference map and estimated the accuracies of the resulted classifications by performing error matrix. We calculated the overall accuracy values based on confusion matrix principle. The calculation provides for a specific test between the classification results and the conditions at ground level [19].

3. Results and Discussion

Table 1 presents the overall accuracy resulted from the experiment when implementing SVM for land cover classification of seven combination of bands. From the results, we can see that the overall accuracies by using DSM and intensity are less than 50 %. Meanwhile, by combining both DSM and intensity data as the input images improved the results to 68%. Furthermore, using nDSM does not improve the accuracy. Nevertheless, combining DSM, nDSM and intensity data can improvement in the overall accuracy of the classification results. We obtained the overall accuracies for about 75-77 %. In conclusion, by using three band combinations produced the best performance of SVM classification by 77.9%.

Figures 3a-g show the classification results for each band combination when we performed SVM in image classifications. For visualization, we compared the classification results with the reference map in Figure 3h. In Figure 3a, only three classes could be identified namely building, trees and low vegetation. Meanwhile in Figure 3b, we can see from the classification result that only vegetation classes such as trees and low vegetation were classified. It might be due to higher intensity value of vegetation classes in the NIR range [12]. Furthermore, in Figure 3c, building features could be identified instead of vegetation classes. By adding more bands, the classification results produced increase accuracy and more features were identified (see Figures 3d-g). From these experiments, the errors in classification results may be due to lack of training data for each classes or maybe due to the similarity in intensity values between land cover classes.

Table 1. The overall accuracy results when performing SVM by using seven band combinations

No.	Band Combination	Overall accuracy
a.	DSM	45.9 %
b.	Intensity	42.7 %
c.	nDSM	48. 8 %
d.	DSM and Intensity	68.1 %
e.	DSM and nDSM	75.2 %
f.	Intensity and nDSM	70.8 %
g.	DSM, Intensity and nDSM	77.9 %

(a)

(b)

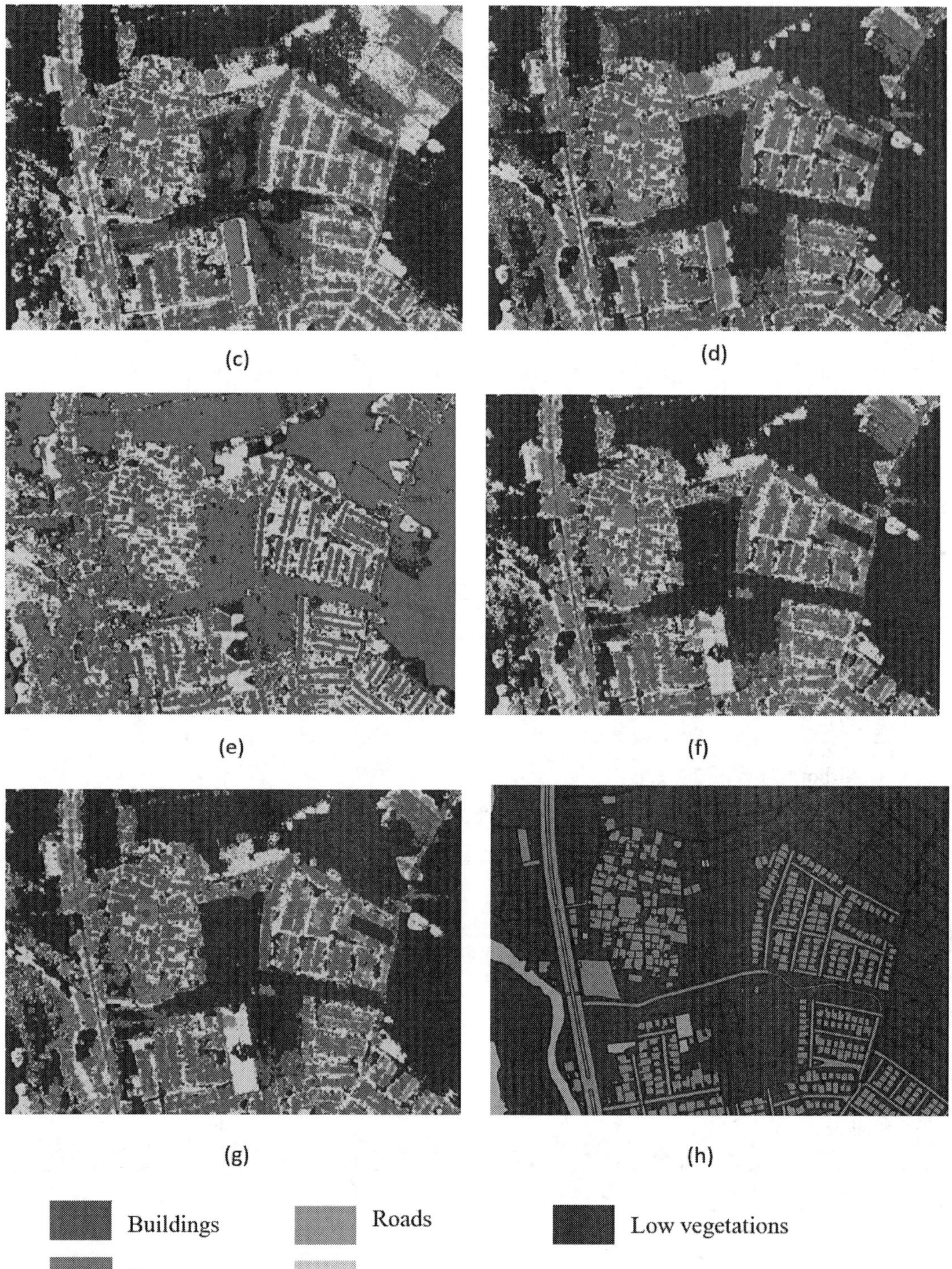

Figure 3. The results land cover classifications by performing SVM using seven band combinations: a) DSM, b) Intensity, and c) nDSM. d) DSM and Intensity, e) DSM and nDSM), f) Intensity and nDSM, g) DSM, Intensity and nDSM, compared with h) the reference map

4. Conclusion

In this study, we elaborated intensity image extracted from Lidar data for land cover classification by using a machine learning algorithm, SVM. The results of the land cover classification show that the classification using individual band (DSM, nDSM, Intensity) has an overall accuracy of less than 50%. Then the classification with a combination of two bands improved the results of accuracy by approximately 20% compare with using individual band. By using combination DTM and nDSM, we only obtained 75.2% accuracy value, however by including the intensity band for combination, we can improve the accuracy of land cover classification up to 77.9%.

Acknowlegdments

We thank Geospatial Infomation Agency of Indonesia for all the materials that has been provided as support for this research.

References

[1] Hariyono M I and Windiastuti R 2018 Classification of LiDAR Data to Generate Digital Terrain Model *39th Asian Conf. Remote Sens* 1743–1754
[2] Meng X, Wang L, and Currit N 2008 Morphology-based Building Detection from Airborne Lidar Data *ASPRS* **75** 437–442
[3] Miliaresis G and Kokkas N 2007 Segmentation and object-based classification for the extraction of the building class from LIDAR DEMs *Comput. Geosci 33 1076–1087*
[4] Nardinocchi C, Forlani G, and Zingaretti P 2003 Classification and filtering of laser data *Int. Arch. Photogramm. Remote Sensing Spat. Inf. Sci* 79–86
[5] Mallet C, Bretar F, Roux M, Soergel U and Heipke C 2011 Relevance assessment of full-waveform lidar data for urban area classification *ISPRS J. Photogramm. Remote Sens.* **66** S71–S84
[6] Chen Z and Gao B 2014 An Object-Based Method for Urban Land Cover Classification Using Airborne Lidar Data *IEEE J. Sel. Top. Appl. Earth Obs. Remote Sens.* 7 4243–4254
[7] Morsy S, Shaker A, and El-Rabbany A 2017 Multispectral LiDAR Data for Land Cover Classification of Urban Areas *Sensors* **17** 5 958–979
[8] Pan S, Guan H, Yu Y, Li J and Peng D 2019 A Comparative Land-Cover Classification Feature Study of Learning Algorithms : DBM , PCA ,and RF Using Multispectral LiDAR Data *IEEE J. Sel. Top. Appl. Earth Obs. Remote Sens.* **12** 4 1314–1326
[9] Song J, Han S, Yu K, and Kim Y 2002 Assessing The Possibility Of Land-Cover Classification Using Lidar Intensity Data *Int. Arch. Photogramm. Remote Sens. Spat. Inf. Sci. 34 3B* 259–262
[10] Flood M 2001 Lidar Activities and Research Priorities in The Commercial Sector *Int. Arch. Photogramm. Remote Sens.* **XXXIV-3/W4** 22–24
[11] Hui L, Di L, Xianfeng H, and Deren L 2008 Laser Intensity Used In Classification Of Lidar Point Cloud Data *IGARSS* **2** 1140–1143
[12] Shaker A and El-ashmawy N 2012 Land cover information extraction using lidar data *Int. Arch. Photogramm. Remote Sens. Spat. Inf. Sci.* **XXXIX-B7** XXII ISPRS Congress167–172
[13] Yan W Y, Shaker A, Habib A and Kersting A P 2012 Improving Classification Accuracy of Airborne LiDAR Intensity Data by Geometric Calibration and Radiometric Correction *ISPRS J. Photogramm. Remote Sens.* **67** 1 35–44
[14] Noi P T and Kappas M 2017 Comparison of Random Forest, k-Nearest Neighbor, and Support Vector Machine Classifiers for Land Cover Classification Using Sentinel-2 Imagery *Sensors* **18** 18–37
[15] Shao Y and Lunetta R S 2012 Comparison of support vector machine , neural network , and CART algorithms for the land-cover classification using limited training data points *ISPRS J.*

Photogramm. Remote Sens. **70** 78–87

[16] Tseng Y H, Wang C K, Chu H J and Hung Y C 2015 Waveform-based point cloud classification in land-cover identification *Int. J. Appl. Earth Obs. Geoinf.* **34** 1 78–88

[17] Qin R A 2015 Mean Shift Vector-Based Shape Feature for Classification of High Spatial Resolution Remotely Sensed Imagery *IEEE Journal of Selected Topics in Applied Earth Observations and Remote Sensing*, **8** 5 1974–1985

[18] Jozdani S E, Johnson B A and Chen D 2019 Comparing Deep Neural Networks, Ensemble Classifiers, and Support Vector Machine Algorithms for Object-Based Urban Land Use Land Cover Classification *Remote Sensing* **11** 1713–1736

[19] Foody G M 2002 Status of land cover classification accuracy assessment *Remote Sensing of Environment* **80** 1 185-201

Synthetic Modeling of Ambient Seismic Noise Tomography Data

Firman Syaifuddin,[1, 2],* Andri Dian Nugraha,[1] Zulfakriza,[1] Shindy Rosalia,[1]

[1] Institut Teknologi Bandung (ITB), Kampus ITB, Jl. Ganesha No.10, Bandung, Jawa Barat, 40132, Indonesia.
[2] Institut Teknologi Sepuluh Nopember (ITS), Kampus ITS, Sukolilo, Jl. Raya ITS, Keputih, Surabaya, Jawa Timur, 60117, Indonesia.

*Corresponding author: firmansyaifuddinakbar@gmail.com

Abstract. Ambient seismic noise tomography is one of the most widely used methods in seismological studies today, especially after a comprehensive Earth noise model was published and noise analysis was performed on the IRIS Global Seismographic Network. Furthermore, the Power Spectral Density technique was introduced to identify background seismic noise in the United States. Many studies have been carried out using the ambient seismic noise tomography method which can be broadly grouped into several groups based on the objectives and research targets, such as to determine the structure of the earth's crust and the upper mantle, to know the thickness of the sedimentary basins, to know the tectonic settings and geological structures, to know volcanic systems and geothermal systems, knowing near-surface geological features and as a monitoring effort the Ambient Noise Tomography method carried out by repeated measurements or time lapse. In this study, we investigate the characteristics of the ambient noise seismic tomography method, both its advantages and limitations of the method by utilizing synthetic data modeling using a simple geological model. Synthetic data is generated based on 1D dispersion curve forward modelling and the forward modeling of surface waves travel time for each period, which is then convoluted with the wavelets of each periods, then doing reverse correlation using a reference signal to produce synthetic recording data. We found that the estimate target depth and vertical resolution depend on the recorded data periods and the synthetic data modeling can be used as a basis in determining the acquisition design.

Keywords: Synthetic modelling, ambient noise seismic tomography

1. Introduction

The Ambient seismic noise tomography is one of the most widely used methods in seismological studies today, especially after a comprehensive Earth noise model was published [1] and noise analysis was carried out on the IRIS Global Seismographic Network [2], then Furthermore, the Power Spectral Density technique was introduced to identify background seismic noise on the United States Plain [3]. The use of cross-correlation techniques began to be used in an effort to extract the Green Function on ambient seismic noise data for a period of 10 to 150 seconds to describe the conditions of the earth's crust and the upper mantle in the United States [4], the same technique was also used to know the changes in seismic wave velocity values around Mount Merapi [5].

Many studies using the ambient seismic noise tomography method have been carried out which in general can be grouped into several groups based on research objectives and objectives, including to

IOP Conf. Series: Earth and Environmental Science **873** (2021) 012096 doi:10.1088/1755-1315/873/1/012096

determine the structure of the earth's crust and the upper mantle, to determine the thickness of the sedimentary basins, knowing tectonic settings and geological structures, knowing volcanic systems and geothermal systems, knowing near-surface geological features, and as an effort to monitor the Ambient Noise Tomography method by repeated measurements or time-lapses.

From the results of the research that has been done, it provides subsurface information according to the desired target, but not many have discussed the limitations of this method such as the relationship between the target depth and the seismic noise data period used in this method. This paper will discuss the relationship between the target depth and the seismic noise data period used so that it is in accordance with the desired resolution with the synthetic modeling approach.

2. Method

2.1. Ambient Seismic Noise Synthetic Modeling

Ambient seismic noise tomography data processing procedures are described by [6] in detail which consists of 4 main stages, data preparation for each station, cross-correlation and temporal stacking, dispersion curve estimation using frequency-time analysis (FTAN) techniques, and data quality control. Referring to the data processing procedure for ambient seismic noise tomography described by [6], the synthetic modeling process for ambient seismic noise data is carried out using reverse flow diagrams described by [6]. The flow chart for synthetic ambient seismic noise data generation is shown in Figure 1.

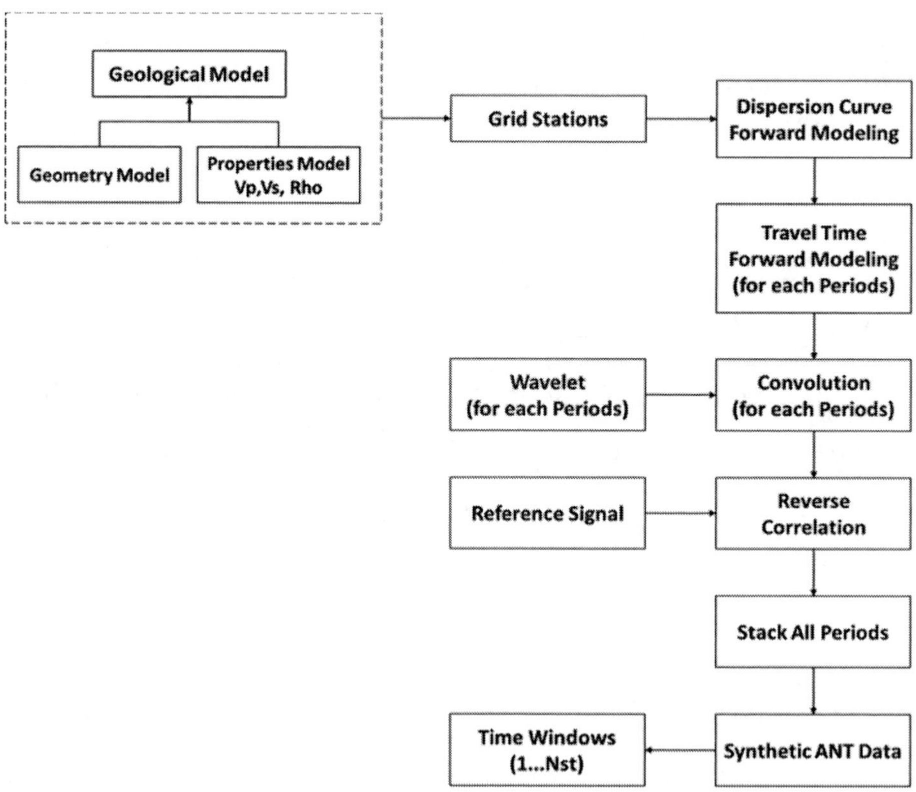

Figure 1. Flow chart for synthetic ambient seismic noise data generation.

2.1.1. Geological Model Building

The synthetic modeling process begins with making a geological model accompanied by physical properties such as P wave velocity, S wave velocity, and the density of the model. In this study, the geological model used is a simple flat layer model with the s wave velocity as the main input based on the equation function Vs (Z) = Vso + k(Z) [7], with Vso = 300 m / s and k = 0.5, where Vso is reference S velocity at surface, K is gradient, and Z is depth, with a depth of 8,000 m and a lateral distance of 30 km. Vp is obtained based on the Vp / Vs relationship of 1.7 and the density is converted from the Vp value. A simple geological model can be seen in Figure 2.

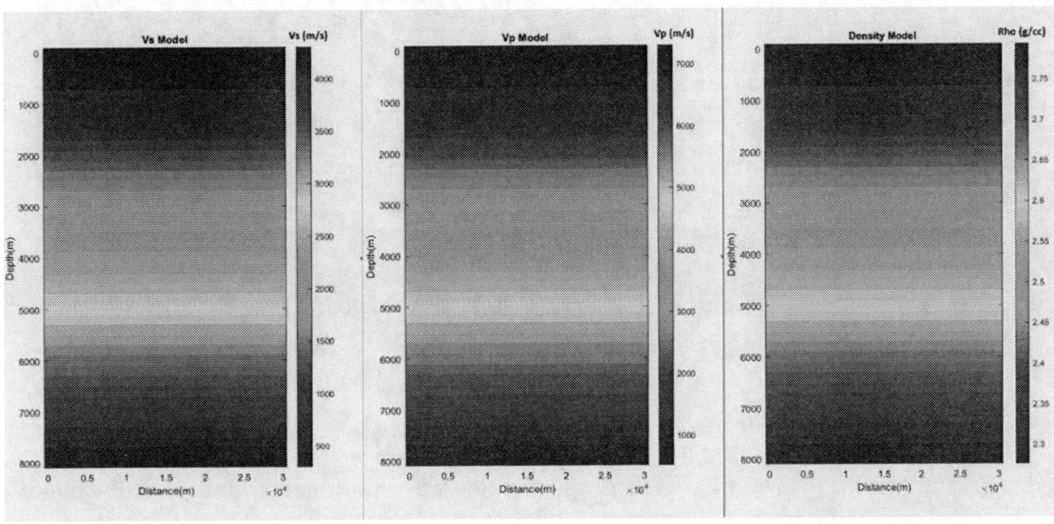

Figure 2. Simple flat layer geological model for Vs, Vp and density models, respectively.

2.1.2. Dispersion Curve Modeling

1D forward modeling dispersion curve is generated using the dispersion curve function obtained by mat_disperse referring to the algorithm developed by [8] and [9]. The 1D curve is modeled at each observation station with intervals of 1000 meters and a period of 0.5, 1, 2, 4, 5, 6, 8, 10, 15, 20, 30, 40, 50 and 100 seconds. The 1D dispersion curve at station 31 can be seen in Figure 3a and in Figure 3b plot the group velocity model at each station as a function of the period.

IOP Conf. Series: Earth and Environmental Science **873** (2021) 012096 doi:10.1088/1755-1315/873/1/012096

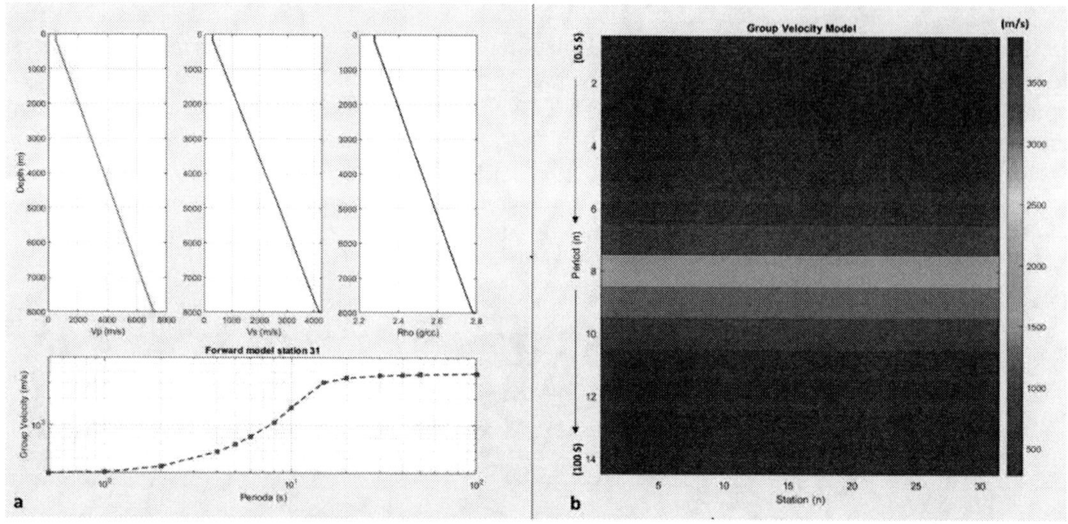

Figure 3. a) 1D forward modelling dispersion curve at station 31 and b) Group velocities as function of periods.

2.1.3. Travel Time Forward Modeling & Wavelet Convolution

Based on the group velocity model, the travel time for surface wave propagation is calculated based on the distance between stations. In Figure 4 is a plot of the wave propagation time as a function of the period.

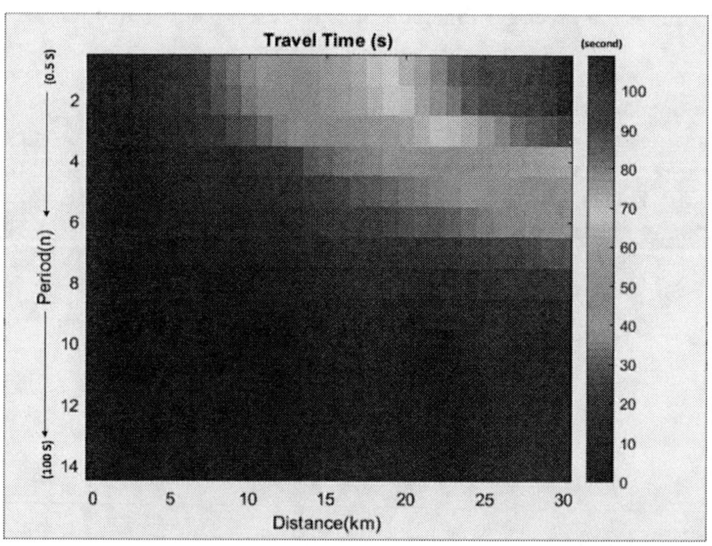

Figure 4. Calculated travel time surface wave propagation in seconds.

The arrival time of the surface waves is calculated based on the travel time surface wave model in each period and a certain distance which is convoluted with the wavelet of each period. Figure 5 shows the wavelets for a period of 100 seconds, 10 seconds, and 1 second, as well as the convolution results for each station.

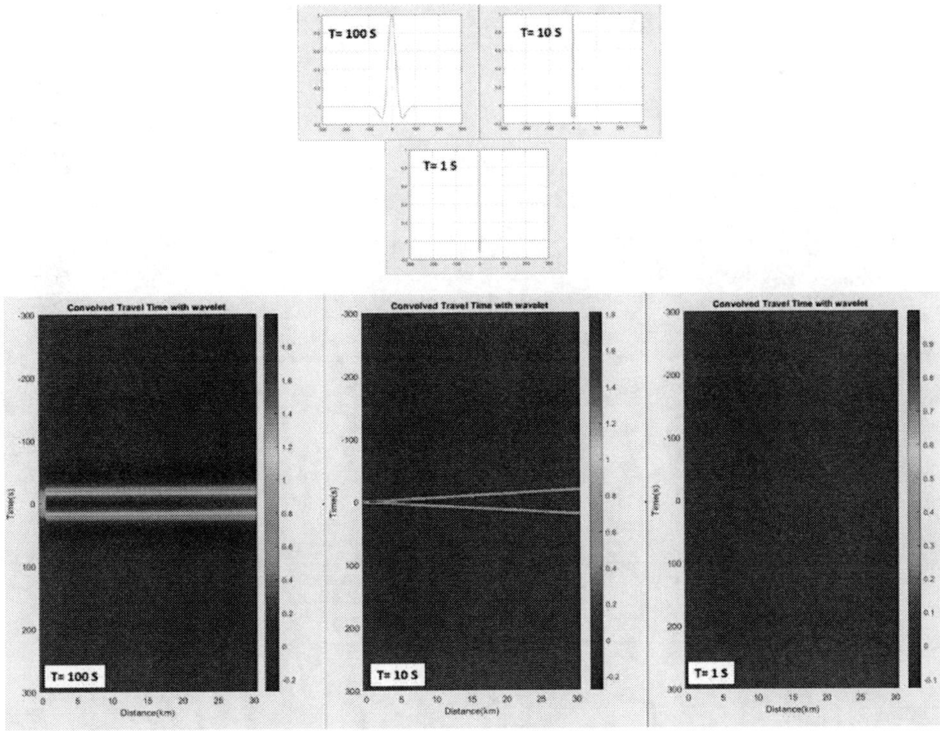

Figure 5. Convoluted wavelets with travel times for periods of 100s, 10s, and 1s, respectively in seconds.

2.1.4. Reverse Correlation and Stacking

The Green function extraction in data processing using the ambient seismic noise tomography method is usually approached by cross-correlation techniques, so that the forward modeling process is carried out in the opposite direction. The reverse cross-correlation process is carried out in the frequency domain by modifying the cross-correlation function in the frequency domain used [10]. In principle, the reverse cross-correlation process is to produce one input between two cross-correlated signals, results of convoluted wavelet with travel time in each period and distance considered as a result of cross-correlation. The reverse cross-correlation resulting desire signal which then considered to be synthetic ambient seismic noise data after stacking in all periods used. Figure 6 shows an example of the reverse cross-correlation results for the 1 second period. Figure 7 is the result of the inverting cross-correlation of each period which is then stacked to produce the desired signal as synthetic ambient seismic noise data.

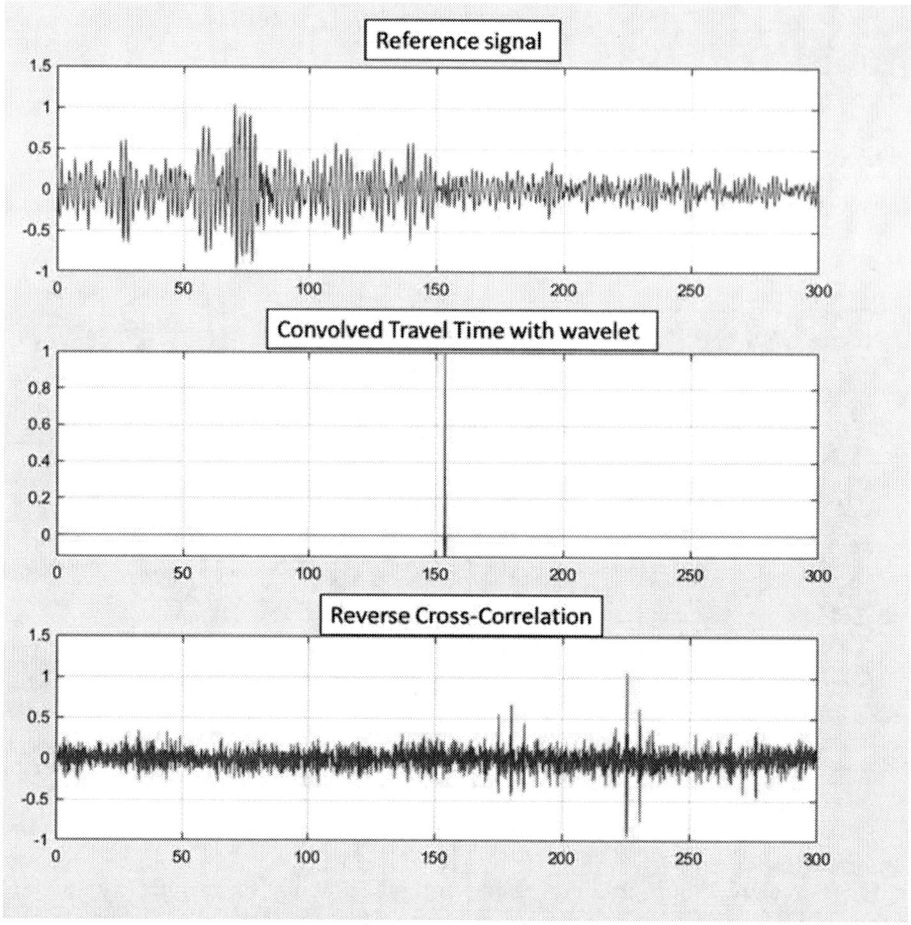

Figure 6. Reverse cross-correlation results using reference signal for period 1 second.

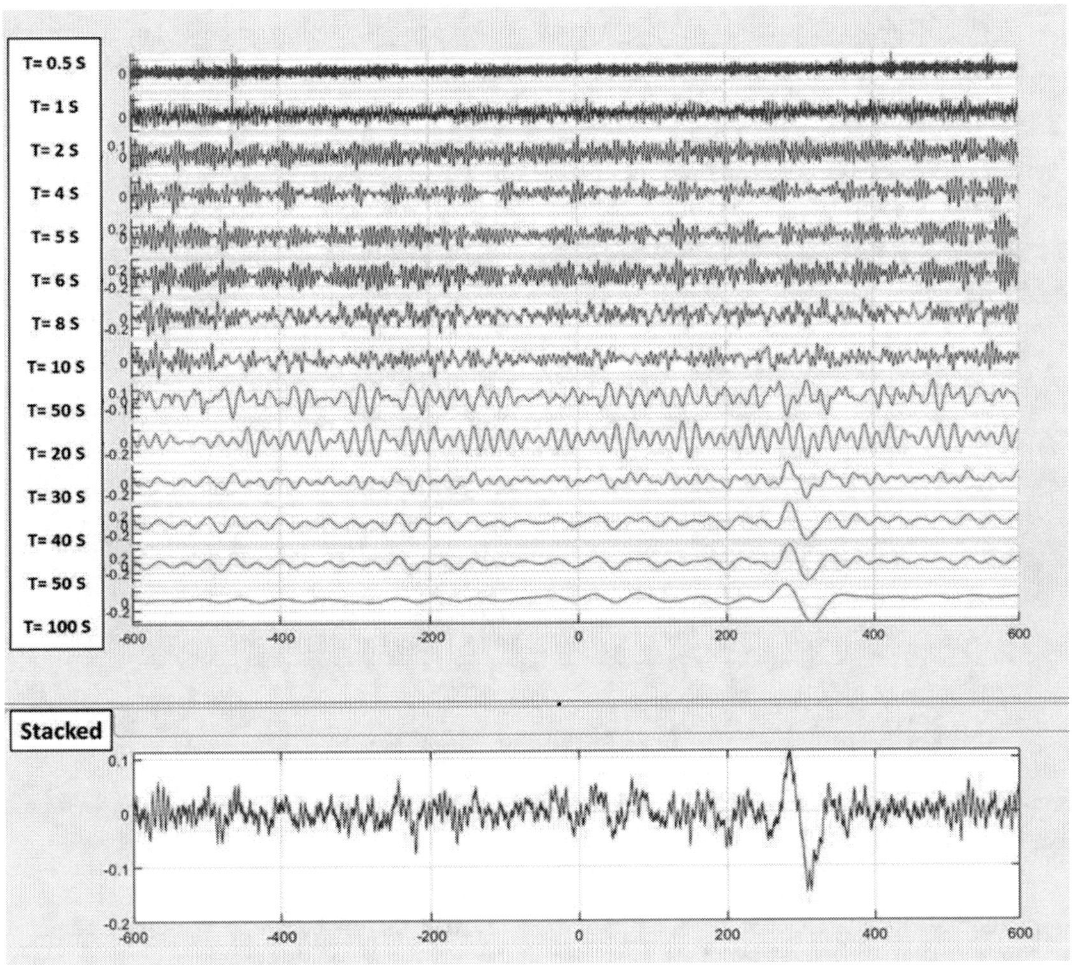

Figure 7. Staked signal all period used as synthetic ambient seismic noise data.

2.2. Vertical and lateral resolution

Both lateral and vertical resolution are important, using a simple analysis of the relationship between velocity and period we can find the lambda or wavelength in the depth domain and distance. Lateral and horizontal resolutions can be used with the 1/4 lambda approach. Table 1 shows the relationship between velocity, period, lambda, and resolution both vertically and horizontally.

IOP Conf. Series: Earth and Environmental Science **873** (2021) 012096 doi:10.1088/1755-1315/873/1/012096

Table 1. The relationship between velocity, periods, lambda, and resolution.

No	T (s)	Group Vel (m/s)	Λ (m)	Λ/4 (m)
1	0.5	275	138	34
2	1	281	281	70
3	2	328	656	164
4	4	482	1930	482
5	5	591	2957	739
6	6	722	4331	1083
7	8	1059	8472	2118
8	10	1579	15788	3947
9	15	3052	45756	11439
10	20	3448	68952	17238
11	30	3667	110106	27527
12	40	3748	149927	37482
13	50	3792	189611	47403
14	100	3872	387186	96796

3. Result

Based on the results of the dispersion curve synthetic modelling, it can be seen that the value of the surface wave group velocity varies according to the period, in a short period it gives a lower group velocity value than in the long period, the change in surface wave group velocity is significant for more than 10 seconds. Group velocity for each period using the mat_disperse algorithm can provide good information about subsurface conditions. Based on the analysis of the group velocity gradient as a function of the existing model, the effective period that provides good information is at 10 seconds, because the larger the 10 s gradient the group velocity is smaller than the 10 second period. The travel time for surface wave variations depends on the period, from the closest to the farthest distance for a period of 0.5 seconds from 0 to 109 seconds while in a 100 second period the travel time ranges from 0 to 8 seconds. The wavelet convolution approach with travel time can be used as a model of surface wave arrival time forwarding, which is a green function approach in processing ambient seismic noise data. The reverse cross-correlation method from the modeling results is promising in the process of producing the desired signal or synthetic data as if it were the result of field recording. Besides the reverse cross-correlation method as an effort to restore the green function, it seems that approaches with reverse deconvolution and reverse cross-coherence should be tried to produce the desired signal in synthetic data modeling. Based on the existing geological model and the maximum station distance of 30 km, the 30 second period becomes the maximum period that can best describe the arrival time of the surface waves from the convolution results with wavelets. A simple approach using the 1/4 wavelength function can direct us in determining survey designs such as the distance between recording stations that can be adjusted to the target depth, as well as determining the maximum period in the extraction of the green function when processing ambient seismic noise data.

IOP Conf. Series: Earth and Environmental Science **873** (2021) 012096 doi:10.1088/1755-1315/873/1/012096

4. Conclusion

The estimate target depth, vertical and lateral resolution depend on the recorded data periods and subsurface velocity model. Effective spacing between stations is strongly influenced by target depth, resolution, data period, and velocity value. Synthetic data modeling or forward modeling can be used as a basis in determining the acquisition design so that data can be recorded optimally in the field.

References

[1] Peterson J 1993 USGS Numbered Series Open-File Report 93-322
[2] Berger J, Davis P, and Göran E 2004 Journal of Geophysical Research: Solid Earth 109 B11
[3] McNamara D E 2004 Bulletin of the Seismological Society of America,.94 4 1517–27
[4] Bensen G, Ritzwoller M, Shapiro N, and Levshin A L 2005 AGU Fall Meeting
[5] Schönfelder S C, and Wegler U 2006 Geophysical Research Letters 33. 21
[6] Bensen G D, Ritzwoller M, Barmin M P, Levshin A L, Lin F, Moschetti M P, Shapiro N M, and Yang Y 2007 Geophysical Journal International 169 3 1239–60
[7] Slotnick M M 1941 Bulletin of the Seismological Society of America 31 2 173–75
[8] Hisada Y 1994 Bulletin of the Seismological Society of America 84 5 1456–72
[9] Lai C G 1998 PhD Georgia Institute of Technology
[10] Kalkan E 2007 MATLAB Central File Exchange

Elastic and Viscoelastic Models of Crust Deformation in Strike-slip Fault

Maulidia Ain Bening[1], David P. Sahara[2], Dian Kusumawati[2], Wahyu Triyoso[2], Sri Widiyantoro[2], Andri Dian Nugraha[2]

Geophysical Department, Faculty of Mining and Petroleum Engineering, ITB, Indonesia
Global Geophysics Research Group, Faculty of Mining and Petroleum Engineerinig, ITB, Indonesia

Corresponding author: maulidia.ain@gmail.com

Abstract. The crust, when viewed over a long period, moves towards one another. Crusts might experience sudden slip on a fault plane and caused fractures or cracks. There are three different types of faults, normal, reverse, and strike-slip faults. Induced stress due to sudden rupture on fault planes capable of creating stress and need to be measured quantitatively to comprehend the earthquake process. To understand the stress that occurs in strike-slip faults in the earth's crust, the previous researchers study the use of elastic materials as the material of the earth's crust, so that the earth crust's deformation is elastic. However, elastic material has linear stress and strain relationship that results in reversible deformations or returns to their original shape. This material is not suitable for modeling the earth's crust's long-term deformation, where the deformation of the earth's crust can be permanent, so a model is needed to solve this problem. In this study, we will compare the stress in the strike-slip fault in the upper crust with elastic materials, while the lower crust and upper mantle have viscoelastic materials compared to purely elastic materials through numerical simulations. This comparison is made to see the comparison between the two approaches with the earth's layers' actual state. The two models is chosen to represent the different failure processes of the earth crust, i.e. the elastic deformation part describes the response to stress in a short period, and the viscous deformation can explain the response over a more extended period. The study of both materials above is based on plate tectonic theory, in which the lithosphere plates will relatively move to each other because the layer material underneath is solid but can flow like a liquid for a long period.

Keywords: Elastic, viscoelastic, numerical simulations, crustal deformation, strike-slip fault

1. Introduction

Displacements on faults occur continuously at tectonic velocities of tens of millimeters per year When the crusts force overcomes the friction of the jagged edges of the fault, the stored up energy is released and will produce slip or sudden rupture on fault planes, causing more displacements to occur on faults during earthquakes and also fractures or cracks. A strike-slip fault is a fault along which the displacement is horizontal, which means no strain in y-direction [Turcotte and Schubert, 1982]. A variant of strike-slip fault plays a significant role on plate tectonics. For example, Great Sumatra Fault

(GSF), a 1650-km-long dextral strike-slip fault, is one of the strike-slip fault located in Indonesia. This fault zone poses major hazards, particularly to the highly populated areas on and around the active fault trace. Sumatra suffered from hundreds of earthquakes with Mw > 4 per year [Rafie et al., 2019]. The structure of strike-slip faults and its association with stress, strain, and displacement is an essential key to understanding intracontinental tectonic, including the process of the earthquake or its prediction and to define the seismic hazard along this fault.

The frictional strength of faults within the crust is an important factor that controls the earthquake mechanics; it governs the dynamics of earthquake nucleation and rupture propagation [Scholz, 2002; Kaneki and Hirono, 2019]. Several parameters affect the fault's frictional strength, such as normal stress, shear stress, rock type, fault orientation, displacement rate, coefficient of friction, and cohesion. Failure can occur along the fault plane when applied shear stress exceeds its resistance, particularly cohesion and coefficient of friction, which is remains the standard conceptual framework of faulting mechanics.

Strike-slip earthquakes and its deformation have been the subjects of intensive studies. Analog models (clay or sandbox experiments) are commonly used to study the long-terms evolution of faults. Still, they can not give quantitative constraints on the 3-D stress needed to understand the cause of crustal deformation [Dooley and Schreurs, 2012].

This study will model the tectonic loading in the strike-slip fault using 3-D numerical finite element code. This fault will be modeled on a crust with an elastic material as the upper crust overlying a visco-elastic material as the lower crust and upper mantle. These two materials are picked based on plate tectonic theory, where the lithosphere plates will relatively move to each other because the layer material underneath is solid but can flow like a liquid for a long period. For the elastic material, the material's behavior will produce reversible deformations or returns to their original shape due to their linear relationship between stress and strain. This kind of deformation can explain the response to stress in a short period. Meanwhile, the viscoelastic material yields viscous deformation that can explain the response over a more extended period [Stüwe, 2007; Bürgmann and Dresen, 2008].

2. Methodology
The process of measuring the perturbation of stress on a fault can be carried out using a numerical modeling method using the finite-element mesh. The finite-element mesh is a method used to divide a model into very small domains, which can be called elements. Element is an area where a series of equations will be applied. This equation represents the equation of the mesh (equation of interest) through a set of polynomial functions assigned to each element. Since these elements are made very small, the calculated solution will approximate the actual solution.

In this research, the 3-D finite-element mesh method is the right method for modeling the movement or deformation of shear faults in the earth's crust. This modeling is made by including several fault model parameters and crust material properties. This modeling is done using Coreform Cubit 2020.2 to generate the mesh model and Pylith [Aagaard *et al.*, 2013, 2017] to produce modeling of the earth's crust with certain parameters.

3. Results and discussion
This research is on-going research. This study has already succeed in modeling the effect of rock strength parameters on the displacement around the fault in elastic materials. The parameters that are being studied are cohesion and coefficient of friction.

The mesh used in this modeling is a hexahedron mesh with a length of 6 km, a width of 6 km, and a height of 4 km (Figure 1). This mesh consists of two layers, namely upper crust and lower crust, and a fault as an interface with model parameters that support this material, such as density of a 2500 kg/m^3, velocity of compressional wave (Vp) 3000 m/s and velocity of shear wave (Vs) 5291.5026 m/s.

Figure 1. Numerical mesh of the finite element model.

The boundary conditions applied to this mesh are quasi-static Dirichlet boundary conditions. Axial displacements are applied on the –x and + x faces to maintain normal stress on the fault, and velocity displacements are applied in –y and + y directions on the same faces to yield a left-lateral sense of shear. This modeling uses a static friction fault with parameters in the form of a friction coefficient of 0.6 [Byerlee, 1978] and cohesion of 2 MPa [Andrews, 2005] to cause slip on the fault. This model was simulated for 200 years. From the result of modeling with applied parameters, it is known that the fault will generate slip at a period of 125 years (Figure 2).

Figure 2 Numerical model of strike-slip fault with coefficient of friction 0.6 and cohesion 2 MPa slip at a period of 125 years.

Observations from this modeling are done by varying the value of the coefficient of friction and cohesion against the strike-slip fault to see the distribution of the displacement resulting from the modeling. From the variation of the coefficient of friction carried out on the crust model, it is known that the higher the coefficient of friction used, the longer the stress accumulates in the fault, which results in slip so that the slip that occurs will also take longer. This is shown in Figure 3, where each model with different coefficient of friction produce slips in different period. Model with friction coefficient 0.6 will slip at period 125 years, model with friction coefficient 0.85 will slip at period 145 years, model with friction coefficient 0.2 will slip at period 85 years and model with friction coefficient 0.4 will slip at period 105 years.

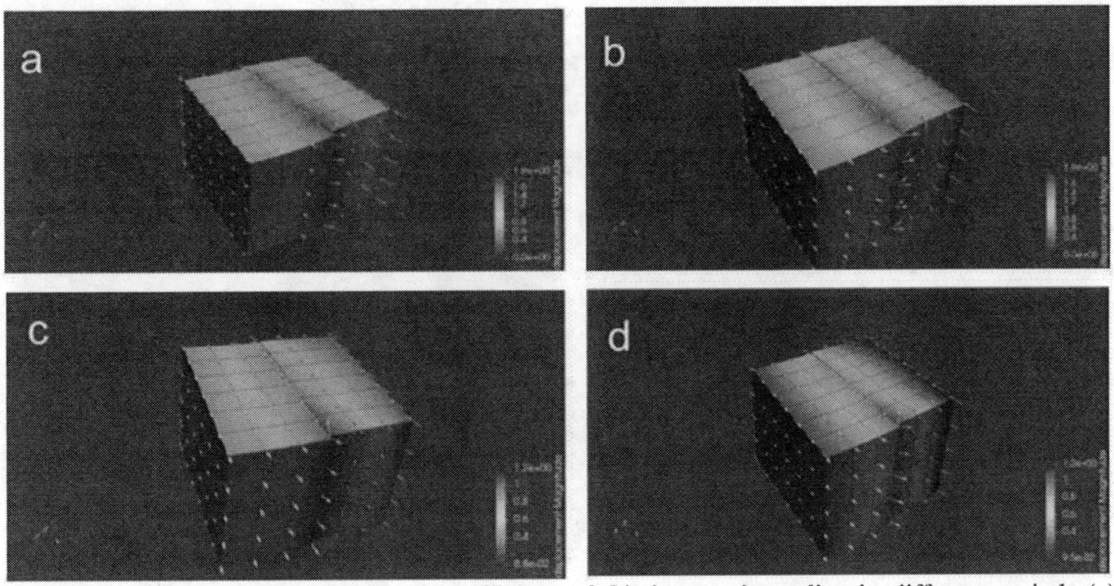

Figure 3. Each model with different coefficient of friction produce slips in different period: (a) Friction coefficient 0.6: 125 years, (b) Friction coefficient 0.85: 145 years, (c) Friction coefficient 0.2: 85 years, (d) Friction coefficient 0.4: 105 years.

In addition, this research also varied the cohesion value used. The higher the cohesion value in a rock, the longer the stress accumulation in the fault can produce slip, so that rocks that have a lower cohesion value will experience slip first than rocks with a higher cohesion value. This is shown in Figure 4, where each model with different cohesion produce slips in different period. Model with cohesion 2 MPa will slip at period 145 years, model with cohesion 5 MPa will slip at period 185 years, and model with cohesion 1 MPa will slip at period 105 years.

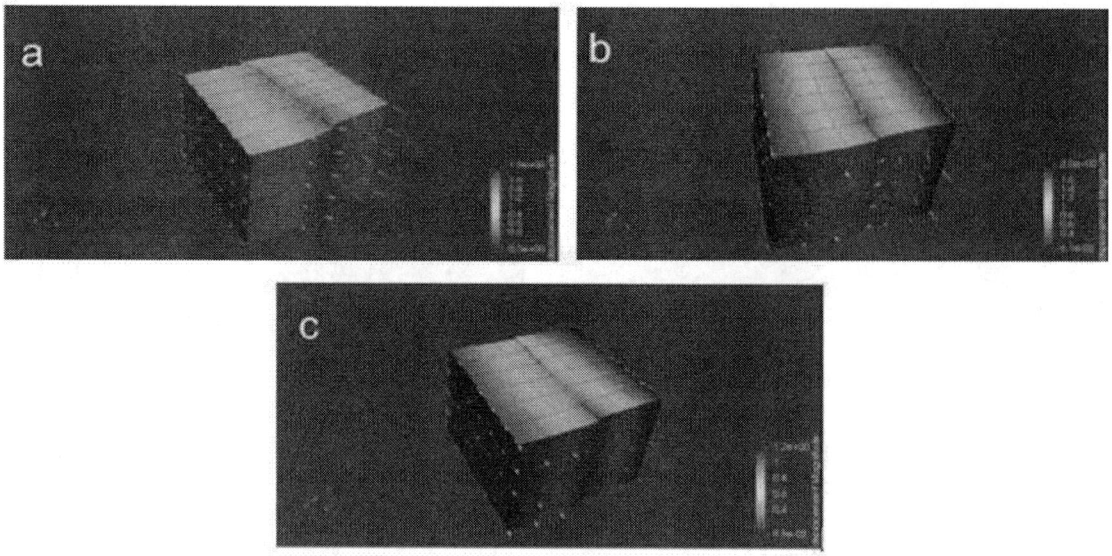

Figure 4. Each model with different coefficient of friction produce slips in different period: (a) Cohesion 2 MPa: 145 years, (b) Cohesion 5 MPa: 185 years, (c) Cohesion 1 MPa: 105 years.

4. Conclusions

We model a simple numerical modeling represent the tectonic load at the model boundary and the frictional strength of the fault. Model of strike-slip fault with coefficient of friction 0.6 and cohesion 2 MPa will slip in the period of 125 years. The variation of frictional strength parameter affects the fault reactivation, the higher the coefficient of friction and cohesion is used, the longer the stress accumulates in the fault which result in the longer slip of the fault will occur. The further study of this research is to apply the modeling concept developed in this study to explain the observed major events along Great Sumatra Fault (GSF).

References

[1] Aagaard B T, Knepley M G, and Williams C A 2013 A domain decomposition approach to implementing fault slip in finite-element models of quasi-static and dynamic crustal deformation *Journal of Geophysical Research: Solid Earth* **118**(6) 3059-3079

[2] Aagaard B, M Knepley, and C Williams 2017 *PyLith v2.2.1* Davis, CA: Computational Infrastructure of Geodynamics. DOI: 10.5281/zenodo.886600.

[3] Andrews D J 2005 Rupture dynamics with energy loss outside the slip zone *Journal of Geophysical Research: Solid Earth* **110**(B1)

[4] Bürgmann R and Dresen G 2008 Rheology of the lower crust and upper mantle: Evidence from rock mechanics, geodesy, and field observations. *Annu. Rev. Earth Planet. Sci.*, **36** 531-567.

[5] Byerlee J 1978 Friction of rocks In *Rock friction and earthquake prediction* pp. 615-626 Birkhäuser, Basel

[6] Dooley T P and Schreurs G 2012 Analogue modelling of intraplate strike-slip tectonics: A review and new experimental results *Tectonophysics* **574-575** 1-71

[7] Kaneki S and Hirono T 2019 Diagenetic and shear-induced transitions of frictional strength of carbon-bearing faults and their implications for earthquake rupture dynamics in subduction zones *Scientific reports* **9**(1) 1-13

[8] Rafie M T, Sahara D P, Widiyantoro S, and Nugraha A D 2019 Impact of the 2004 sumatra-andaman earthquake to the stress heterogeneity and seismicity pattern in Nothern Sumatra,

Indonesia In *IOP conference series: earth and environmental science* (IOP Publishing) **318** No 1 p 012010
[9] Scholz C H Te 2002 *Mechanics of Earthquakes and Faulting* (Cambridge university press)
[10] Stüwe K 2007 *Geodynamics of the lithosphere: an introduction*, Springer Science & Business Media.
[11] Turcotte D L, & Schubert G 2002 *Geodynamics* (Cambridge university press)
[12] Wang H, Liu M, Ye J, Cao J, Jing Y 2017 Strain partitioning and stress perturbation around stepovers and bends of strike-slip faults: numerical results *Tectonophysics* **721** 211-226

Hypocenter Determination of Volcano-Tectonic (VT) Earthquake around Agung Volcano in Period October-December 2017 Using a Non-Linear Location Method: A Preliminary Result

P P Rahsetyo[1], D P Sahara[2], A D Nugraha[2], D K Syahbana[3], Zulfakriza[2], Ardianto[1], Martanto[3], H Afif[3], A W Baskara[1], I B Hermawan[1] and A D Sasongko[1]

[1]Geophysical Engineering Study Program, Faculty of Mining and Petroleum Engineering, Bandung Institute of Technology, Bandung Indonesia
[2]Global Geophysics Research Group, Faculty of Mining and Petroleum Engineering, Bandung Institute of Technology, Bandung Indonesia
[3]Center for Volcanology and Geohazard Mitigation, Geological Agency, Ministry of Energy and Mineral Resources, Bandung Indonesia

E-mail: puputrahsetyo@gmail.com

Abstract. Agung is one of active volcanoes in Indonesia, located on island of Bali. Since 1963, Agung has not had significant activity, until in September 2017 the volcano was active again which was marked by increased seismic activity and eruptions in November 2017. Therefore, to analyze the dynamics and processes of active volcanic eruptions requires an understanding of the structure of the volcano, especially the position of the magma reservoir and its path. The depiction of the structure of this volcano can be analyzed by determining the location of the earthquake due to volcanic activity, especially Volcano-Tectonic (VT) earthquake. In this study, we determined the location of the hypocenter around the Agung using the non-linear location method. VT earthquakes have similar characteristics to tectonic earthquakes so this method can be used to determine the initial hypocenter. The data used in this study came from 8 PVMBG seismographs from October to December 2017. We manually picking arrival time of P- and S-waves from the 3948 VT events found. Pair of P and S wave phases with 18741 P-wave phases and 17237 S-wave phases, plotted in a wadati diagram resulting in a vp/vs ratio of 1.7117. We use 1D velocity models derived from Koulakov with the assumption that the geology of the study area is not much different from the volcanoes in Central Java. The resulting hypocenter distribution shows a very random location and has uncertain X, Y, and Z directions from a range of 0 to 91 km. This study limits this uncertainty to 5 km resulting in a more reliable earthquakes distribution of 3050 events. The results indicate 2 clustered events, a swarm of VT events that occur every month at a depth of 8 to 15 km and there are 2 paths that lead to the top of Agung and SW of that swarm. These preliminary results will be used to update 1D velocity model and relocate the events beneath Agung region for further studies.

1. Introduction

Agung is one of active volcanoes in Indonesia, located on Bali Island, and became active in September 2017 and erupted again in November 2017 after dormant since 1963. The eruption in 1963-1964 was one of the biggest eruptions in Indonesian history after the Krakatoa eruption in 1883, especially since the eruption affected the global climate [1–5]. The explosive eruption column reached more than 20 km above the summit of Agung to penetrate the tropopause and cause injection of ash and sulfur gas in the stratosphere [5]. Influence of eruption results in this stratosphere can reduce climate temperatures around -0.3° C[4]. Agung eruption can occur as large as in 1963 if the composition of magma and eruption type is still the same [4]. However, the eruption in 1963 did not have specific data information about its eruption, due to limited tools. Therefore, Agung needs to be studied more deeply about the nature and process of its dynamics apart from the previous eruption comparison.

To understand the dynamics and eruption of active volcanoes requires an understanding of the structure of volcanoes, especially the position of the storage and the path of magma. It has been obtained several studies on this volcano such as research on tomography[6], geology[7], satellite imaginary [8], activity reports[9], as well as determining the location of the hypocenter swarm with regional network[10]. However among the previous studies did not explain in detail for the seismicity during the initial activity in 2017 on this volcano. So that in this study we want to determine the location of the hypocenter of the seismicity that occurred before and during the 2017 eruption using the local network in period October - December 2017. Determination of the initial hypocenter using the non-linear location method[11].

2. Data and Methods: Non-Linear Location Determination

The data used in this study are earthquake waveform data recorded by 8 seismometers from the Center for Volcanology and Geological Disaster Mitigation (PVMBG) in the period 18 October- 31 December 2017. The distribution of this seismometer is around Agung (Figure 1). The earthquake used in this study was volcanic earthquake type A (VTA) and B (VTB).

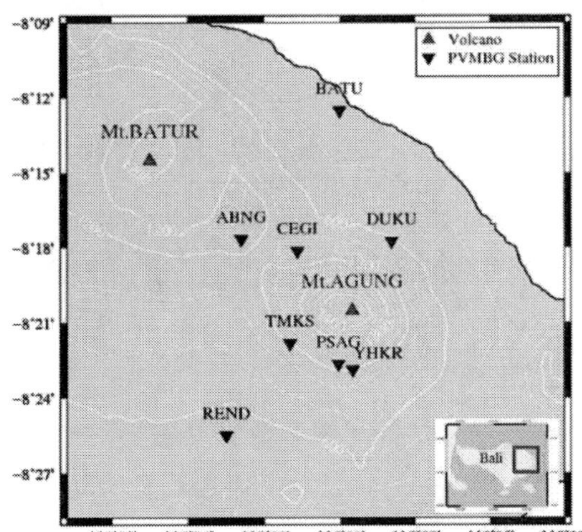

Figure 1. Distribution of seismic stations around Agung

In this study, we manually picked arrival time of P and S-waves uses Seisgram2k [12] and determination of the hypocenter location uses the non-linear location method with the NonLinLoc (NLLoc) program [11,12] and uses a 1D Koulakov seismic velocity model [13]. NLLoc implements a probabilistic inversion from Tarantola and Valette [14] and produce Posterior Density Function (PDF) using Oct-Tree sampling algorithm to estimates for hypocenter locations [11]. Oct-Tree works by dividing the data area in several three-dimensional (3D) cells recursively, then the hypocenter position

IOP Conf. Series: Earth and Environmental Science **873** (2021) 012098 doi:10.1088/1755-1315/873/1/012098

is determined by calculating the PDF value of each cell[12]. The largest PDF value of the cells is then taken, and then divided again into several small cells, until finally obtained the greatest probability at a hypocenter location solution. We use a grid size of 1520 x 1520 x 420 with the node for each grid of 0.25.

3. Results, Discussions and Conclusions

The initial step of this study is to look for VT events that occurred during the period 18 October to 31 December 2017. VT events have the same type as tectonic earthquakes so that P- and S- waves can be determined. From the determination of this event, manual picking will be carried out to obtain 3948 VT events with a VTA that has 17,064 pairs of P and S that can be plotted in the Wadati diagram (Figure 2). This diagram is used to examine the linear relationship and time quality P and S, the mean Vp/Vs of the earth's crust is $\sqrt{3}$ or close to 1.73. Based on the Wadati diagram obtained, 1.71, it shows that the arrival times that we observed are good enough to locate the hypocenters.

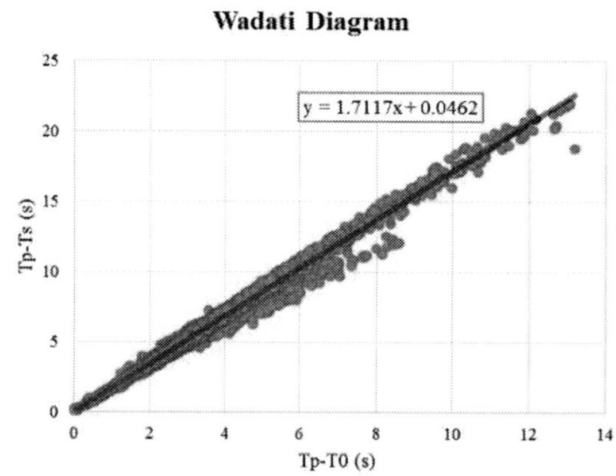

Figure 2. Wadati diagram of the picked phase of each event that has been obtained. The black line indicates Vp/Vs ratio.

The results of determining the location of the hypocenter with NLLoc provide a location with various kinds of errors and uncertainty based on the probability used. The results of the uncertainty show a far range, from 0.1058 to 91.39891 km (Figure 3). This uncertainty can be caused because the search for grid locations is still very far away so that the PDF will be very low and the uncertainty will be very high. Therefore, in order to reduce the error in determining the location, we will select data with uncertainty below 5 km to get the distribution of the number of VT events over time in Figure 4.

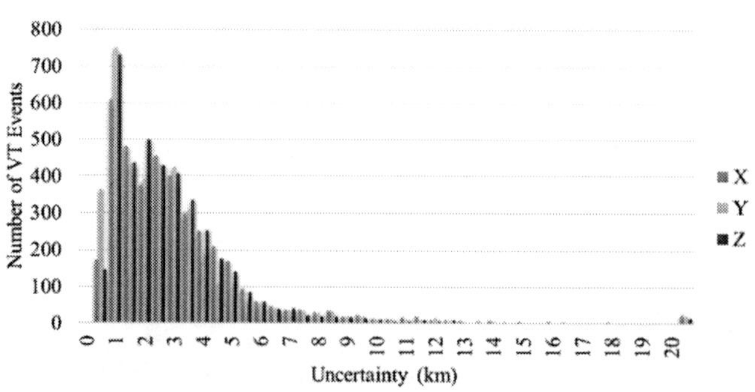

Figure 3. Uncertainty histogram for determining Nonlinloc locations based on X (green), Y (yellow), and Z (blue) directions.

Apart from uncertainty, this 3050 events errors in determining the location can also be based on the residual travel time (Figure 5). It can be seen in the figure that all events have a very large travel-time residual range, and reduced in events with uncertainty below 5 km. The travel-time residual above 0.2 s and below 0.2 s the amount is reduced. So that errors in determining the location are also reduced.

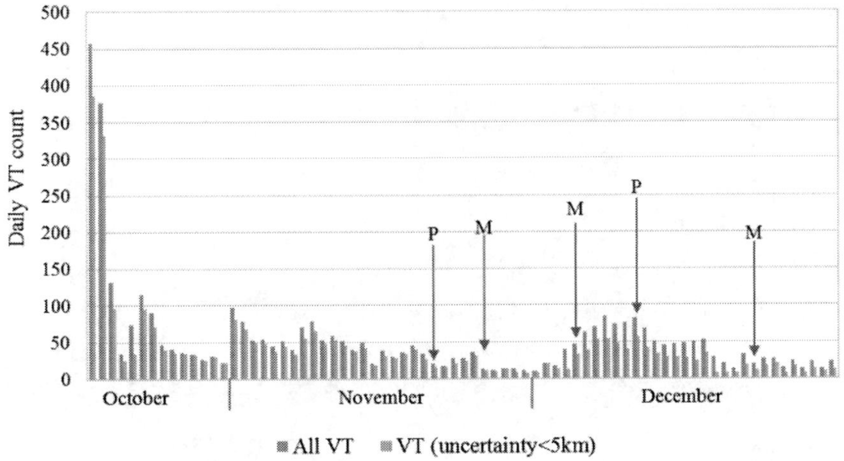

Figure 4. Distribution of the number of VT events in a time function. All VTs in blue, for VTs under 5 km it is yellow. P is a phreatic eruption and M is a magmatic eruption.

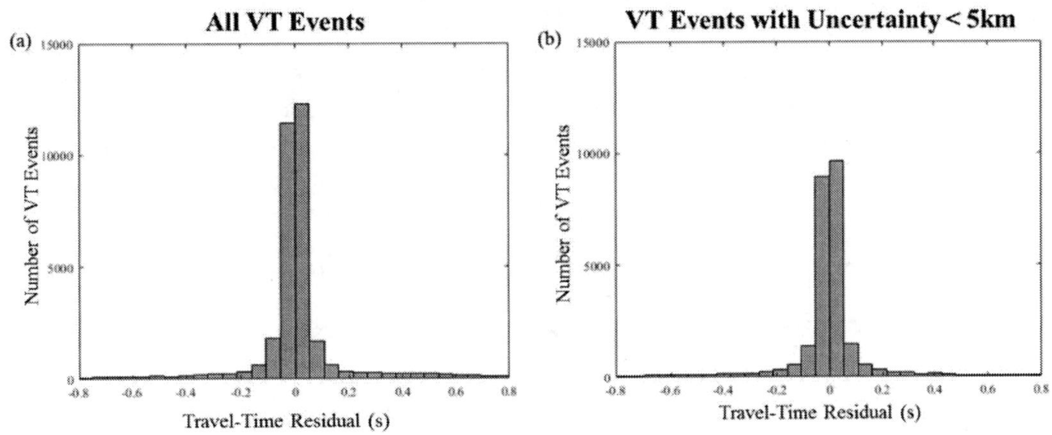

Figure 5. Travel-time residual from hypocenter determination with NLLoc for all VT events (a) and VT events with uncertainty below 5 km (b).

Figure 6 shows the distribution of VT events in Agung during the period 18 October to 31 December 2017. These events are grouped by month, which is clear in October, it can be seen that swarm in the west of Agung. This month is the month before the eruption, which is the opening of the way for magma to rise to the surface. But this month not only happen to the top of Agung, but also located on the west side of Agung, which has a direction of SW (Figure 7).

The results of this study obtained a hypocenter distribution which adequately describes the displacement of magma from October to December 2017 (Figure 7). It is seen that there is a swarm of VT earthquakes at a depth of 8 to 15 km which every month always occurs in that area. And there are 2 different paths, the first one that goes to the top and the next one that goes to the SW direction of that swarm.

Based on our study, we identified the hypocenter location of 3050 events around Agung by applying a non-linear method with uncertainty under 5 km that distributed in 2 clustered, a swarm of VT events that occur every month at a depth of 8 to 15 km and there are 2 paths lead to the top of Agung and SW of that swarm. However, this result is still a lot of earthquakes that scatter, especially in November and December. So that further research is needed to get a more definite hypocenter. This preliminary result will be used to update 1D velocity model and relocate the events beneath Agung region for further studies.

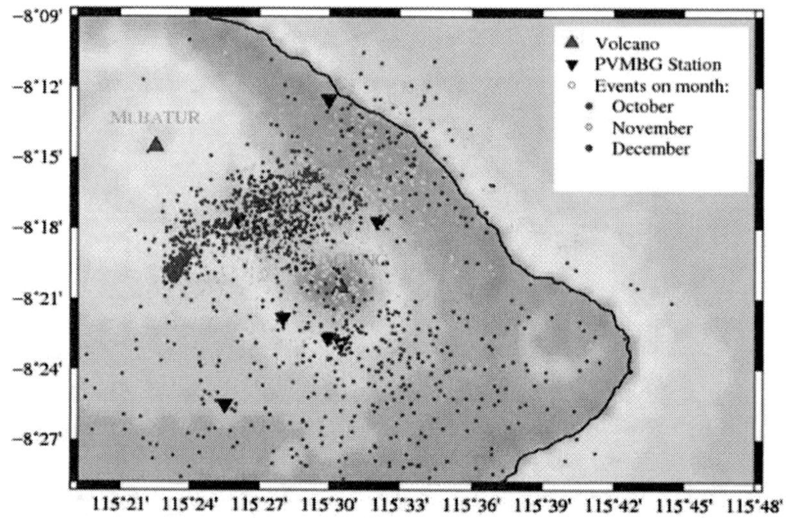

Figure 6. Distribution of VT events during the period 18 October to 31 December 2017. Red dots are VT events in October, yellow dots are events in November and blue dots are events in December. Red triangles are volcanoes and black inverted triangles are PVMBG stations.

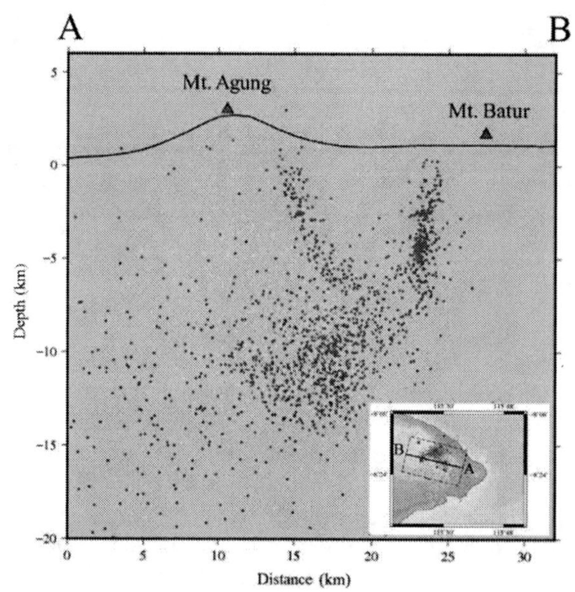

Figure 7. VT events distribution around Agung on the A-B cross line. Red dots are VT events in October, yellow dots are events in November and blue dots are events in December. Red triangles are volcanoes.

Acknowledgments
The authors wish to thank the Center for Volcanology and Geological Hazards Mitigation (PVMBG) for their permission to uses this data. Thanks to Bandung Institute of Technology (ITB) and Partnerships for Enhanced Engagement in Research (PEER) team for their support and assistance so far. Thank to Prof. Anthony Lomax for NLLoc and Seisgram2k, the tools we use to process data. And we like to thank Paul Wessel and Walter H.F. Smith for the Generic Mapping Tools (GMT) that we use the program to made all the figure in this study.

References

[1] Cadle R D, Kiang C S and Louis J-F 1976 The global scale dispersion of the eruption clouds from major volcanic eruptions *J. Geophys. Res.* **81** 3125–32

[2] Hansen J E, Wang W-C and Lacis A A 1978 Mount Agung Eruption Provides Test of a Global Climatic Pertubation *Science (80-.).* **199** 1065–8

[3] Rampino M R and Self S 1982 Historic eruptions of Tambora (1815), Krakatau (1883), and Agung (1963), their stratospheric aerosols, and climatic impact *Quat. Res.* **18** 127–43

[4] Self S, Rampino M R and Barbera J J 1981 The possible effects of large 19th and 20th century volcanic eruptions on zonal and hemispheric surface temperatures *J. Volcanol. Geotherm. Res.* **11** 41–60

[5] Self S and Rampino M R 2012 The 1963-1964 eruption of Agung volcano (Bali, Indonesia) *Bull. Volcanol.* **74** 1521–36

[6] Zulfakriza Z, Nugraha A D, Widiyantoro S, Cummins P R, Syahbana D K, Priambodo I C, Martanto M, Ardianto A, Oliver J and Hammond S 2020 Tomographic Imaging of the Agung-Batur Volcano Complex , Bali , Indonesia , From the Ambient Seismic Noise Field **8** 1–11

[7] Fontijn K, Costa F, Sutawidjaja I, Newhall C G and Herrin J S 2015 A 5000-year record of multiple highly explosive mafic eruptions from Gunung Agung (Bali, Indonesia): implications for eruption frequency and volcanic hazards *Bull. Volcanol.* **77**

[8] Albino F, Biggs J and Syahbana D K 2019 Dyke intrusion between neighbouring arc volcanoes responsible for 2017 pre-eruptive seismic swarm at Agung *Nat. Commun.* **10**

[9] Syahbana D K, Kasbani K, Suantika G, Prambada O, Andreas A S, Saing U B, Kunrat S L, Andreastuti S, Martanto M, Kriswati E, Suparman Y, Humaida H, Ogburn S, Kelly P J, Wellik J, Wright H M N, Pesicek J D, Wessels R, Kern C, Lisowski M, Diefenbach A, Poland M, Beauducel F, Pallister J, Vaughan R G and Lowenstern J B 2019 The 2017–19 activity at Mount Agung in Bali (Indonesia): Intense unrest, monitoring, crisis response, evacuation, and eruption *Sci. Rep.* **9** 1–17

[10] Gunawan M T, Kusnandar R, Supendi P, Nugraha A D, Puspito N T and Sahara D P 2020 Analysis of swarm earthquakes around Mt . Agung Bali , Indonesia prior to November 2017 eruption using regional BMKG network *Geosci. Lett.*

[11] Lomax A and Curtis A 2001 Fast, probabilistic earthquake location in 3D models using Oct-Tree Importance sampling *Geophys. Res. Abstr* **3** 955

[12] Lomax A, Virieux J, Volant P and Berge-Thierry C 2000 Probabilistic Earthquake Location in 3D and Layered Models 101–34

[13] Koulakov I, Jakovlev A and Luehr B G 2009 Anisotropic structure beneath central Java from local earthquake tomography *Geochemistry, Geophys. Geosystems* **10**

[14] Tarantola A and Valette B 1982 Inverse Problems = Quest for Information *J. Geophys.* **50** 159–70

Analysis of Hypocenter Relocation by Using Double Difference Method in a Deep Underground Mining

Wahyu Hidayat[1,4], **David P. Sahara**[3], **Sri Widiyantoro**[3], **I Putu Raditya Ambara Putra**[2], **Nabiel H. Shihab**[2], **Rizka Amalia**[2], **Azhar Harisandi**[2], and **Suharsono**[4]

[1]Doctoral Program of Geophysical Engineering, Faculty of Mining and Petroleum Engineering, Institut Teknologi Bandung
[2]Master Program of Geophysical Engineering, Faculty of Mining and Petroleum Engineering, Institut Teknologi Bandung
[3]Global Geophysics Research Group, Faculty of Mining and Petroleum Engineering, Institut Teknologi Bandung
[4]Geophysical Engineering, Facultyof Mineral Technology, UPN "Veteran" Yogyakarta

Corresponding author: hidayat18@gmail.com

Abstract. Hypocenter relocation is one of the keys to success in the analysis of seismicity induction in underground mines. Overburden thickness, topography, geological complexity, and mining activities can result in newly induced seismicity that can endanger the safety of underground mine workers. The relatively narrow underground mine area requires the most accurate hypocenter location information possible. The double-difference algorithm approach is one of the keys to overcoming this problem. The double-difference method is a relative location method that tries to minimize the residuals between the observed and calculated travel time differences for pairs of microseismic events at each station, by adjusting the differences between all pairs of events at each station repeatedly. In this study, we utilized microseismic measurement data in the deepest underground mine in Indonesia. A total of 1783 seismic events were successfully relocated. The relocation results show the rock mass stress which is illustrated by the distribution of events around the cave, especially the abutment area and underground mining tunnels.

Keywords: Induced Seimicity, Double-difference method, Underground mining

1. Introduction

Microseismic monitoring is a method of underground seismic investigation based on the detection of the location of earthquakes with small magnitudes that occur in rocks due to natural or artificial processes [2]. Underground mining seismic monitoring system can explain the process of damage to rock mass and rock failure due to mining-induced and geological structures as well as monitoring tunnel development evolution and cave evolution [1]. Monitoring of seismic activity in underground mining areas is very necessary so that the evacuation and handling process carried out when a seismic event that is dangerous to the safety of workers can be carried out quickly and measured.

The determination of the hypocenter location is the first step in seismological studies [2]. Determination of the location of an earthquake hypocenter with high accuracy is required to analyze velocity structures, identify fault zones, microfracture distribution, and orientation as well as to analyze both global and local seismicity. For these purposes, the relocation of an earthquake hypocenter is very important. To get an accurate location, especially a local scale in underground mines, an efficient and accurate location determination algorithm is required. The Double - Difference method is a method that can relocate earthquakes simultaneously [4] [5] [6]. This method works by pairing earthquakes and utilizing the time difference between the two events to obtain a more accurate hypocenter location. The principle of this method is that if the distance between the two events (a pair of events) that are paired is relatively small compared to the distance to the observing station, then the ray path and waveform of the two events can be considered almost the same. With this assumption, the difference in travel time between the two events recorded at the same station is considered a function of the distance between the two hypocenters [6].

2. Data and methodology

The data used in this research is microseismic catalog data for one month (January 2019) that was recorded by 74 stations from the underground mining in Indonesia. The microseismic catalog data consists of the date, time of occurrence, station, magnitude, depth, and uncertainty of each of these parameters. Total microseismic events were 1783 microseismic events with magnitudes from -1.3 to 1.5 Mw. The microseismic catalog data then calculated the travel time and determined the pair event using ph2dt. Based on the results of ph2dt, 1765 events were selected with a total phase of 2994. The P-wave phase selected pairs as many as 49910 out of 57991 or 86%, while for the S-wave phase, 24943 pairs were selected out of 28752 (86%). There is 1 cluster, the microseismic event is around the mine operation (cave). The initial velocity model used is a 3-Dimensional velocity model. 3-Dimensional velocity model is obtained from geological information. The distribution of the magnitude of the research area is shown in Figure 1.

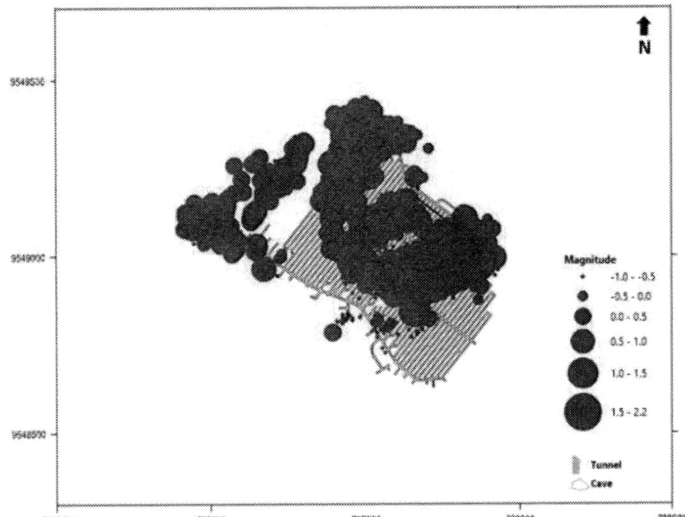

Figure 1. Distribution of the magnitude of the seismic event at research area

The double-difference method assumes that if the distance between two adjacent event sources is much smaller than the distance to the station and has a heterogeneity of velocity which is considered the same, then the waves from both sources are considered almost the same [4]. Figure 2 shows an illustration of two event sources *i* and *j* which are close and relatively far from the recording sensors *k* and *l*.

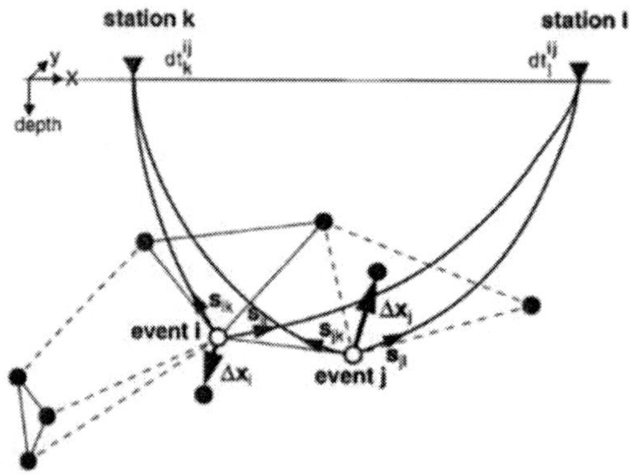

Figure 2. Double-difference methode ilsutration [4].

The arrival time of the *T* wave from the event source *i* to an event station is expressed by the wave-ray theory,

$$T_k^i = \tau^i + \int_i^k u \, ds \tag{1}$$

where τ^i is the event occurrence time from the source *i*, *u* is the slowness vector, and ds is the partition of the wave raypath. The relationship between the arrival time and the source location is not very linear, therefore linearization is carried out by calculating the misfit between the observed data and the predicted arrival timer_k^i.

$$r_k^i = \sum_{l=1}^{3} \frac{\partial T_k^i}{\partial x_l^i} \Delta x_l^i + \Delta\tau^i + \int_i^k \delta u \, ds \tag{2}$$

Equation 2 is substituted with the occurrence of source *j* received by station *k*, then it is obtained.:

$$r_k^i - r_k^j = \sum_{l=1}^{3} \frac{\partial T_k^i}{\partial x_l^i} \Delta x_l^i + \Delta\tau^i + \int_i^k \delta u \, ds - \sum_{l=1}^{3} \frac{\partial T_k^j}{\partial x_l^j} \Delta x_l^j - \Delta\tau^j - \int_i^j \delta u \, ds. \tag{3}$$

By assuming the two sources are so close that the light lines of the waves are nearly equal and cancel each other out, equation 3 can be simplified to:

$$r_k^i - r_k^j = \sum_{l=1}^{3} \frac{\partial T_k^i}{\partial x_l^i} \Delta x_l^i + \Delta\tau^i - \sum_{l=1}^{3} \frac{\partial T_k^j}{\partial x_l^j} \Delta x_l^j - \Delta\tau^j \tag{4}$$

where $r_k^i - r_k^j$ is a double – difference [4]. Double - difference is the time difference between the observed data and the calculated data between two adjacent event sources which can be written as [6]:

$$r_k^i - r_k^j = (T_k^i - T_k^j)^{obs} - (T_k^i - T_k^j)^{calculation}. \tag{5}$$

The difference in arrival time $(T_k^i - T_k^j)^{obs}$ can be obtained through waveform cross-correlation or from absolute time difference from event catalog data.

3. Results and discussion

The data used in data processing in this study include data absolute travel time at each station, difference time data for each pair of events (cross-correlation difference time or catalog difference time), station coordinate data, earthquake catalog data, and initial velocity model. [5] [6]. Determination of the initial velocity model is carried out by utilizing geological data information based on the results of logging and coring of rocks in underground mines vertically and horizontally in 3 dimensions. Data selection is done by making a plot between the travel time and the distance traveled and based on rock velocity information from rock test results in the laboratory. The difference time data for each pair of event events is in the form of cross-correlation difference times or catalog difference times which are then used in the inversion process using the TomoDD program package [6]. Figure 3a is a histogram of the residual travel time before being relocated, while figure 3b is a histogram that has been relocated. The residual RMS distribution before being relocated was in the range of -3 ms to 2 ms. The histogram after relocation shows a high-frequency distribution value and is close to 0 with a value range of -2 ms to 2 ms, which shows that the microseismic relocation is statistically better than before the relocation. To determine the distance of the shifting of the microseismic before and after being relocated, a histogram of the shift versus the amount of data and a rosette diagram was made to determine the dominant direction of distribution (Figures 4 and 5).

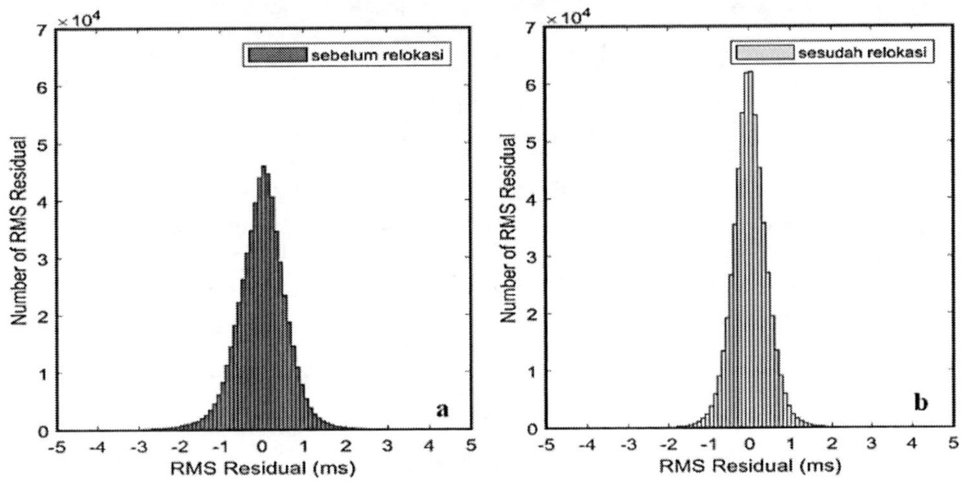

Figure 3. a. Histogram of microseismic residual RMS before being relocated, b. after being relocated

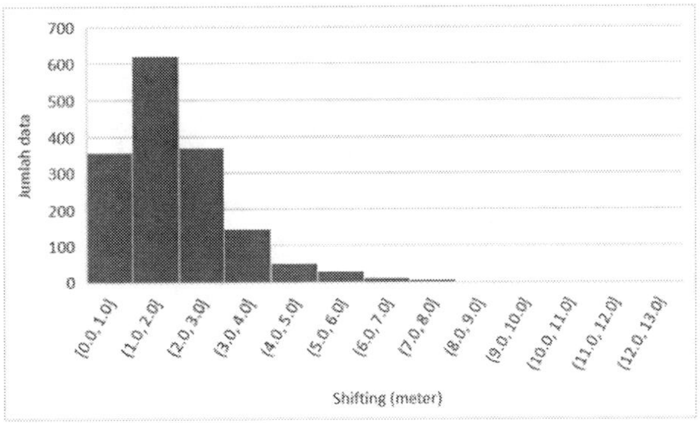

Figure 4. Histogram of shifting before and after being relocated.

The shift of the hypocenter before and after being relocated shifted from 0.01 to 13 meters (Figure 4). This shows that the event relocation in the mine needs better accuracy because the mining area has a relatively narrow supporting utility. The distance between panels and certain levels is in the order of meters. The result of the event shift histogram was then made a reset diagram. The distribution of the microseismic shown by the rosette diagram spreads in all directions, but dominantly in the NE-SW direction (Figure 5).

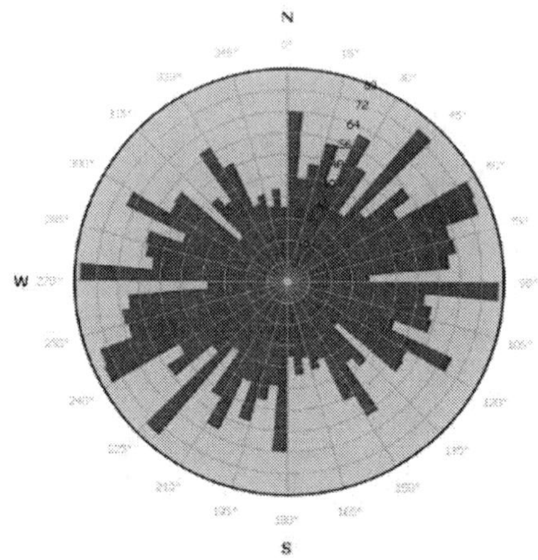

Figure 5. Rosette diagram.

The distribution of the hypocenter of the event using the Double - Difference method before relocation and after the relocation is shown in Figure 6. The distribution of the hypocenter of the event before the relocation (blue color) is more spread out, on the other hand, after the relocation of the hypocenter is more congested (orange color). The relocation results show the rock mass stress which is illustrated by the distribution of events around the cave, especially the abutment area and underground mining tunnels.

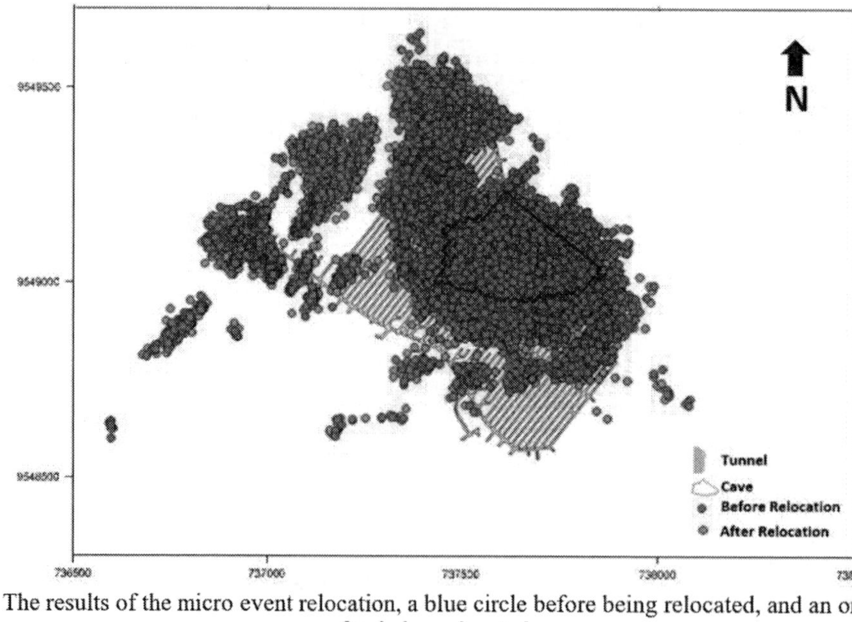

Figure 6. The results of the micro event relocation, a blue circle before being relocated, and an orange color after being relocated

4. Conclusion

The double-difference method has succeeded in increasing the accuracy of local scale microseismic relocation. The results of the relocation indicated a shift in the location as far as an average of ± 0.01 to 13 meters from the initial location.

5. References

[1] Flores, G., and Catalan., A. (2019): A transition from a large open pit into a novel "macroblock variant" block caving geometry at Chuquicamata mine, Codelco Chile. *Journal of Rock Mechanics and Geotechnical Engineering, Pages 549-561.* https://doi.org/10.1016/j.jrmge.2018.08.010

[2] Gong SY., Jing Lia, Feng Ju, Lin-ming Dou, Jiang He, Xin-yuan Tian. (2019): Passive seismic tomography for rockburst risk identification based on *adaptive-grid* method. *Tunnelling and Underground Space Technology, Pages 198–208.*

[3] Nugraha, DA. (2017): *Tomografi Seismik.* Penerbit ITB Press 2017, ISBN 978-602-5417-48-1.

[4] Waldhauser, F., and Ellsworth, W.L. (2000): A Double-Difference Earthquake Location Algorithm: Method and Application to the Northern Hayward Fault, California. *Bulletin of the Seismological Society of America 90*, 1353–1368. https://doi.org/10.1785/0120000006

[5] Waldhauser, F. (2001): *HypoDD: A Computer Program to Compute Double-difference Earthquake Location*, U. S. Geol. Surv. Open-file report, 01-113, Menlo Park, California.

[6] Zhang, H.J., and Thurber, C. (2005): *Adaptive mesh* seismic tomography based on tetrahedral and Voronoi diagrams: Application to Parkfield, California. *J. Geophys. Res. 110 (B4), 225–243*

IOP Conf. Series: Earth and Environmental Science **873** (2021) 012100 doi:10.1088/1755-1315/873/1/012100

Maluku Sea Plate Faulting Regime Analysis: A Preliminary Study

Jonathan Adii[1], Dian Kusumawati[1*], Cevian Falevi[1] and David P Sahara[1]

[1]Global Geophysics Research Group, Institut Teknologi Bandung

dianksurya@gmail.com

Abstract. Present day Molucca or Maluku sea plate in the eastern of Indonesia possesses a complex tectonic setting. This complex tectonic setting has been formed due to the collision of an actively moving Eurasian plate and Philippine sea plate toward the Maluku sea plate. At the west, Maluku sea plate is subducting beneath Sangihe arc, which began in the early Miocene. While at the east, Maluku sea plate is subducting under Halmahera arc, since in the middle Miocene. These subduction processes take place up to the present. Therefore, it has formed Maluku sea plate into an inverted U-shape slab under a thickening accretionary complex. Seismicity distribution has clearly shown the U-shape slab. Earthquake events take place on the subducting slab, and interestingly on the above accretionary complex as well. Maluku sea plate might pose hazards to surrounding islands: northern Sulawesi, Halmahera island, Sangihe island and Talaud island. The possible hazard, for instance, a thrusting earthquake which may generate tsunami to the nearby islands. Hence, understanding its tectonic and seismicity signature, especially at the shallow part, are indeed important in the Maluku sea region. Faulting regime could be analyzed using focal mechanism ternary diagram analysis, by categorizing the focal mechanisms' strike, dip and rake values. Thus, in this study we aim to analyze faulting regime and hazard potential in the complex. Maluku setting using ternary diagram analysis.

Keywords: *Maluku sea plate, ternary diagram, faulting regime, double subduction zone*

1. Introduction

Present day Molucca or Maluku sea possesses a complex tectonic setting. Its location is at the triple junction of three major plate, i.e., Philippine-sea, Eurasian, and Australian plate [1]. It separates volcanic Sangihe arc in the west and volcanic Halmahera arc in the east, forming an arc-arc collision or double subduction in the Maluku sea plate [2]. The ongoing movement of plates and the colliding fore arcs at both sides, are surpassing the Maluku sea plate. At the present, the Maluku sea plate has been almost eradicated by these subductions [3, 4], forming an inverted U-shape beneath colliding fore arcs [2]. Though this collision complex area has low gravity [2], many earthquake events took place in this area: in the subducting slab (depth > 25 km) and above the accretionary complex/ collision complex area (0 km < depth < 25 km) [2]. Indeed, the Maluku sea plate might pose earthquake or tsunami hazards to surrounding islands.

Earthquake focal mechanisms can give us a vivid representation of the faulting mechanism of an event: their type and parameter (strike, dip rake, B, P and T axis). Anderson [5] categorized the type of earthquake faulting based on the stress regime into normal fault, strike slip fault and thrust fault. By utilizing the P/T and B axis parameter, these focal mechanism's types could be further classified in the Ternary diagram.

As we know that dip slip mechanism, such as thrust fault and normal fault, in the offshore could displace sea columns and trigger tsunamis. Hence, analysing focal mechanisms in the Maluku region is important. In this study we aim to analyse faulting regime and signature in the complex Maluku setting using focal mechanism and Ternary diagram analysis.

2. Data and Methodology

2.1. Ternary Diagram

Ternary diagram categorized focal mechanism into their faulting mechanism, i.e., strike slips, normal or reverse. The diagram utilizes the B, T and P axis plunge of the focal mechanism for the classification. The result is plotted in the diagram using the projection concept of a point in a spherical octant over a planar surface. Frohlich and Apperson [6] represent the spherical octant as a triangle, which can classify the focal mechanism into pure strike slip, pure normal and pure reverse faulting. Kaverina et al. [7] and Kagan [8] further modified the projection and used Lambert Azimuthal Equal-Area. Álvarez-Gómez [9] adopts the modified projection in the Ternary diagram so it could also plot the oblique faulting. Hence, there are seven categories of focal mechanism in the Ternary diagram: normal, normal-strike slip, strike slip-normal, strike slip, strike slip-reverse, reverse-strike slip and reverse. In this study we used the FMC, a Python program for classifying and plotting focal mechanism data [9].

2.2. Data

We used focal mechanisms data from the Global Centroid Moment Tensor Catalog [10, 11] to analyse the earthquake event in our study area. We took events which occurred between January 1st, 1976, to September 30th, 2020, with depth between 0 km to 100 km and with minimum magnitude of Mw 4. The given criteria generate a maximum magnitude of Mw 8 in our study area. There are approximately 999 events. These events are plotted in figure 1.

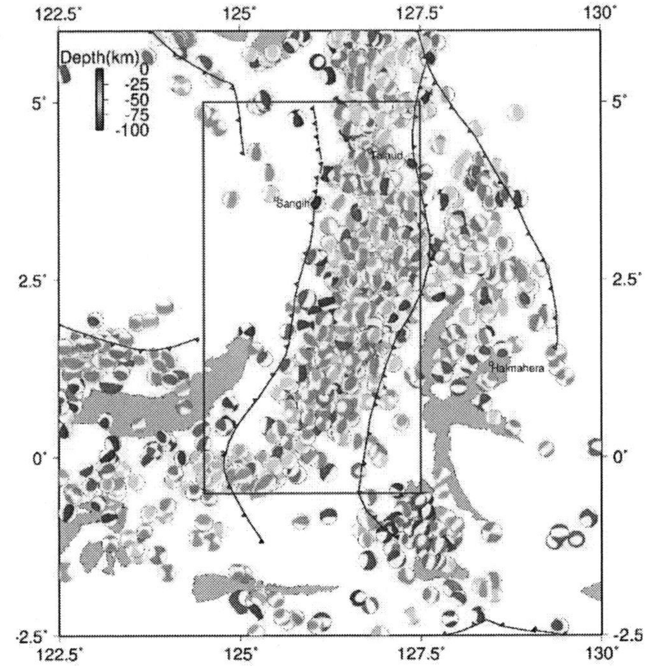

Figure 1. Event's focal mechanisms distribution in Maluku sea plate and surrounding area. The events are between January 1st, 1976, to September 30th, 2020, with depth between 0 km to 100 km and with minimum and maximum magnitude of Mw 4 and Mw 8. The data was taken from Global Centroid Moment Tensor Catalog [10, 11]. Solid blue square referring to events plotted in P/T axis diagram of figure 5.

3. Result and Discussion

We plotted all the events in figure 1 to the Ternary diagram in figure 2. The focal mechanisms are classified into pure strike slip, pure normal, pure reverse, and oblique mechanisms. In this study we focus only on the pure mechanisms which we refer as strike slip, normal and reverse faulting mechanisms.

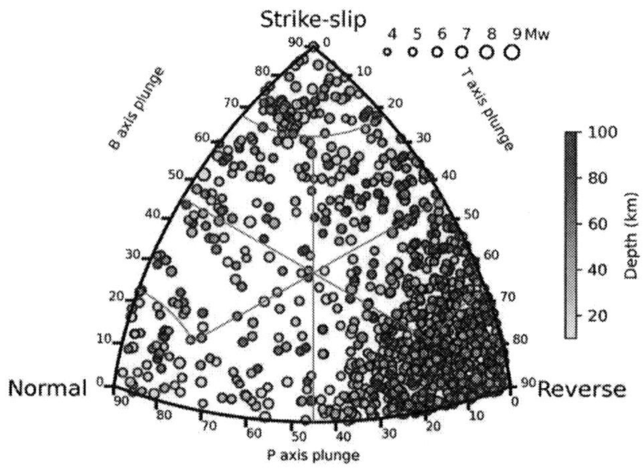

Figure 2. Ternary diagram of focal mechanisms in Maluku sea plate and surrounding area.

From the Ternary diagram classification, there are a few strike slip mechanisms (64 events) which are mostly clustered in the inland (0 km – 25 km) of the northern Halmahera Island (figure 3 (a)). For the reverse mechanism, there are 513 events (figure 3 (b)) which are mostly located in the 0 to 50 km. A rather shallow depth of the Maluku sea plate double subduction zone. Indeed, reverse faulting is a common mechanism in the subduction zone [12]. There are several normal mechanisms (80 events) with varied depth in the double subduction zone area. Normal mechanisms are commonly found at the deep part of the subduction system [12]. However, in the Maluku sea plate we found normal events at shallow depth. The shallow normal events might be related to complex collision; and might pose a tsunami hazard [13]. We plotted the normal mechanisms in a map view and cross-section in figure 4.

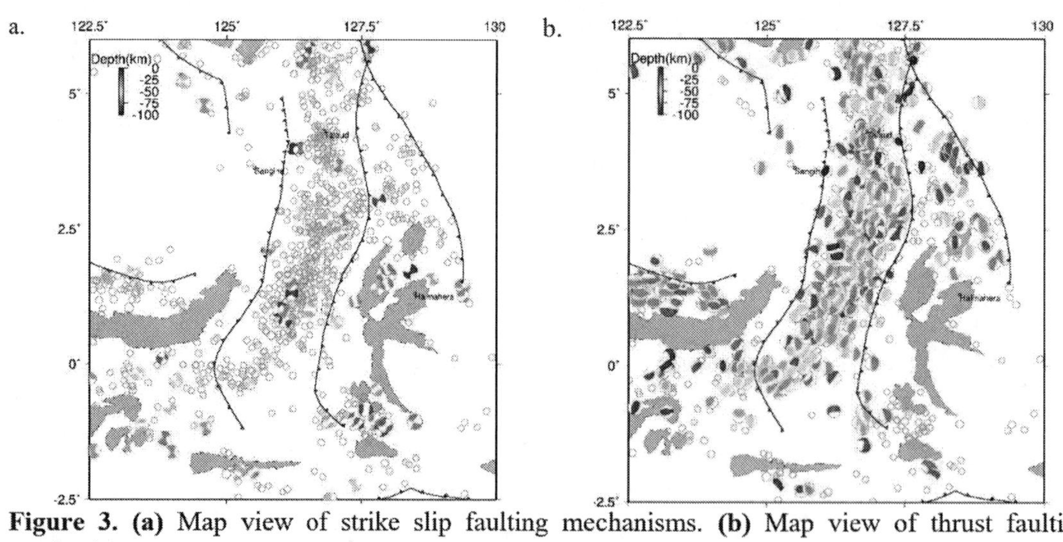

Figure 3. (a) Map view of strike slip faulting mechanisms. **(b)** Map view of thrust faulting mechanisms.

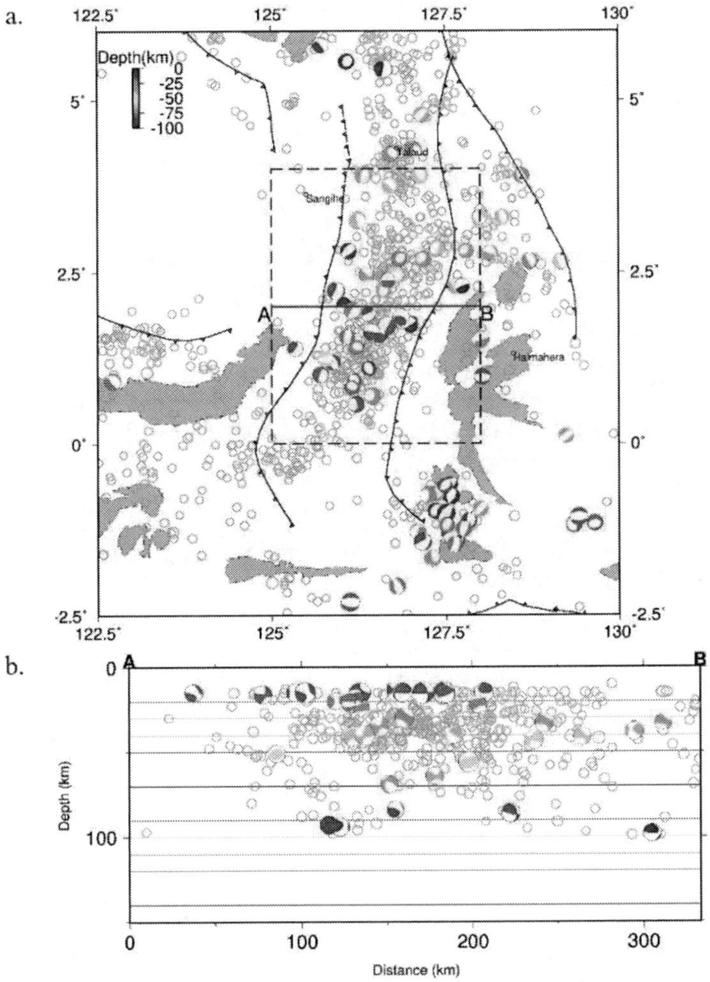

Figure 4. (a) Map view of normal faulting mechanisms. Dash blue square and red line are cross section area and line. **(b)** Cross section of normal focal mechanisms which lie in cross section area. Gray circles are showing the non-normal faulting mechanisms event distribution.

We further assessed the stress direction in the Maluku sea plate region. We plotted the focal mechanisms P/T axes inside the area bounded by solid blue square in figure 1 to P/T axes diagram on figure 5 (a) for 0 to 25 km depth and (b) for 25 km to 50 km depth. We used the STRESSINVERSE package [14] to plot the P/T axes diagram. There are 136 and 291 focal mechanisms in the depth ranges of 0 to 25 km and 25 to 50 km, respectively. The depth division follows the interpreted complex collision and the upper double subduction boundary [15]. In the 0 to 25 km depth the P/T axes are scattered. The stress orientation is less consistent, which might be related to the complex collision process. While in the deeper depth (25 to 50 km), we found focused P/T axes which show reverse faulting regime. The reverse stress regime is related to the double subduction process in the Maluku sea plate.

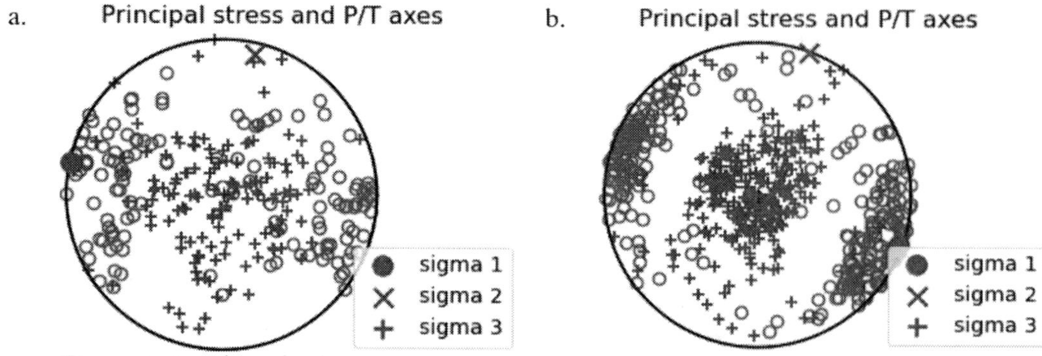

Figure 5. Focal mechanisms plotted in P/T axes for: **(a)** 0 to 25 km depth (136 events) and; **(b)** 25 to 50 km depth (291 events). The events are those which located inside the solid blue square in Figure 1. The P/T axes diagrams are plotted using STRESSINVERSE package [14].

4. Conclusion

Focal mechanisms distribution shows that Maluku sea region has dominantly reverse mechanism and several normal mechanisms with varied depth. The shallow normal mechanisms might be related to complex collisions. Both shallow reverse and normal mechanisms could possibly pose a tsunami hazard. At 0-25 km depth, the stress orientation is scattered due to the complex collision process. While consistent/ focused stress orientation found at 25-50 km depth, which is related to the dominant reverse mechanism in the subduction system.

In the future we will divide the depth section more detail and assess the focal mechanisms, especially the thrust mechanism, in each depth; to get the clear and detail picture of the Maluku sea plate thrusting regime.

References

[1] Widiwijayanti C, Tiberi C, Deplus C, Diament M, Mikhailov V and Louat R 2004 Geodynamic evolution of the northern Maluku Sea area (Eastern Indonesia) constrained by 3-D gravity field inversion. *Tectonophysics 386*(3-4) pp 203-222

[2] Hall R 1987 Plate boundary evolution in the Halmahera region, Indonesia. *Tectonophysics 144*(4) pp 337-352

[3] Cardwell R K, Isaacks B L and Karig D E 1980 The spatial distribution of earthquakes, focal mechanism solutions, and subducted lithosphere in the Philippine and northeastern Indonesian islands *GMS 23* pp 1-35

[4] McCaffrey R, Silver E A and Raitt R W 1980 Crustal structure of the Maluku Sea collision zone, Indonesia in: D.E. Hayes (Editor) The Tectonic and Geologic Evolution of South-East Asian Seas and Islands *Am. Geophys Union* Monogr. 23: 161-177.

[5] Anderson E M 1951 The Dynamics of Faulting and Dyke Formation with Applications to Brittan, Edinburgh, Oliver and Boyd.

[6] Frohlich C and Apperson K D 1992 Earthquake focal mechanisms, moment tensors, and the consistency of seismic activity near plate boundaries *Tectonics* 11 (2) 279–296.

[7] Kaverina A N, Lander A V and Prozorov A G 1996 Global creepex distribution and its relation to earthquake-source geometry and tectonic origin *Geophysical Journal International 125*(1) pp 249-265

[8] Kagan Y Y 2005 Double-couple earthquake focal mechanism: random rotation and display. *Geophysical Journal International 163*(3) pp 1065-1072

[9] Álvarez-Gómez J A 2014 FMC: a one-liner Python program to manage, classify and plot focal mechanisms *EGUGA* p 10887.

[10] Dziewonski A M, Chou T A and Wondhouse J H 1981 Determination of earthquake source parameters from waveform data for studies of global and regional seismicity *J. Geophys Res 86*(2) pp 825-2

[11] Ekström G, Nettles M and Dziewonski A M 2012 The global CMT project 2004-2010: Centroid-moment tensors for 13,017 earthquakes *Phys. Earth Planet. Inter.* 200-201 1-9 2012 doi: 10.1016/j.pepi.2012.04.002

[12] Stein S and Wysession M 2009 *An introduction to seismology, earthquakes, and earth structure*. John Wiley & Sons.

[13] Cummins P R, Pranantyo I R, Pownall J M, Griffin J D, Meilano I and Zhao S 2020 Earthquakes and tsunamis caused by low-angle normal faulting in the Banda Sea, Indonesia *Nature Geoscience 13*(4) pp 312-318

[14] Vavryčuk V 2014 Iterative joint inversion for stress and fault orientations from focal mechanisms *Geophysical Journal International 199*(1) pp 69-77

[15] Zhang Q, Guo F, Zhao L and Wu Y 2017 Geodynamics of divergent double subduction: 3-D numerical modeling of a Cenozoic example in the Maluku Sea region, Indonesia *Journal of Geophysical Research: Solid Earth 122*(5) pp 3977-3998

IOP Conf. Series: Earth and Environmental Science **873** (2021) 012101 doi:10.1088/1755-1315/873/1/012101

Determination of Shear Wave Splitting Parameters in 2018 Lombok Earthquake Using Rotation Correlation Method: Preliminary Result

Annisa Trisnia Sasmi*[1], Andri Dian Nugraha[2], Muzli Muzli[3,4], Sri Widiyantoro[2,5], Zulfakriza Zulfakriza[2], Shengji Wei[3], David P Sahara[2], Nanang T Puspito[2], Awali Priyono[2], Haunan Afif[6], Pepen Supendi[1,7], Daryono Daryono[4], Ardianto Ardianto[1], Devy Kamil Syahbana[6], Yayan Mi'rojul Husni[8], Billy S Prabowo[8], Kadek Hendrawan Palgunadi[9], Achmad Fajar Narotama Sarjan[1]

[1] Geophysical Engineering, Faculty of Mining and Petroleum Engineering, Institut Teknologi Bandung, Bandung, Indonesia
[2] Global Geophysics Research Group, Faculty of Mining and Petroleum Engineering, Institut Teknologi Bandung, Bandung, Indonesia
[3] Earth Observatory of Singapore/Asian School of the Environment, Nanyang Technological University, Singapore
[4] Meteorology, Climatology, and Geophysics Agency (BMKG), Jakarta, Indonesia
[5] Faculty of Engineering, Maranatha Christian University, Bandung 40164, Indonesia
[6] Center for Volcanology and Geological Hazard Mitigation (PVMBG), Bandung, Indonesia
[7] Meteorology, Climatology, and Geophysics Agency (BMKG), Bandung, Indonesia
[8] Laboratory of Volcanology and Geothermal, Faculty of Mining and Petroleum Engineering, Institut Teknologi Bandung, Bandung, Indonesia
[9] Physical Science and Engineering, King Abdullah University of Science and Technology, Saudi Arabia

Corresponding author's email: trisnia24@gmail.com

Abstract. Shear-wave splitting (SWS), or the propagation of two independent shear waves, can be used as an indicator of seismic anisotropy. In this study, we utilize this concept using aftershock data of the 2018 Lombok earthquake which had been acquired in period of August 4 – September 9, 2018. The goal of this research is to better understand the crack distribution related to the rupture zone of the 2018 Lombok earthquake. After applying instrument correction to the data, the waveform data were then windowed in each P and S arrival time. To determine the SWS parameters, we performed rotation in each horizontal seismogram components. The horizontal components were rotated from azimuth 0° to 180° with an increment of 1°. Cross-correlation coefficient (CCC) was determined for each rotation angle. The polarization direction and the SWS delay time were chosen from the parameters shown in the highest value of CCC.

IOP Conf. Series: Earth and Environmental Science **873** (2021) 012101 doi:10.1088/1755-1315/873/1/012101

Keywords: Shear Wave Splitting, 2018 Lombok Earthquake, Seismogram Rotation, Cross Correlation

1. Introduction

Lombok Island is one of the islands in Indonesia, which is located in the border area between the Sunda Arc and the Banda Arc. The seismicity condition of Lombok Island is controlled by two tectonic settings around the area, including: the megathrust trench in the south of Lombok Island [1,2], and the Flores Back Arc Thrust / FBT system in the north [1,2,3]. In addition, Lombok Island is flanked by the Lombok Strait Strike Slip Fault in the west, and the Sumbawa Strait Strike Slip Fault in the east [4]. Some researchers also suggest a submerging structure in the north of Lombok Island called Flores Oceanic Crust / FOC [4, 5, 6, 7, 8]. The implication of the complexity of this tectonic feature is the vulnerability of Lombok Island to earthquakes. Seismic activity on the island of Lombok also has been recorded since 1856 [9]

On July 29 2018, an earthquake measuring M 6.4 hit Lombok Island [10]. A series of significant earthquakes reoccurred on August 5 (M 7.0), August 9 (M 5.9) and August 19 (M 6.3 and M 6.9), with earthquake intensity on the VIII-IX MMI scale [11]. BMKG [10] reports that the hypocenter of these significant earthquakes is mostly located in the north of Lombok Island. This disaster caused 564 causalities, with the most victims were reported in the North Lombok Regency area [12].

A number of studies have been conducted to explain the mechanism of the 2018 Lombok earthquake. Based on the location of the hypocenter, the main earthquakes on August 5, August 9 and August 19, were shallow earthquakes with a depth of less than 35 km [7, 13]. The significant earthquakes that had occurred several times were thought to be the result of segmentation of the rupture that was formed before the emergence of the earthquakes with strong magnitudes [7, 14]. As a result of the 2018 series of earthquakes in Lombok, a deformation occurred in the north of Lombok Island [15], causing an uplift of 70 cm [16]. However, more detailed information regarding the rupture of these significant earthquakes (such as the orientation and distribution of the crack in the rupture zone, as well as the direction of slip movement) is still not well defined.

The presence and orientation of cracks in an area can be analyzed using the SWS concept. Splitting or polarization of shear waves occurs when shear waves propagate through an anisotropic medium, one of the causes is a fractured zone. This process produces orthogonal polarization of shear fast and slow waves which are separated based on the characteristics of the delay time. SWS using local station network data has been applied to study the faulting system in the 1995 Kobe earthquake [17]. Parameters that are also generated using the SWS method are shear wave anisotropy parameters in rocks (K). Seismic anisotropy is defined as the variation in seismic wave velocity with respect to direction [18]. The anisotropy properties are caused by 3 things: (i) laminations that occur in rocks, (iii) orientation of minerals, (iii) fractures and microcracks [19]. The SWS tomography study has been applied by [20] to determine the distribution of anisotropy (K) values in the active fault region.

In this study, to explore the explanation of the 2018 Lombok earthquake rupture, SWS was conducted by utilizing data from 16 local seismographic stations which had been installed in the Lombok Island region on August 4 to September 9 2018. This study is expected to provide detailed information about the crack orientation in the Lombok rupture zone and its spatial distribution.

2. Data and Method

To perform the Shear Wave Splitting (SWS) analysis, the earthquake waveforms used in this study were recorded via 16 temporal seismographic network from ITB and EOS / NTU. The seismographic network was installed on June 5 to September 9 2018 with the configuration as shown in Figure 1. The location of earthquake hypocenters had previously been determined [7].

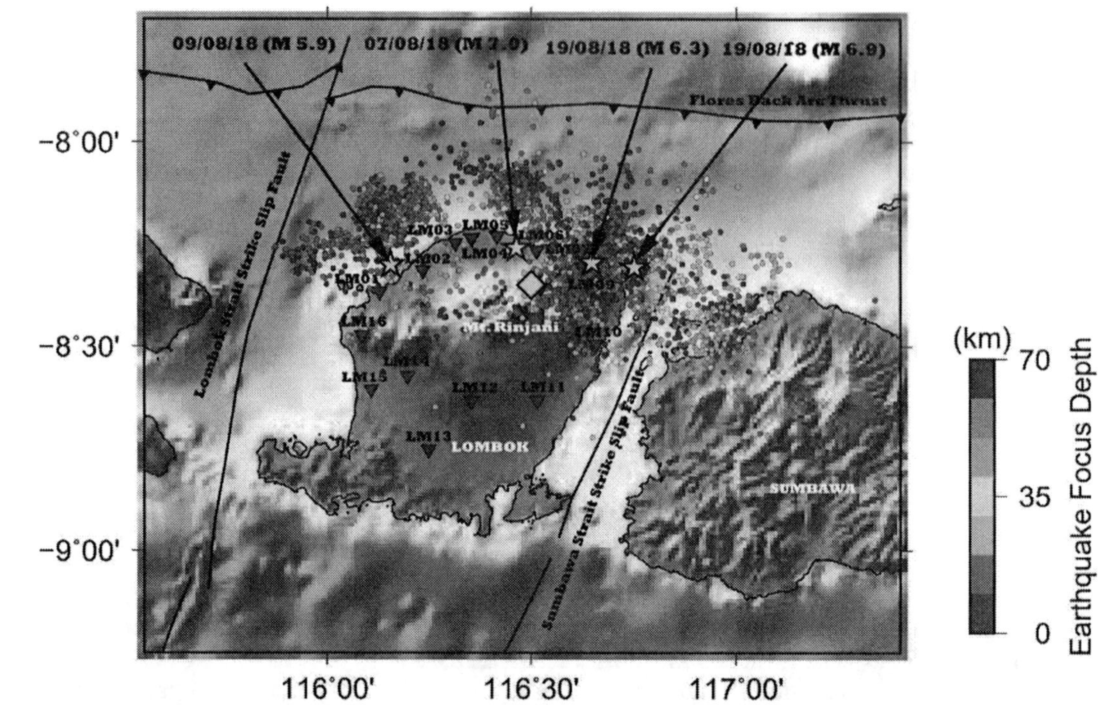

Figure 1. The location of station distribution (inverted blue triangles) and earthquake hypocenters used in this study. The yellow stars show the location of the significant earthquakes from [7]: (August 9 [M 5.9], August 5 [M 7.0], and August 19 [M 6.3 and M 6.9]); The diamond symbol depicts the mainshock of July 29 [M 6.9] event. The color scale represents the earthquake focal depth.

The SWS technique is carried out to determine the propensity of the cracks distribution which were formed during the earthquake series. This process is applied to determine the delay time (dt) and polarization angle (φ) between the two pairs of S-phase wave, that is, S_{fast} and S_{slow}. This process is applied to the local network data. There are several stages of routine data processing that have to be prepared before performing a SWS analysis, including the selection of waveform data, equalization of sampling frequency, and instrument correction. In the data selection process, we filter the data as following: the data must be recorded by a minimum of three seismometer stations; the data has a complete seismometer component (vertical, E-W, N-S); and has a high signal-to-noise ratio on S-phase waveform. The S_{fast} and S_{slow} dominant frequency were determined by using the Short Time Fourier Transform (STFT). The next process is determining the SWS parameters by applying the rotation-correlation method [21].

3. Result and Discussion

A total of 2,619 aftershocks events have been identified for the S phase for further SWS analysis. From these events, the number of S phases that met the criteria was 14,249. The number of phases read at each station can be seen in Figure 2. The windowing process using STFT had been applied to determine the dominant frequency of each S_{fast} and S_{slow} phase. This process was applied to determine the dominant frequency of the 0.4 s wavelength sample picking area. Then, the dominant frequency was used to set the length of the signal period. This periode length then will be used to perform the cross-correlation between two signal, S_{fast} and S_{slow}.

IOP Conf. Series: Earth and Environmental Science **873** (2021) 012101 doi:10.1088/1755-1315/873/1/012101

The determination of the polarization angle and SWS delay time was performed using cross-correlation between the two waveforms that have been rotated into radial and transversal directions (Figure 3). We vary the rotation angle between 0° and 180° with an increment of 1°. Meanwhile, the value of the delay time used in this study has a range from 0 to 0.2 s with an increment of 0.01 s. The reason for using this increment is that the sampling rate used in this study was 0.01 s. Based on these two parameters, a correlation contour will be formed as shown in Figure 4. The value of delay time and the angle of polarization can be found in the highest contour of every pair of S_{fast} and S_{slow}. This process is still continuing to obtain all of the *dt* and the polarization angle value of every S event.

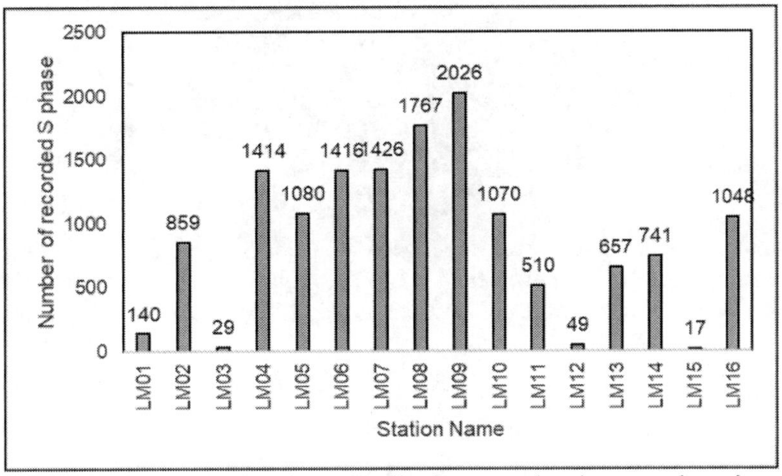

Figure 2. Number of S phase distribution at each seismograph station

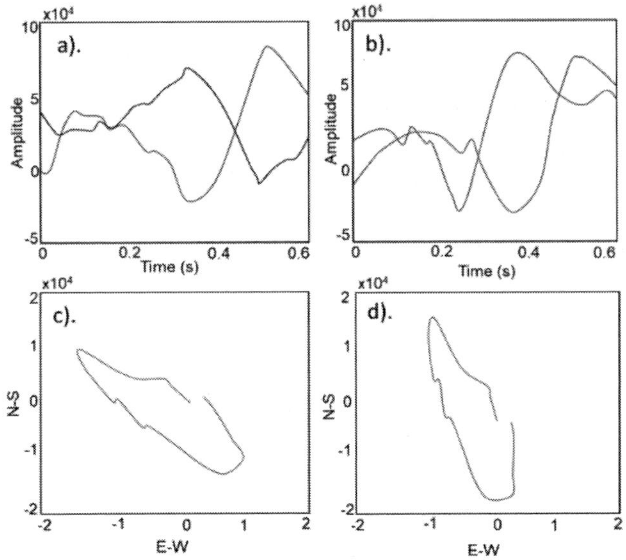

Figure 2. a). An example of cropped waveform (from LM09 station) before rotation; b). Waveform after rotation from station LOM-09; c). Hodogram before rotation; d). Hodogram after rotation.

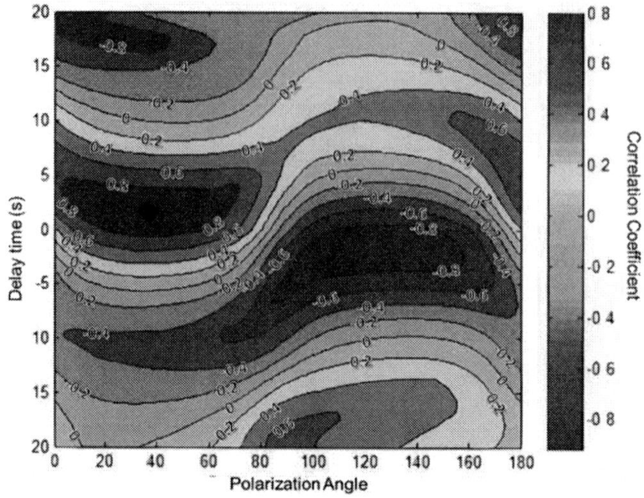

Figure 3. Contour of rotation-correlation between a pair of S phase in LM09 station. The color scale bar depicts the cross-correlation coefficient (CCC). The highest coefficient value is depicted by the black diamond symbol, and shows that the polarization angle of the event pair is 38° with a delay time of 2.4 s.

4. Conclusions

A number of 14,249 S-phase events of 2,619 Lombok aftershock events had been determined to conduct SWS analysis. The windowing process of S_{fast} and S_{slow} had been implemented using STFT as a benchmark in rotation-correlation process. The rotation-correlation process was applied using a range of 1°-180° polarization angle, and the highest correlation value shows the delay time and polarization angle of a pair of S phase event. This process will be applied to all of the data in the future in order to investigate the anisotropy of the study area in detail.

Acknowledgements

We would like to thank our colleagues at Institut Teknologi Bandung (ITB), the Earth Observatory Singapore (EOS), the Agency for Meteorology, Climatology, and Geophysics of Indonesia (BMKG), and the Center for Volcanology and Geological Hazard Mitigation of Indonesia (PVMBG) for their collaboration. We also express our gratitude for the PMDSU Batch IV scholarship program (2018) established by the Indonesian Ministry of Research, Technology, and Higher Education and for the support and research opportunities given to A.T. Sasmi, "Hibah Riset Penelitian Dasar Unggulan Perguruan Tinggi, Kemenristek/BRIN 2019 – 2021" and "Hibah Riset Institut Teknologi Bandung 2019-2020" DIKTI research funding 2019 awarded to A. D. Nugraha & N. T. Puspito, and the World Class Research funding to S. Widiyantoro.

References

[1] W B Hamilton 1977 Subduction in the Indonesian region *Am. Geophys. Union* Maurice Ewing Series **1** 15–31

[2] W B Hamilton 1979 Tectonics of the Indonesian region *Tech. Rept.* **1078** U.S. Govt. Print. Off., Wash

[3] R McCaffrey and J Nabelek 1984 The geometry of back arc thrusting along the eastern Sunda Arc, Indonesia: constraints from earthquake and gravity data *J. Geophys. Res.* **89** no B7 6171–6179

[4] M Irsyam, S Widiyantoro, D H Natawidjaya, I Meilano, A Rudyanto, S Hidayati, W Triyoso, N

R Hanifa, D Djarwadi, L Faizal, and S Sunarjito 2017 Peta sumber dan bahaya gempa Indonesia tahun 2017 *Pusat Penelitian dan Pengembangan Perumahan dan Permukiman, Kementerian Pekerjaan Umum dan Perumahan Rakyat* (In Indonesian)

[5] S Widiyantoro, J D Pesicek, and C H Thurber 2011 Subducting Slab Structure below the eastern Sunda arc inferred from non-linear tomography imaging *In the SE Asian Gateway: History and Tectonics of the Australian-Asia Collision* (Special Publication) ***139-155*** *Geol. Soc. Lond. Mem.*

[6] T Zubaidah, M Korte, M Mandea, and M Hamoudi 2014 New insight into regional tectonics of the Sunda-Banca Arcs region from integrated magnetic and gravity modelling *J Asia Earth Sci* **80** 172-184

[7] A T Sasmi, A D Nugraha, M Muzli, S Widiyantoro, Z Zulfakriza, S Wei, D P Sahara, A Riyanto, N T Puspito, A Priyono, T Greenfield, H Afif, P Supendi, D Daryono, A Ardianto, D K Syahbana, Y M Husni, B S Prabowo, and A F N Sarjan 2020 Hypocenter and Magnitude Analysis of Aftershocks of the 2018 Lombok Earthquakes, Indonesia, Using Local Seismographic Networks *Seismol. Res. Lett.* **XX** 1-11 doi: 10.1785/0220190348

[8] H Afif, A D Nugraha, M Muzli, S Widiyantoro, D P Sahara, A Riyanto, T Greenfield, N T Puspito, A Priyono, A T Sasmi, P Supendi, A Ardianto, D K Syahbana, S Rosalia, A Cipta, Y M Husni 2021 Local Earthquake Tomography of The Source Region of The 2018 Lombok Earthquake Sequence, Indonesia *Geophys. J. Int.*

[9] A Wichmann 1918 The Earthquakes of The Indian Archipelago Until The Year 1857 *Natuurkunde* **2** Sect Deel 20 No 4 and Deel 22 No 5

[10] Badan Meteorologi, Klimatologi, dan Geofisika (BMKG) 2019 Katalog Gempa Bumi Signifikan dan Merusak Tahun 1821-2018 Pusat Gempa dan Tsunami BMKG *https://www.bmkg.go.id/gempabumi/katalog-gempabumi-signifikan.bmkg* Accessed at September 14, 2020

[11] R Robiana, P A Minarno, S Hidayati, S Supartoyo, and F Nurfalah 2018 Gempa Lombok Tanggal 29 Juli 2018 dan Dampaknya—Kajian Rangkaian Gempa Lombok Provinsi Nusa Tenggara Barat *Pusat Penelitian dan Pengembangan Perumahan dan Pemukiman, Badan Penelitian dan Pengembangan, Kementerian Pekerjaan Umum dan Perumahan Rakyat* (In Indonesian)

[12] Badan Nasional Penanggulangan Bencana (BNPB) 2019 Gempa Lombok https://www.bnpb.go.id/uploads/24/info-bencana-agustus-2018-1.pdf Accessed at June 30 2020

[13] X Yang, S C Singh, and A Tripathi 2020 Did The Flores Backarc thrust rupture offshore during the 2018 Lombok Earthquake sequence in Indonesia? *Geophys. J. Int.* **221** 758-768

[14] C Wang, X Wang, W Xiu, B Zhang, G Zhang, and P. Liu 2020 Characteristics of the Seismogenic Faults in the 2018 Lombok, Indonesia, Earthquake Sequence as Revealed by Inversion of InSAR Measurements *Seismol. Res. Lett.* **XX** 1–12 doi: 10.1785/0220190002

[15] S Susilo, A Wijanarto, S Wibowo, and H Z Abidin 2018 GPS/GNSS Analysis On Lombok Earthquakes: Co-Seismic Deformation *Experimental Findings* doi: 10.13140/RG.2.2.34871.78242

[16] R Agustan, N Hanifa, Y Anantasena, M Sadly,. and T Ito 2019 Ground Deformation Identification related to 2018 Lombok Earthquake Series based on Sentinel-1 Data *Earth and Environ. Sci.* **280** (2019) 012004 doi:10.1088/1755-1315/280/1/012004

[17] K Tadokoro and M Ando Evidence for rapid fault healing derived from temporal changes in S wave splitting *Geophys. Res. Lett.* **29** no 4 1047 10.1029/2001GL013644, 2002

[18] S Crampin 1984 Effective anisotropic elastic constants for wave propagation through cracked solids *Geophys. J. Int.* **79**(1) 135-145 https://doi.org/10.1111/j.1365-246X.1984.tb05029.x

[19] L Vernik and A Nur 1992 Ultrasonic velocity and anisotropy of hydrocarbon source rocks *Geophys.* **57 (5)** https://doi.org/10.1190/1.1443286

[20] H Zhang, Y Liu, C Thurber, and S Roecker 2007 Three-dimensional shear-wave splitting tomography in the Parkfield, California, region *Geophys. Res. Lett.* **34** L24308 doi:10.1029/2007GL031951, 2007
[21] L Vecsey, J Plomerová, and V Babuška 2008 Shear-wave splitting measurements — Problems and solutions *Tectonophysics* **462** (2008) 178–196

IOP Conf. Series: Earth and Environmental Science **873** (2021) 012102 doi:10.1088/1755-1315/873/1/012102

Prior Analysis in Determination of TI Anisotropy Type through Well-Based Modelling: A Case Study on the Deep Water Environment

Madaniya Oktariena[1], Wahyu Triyoso[2], Fatkhan Fatkhan[2], Sigit Sukmono[2], Erlangga Septama[3], Pongga D Wardaya[3], Befriko S Murdianto[4], Rusalida Raguwanti[3], Vida I Rossa[3]

[1] Postgraduate Program of Geophysical Engineering, Faculty of Mining and Petroleum Engineering, Institut Teknologi Bandung, Jalan Ganesha 10, Bandung 40132, Indonesia
[2] Faculty of Mining and Petroleum Engineering, Institut Teknologi Bandung, Jalan Ganesha 10, Bandung 40132, Indonesia
[3] PT Pertamina (Persero) Upstream Research & Technology Innovation, SopoDel Office Tower 50th Floor, Jalan Mega Kuningan Barat III Jakarta 12950, Indonesia
[4] Pertamina Hulu Kalimantan Timur, CIBIS NINE Tower 9th Floor, Jalan TB Simatupang 2, Jakarta Selatan 12560, Indonesia

madaniyaoktariena@outlook.com

Abstract. The existence of anisotropy phenomena in the subsurface will affect the image quality of seismic data. Hence a prior knowledge of the type of anisotropy is quite essential, especially when dealing with deep water targets. The preliminary result of the anisotropy of the well-based modelling in deep water exploration and development is discussed in this study. Anisotropy types are modelled for Vertical Transverse Isotropy (VTI) and Horizontal Transverse Isotropy (HTI) based on Thomsen Parameters of ε and γ. The parameters are obtained from DSI Logging paired with reference δ value for modelling. Three initial conditions are then analysed. The first assumption is isotropic, in which the P-Wave Velocity, S-Wave Velocity, and Density Log modelled at their in-situ condition. The second and third assumptions are anisotropy models that are VTI and HTI. In terms of HTI, the result shows that the model of CDP Gather in the offset domain has a weak distortion in Amplitude Variation with Azimuth (AVAz). However, another finding shows a relatively strong hockey effect in far offset, which indicates that the target level is a VTI dominated type. It is supported by the geomechanical analysis result in which vertical stress acts as the maximum principal axis while horizontal stress is close to isotropic one. To sum up, this prior anisotropy knowledge obtained based on this study could guide the efficiency guidance in exploring the deep water environment.

1. Introduction

Seismic imaging in an anisotropy media in hydrocarbon exploration is needed. It is because of the fact of the subsurface is anisotropy media. Previous elastic theories that assume rock as an isotropic media is slowly developing into adaptation to the real subsurface phenomenon. The complex geological structure, mineral alignment during sedimentation, and compaction related to overburden are among the

IOP Conf. Series: Earth and Environmental Science **873** (2021) 012102 doi:10.1088/1755-1315/873/1/012102

factors triggering anisotropy within rock [1] [2]. Transverse Isotropic (TI) is commonly used to simplify the anisotropy in sedimentary rock [3]. TI Anisotropy assumes that a media is isotropic in one direction while anisotropic in its perpendicular plane [4]. It fits the sand/shale lamination reservoir's description representing Vertical Transverse Isotropy (VTI) condition and the fractured clastic reservoir that represents Horizontal Transverse Isotropy (HTI) condition.

The existence of anisotropy in subsurface conditions could influence the subsurface image as captured by seismic reflection data. Both VTI and HTI cases have a different impact on the recorded seismic reflection data [5]. A deep water setting might further enhance its complexity [6]. A thorough exploration planning is necessary to obtain the best data quality for subsurface imaging at a significant depth with the most effective cost. Prior knowledge of TI anisotropy type and degree would be quite essential in deep water exploration. A better understanding of anisotropy will help a lot in the exploration plan to design the acquisition parameter the proper seismic processing flow. It could also become the first clue in development strategies as guidance for the azimuth of directional drilling and maximum offset determination in the seismic 3D acquisition.

2. Methods

Well-based synthetic modeling is a necessary preliminary analysis to further explore or develop decisions in the deep water environment. In the deep water reservoir, Well SAD-4 is part of Sadewa Field discovered in East Kalimantan, drilled to penetrate a clastic reservoir at about 11000ft. Clastic reservoirs evaluated in this research are quartz-rich, moderately to highly porous, fairly compacted, and upper very fine sand to upper fine sand grain size with moderate sorting.

The anisotropy effect is to be modeled to obtain a preliminary understanding of the target's TI Anisotropy type. Parameter ε and γ of Thomsen's Anisotropy are obtained from DSI Logging paired with reference δ value as the input for well-based synthetic modeling. Thomsen's Anisotropy Parameter ε and γ range from -0.01 – 0.04 in which considered as weak anisotropy. The δ value refers to the sandstone value of the Typical Thomsen Parameter Value table [7] with adjustment accordingly to the SAD-4 Thomsen's Anisotropy Parameter magnitude range. Weak anisotropy has a magnitude of ε and δ that less than 0.2 [8].

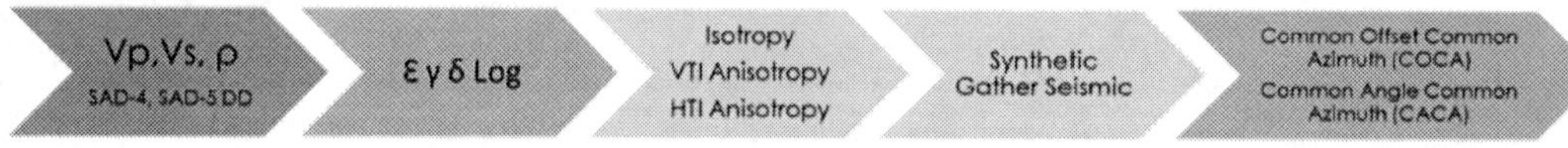

Figure 1. A workflow diagram of preliminary TI anisotropy determination through well-based modeling.

Well-based modeling applies three initial conditions of anisotropic assumption in creating the synthetic seismic gather. The first assumption is the Isotropic Media, where the P-Wave Velocity, S-Wave Velocity, and Density Log would be modeled at their in-situ condition. The second assumption is Vertical Transverse Isotropic (VTI), and the third assumption is Horizontal Transverse Isotropic (HTI). Azimuth information is also added to the synthetic modeling. The synthetic seismic is divided into four azimuthal groups centered at N70^0, N160^0, N250^0, and N340^0. The modeling result will be sorted into Common Offset Common Azimuth (COCA) and Common Angle Common Azimuth (CACA). The maximum offset of COCA is 7000 meters to provide a sufficient target level coverage, while the ultra-far angle of CACA is 62^0 following a preliminary angle analysis. These seismic gather sorting allow us to see both the VTI & HTI effect in a single seismic image. Ray tracing technique is applied for the offset to angle mapping of the synthetic seismic gather. We make use of the Sonic Log in SAD-4 as the replacement velocity to perform the tracing from the surface.

IOP Conf. Series: Earth and Environmental Science **873** (2021) 012102 doi:10.1088/1755-1315/873/1/012102

In the isotropic assumption, the reflectivity estimation is based on the Zoeppritz equation. It is addressed to ensure a more accurate estimate of the event amplitude of reflection from the target interface compared to the linearized one. In the TI anisotropy assumption, the Schoenberg-Protázio Extension to Anisotropic Zoeppritz Theory is applied. Schoenberg-Protázio extended the Zoeppritz equations to anisotropy by introducing the (2 x 2) block matrices (see Eq. 1), providing the solution to four components of both reflected and transmitted waves [9].

$$X = \begin{bmatrix} Vp_{S1} & Vs_{S3_S} \\ -\rho Vp\Gamma & 2\rho Vs^3{}_{S1\,S3_S} \end{bmatrix}; Y = \begin{bmatrix} -2\rho VpVs^2{}_{S1\,S3_P} & -\rho Vs\Gamma \\ Vp_{S3_P} & Vs_{S1} \end{bmatrix} \quad \text{(Eq. 1)}$$

S1 is horizontal slowness, $S3_P$ is vertical P-wave slowness, $S3_S$ is vertical S-wave slowness, and Γ is equal to $(1 - 2V_s^2 s_1^2)$. The synthetic gather is conducted using the Azimuthal Plane Wave technique for the synthetic to have appeared as NMO corrected seismic gather. The synthetic gather of the VTI and HTI of the well-based modeling results are then analyzed. The primary purpose is to better understand anisotropy's nature on seismic reflectivity response. In turn, to understand better relating the TI anisotropy role to the deep water environment.

3. Results

The result of well-based modelling with isotropic assumption using in-situ log condition shows an indication of hockey stick effect at the synthetic gathers seismic far offset. Thomsen's Anisotropy Parameter value hasn't been added to the original P-Wave, S-Wave, and Density log data at this in-situ condition.

In terms of HTI, the result showcases that the synthetic Common Offset Common Azimuth seismic gather shows a weak response with minimal distortion in Amplitude Variation with Azimuth (AVAz). Another finding through VTI assumption shows a relatively strong hockey effect in the far offset of the synthetic Common Offset Common Azimuth seismic gather. With additional anisotropy data calculated from Dipole Shear Sonic Imager (DSI) Logging, the hockey stick effect is still present and further enhanced (see Fig. 2a).

However, the strong seismic response on anisotropic is showed by the Common Angle Common Azimuth seismic gather. The hockey stick effect is smoother without any sudden jump in value. The distortion in AVAz is also more pronounced (see Fig. 2b).

4. Discussion

Anisotropy, well-based modelling at the isotropic assumption showcases that the in-situ log data, DTCO, DTSM, RHOB, contains the record of the anisotropic tendency of rocks. It confirms the theory of Li & Pickford [2] stated that most of the rocks are considered weakly anisotropic. The Thomsen's Anisotropy Parameter of SAD-4 calculated from DSI Logging also falls into the weak anisotropy type with a value of ε and γ below 0.2 in magnitude. The use of the anisotropy parameter in the modelling will enhance the anisotropy's effect in the synthetic seismic gather.

Well-based modelling with the HTI assumption shows a weak distortion in AVAz. It could represent a hairline fracture resulting from overburden and compaction commonly found in the clastic sandstone reservoir. The VTI assumption's well-based modelling shows a clear and strong hockey stick effect in the synthetic Common Angle Common Azimuth seismic gather. Analysis into the Pre-SDM seismic gather further confirms this synthetic finding.

This study shows that the well-based modelling concludes that the target level is VTI anisotropy dominated. The work is aligned with Geomechanical MEM Analysis on SAD-4 that stated vertical stress acts as the maximum principal axis. Meanwhile, the magnitude contrast of maximum horizontal stress and mini-mum horizontal stress is close to the isotropic one.

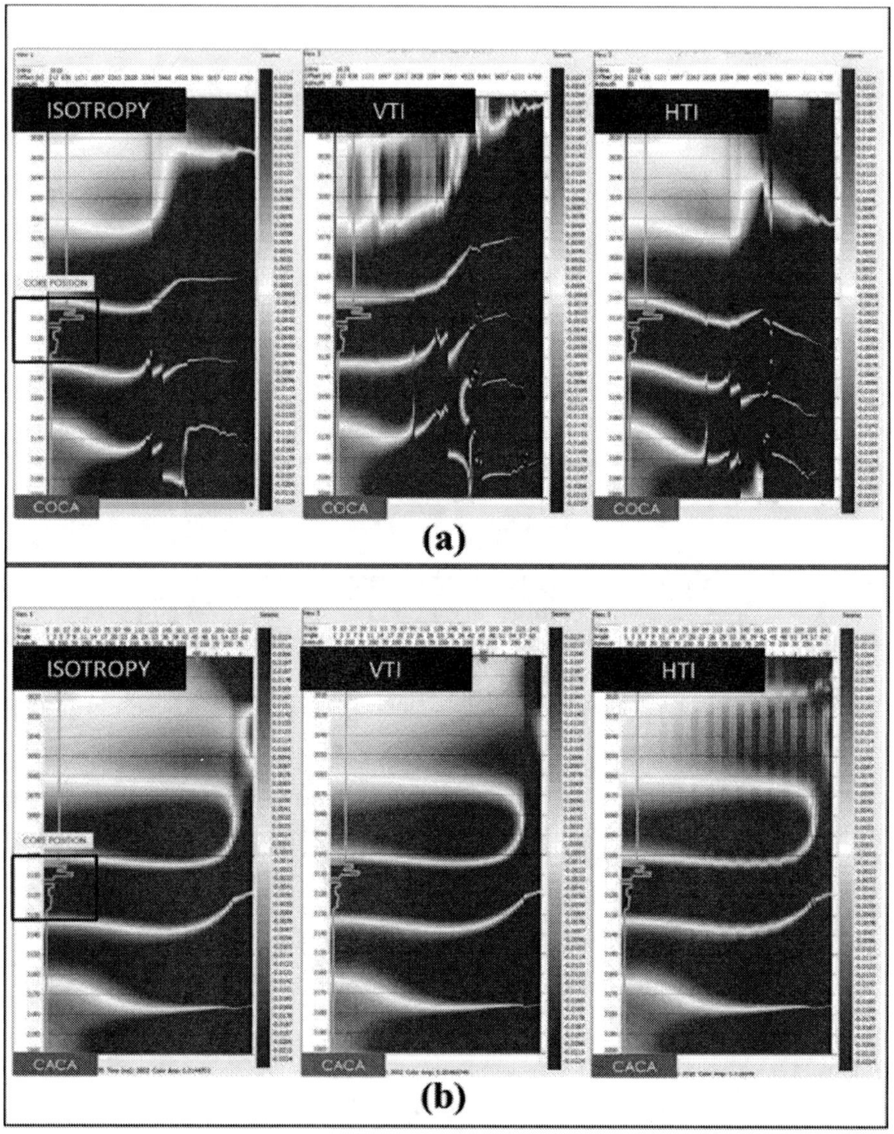

Figure 2. Synthetic seismic gathers result from anisotropy, well-based modeling with in-situ, VTI, and HTI assumption sorted into (a) Common Offset Common Azimuth (b) Common Angle Common Azimuth gather.

5. Conclusions

The deep water environment has been posing as a challenge in hydrocarbon exploration. Prior knowledge about the subsurface anisotropy condition would benefit in more sufficient planning of seismic exploration and further development. The well-based modelling could act as a first step in identifying the anisotropy type before a more comprehensive study is being conducted. The VTI is found in SAD-4 related to the determination of far offset design in the seismic acquisition. This prior anisotropy knowledge obtained from a simple preliminary analysis in the form of well-based synthetic modelling could guide a more efficient exploration in a deep water environment.

References

[1] Liu X 1994 Non-linear elasticity, seismic anisotropy, and petrophysical properties of reservoir rocks *PhD Thesis* Stanford University, California
[2] Li Y 2002 *SEG Technical Program Expanded Abstracts*
[3] Saberi M R, Ting J. 2016 *First break* **34** 41-48
[4] Levin F K 1979 *Geophysics* **44** 918-936
[5] Alkhalifah T, Tsvankin I 1995 *Geophysics* **60** 1550-1566
[6] Hall M 2003 *CSEG Recorder* **28** 1-3
[7] Tsvankin I 1997 *Geophysics* **62** 1292-1309
[8] Xiao C 2006 *MSc Thesis* University of Calgary
[9] Schoenberg M and Protazio J 1992 *The Journal of the Acoustical Society of America* **88** 125-144

AUTHOR INDEX